DATE DUE

APR 1 0 1996	SEP 2 6 2003
SEP 2 6 1996	
Oct 10	
~~NOV 19~~	
~~OCT 2 1 1996~~	
Dec 4	
APR - 2 1997	
JAN - 5 1998	
APR - 3 1998	
APR - 9 1998	
APR 2 7 1998	
FEB 1 2 1999	
MAR - 6 2000	
NOV 1 5 2000	
NOV 2 9 2000	
APR - 9 2002	
OCT 1 1 2002	

BRODART.   Cat.

MICROBIAL

TRANSFORMATION AND

DEGRADATION OF TOXIC

ORGANIC CHEMICALS

WILEY SERIES IN
ECOLOGICAL AND APPLIED MICROBIOLOGY

SERIES EDITOR, Ralph Mitchell, Division of Applied Sciences, Harvard University

RECENT TITLES

THERMOPHILES: General,
Molecular, and Applied Microbiology
Thomas D. Brock, Editor, 1986

INNOVATIVE APPROACHES
TO PLANT DISEASE CONTROL
Ilan Chet, Editor, 1987

PHAGE ECOLOGY
Sagar M. Goyal, Charles P. Gerba,
and Gabriel Bitton, Editors, 1987

BIOLOGY OF ANAEROBIC
MICROORGANISMS
Alexander J.B. Zehnder, Editor, 1988

THE RHIZOSPHERE
J.M. Lynch, Editor, 1990

BIOFILMS
William G. Characklis and
Kevin C. Marshall, Editors, 1990

ENVIRONMENTAL MICROBIOLOGY
Ralph Mitchell, Editor, 1992

BIOTECHNOLOGY IN PLANT
DISEASE CONTROL
Ilan Chet, Editor, 1993

ANTARCTIC MICROBIOLOGY
E. Imre Friedmann, Editor, 1993

WASTEWATER MICROBIOLOGY
Gabriel Bitton, 1994

EFFECTS OF ACID RAIN ON
FOREST PROCESSES
Douglas L. Godbold and Aloys Hüttermann,
Editors, 1994

MICROBIAL TRANSFORMATION
AND DEGRADATION OF
TOXIC ORGANIC CHEMICALS
Lily Y. Young and Carl E. Cerniglia,
Editors, 1995

MICROBIAL TRANSFORMATION AND DEGRADATION OF TOXIC ORGANIC CHEMICALS

Edited by

LILY Y. YOUNG

Center for Agricultural Molecular Biology
Cook College, Rutgers University
New Brunswick, New Jersey

CARL E. CERNIGLIA

National Center for Toxicological Research
U.S. Food and Drug Administration
Jefferson, Arkansas

WILEY-LISS

A JOHN WILEY & SONS, INC., PUBLICATION
New York • Chichester • Brisbane • Toronto • Singapore

Address All Inquiries to the Publisher
Wiley-Liss, Inc., 605 Third Avenue, New York, NY 10158-0012

Library of Congress Cataloging-in-Publication Data
Microbial transformation and degradation of toxic organic chemicals /
    edited by Lily Y. Young, Carl E. Cerniglia.
        p.   cm. — (Wiley series in ecological and applied
microbiology)
    Includes bibliographical references and index.
    ISBN 0-471-52109-4 (alk. paper)
    1. Bioremediation. 2. Sanitary microbiology. I. Young, Lily Y.
II. Cerniglia, Carl E. III. Series.
TD192.5.M326 1995
628.5'2—dc20   95-15401
                                                                     CIP

The text of this book is printed on acid-free paper.

CONTENTS

CONTRIBUTORS   vii

PREFACE   xi
Lily Y. Young and Carl E. Cerniglia

PART I. THE ISSUES

1. MICROBIAL VERSATILITY   3
Norberto J. Palleroni

2. CHEMICAL CONTAMINATION OF THE
ENVIRONMENT: SOURCES, TYPES, AND FATE OF
SYNTHETIC ORGANIC CHEMICALS   27
Norbert G. Swoboda-Colberg

PART II. THE MICROBIOLOGY

3. CLEANUP OF PETROLEUM HYDROCARBON
CONTAMINATION IN SOIL   77
Ingeborg D. Bossert and Geoffrey C. Compeau

4. MICROBIAL REDUCTIVE DECHLORINATION
OF POLYCHLORINATED BIPHENYLS   127
Donna L. Bedard and John F. Quensen III

5. BACTERIAL CO-METABOLISM OF HALOGENATED
ORGANIC COMPOUNDS   217
Lawrence P. Wackett

6. THE MICROBIAL ECOLOGY AND PHYSIOLOGY
OF ARYL DEHALOGENATION REACTIONS AND
IMPLICATIONS FOR BIOREMEDIATION   243
Joseph M. Suflita and G. Todd Townsend

7. MECHANISMS OF POLYCYCLIC AROMATIC
HYDROCARBON DEGRADATION   269
*John B. Sutherland, Fatemeh Rafii, Ashraf A. Kahn,
and Carl E. Cerniglia*

v

8. ANAEROBIC DEGRADATION OF NONHALOGENATED
HOMOCYCLIC AROMATIC COMPOUNDS COUPLED
WITH NITRATE, IRON, OR SULFATE REDUCTION   307
Patricia J. S. Colberg and Lily Y. Young

9. ORGANOPOLLUTANT DEGRADATION
BY LIGNINOLYTIC FUNGI   331
Kenneth E. Hammel

**PART III. APPLICATIONS IN CLEANUP AND
BIOREMEDIATION**

10. MICROBIOLOGICAL TREATMENT OF CHEMICAL
PROCESS WASTEWATER   349
Laurence E. Hallas and Michael A. Heitkamp

11. BIOREMEDIATION OF CHLOROPHENOL WASTES   389
Max M. Häggblom and Risto J. Valo

12. BIOLOGICAL TREATMENT OF CHLORINATED
ORGANICS   435
Peter Adriaens and Timothy M. Vogel

13. FIELD TREATMENT OF COAL TAR-CONTAMINATED
SOIL BASED ON RESULTS OF LABORATORY
TREATABILITY STUDIES   487
Margaret Findlay, Samuel Fogel, Lee Conway,
and Arthur Taddeo

14. *IN SITU* PROCESSES FOR BIOREMEDIATION
OF BTEX AND PETROLEUM FUEL PRODUCTS   515
Gene F. Bowlen and David S. Kosson

PART IV. FUTURE TRENDS

15. DEGRADATIVE GENES IN THE ENVIRONMENT   545
Tamar Barkay, Sylvie Nazaret, and Wade Jeffrey

16. RISK ASSESSMENT FOR TOXIC CHEMICALS
IN THE ENVIRONMENT   579
David W. Gaylor

17. HAZARDOUS CHEMICALS AND BIOTECHNOLOGY:
PAST SUCCESSES AND FUTURE PROMISE   603
Burt D. Ensley and Mary F. DeFlaun

INDEX   631

CONTRIBUTORS

PETER ADRIAENS, Department of Civil and Environmental Engineering, University of Michigan, Ann Arbor, MI 48109 **[435]**

TAMAR BARKAY, Microbial Ecology and Biotechnology, U.S. Environmental Protection Agency, Gulf Breeze, FL 32561 **[545]**

DONNA L. BEDARD, Bioremediation Research Program, Environmental Laboratory, GE Corporate Research and Development, Schenectady, NY 12301 **[127]**

INGEBORG D. BOSSERT, Research and Development Center, Texaco, Inc., Beacon, NY 12508 **[77]**

GENE F. BOWLEN, Envirogen, Inc., Lawrenceville, NJ 08648; present address: Accutech Remedial Systems, Keyport, NJ 07735 **[515]**

CARL E. CERNIGLIA, Division of Microbiology, National Center for Toxicological Research, U.S. Food and Drug Administration, Jefferson, AR 72079 **[269]**

PATRICIA J.S. COLBERG, Department of Zoology and Physiology, University of Wyoming, Laramie, WY 82071 **[307]**

GEOFFREY C. COMPEAU, Enviros Incorporated, Kirkland, WA 98033 **[77]**

LEE CONWAY, Northeast Utilities Service Company, Hartford, CT 06141 **[487]**

MARY F. DEFLAUN, Envirogen, Inc., Lawrenceville, NJ 08648 **[603]**

BURT D. ENSLEY, Envirogen, Inc., Lawrenceville, NJ 08648; present address: Phytotech, Inc., Monmouth Junction, NJ 08852 **[603]**

MARGARET FINDLAY, Bioremediation Consulting, Inc., Newton, MA 02159 **[487]**

SAMUEL FOGEL, Bioremediation Consulting, Inc., Newton, MA 02159 **[487]**

DAVID W. GAYLOR, National Center for Toxicological Research, U.S. Food and Drug Administration, Jefferson, AR 72079 **[579]**

The numbers in brackets are the opening page numbers of the contributors' articles.

MAX M. HÄGGBLOM, Center for Agricultural Molecular Biology, Cook College, Rutgers University, New Brunswick, NJ 08903 **[389]**

LAURENCE E. HALLAS, Wastewater Solutions Division, LMII Corporation, St. Louis, MO 63131 **[349]**

KENNETH E. HAMMEL, Forest Products Laboratory, U.S. Department of Agriculture, Madison, WI 53705 **[331]**

MICHAEL A. HEITKAMP, Environmental Sciences Center, Monsanto Company, St. Louis, MO 63166 **[349]**

WADE JEFFREY, Center for Environmental Diagnostics and Bioremediation, University of West Florida, Pensacola, FL 32514 **[545]**

ASHRAF A. KHAN, Division of Microbiology, National Center for Toxicological Research, U.S. Food and Drug Administration, Jefferson, AR 72079 **[269]**

DAVID S. KOSSON, Department of Chemical and Biochemical Engineering, Rutgers, The State University of New Jersey, Piscataway, NJ 08855 **[515]**

SYLVIE NAZARET, Laboratoire d'Ecologie Microbienne du Sol, U.A. CNRS, Université Lyon I, Villeurbanne, F-69622, France **[545]**

NORBERTO J. PALLERONI, Center for Agricultural Molecular Biology, Cook College, Rutgers University, New Brunswick, NJ 08903 **[3]**

JOHN F. QUENSEN III, Department of Crop and Soil Sciences, Michigan State University, East Lansing, MI 48824 **[127]**

FATEMEH RAFII, Division of Microbiology, National Center for Toxicological Research, U.S. Food and Drug Administration, Jefferson, AR 72079 **[269]**

JOSEPH M. SUFLITA, Department of Botany and Microbiology, University of Oklahoma, Norman, OK 73019 **[243]**

JOHN B. SUTHERLAND, Division of Microbiology, National Center for Toxicological Research, U.S. Food and Drug Administration, Jefferson, AR 72079 **[269]**

NORBERT G. SWOBODA-COLBERG, Department of Geology and Geophysics, University of Wyoming, Laramie, WY 82071 **[27]**

ARTHUR TADDEO, Unitec Environmental Consultants, Inc., Honolulu, HI 96819 **[487]**

G. TODD TOWNSEND, Department of Botany and Microbiology, University of Oklahoma, Norman, OK 73019 **[243]**

RISTO J. VALO, Bioteam Corp., Maininkitie 21 A6, Espoo 02320, Finland; present address: Soil and Water, Ltd., Helsinki 00210, Finland **[389]**

TIMOTHY M. VOGEL, Department of Safety and the Environment, Rhone-Poulenc Industrialisation, 69153 Decines-Charpieu, France **[435]**

LAWRENCE P. WACKETT, Department of Biochemistry and Institute for Advanced Studies in Biological Process Technology, University of Minnesota, St. Paul, MN 55108 **[217]**

LILY Y. YOUNG, Center for Agricultural Molecular Biology, Cook College, Rutgers University, New Brunswick, NJ 08903 **[307]**

PREFACE

Recent high interest in using microorganisms for environmental cleanup began after the Exxon Valdez ran aground and released ten million gallons of oil into Prince William Sound, Alaska, in March 1989. Although oil is a natural substance, quantities of this magnitude were clearly a hazard to the environment, and the bioremediation efforts that were part of the overall cleanup sparked a renewed interest in using microbial processes for removing hazardous wastes from the environment. Microbiologists and environmental engineers have long known that microbial biodegradation and bioremediation are not new concepts and that waste treatment, composting, and natural attenuation, for example, all rely on microorganisms for removal of undesirable organic compounds. Both indigenous and genetically modified microorganisms are now being studied for their ability to degrade hazardous chemicals such as polyaromatic hydrocarbons (PAHs), benzene, toluene, xylenes (BTXs), polychlorinated biphenyls (PCBs), and chlorinated solvents such as perchlorethene (PCE) and trichlorethene (TCE). The potential benefits that would be derived from new biotechnological processes and the use of bioremediation has stimulated research worldwide. This volume provides current perspectives from both laboratory and field studies on the biodegradation of toxic chemicals. The contributors to this book are recognized experts in their field and represent academic, government, and industrial laboratories.

The book is organized into four parts, with the authors giving their unique research experiences and perspectives in the following areas: Part I, **The Issues,** which examines the diversity and versatility of microorganisms involved in the degradation of hazardous chemicals, and the sources, types, and fate of synthetic organic chemicals in the environment; Part II, **The Microbiology,** which explores the basic pathways and mechanisms of microbially mediated biodegradation of a variety of hazardous chemicals found in the environment; Part III, **Applications in Cleanup and Bioremediation,** whereby biodegradation principles are applied in managed processes and *in situ* treatment for biotreatment of toxic chemicals such as solvents, chlorinated aromatics, and fuel-derived waste; and Part IV, **Future Trends,** which looks at the new strategies available to us with molecular biotechnology approaches and examination of risk assessment. We believe this volume can serve as a resource on the current state of the art in our understanding of these issues and that both advanced students and practitioners will find it useful.

We thank Dr. William Curtis of Wiley-Liss for his unflagging support, Dr. Ralph Mitchell for his initiation of this volume, and the contributing authors for their effort and cooperation.

LILY Y. YOUNG
CARL E. CERNIGLIA

PART I. THE ISSUES

MICROBIAL VERSATILITY

NORBERTO J. PALLERONI

Center for Agricultural Molecular Biology, Cook College, Rutgers University, New Brunswick, New Jersey 08903

1. INTRODUCTION IN THE WAY OF AN APOLOGY

Anyone planning to write on microbial versatility cannot fail to recall the superb reflexions on this fascinating subject contained in a number of publications by A.J. Kluyver and C.B. van Niel, particularly the chapters of the charming little book *The Microbe's Contribution to Biology* that they wrote in collaboration (Kluyver and van Niel, 1956).

These two men had an invaluable participation in the development of general microbiology in this century. Kluyver succeeded Beijerinck in the prestigious chair of microbiology at Delft, after a few years spent in Java. From then on he remained in his country, which meant that he had to find the hard way of surviving with dignity through the darkness of German occupation. His work dealt with carbohydrate biochemistry, biochemical activities of bacteria (methanogens, hydrogen utilizers, lactic acid producers, nitrogen fixers, denitrifiers), and yeasts and other fungi, including industrially important fermentations. He directed numerous doctor's theses, one of which was that of van Niel, a fundamental work on propionic acid bacteria. His life and work are reviewed in the book edited by Kamp et al. (1959), which includes some of his most important writings originally published in Dutch or German, but mercifully translated for the book into English by van Niel. Kluyver's clear insight in microbial metabolism, together with an immaculate human quality, left in all who were honored with his friendship a truly unique and indelible impression.

After his association with Kluyver in their native country, van Niel emigrated to America and settled in Pacific Grove, California, until his death in 1985. There,

Microbial Transformation and Degradation of Toxic Organic Chemicals, pages 3–25

aside from his original contributions to microbiology, his teachings became legendary. Those who were fortunate to be able to attend one of his summer courses in a laboratory where the modest equipment made more evident than anywhere else that the best instrument is by far the human brain, will never forget his classes, a balanced mixture of solid science and histrionics that pointed to the path of total conversion. Many of the converts, by the way, became the best microbiologists that America ever had.

Simply updating the material contained in the chapters of *The Microbe's Contribution to Biology* and in many of their papers is not an infalible antidote against comparisons with these masters' inspiration. For what it is worth, this chapter is dedicated to their dear memory.

2. GENERAL CONSIDERATIONS ON MICROBIAL VERSATILITY

Some three centuries ago, scientists became aware of the existence of a living world invisible to the naked eye, but for some one hundred and fifty years thereafter, they did not have a clear idea about the meaning of these small living forms in the economy of nature. The explosive development of microbiology in the second half of the nineteenth century hinges on the personality of a single man, Louis Pasteur, who envisioned more precisely than other scientists the role of microorganisms in natural processes. His clear insight brought to an end the theory of spontaneous generation, which only had the support of objectionable experiments. This, in turn, implied that, according to Pasteur, microbes are not the consequence of fermentation, putrefaction, or disease, but the direct cause of such complex processes. *Cause* and *consequence* were not in the proper order, and once again a basic concept had emerged from an examination of familiar facts from a novel, unfamiliar perspective.

In truth, Pasteur was not the first to propose the idea that microbes are causative agents of transformations. In 1845 the potato disease then commonly known as *murrain* broke out with devastating effects in various European countries, and the correct explanation for the course of events was given by the Reverend M.J. Berkeley, who, in lively arguments against the opinion of Dr John Lindley, insisted that the fungus observed in the plant tissues was in all likelihood the causal agent of the disease and not the consequence of the weather conditions (Large, 1940). This happened a few decades before Pasteur's work, which goes to teach us an important lesson, namely, that great ideas often depend on other factors (such as right timing) to have a lasting effect in the history of scientific endeavor.

From Pasteur's time, the world of *microbiology* was enormously enriched with the discovery of a vast variety of microbial types and activities. Application of the techniques of pure culture developed in Koch's laboratory was a decisive factor in helping to establish the identity of newly discovered microorganisms and in the determination of specificity of pathogenic bacteria. At the same time, by application of ecological principles, Sergei Winogradsky and Martinus Beijerinck became the principal protagonists in the search for and discovery of microbial activities that were not only unique, but also essential for maintaining the balance of nature. One of the most important innovations to biological knowledge was made by Winogradsky when he demonstrated that autotrophy, i.e., the capacity to live at the expense of CO_2 as the sole carbon source, was not necessarily linked to the use of light energy. During the first half of the twentieth century, a large inventory of biochemical pathways, both biosynthetic and degradative, was assembled, and microorganisms became prime examples of life's versatility. It is only among microbes that we observe examples of tolerance or preference for extreme habitats, ranging in temperature from the freezing to the boiling points of water, or from one end to the other in the scales of pH or redox potentials.

Microbial versatility is outstanding in the tolerance for extreme conditions as well as in the rapidity with which microbes adapt and modify their enzymatic machinery to face new challenges in the environment. It is among microorganisms that we can find catabolic activities of great interest, which allow bioremediation of environments heavily contaminated with recalcitrant organic compounds that are toxic for higher forms of life. The list of electron donors and acceptors used by various microbial groups includes examples that do not find parallel in the biology of higher organisms. The variety of reactions that can be carried out by microorganisms is much greater than that found in higher forms of life. Some of them were developed under the conditions prevailing in our planet before the advent of oxygen-generating photosynthesis, and later some of the biochemical pathways of an essentially anaerobic nature were supplemented with the necessary links to amospheric oxygen as final electron acceptor. The examples have not only enriched our "inventory," but in addition they have helped us to understand metabolism in general and to grasp the unity underlying life's complexity.

In spite of the many variations of biochemical pathways that we find among microorganisms, there is a basic unity in the biochemical processes carried out by living systems. In the biological transformation of organic compounds, hydrogen appears so intimately associated with carbon that its importance as the real fuel for the cells appears overshadowed. Energy-yielding processes are linked to electron transport, and this in turn is associated with the transport of protons. The energy liberated in discrete steps is stored in high energy compounds for use when needed in energy-requiring processes such as biosynthesis. One aspect of the unity of life is represented by the use of the same energy coinage (ATP). The general strategies for use of energy and carbon sources are very similar in all organisms. However, this underlying unity is often hidden behind an enormous complexity in the details of biochemical pathways, to which microorganisms contribute to a high degree.

3. DEVELOPMENT OF IDEAS ON MICROBIAL VARIABILITY

Microbes are very small, which means that the surface to volume ratio is enormous, and therefore they are highly exposed in nature to many mutational agents and selective pressures that tend to alter their genetic constitution. The natural challenges are more critical to the individual in the case of unicellular microorganisms.

The genetic changes provoked by different environmental factors are readily expressed in prokaryotes due to the haploid nature of their genomes. These characteristics of microorganisms must have played a key role during evolution in the production of the large variety of physiological types that we know today. At the same time, by handling microorganisms in the laboratory, the microbiologist is well aware of the capacity for change of isolated strains and can mimic evolution by applying defined environmental conditions and strong selective forces. At present we have an imprecise idea of the mechanisms by which the enormous variety of microorganisms found in nature have evolved to their present state. In contrast, our knowledge of factors and mechanisms of variation of pure cultures under laboratory conditions is considerable.

The discussion in this chapter refers mostly to prokaryotic organisms, that is, bacteria. It is also mandatory to emphasize at this point the importance of avoiding the temptation of describing bacterial activities that also occur in eukaryotes but are better known in bacteria simply because these organisms are easier to study under controlled laboratory conditions. A clear-cut distinction is, however, not possible, and some occasional reference will be made to activities that have been observed in both bacteria and higher organisms. Many biochemical and genetic properties that seem now to be uniquely bacterial in nature may eventually be identified in eukaryotes with the help of bacterial models.

The development of ideas about the origin and mechanism of microbial variation was very uneven. On the one hand, Mendelian genetics of eukaryotic microorganisms placed them close to plants and animals, and some of them (*Neurospora, Saccharomyces, Paramecium*) soon were choice subjects for genetic research. In the case of bacteria, for several decades our ideas were very elementary, and only in the 1940s was the subject attacked for the first time on solid scientific grounds. The demonstration by Luria and Delbrück (1943) of the occurrence of spontaneous mutations among cells not yet exposed to the selecting factor introduced bacteria as subjects for statistical studies in the genetics of populations, and the discovery of conjugation in *Escherichia coli* by Lederberg and Tatum (1946) elevated this species to a position of preeminence in genetic research. Bacteriologists could, from then on, speak about mutation and selection on firm bases, a right that thus far had been denied to them on account of lack of solid genetic evidence.

Soon after these discoveries were made, a lively polemic took place between two schools of thought, represented, on the one hand, by bacterial geneticists, and on the other, by those who were inclined to attribute to physicochemical causes the reasons for bacterial variations. The main representative of this last tendency was a distinguished kineticist, Cyril Hinshelwood. In trying to explain the response of *Aerobacter aerogenes* to the utilization of D-arabinose, which occurs after a considerable lag, Hinshelwood (1946) thought that the change happened in all the cells of the population. When he tried to give a similar explanation for the origin of resistance to antibacterial compounds, the opposition manifested by geneticists who attributed both types of changes to mutation and selection became very lively.

Hinshelwood maintained that low concentrations of some drugs were likely to shift the normal balance of some cellular reactions, reaching a new equilibrium where interaction with the drug was less effective. This change could be perpetuated in the cells' cytoplasm for many generations afterwards (Hinshelwood, 1946; Woof and Hinshelwood, 1960). In turn, the geneticists pointed to the fact that, even under very unfavorable conditions, bacteria are capable of some growth, and the populations arising from single cells could reach a sufficient size to include one or more mutants capable of proliferating under the new conditions.

Even though Hinshelwood knew the work of the bacterial geneticists, and did not deny the appearance of mutants under certain circumstances, the emphasis he placed on physicochemical causes resulted in an almost universal rejection of his explanations by the bacteriology community. Recently the episode was revived (Watson, 1991), but the remarks, clearly unfavorable to Hinshelwood's reputation, were opposed by Rubin (1991), who sustained that, even though Hinshelwood's reasoning was focused on physicochemical arguments, his observations revealed potential changes in the behavior of all cells in a population by interaction with external conditions. In fact, the study of these properties appears to have been delayed by the immense popularity of the approaches based on mutation and selection heralded by the bacterial geneticists.

In later sections of this chapter, attention is paid to the various factors and mechanisms underlying microbial versatility. Some of them may occur in a very small number of cells of the population, and the basic mechanism is a change in the composition of the microbial genome, while others are the result of alterations in the behavior of all cells due to changes in the structure of the genome, but not to changes in its composition.

4. MUTATIONAL CHANGES

Mutations in many genes can be observed to appear spontaneously with low frequency, which usually can be increased by several orders of magnitude through the action of mutagenic agents. Presumably similar mechanisms are involved in the production of mutants in all living organisms, but a brief discussion here is justified because the use of prokaryotes as subjects for research has been crucial for a deeper understanding of mutational events.

To a microbiologist, a pure culture is a population of microorganisms derived from the multiplication of a single cell. The most convincing way of obtaining such a culture is by micromanipulation, but in practice micromanipulators are seldom used for this purpose, and simpler procedures, such as streaking a suspension on the surface of an agar medium, give a fair assurance, after one or more repeats, of satisfactory microbiological purity.

However, even with the absolute certainty of single cell origin, a culture is never pure from the genetic point of view. Variants of many different types constantly appear during growth of the population, and application of a selective pressure results in the establishment of a new population, which displaces the original one. In turn, further selection exerted on the new population easily shows its genetic heterogeneity so that genetic stability is a goal that can never be achieved in practice. As pointed out by Bryson and Szybalski (1952), if the number of characteristics under examination is large, each cell in the population may well be unique.

Many of the changes that a "pure culture" undergoes are simple mutations. As previously mentioned, by the use of the so-called fluctuation test, Luria and Delbrück (1943) demonstrated that the appearance of mutants occurs independently of the selective agent. Recently this genetic concept was challenged on the basis of observations showing different distributions of mutants depending on the nature of the selective agent (Cairns et al., 1988, Stahl, 1988). Be as it may, selection preserves certain mutations, and it is the main driving force in the changes occurring during evolution. The bacteriologist also makes use of this principle to select mutants capable of accomplishing certain useful biochemical tasks. Use of mutants in research has been instrumental in studies on many biochemical pathways and in genetic investigations leading to the construction of chromosomal maps.

Resistance to high concentrations of most antimicrobial agents (among them, the penicillins) can be built in microbial populations in a stepwise manner, that is, by exposure to increasing concentrations of the drugs. Of course, at each step, the sensitive cells, always in the majority, are eliminated. This has been named *obligatory multiple-step resistance* (Hsie and Bryson 1952; Bryson and Szybalski, 1952; Szybalski and Bryson, 1952). In other cases, resistance to high concentrations can develop in a single step or through the use of a progressively increasing series of drug concentrations, although single-step mutants resistant to any of the concentrations may appear at any time, even in drug-free media. This process is less common than the former, and it is called *facultative multiple-step resistance.* Acquisition of resistance to streptomycin follows this pattern. Indeed, stepwise acquisition of drug resistance seems to argue in favor of Hinshelwood's kinetic theory. However, Demerec (1945, 1948) demonstrated the genetic bases of such phenomena, showing that different genes are responsible for the various levels of antibiotic resistance.

5. GENE REARRANGEMENTS

The stability of the bacterial chromosome is quite remarkable, and many microbial strains conserve their basic properties for considerable lengths of time, even after innumerable transfers in artificial culture media under laboratory conditions. However, the genome is far from an immutable structure. A striking example of gene reassortment taken from the eukaryotic world is the redistribution of gene fragments in various combinations that give the enormous diversity of immunoglobulin genes of animals (Yankopoulos et al., 1986). Aside from mutations, there are other variations at the molecular level that have been studied in prokaryotes, although some of them also operate in eukaryotes, where research often has been facilitated by reference to bacterial models. These variations include changes in the copy number of genes or in the control of expression of genes coding for the same function.

These changes are globally known as *gene rearrangements,* and they can be classified, according to Borst and Greaves (1987), as follows. First, there are "incidental" rearrangements that originate in errors in the replication, repair, or recombination of DNA and also from insertion of DNA segments (insertion elements, transposons, plasmids, viral DNA) in the genome. Second, there are "programmed" gene rearrangements, whose effects are predictable, resulting from the action of specific enzymes and the respective regulatory mechanisms. In this category are placed changes in gene dosage caused by deletions and amplifications and mechanisms such as inversions and transpositions that do not alter the gene dosage but modify the expression of genes.

In a later section, changes in gene expression due to environmental factors such as the nutrient composition of the medium, its osmolarity, oxygen tension, temperature, or radiation are mentioned. Most of the cells in the population may be involved in the reaction to these factors, which in particular cases may cause changes in the structure of the genome.

5.1. Insertions

Insertion elements and transposons are segments of DNA capable of insertion in different places in the genome, with which they do not share homology, i.e., their insertion is independent of homologous recombination. Insertion elements may be reduced to the components required for the process of moving from place to place, while transposons also carry extra genes that often permit their identification. Frequently these are genes of resistance to antibiotics. The insertion sequences flanking the insertion elements are required for transposition, which means that in principle any segment of DNA that becomes blanked by these sequences may be converted into a transposable element (Hardman and Gowland, 1985).

One of the striking consequences of transposition is the mutational effect that results from the interruption of the continuity of the gene by the insertion element or transposon. In this case the insertion element blocks the transcription and the gene ceases to be expressed. The silencing effect can be due to the presence of termination signals (Kleckner, 1981), which may also be manifested at some distance from the point of insertion (Stokes and Hall, 1984). In other instances, the transposable element carries a promoter capable of activating inactive genes. This results in an enormous potential for variation and is in part responsible for the versatility of microorganisms (Kleckner, 1981). Transposons are also very effective in transmitting resistance to antibiotics among bacteria (Datta, 1988).

A particularly interesting case has been studied by Professor Tom Lessie and his collaborators. Their experiments point to a possible mechanism by which bacteria as versatile as many of the species of aerobic pseudomonads evolve through the recruitment of foreign genes. For most of the experiments they have used one strain of *Pseudomonas cepacia*. The introduction of foreign genes into this strain by plasmid transfer sometimes results in poor expression because of lack of appropriate promoter. However, upon repeated transfers under selective conditions, expression may improve considerably, and the mechanism underlying the change involves insertion elements that move from the chromosome to the neighborhood of the foreign genes (Lessie et al., 1990; Lessie and Gaffney, 1986; Drlica and Rouviere-

Yaniv, 1987). *P. cepacia* has an outstanding nutritional versatility (Palleroni and Holmes, 1981), and perhaps mechanisms of the type just described may have been responsible for the permanent incorporation by transposition of imported genes into the bacterial genome.

The possibilities for variation through this type of genetic change are enormous and include activation of silent pathways, one of which is particularly interesting. Transfer of the lactose gene (*lacZ*) from *Escherichia coli* to *P. cepacia* resulted in poor expression, but eventually, by the influence of insertion elements carrying the right promoter, the expression improved, and in one case the insertion of one of these elements within *lacZ* allowed growth of the cells on lactulose, which differs from lactose in having in its structure fructose instead of glucose (Lessie and Gaffney, 1986). In *P. cepacia* transposition of insertion elements can use as target both resident and foreign plasmids. In *Streptococcus,* in addition, transposition of Tn*916* can be intercellular through conjugation (Clewell et al., 1988).

Cases of transposition in eukaryotic organisms have been discussed by Kingsman et al. (1988).

5.2. Plasmids and Microbial Versatility

Whether or not they become integrated in the structure of the bacterial genome, plasmids are important reservoirs of genetic material that can expand considerably the versatility of host cells. Plasmids are replicons of smaller size than chromosomes and carry genes that are dispensable, except under certain special conditions. Because of their practical interest, plasmids carrying markers conferring resistance to antibiotics ("R factors") were responsible for initiation of intensive research in these extrachromosomal elements. The antibiotic-resistance markers are now extensively used in research for the purpose of plasmid identification in laboratory experiments.

Plasmids seem to have acquired resistance markers within a very short time, and this is another example of striking microbial versatility. Datta (1985) mentions the observation that of the strains of Enterobacteriaceae of the "preantibiotic" era in Professor Murray's collection, many contained conjugative plasmids, but none had antibiotic-resistance markers. She concluded that "antibiotic-resistance genes are those most evidently acquired and lost by plasmids within our working experience. Selection for drug resistance, especially in hospitals and farms, has been intense, but it still seems almost incredible that resistance genes to so many unrelated antibiotics should have been accumulated in so short a time."

Aside from resistance genes, the range of properties that plasmids can confer to the host cells is very large. They can act as fertility factors or as inhibitors of fertility, and they can carry genes of resistance to various chemical or physical deleterious agents or genes coding for enzymes of degradative pathways. The sizes of plasmids vary within a wide range, and the largest may exceed 10% of the size of the host chromosome. Since they have a low copy number, for all practical purposes they can be considered as small chromosomes. Some plasmids can be transmitted from cell to cell through conjugation, while others can only be transferred with the help of conjugative plasmids or are liberated only after lysis of the host cells and enter other cells by transformation.

As pointed out by Nakatsu et al. (1991), there are two very effective selective pressures in nature: the antibiotics and the organic compounds that pollute the environment (xenobiotics). Surprisingly, the molecular bases for resistance to these agents or for their degradation are usually located in plasmids. Catabolic plasmids represent a valuable reservoir of biochemical potentialities for bacteria and will be mentioned again in a later section. Under nonselective conditions, they usually can survive in a few cells of the population. When the selective agent is present, these few individuals can be crucial for the perpetuation of the population.

5.3. Amplifications and Deletions

The normal modulators of the synthesis of enzymes and other gene products are repressors and inducers, and they normally satisfy the physiological requirements of the cells. But when these face the need to produce a higher concentration of a product, for instance, an enzyme, mutation is not the choice solution. Rather than mutation, the common response is to modify the gene dosage by means of duplications or amplifications. This is observed, for instance, in cases of the development of resistance to the analog of a compound required by the cells for growth. To compete successfully with the incorporation of the analog, more normal compound is needed, a challenge that the cells face by increasing the dosage of the gene(s) involved in its synthesis. Gene amplifications can also be the solutions chosen by revertants of mutants that produce less effective enzyme (Folk and Berg, 1970). Likewise, resistance to antibiotics can increase by duplication of resistance genes (Yagi and Clewell, 1976).

Perhaps the most dramatic case of enzyme amplification ever recorded was that of selection in the laboratory for strains of *E. coli* producing high amounts of β-galactosidase. This was achieved by Novick and Horiuchi (1961) by growing the cells in the chemostat in the presence of limiting concentrations of lactose. A mutation in the regulatory gene to constitutivity was followed by gene amplification, and β-galactosidase was produced at levels representing up to 25% of the total cell protein.

Gene dosage increases by amplification, but this not always results in an increase in the expression of the amplified genes. A familiar example of amplification is that of the ribosomal RNA genes in *E. coli* and, very possibly, in many other bacterial species. Seven transcriptional units are present in the chromosome, which may only represent a safety device against the risk of deletion of such important genes. Deletion of one of the copies has been achieved experimentally and has had no harmful effect (Ellwood and Nomura, 1980). The sets of the genes for the three ribosomal components (in the order 16S, 23S, and 5S) are not adjacent (which reduces the danger of total deletion), the three genes are transcribed together, and the messenger RNA is processed after transcription (Dunn and Studier, 1973).

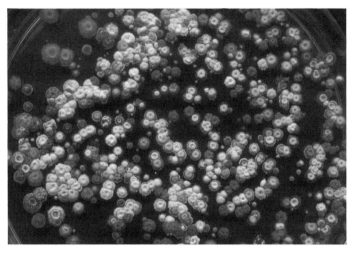

Figure 1.1.   Colonies of *Streptomyces lividans* strain TK20. Colonies of different types breed true for a few transfers, but gradually revert to the heterogeneous appearance of the culture shown here.

Amplification of the genes coding for ribosomal proteins does not result in an increase in gene products. More mRNA is produced, but most of the excess is degraded (Jinks-Robertson and Nomura, 1987).

Microbiologists who have worked with streptomycetes are well aware of the phenotypic variability of cultures of these organisms. Streaking on agar may result in a mixture of colony types (Fig. 1.1), and isolation of single colonies seldom result in lines that "breed true" in their diverse phenotypic characteristics.

Genes coding for enzymes of biosynthetic pathways leading to antibiotic production by antinomycetes frequently are amplified under the selective conditions imposed on these organisms in industrial practice, although in a number of examples mentioned in the literature amplification seems to have taken place for no apparent reason, as in the synthesis of a secondary metabolite in the anthracycline-producing strains of *Streptomyces peuceticus* studied by Stutzman-Engwall and Hutchinson (1989). Amplifications in the streptomycetes can attain very high levels. At the same time, the total genome size tends to remain unchanged, which can only be achieved by a combination of amplifications and deletions. In fact, the notorious instability of *Streptomyces* strains seems to derive from these phenomena, and loss of properties, sometimes attributed to plasmid loss, actually may be due to deletions.

The size of the regions affected by amplification and deletions can reach values higher than those ever reported for other bacteria. Analyses of *Streptomyces glaucescens* have defined a map of the instability region of a size of more than 900 kilobase (kb) pairs, where large genomic rearrangements occur (Ono et al., 1982; Hasegawa et al., 1985; Birch et al., 1990). Deletions can reach more than 800 kb

(Birch et al., 1989), and amplifications have been observed that include tandem repeats of up to 1,500 kb that extend to 30% of the genome (Häusler et al., 1989).

Normally deletions have a negative consequence, that is, they correlate with loss of function, but there are rare instances when the effect is just the opposite. Activation of the *nifD* gene in the chromosome of *Anabaena variabilis* heterocysts depends on the deletion of an 11 kb fragment; otherwise, nitrogen fixation does not occur (Brusca et al., 1989).

Amplifications have also been studied in eukaryotic organisms, and they share some of the characteristics observed in bacteria (Schimke, 1982). Recently, studies on the amplification of a locus implicated in cases of human sarcomas has provoked a great deal of excitement (Lane, 1992).

5.4. Inversions and Translocations

In *Salmonella,* the alternative expression of two genes, H_1 and H_2, coding for different flagellins, is the basis for a phenomenon that has been studied in considerable detail (Simon and Silverman, 1983). A DNA segment undergoes inversion, catalyzed by the product of a recombinase gene. In one of the orientations of the fragment, H_2 and its promoter become close to one another, and the gene is transcribed, as well as a gene whose product represses H_1. In the other orientation, H_1 is expressed instead, but H_2 and the repressor gene are not. Thus, the key component in the mechanism of flagellar variation is the coupling or uncoupling of H_2 with its promoter. Orientation effect is also the basis for the expression of genes coding for the tail fibers of phage Mu, which determine the host specificity (Kamp et al., 1979).

Products of genes involved in inversion have been called *invertases* (although, obviously, they have nothing to do with the enzyme of the same name). Inversions observed in these and other instances may have originated in a similar manner (Enomoto and Stocker, 1975; Enomoto et al., 1983; Szekely and Simon, 1983).

Variations in the antigenic properties of surface components, by which pathogenic bacteria can evade the host defenses, may be achieved by mechanisms other than inversions. In *Neisseria gonorrhoeae* the antigenic variation of the fimbriae by which bacteria attach to host cells involves a complex mechanism that will be briefly outlined here. In the genome there are one or two expression loci (*pilE1* and *pilE2*), depending on the strain, and up to seven silent loci (pilS1 to pilS7), with incomplete genes and no promoters. Recombination between one of the silent genes and *pilE* occurs in a nonreciprocal manner that has been called *gene conversion* by analogy with a nonreciprocal recombination phenomenon that has been observed in fungi. Silent pilin genes can be expressed in the new location, and therefore they represent reserve genetic material to be used for the formation of recombinant pilin molecules (Meyer and Haas, 1988). The net result is an enormous potential to produce variant pili in this species. The interpretation of this phenomenon as a case of gene conversion exclusively has been disputed by Seifert et al. (1988), who found that under certain conditions transformation with DNA liberated from dead cells can be an additional mechanism generating variability of the competent *Neisseria gonorrhoeae* cells.

6. VARIATIONS IN GENE EXPRESSION DUE TO CHANGES IN THE STRUCTURE OF THE GENOME

The DNA in the bacterial nucleoid and plasmids is negatively supercoiled, a conformation for which the enzyme gyrase is responsible (Gellert et al., 1976). Since negative supercoiling (or superhelicity) results from twisting in the direction of unwinding the DNA, one effect of increasing the negative supercoiling should be an enhancement in gene expression, because transcription requires separation of the DNA strands. In contrast to prokaryotes, the DNA of eukaryotic organisms is essentially relaxed in vivo, a difference that is attributed to the proteins that are associated with the DNA in each case. The extent of supercoiling or superhelicity can be estimated in vitro by dye titration (Drlica, 1987), but it is thought that the degree of superhelicity in vivo is probably of a different magnitude (Peck and Wang, 1985).

Environmental factors can modify the degree of supercoiling or superhelicity of the bacterial DNA, and this in turn is reflected in changes in gene expression. Much evidence is growing in support of the hypothesis that torsional stress may be an important element in global regulatory mechanisms (Thompson et al., 1990). Reports describing variations in gene expression refer to the effect of media composition (Balke and Gralla, 1987), temperature (Goldstein and Drlica, 1984), oxygen tension (Dorman et al., 1988), or osmolarity (Higgins et al., 1987, 1988) on the superhelicity of the genome. For instance, an increase in the expression of the gene *proU*, which codes for a transport system with high affinity for the osmoprotectant betaine, can be provoked by growth in a medium of high osmolarity (Cariney et al., 1985). The idea that this may be due to an increase in DNA supercoiling is supported by the fact that inhibitors of DNA gyrase (such as novobiocin and nalidixic acid) substantially reduce the expression of *proU* in *Salmonella typhimurium* (Higgins et al., 1988).

By the same token, variations in intracellular conditions also must have an influence on the genome structure and, consequently, on gene expression, although here the experimentation is more indirect and the results more difficult to interpret. In the *E. coli* composition one basic protein (protein HU) that resembles the histones of eukaryotes is very abundant (Drlica and Rouviere-Yaniv, 1987), and its interaction with DNA may cause looping and supercoiling, which in turn may have an effect on gene expression (Broyles and Pettijohn, 1986). Another endogenous factor that locally modifies the degree of DNA supercoiling is transcription. An activation of transcription increases the level of negative superhelicity, mainly because of differential swiveling on opposite sites of the involved region (Figueroa and Bossi, 1988). As mentioned before, strand separation is an essential step in transcription and replication, but the natural negative supercoiling of DNA may not be indispensable for these processes to occur normally (Drlica, 1984), and transcription of different genes is affected differently by variations in the degree of superhelicity (Brahms et al., 1985; Borowiec and Gralla, 1987).

Information can be transmitted at considerable distances along the DNA molecule (Schleif, 1988). This "action at a distance" is facilitated by the formation of loops that bring the two components within close proximity. Protein molecules with affinity for the two sites are involved (Wang and Giaever, 1988). Helicases, important enzymes in the process of replication that bind to DNA and search for particular sequences that they are able to cleave, are an example (Baker et al., 1986, 1987).

High quality electron micrographs of DNA loops have been published (Krämer et al., 1987). According to Schleif (1988), looping is quite a rational mechanism to regulate gene activity, since, among other things, it avoids overcrowding the site at which various factors are needed for interaction with RNA polymerase during transcription.

7. OTHER RESPONSES TO ENVIRONMENTAL FACTORS

Changes in environmental factors can have other definite effects on gene expression, particularly those known to subject the cells to stressful conditions. Changes in temperature, presence of heavy metals or hydrogen peroxide, or a high pH induce in the cells the production of proteins that have been globally named *heat shock* proteins (Neidhardt and VanBogelen, 1987; VanBogelen et al., 1987; Taglicht et al., 1987; Morgan et al., 1986). The response may also originate in changes of DNA superhelicity (Travers and Mace, 1982), but in most cases there is no evidence in support of this hypothesis.

In recent years considerable progress has been made in genetic studies of the reaction of bacteria to stress and in particular in the induction of the so-called SOS functions. This complex response is expressed in reactions directed to the repair of the damage done by radiation or by some chemicals to DNA, but it also has some consequences that are not so desirable, such as enhanced mutagenesis, the induction of resident prophages to the lytic phase, and the inhibition of cell division. Several processes are induced in a coordinated manner in the SOS response, involving the regulation of transcription of 20 or more unlinked genes that are part of the SOS regulon. The SOS functions have been aptly summarized by Thliveris et al. (1990). The chain of reactions starts by activation of the protein RecA by single-stranded DNA resulting from the damage. The protein, bound to DNA near the damaged site, interacts with a repressor (the LexA protein), destroying its capacity for dimer formation, an indispensable condition for its repressor action. The alteration of LexA to an inactive form causes transcription of the SOS genes.

The case of amplification of eukaryotic genes previously mentioned (Lane, 1992) has features reminiscent of the prokaryotic SOS response.

8. CHANGES IN GENE EXPRESSION
DURING DEVELOPMENTAL CYCLES

The regulation of gene expression in the complex developmental cycles of some bacteria deserve mention here as examples of the enormous diversity in this group of organisms. Particular cases include the cycles observed in *Bacillus, Caulobacter,* the myxobacteria, and the actinomycetes, where sets of genes can be expressed sequentially, in a sort of "cascade" effect, as an integral part of developmental programs. The genetic bases of this most interesting aspect of microbial diversity have been covered for several important bacterial groups in the book edited by Hopwood and Chater (1989).

Some stress responses lead to differentiation, sporulation being a typical example (Smith et al., 1990). Streptomycetes and bacilli are two types of organisms that respond to nutrient exhaustion by undergoing sporulation. The spores produced by the two groups are different in the process of their formation and in fundamental morphological and physiological characteristics (in fact, *spore* may be one of the most abused terms in microbiology!), but both represent forms of resistance to unfavorable conditions.

Formation of spores by bacilli has been the subject of intensive research for many years. Approximately 50 genes are involved in the control of sporulation in *Bacillus subtilis* (Losick et al., 1986; Errington, 1988). Stages in the process include separation of the region to be occupied by the spore by formation of a septum, growth of a membrane surrounding the forespore, and formation of a cortex that will be part of the spore coat. The bacillar endospore, finally released from the mother cell, is an example of the most resistant structure produced by any organism. The complex events that occur at the initial stage of sporulation involve the activation of genes triggered by a sensor sensitive to nutrient stress. These events are clearly described by Smith et al. 1989, 1990).

9. A CASE HISTORY: THE AEROBIC PSEUDOMONADS
AS AN EXAMPLE OF NUTRITIONAL VERSATILITY

Pseudomonas species became prominent examples of nutritional versatility at the beginning of this century through the work of den Dooren de Jong (1926). The well-deserved reputation of these gram-negative organisms that emerged from this fundamental piece of research stood the test of time, and members of the genus are still counted among the bacterial species capable of living in very simple media at the expense of the largest variety of organic compounds (Stanier et al., 1966). Variations in the nutritional spectra of different species, as predicted by den Dooren de Jong, have important taxonomic implications and can be used for species differentiation and identification (Palleroni and Doudoroff, 1972; Palleroni, 1984).

Up to the 1960s, the biochemical literature was enriched with the description of many catabolic pathways performed by aerobic pseudomonads isolated from enrichment cultures using many different carbon sources of low molecular weight.

The genus *Pseudomonas* is now restricted to the species of Group I as defined by the ribosomal RNA–DNA reassociation experiments of Palleroni et al. (1973), but still some of its members are among the most versatile of gram-negative bacteria. However, the most remarkable example is placed in Group II, which is phylogenetically distant from "true" *Pseudomonas*. This is the case of *"Pseudomonas" cepacia* (Palleroni and Holmes, 19981; Palleroni, 1984). Strains of this species have the capacity to live at the expense of more than 100 different carbon sources, some of which may be degraded by the same strain using more than one catabolic pathway. We have already mentioned some of the experiments performed on this species in Professor Lessie's laboratory.

A substantial amount of information is available today on pathways of degradation of aromatic compounds by *Pseudomonas* species, particularly *P. putida* and *P. aeruginosa* (Clarke and Richmond, 1975; Galli et al., 1992). Many enrichments under aerobic conditions using these compounds lead to the isolation of pseudomonads. In a number of instances, the genes coding for enzymes of these catabolic pathways are carried by catabolic plasmids, of which those responsible for the degradation of camphor, toluene, naphthalene, or octanol are well known (Boronin, 1992). Of these, plasmid TOL (with genes for the degradation of toluene and derivatives) has been the subject of a great deal of genetic and biochemical research (Assinder and Williams, 1990).

Pseudomonas species are a rich source of different types of plasmids, of which a substantial proportion carry genes conferring resistance to antibiotics. Much of the research in the genus has been focused on the type species *P. aeruginosa,* a well-known opportunistic human pathogen that can be readily isolated from soil, water, a number of other natural sources, and clinical specimens. Plasmids carrying multiple drug resistance markers are often found in strains of the species and are cause of much medical concern (Doggett, 1979).

The species has been the subject of intensive genetic research, and approximately 250 markers have been located in the genetic map of one strain, PAO (Holloway et al., 1990). A physical map of this strain has been constructed by Roemling et al. (1989).

Strains of other species are being introduced to genetic research. For one of them, strain PPN of *P. putida,* about 70 markers have been mapped. In this case, as in *P. aeruginosa* PAO, a combination of conjugation and transduction has been used, and a comparison of gene arrangement has been possible. The analysis has shown that, contrary to the case of the enterics, genes of biosynthetic patways do not tend to be contiguous. On the other hand, the genes of catabolic pathways generally are clustered. Differences in gene arrangement in the chromosomal maps of *P. aeruginosa* and *P. putida* can be explained by processes such as inversions and translocations, which may have occurred since the two species became differentiated in evolution from a common ancestor (Holloway et al., 1990). Obviously, as indicated in another section, Lessie's hypothesis on recruitment of foreign genes in *P. cepacia* could have been operative in these two fluorescent species during evolution. Their chromosomes may have increased in size mainly by incorporation of catabolic genes recruited from other sources.

The extensive nutritional characterization of many *Pseudomonas* species (Stanier et al., 1966) has stimulated a great deal of biochemical research, and the internal subdivision of the genus into five groups often has been taken as a useful frame of reference for comparative studies. The volume of work performed in many laboratories is enormous, and no attempt will be made to summarize it here. However, work performed by Professor P.H. Clarke and her collaborators on *P. aeruginosa* aliphatic acetamidase deserves some comments, since it represents a clear demonstration of the versatility of bacteria to change their biochemical attributes when subjected to the strong selective pressures that can be created under laboratory conditions. The amidase of wild-type *P. aeruginosa* hydrolyzes several aliphatic amides, of which acetamide is the best inducer. Mutants having amidases with altered substrate specificities, regulatory properties, and thermal stabilities could be obtained, and families of enzymes with novel properties could be constructed by means of sequential mutations. Some mutants were able to hydrolyze the aromatic phenylacetamide; in other words, by a process of evolution in the laboratory, acetamidase could be changed to phenylacetamidase, and the acetamidase activity of the original enzyme was largely lost.

Many other interesting aspects of the evolution of structural and functional properties of *Pseudomonas* enzymes have been ably reviewed by Clarke and Slater (1986).

10. EPILOGUE

The examples discussed in the foregoing sections have been chosen out of a multitude of mechanisms developed by microbes to face a variety of chemical and physical challenges in their natural habitats. This remarkable versatility is not surprising in organisms that, as far as we can guess, are direct descendants of the oldest living systems that appeared in the planet about three and a half billion years ago, during which time they had to face many environmental changes.

Some of the responses of microorganisms to factors in the environment where they live occur in the majority of the cells in the population through the agency of inducible systems, which are economical solutions to variable conditions since they can be made fully operative only when needed. Other changes involve a minority of members of the microbial population and allow them to multiply under selective conditions. Several examples of reversible alterations in the genome structure by environmental factors have been mentioned in the preceding sections, and it is not too far fetched to think that some natural or synthetic organic compounds, as well as other changes in ecological conditions, may also participate in provoking changes in the genome structure and composition resulting in variations in gene expression. This is an interesting avenue of research in microbial ecology and its environmental implications.

The remarkable versatility of microorganisms justifies our reliance on their seemingly inexhaustible capacity for coping with unusual conditions. Various aspects of the principles underlying the methods and applications of microbial ecology have been recently reviewed in a book edited by Levin et al. (1992). The natural capabilities of the indigenous microflora can be favored by proper management of the natural habitats, in addition to which a substantial amount of effort is devoted today to the construction of genetically engineered microorganisms (GEMs) with desirable characteristics for the solution of specific environmental problems. Progress has been substantial, but much more work will be required to ensure success in practical applications. Even though research on the development of vectors for the expression under field conditions is very active, it must be admitted that so far "procedures to ensure the ecological predictability of GEMs remain in their infancy" (De Lorenzo and Timmis, 1992).

Before ending this brief account of microbial versatility, I would like to evaluate from a new perspective some of the consequences of the remarkable flexibility of bacteria to face, evade, resist, or modify adverse conditions. It is my impression that, far from being pushed to the verge of extinction, prokaryotes will not only survive, but in addition, their diversity *may actually* increase due to their capacity to generate novel systems to counteract unfavorable environmental conditions. Induction of enzymatic systems, creation of new pathways of mutation and selection, and horizontal transfer of catabolic genes among unrelated species are all factors that tend to increase microbial diversity.

The impact of human activities on the environment range from mild to catastrophic, but the consequences are usually measured in terms of the damage inflicted upon plants and animals. In his marvelous account of life's diversity, Wilson (1992) comments on the factors that have forced populations of freshwater fishes into decline. These are 1) destruction of physical habitat, 2) displacement by introduced species, 3) alterations of habitats by chemical pollutants, 4) hybridization with other species and subspecies, and 5) overharvesting. These factors affect other higher forms of life, but in the case of microorganisms the effect of chemical pollutants probably is the most important.

Since microbes are invisible to the naked eye, and our inventory of microbial species is far from complete, it seems senseless to argue here about "species extinction" or to suggest inclusion of some of them in an "endangered species" list. Moreover, microbial species are widely dispersed, and there are many sources from which the same species can repopulate mistreated areas once conditions have returned to normal. On the other hand, in response to strong selective pressures, prokaryotes quickly "learn" to tolerate and even counteract the impact of many abnormal conditions in the medium. Having survived after a long history of changes in their natural habitats, these organisms are best endowed to face the challenge of deteriorated chemical conditions in the environment.

REFERENCES

Assinder SJ, Williams PA (1990): The TOL plasmids: Determinants of the catabolism of toluene and the xylenes. Adv Microb Physiol 31:1–69.

Baker TA, Funnell BE, Kornberg A (1987): Helicase action of DnaB protein during replication from the *Escherichia coli* chromosomal origin in vitro. J Biol Chem 262:6877–6885.

Baker TA, Sekimizu K, Funnell BE, Kornberg A (1986): Extensive unwinding of the plasmid template during staged enzymatic initiation of DNA replication from the origin of the *Escherichia coli* chromosome. Cell 45:53–64.

Balke VL, Gralla JD (1987): Changes in linking number of supercoiled DNA accompany growth transitions in *Escherichia coli*. J Bacteriol 169:4499–4506.

Birch A, Häusler A, Hütter R (1990): Genome rearrangement and genetic instability in *Streptomyces* spp. J Bacteriol 172:4138–4142.

Birch A, Häusler A, Vögtli M, Krek W, Hütter R (1989): Extremely large chromosomal deletions are intimately involved in genetic instability and genomic rearrangements in *Streptomyces glaucescens*. Mol Gen Genet 217:447–458.

Boronin AM (1992): Diversity and relationships of *Pseudomonas* plasmids. In Galli E, Silver S, Witholt B (eds): *Pseudomonas*. Molecular Biology and Biotechnology. Washington, DC: American Society for Microbiology, pp 329–340.

Borowiec JA, Gralla JD (1987): All three elements of the *lac* p^s promoter mediate its transcriptional response to DNA supercoiling. J Mol Biol 195:89–97.

Borst P, Greaves DR (1987): Programmed gene rearrangements altering gene expression. Science 235:658–667.

Brahms JG, Dargouge O, Brahms S, Ohara Y, Vagner V (1985): Activation and inhibition of transcription by supercoiling. J Mol Biol 181:455–465.

Broyles S, Pettijohn D (1986): Interaction of the *Escherichia coli* HU protein with DNA. J Mol Biol 187:47–60.

Brusca JS, Hale MA, Carrasco CD, Golden JW (1989): Excision of an 11-kilobase-pair DNA element from within *nifD* gene in *Anabaena variabilis* heterocysts. J Bacteriol 171:4138–4145.

Bryson V, Szybalski W (1952): Microbial selection. Science 116:45–51.

Cairns J, Overbaugh J, Miller S (1988): The origin of mutants. Nature 335:142–145.

Cariney J, Booth IR, Higgins CF (1985): Osmoregulation of gene expression in *Salmonella typhimurium: proU* encodes an osmotically induced betaine transport system. J Bacteriol 164:1224–1232.

Clarke PH, Richmond MH (eds) (1975): Genetics and Biochemistry of *Pseudomonas*. London: John Wiley & Sons, pp 1–366.

Clarke PH, Slater JH (1986): Evolution of enzyme structure and function in *Pseudomonas*. In Sokatch JR (ed): The Bacteria. A Treatise on Structure and Function, vol X. Orlando: Academic Press, pp 71–144.

Clewell DB, Senghas E, Jones JM, Flannagan SE, Yamamoto M, Gawron-Burke C (1988): Transposition in *Streptococcus:* Structural and genetic properties of the conjugative transposon Tn916. In Kingsman AJ, Chater KF, Kingsman SM (eds): Transposition. Cambridge, England: Society for General Microbiology, Cambridge University Press, pp 43–58.

Datta N (1985): Plasmids as organisms. In Helinski DR, Cohen SN, Clewell DB, Jackson DA, Hollaender A (eds): Plasmids in Bacteria. New York: Plenum Press, pp 3–16.

Datta N (1988): Introduction. In Kingsman AJ, Chater KF, Kingsman SM (eds): Transposition. Cambridge, England: Cambridge University Press, pp 1–3.

De Lorenzo V, Timmis KN (1992): Specialized host–vector systems for the engineering of *Pseudomonas* strains destined for environmental release. In Galli E, Silver S, Witholt B (eds): *Pseudomonas*. Molecular Biology and Biotechnology. Washington, DC; American Society for Microbiology, pp 415–428.

Demerec M (1945): Production of *Staphylococcus* strains resistant to various concentrations of penicillin. Proc Natl Acad Sci USA 31:16–24.

Demerec M (1948): Origin of bacterial resistance to antibiotics. J Bacteriol 56:63–74.

den Dooren de Jong LE (1926): Bijdrage tot de kennis van het mineralisatieprocess. Rotterdam: Nijgh & Van Ditmar, pp 1–200.

Doggett RG (ed) (1979): *Pseudomonas aeruginosa*. Clinical Manifestations of Infection and Current Therapy. New York: Academic Press, pp 1–504.

Dorman CJ, Barr GC, NiBhriain N, Higgins CF (1988): DNA supercoiling and the anaerobic and growth phase regulation of *tonB* gene expression. J Bacteriol 170:2816–2826.

Drlica K (1984): Biology of bacterial deoxyribonucleic acid topoisomerases. Microbiol Rev 48:273–289.

Drlica K (1987): The nucleoid. In Neidhardt FC, Ingraham JL, Low KB, Magasanik B, Schaechter M, Umbarger HE (eds): *Escherichia coli* and *Salmonella typhimurium*. Cellular and Molecular Biology. Washington, DC: American Society for Microbiology, pp 91–103.

Drlica K, Rouviere-Yaniv J (1987): Histone-like proteins of bacteria. Microbiol Rev 51: 301–319.

Dunn JJ, Studier FW (1973): T7 early RNAs and *Escherichia coli* ribosomal RNAs are cut from large precursor RNAs in vivo by ribonuclease III. Proc Natl Acad Sci USA 70:3296–3300.

Ellwood M, Nomura M (1980): Deletion of a ribosomal ribonucleic acid operon in *Escherichia coli*. J Bacteriol 143:1077–1080.

Enomoto M, Oosawa H, Momotu H (1983): Mapping of the *pin* locus for a site-specific recombinase that causes flagellar phase variation in *E. coli* K-12. J Bacteriol 156: 663–668.

Enomoto M, Stocker BAD (1975): Integration at *hag* or elsewhere of *H2* genes transduced from *Salmonella* to *Escherichia coli*. Genetics 81:595–614.

Errington J (1988): Regulation of sporulation. Nature 333:399–400.

Figueroa N, Bossi L (1988): Transcription induces gyration of the DNA template in *Escherichia coli*. Proc Natl Acad Sci USA 85:9416–9420.

Folk WR, Berg P (1970): Isolation and partial characterization of *Escherichia coli* mutants with altered glycyl transfer ribonucleic acid synthetases. J Bacteriol 102:193–203.

Galli E, Silver S, Witholt B (eds) (1992): *Pseudomonas*. Molecular Biology and Biotechnology. Washington, DC: American Society for Microbiology, pp 1–443.

Gellert M, Mizuuchi K, O'Dea MH, Nash H (1976): DNA gyrase: An enzyme that introduces superhelical turns into DNA. Proc Natl Acad Sci USA 73:3872–3876.

Goldstein E, Drlica K (1984): Regulation of bacterial DNA supercoiling: Plasmid linking numbers vary with growth temperature. Proc Natl Acad Sci USA 84:4046–4050.

Hardman DJ, Gowland PC (1985): Large plasmids in bacteria. Part 2. Genetics and evolution. Microbiol Soc 2:184–190.

Hasegawa M, Hintermann G, Simonet JM, Crameri R, Piret J, Hütter R (1985): Certain chromosomal regions in *Streptomyces glaucescens* tend to carry amplifications and deletions. Mol Gen Genet 200:375–384.

Häusler A, Birch A, Krek W, Piret J, Hütter R (1989): Heterogeneous genomic amplification in *Streptomyces glaucescens:* Structure, location and junction sequence analysis. Mol Gen Genet 217:437–446.

Higgins CF, Cairney J, Stirling DA, Sutherland L, Booth IR (1987): Osmotic regulation of gene expression: Ionic strength as an intracellular signal? Trends Biochem Sci 12: 339–344.

Higgins CF, Dorman CJ, Stirling DA, Waddell L, Booth IR, May G, Bremer E (1988): A physiological role for DNA supercoiling in the osmotic regulation of gene expression in *S. typhimurium* and *E. coli.* Cell 52:569–584.

Hinshelwood CN (1946): The Chemical Kinetics of the Bacterial Cell. London: Oxford University Press (Clarendon).

Holloway BW, Dharsmsthiti S, Krishnapillai V, Morgan A, Obeyesekere V, Ratnaningsih E, Sinclair M, Strom D, Zhang C (1990): Patterns of gene linkages in *Pseudomonas* species. In Drlica K, Riley M (eds): The Bacterial Chromosome. Washington, DC: American Society for Microbiology, pp 97–105.

Holloway BW, Roemling U, Tuemmler B (1994):Genomic mapping of *Pseudomonas aeruginosa.* J Gen Microbiol 140:2907–3189.

Hopwood DA, Chater KE (eds) (1989): Genetics of Bacterial Diversity. London: Academic Press.

Hsie J-Y, Bryson V (1952): Genetic studies on the development of resistance to neomycin and dihydrostreptomycin in *Mycobacterium ranae.* Am Rev Tuberc 62:286–299.

Jinks-Robertson S, Nomura M (1987): Ribosomes and tRNA. In Neidhardt FC, Ingraham JL, Low KB, Magasanik B, Schaechter M, Umbarger HE (eds): *Escherichia coli* and *Salmonella typhimurium.* Cellular and Molecular Biology. Washington, DC: American Society for Microbiology, pp 1358–1385.

Kamp AF, La Rivière JWM, Verhoeven W (eds) (1959): Albert Jan Kluyver. His Life and Work. Amsterdam: North-Holland, pp 1–567.

Kamp D, Chow LT, Broker TR, Kwoh D, Zipser D, Kahman R (1979): Site-specific recombination in phage Mu. Cold Spring Harbor Symp Quant Biol 13:1159–1167.

Kingsman AJ, Chater KF, Kingsman SM (eds) (1988): Transposition. Cambridge, England: Cambridge University Press, pp 1–375.

Kleckner N (1981): Transposable elements in prokaryotes. Annu Rev Genet 15:341–404.

Kluyver AJ, van Niel CB (1956): The Microbe's Contribution to Biology. Cambridge, MA: Harvard University Press, pp 1–182.

Krämer H, Niemöller M, Armouyal M, Revet B, v Wilken-Bergmann B, Müller-Hill B (1987): *lac* repressor forms loops with linear DNA carrying two suitably spaced *lac* operators. EMBO J 6:1481–1491.

Lane DP (1992): p53, guardian of the genome. Nature 358:15–16.

Large EC (1940): The Advance of Fungi. New York: Henry Holt, pp 1–488.

Lederberg J, Tatum E (1946): Gene recombination in *E. coli.* Nature 158:558.

Lessie TG, Gaffney T (1986): Catabolic potential of *Pseudomonas cepacia.* In Sokatch JR (ed): The Bacteria, vol X, The Biology of *Pseudomonas*. Orlando: Academic Press, pp 439–481.

Lessie TG, Wood MS, Byrne A, Ferrante A (1990): Transposable gene-activating elements in *Pseudomonas cepacia*. In Silver S, Chakrabarty AM, Iglewski B, Kaplan S (eds): *Pseudomonas: Biotransformation, Pathogenesis and Evolving Biotechnology*. Washington, DC: American Society for Microbiology, pp 279–291.

Levin MA, Seidler RJ, Rogul M (eds) (1992): Microbial Ecology. Principles, Methods, and Applications. New York: McGraw-Hill, pp 1–945.

Losick R, Youngman P, Piggot PJ (1986): Genetics of endospore formation in *Bacillus subtilis*. Annu Rev Genet 20:625–669.

Luria SE, Delbrück M (1943): Mutations of bacteria from virus sensitivity to virus resistance. Genetics 28:491–511.

Meyer TF, Haas R (1988): Phase and antigenic variation by DNA rearrangements in procaryotes. In Kingsman AJ, Chater KF, Kingsman SM (eds): Transposition. Cambridge, England: Cambridge University Press, pp 192–219.

Morgan RW, Christman MF, Jacobsen FS, Storz G, Ames BN (1986): Hydrogen peroxide-inducible proteins in *Salmonella typhimurium* overlap with heat shock and other stress proteins. Proc Natl Acad Sci USA 83:8059–8063.

Nakatsu C, Ng J, Singh R, Straus N, Wyndham C (1991): Chlorobenzoate catabolic transposon Tn*5271* is a composite class I element with flanking class II insertion sequences. Proc Natl Acad Sci USA 88:8312–8316.

Neidhardt FC, VanBogelen RA (1987): Heat shock response. In Neidhardt FC, Ingraham JL, Low KB, Magasanik B, Schaechter M, Umbarger HE (eds): *Escherichia coli* and *Salmonella typhimurium:* Cellular and Molecular Biology. Washington, DC: American Society for Microbiology, pp 1334–1345.

Novick A, Horiuchi T (1961): Hyper-production of β-galactosidase by *Escherichia coli* bacteria. Cold Spring Harbor Symp Quant Biol: 239–245.

Ono H, Hintermann G, Crameri R, Wallis G, Hütter R (1982): Reiterated DNA sequences in a mutant strain of *Streptomyces glaucescens* and cloning of the sequence in *Escherichia coli*. Mol Gen Genet 186:106–110.

Palleroni NJ (1984): Genus I. *Pseudomonas*. In Krieg NR, Holt JG (eds): Bergey's Manual of Systematic Bacteriology. Baltimore: Williams & Wilkins, pp 1441–199.

Palleroni NJ, Doudoroff M (1972): Some properties and subdivisions of the genus *Pseudomonas*. Annu Rev Phytopathol 10:73–100.

Palleroni NJ, Holmes B (1981): *Pseudomonas cepacia* sp. nov. nom. rev. Int J Syst Bacteriol 31:479–481.

Palleroni NJ, Kunisawa R, Contopoulou R, Doudoroff M (1973): Nucleic acid homologies in the genus *Pseudomonas*. Int J Syst Bacteriol 23:333–339.

Peck LJ, Wang JC (1985): Transcriptional block caused by a negative supercoiling induced structural change in an alternating CG sequence. Cell 40:129–137.

Roemling U, Grothues D, Bautsch W, Tuemmler B (1989): A physical genome map of *Pseudomonas aeruginosa* PAO. EMBO J 8:4081–4089.

Roemling U, Tuemmler B (1991): The impact of two-dimensional pulsed-field gel electrophoresis techniques for the consistent and complete mapping of bacterial genomes: Refined physical of *Pseudomonas aeruginosa* PAO. Nucleic Acids Res 19:3199–3206.

Rubin H (1991): Hinshelwood in the Pantheon? Nature 351:600.

Schimke RT (1982): Summary. In Schimke RT (ed): Gene Amplification. Cold Spring Harbor, NY: Cold Spring Harbor Laboratory, pp 317–333.

Schleif R (1988): DNA looping. Science 240:127–128.

Seifert HS, Ajioka RS, Marchal C, Sparling PF, So M (1988): DNA transformation leads to pilin antigenic variation in *Neisseria gonorrhoeae*. Nature 336:392–395.

Simon MI, Silverman M (1983): Recombinational regulation of gene expression in bacteria. In Beckwith J, Davies J, Gallant JA (eds): Gene Function in Prokaryotes. Cold Spring Harbor, NY: Cold Spring Harbor Laboratory, pp 211–227.

Smith I, Dubnau E, Gaur N, Lewandoski M, Weir J, Cabane K, Nair G (1990): Sporulation: A comprehensive stress response which leads to differentiation. In Drlica K, Riley M (eds): The Bacterial Chromosome. Washington, DC: American Society for Microbiology, pp 389–403.

Smith I, Slepecky RA, Setlow P (eds) (1989): Regulation of Procaryotic Development. Washington, DC: American Society for Microbiology.

Stahl FW (1988): A unicorn in the garden. Nature 335:112–113.

Stanier RY, Palleroni NJ, Doudoroff M (1966): The aerobic pseudomonads: A taxonomic study. J Gen Microbiol 43:159–271.

Stokes HW, Hall BG (1984): Topological repression of gene activity by a transposable element. Proc Natl Acad Sci USA 81:6115–6119.

Stutzman-Engwall KJ, Hutchinson CR (1989): Multigene families for anthracycline antibiotic production in *Streptomyces peuceticus*. Proc Natl Acad Sci USA 86:3135–3139.

Szekely L, Simon MI (1983): The DNA sequence adjacent to flagellar genes and the evolution of flagellar phase variation. J Bacteriol 155:74–81.

Szybalski W, Bryson V (1952): Bacterial resistance studies with derivatives of isonicotinic acid. Am Rev Tuberc 65:768–770.

Taglicht D, Padan E, Oppenheim AB, Schuldiner S (1987): An alkaline shift induces heat shock response in *Escherichia coli*. J Bacteriol 169:885–887.

Thliveris AT, Ennis DG, Lewis LK, Mount DW (1990): SOS functions. In Drlica K, Riley M (eds): The Bacterial Chromosome. Washington, DC: American Society for Microbiology, pp 381–387.

Thompson RJ, Davies JP, Lin G, Mosig G (1990): Modulation of transcription by altered torsional stress, upstream silencers, and DNA-binding proteins. In Drlica K, Riley M (eds): The Bacterial Chromosome. Washington, DC: American Society for Microbiology, pp 227–240.

Travers AA, Mace HAF (1982): The heat shock phenomenon in bacteria—a protection against DNA relaxation? In Schlessinger M, Ashburner M, Tissieres A (eds): Heat Shock from Bacteria to Man. Cold Spring Harbor, NY: Cold Spring Harbor Laboratory, pp 127–130.

VanBogelen RA, Kelley PM, Neidhardt FC (1987): Differential induction of heat shock, SOS, and oxidation stress regulons and accumulation of nucleotides in *Escherichia coli*. J Bacteriol 169:26–32.

Wang JC, Giaever GN (1988): Action at a distance along a DNA. Science 240:300–304.

Watson JD (1991): Salvador E. Luria (1912–1991). Nature 350:113.

Wilson EO (1992): The Diversity of Life. Cambridge, MA: Harvard University Press, pp 1–424.

Woof JB, Hinshelwood CN (1960): Chloramphenicol resistance of *Bacterium lactis aero-genes (Aerobacter aerogenes)*. I. Adaptive and lethal processes in liquid media and on agar plates. Proc R Soc Lond B Biol Sci 153:321.

Yagi Y, Clewell DB (1976): Plasmid-determined tetracycline resistance in *Streptococcus faecalis:* Tandemly repeated resistance determinants in amplified forms of pAM α DNA. J Mol Biol 102:583–600.

Yancopoulos G, Alt FW (1986): Regulation of the assembly and expression of variable-region genes. Annu Rev Immunol 4:339–368.

Zhang Y, Heym B, Allen B, Young D, Cole S (1992): The catalase–peroxidase gene and isoniazid resistance of *Mycobacterium tuberculosis*. Nature 358:591–593.

2

CHEMICAL CONTAMINATION OF THE ENVIRONMENT: SOURCES, TYPES, AND FATE OF SYNTHETIC ORGANIC CHEMICALS

NORBERT G. SWOBODA-COLBERG

Department of Geology and Geophysics, University of Wyoming, Laramie, Wyoming 82071

1. SOURCES OF ORGANIC CHEMICALS

1.1. Introduction

The number, diversity, and complexity of organic chemicals being produced at present are overwhelming. Organic chemicals are ubiquitous and affect every aspect of modern life. Over 100,000 chemicals are produced commercially, yet good information on their environmental fate or impact on human health exists for only a small fraction of these—perhaps 100 compounds (Englande and Guarino, 1992). This chapter presents an overview of major sources of synthetic organic chemicals. It is cursory in some respects, with certain groups of highly specialized chemicals treated only very summarily (e.g., drugs, polymers and plastics, chemicals used in the electronics industry). I have attempted to choose compounds according to one or more of the following selection criteria: the chemical is either produced or released in large quantities, and/or is of widespread use, and/or is of health or environmental concern.

Microbial Transformation and Degradation of Toxic Organic Chemicals, pages 27–74
© *1995 Wiley-Liss, Inc.*

A major goal of this chapter is to survey the vast multitude of compound classes represented and the enormous number of functional groups and their possible combinations and arrangements. While the complete list of organic compounds in use today would fill several volumes by itself, an attempt was made to include all of the major types of compounds without any special regard to their chemical stability or biodegradability.

The selection of representative compounds was especially difficult for the pesticides. As of 1986, 561 products were in use (Parry, 1989). The last 10 years have seen an explosion of new products, and the total number of registered products in the United States is 1,200, although only 850 are currently in production (Ware, 1989). Many older pesticides were banned from use in the United States by the EPA in the 1970s and 1980s, but, since some of them were once widely used, they are still of environmental concern. The pesticides included in Table 2.1 are either "interesting" structural representatives or pesticides that are scrutinized by EPA regulations concerning air, soil, and water pollution.

Some groups of organic compounds that are used "in bulk" by a wide variety of industries are discussed under the topic of "petrochemicals." In the following sections, contributions from specific industries are discussed in more detail.

1.2. Petrochemical Industry

The petrochemical industry is the source of almost all of the chemical production as well as the largest producer by volume. The petrochemical industry includes the oil/gas industry, refineries, and the production of basic chemicals such as vinyl-chloride (s1)[1] and benzene (s2).

vinylchloride
(s1)

benzene
(s2)

The most immediately obvious output of the petrochemical industry is refined petroleum products such as gasoline. In addition to the saturated aliphatic components that make up approximately 70% of average gasoline, there are a number of aromatic hydrocarbons that account for the remaining 30%. In order of abundance (approximate percentage), they are xylenes (10%), toluene (4.7%), 1,2,4-trimethylbenzene (3.3%), 1-methyl-3-ethylbenzene (2.7%), benzene (2.3%), ethylbenzene (1.6%), C_{11}-alkylbenzenes (1.6%), 1,3,5-trimethylbenzene (1.2%) (and others) (Mehlman, 1992), and methyl *tert*-butyl ether (Hartley and Englande, 1992).

---

[1] Chemical structures are given for all but the very simple compounds. Structures are numbered for easy reference, and numbers s1–s90 refer to structures in text. Chemical structures a1–a116 refer to structures given in Appendix 1.

TABLE 2.1.   List of Synthetic Organic Chemicals in the Environment[a]

Compound (Other Names or Descriptions in Parentheses)	Chemical Structure[b]	Source or Application	Reference
Low-molecular-weight, Nonaromatic compounds			
Methane and related compounds			
Methylene chloride (CH_2Cl_2)[c,d]		Fumigant	Metcalf (1971)
Carbon tetrachloride (CCl_4)[c–e]		Fumigant	Metcalf (1971)
Methylbromide (CH_3Br)[c]		Fumigant, nematicide	Ware (1989)
CFC-11, Freon-11 (CCl_3F)	s28	Solvent	
Chloropicrin (CCl_3NO_2)		Soil fumigant, nematicide	Ware (1989)
Ethane and related compounds			
1,2-Dibromoethane[c,d]		Nematicide	Metcalf (1971)
Chlorothene (1,1,1-trichloroethane)[c,d]	s24	Solvent, fumigant	Ware (1989)
1,1,2-Trichloroethane[c,d]			
1,1,2,2-Tetrachloroethane[c,d]			
Hexachloroethane[d]			
CFC-113 ($CCl_2F-CClF_2$)	s29		
CH_3-CCl_2F (HCFC)	s30		
1,1-Dichloro-1-nitroethane		Fumigant	Metcalf (1971)
Ethylene and related compounds			
Ethylene	s8	Bulk chemical	Economic Trends (1993)
1,2-Dichloroethylene[c,d]		Fumigant	Ware (1989)
1-Bromo-2-chroroethylene		Soil fumigant	Metcalf (1971)
1,2-Dibromoethylene[f]		Soil fumigant	Ware (1989)
Trichloroethylene[c,d]	s22	Fumigant, solvent	Metcalf (1971)
Tetrachloroethylene[c,d]	s23	Bulk chemical	Castaldi and Ford (1992)
Propane, butane, and related compounds			
Propylene	s9	Bulk chemical	Economic Trends (1993)
Butadiene		Bulk chemical	
1,2-Dichloropropane[c,d]		Soil fumigant, nematicide	Metcalf (1971)
1,2-Dibromo-3-chloropropane[c]		Soil fumigant, nematicide	Metcalf (1971)
Vinylchloride[c,d]	s1	Bulk chemical	Economic Trends (1993)
Propargyl bromide		Soil fumigant, nematicide	Metcalf (1971)
Isoprene (2-methyl-1,3-butadiene)[f]			

(*continued*)

TABLE 2.1. *(Continued)*

Compound (Other Names or Descriptions in Parentheses)	Chemical Structure[b]	Source or Application	Reference
Telone II® (1,3-dichloropropene)		Soil fumigant, nematicide	Ware (1989)
Alcohols			
Methanol		Bulk chemical	
2-Chloroethanol[e]			
Isopropanol	s13	Solvent, bulk chemical	
Ethylene glycol (1,2-ethanediol)		Bulk chemical	
Propylene glycol (1,3-propanediol)	s10	Bulk chemical	Economic Trends (1993)
1,3-Dichloropropan-2-ol[e]			
Butanol		Bulk chemical	
1,4-Butanediol		Petrochemical, major raw product	Mitsubishi Kasei Corporation (1988)
Octanol		Bulk chemical	
Acids			
Acetic acid		Bulk chemical	Economic Trends (1993)
Sorbic acid	s78	Cosmetics	de Kruijf et al. (1989)
Chloroacetic acid[e]			
Sodium fluoroacetate		Rodenticide	Ware (1989)
Sodium trichlooroacetate (TCA)		Herbicide	Ware (1989)
Dalapon (2,2-dichloropropanoic acid)		Herbicide	Ware (1989)
Fenac (Fenatrol®; 2,3,6-trichlorophenylacetic acid)		Herbicide	Ware (1989)
Aldehydes, ketones			
Acetaldehyde[f]		Bulk chemical	
Formaldehyde[f]	s12	Bulk chemical	Andrews and Reinhardt (1989)
Acrolein (2-propenal)[d,g]		Herbicide	Ware (1989)
Acetone (dimethyl ketone)		Bulk chemical	
Methyl ethyl ketone (MEK)[c]	s40	Petrochemical	Castaldi and Ford (1992)
4-Methyl-2-pentanone		Petrochemical	Castaldi and Ford (1992)
Acetophenone (methyl phenyl ketone)[g]		Petrochemical	
Methyl isobutyl ketone	s64	Paint solvent	Reisch (1993)

(continued)

TABLE 2.1. (*Continued*)

Compound (Other Names or Descriptions in Parentheses)	Chemical Structure[b]	Source or Application	Reference
Esters			
Ethyl formate		Fumigant	Metcalf (1971)
Methyl formate		Fumigant	Metcalf (1971)
Butyl acetate[f]			
Pentyl acetate[f]			
Vinyl acetate[f]		Bulk chemical	
Methylmethacrylate[f,g]	s48	Polymer production	Min et al. (1994)
Sta-Way®	a1	Insect repellant	Ware (1989)
Ethers, epoxides			
Methyl *tert*-butyl ether	s7	Gasoline	Hartley and Englande (1992)
β-(2-Chloroethyl)ether[d]			
Metaldehyde	a2	Molluscicide	Ware (1989)
β,β′-Dichlorodiethyl ether		Soil fumigant	Metcalf (1971)
1,2-Diethoxyethane	s19	Plastics	Sheldon and Hites (1979)
1-Chloro-2-(2-[*p*-1′,3′-tetramethylbutylphenoxy]ethoxy)-ethane	a3	Plasticizer	Sheldon and Hites (1979)
Ethylene oxide ([CH$_2$]$_2$O)		Fumigant, bulk chemical	Ware (1989)
Propylene oxide[f]		Fumigant	Metcalf (1971)
Amines			
Methylamine[f]			
Butylamine[f]			
Diethylamine[f]			
Triethylamine[f]			
Trimethylamine[f]			
Ethylenediamine[f]			
Nitriles, isothiocyanates			
Acetonitrile[g]		Solvent	
Acrylonitrile[c,d]	s11	Fumigant, bulk chemical	Metcalf (1971)
Vorlex® (methylisothiocyanate)		Nematicide	Ware (1989)
Dazomet (Basamid®)	a4	Nematicide	Ware (1989)
Cyclic, nonaromatic			
Cyclohexane[f]		Petrochemical, bulk chemical	Taylor and McLean (1992)
d-Limonene	s27	Insecticide, solvent	Ware (1989)

(*continued*)

TABLE 2.1. (*Continued*)

Compound (Other Names or Descriptions in Parentheses)	Chemical Structure[b]	Source or Application	Reference
Lindane[d,g,h]	s69	Insecticide	Ware (1989)
Caprolactam	a5	Bulk chemical	Economic Trends (1992)
Toxaphene (terpene)[d,g]	a6	Insecticide	Ware (1989)
α-Pinene (terpene)	s25	Solvent	
Hexachlorocyclopentadiene[d,g]			
Dioxane		Solvent	
Simple aromatic compounds			
Benzene and related compounds			
Benzene[c,d]	s2	Solvent, bulk chemical	Castaldi and Ford (1992)
Toluene[c,d]	s3	Solvent, bulk chemical	Castaldi and Ford (1992)
Xylenes[c,f]	s4	Solvent, bulk chemical	Castaldi and Ford (1992)
Ethylbenzene[c-e]	s6	Solvent, bulk chemical	Castaldi and Ford (1992)
Styrene[c,f]	s14	Bulk chemical	Castaldi and Ford (1992)
Cumene (isopropylbenzene)		Bulk chemical	Economic Trends (1992)
1,2,4-Trimethylbenzene	s5	Gasoline	Mehlman (1992)
Chlorobenzene[c,d]		Petrochemical	Castaldi and Ford (1992)
1,2-Dichlorobenzene[c,d]		Petrochemical, wood treatment	Metcalf (1971)
1,4-Dichlorobenzene[d,e]	s71	Moth proofing	Ware (1989)
1,2,3-Trichlorobenzene[g,h]		Wood treatment	Metcalf (1971)
1,2,4-Trichlorobenzene[d,g,h]		Wood treatment	Metcalf (1971)
Pentachlorobenzene[g]			
Hexachlorobenzene[d,g,h]		Petrochemical, fungicide	Quirijns et al. (1979), Ware (1989)
Chlorocymenes (methyl isopropyl benzenes)	s74	Pulp and paper	Bjørseth et al. (1979)
Polyaromatic hydrocarbons (PAHs)			
Naphthalene[f,g]	s18	Moth proofing, petrochemical	Metcalf (1971), Castaldi and Ford (1992), Ware (1989)
2-Methylnaphthalene[g]		Petrochemical	Castaldi and Ford (1992)

(*continued*)

TABLE 2.1. *(Continued)*

Compound (Other Names or Descriptions in Parentheses)	Chemical Structure[b]	Source or Application	Reference
2-Chloronaphthalene[d,g]			
Acenaphthene[f,g]	s41	Petrochemical	Castaldi and Ford (1992)
Anthracene[e–g]	s42	Petrochemical	Castaldi and Ford (1992)
Acenaphthylene[f,g]	a8		
Fluorene[f,g]	a9	Petrochemical	Castaldi and Ford (1992)
Phenanthrene[f,g]	s43	Petrochemical	Castaldi and Ford (1992)
Chrysene[f,g]	s44	Petrochemical	Castaldi and Ford (1992)
Fluoranthene[f,g]	a10	Petrochemical	Lau et al. (1993), Castaldi and Ford (1992)
Benzo[a]anthracene[f,g]		Petrochemical	Marks et al. (1992)
Dibenz[a,h]anthracene		Petrochemical	
7,12-Dimethylbenz[a]anthracene[g]		Petrochemical	
Benzo[b]fluoranthene[f,g]	s46	Petrochemical	Lau et al. (1993), Castaldi and Ford (1992)
Benzo[k]fluoranthene[f,g]		Petrochemical	Lau et al. (1993), Castaldi and Ford (1992)
Pyrene[f,g]	s45	Petrochemical	Castaldi and Ford (1992)
1,2-Benzo(a)pyrene[f,g]			Lau et al. (1993)
Benzo(ghi)perylene[f,g]	a11		Lau et al. (1993)
Indeno(123-cd)pyrene[f,g]	a12		Lau et al. (1993)
"Creosote"		Wood treatment	Metcalf (1971)
Isophorone[f,g]	a13	Petrochemical wood preservation	Castaldi and Ford (1992), Metcalf (1971)
Biphenyls			
Biphenyl[e]			
Benzidine[d]	a14		
3,3'-Dichlorobenzidine[d,g]			
PCBs[d,g,h]	s37	Petrochemical	Lau et al. (1993), Castaldi and Ford (1992)

(continued)

TABLE 2.1. (*Continued*)

Compound (Other Names or Descriptions in Parentheses)	Chemical Structure[b]	Source or Application	Reference
DDT relatives			
DDT (4,4'-DDT)[d,g,h]	s54	Insecticide	Lau et al. (1993), Metcalf (1971)
TDE (4,4'-DDD)[d,f,g]	a15	Insecticide	Ware (1989)
4,4'-DDE[d,g]	a16	Insecticide	Metcalf (1971)
Ethylan (Perthane®)	a17	Insecticide	Ware (1989)
Methoxychlor[f,g]	a18	Insecticide	Ware (1989)
DMC	a19	DDT synergist, acaricide	Metcalf (1971)
Chlorobenzilate[g]	a20	Acaricide	Ware (1989)
Dicofol (Kelthane®)[f]	a21	Acaricide	Ware (1989)
N,N-Dibutyl-p-chlorobenzenesulfonamide	a22	DDT "anti-resistant"	Metcalf (1971)
Higher molecular weight, primarily oxygen functional groups			
Cyclodienes			
Chlordane[d,g]	a23	Insecticide	Ware (1989)
Heptachlor[d,g]	a24	Insecticide	Ware (1989)
Aldrin[d,g,h]	s49	Insecticide	Quirijns et al. (1979), Ware (1989)
Mirex	a25	Insecticide	Ware (1989)
Dieldrin[d,g,h]	s70	Insecticide	Lau et al. (1993), Quirijns et al. (1979), Ware (1989)
Endrin (isomer of di-eldrin)[d,g,h]		Insecticide, rodenticide	Quirijns et al. (1979), Ware (1989)
Endosulfan (Thiodan®)[d,g,h]	a26	Insecticide	Ware (1989)
Chlorodecone (Kepone®)[f,g]	a27	Insecticide	Ware (1989)
Substituted benzenes			
Benzyl alcohol[g]			
Benzyl chloride[f]			
Aniline[f]	s15		
Nitrobenzene[d,g]	s16		
1,3-Dinitrobenzene[d,g]			
1,3,5-Trinitrobenzene[g]			
Chloronitrobenzenes		Fungicide	Ware (1989)
Pentachloronitrobenzene[g]			
Azobenzene[d]	a28	Acaricide	Metcalf (1971)
Trifenmorph (Frescon®)	a29	Molluscicide	Ware (1989)
Chlorinated dioxins	s38	Byproduct of combustion	Choudhary et al. (1983)

(*continued*)

TABLE 2.1. *(Continued)*

Compound (Other Names or Descriptions in Parentheses)	Chemical Structure[b]	Source or Application	Reference
Substituted toluenes			
Nitrotoluene[f]			
2,4-Dinitrotoluene[d,g]			
2,6-Dinitrotoluene[d,g]			
4-Chloro-2-nitrotoluene[e]			
o-Toluidine (2-aminotoluene)			
5-Nitro-o-toluidine[g]			
p-Toluic acid		Petrochemical	Macarie et al. (1992)
Toluene diisocyanate		Bulk chemical	Economic Trends (1992)
Chloramine-T	s36	Disinfectant	Ware (1989)
Phenols			
Phenol[d,g]	s31	Petrochemical	Castaldi and Ford (1992)
2,4-Dimethylphenol[d,g]		Petrochemical	Castaldi and Ford (1992)
2-Chlorophenol[d,g]		Petrochemical	
2,4-Dichlorophenol[d,g]		Petrochemical	Castaldi and Ford (1992)
2,4,5-Trichlorophenol[g]		Fungicide	Metcalf (1971)
2,4,6-Trichlorophenol[d,g]	s66	Fungicide	Metcalf (1971)
2,3,4,6-Tetrachlorophenol[g]		Fungicide	Metcalf (1971)
Pentachlorophenol[d,g,h]		Fungicide, wood preservation, desiccant	Metcalf (1971)
2-Amino-4-chlorophenol[e,g]			
2-Nitrophenol[d,g]			
4-Nitrophenol[d,g]			
2,4-Dinitrophenol[d,g]		Fungicide	Ware (1989)
Vancide BL	a30	Fungicide	Metcalf (1971)
Dinocap (Karathane®)	a31	Fungicide	Ware (1989)
Dinoseb[g]	s50	Fungicide, herbi-cide, insecti-cide, desiccant	Ware (1989)
Ethyl benzyl ether	s21	Plastics	Sheldon and Hites (1979)
2-Phenoxyethanol	s20	Plastics	Sheldon and Hites (1979)
o-Phenylphenol	s33	Disinfectant	Ware (1989)
Dichlorophene	a32	Fungicide	Metcalf (1971)
Hexachlorophene	s34	Disinfectant	Ware (1989)
Cresols (methylphenols)[f,g]	s32		

(continued)

TABLE 2.1. (*Continued*)

Compound (Other Names or Descriptions in Parentheses)	Chemical Structure[b]	Source or Application	Reference
4-Chloro-*m*-cresol[d,g]		Cosmetics	de Kruijf et al. (1989)
4,6-Dinitro-*o*-cresol (DNOC)[d,g]		Fungicide, herbicide	Ware (1989)
Xylenols (dimethylphenols)[f]			
Resorcinol (1,3-dihydroxybenzene)[f]			
Chlorocatechols	s75	Pulp and paper	Bjørseth et al. (1979)
Chloroguaiacols	s76	Pulp and paper	Brezny et al. (1992)
Chloroveratroles (1,2-dimethoxybenzenes)		Pulp and paper	Brezny et al. (1992)
Benzoic acid and related compounds			
Benzoic acid		Petrochemical	Macarie et al. (1992)
Salicylic acid	s77	Preservative	de Kruijf et al. (1989)
4-Formylbenzoic acid		Petrochemical	Macarie et al. (1992)
p-Methylbenzoic acid	s79	Cosmetics	de Kruijf et al. (1989)
p-Benzylbenzoic acid	s80	Cosmetics	de Kruijf et al. (1989)
Alkyl benzoates		Cosmetics	de Kruijf et al. (1989)
Benzyl benzoate		Cosmetics	de Kruijf et al. (1989)
Dicamba (Banvel®)[f]	s51	Herbicide	Ware (1989)
Halazone	s35	Disinfectant, Cl_2-reagent	Ware (1989)
Benzonitriles			
Benzonitrile[f]			
Diphenatrile	a33	Herbicide	Metcalf (1971)
Bromoxynil (Brominal®; 3,5-dibromo-hydroxybenzonitrile)		Herbicide	Ware (1989)
Chlorothalonil (Bravo®; 2,4,5,6-tetrachloro-1,3-dicyanobenzene		Fungicide	Ware (1989)

(*continued*)

TABLE 2.1. *(Continued)*

Compound (Other Names or Descriptions in Parentheses)	Chemical Structure[b]	Source or Application	Reference
Dichlobenil (Casaron®; 2,6-dichlorobenzonitrile)		Herbicide	Ware (1989)
Phthalic acid and related compounds			
Dimethylphthalate[d,g]	s17	Insect repellent	Ware (1989)
Bis(2-ethylhexyl)phthalate[d]		Petrochemical	Castaldi and Ford (1992)
Butyl benzyl phthalate[d,g]		Petrochemical	Castaldi and Ford (1992)
Di-*n*-butylphthalate[d,g]		Petrochemical	Castaldi and Ford (1992)
Dioctylphthalate		Bulk chemical	
Endothall (Endothal®)	s53	Herbicide	Ware (1989)
Captan[f]	a34	Fungicide, bird repellent	Ware (1989)
Folpet	a35	Fungicide	Ware (1989)
Terephthalic acid	s47	Petrochemical	
DCPA (chlorothal, dimethyl-tetrachloroterephthalate)		Herbicide	Ware (1989)
Phenoxy acids			
2,4-D[e–g]	a36	Herbicide	Ware (1989)
MCPA	a37	Herbicide	Ware (1989)
2,4,5-T[f,g]	a38	Herbicide	Ware (1989)
Silvex (2,4,5-TP)[f,g]	a39	Herbicide	Ware (1989)
Oxyphenoxy acid esters			
Fluazifop-butyl (Fusilade®)	a40	Herbicide	Ware (1989)
Quizalofop-ethyl (Assure®)	a41	Herbicide	Ware (1989)
Quinones			
Chloranil (Spergon®; 2,3,5,6-tetrachlorobenzoquinone)		Fungicide	Ware (1989)
1,4-Naphthoquinone[g]		Fungicide	Metcalf (1971)
Dichlone (Phygon®; 2,3-dichloro-1,4-naphthaquinone)[f]	s52	Fungicide	Ware (1989)
Phenanthraquinones		Fungicide	Metcalf (1971)
Anthraquinones		Fungicide	Metcalf (1971)
Acenaphthoquinones		Fungicide	Metcalf (1971)
Indandiones			
Pindone (Pival®)	a42	Rodenticide	Ware (1989)
Diphacinone (Diphacin®)	a43	Rodenticide	Ware (1989)
Chlorophacinone (Rozol®)	a44	Rodenticide	Ware (1989)
Furanes			
Tetrahydrofuran		Solvent, petro-chemical	Taylor and McLean (1992)

(continued)

TABLE 2.1. (*Continued*)

Compound (Other Names or Descriptions in Parentheses)	Chemical Structure[b]	Source or Application	Reference
Furfural[f]	a45		
Dibenzofuran[g]		Petrochemical	Castaldi and Ford (1992)
Chlorinated dibenzofuranes	s39	Combustion processes	
MGK® Repellent 11	a46	Insect repellent	Ware (1989)
Hydroxycoumarins			
Dicumarol	a47	Rodenticide	Ware (1989)
Warfarin	a48	Rodenticide	Ware (1989)
Coumachlor	a49	Rodenticide	Ware (1989)
Coumafuryl	a50	Rodenticide	Ware (1989)
Brodifacoum (Talon®)	a51	Rodenticide	Ware (1989)
Diphenyl ethers			
Nitrofen	a54	Herbicide	Ware (1989)
Fomesafen (Flex®)	a55	Herbicide	Parry (1989), Ware (1989)
4-Bromophenyl phenyl ether[d,g]			
Miscellaneous compounds			
GA3 (Gibberellin®)	a52	Growth promoter	Ware (1989)
Bromethalin (Vengeance®)	a53	Rodenticide	Ware (1989)
Higher molecular weight, primarily nitrogen functional groups			
Diphenylamines			
Diphenylamine[g]			
N-nitrosodiphenylamine[d]			
Amides			
Diphenamid (Dymid®)	a56	Herbicide	Ware (1989)
Napropamide (Devrinol®)	a57	Herbicide	Ware (1989)
Pronamide (Kerb®)[g]	a58	Herbicide	Ware (1989)
Furalaxyl (Fongarid®)	a59	Fungicide	Ware (1989)
NAD (naphthaleneacetamide)		Herbicide, growth regulator	Ware (1989)
Niclosamide (Bayluscid®)	a60	Molluscicide	Ware (1989)
Acetanilides and related compounds			
Alachlor (Lasso®)	s62	Herbicide	Ware (1989)
N-methyl-2-pyrrolidone	s26	Solvent	Wolf et al. (1991)
MGK 264®	a61	Insecticide synergist	Ware (1989)
Procymidone (Sumilex®)	a62	Fungicide	Ware (1989)
Amitraz (Mitac®)	a63	Insecticide	Ware (1989)

(*continued*)

TABLE 2.1. (*Continued*)

Compound (Other Names or Descriptions in Parentheses)	Chemical Structure[b]	Source or Application	Reference
Nitroanilines			
Nitroanilines[g]			
Trifluralin (Treflan®)[h]	a64	Herbicide	Ware (1989)
Oryzalin (Surflan®)	a65	Herbicide	Ware (1989)
Phenylureas, uracils			
Fluometuron (Cotoran®)	s60	Herbicide	Ware (1989)
Diuron (Karmex®)[f]	a66	Herbicide	Ware (1989)
Linuron (Lorox®)[e]	a67	Herbicide	Ware (1989)
Bromacil (Hyvar®)	a68	Herbicide	Ware (1989)
Carbamates			
Carbamates	s72	Textile finishing	Andrews and Reinhardt (1989)
Monoalkyl dimethylolcarbamate		Textile finishing	Andrews and Reinhardt (1989)
Aldicarb (Temik®)	a70	Insecticide, molluscicide, nematicide	Ware (1989)
Thiodicarb (Larvin®)	a69	Insecticide, molluscicide	Ware (1989)
Carbaryl (Sevin®)[f]	a71	Insecticide, molluscicide	Ware (1989)
Carbofuran (Furadan®)[f]	s61	Insecticide, molluscicide, nematicide	Ware (1989)
Methiocarb (Mesurol®)[f]	a72	Insecticide, molluscicide	Ware (1989)
Bendiocarb (Ficam®)	a73	Insecticide, molluscicide	Ware (1989)
Mexacarbate (Zectran®)[f]	a74	Insecticide, molluscicide	Ware (1989)
Barban (Carbyne®)	a75	Herbicide	Ware (1989)
Asulam (Asulox®)	a76	Herbicide	Ware (1989)
Pyridines and related compounds			
Pyridine	a77	Petrochemical	Taylor and McLean (1992)
4-Aminopyridine (Avitrol®)		Bird repellent	Ware (1989)
MGK® repellent 326	a78	Insect repellent	Ware (1989)
Picloram (Tordon®; 4-amino-2-carboxy-3,5,6-trichloropyridine)		Herbicide	Ware (1989)

(*continued*)

TABLE 2.1. *(Continued)*

Compound (Other Names or Descriptions in Parentheses)	Chemical Structure[b]	Source or Application	Reference
Quinoline[f]	a79		
8-Hydroxyquinoline	s82	Wood preservation	Metcalf (1971)
Diquat[f]	s55	Herbicide	Ware (1989)
Paraquat	a80	Herbicide	Ware (1989)
Nicotine	a81	Insecticide, fumigant	Ware (1989)
Nicotine relatives		Insecticide	Ware (1989)
Fluridone (Sonar®)	a82	Herbicide	Ware (1989)
Pyrithion	s81	Cosmetics	de Kruijf et al. (1989)
Dimethirimol (Milcurb®)	s56	Fungicide	Ware (1989)
Chloridazon, pyrazon (Pyramin®)[e]	a83	Herbicide	Ware (1989)
Triazines			
Atrazine (Aatrex®)[h]	s57	Herbicide	Ware (1989)
Simazine (Princep®, atrazine analog)[h]		Herbicide	Ware (1989)
Anilazine (Dyrene®)	a84	Fungicide	Ware (1989)
Cyromazine (Larvadex®)	a85	Insect repellent	Ware (1989)
ICA (isocyanuric acid)[e]	a86	Algicide	Ware (1989)
TICA (trichloro-ICA)		Algicide	Ware (1989)
Triazoles, imidazoles			
Amitrole	a87	Herbicide	Ware (1989)
Bitertanol (Baycor®)	a88	Fungicide	Ware (1989)
Imazapyr	a89	Herbicide	Ware (1989)
Thiabendazole (TBZ)	a90	Fungicide	Ware (1989)
Special C/N compounds			
Kinetin (purin)	a91	Growth promoter	Ware (1989)
Azoadamantane	s67	Preservative	Summers (1992)
TNT	s85	Explosive	Borman (1994)
TNAZ	s88	Explosive	Borman (1994)
RDX	s87	Explosive	Borman (1994)
HMX	s86	Explosive	Borman (1994)
CL-20	s89	Explosive	Borman (1994)
Organophosphates			
Nonaromatic			
Glyphosate (Roundup®)	a92	Herbicide	Ware (1989)
TEPP (tetraethylpyrophosphate)		Insecticide	Ware (1989)
Malathion[f,h]	s58	Insecticide	Ware (1989)
Dimethoate (Cygon®)[e]	a93	Herbicide	Ware (1989)
Disulfoton (Di-Syston®)[f,g]	a94	Herbicide	Ware (1989)

(continued)

TABLE 2.1. *(Continued)*

Compound (Other Names or Descriptions in Parentheses)	Chemical Structure[b]	Source or Application	Reference
Phorate (Thimet®)[g]	a95	Nematicide	Ware (1989)
Dichlorvos (Vapona®)[f,h]	a96	Insecticide	Ware (1989)
Mevinphos (Phosdrin®)[e,f]		Insecticide	Ware (1989)
Ethion (Nialate®)[f]	a97	Insecticide	Ware (1989)
Naled (Dibrom®)[f]	a98	Insecticide	Ware (1989)
Aromatic			
Parathion, methyl[e–g]	s59	Insecticide	Ware (1989)
Parathion, ethyl[e–g]		Insecticide	Ware (1989)
Tetrachlorvinphos (Gardona®)		Insecticide	Ware (1989)
Crufomate (Ruelene®)		Insecticide	Ware (1989)
Famphur (Bash®)[g]	a99	Insecticide	Ware (1989)
Fenitrothion (Accothion®)[h]		Insecticide	Ware (1989)
Fenthion (Baytex®)[e]		Insecticide	Ware (1989)
Diazinon[f]	a100	Insecticide	Ware (1989)
Azinophosmethyl (Guthion®)[f,h]	a101	Insecticide, molluscicide	Ware (1989)
Chlorpyrifos (Dursban®)[f]	a102	Insecticide	Ware (1989)
Organosulfates and related compounds			
Tetrasul	a103	Acaricide	Ware (1989)
Ovex (Ovotran®)	a104	Acaricide	Ware (1989)
Propargite (Omite®)[f]	a105	Acaricide	Ware (1989)
Methyl methanesulfonate		Petrochemical	Castaldi and Ford (1992)
Thiocarbamates, dithiocarbamates			
EPTC	a106	Herbicide	Ware (1989)
Thiobencarb (Bolero®)	a107	Herbicide	Ware (1989)
Diallate[g]	a108	Herbicide	Ware (1989)
Thiram	a109	Fungicide	Ware (1989)
Ferbam (ferric dimethyldithio-carbamate)		Fungicide	Ware (1989)
Zineb (zinc ethylenebisdithio-carbamate)		Fungicide, molluscicide	Ware (1989)
Maneb (manganese ethylene-bisdithiocarbamate)		Fungicide	Ware, (1989)
Organothiocyanates, thioureas, isothiazolones			
Thanite® (isobornyl thiocyano-acetate)		Insecticide	Ware (1989)
α-Naphthylthiourea (Antu®)		Rodenticide	Ware (1989)
Tricyclazole (Beam®)	a110	Fungicide	Ware (1989)
Bentazon	a111	Herbicide	Ware (1989)
Methylisothiazolone	s83	Preservative, paints	Law and Lashen (1991)

(continued)

TABLE 2.1. (*Continued*)

Compound (Other Names or Descriptions in Parentheses)	Chemical Structure[b]	Source or Application	Reference
Methylchloroisothiazolone	s84	Preservative, paints	Law and Lashen (1991)
Promexal W50	s68	Preservative, paints	Eacott and Linley (1991)
Complex organic chemicals			
Pyrethroids			
Pyrethrin I[f]	a115	Insecticide	Ware (1989)
Permethrin (Ambush®)[f]		Insecticide	Ware (1989)
Resmethrin (Synthrin®)[f]		Insecticide	Ware (1989)
Tetramethrin (Neo-Pynamin®)[f]		Insecticide	Ware (1989)
Tefluthrin (Force®)[f]		Insecticide	Ware (1989)
Others			
Carboxin (Vitavax®, oxathiin)	a112	Fungicide	Ware (1989)
Red squill (Scilliroside, glycoside)	a113	Rodenticide	Ware (1989)
Strychnine (alkaloid)[f]	a114	Rodenticide	Ware (1989)
Rotenone	a116	Insecticide	Ware (1989)
Metalorganic compounds			
Potassium antimonyl tartarate		Molluscicide	Metcalf (1971)
PMA (phenylmercury acetate)		Fungicide	Ware (1989)
Aryloxy-mercuric chlorides		Fungicide	Metcalf (1971)
Alkyl-mercuric chlorides		Fungicide	Metcalf (1971)
Ceresan	s65	Fungicide	Ware (1989)
Tributyl tin[h]		Fungicide, molluscicide	Lau et al. (1993), Metcalf (1971)
Triphenyl tin salts[h]	s63	Fungicide, molluscicide	Metcalf (1971)

[a]Compounds are grouped by compound classes and by increasing complexity. Many of the more complex compounds could be assigned to a number of compound classes. It was attempted to assign them according to their dominant features. Compound classes are grouped as follows: Simple nonaromatic compounds, simple monoaromatic and polyaromatic compounds, more complex compounds with mainly oxygen functional groups, compounds with mainly nitrogen functional groups, organophosphorous compounds, organosulfur compounds, and very complex compounds. References are given where appropriate. Some chemicals are of general use, and a specific reference is not given.

[b]Structures marked "s" are given in the text, in sequence of their appearance. Structures marked "a" are listed in Appendix 1, in the sequence of their appearance in this table.

[c]Specified for groundwater detection monitoring programs by the U.S. EPA (U.S. Government, 1993f).

[d]Specified for effluent testing for many industries by the U.S. EPA (U.S. Government, 1993b).

[e]Specified on the "Black List" (of potential future concern) of toxic chemicals in the environment, United Kingdom (McClure et al., 1991).

[f]Specified for effluent testing if suspected present for many industries by the U.S. EPA (U.S. Government, 1993b).

[g]Specified for groundwater assessment monitoring programs by the U.S. EPA (U.S. Government, 1993f).

[h]Specified on the "Red List" (of immediate concern) of toxic chemicals in the environment, United Kingdom (McClure et al., 1991).

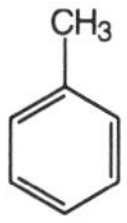

toluene
(s3)

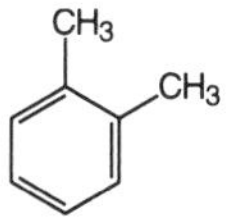

o-xylene
(s4)

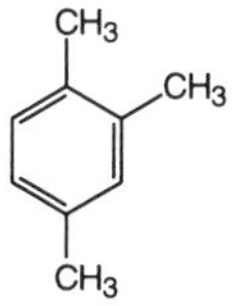

1,2,4-trimethylbenzene
(s5)

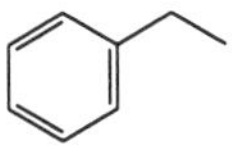

ethylbenzene
(s6)

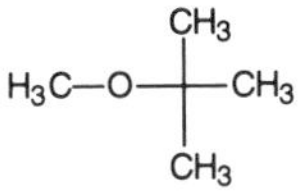

methyl *tert*-butyl ether
(s7)

The next most important class of products of the petrochemical industry are the bulk chemicals that are subsequently used in further organic synthesis. Such "primary" chemicals include ethylene (s8), propylene (s9), propylene glycol (s10), isopropyl alcohol (s13), acetic acid, formaldehyde (s12), acrylonitrile (s11), benzene (s2), toluene (s3), xylenes (s4), and styrene (s14) (Economics Trends, 1993).

ethylene
(s8)

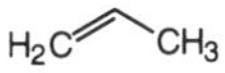

propylene
(s9)

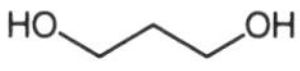

propylene glycol
(s10)

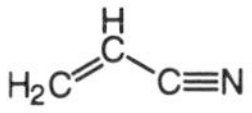

acrylonitrile
(s11)

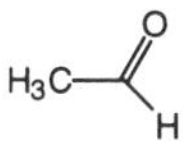

formaldehyde
(s12)

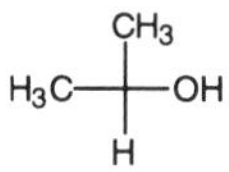

isopropyl alcohol
(isopropanol)
(s13)

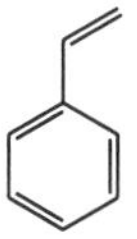

styrene
(s14)

More complex production intermediates include the chlorobenzenes, aniline (s15), nitrobenzene (s16), naphthalene (s18), phthalates (s17), and innumerable others.

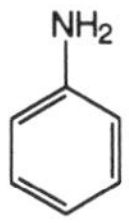

aniline
(s15)

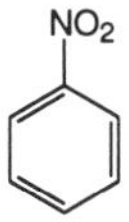

nitrobenzene
(s16)

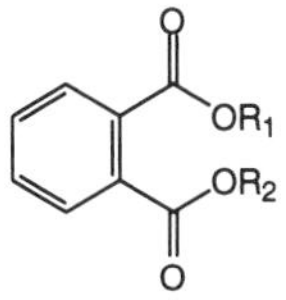

phthalates
(s17)

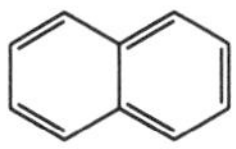

naphthalene
(s18)

All of these compounds are produced and handled in large quantities in a number of industries, such as in the production of solvents, pesticides, aluminum, synthetic rubber, and plastics (Fishbein, 1979a–d). By virtue of the enormous quantities of these chemicals being handled, the relatively small fraction that escapes the production process during handling or as waste still amounts to large quantities of these hydrocarbons entering the environment.

A large group of compounds in itself is produced from ethylene glycol, including plasticizers, copolymers, cross-linking agents, de-icing chemicals, textile finishing reactants, industrial solvents, surfactants, and detergents (Sheldon and Hites, 1979). Compounds with up to 10 ethylene glycol units are being produced, and, due to their widespread use, a number of these compounds are found in surface waters (Sheldon and Hites, 1979). Typical compounds are 1,2-diethoxyethane (s19), 2-phenoxyethanol (s20), and ethyl benzyl ether (s21). A long-chained tetramethylbutylphenoxy derivative (Table 2.1, a3) of ethylene glycol was found in the Delaware River and probably originated from the plastics industry in the area (Sheldon and Hites, 1979).

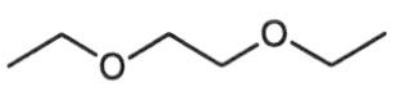

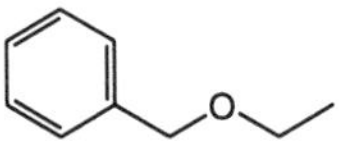

| 1,2-diethoxyethane | 2-phenoxyethanol | ethyl benzyl ether |
| (s19) | (s20) | (s21) |

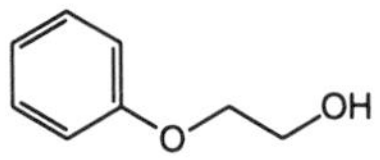

Degreasing solvents are very widely used and are the major sources of trichloroethylene (s22), tetrachloroethylene (s23), trichloroethane (s24), methanol, and chlorofluorocarbons (CFCs; s28, s29). New alternatives are being sought. Candidates include isopropyl alcohol (s13), terpenes (s25), N-methyl-2-pyrrolidone (s26), and hydrochlorofluorocarbons (HCFCs) (Wolf et al., 1991). d-Limonene (s27) has for some time been considered as another alternative, but is still being evaluated (Wolf et al., 1991; Reisch, 1993).

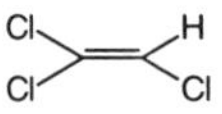

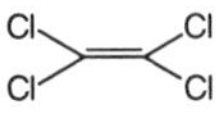

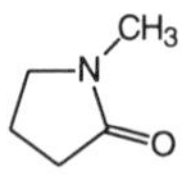

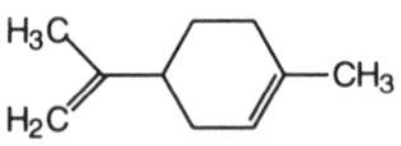

| trichloroethylene | tetrachloroethylene | 1,1,1-trichloroethane |
| (s22) | (s23) | (s24) |

α-pinene	N-methyl-	d-limonene
(turpentine oil,	2-pyrrolidone	(s27)
a terpene)	(s26)	
(s25)		

CFC-11	CFC-113	dichlorofluoroethane
(Freon-11)	(s29)	(an HCFC)
(s28)		(s30)

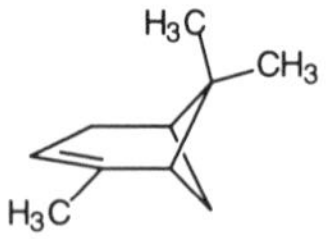

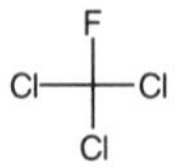

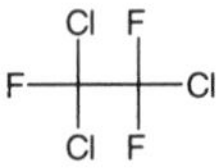

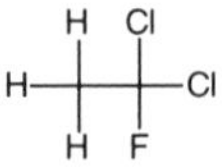

Disinfectants are found in many commercial and household products. Typical compounds include phenol (s31), *o*-cresol (s32), *o*-phenylphenol (s33), hexachlorophene (s34), and sulfonic acid relatives like halazone (s35) and chloramine-T (s36) (Ware, 1989). In 1982, the EPA had 600 compounds registered as active ingredients in disinfectants, while there were only 250 remaining in 1988. Major places of disinfectant use include beverage and food processing plants, eating establishments, barber and beauty shops, hospitals, laundry and dry cleaning operations, swimming pools, carpets, and maintenance of commercial and private property (Ware, 1989).

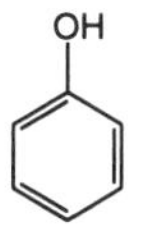

phenol (s31) *o*-cresol (s32) *o*-phenylphenol (s33) hexachlorophene (s34)

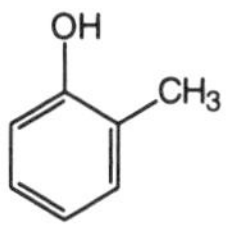

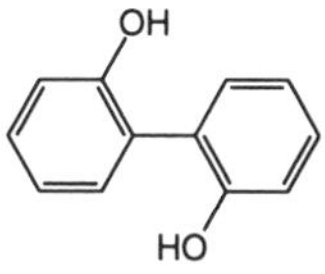

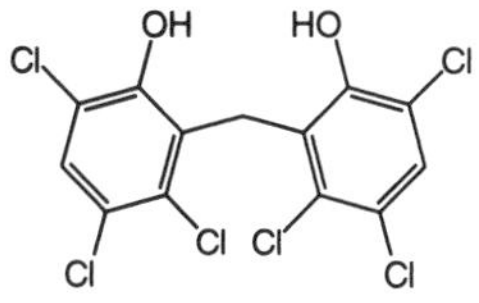

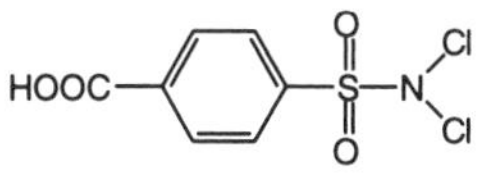

halazone (s35) chloramine-T (s36)

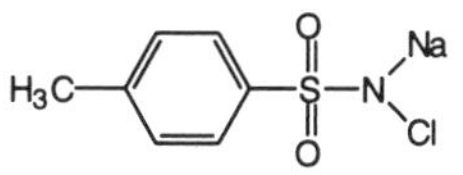

Another class of compounds that has been used in large quantities and is of great environmental concern are the polychlorinated biphenyls (PCBs; s37). They have been produced in several countries and are marketed and used worldwide. PCBs have been used in large quantities as heat transfer fluids, hydraulic fluids, and solvent extenders. PCBs are chemically inert, resist microbial degradation, and tend to accumulate in the food chain due to their lipophilic nature. Large quantities (up to one-third of the total U.S. production) are estimated to have entered the environment and, because of their extreme stability, PCBs have been transported throughout the biosphere. There is an abundant literature on various aspects of PCBs. More detailed discussions of PCBs and their toxicological properties are given by Safe (1987) and D'Itri and Kamrin (1983).

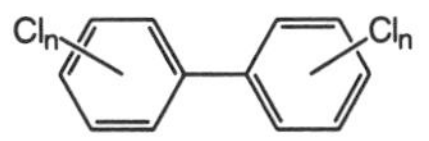

polychlorinated biphenyls (PCBs) (s37)

An extremely diverse class of compounds that are also of extreme environmental concern are the chlorinated dioxins (s38) and the related dibenzofurans (s39). They are primarily products of combustion processes (e.g., municipal and industrial garbage incineration) and impurities in chlorophenol products (Choudhary et al., 1983). Dioxins and dibenzofurans can be both mono- and polychlorinated, which gives rise to a large number of possible isomers. An extensive review of dioxins and dibenzofurans and their toxicity is given by Choudhary et al. (1983).

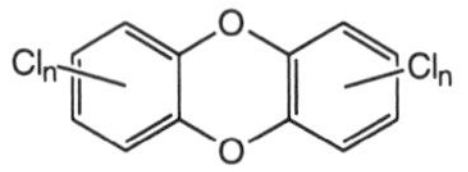

chlorinated dioxins
(s38)

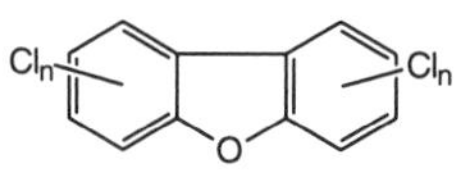

chlorinated dibenzofuranes
(s39)

An important way of releasing organic compounds into the environment is through petrochemical waste sludges (Castaldi and Ford, 1992). Compound classes of major concern (those on the EPA priority list) are volatiles, PAHs, substituted phthalate esters (s17), dibenzofuranes (s39), phenols (s31) and PCBs (s37), just to mention the most important categories (Castaldi and Ford, 1992). Volatiles include both chlorinated and nonchlorinated organics like benzene, toluene (s3), ethylbenzene (s6), xylenes (s4), styrene (s14), methyl ethyl ketone (s40), other ketones, and chlorobenzenes. Typical PAHs are naphthalenes (s18), acenaphthene (a41), anthracene (s42), phenanthrene (s43), benzofluoranthenes (s46), chrysene (s44), and pyrene (s45). For a complete listing, see the pertinent EPA regulations that are published in the *Code of Federal Regulations* (U.S. Government, 1993a–f). Many other countries have compiled their own priority lists, which include a similar selection of priority compounds (Taylor and McLean, 1992).

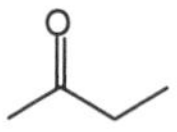

methyl ethyl
ketone
(s40)

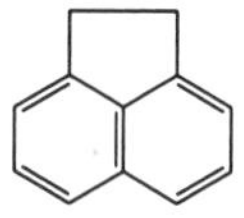

acenaphthene
(s41)

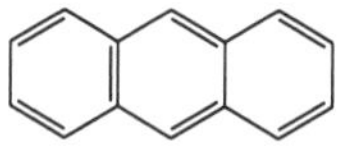

anthracene
(s42)

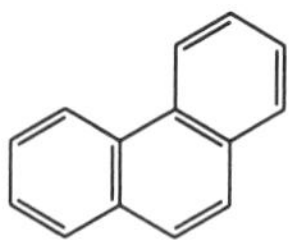

phenanthrene
(s43)

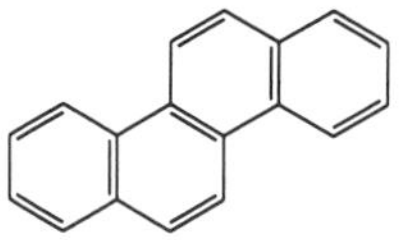

chrysene
(s44)

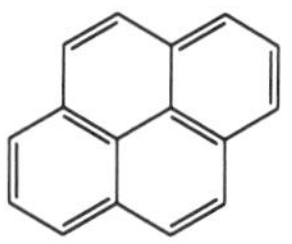

pyrene
(s45)

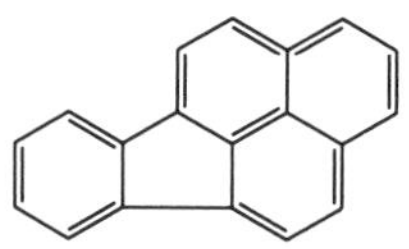

benzo(3,4)fluoranthene
(s46)

1.3. Plastics Industry

The plastics industry, which is closely related to the petrochemical industry, is enormous in itself, and its arsenal of chemicals is just as impressive. The reader may get an impression of the chemistry involved from Järvisalo et al. (1984). In addition to bulk raw materials like styrene (s14), vinylchloride (s1), aniline (s15), terephthalic acid (s47), or methyl methacrylate (s48) and solvents like toluene (s3), benzene (s2), and methanol, the plastics industry also uses a number of highly specialized, complex organic compounds such as antioxidants, plasticizers, cross-linking agents, and reagents to modify resins and improve fiber characteristics. A detailed discussion of all the chemicals used in the plastics industry is not possible in this chapter, but examples have been published by Martakis and Niaounakis (1992), Mathur and Varma (1992), Bodmeier and Paeratakul (1994), and Min et al. (1994).

terephthalic acid
(s47)

methyl methacrylate
(s48)

1.4. Pesticide Industry

Besides the plastics industry, the pesticide industry is probably the most important consumer of petrochemical products as well as producer of organic compounds. A large volume of bulk chemicals goes into pesticide synthesis, and this industry is, therefore, responsible for a large share of organic wastes produced, including primary reactants, solvents, intermediates, and final products. Table 2.1 documents the wide variety of compounds produced by this industry.

Pesticides are given a relatively high priority in this chapter (as documented by their abundance in Table 2.1) for several reasons. For one, pesticides serve as examples of the extreme diversity of synthetic organic chemicals being synthesized on a large scale today. In addition, pesticides are being introduced into the environment on an equally large scale. Finally, consistent with their use, they are designed to have biocidal properties and thus are *a priori* of environmental concern. Despite the fact that modern pesticides tend to be highly specific and selective, pesticides will remain an environmental concern for some time to come. Testing of newly introduced pesticides is mainly restricted to determining acute and perhaps chronic toxicity to humans, but usually there is very little information available on the fate of a particular pesticide in the environment. Also, many older pesticides that have been taken off the market in the United States and in European countries are still in use today in Third World countries, DDT being the prime example (Ware, 1989).

A detailed discussion of individual pesticides is out of place in the context of this chapter, and the reader is referred to the many books on pesticides that discuss their history, use, and biological action (Kearny and Kaufman, 1976, 1988; Ware, 1989; Hassall, 1990). The compilation found in White-Stevens (1971a,b) is somewhat older, but gives detailed information about the chemical and physical properties of pesticides, including solubility, vapor pressure, and chemical reactivity.

For the purposes of this chapter, suffice it to say that there is virtually no limitation to what basic compound classes and functional groups may be found in pesticides. Upon examination of Table 2.1, the reader may notice that there is a pesticide representative in virtually every class and subclass of compounds listed. It should also be emphasized that the list is far from complete and that usually only one, or at most a few, typical representatives are included in the list.

To summarize, the most commonly found central structures in pesticides are benzene and benzene derivatives (cyclodienes, s49; phenols, s50; benzoic acids, s51; quinones, s52; phthalates, s53; diphenyls, s54), often chlorinated and often heterocyclic (i.e., pyridines, s55; pyrimidines, s56; pyrimidazoles; triazines, s57). Also very common are organophosphorous compounds (organophosphates and organothiophosphates, s58, s59), ureas (s60), carbamates (s61), acetanilides (s62), and organometal compounds (s63). Common functional groups are halogens and hydroxy, alkoxy, trihalomethyl (including CF_3), aryl, nitrile, nitro, and amino groups.

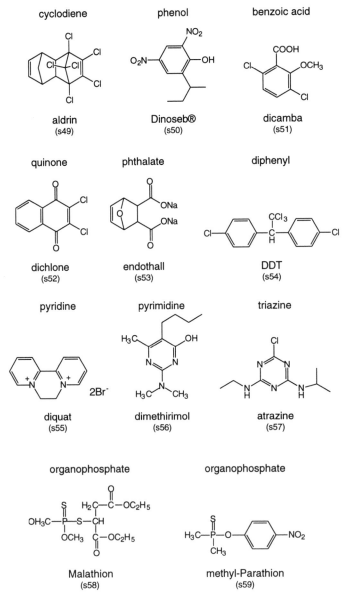

urea　　　　　carbamate　　　　　acetanilide

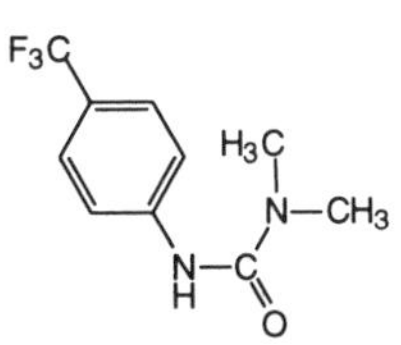

fluometuron
(s60)

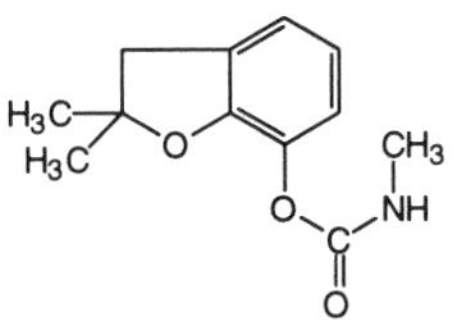

carbofuran
(s61)

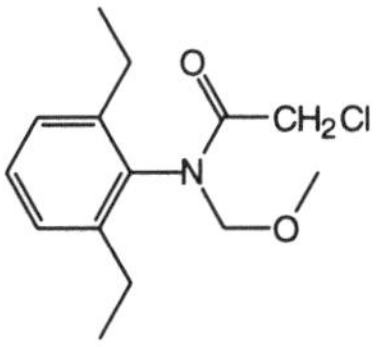

alachlor
(s62)

organometal

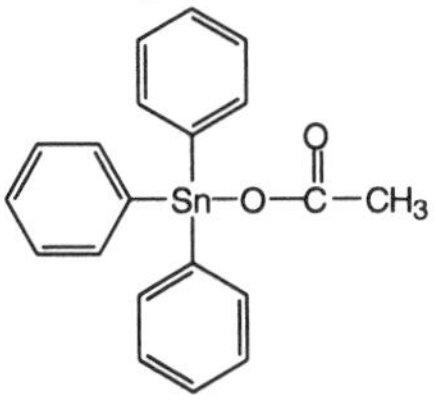

triphenyltin acetate
(s63)

1.5. Paint Industry

Solvents are a major ingredient in paint formulations and include xylene (s4), toluene (s3), methyl ethyl ketone (s40), and methyl isobutyl ketone (s64) (Reisch, 1993). Since the release of these solvents into the atmosphere is becoming more and more restricted, the paint industry is looking for alternatives. Candidates include various alkyl acetates, ketones, and polyglycol ethers (Reisch, 1993).

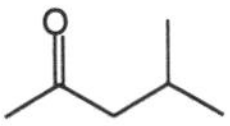

methyl isobutyl ketone (s64)

Preservatives are also an important component in paints. A number of preservatives are currently in use, including organotins (s63), mercury-containing compounds (s65), chlorinated phenolics (s66), and formaldehyde-releasing chemicals. All of these preservatives have problems with regard to their fate in the environment or their efficiency as biocides. Some newer developments are, for example, azoadamantane-related compounds (Summers, 1992). Azoadamantane (or tetra-aza-adamantane; s67) itself is of limited efficiency as a preservative, but N-methyl- and N-chloropropenyl-substituted relatives are used in paints and cosmetics. As a curiosity it may be noted here that a very close relative, hexanitrohexa-aza-adamantane, is used as an explosive (Liebenberg et al., 1993). Another relatively new compound that is designed to be used primarily in paints is Promexal W50 (s68), an isothiazolin derivative (Eacott and Linley, 1991). It exhibits better antimicrobial activity than other products and is very water soluble, and thus easily added to paint formulations.

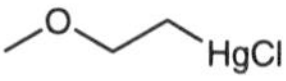

ceresan	2,4,6-trichloro-phenol	azo-adamantane	Promexal W50
(s65)	(s66)	(s67)	(s68)

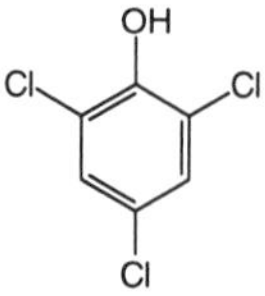

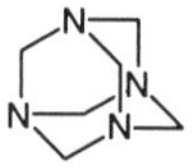

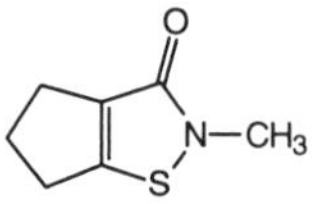

1.6. Household Use

A large number of compounds described in other sections are designed to be used in the home, such as paints, cosmetics, personal care products, cleaning and disinfection products, and even foods. Pesticides are used on house plants, in the garden, and on pets.

Constituents and additives in paints and cosmetics have been discussed in previous sections. An extensive list of cleaning and polishing product formulations is given by Flick (1986). The dominant active ingredients are surfactants, which are represented by a number of substituted alkyl and aryl sulfonates and a large variety of ethylene glycol derivatives of various lengths. There is only limited use of aromatic solvents in this class of products.

Pesticides are used very commonly and in a number of applications, both in hidden ways through the use of paints and cosmetic products and directly as insect repellents and for pest control on plants and animals. Awareness of the potential danger of pesticides to human health has greatly increased over the past decade, though not long ago pesticides were used rather naively. Significant amounts of now strictly regulated or even banned pesticides have found their way into the environment. For example, lindane (s69) was used in household fumigants in the form of pellets attached to light bulbs and electric wall vaporizers (Ware, 1989). Other examples are DDT (s54) and dieldrin (s70), which have been widely used to treat fabrics against moths; p-dichlorobenzene (s71) and naphthalene (s18), which are used as space fumigants for moth proofing in storage spaces (Metcalf, 1971); and dichlorobenzenes, which are found as disinfectants in toilet bowls.

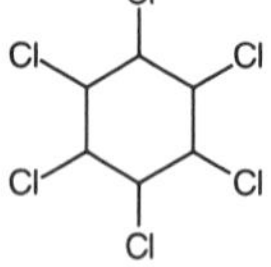

lindane	dieldrin	p-dichloro-benzene
(s69)	(s70)	(s71)

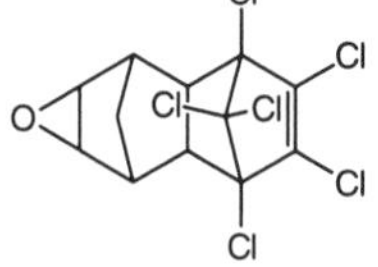

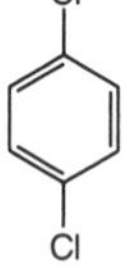

1.7. Electronics Industry

Apart from being one of the major users of chlorinated and nonchlorinated solvents such as methylene chloride, tetrachloromethane, trichloroethylene (s22), perchloroethylene (s23), toluene (s3), xylenes (s4), and CFCs (s28, s29), the electronics industry is also ever-increasingly involved in the use of highly specialized chemicals in the production of electronic components. The production of electronic circuits involves the use of "resist" layers on silicon dioxide in which the circuit patterns are etched. Chemicals used are typically compounds that form resins and are photosensitive. It is far beyond the scope of this chapter to explore all of the different kinds of compounds in use in this field, but a recent review is given by Willson and Bowden (1989). A more complete discussion of synthetic organic chemicals used in "high tech" industries is given by Sherry (1985).

1.8. Textile Industry

In addition to use of some pesticides for insect/moth control, the textile industry uses organic solvents both in the manufacturing and in the cleaning process. The almost "universal" solvent for cleaning/dry cleaning purposes is tetrachloroethylene (s23), while methanol sometimes is used in a mixture with tetrachloroethylene (Carr and Weedall, 1989; Gulyas and Hemmerling, 1990).

In textile manufacturing, fabrics are often "finished" to improve their resistance against decoloration, wear, and knitting. The finishing process involves the application of alkylcarbamates (s72), which form very stable esters with the cellulose matrix. In the process, formaldehyde is released and has to be removed before the product can be marketed. Substituted ureas (s73) are also being used (Andrews and Reinhardt, 1989).

carbamate
(s72)

urea
(s73)

A large portion of the textile industry uses synthetic fibers. As in the plastics industry, the production of synthetic fibers involves a number of primary monomeric compounds such as vinylchloride (s1) and propylene (s9), as well as a whole host of reagents controlling the polymerization reactions and surface properties of the resulting fibers.

1.9. Pulp and Paper Industry

The most significant production of synthetic organic chemicals in the pulp and paper industry occurs during the chlorine bleaching of pulp. All in all, over 200 chlorinated compounds have been found in effluents from pulp mills, ranging from low-molecular-weight materials like chlorinated phenolic compounds (e.g., s66) or carboxylic acids to an ill-defined class of compounds known as chlorolignin with molecular weights of 1,000 g/mol and higher (Häggblom and Salkinoja-Salonen, 1991). Other components that have been identified in pulp bleaching effluents include chlorinated cymenes (s74), chlorinated catechols (s75), hexachlorobenzene, and chloroguaiacols (s76) (Bjørseth et al., 1979).

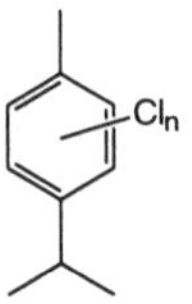

| chlorinated cymenes (s74) | chlorinated catechols (s75) | chlorinated guaiacols (s76) |

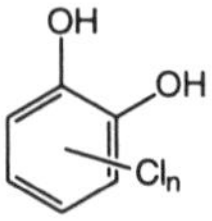

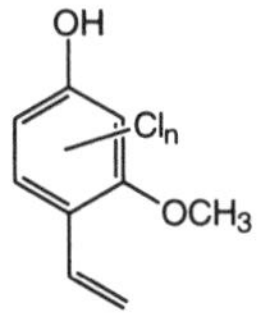

1.10. Cosmetics Industry and Medical/Pharmaceutical Industry

A large number of active ingredients are used in the general sector of cosmetics and drug industries. It is hardly possible to give justice to a class of compounds as varied and as numerous as this one within the framework of a short review; organic chemicals related to the medical field have been omitted altogether. Further justification for this exclusion was taken from the fact that compounds used in the medical field often are naturally occurring substances (or close relatives). Relatively small amounts of these compounds are being produced, and only small quantities are expected to be released into the environment.

Apart from medically active ingredients, preservatives in cosmetic products or drugs are one of the more important classes of compounds. Preservatives that are typically used are simple compounds like salicylic acid (s77) or sorbic acid (s78), but many benzene and benzoic acid derivatives, particularly 4-alkyl benzoic acids (s79) and benzyl benzoic acids (s80), are also employed. Frequently used as well are xylenols, cresols, pyridine relatives (e.g., pyrithion, s81; 8-hydroxyquinoline, s82), and others (de Kruijf et al., 1989). A short survey of colorings and preservatives in drugs is given by Pollock et al. (1989). A more complete description of chemicals used in cosmetics and drugs is given by Smolinske (1992).

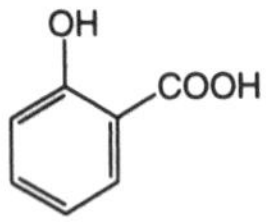

| salicylic acid (s77) | sorbic acid (s78) | 4-methylbenzoic acid (s79) |

| 4-benzylbenzoic acid (s80) | pyrithion (s81) | 8-hydroxyquinoline (s82) |

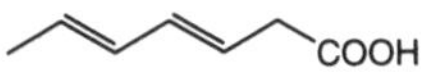

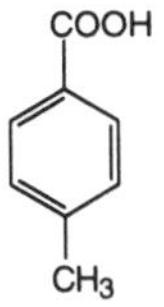

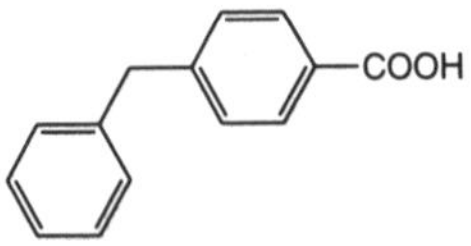

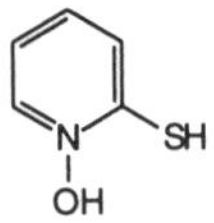

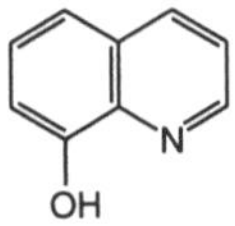

1.11. Metals Industry

While the metal finishing industry is mostly concerned with inorganic, metal-containing wastes, solvents such as CCl_4 (tetrachloromethane) or 1,1,1-trichlorethane (s24) are used for degreasing and cleaning, although their use appears to be on the decline, with alkaline cleaning solutions replacing organic solvents. A recent review of hazardous waste characteristics in the metals industry can be found in PRC Environmental Management (1989).

A minor area of synthetic chemical application in the metal industry is the use of preservatives in metalworking fluids. The compounds used here are similar to those used in paints and in some cosmetics. Among the more popular compounds are isothiazolones, especially MCI, a mixture of methylisothiazolone (s83), and methylchloroisothiazolone (s84) (Law and Lashen, 1991).

methylisothiazolone
(s83)

methylchloroisothiazolone
(s84)

1.12. Wood Preservation

Most of the compounds used for wood preservation have already been introduced. The primary compound categories are chlorinated benzenes (dichlorobenzenes, trichlorobenzenes) and chlorinated phenols (especially pentachlorophenol; McNeill, 1989). A complex mixture of polyaromatic hydrocarbons and substituted monoaromatics (phenols, s31; xylenes, s4; cresols, s32) known as *creosote* was widely used as a preservative for railroad ties (Arvin and Flyvbjerg, 1992). Other compounds include fungicides such as tributyltin, 8-hydroxyquinoline (s82), and dieldrin (s70; McNeill, 1989).

1.13. Explosives Industry

The explosives industry produces a very interesting class of compounds. Generally speaking, explosives are relatively simple C-N-O compounds, often cyclic, with the nitrogen atoms in azo functionality or as part of nitro groups. Probably the most famous representative of this class of compounds is TNT (s85), or trinitrotoluene. Other popular explosives are HMX (s86) and RDX (s87), while some newer explosives include CL-20 (s88) and TNAZ (s89) (Borman, 1994). Still hypothetical but of great interest because of their potential density:energy ratio are the aza relatives of cubane (C_8H_8, s90), C/N cubanoids with varying numbers of nitrogen-replacing carbon atoms (Engelke, 1993). A more complete compilation and description of explosives is given by Liebenberg et al. (1993).

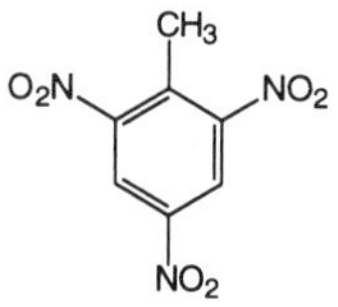

TNT
(s85)

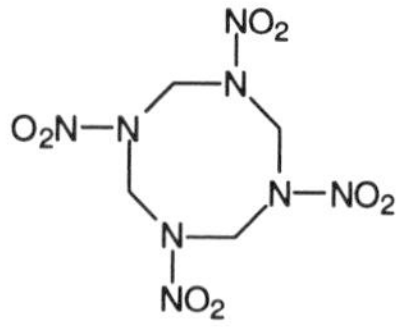

HMX
(s86)

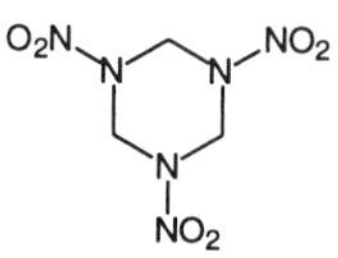

RDX
(s87)

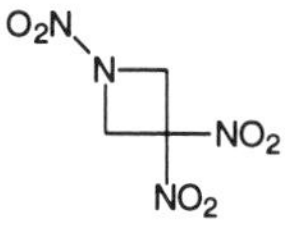

TNAZ
(s88)

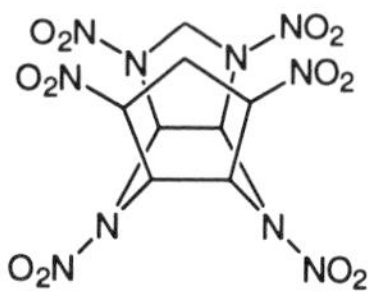

CL-20
(s89)

cubane
(not an explosive)
(s90)

To the author's knowledge, little is known about the chemistry and fate of most of these compounds, partly due to the secretive nature of their development and application. It can be assumed that numerous related compounds are either used as intermediates or produced as wastes.

1.14. Energy Industry/Combustion of Fossil Fuels

Although the energy/combustion area is closely related to the petrochemical industry, emphasis in the energy sector is on the combustion of fossil fuels rather than on their use for synthetic processes. The common ground, of course, is the handling of bulk chemicals in large quantities (such as gasoline or diesel fuel) and thus the inadvertent release of a number of relatively simple hydrocarbons into the environment. Gasoline, for example, typically contains approximately 70% aliphatic hydrocarbons and up to 30% aromatic hydrocarbons such as toluene, ethylbenzene, and others (see section on petrochemicals). Leaking underground storage tanks are but one very common source, with many other ways possible for such a commonly handled chemical to find its way into the environment.

Combustion, on the other hand, gives rise to an entirely new class of compounds, mainly nitrated forms of the hydrocarbons found in gasoline and their reaction products. The exact nature of the combustion process determines which products are formed. For example, hydrocarbon emissions from engines using unleaded gasoline have been virtually eliminated due to the use of catalytic converters, but diesel engines still produce significant amounts of hydrocarbons and nitrated hydrocarbons, as do combustion processes used in power plants. A class of compounds of particular concern are the nitrated polyaromatic hydrocarbons because of their carcinogenic potential. For a more detailed discussion of this compound class, see White (1985).

2. SYNTHETIC ORGANIC CHEMICALS: AMOUNTS PRODUCED

This section discusses both the production volume of bulk chemicals and some selected compound classes and the volume of waste products generated by some industries. The production volume does not have any direct correlation with the amounts of synthetic organic chemicals entering the biosphere, but it may serve as an indicator of the overall production activity and of the volumes of chemicals being handled. Relatively good information is available on the quantities of bulk chemicals produced worldwide; the more specialized the chemical is, the less available the information. Very little information is available on the quantities of synthetic organic chemicals that are released into the environment by accident, by discharge, or as byproducts.

2.1. Petrochemical Industry

The U.S. output of bulk chemicals in 1991 and 1993, together with some data on production volumes in Japan, is summarized in Table 2.2. Benzene, ethylene, and propylene are clearly the compounds produced in the largest quantities. Year-to-year variations are generally very small (a few percent), and major shifts in production volumes of individual compounds occur over a period of several years. In order to set the U.S. output into perspective, the worldwide production volume of BTEX compounds and gasoline is given in Table 2.3. The United States, Western Europe, and the Far East clearly dominate the picture, a fact that is also reflected in statistics on the worldwide consumption of synthetic materials (Table 2.4).

The demand for some widely used chlorinated solvents, listed by different applications, is summarized in Table 2.5, together with some data on total production volumes. Other important production sectors are nitroaromatics (Table 2.6) and terephthalic acid, which is used in polyester fibers and polyethylene terephthalate bottles. In 1990, 6,520 ktons of terephthalic acid were produced in the United States alone (Macarie et al., 1992).

2.2. Agrochemical Industry

The total agrochemical market worldwide in 1985 was $16 billion, of which $7 billion were for herbicides. The U.S. market in 1987 amounted to $3.5 billion, with herbicides contributing 70% of the total (Parry, 1989). This information is not easily translated into pesticide production volumes, but it gives at least an impression of the order of magnitude of pesticide production.

Major applications for herbicides are in maize, soya, cotton, and grain production, as well as in miscellaneous vegetative control (Parry, 1989). The production of herbicides broken down into individual categories is given in Table 2.7. Very few numbers are available on the production volumes of individual pesticides. A small selection of herbicides is listed in Table 2.8. Considering that a large proportion of pesticides is designed for agricultural application, a significant amount of all the

TABLE 2.2. Production of Bulk Chemicals in the United States and Japan

Compound	U.S. 1991[a] (ktons/year)	U.S. 1993[b] (ktons/year)	Japan 1987[c] (ktons/year)
Acetaldehyde	n.a.	n.a.	285
Acetic acid	1,680	1,587	367
Acetone	n.a.	n.a.	260
Acrylonitrile	1,260	1,258	573
Benzene	6,047	6,246	2,224
Butadiene	n.a.	n.a.	707
Butanol	n.a.	n.a.	259
Caprolactam	578	576	470
Cumene	2,479	2,177	n.a.
Cyclohexane	n.a.	n.a.	557
Dioctylphthalate	n.a.	n.a.	261
Ethylene	18,723	18,552	4,585
Dichloroethylene	n.a.	n.a.	2,051
Tetrachloroethylene	112	130	n.a.
Ethylene glycol	n.a.	n.a.	313
Ethylene oxide	2,318	1,761	484
Formaldehyde	3,090	3,231	n.a.
Isopropanol	542	583	n.a.
Methanol	3,206	3,580	181
Octanol	n.a.	n.a.	257
Phenol	n.a.	n.a.	308
Phthalic anhydride	n.a.	n.a.	172
Propylene	9,924	9,818	3,370
Propylene glycol	n.a.	n.a.	236
Sytrene	3,793	3,861	1,559
Terephthalic acid	n.a.	n.a.	1,047
Toluene	3,784	3,160	881
Toluene diisocyanate	346	318	91
Urea	n.a.	n.a.	729
Vinylacetate	1,163	1,243	n.a.
o-Xylene	500	446	1,767[d]
p-Xylene	2,653	2,580	n.a.

[a]Based on first quarter (Economic Trends, 1992).
[b]Based on first quarter (Economic Trends, 1993).
[c]Kiefer (1988).
[d]Total xylenes.

pesticides produced is directly released into the environment, and another significant percentage of pesticides will find their way into the environment due to their outdoor use, improper disposal, or as byproducts from the production process.

One of the few pesticides for which some specific information is available is DDT. Its use in the United States was outlawed in 1973, but 4 billion pounds had been used worldwide since 1940, with 80% of that amount used in agriculture.

TABLE 2.3. BTX and Gasoline Demand in 1989[a]

Countries	Benzene	Toluene[b]	Xylenes[b]	gasoline[b,c]
United States	6.3	4.1	4.8	23.28
Western European	6.0	2.8	2.6	n.a.
Far Eastern	4.5	2.1	3.7	n.a.
Other	5.3	2.2	2.0	n.a.
World	22.1	11.2	13.1	n.a.

[a]Adapted from Acosta and Barry (1990).
[b]Mtons/year.
[c]Converted from gallons.

TABLE 2.4. Worldwide Consumption of Synthetic Materials[a]

Countries	Plastics	Rubbers	Fibers
North American	15.2	9.0	10.1
Japan	13.9	8.0	6.0
Western European	13.2	5.0	10.1
Latin American	9.0	1.0	2.0
Asian	3.0	0.7	2.0
African	2.0	0.5	n.a.

[a]Adapted from Vergara (1989). Values are kg/capita.

TABLE 2.5. The Use of Chlorinated Solvents in the United States (1988)[a]

Application	TCE	PERC	METH	TCA	CFC-113
Vapor degreasing	47.1	18.1	5.8	106.0	17.7
Dry cleaning	—	120.0	—	—	2.0
Intermediate	7.0	80.0	—	22.5	5.3
Cold cleaning	14.1	6.7	17.2	48.0	4.2
Electronics	3.2	1.3	16.9	17.0	40.2
Aerosols	—	3.0	20.0	40.8	0.6
Paint stripping	—	—	50.0	—	—
Adhesives	—	—	5.0	26.0	—
Coatings	—	7.0	—	17.2	—
Flexible foam	—	—	23.2	—	—
Pharmaceuticals	—	—	14.4	—	—
Textiles	1.0	2.0	—	7.0	0.5
Food processing	—	—	4.2	—	—
Pesticides	—	—	1.0	3.0	—
Other	—	20.0	49.3	10.5	7.5
Total demand	72.0	258.0	207.0	298.0	78.0
Total production	82.0	226.0	229.0	328.0	78.0
Total production 1990[b]	148.0	345.0	244.0	476.0	n.a.
Worldwide production in 1976[c]	n.a.	n.a.	500.0	n.a.	n.a.

[a]TCE, trichloroethylene; PERC, perchloroethylene, tetrachloroethylene; METH, methylenechloride; TCA, trichloroethane; CFC-113, CCl_2F-$CCIF_2$. Values are ktons/year. Adapted from Wolf et al. (1991).
[b]From Shariff and Rotman (1991).
[c]From Fishbein (1979b).

TABLE 2.6. Production of Nitroaromatics (1980)[a]

Product Category	Market Size (Mtons/year)
Polymers[b]	572
Rubber chemicals	101
Dyes	21
Pharmaceuticals	13
Pesticides	69
Other	32

[a]Production volumes include derivatives of benzene, chlorobenzene, and toluene. Adapted from Rickert (1985).
[b]Mostly polyurethane produced from aniline.

TABLE 2.7. Worldwide Herbicide Market (1986)[a]

Product Category	No. of Products	Market Value (M$)	Typical Product
Triazines	29	1,425	Atrazine
Amides (Haloacetanilides)	24	940	Alachlor
Carbamates, thiocarbamates	34	765	EPTC
Toluidines	22	660	Treflan
Ureas	37	670	Linuron
Hormones	34	435	2,4-D
Diphenyl ethers	29	345	Flex
Others	≈350	1,835	Paraquat, diazines

[a]Adapted from Parry (1989).

TABLE 2.8. Worldwide Production of Herbicides (1984)[a]

Product	Production (ktons/year)
Atrazine	39
Cyanazine	13
Diuron	3
2,4-D	30
Alachlor	50
Glyphosate	12

[a]Adapted from Parry (1989).

Some countries still use DDT today for mosquito control and other purposes, and it can find its way into the United States through agricultural products from other countries.

Two other pesticides, namely, alachlor and atrazin, are of particular interest because of their toxicity. The U.S. EPA estimated that in 1987 100 million pounds of each of these two pesticides were used in the United States mostly on corn, cotton, and soybean crops (Ware, 1989).

2.3. Paint Industry

Total paint production in the United States in 1992 was close to 1 billion gallons (Reisch, 1993). With an average formulation of 10% – 30% volatile organic compounds (VOCs), this amounts to a significant volume of VOCs being released into the atmosphere. VOC emissions of 990 million pounds were estimated for 1990, mainly originating from architectural and maintenance coatings (Reisch, 1993).

2.4. Waste Production

As an indicator of the order of magnitude of waste production, total hazardous waste volumes for some European countries are given in Table 2.9. A fraction of these wastes was transported to a country other than the producing country, as shown by the import/exported statistics (Table 2.9). Hazardous wastes exported to other countries were mostly oil-polluted soils and sludges, hydroxide sludges, acidic and alkali solutions, and wastes containing halogens, pesticides, and organic solvents. A large volume of the exported wastes was disposed of in the ocean via Belgium. With production levels of synthetic organic chemicals in the United States

TABLE 2.9.   Hazardous Waste Production in Europe[a]

Country	Quantity Produced	Imported	Exported
West Germany	4,900	n.a.	n.a.
East Germany	n.a.	0.0	345.6
France	2,000	4.9	21.5
England	1,500	n.a.	n.a.
Sweden	550	0.8	0.0
The Netherlands	280	13.4	1.9
Norway	120	n.a.	n.a.
Switzerland	100	10.9	11.1
Denmark	63	0.7	0.0
Finland	87	n.a.	n.a.
Italy	n.a.	0.4	4.1
Austria	n.a.	0.4	1.7
Belgium	n.a.	1.6	917.3

[a]Production is given for 1984 and includes all types of hazardous waste. Values are ktons/year. Adapted from Shin (1992).

being comparable to production levels in Western Europe, it can be assumed that the volume of hazardous wastes produced in the United States is of a similar magnitude.

Typical quantities of hazardous wastes produced in West Germany in 1986 are given in Table 2.10. Approximately 20% of the total volume is organic material of various kinds. A significant amount of hazardous wastes falls into the category "others" (27% of total), which may contain more organic material. The category "sulfur containing" is not defined, but could potentially also include significant amounts of organic compounds.

Only erratic information exists on the waste production of specific industries. Petrochemical plants, for example, can produce a massive output of wastes containing organic chemicals. Basic compounds like benzene, simple chlorinated benzenes, and hexachlorobenzene are produced in large quantities, and the corresponding waste volumes can be expected to be rather large. PAHs, for example, can be found in concentrations ranging from 50 to 1,000 mg/kg of sludge (Marks et al., 1992).

Hexachlorobenzene (HCB) wastes produced in the United States in 1970 amounted to 1,533 tons, and another 19,164 tons were accumulated as HCB-containing waste, which is defined by lower concentrations of HCB (Fishbein, 1979d). Source industries in the case of HCB were mostly manufacturers of chlorinated solvents, with some contributions from pesticide plants and electrolytic chlorine manufacturers. Wastes containing high levels of HCB were disposed of in landfills, by deep well injection, or incineration (approximately 50% of these wastes being incinerated). Up to 80% of the total volume of wastes containing lower levels were injected into deep wells (Fishbein, 1979d).

The pulp and paper industry is an important source of organic chemical wastes. Information on waste volumes is not easily accessible, but it has been estimated that chlorinated organic compounds, collectively classified as adsorbable organic halide (AOX) amount to about 2.5 kg per ton of pulp produced (Bryant et al., 1992).

TABLE 2.10.   Hazardous Waste Production in West Germany[a]

Waste Category	Production (ktons/year)
Sulfur-containing waste	2,160
Oil-containing waste	490
Paint sludges	250
Organic solvents (halogenated)	223
Organic solvents (not halogenated)	90
Galvanic waste	190
Salt sludge	125
Others	1,342

[a]Production for different waste categories is given for 1986. Adapted from Shin (1992).

3. FATE OF SYNTHETIC ORGANIC CHEMICALS
IN THE ENVIRONMENT

There is an abundant literature on the fate of synthetic organic chemicals in the environment. It is a very complex topic and well beyond the scope of this chapter to review the fate of individual compound classes in detail. The discussion will, therefore, be limited to some general principles, highlighting a few selected aspects. For more extensive treatment, the reader is referred to the literature on specific topics such as transport, chemical reactivity, and biodegradation. A small selection of books and review articles treating the general topic may serve as a starting point, especially those on fate in general (Ney, 1990); long-range atmospheric transport (Knap, 1990); transport in general (Albaigés and Frei, 1986; Iwata et al., 1993); bioaccumulation (Connell, 1990); nitroaromatics (Rickert, 1985); and dioxins (Choudhary et al., 1983).

A few selected papers on more specific topics may serve as an illustration of the information that has become available in recent years, especially those on solvents (Longstaff et al., 1992); wood preservation chemicals (McNeill, 1989; Arvin and Flyvbjerg, 1992); chorinated benzene relatives (Remberger et al., 1993); PCBs and PAHs (Cooke et al., 1979; Arvin and Flyvbjerg, 1992; Christensen and Zhang, 1993; Bergen et al., 1993; Herbert et al., 1993; Järnberg et al., 1993); and pesticides (Braun and Frank, 1980; Frank and Sirons, 1980; Croll, 1991; Gomme et al., 1992; Achman et al., 1993; Brown et al., 1993; Kolpin and Kalkhoff, 1993).

3.1. The Fate of Organic Chemicals in the Environment:
A General Picture

A simplified schematic representation of the fate of many organic compounds in the environment is presented in Figure 2.1. It is meant to serve as an overview of the most important parameters and processes controlling the fate of synthetic organic chemicals in the environment. Parameters include the physical and chemical characteristics of a compound, its behavior with respect to chemical reactions or microbial degradation, and other physical conditions such as temperature, availability of water, light, and oxygen. The major processes responsible for distributing synthetic organic chemicals throughout the biosphere are volatilization and atmospheric transport, transport in groundwater and surface waters in soluble form or adsorbed to particles, or movement through the food chain. The major sinks are the atmosphere, soils, sediments, oceans, and the highest members of a given food chain.

The scheme presented in Figure 2.1 is obviously overly simplified since there are many subtle distinctions and unknowns that are not included. For example, probably no compound is totally insoluble in water. Even very hydrophobic compounds like chlorinated solvents or PCBs are found in aquatic environments, and transport in surface and ocean waters are a major component in PCB distribution (Iwata et al., 1993). Other important mechanisms determining the fate of synthetic organic chemicals are chemical and microbial transformations or degradation. In both chemical and especially in microbial processes, it is not necessarily enough that a compound

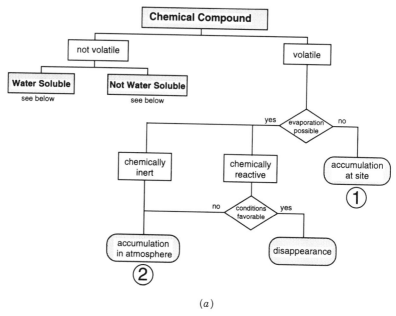

(a)

Figure 2.1.   A simplified diagram illustrating the fate of a synthetic chemical compound (a), both water soluble (b) and nonwater soluble (c), in the environment.

is amenable to degradation; the environmental conditions also must be met. For example, CFCs are rather stable compounds and only undergo photochemical reactions in specific regions of the stratosphere. Another class of processes that are highly dependent on all environmental factors being favorable are microbial transformations. Controlling factors are bioavailability, presence or absence of oxygen and nutrients, presence of water, presence of indigenous microbial communities, temperature, and pH.

3.2. Special Remarks

Some more specialized aspects concerning the fate of synthetic organic chemicals should also be mentioned here:

1. Toxic intermediates formed during microbial transformations. During microbial (or, for that matter, chemical) degradation, one has to keep in mind that in many cases degradation involves a number of transformations prior to mineralization. There is always the potential for the formation of toxic intermediates. The formation of vinylchloride during the degradation of perchlorethylene is a well known example (Vogel and McCarty, 1985). The formation

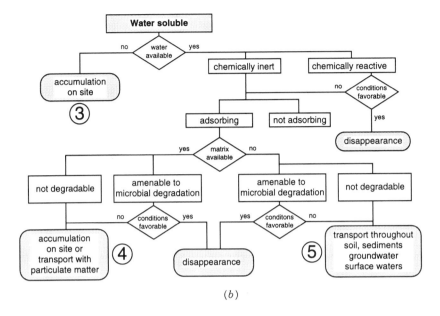

(b)

Figure 2.1. (*Continued*)

of the very potent carcinogen 7,8-dihydrol-9,10-epoxide of benzo[*a*]pyrene during benzo[*a*]pyrene metabolism is another example (Marks et al., 1992), as is the formation of dioxins during the combustion of organic wastes (Choudhary et al., 1983).

2. Complexity of chemical precludes detailed study of fate in environment. Toxaphene, a chlorinated terpene used as an insecticide, is actually not a single compound, but a mixture of up to 200 polychlorinated isomers with variable chlorine contents, where no single component accounts for more than a few percent of the total mixture. The toxicology and fate in higher or lower organisms are extremely difficult to assess because of the complexity of the mixture (Ware, 1989).

3. Predictability of fate from chemical structure. Although many attempts have been made to correlate structure and amenability to chemical and/or micro-bial degradation (Golberg, 1983; Wolfe et al., 1980; Paris et al., 1982; Paris and Wolfe, 1987), only very limited statements can be made, and ultimately the fate of synthetic organic chemicals has to be examined for each individual compound. The chemical structure of a compound can give certain clues as to the reactivity, but very similar compounds can still exhibit highly different degradation rates. For example, many of the organophosphorous pesticides have very similar structures, but show very different reactivities (Ware, 1989).

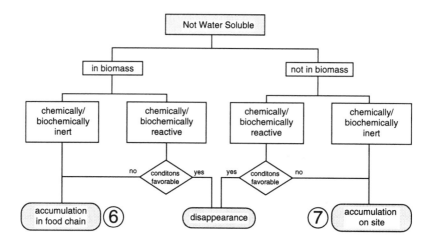

Examples:

① Chlorinated solvents in landfills; although they are volatile they can be trapped; also, despite the fact that they are very sparingly water soluble, they are present in water at low concentrations, but high enough to be of concern.

② CCl₄, CFC's "unreactive" except in specific parts of the atmosphere

③ Water soluble pesticides in arid climates

④ Chlorinated benzoic acids, some pesticides

⑤ Many pesticides, petrochemicals

⑥ DDT, PCBs etc.

⑦ Coal tar, PAHs

(c)

Figure 2.1. (Continued)

ACKNOWLEDGEMENTS

I would like to thank most of all Lily Young for her extreme patience and strong support over the long period of time it took to complete this chapter. I would also like to thank Patricia Colberg, my partner in crime, for putting up with my working weekends, late nights, and early mornings; Linda Mickley for an excellent and speedy job of editing the manuscript; and last, but not least, the staff at John Wiley & Sons for not giving up on the project.

REFERENCES

Achman DR, Hornbuckle KC, Eisenreich SJ (1993): Volatilization of polychlorinated biphenyls from Green Bay, Lake Michigan. Environ Sci Technol 27:75–87.

Acosta T, Barry WP (1990): The changing aromatics marketplace of the 1990s. Chem Ind 7:220–223.

Albaigés J, Frei RW (1986): Fate of Hydrocarbons in the Environment: An Analytical Approach. New York, NY: Gordon and Breach Science Publishers.

Andrews BAK, Reinhardt RM (1989): Reaction equilibria and textile performance from low formaldehyde release carbamate finishing. Text Res J 59:53–60.

Arvin E, Flyvbjerg J (1992): Groundwater pollution arising from the disposal of creosote waste. J Inst Water Environ Manage 6:646–652.

Bergen BJ, Nelson WG, Pruell RJ (1993): Partitioning of polychlorinated biphenyl congeners in the seawater of New Bedford Harbor, Massachusetts. Environ Sci Technol 27: 938–942.

Bjørseth A, Carlberg GE, Møller M (1979): Determination of halogenated organic compounds and mutagenicity testing of spent bleach liquors. Sci Total Environ 11:197–211.

Bodmeier R, Paeratakul O (1994): The distribution of plasticizers between aqueous and polymer phases in aqueous colloidal polymer dispersions. Int J Pharma 103:47–54.

Borman S (1994): Advanced energetic materials emerge for military and space applications. Chem Eng News 72:18–22.

Braun HE, Frank R (1980): Organochlorine and organophosphorus insecticides: Their use in eleven agricultural watersheds and their loss to stream waters in southern Ontario, Canada, 1975–1977. Sci Total Environ 15:169–192.

Brezny R, Joyce TW, Gonzalez B (1992): Biotransformation in soil of chloroaromatic compounds related to bleach plant effluents. Water Sci Technol 26:397–406.

Brown MA, Petreas MX, Okamoto HS, Mischke TM, Stephens RD (1993): Monitoring of malathion and its impurities and environmental transformation products on surfaces and in air following an aerial application. Environ Sci Technol 27:388–397.

Bryant CW, Avenell JJ, Barkley WA, Thut RN (1992): The removal of chlorinated organics from conventional pulp and paper wastewater treatment systems. Water Sci Technol 26:417–425.

Carr CM, Weedall PJ (1989): Effect of solvent treatments on the handle of wool fabrics. Text Res J 59:45–48.

Castaldi FJ, Ford DL (1992): Slurry bioremediation of petrochemical waste sludges. Water Sci Technol 25:207–212.

Choudhary G, Keith LH, Rappe C (1983): Chlorinated Dioxins and Dibenzofurans in the Total Environment. Boston, MA: Butterworth.

Christensen ER, Zhang X (1993): Sources of polycyclic aromatic hydrocarbons to Lake Michigan determined from sedimentary records. Environ Sci Technol 27:139–146.

Connell DW (1990): Bioaccumulation of Xenobiotic Compunds. Boca Raton, FL: CRC Press.

Cooke M, Nickless G, Povey A, Roberts DJ (1979): Poly-chlorinated biphenyls, polychlorinated naphthalenes and polynuclear aromatic hydrocarbons in Severn Estuary (U.K.) sediments. Sci Total Environ 13:17–26.

Croll BT (1991): Pesticides in surface waters and groundwaters. J Inst Water Environ Manage 5:389–395.

D'Itri FM, Kamrin MA (1983): PCBs: Human and Environmental Hazards. Boston, MA: Butterworth.

de Kruijf N, Schouten A, Rijk MAH, Pranoto-Soetardhi LA (1989): Determination of preservatives in cosmetic products II. High-performance liquid chromatographic identification. J Chromatog 469:317–328.

Eacott C, Linley C (1991): A new preservative for paint. Mfg Chem 62:19–21.

Economic Trends (1992): Quarterly chemical output report. Chem Week 151:37.

Economic Trends (1993): Quarterly chemical output report. Chem Week 152:29.

Engelke R (1993): Ab initio calculations of ten carbon/nitrogen cubanoids. J Am Chem Soc 115:2961–2967.

Englande AJ Jr, Guarino CF (1992): Toxics management in the chemical and petrochemical industries. Water Sci Technol 26:263–274.

Fishbein L (1979a): Potential halogenated industrial carcinogenic and mutagenic chemicals I. Halogenated unsaturated hydrocarbons. Sci Total Environ 11:111–161.

Fishbein L (1979b): Potential halogenated industrial carcinogenic and mutagenic chemicals II. Halogenated saturated hydrocarbons. Sci Total Environ 11:163–195.

Fishbein L (1979c): Potential halogenated industrial carcinogenic and mutagenic chemicals III. Alkane halides, alkanols and ethers. Sci Total Environ 11:223–257.

Fishbein L (1979d): Potential halogenated industrial carcinogenic and mutagenic chemicals IV. Halogenated aryl derivatives. Sci Total Environ 11:259–278.

Flick EW (1986): Household and Automotive Cleaners and Polishes. Park Ridge, NJ: Noyes Publications.

Frank R, Sirons GJ (1980): Chlorophenoxy and chlorobenzoic acid herbicides; their use in eleven agricultural watersheds and their loss to stream waters in southern Ontario, Canada 1975–1977. Sci Total Environ 15:149–167.

Golberg L (1983): Structure–Activity Correlations as a Predictive Tool in Toxicology: Fundamentals, Methods, and Applications. Washington, DC: Hemisphere Publishing Corp.

Gomme J, Shurvell S, Hennings SM, Clark L (1992): Hydrology of pesticides in a Chalk catchment: Groundwaters. J Inst Water Environ Manage 6:172–178.

Gulyas H, Hemmerling L (1990): Tetrachloroethene air pollution originating from coin-operated dry cleaning establishments. Environ Res 53:90–99.

Häggblom M, Salkinoja-Salonen M (1991): Biodegradability of chlorinated organic compounds in pulp bleaching effluents. Water Sci Technol 24:161–170.

Hartley WR, Englande AJ Jr (1992): Health risk assessment of the migration of unleaded gasoline: A model for petroleum products. Water Sci Technol 25:65–72.

Hassall KA (1990): The Biochemistry and Uses of Pesticides, 2nd ed. New York: VCH Publishers.

Herbert BE, Bertsch PM, Novak JM (1993): Pyrene sorption by water-soluble organic carbon. Environ Sci Technol 27:398–403.

Iwata H, Tanabe S, Sakai N, Tatsukawa R (1993): Distribution of persistent organochlorines in the oceanic air and surface seawater and the role of ocean on their global transport and fate. Environ Sci Technol 27:1080–1098.

Järnberg U, Asplund L, de Wit C, Grafström A-K, Haglund P, Jansson B, Lexén K, Strandell M, Olsson M, Jonsson B (1993): Polychlorinated biphenyls and polychlorinated naphthalenes in Swedish sediment and biota: Levels, patterns, and time trends. Environ Sci Technol 27:1364–1374.

Järvisalo J, Pfäffli P, Vainio H (1984): Industrial Hazards of Plastics and Synthetic Elastomers. New York: Alan R. Liss.

Kearny PC, Kaufman DD (1976): Herbicides. New York: Marcel Dekker.

Kearny PC, Kaufman DD (1988): Herbicides. New York: Marcel Dekker.

Kiefer DM (1988): Chemical business will be brisk despite slower growth. Chem Eng News 66:28–45.

Knap AH (1990): The Long-Range Atmospheric Transport of Natural and Contaminant Substances. Boston, MA: Kluwer Academic Publishers.

Kolpin DW, Kalkhoff SJ (1993): Atrazine degradation in a small stream in Iowa. Environ Sci Technol 27:134–139.

Lau MM, Rootham RC, Bradley GC (1993): A strategy for the management of contaminated dredged sediment in Hong Kong. J Environ Manage 38:99–114.

Law AB, Lashen ES (1991): Microbicidal efficacy of a methylchloro/methylisothiazolone/ copper preservative in metalworking fluids. Lubr Eng 47:25–30.

Liebenberg DH, Armstrong RW, Gilman JJ (1993): Structure and Properties of Energetic Materials. Pittsburgh, PA: Materials Research Society.

Longstaff SL, Aldous PJ, Clark L, Flavin RJ, Partington J (1992): Contamination of the Chalk aquifer by chlorinated solvents: A case study of the Luton and Dunstable area. J Inst Water Environ Manage 6:541–550.

Macarie H, Noyola A, Guyot JP (1992): Anaerobic treatment of a petrochemical wastewater from a terephthalic acid plant. Water Sci Technol 25:223–235.

Marks RE, Field SD, Wojtanowicz AK, Britenbeck GA (1992): Biological treatment of petrochemical wastes for removal of hazardous polynuclear aromatic hydrocarbon constituents. Water Sci Technol 25:213–220.

Martakis N, Niaounakis M (1992): Comparison of two distinct phenolic antioxidants in gamma-sterilized NR compounds used as rubber pistons in syringes for single use. J Appl Polymer Sci 46:1737–1748.

Mathur A, Varma IK (1992): Tris(3-aminophenyl)phosphine oxide-based nadimide resins. II. J Appl Polymer Sci 46:1749–1757.

McClure NC, Fry JC, Weightman AJ (1991): Genetic engineering in wastewater treatment. J Inst Water Environ Manage 5:608–616.

McNeill A (1989): The effects of a timber preservative spillage on the ecology of the River Lossie. J Inst Water Environ Manage 3:496–504.

Mehlman MA (1992): Dangerous and cancer-causing properties of products and chemicals in the oil refining and petrochemical industry. Environ Res 59:238–249.

Metcalf RL (1971): The chemistry and biology of pesticides. In White-Stevens R (ed): Pesticides in the Environment. New York: Marcel Dekker, pp 1–144.

Min KE, Lim JC, Seo WY, Kwon HK (1994): Plasticizer effect on miscibility of isotactic poly(methyl methacrylate)-poly(vinyl chloride) blends. J Appl Polymer Sci 51:1521–1525.

Mitsubishi Kasei Corporation (1988): 1,4-Butanediol/tetrahydrofuran production technology. Chemtech 18:759–763.

Ney RE Jr (1990): Where Did That Chemical Go? A Practical Guide to Chemical Fate and Transport in the Environment. New York: Van Nostrand Reinhold.

Paris DF, Wolfe NL (1987): Relationships between properties of a series of anilines and their transformation by bacteria. Appl Environ Microbiol 53:911–916.

Paris DF, Wolfe NL, Steen WC (1982): Structure–activity relationships in microbial transformation of phenols. Appl Environ Microbiol 44:153–158.

Parry KP (1989): Herbicide use and invention. In Dodge AD (ed): Herbicides and Plant Metabolism. Cambridge: Cambridge University Press, pp 1–20.

Pollock I, Young E, Stoneham M, Slater N, Wilkinson JD, Warner JO (1989): Survey of colourings and preservatives in drugs. Br Med J 299:649–651.

PRC Environmental Management Inc (1989): Hazardous Waste Reduction in the Metal Finishing Industry. Park Ridge, NJ: Noyes Data Corporation.

Quirijns JK, van der Paauw CG, ten Noever de Brauw MC, de Vos RH (1979): Survey of the contamination of dutch coastal waters by chlorinated hydrocarbons, including the occurrence of methylthio-pentachlorobenzene and di-(methylthio)tetrachlorobenzene. Sci Total Environ 13:225–233.

Reisch MS (1993): Paints & coatings. Chem Eng News 71:34–61.

Remberger M, Hynning P-Å, Neilson AH (1993): Release of chlorocatechols from a contaminated sediment. Environ Sci Technol 27:158–164.

Rickert DE (1985): Toxicity of Nitroaromatic Compounds. Washington, DC: Hemisphere Publishing.

Safe S (1987): Polychlorinated Biphenyls (PCBs): Mammalian and Environmental Toxicology. New York: Springer-Verlag.

Shariff S, Rotman D (1991): Chlorinated solvents are still hanging on. Chem Week 148:11.

Sheldon LS, Hites RA (1979): Environmental occurrence and mass spectral identification of ethylene glycol derivatives. Sci Total Environ 11:279–286.

Sherry S (1985): High Tech and Toxics: A Guide for Local Communities. Sacramento, CA: Golden Empire Health Planning Center.

Shin IKC (1992): The situation and the problems of hazardous waste treatment in Germany. Water Sci Technol 26:31–40.

Smolinske SC (1992): Handbook of Food, Drug, and Cosmetic Excipients. Boca Raton, FL: CRC Press.

Summers WR (1992): Characterization of azoniaadamantane-based preservatives by combined reversed-phase cation-exchange chromatography with suppressed conductivity detection. Anal Chem 64:1096–1099.

Taylor MRG, McLean RAN (1992): Overview of clean-up methods for contaminated sites. J Inst Water Environ Manage 6:408–417.

U.S. Government (1993a): National emission standards for hazardous air pollutants for source categories. In Code of Federal Regulations. Title 40 Part 63 (40 CFR 63), pp 318–319.

U.S. Government (1993b): EPA administered permit programs: The national pollutant discharge elimination system. In Code of Federal Regulations. Title 40 Part 122 (40 CFR 122), pp 128–133.

U.S. Government (1993c): Toxic pollutant effluent standards. In Code of Federal Regulations. Title 40 Part 129 (40 CFR 129), pp 246–257.

U.S. Government (1993d): Water quality standards. In Code of Federal Regulations. Title 40 Part 131 (40 CFR 131), pp 271–295.

U.S. Government (1993e): National primary drinking water regulations. In Code of Federal Regulations. Title 40 Part 141 (40 CFR 141), pp 593–691.

U.S. Government (1993f): Criteria for municipal solid waste landfills. In Code of Federal Regulations. Title 40 Part 258 (40 CFR 258), pp 350–386.

Vergara W (1989): World petrochemicals in the coming decade. Chem Eng Prog 85:24–32.

Vogel TM, McCarty PL (1985): Biotransformation of tetrachloroethylene to trichloroethylene, dichloroethylene, vinyl chloride, and carbon dioxide under methanogenic conditions. Appl Environ Microbiol 49:1080–1083.

Ware GW (1989): The Pesticide Book. Fresno, CA: Thomson Publications.

White CM (1985): Nitrated Polycyclic Aromatic Hydrocarbons. New York: Dr. Alfred Huethig Verlag.

White-Stevens R (1971a): Pesticides in the Environment, vol 1, part I. New York: Marcel Dekker.

White-Stevens R (1971b): Pesticides in the Environment, vol 1, Part II. New York: Marcel Dekker.

Willson CG, Bowden MJ (1989): Resist materials for microelectronics. Chemtech 19: 102–110.

Wolf K, Yazdani A, Yates P (1991): Chlorinated solvents: Will the alternatives be safer? J Air Waste Manage Assoc 41:1055–1061.

Wolfe NL, Paris DF, Steen WC, Baughman GL (1980): Correlation of microbial rates with chemical structure. Environ Sci Technol 14:1143–1144.

APPENDIX 1: ADDITIONAL CHEMICAL STRUCTURES

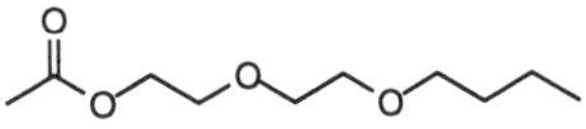

a1
Sta-Way®

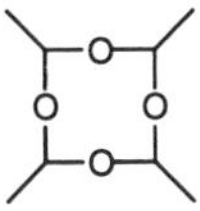

a2
metaldehyde

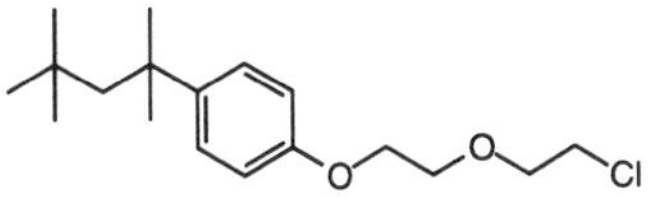

a3
plasticizer

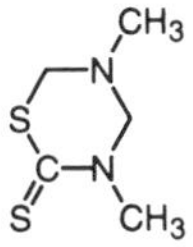

a4
dazomet

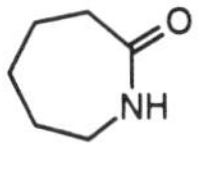

a5
caprolactam

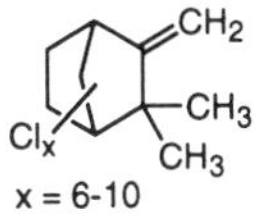

a6
toxaphene

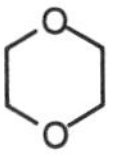

a7
dioxane

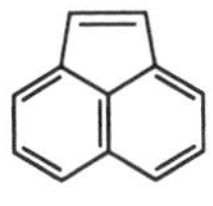

a8
acenaphthylene

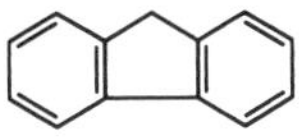

a9
fluorene

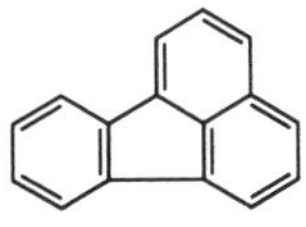

a10
fluoranthene

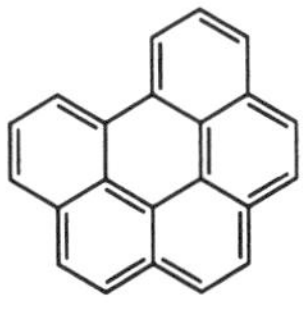

a11
benzo(ghi)perylene

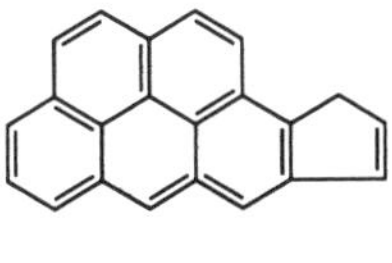

a12
indeno(123-cd)pyrene

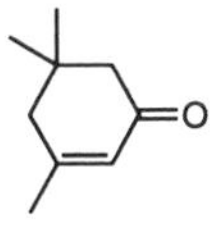

a13
isophorone

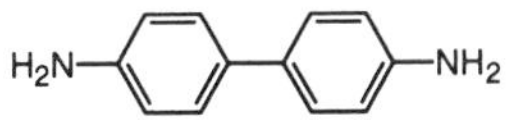

a14
benzidine

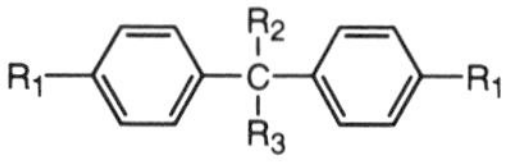

	R₁	R₂	R₃	
a15	Cl-	CHCl₂-	H-	TDE
a17	C₂H₅-	CHCl₂-	H-	ethylan
a18	CH₃O-	CCl₃-	H-	methoxychlor
a19	Cl-	CH₃-	OH-	DMC
a20	Cl-	C₂H₅OOC-	OH-	chlorobenzilate
a21	Cl-	CCl₃-	OH-	dicofol

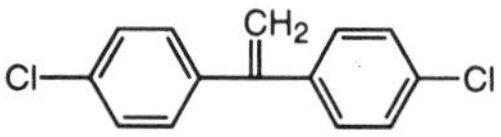

a16
4,4'-DDE

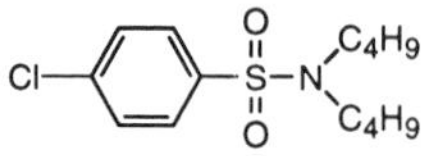

a22
DDT "antiresistant"

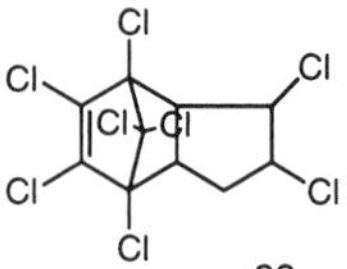

a23
chlordane

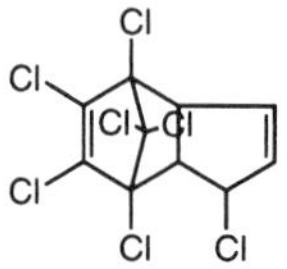

a24
heptachlor

a25
mirex

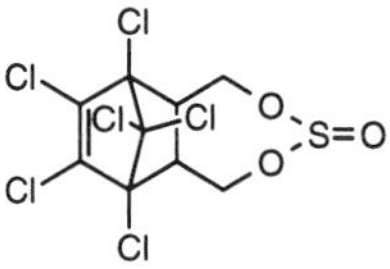

a26
endosulfan

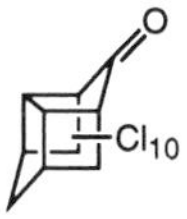

a27
chlorodecone

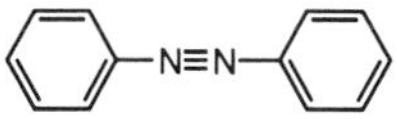

a28
azobenzene

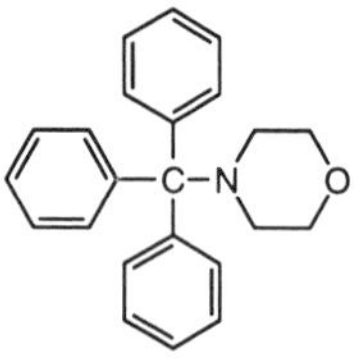

a29
trifenmorph

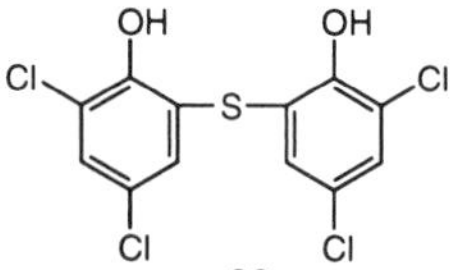

a30
vancide BL

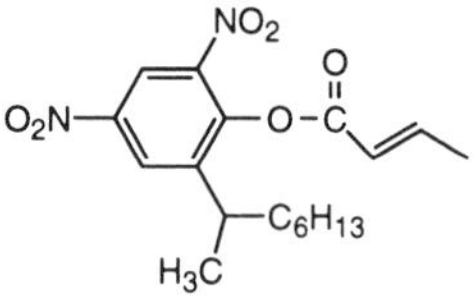

a31
dinocap

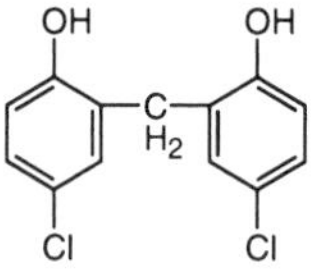

a32
dichlorophene

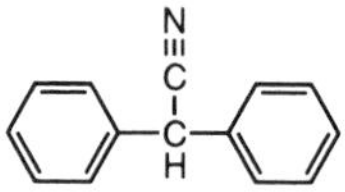

a33
diphenatrile

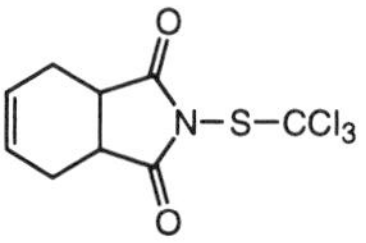

a34
captan

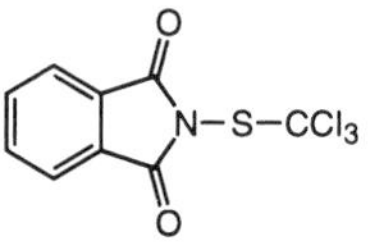

a35
folpet

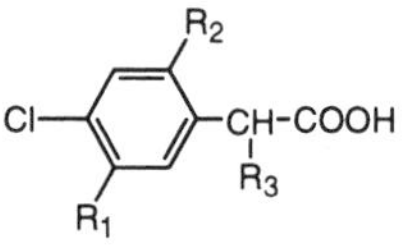

	R₁	R₂	R₃	
a36	-	Cl-	H-	2,4-D
a37	-	CH₃-	H-	MCPA
a38	Cl-	Cl-	H-	2,4,5-T
a39	Cl-	Cl-	CH₃-	2,4,5-TP

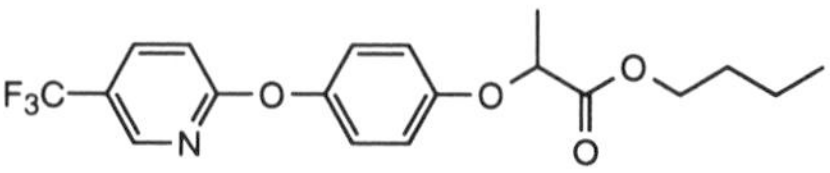

a40
fluazifop-butyl

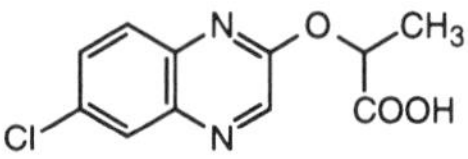

a41
quizalofop-ethyl

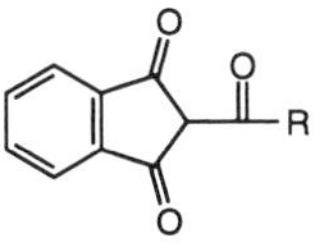

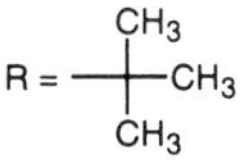

a42
pindone

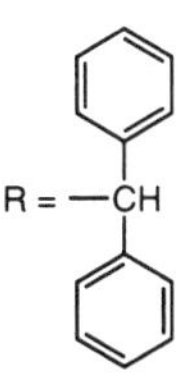

a43
diphacinone

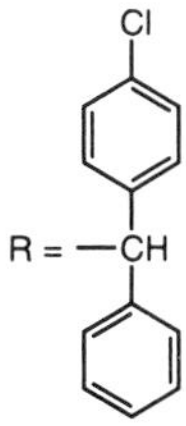

a44
chlorophacinone

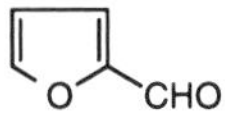

a45
furfural

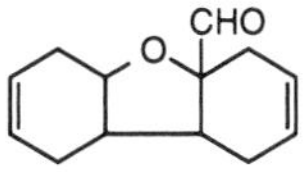

a46
MGK® repellent II

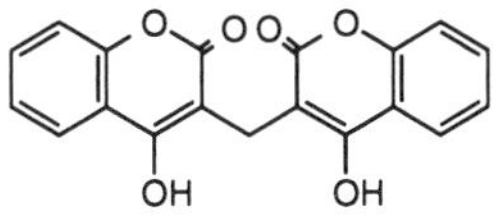

a47
dicumarol

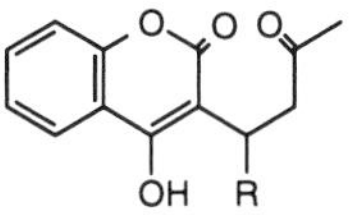

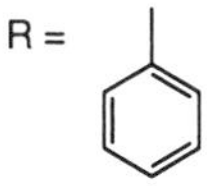

R =

a48
warfarin

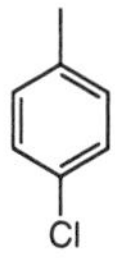

a49
coumachlor

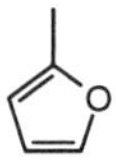

a50
coumafuryl

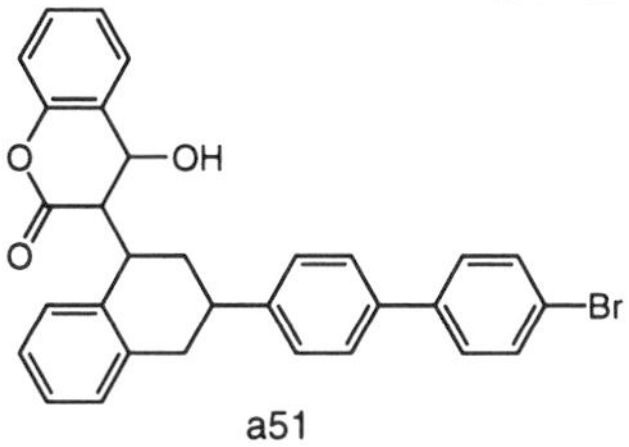

a51
brodifacoum

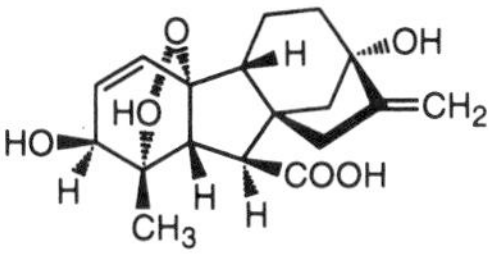

a52
GA3

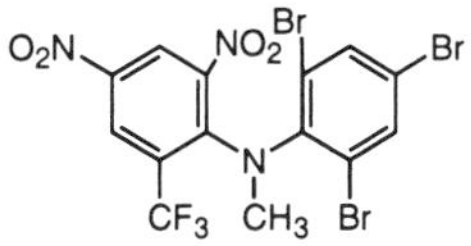

a53
bromethalin

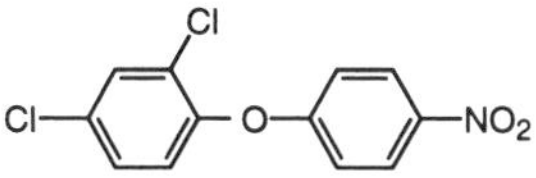

a54
nitrofen

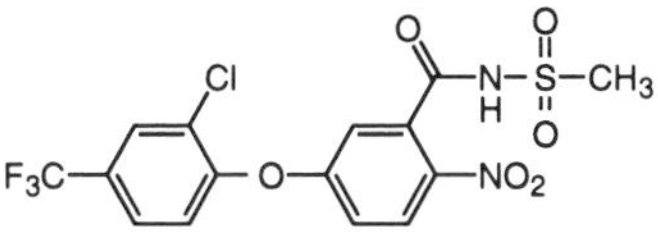

a55
fomesafen

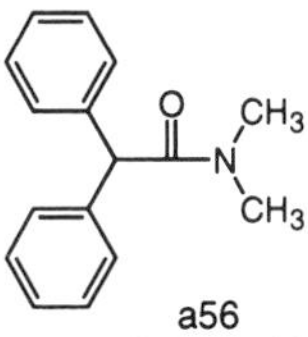

a56
diphenamid

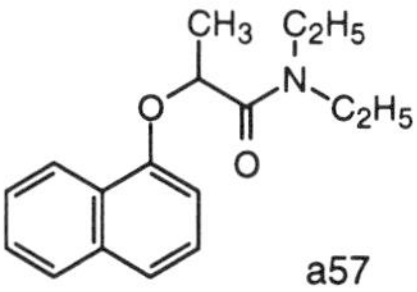

a57
napropamide

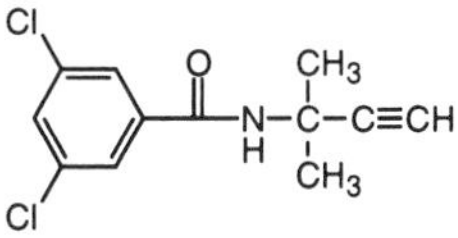

a58
pronamide

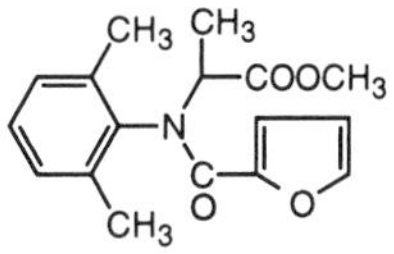

a59
furalaxyl

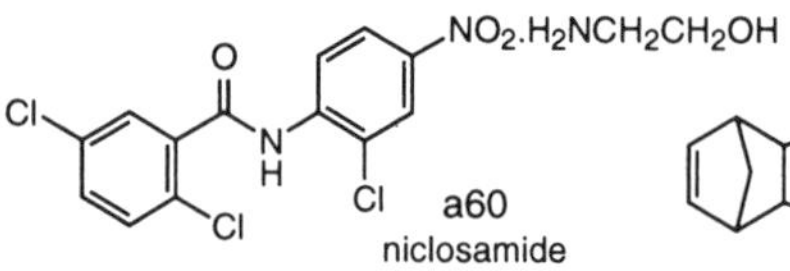

a60
niclosamide

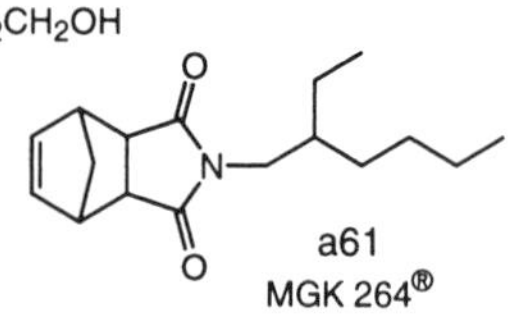

a61
MGK 264®

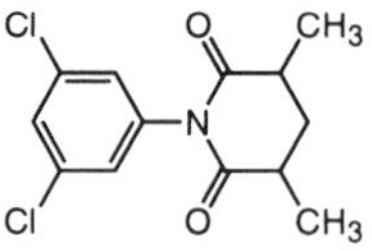

a62
procymidone

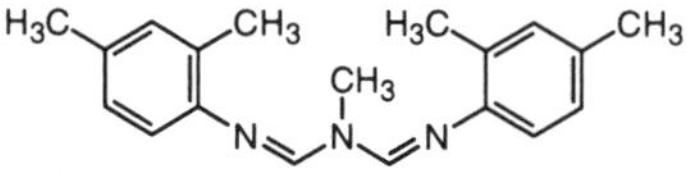

a63
amitraz

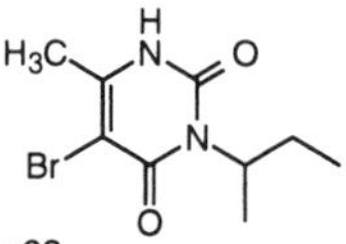

a68
bromacil

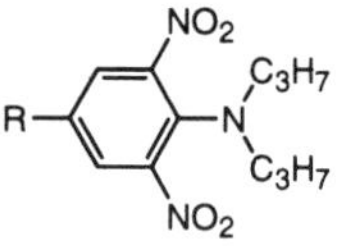

$-CF_3$ a64 trifluralin

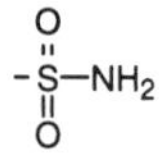

a65 oryzalin

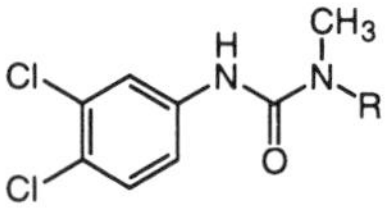

-CH3 a66 diuron

-OCH3 a67 linuron

a69
thiodicarb

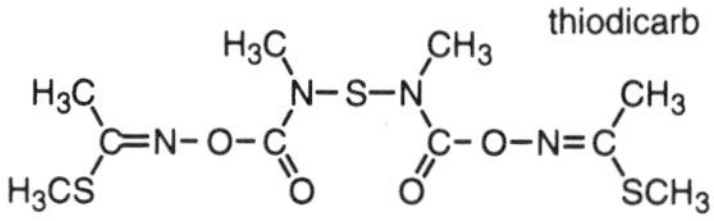

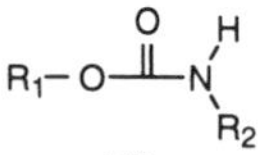

	R1	R2
a70 aldicarb	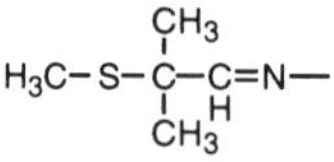	-CH3
a71 carbaryl	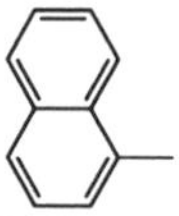	-CH3
a72 methiocarb	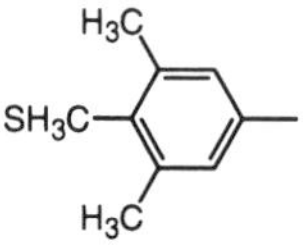	-CH3
a73 bendiocarb	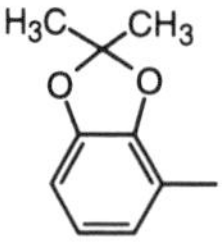	-CH3
a74 mexacarbate	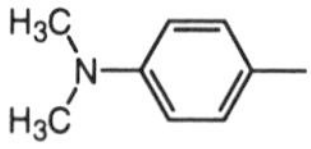	-CH3
a75 barban	$ClH_2C-C\equiv C-CH_2$-	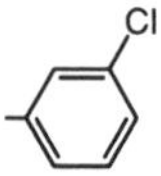
a76 asulam	-CH3	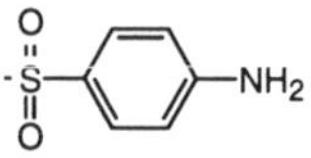

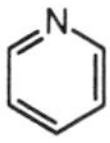

a77
pyridine

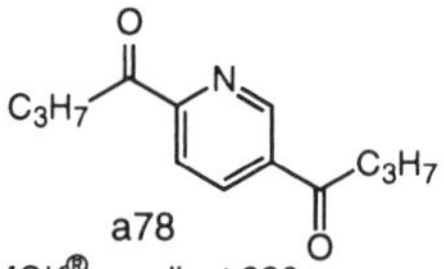

a78
MGK® repellent 326

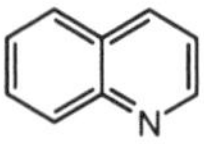

a79
quinoline

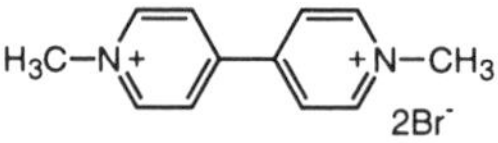

a80
paraquat

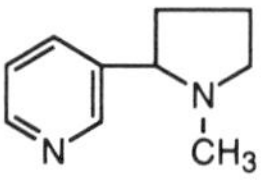

a81
nicotine

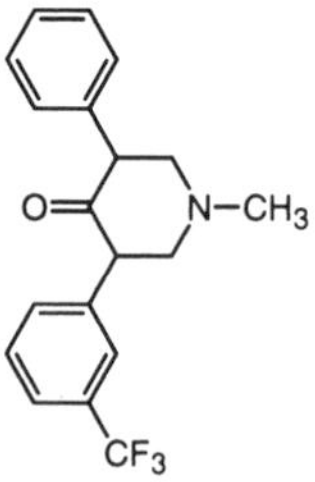

a82
fluridone

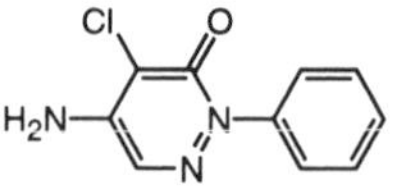

a83
pyrazon
chloridazon

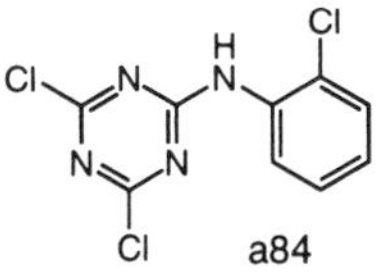

a84
anilazine

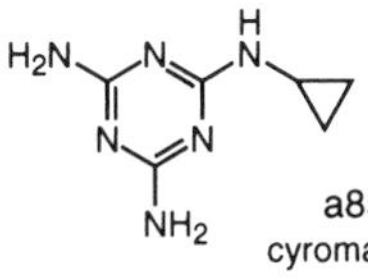

a85
cyromazine

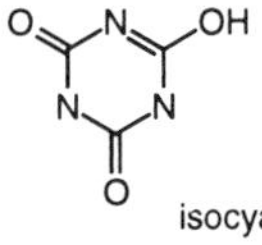

a86
isocyanuric acid

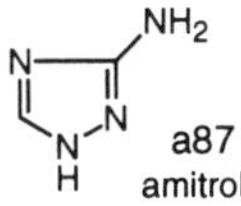

a87
amitrole

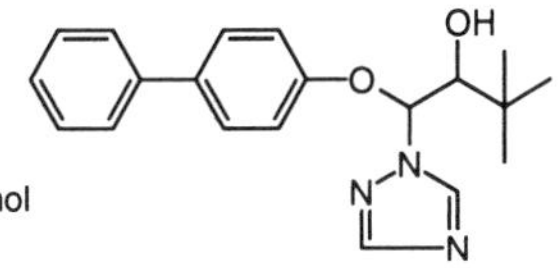

a88
bitertanol

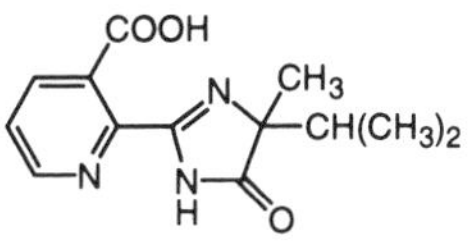

a89
imazapyr

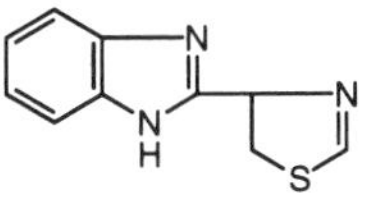

a90
thiabendazole (TBZ)

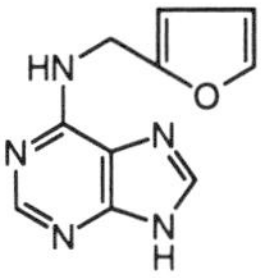

a91
kinetin

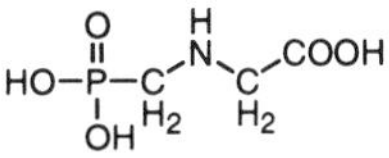

a92
glyphosate

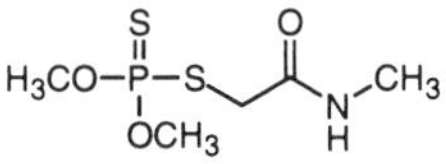

a93
dimethoate

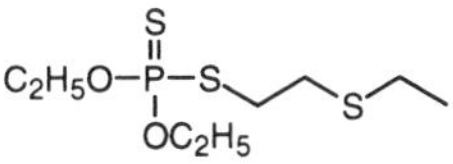

a94
disulfoton

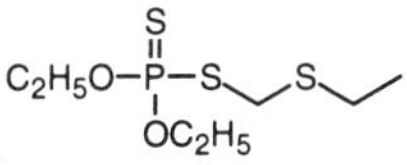

a95
phorate

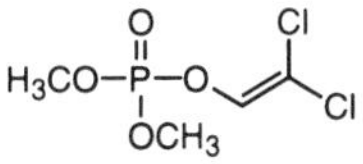

a96
dichlorvos

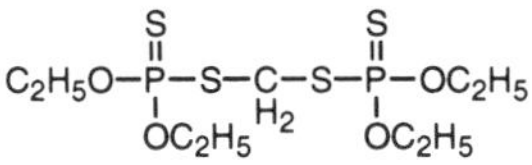

a97
ethion

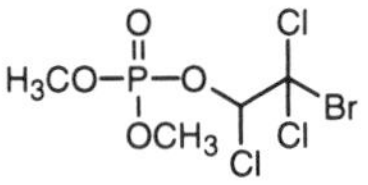

a98
naled

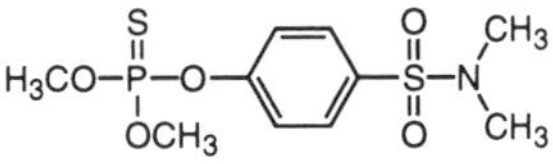

a99
famphur

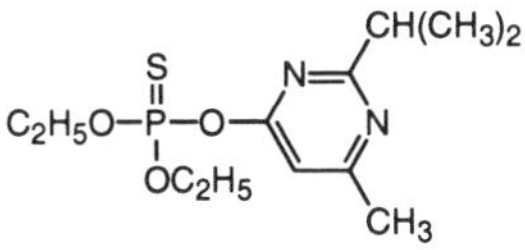

a100
diazinon

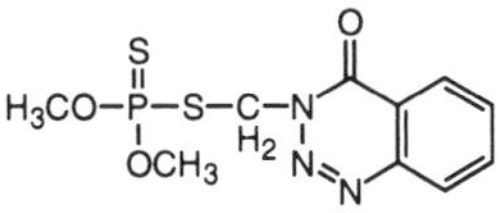

a101
azinphosmethyl

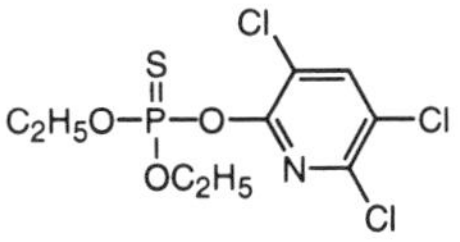

a102
chlorpyrifos

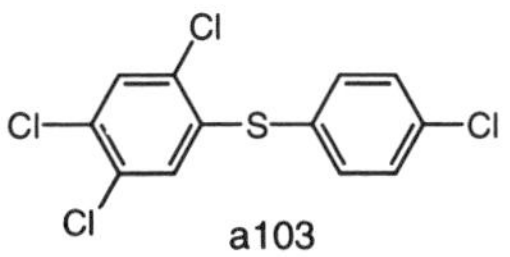

a103
tetrasul

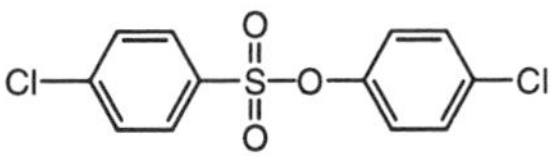

a104
ovex

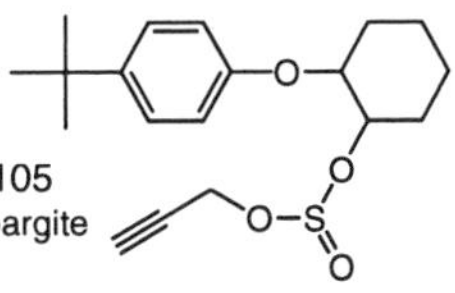

a105
propargite

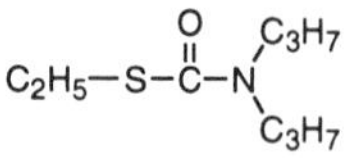

a106
EPTC

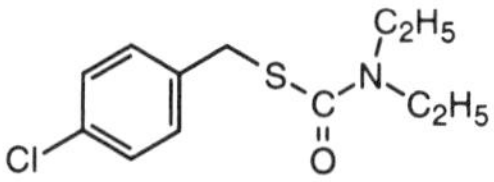

a107
thiobencarb

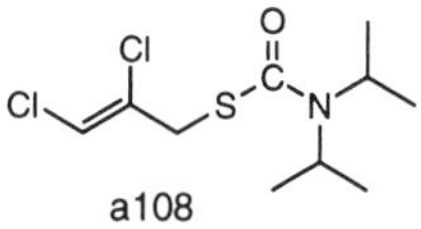

a108
diallate

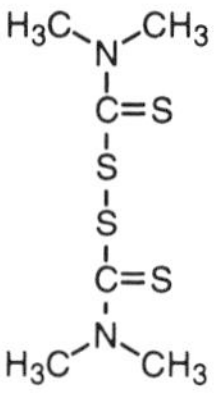

a109
thiram

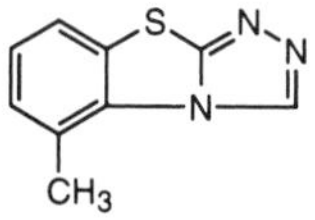

a110
tricyclazole

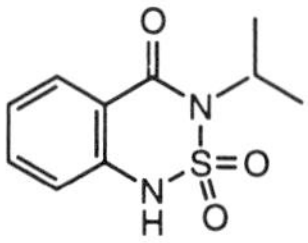

a111
bentazon

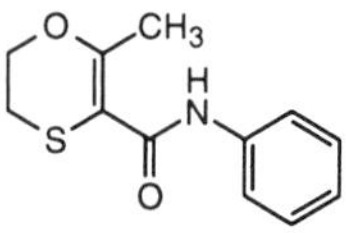

a112
carboxin

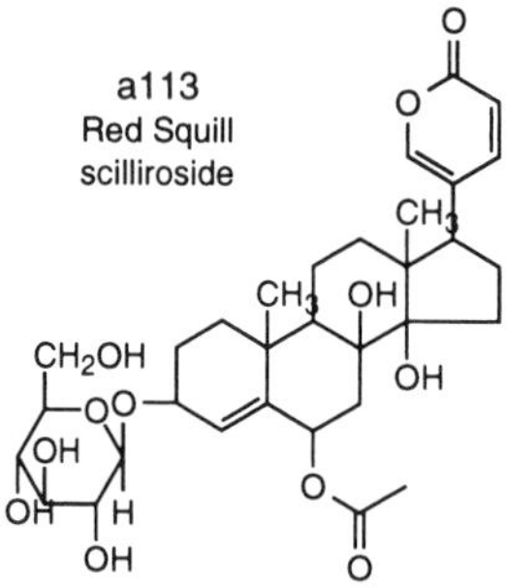

a113
Red Squill
scilliroside

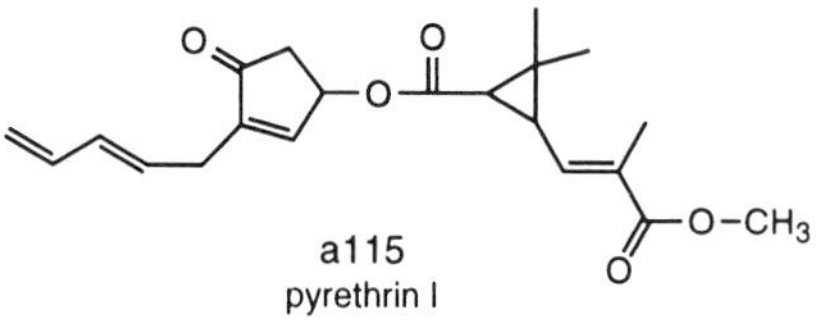

a115
pyrethrin I

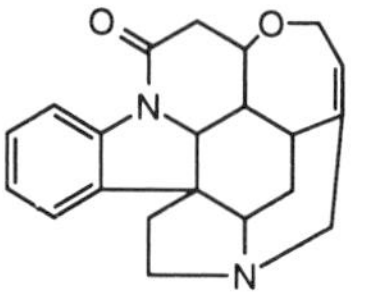

a114
strychnine

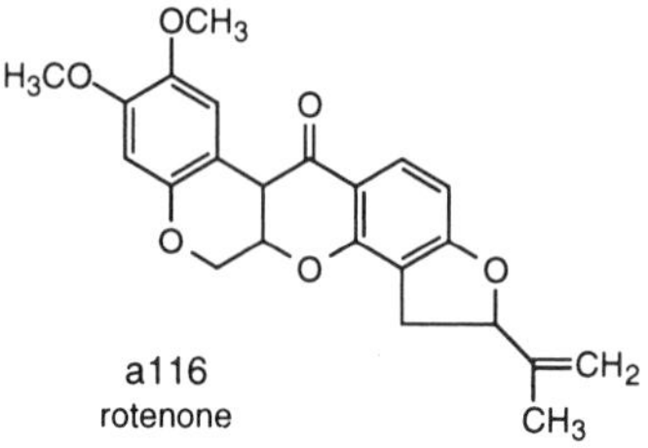

a116
rotenone

PART II. THE MICROBIOLOGY

3

CLEANUP OF PETROLEUM HYDROCARBON CONTAMINATION IN SOIL

INGEBORG D. BOSSERT

Research and Development Center, Texaco, Inc., Beacon, New York 12508

GEOFFREY C. COMPEAU

Enviros Incorporated, Kirkland, Washington 98033

1. INTRODUCTION

Carbon is distributed in the environment as chemical compounds that range in form from gases, e.g., methane or carbon dioxide; to liquids, e.g., plant oils or petroleum; to solids, e.g., simple and polymeric sugars or heavy tars. The degradation and removal of many of these compounds from the environment occurs through natural processes. Primary mediators for these transformations are microorganisms, in particular naturally occurring fungi and bacteria. Much has already been written about the microbiology of carbon cycling in the environment, in particular the biodegradation of organic contaminants. This review focuses on the fate and biodegradation of petroleum and its products in the soil environment and on technologies for the biological treatment of contaminated soils.

2. MICROBIAL DEGRADATION OF HYDROCARBONS

As a class of compounds, hydrocarbons comprise some of the more biodegradable contaminants of concern in the environment. In the current jargon, the term *hydro-*

Microbial Transformation and Degradation of Toxic Organic Chemicals, pages 77–125

77

carbon loosely refers to organic compounds that consist primarily of hydrogen and carbon, but may contain other substituents as well. For example, phenolics, organic acids, and even halogenated organics may sometimes be referred to as hydrocarbons. For the purposes of this discussion, and unless otherwise noted, the term *hydrocarbons* is used in its strictest sense, referring to petroleum-derived compounds consisting of only hydrogen and carbon atoms. As such, hydrocarbons are ancient, naturally occurring compounds (Berger and Anderson, 1981). The presence of hydrocarbons in the environment for such a long time has provided a selective pressure for microorganisms and enzyme systems that degrade these substrates. This concept, the "principle of microbial infallibility," was first introduced by Alexander (1965). Not all hydrocarbons are equally biodegradable, however. Many different factors dictate the metabolism and ultimate environmental fate of hydrocarbon contaminants. Here, we first define the substrate, then discuss in detail the factors that contribute to the biodegradation of hydrocarbons in soils, and, finally, how these contaminated environments can be cleaned up using bioremediation technologies.

2.1. Undesired Biodegradation (Microbial "Spoilage")

The biodegradation of hydrocarbons is not always desirable. Petroleum hydrocarbons, by their nature, are generally biodegradable, including those stored in natural reservoirs where they were first formed eons ago. Oil (or hydrocarbon)–degrading microorganisms are ubiquitous to surface and near-subsurface soils (Beerstecher, 1954; Bossert and Bartha, 1984) and have been detected also in deep formations (Balkwill, 1989; Fredrickson et al., 1991). Those microorganisms that are associated with oil-bearing formations are indirectly responsible for substantial economic losses during production (Herbert, 1985).

The fact that petroleum deposits are not more degraded in their natural state is in large part due to the prevailing environmental conditions, e.g., little or no oxygen and available water, high temperatures and pressures, and high salinity. These environmental constraints adversely impact *in situ* microbial growth and activity. When one or more of these factors are altered to conditions more favorable for microbial growth, the biodegradation of significant amounts of the source petroleum can result. For example, correlations between freshwater intrusion and/or depth and the degradation of petroleum in a reservoir have been reported (Bailey et al., 1973; Milner et al., 1977; Price, 1980).

In addition to overall losses, the *quality* of a petroleum reservoir can be changed by microbial activity. Sulfate-reducing bacteria (SRB) have long been implicated in petroleum reservoir deterioration. In general, the anaerobic SRB possess a very limited substrate range. In-reservoir petroleum biodegradation occurs primarily at oxic–anoxic interfaces, where the synergistic activities of both aerobic and anaerobic bacteria bring about the biodegradation of the oil; the products of aerobic metabolism serve as substrates for the SRB (Jobson et al., 1979; Vance and Brink, 1994).

At the other end of the sulfur-utilizing spectrum are the aerobic, sulfur-oxidizing bacteria. It has been suggested by some that these can be used to advantage in recovering or upgrading crude oil, e.g., the *in situ* removal of sulfur compounds or the solubilization of heavier components (Gallagher et al., 1993; Monticello and Finnerty, 1985; Kim et al., 1990; Fedorak, 1990; Finnerty et al., 1983). In most cases, however, the desirable components of the crude oil, e.g., straight-chain aliphatics, are preferentially biodegraded over heavier fractions. Control of metabolic processes that preferentially biodegrade lighter hydrocarbons is a key element to developing a directed biological treatment, or "biobenefication," process for upgrading crude oils and coal to a higher valued product. The selective biodegradation of heavier, heteroatom components presents a challenge to current research efforts focused on the upgrading of fossil fuel through biobenefication processes.

Several activities during the production of petroleum crude oil provide conditions that favor microbial growth and corrosion, which may result in the degradation of field equipment and the produced crude. Under field conditions, indigenous microorganisms, including both aerobic and anaerobic species, can attack the metal surfaces of equipment and depolarize the surface, creating a concentration cell, sometimes causing severe corrosion problems and/or the production of toxic hydrogen sulfide gas (Iverson and Olson, 1984; Obuekwe et al., 1983). For a more detailed discussion of microbially induced corrosion and methods for its mitigation, the reader is referred to Iverson and Olson (1984) and Hamilton (1985).

Other aspects of misplaced or unwanted hydrocarbon biodegradation occur in stored fuel products, which can result in economic losses as well as hazardous situations. For example, biofouling during fuel storage results in biomass production that can introduce impurities and particulate contamination to the fuel. The net result is a lowering of the quality of stored fuels. This and other types of detrimental biodegradation of hydrocarbons is beyond the scope of this chapter and can be found elsewhere (Hill, 1984).

2.2. Hydrocarbon Biodegradation in Soil

Aside from their natural, contained niche within rock reservoirs, crude petroleum hydrocarbons and refined products can be displaced and enter the environment. Hydrocarbon contamination can occur as combined or singular components of crude oil or refined petroleum, and may impact a variety of environmental matrices. Discussion here will be limited to the contamination of unsaturated soils and factors affecting hydrocarbon bioremediation.

Introduction of hydrocarbon contaminants into the environment may be naturally occurring, as in natural oil seeps, or anthropogenic, as in the case of accidental spills and leakages. Unlike many other organic compounds, however, petroleum and its products generally are readily biodegradable contaminants, given the proper environmental conditions.

Aside from unintentional contamination, hydrocarbons have been sometimes deliberately applied to soils. In the past, oils were applied to roads as a means for dust prevention. More recently, soils have, in effect, served as "solid-state bioreactors" for treating petroleum wastes. By applying hydrocarbonaceous wastes to soils in a controlled manner, the materials are biodegraded, i.e., destroyed, by soil micro-organisms, thereby proving a permanent, cost-effective disposal option. Landfarming, or land treatment of oily wastes, has been successfully used for years by the petroleum and related industries to clean up contaminated soils and/or to treat wastes (Hosler et al., 1992; Huesemann and Moore, 1993; Bartha and Bossert 1984; Bossert, et al., 1984). Variations of this technology and other methods for soil bioremediation are discussed later in this chapter.

2.2.1. The Substrate

Petroleum is a complex mixture of hydrocarbons and related compounds. Its con-stituent hydrocarbons span a wide range of molecular sizes, ranging from methane to polynuclear aromatics and asphaltenes. In addition to their molecular size, hydro-carbons occur as a variety of molecular structures in both unsaturated and saturated forms: straight and branched chains, singular and condensed rings, and combina-tions of substituted (aromatic) rings. Depending on its source material, petroleum can also contain varying amounts of S-, N-, and O-heterocyclic compounds and heavy metals (Gary and Handwerk, 1984; Yoshiba et al., 1988; Sullivan and John-son, 1993; Monticello and Finnerty, 1985; Morgan and Watkinson, 1989). These structural and compositional variations affect biodegradation processes, depending on how easily the hydrocarbons can be metabolized by microorganisms, either singly or as a mixture (Fedorak and Westlake, 1984; Fedorak et al., 1983). Figure 3.1 shows the composition of two crude oils from nearby fields in California, and Table 3.1 presents the sulfur content of some "typical" crude oils. Both presenta-tions illustrate how composition can vary, even within similar geographical regions.

The basic structural compounds found in crude oil do not change during refining.

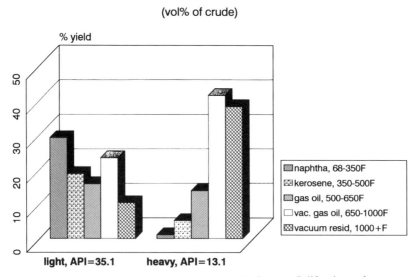

Figure 3.1. Comparison of wide cut yields for two California crudes.

TABLE 3.1.   Organic Sulfur Content of Some Crude Oils[a]

Source	Percent Sulfur
Kuwait	2.6
Middle East	1.5–3.0
Venezuela	1.7
Mississippi	1.6
West Texas	0.05–5.0
California	1.0
Canada	0.44
East Texas	0.26
North Africa	0.18
Far East	0.10

[a]Reproduced from Monticello and Finnerty (1985), with permission of the publisher.

Through cracking, hydrogenation, reforming, and related processes, the refining process primarily shifts the relative makeup of the crude oil into proportions that better match marketing needs. An example of the catalytic upgrading of a naphtha stream is presented in Figure 3.2. During this refining process, less desirable naphthenes (primarily cycloalkanes) are dehydrogenated to aromatic compounds. With a makeup that is more defined, refined petroleum provides the same substrates as crude oil, but generally in vastly different proportions. As with crude oils, different refined components exhibit varying rates of biodegradability (Song and Bartha, 1990).

2.2.2. The Physical (Abiotic) Fate of Hydrocarbons in the Soil Environment

Soils vary widely with regard to geology, hydrology, climate, fertility, and other physical attributes; for a detailed discussion of soils, the reader is referred to a soil science text (Tan, 1982; Buchman and Brady, 1969). The geophysiochemical prop-

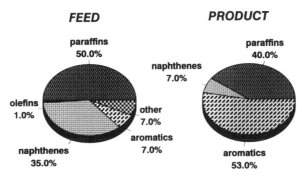

Figure 3.2.   Catalytic refining of crude naphthas. (Adapted from Gary and Handwerk, 1984.)

erties of a soil are instrumental in determining the fate of a contaminant. By virtue of complex matrix interactions, soils often mitigate the potentially toxic effects of a contaminant, through binding and sorption phenomena, while also providing a solid, physical support to help protect and stabilize microorganisms and their cellular components. In addition, soils often define the physiological constraints in a particular environment. These, in turn, impact microbial activity, i.e., biodegradation, and are discussed in more detail later on.

Most soils are multiphasic systems, containing an ionic solid mineral matrix and some associated organic matter, which is surrounded by a water film, as is schematically shown in Figure 3.3. In unsaturated soils, generally referred to as the *vadose zone,* a gas phase permeates the pore spaces; in saturated soils, pore spaces are part of the aqueous phase. When hydrocarbons are introduced onto the surface of a soil, a number of physical phenomena impact their removal or fate in the environment

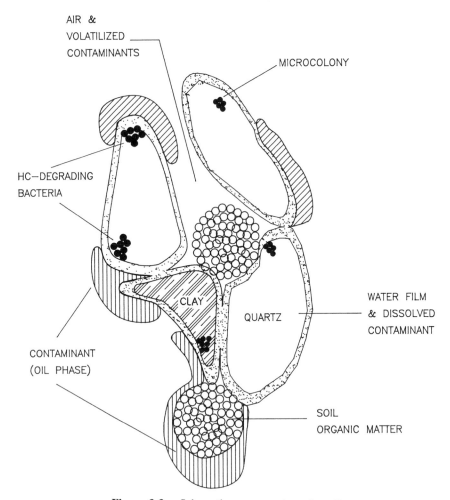

Figure 3.3. Schematic representation of a soil.

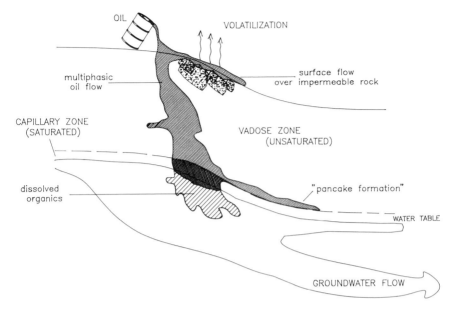

Figure 3.4. Representation of various methods by which hydrocarbons can be distributed through a soil matrix. (Adapted from Bossert and Bartha, 1984.)

(Bossert and Bartha, 1984; McGill et al., 1981). For fresh spills of light hydrocarbons, volatilization may play an important role in removing the material from the soil surface, particularly on less permeable surfaces. For heavier hydrocarbons, auto-, thermal-, and photo-oxidation mechanisms, in addition to biological degradation, may partially oxidize the contaminants on the soil surface, making them more water soluble and perhaps more bioavailable. More polar intermediates also exhibit greater movement through the underlying unsaturated soil, or vadose zone, eventually to the water table. This may not always be the case; laboratory studies conducted by Miller and coworkers (1988) have demonstrated that photoproducts of benzo[a]pyrene photolysis are rapidly detoxified, most likely through mineralization or binding to the soil organic matter.

Water-immiscible hydrocarbons can also move down through the soil, albeit more slowly, thereby introducing an additional oil phase to a multiphasic matrix, which may change the wetting properties (and the water holding capacity) or a soil. Figure 3.4 gives a representation of the various modes in which hydrocarbons can distribute through a soil. Note that this illustrates movement through an idealized, permeable soil. For less permeable, more heterogeneous and/or fissured matrices, channelling can occur.

2.2.3. The Biological Fate of Hydrocarbons in the Soil Environment

Hydrocarbon-degrading microorganisms add yet another dimension to an already complex soil system. The biological fate of contaminants can follow a number of routes. Under ideal conditions, the hydrocarbons are completely mineralized to carbon dioxide and water, with some biomass production. More often, however, biodegradation is not complete. Metabolism to partially oxidized products also serves to remediate a soil by fixing and stabilizing potentially hazardous materials. Partially degraded, i.e., oxidized, hydrocarbons may become incorporated into the soil organic matter or humified (Shannon and Unterman, 1993; Schwarzenbach et al., 1993; Tate, 1987; McGill et al., 1981).

Hydrocarbons are water-immiscible compounds, yet microorganisms generally reside in the aqueous phase. For effective biodegradation to occur, it is therefore essential that the contaminant substrate be "bioavailable" to the degrading microbial communities (Gerson, 19985). Many microorganisms possess an ability to overcome partitioning effects and utilize water-insoluble substrates. In some cases, the production of extracellular, surface active agents are produced that may solubilize hydrocarbons into the aqueous phase (Miller, 1994). In other cases, hydrophobic cell walls may actually partition with hydrocarbons present in the soil or attach to water–hydrophobe interfaces (Efroymson and Alexander, 1991). The mechanism(s) for hydrocarbon transport into, and assimilation by, microorganisms is not entirely understood (Britton, 1984). Further discussion of substrate bioavailability is presented later in this chapter.

2.3. Types and Numbers of Microbial Mediators for Biodegradation

Soils are microbially rich environments. They contain a large variety of microorganisms, ranging in complexity from prokaryotic bacteria and cyanobacteria to eukaryotic algae, fungi, and protozoa. As the results of more and more studies become available, it becomes increasingly clear how broad the taxonomic diversity and metabolic spectra are that comprise hydrocarbon biodegradation. A survey of the literature conducted by Bossert and Bartha (1984) showed 22 bacterial and 31 fungal isolates from soil utilized hydrocarbons as a growth substrate. Combined with subsequently published articles and reviews (Magor et al., 1986; Leahy and Colwell, 1990; Cerniglia, 1993; Sutherland, 1992), the reported number of hydrocarbon-utilizing, bacterial genera isolated has grown to at least 28, with the gram-negative, nonfermentative bacteria, as well as coryneforms, appearing to be the predominant, most often isolated, hydrocarbon-degrading groups. The number of identified fungal genera has grown to 66. The relatively large increase in fungal isolates most probably is due to a growing interest in this eukaryotic sector of the soil microbiota, which has been less scrutinized in the past. The large increase in fungal isolates also reflects different selection criteria; the more recent tallies include genera, e.g., *Phanerochaete* and *Cunninghamella,* that may only co-metabolize or partially oxidize hydrocarbons. Although unable to grow on their reported substrates, these may play a significant role in the biodegradation of residual hydrocarbons in mixed, soil microbial populations.

In addition to bacteria and fungi, a number of cyanobacteria and algae have also been implicated in hydrocarbon biodegradation (Cerniglia, 1993). Owing to their photosynthetic nature, it is assumed that they play a lesser role in deeper soil environment; this is reflected in their relatively lower numbers (Sinclair and

Ghiorse, 1989). Protozoa are yet another component of the soil microbiota. They are found in most soils, and as predators they fill an important ecological niche (Sinclair and Ghiorse, 1989; Madsen, 1991; Sinclair et al., 1993). In the past, however, little information on their role in hydrocarbon biodegradation was available. Rogerson and Berger (1981) reported that protozoa, grazing on hydrocarbon-degrading bacteria and yeasts, may serve as a conduit for hydrocarbon assimilation into the food chain. Moreover, their studies demonstrated that crude oil in their culture medium did not adversely affect laboratory cultures and could be detected in food vacuoles of the protozoa. In another study, Rogerson and Berger (1983) demonstrated that a ciliated protozoan stimulated crude oil biodegradation. Recent reports by Madsen and coworkers (1991) have extended these earlier findings to propose a novel application for assessing of *in situ* bioremediation of hydrocarbon-contaminated soils. In their study of an aquifer sediment contaminated with coal tar, they demonstrated a direct correlation between contaminant concentration, numbers of PAH-degrading bacteria, and numbers of protozoans grazing on the (PAH-degrading) bacteria in subsurface soils. The same pattern of microbial distribution, but reflecting opposing conditions, was also evident in samples collected outside the area of contamination, i.e., no hydrocarbon substrate, low numbers of bacteria and actinomyces, and concomitantly, low numbers of protozoa. It is noteworthy that the distribution of fungi at this site remained constant, regardless of the soil conditions.

Other soil organisms, even members of the macrobiota, have been implicated in hydrocarbon biodegradation, albeit indirectly. For example, Gardner and coworkers (1979) reported that inoculation of a polychaete, *Capitella capitata,* marginally increased rates of PAH biodegradation in crude oil–contaminated sediments. The slight increase in activity in this particular study was attributed to better mixing from work activity; however, they also claimed that some of their earlier studies demonstrated the ability of the polychaete to metabolize hydrocarbons.

2.3.1. Microbial Ecology

The effect of hydrocarbon contamination on microbial populations in soils can vary considerably, depending on the type, amount, and age of the contaminant, as well as the prevailing environmental conditions. Unlike surface bodies of water or groundwater, which will diffuse the effects of contamination through dilution and migration, the bulk of contamination in unsaturated soils will remain localized, potentially exerting pronounced effects on the immediate soil microbiota. However, by virtue of their physical matrix and chemistry, most soils are actually good mitigators for toxic effects. By (ionic or covalent) binding, and/or sorption onto the soil organic material, contaminants quickly become immobilized and less bioavailable. Moreover, the solid soil structure aids in maintaining cell membrane integrity in the presence of contaminants exhibiting solvation effects.

Very little information is available on the microbial ecology of soils, in particular, the microbial succession and how different groups of microorganisms interact with one another during hydrocarbon biodegradation. There are numerous reasons for this. First, soils and their associated populations vary greatly, depending on geology, climate, and contamination; the ecology of a site may vary widely from site to site. Second, soils are extremely heterogeneous; sampling to obtain a representative sample for evaluation (and/or comparison) is nearly impossible. Third, it is very difficult, if not impossible, to determine *in situ* numbers and types accurately, especially when evaluating mixed populations; only those microorganisms that can be cultured in the laboratory have been extensively studied and identified. Finally, enumeration methodologies range widely and can provide, at best, only an estimate of microbial numbers and types. For these reasons, it is better to look at trends rather than compare absolute numbers of microorganisms in soils.

Compared with pristine soils, hydrocarbon-contaminated soils possess relatively large numbers of microorganisms; however, their microbial diversity is usually diminished and population shifts have been observed (Bossert and Bartha, 1984; Morgan and Watkinson, 1989; Leahy and Colwell, 1990; Song and Bartha, 1990; Atlas, 1991; Thomas and Ward, 1992). This is to be expected, since hydrocarbons, in general, provide a large source of carbon substrate for growth that provides a selective advantage to hydrocarbon-utilizing bacteria; these become selectively "enriched" under otherwise nonlimiting conditions for growth.

2.3.2. *Fungi Versus Bacteria*

Although it is widely accepted that bacteria and fungi are primary mediators in hydrocarbon biodegradation, little is known about their respective contributions. Filamentous fungi and antinomycetes may play the more important role during the early colonization of a contaminated soil. Hyphal structures and increased surface area allow for better penetration into, and contact with, the hydrocarbons; their extracellular enzymes, e.g., oxidases, may further extend their activity into the soil. Moreover, fungi tend to withstand greater fluctuations in environmental conditions, e.g., moisture and pH. Bacteria, on the other hand, exhibit a more versatile metabolism and therefore may play a greater role during the latter stages of biodegradation, when the remaining substrate is more chemically varied. These generalizations are based on what is currently known about the respective characteristics of soil bacteria and fungi. To date, only a few limited studies have been conducted to define more clearly the relative degradative roles of bacteria and fungi in soils.

In an early study using antibiotic inhibitors to determine the respective roles of glucose utilization in a variety of soils, Anderson and Domsch (1975) reported that microbial activity in the soils tested was predominantly due to the fungal component. Using a similar inhibition technique, Song and coworkers (1986) reported that the bulk of *n*-hexadecane mineralization (82%) in a previously uncontaminated soil was attributable to the bacterial population; only 13% of the hydrocarbon-degrading activity was mediated by the fungi. Glucose utilization in their study appeared to be more evenly divided between fungal and bacterial populations, as in the Anderson and Domsch study. More recently, Carmichel and Pfaender (1994) reported on similar results with an aromatic substrate. Further studies are needed to develop a better understanding of how microbial populations interact and shift under different and changing scenarios, e.g., chronic or prolonged exposure to hydrocarbons, varying environmental conditions, and different hydrocarbon substrates or mixtures of hydrocarbons whose components change over time.

2.4. Biodegradation Pathways

Biodegradation is a biologically driven process that relies on microorganisms to metabolize a hydrocarbon substrate. In general, the hydrocarbon substrate serves as the food source for microbial growth and/or energy. In some cases, especially for more recalcitrant compounds, a single organism lacks the full complement of uptake and/or enzymatic capabilities to metabolize a particular substrate. *Co-metabolism* of the hydrocarbon substrate may occur, but only in the presence of other metabolizable substrates. The phenomenon describes a mechanism whereby a hydrocarbon substrate is only partially metabolized in the presence of other substrates and cannot itself provide carbon for energy or growth to the microorganism. It is often evident as syntrophic or commensalistic interactions within mixed microbial communities, (Atlas and Bartha, 1993; Liu and Suflita, 1993) and may in large part be responsible for the biodegradation of persistent hydrocarbon residues in soils. Another metabolic phenomenon observed for hydrocarbon substrates is *analog enrichment,* which describes *cross-acclimation* (Leahy and Colwell, 1990) between compounds of similar structure (Bauer and Capone, 1988).

2.4.1. Anaerobic Versus Aerobic Metabolism

Metabolic steps in the biodegradation of hydrocarbons follow two major strategies: oxidation and/or reduction. Because hydrocarbons are already chemically reduced and stable compounds (a practical demonstration of this is the longevity of petroleum reservoirs), further reduction, while thermodynamically possible, is not a primary mode for biodegradation, even under strict anaerobic conditions. A number of reports (Aeckersberg et al., 1991; Berry et al., 1987; Ward and Brock, 1978; Grbic-Galic and Vogel, 1987; Vogel and Grbic-Galic, 1986; Schink, 1985) have demonstrated that toluene, benzene, and a variety of alkanes can be biodegraded under the strictest of anaerobic conditions by sulfidogenic and methanogenic cultures. In these well-documented cases, anaerobic metabolism still follows an oxidative strategy. In the absence of molecular oxygen, water-derived oxygen serves as a reactant, while carbon dioxide or sulfate serve as the electron acceptors for anaerobic oxidation of the substrates to hydroxylated aromatics or fatty acids, respectively; further metabolism can then follow one of several established routes, such as ring cleavage and β-oxidation. For more information on anaerobic aromatic metabolism, the reader is referred to Chapter 8 (Colberg and Young).

Hydrocarbon biodegradation under anaerobic, denitrifying conditions also follows an oxidative strategy. In the presence of nitrate, hydrocarbon substrates, e.g., toluene, are metabolized to oxidized intermediates prior to further biodegradation (Evans et al., 1992; Flyvbjerg et al., 1993; Zeyer et al., 1986). In a series of well-documented laboratory and field studies on the biodegradation of BTEX contaminants in a nitrate-amended, subsurface aquifer soil, Hutchins and coworkers (1992,

1991) and Barbaro et al. (1992) reported that the substituted aromatics, toluene, ethylbenzene, and xylenes were biologically removed from the soil under denitrifying conditions. In the same studies, benzene levels dropped only after small amounts of molecular oxygen were provided, presumably to aid in the initial ring oxidation of the molecule by oxygenases. For unsubstituted aromatics, oxygenation may help to "prime" the molecule for further attack by destabilizing the aromatic ring via delocalization of its π electrons.

This "priming" phenomenon has practical implications for the field. Reduced substrates will, at best, undergo only very slow biodegradation in anaerobic or oxygen-limited subsoils. However, oxidized metabolites produced near the surface may migrate down into the lower depths and be more readily degraded in the oxygen-limited subsurface, where alternate electron acceptors such as sulfate or nitrate predominate. In addition, partially oxidized metabolites are generally more water-soluble and may partition more readily into the aqueous phase.

Although there is a growing amount of evidence that reduced substrates such as hydrocarbons can indeed be biodegraded in the absence of molecular oxygen, biodegradation proceeds most rapidly and efficiently under nonlimiting, aerobic conditions, where oxygen is available to serve both as reactant and electron acceptor in metabolism. The higher biodegradation rates observed in aerobic environments often are indicative of faster aerobic growth rates and may reflect a greater net production of energy during oxidative phosphorylation and electron transport. Biodegradation under aerobic conditions is usually more "complete," resulting in greater rates of mineralization of the hydrocarbon contaminant to its ultimate endproducts, carbon dioxide and water. This apparent effect on the "quality" of biodegradation also has important implications for the field. To minimize future liabilities from a contaminated site, it is essential that a bioremedy be conducted to minimize the formation of undesirable intermediary metabolites, which may be more mobile and/or toxic while maximizing biodegradation of the contaminants to their ultimate and harmless endproducts.

In general, aerobic metabolism of hydrocarbons requires oxygenase enzymes, which incorporate molecular oxygen into the reduced substrate. For aliphatic hydrocarbons, alcohols are initially produced; these are oxidized sequentially, via dehydrogenases, to carboxylic acids, which then undergo β-oxidation. In the case of aromatic substrates, as well as condensed ring compounds such as polyaromatic hydrocarbons (PAHs), hydroxylation of the ring occurs via mono- or dioxygenase enzymes in eukaryotes and prokaryotes, respectively (Cerniglia, 1993). After diol formation, the ring is cleaved, then further degraded. Of course, many variations to these metabolic schemes exist and are discussed in much greater detail elsewhere (Cerniglia, 1984a,b; Chapter 7 of this book).

2.4.2. Effect of Substrate Structure on Biodegradation

The chemical structure of a contaminant has both direct and indirect impacts on how well the substrate will be metabolized, i.e., biodegraded. First, metabolic or physiological constraints by microorganisms will directly impact how readily a substrate can be degraded. For example, light aliphatic hydrocarbons, approximately C3 to C8 in length, may have a solvation effect on cellular lipids and membranes, in effect destroying cellular integrity. Moreover, their increased water solubility may increase toxicity (Britton, 1984). For other *n*-alkanes, chain length often determines their biodegradability by a specific microorganism. A singular type of culture usually metabolizes a specific range of chain-length preferentially, perhaps due to enzyme specificity or to restricted uptake mechanisms. Furthermore, the metabolism of branched aliphatic hydrocarbons may be sterically hindered and create metabolic bottlenecks at branch points. Angular ring arrangement in polynuclear aromatic hydrocarbons may also impact biodegradability and may even determine points of metabolic attack, such as the "bay region" in benzol[α]pyrene (Cerniglia, 1984b).

In general, the biodegradability of hydrocarbons follows the expected trend: Straight-chain aliphatics are more readily biodegraded than branched-chain hydrocarbons, and simple aromatics biodegrade faster than condensed, polyaromatic, or alkylated aromatics.

Second, type and size of chemical structure may indirectly affect biodegradability by altering the bioavailability of the contaminant to the degrading microorganisms. Because hydrocarbons are hydrophobic, water-insoluble compounds, bioavailability is perhaps key to determining a contaminant's biodegradation potential. Biodegradability of hydrocarbons in soils has been demonstrated to correlate to their water solubilities, which are generally inversely proportional to their respective molecular weights (Bossert and Bartha, 1986; Klevens, 1950). Other structural attributes, such as degree of unsaturation, can affect water solubility and ultimately uptake and availability to the degrading microorganisms. Given the same, or similar molecular weight, unsaturated and aromatic compounds exhibit greater water solubility. Bioavailability and how it affects biodegradation in soils is discussed below.

2.5. Physiological Factors Affecting Biodegradation

Physiological constraints on biodegradation encompass two major areas: 1) nutrient or metabolic limitation and 2) bioavailability of the contaminant substrate. Each can directly impact cellular metabolism and therefore biodegradation.

2.5.1. Nutrient Status

The nutrient status of a soil directly impacts microbial activity and biodegradation. Whereas organic contaminants provide the major source of carbon for microbial growth and energy, other mineral nutrients are also essential for good activity. In particular, nitrogen, and to a lesser extent phosphorus, are necessary for cellular metabolism. Bioavailable forms of these ancillary nutrients may not be present in sufficient concentrations in contaminated soils. In addition to inorganic macronutrients, trace amounts of metals and other elements are required for proper microbial nutrition. Because these are generally present in most soils, discussion here is limited to the macronutrients: nitrogen and phosphorus.

Nitrogen and phosphorus are necessary cellular components, but are often not present at adequate levels, owing to the heavy carbon load in contaminated soils. Furthermore, the form, or species, in which they occur may not be in a biologically available form to sustain optimal rates of metabolism, i.e., biodegradation. Under these circumstances, nitrogen and/or phosphorus fertilizers may be added to a soil to improve biodegradation rates. Responses to fertilizer amendments appear site specific and can vary widely, even within a given site (Bossert et al., 1991; Bossert and Bartha, 1984; Leahy and Colwell, 1990). In most cases, responses to fertilization are almost immediate (Hinchee and Arthur, 1991), whereas other soils show a mixed, or no response to fertilization (Morgan and Watkinson, 1992; Manilal and Alexander, 1991). The latter may be due to sufficient nutrient levels already present in some soils or to nitrogen fixation by soil microorganisms and/or to the inherent heterogeneity of soils (Bossert and Bartha, 1984).

Numerous field and laboratory studies report a wide application range for nitrogen and/or phosphorus supplementation to soil. On a C/N molar basis, reports for optimal fertilizer applications range from approximately 2 to 200 for nitrogen (Morgan and Watkinson, 1989; McMillen et al., 1993). It should be noted, however, that excessively high nitrogen loadings, e.g., C/N ratios less than 20, may result in an inhibition of soil activity, possibly due to nitrite toxicity. C/P ratios usually range at least one order of magnitude less. These values should be used strictly as guidelines; each soil and its microbiota are unique and therefore have different requirements. Before any bioremediation of a contaminated soil begins, treatability studies should be conducted on the respective soils to determine what deficiencies may be present and what the nutritional requirements are for the degrading microbial populations; these are discussed in greater detail later in the chapter.

The chemical form in which a fertilizer is applied may also impact biodegradation of hydrocarbon contaminants. There is a growing body of evidence that chemical speciation of nitrogen (and phosphorus) can affect microbial activity, particularly in soils (Rasiah et al., 1992a). The effect may be twofold, and may be especially critical for *in situ* bioremediations.

First, interaction of the fertilizer with a soil will determine how an amendment distributes through the matrix. For nitrogen amendments, neutral or anionic species will generally move unhindered through the matrix, whereas the movement of cationic species can be restricted by the cationic exchange capacity of a soil (Fetter, 1993). Phosphorus amendments, usually added as phosphate salts, readily complex with soil minerals, producing insoluble, immobile precipitates. The use of tripolyphosphates is less likely to result in unwanted precipitation but can instead disrupt soil structure (Morgan and Watkinson, 1992).

Second, the chemical speciation of fertilizers may have direct effects on uptake and utilization. In the case for nitrogen, studies using the three most widely used forms of the fertilizer, urea, ammonia, and/or nitrate, have demonstrated mixed results on microbial activity. In one laboratory study using core material from a diesel-contaminated soil, all three nitrogen forms increased rates of hydrocarbon mineralization, with urea showing up to an additional 35% increase in the mineralization rate compared with the next most effective fertilizer treatment, NH_4NO_3, (Bossert et al., 1991). In another study, Harder and coworkers, (1991) observed little difference in hydrocarbon losses or oxygen uptake when contaminated laboratory soils were treated with either nitrate or ammonia fertilizers. The lack of response in these studies most probably indicates no nitrogen limitation in the initial soil.

Phosphorus speciation also appears to affect microbial activity. In an aqueous system, Robertson and Alexander (1992) concluded that speciation of phosphate salts, as influenced by pH and conjugate ions, impacted both solubility and availability during biodegradation of a variety of organic compounds. In another aqueous system (Jones and Alexander, 1988), high phosphate levels required for *p*-nitrophenol biodegradation by a *Flavobacterium* species were probably due to chemical precipitation.

2.5.2. Bioavailability

The bioavailability of substrate to the degrading microbial community is also a critical factor in determining the fate of a contaminant. A recent review (Mihelcic et al., 1993) points out the difficulty in establishing unifying principles on how bioavailability affects biodegradation, especially in soils. Studies performed on highly variable systems, such as soils, can provide results that are mixed and often difficult to compare. In fact, they can even be contradictory, owing to differences in matrix (soil type), contaminant, equilibration time, and, in cases where surfactants are used to solubilize the contaminant, surfactant type and/or sorption and concentration (Van Dyke et al., 1993b).

In a comprehensive review on surfactants and solubilization, Miller (1994) reports that the uptake and utilization of gaseous and liquid hydrocarbons may occur either in the dissolved state, or directly by surface (interfacial) contact. The microbial utilization of solid, hydrophobic substrates requires solubilization, or emulsification, prior to uptake and metabolism (Wodzinski and Coyle, 1974). Others (Volkering et al., 1993) have determined that growth on crystalline substrates, e.g., naphthalene, results in linear growth rates, indicating that partitioning, i.e., solubilization, of the substrate is rate limiting to biodegradation. According to a review by Britton (1984; see also Rosenberg et al., 1983), uptake of hydrocarbons most likely occurs by attachment, then incorporation into the cystoplasmic membrane. Alternately, transport occurs by passive or facilitated diffusion in the presence of solubilizing agents; intracellular transport is probably coordinated with enzymatic oxidation.

For example, some researchers have shown enhanced metabolism of polycyclic aromatics in the presence of a nonionic surfactant (Aronstein and Alexander, 1993; Tiehm and Zumft, 1992), whereas others have demonstrated that surfactants may actually inhibit metabolism during micellization (Laha and Luthy, 1992). The latter hypothesize that the micelles undergo a form of reversible, physiological interaction with cell membranes, thereby temporarily inhibiting biodegradation.

The use of surfactants for overcoming bioavailability limitations has received much recent attention, even though their potential to improve biodegradation had been recognized earlier (Mulkin-Phillips and Stewart, 1975; Atlas and Bartha, 1973). Because hydrocarbons are hydrophobic compounds, and microorganisms require an aqueous environment for optimal growth and activity, biodegradation occurs in (at least) a biphasic system, comprised of immiscible components. Surface-active agents, containing both hydrophobic and hydrophilic moieties, provide a means for decreasing interfacial tensions and enhancing the miscibility of two or more phases. Commercially available surfactants, both ionic and nonionic in nature (Laha and Luthy, 1992; Haüsler et al., 1992; Pennell et al., 1993; Thai and Maier, 1992), as well as biosurfactants and biosurfactant-producing bacteria, have been investigated for their ability to increase bioavailability (Miller, 1994; Scheibenbogen et al., 1994; Van Dyke et al., 1993a,b; Volkering et al., 1993; Zhang and Miller, 1992). Although results regarding their efficacy are mixed, it appears that one of the more effective applications is their use in soils where contaminants are sorbed to the matrix (Volkering et al., 1993).

Other methods for increasing bioavailability may also enhance the biodegradation of contaminants in a soil. For example, physical disruption of soil aggregates using sonication has been reported to increase biodegradation rates effectively in a landfarm soil (Rasiah et al., 1992b). Others have demonstrated that soil constituents may significantly impact the bioavailability of contaminants (Weissenfels et al., 1992). In the latter study, two contaminated soils with similar contamination histories demonstrated very different biodegradation profiles. Under the same conditions, one presented high PAH-degradation rates, relying on its native microbial populations, whereas the other demonstrated no PAH biodegradation, even after inoculation with known PAH degraders. After ruling out toxicity effects, the lack of activity in the latter was attributed to differences in bioavailability within the two soil matrices. This soil had a higher soil organic matter content, which more tightly bound the contaminants. Similar results have been reported by others (Manilal and Alexander, 1991), where mineralization rates of a contaminant, e.g., phenanthrene, are lower in soils with a high organic matter content, which readily sorbs hydrophobic compounds. Soluble humic substances, in particular humic and fulvic acids, appear to be major binding sites; their binding potential can be attentuated by mineral soil components, as well as pH and salt concentrations (Schlautman and Morgan, 1993). Weathering, or the age of contamination, may also affect bioavailability by physically trapping, hindering, and/or slowing desorption of contaminants from the soil (Connaughton et al., 1993).

2.6. Environmental Factors Affecting Biodegradation

In addition to those parameters already discussed, a number of other environmental factors have direct impact on microbial activity in soils, i.e., hydrocarbon biodegradation. Soil moisture and water potential, temperature, soil reaction (pH), and porosity (mass transfer and aeration) are additional soil/climatic properties that have direct bearing on hydrocarbon biodegradation. Also, high levels of heavy metals, salts, or other inhibitory compounds will affect the soil microbial activity. These topics have all been extensively reviewed elsewhere and are not included in this discussion. For further information, the reader is referred to a number of reviews in the literature (Shannon and Unterman, 1993; Leahy and Colwell, 1990; Morgan and Watkinson, 1989; Bossert and Bartha, 1984; McGill et al., 1981).

While it is important to understand and identify environmental and/or physiological constraints that affect biodegradation, ultimately it is necessary to measure and assess microbial activity in impacted soils. Owing to the complex and varied nature of soils, it is often difficult to observe and measure directly the activity of their microbiota, yet such information is vital for well-designed and -implemented bioremediation efforts.

3. MEASUREMENT AND INDICATORS OF BIODEGRADATION ACTIVITY IN SOILS

Numerous strategies are available for evaluating biodegradation in soils. Measurement of reactants, i.e., hydrocarbon contaminants, and/or their endproducts provides a good means for obtaining a mass balance and determining the efficacy of biodegradation. Mass balance studies directly quantify the routes of hydrocarbon loss in the soil environment; these include (bio)degradation and (bio)transformation, as well as partitioning and transport within the matrix (Grady, 1985). Other approaches provide less direct evidence, yet also impart valuable information for assessing biodegradation. The enumeration of microorganisms and/or measurement of their *in situ* metabolic activities serve as good indicators for the biodegradation potential of a contaminated soil. The specific analytical methods for measuring substrate and indicators of microbial activity will change depending on the needs of the experiment or project.

The following sections summarize practical considerations in the use and application of analytical techniques for the evaluation of hydrocarbon biodegradation processes, with implementation ranging from laboratory studies to full-scale bioremediation in the field.

3.1. Hydrocarbon Analysis

Direct chemical analysis of a contaminant and its metabolites provides unequivocal proof to establish remediation efficacy. An important criterion for any treatment technology is the analytical method chosen to verify the treatment process. First, samples to be analyzed must be representative of the matrix under study. Given the heterogeneity of soils, this often poses a sizable challenge to the investigator. A number of strategies can be followed to minimize sampling error and provide greater representation of *in situ* conditions. For example, multiple and random sampling, thorough mixing and/or compositing of the collected samples, and increased aliquot sizing for analysis will improve the reproducibility and precision of analytical results, ultimately affording better extrapolation of analytical results to the field.

Assuming proper collection and preparation, the analysis of environmental samples can be divided into three interdependent processes: extraction, separation, and detection. For *extraction,* efficiencies for petroleum-derived, organic compounds can vary greatly with respect to their relative solubilities in a solvent; this is further compounded by their varying interactions with different soil matrices. Likewise, *separation*, e,g., chromatographic techniques, rely on differences in chemical structure. Behavior of contaminants during separation procedures is directly affected by the physical and chemical properties of both analyte and the separation matrix and by their interactions. Choice of *detection* method also imparts a potential bias, depending on the type of analytical information sought. When selecting an appropriate detection method, it is necessary to consider post-analysis needs. For example, some methods, e.g., spectrophotometry, enable recovery of the analyte, whereas others, e.g., flame ionization, are a destructive means to an end. These issues must be given careful consideration when choosing an analytical methodology.

3.1.1. Recovery From the Matrix

Hydrocarbon losses are commonly measured in the laboratory and field by the periodic extraction of the contaminants from replicate soil samples. The most frequently used solvents for extraction of hydrocarbons include freon, carbon tetrachloride, and methylene chloride; other more polar solvents may be used to preferentially extract partially oxidized intermediates. Though not yet widely in use, recent efforts are toward the utilization of nonchlorinated species or supercritical fluids (Akgerman et al., 1992; Andrews et al., 1990; Dooley et al., 1990, Tehrani, 1993).

Liquid extraction is generally conducted in a Soxhlet apparatus or by mechanical grinding and sonication. Volatile aromatic hydrocarbons may be analyzed from solvent extracts or directly from a solid matrix by "purge and trap" techniques in which volatile organic compounds are desorbed from soil samples by a inert gas stream. However, the efficacy of purge and trap methods has been called into question by some (Travis and Macinnis, 1992). The evidence from these studies indicate that purge and trap methods greatly underestimate soil contamination. This is especially the case for weathered contaminants that have been in contact with a soil for a long period, effectively entering its matrix.

3.1.2. Separation and Detection

Following extraction from the soil, hydrocarbon contaminants are detected and quantified by gravimetric, spectroscopic, or ionic methods. When the contaminants are present as a mixture of compounds, it is often necessary to separate the analytes by chromatographic methods. The combination and choice of methods used will depend on the amount, type, and mixture of analyte(s) present, the facilities available, and regulatory requirements. Many variations in methodology exist, and it is important to verify that appropriate quality assurance and quality control issues are addressed before accepting the data generated by a particular method. For example, a background control of uncontaminated soil should be included to evaluate background contributions to the measured hydrocarbon concentration.

3.1.3. Methods for Analysis

Numerous methods exist for the chemical analysis of hydrocarbons. Discussion here will be limited to an overview of those methods that represent the most widely used for biodegradation studies. A more detailed account of methods and their modifications can be found in Greenberg et al. (1992).

A growing number of contract laboratories are now available to conduct EPA Solid Waste Methods (SW-846). Total petroleum hydrocarbons (TPHs) may be quantified by methods 418.1 and 8015; aromatic compounds by methods 8020 and 8240; and semivolatiles and PAHs by methods 8310 and 8270. The latter methods (for aromatic compounds and PAHs), which rely on chromatographic procedures with detection of specific compounds, are less ambiguous than those for TPH determination. Whereas chromatographic methods are calibrated to individual compounds with specific compound standards, TPH analyses do not distinguish between the thousands of discreet compounds and provides only limited analysis and interpretation.

TPHs are commonly quantified by method 418.1 (with either gravimetric or infrared spectroscopic detection). Because the method was originally developed for water analysis, a variety of modifications (included in Method 5520, Greenberg et al., 1992) are currently in use for soil applications. Widely used because of its relative simplicity and cost-effectiveness, the method is generally more accurate for low concentrations. Although intended to be specific for hydrocarbons as a class, it cannot discriminate between types of hydrocarbon or individual compounds present in a sample. For example, gasoline, diesel fuels, and heavy machine oils are generally not resolved by this method. Other limitations include interferences by natural materials (resins, oils, humic materials). Moreover, during measurement of carbon–hydrogen stretching in the infrared spectrum, the use of inappropriate analytical standards may not adequately reflect the composition of hydrocarbons in the test sample (Puttman, 1988). While an understanding of these limitations appears to be growing, the method remains a "work horse" method for determination of total hydrocarbons. The results obtained by the method continue to be the accepted, regulatory endpoint for environmental clean ups in many states.

Gas chromatographic (GC) methods are widely used to determine quantitative changes in individual hydrocarbon concentrations, as well as qualitative changes within classes of hydrocarbons. Many states have adopted analytical requirements based on this method, but generally on a more selective basis than TPH. Using GC methods, gasoline, diesel, and other volatile and semivolatile distillate fractions can be quantified and the net changes evaluated over time. If hydrocarbons of greater than C35 (low volatility) are of interest, high performance liquid chromatography (HPLC) can also be used. However, HPLC methodologies are not typically performed by contract laboratories and are not specified by state regulations. The primary issue regarding appropriate use of GC techniques is chromatogram integration. Typically, quantification calls for complete baseline integration; however, many laboratories carry out "peak skimming" integration, which only accounts for resolved hydrocarbons rather than the entire area represented by resolved and unresolved hydrocarbon.

3.1.4. Interpretation of Data

Analytical testing is critically important to the evaluation of hydrocarbon biodegradation test results. Although the degradation of hydrocarbons is well understood, many reports appear in trade publications and product literature concerning the successful biodegradation of heavy weight hydrocarbons and other petroleum-based products, such as motor oil and hydraulic fluids. Claims such as these must be carefully scrutinized. The reported data often do not reflect a real reduction in high molecular weight hydrocarbons but rather can show misleading results obtained through the use of inappropriate and nonapplicable analytical techniques. By employing an accepted GC method with regulatory standards designed for diesel-range hydrocarbons, heavier, less-volatile hydrocarbons, e.g., greater than C30, are not quantified. This scenario, as well as just poor-quality data obtained without adequate quality controls, can lead to an erroneous assessment of biodegradation activity. On two occasions in the past year, we have found faulty analytical data to be the cause of claimed "heavy hydrocarbon degradation." Split confirmation samples sent to an audited laboratory had petroleum concentrations nearly 10-fold higher than vendor claim due to inconsistencies in the quantification of gas chromatograms (G.C. Compeau, unpublished data).

3.2. Indicators and Measurement of Biodegradation Activity in Soils

The definitive means for evaluating the clean up of contaminated soils is a direct measure of hydrocarbon losses, or the residuum. However, the success of a bioremediation can often be estimated, or predicted, by measuring the biodegradability of a contaminant. Because biodegradation is a microbially driven process, it can be indirectly assessed by measuring the microbial populations and their activities in the impacted environment. These can be evaluated by following a number of different strategies, including the measurement of microbial numbers, biomass, and/or activity. Ideally, the methods used to quantify these parameters should not disturb the matrix or metabolic processes that they attempt to measure. As a screening tool, they should exhibit ease and simplicity of use. For soils, this is often a difficult challenge. Methods now exist that provide a number of options to the investigator for the measurement of microbial, i.e., biodegradative, activity in the soil environment. Proper application of the strategies and techniques that follow requires careful interpretation of results and extrapolation to the field.

3.2.1. Enumeration

Typically, the analysis of microbial numbers is ancillary to the disappearance of a target compound and evaluation of key metabolites. At best, numbers of microorganisms in a given environment can serve as an indicator of how well the environment can support microbial growth and, *indirectly,* its potential for biodegradation. Classic and novel methods are available for detecting, identifying, and enumerating microorganisms important to biodegradation processes. Unfortunately, most of these methods exert some selectivity and do not always reflect *in situ* conditions. As long as their individual limitations and biases are recognized, enumeration methods provide useful information on the "microbial vitality" of a soil and help to monitor population dynamics during the course of remediation.

A key limitation of enumeration techniques in general and especially with respect to bioremediation is that they are exclusive and may not account for all degradative organism types present, detecting only those that comprise a culturable, or observable, portion of the population. Culture techniques are routinely used to enumerate total heterotrophic as well as substrate-specific microorganisms (Gerhardt et al., 1981). Plate counts and most probable number (MPN) determinations, using either undefined or selective media, are two of the most widely used methods for both laboratory studies and the assessment of contaminated soils for full-scale, field bioremediations (Beliaeff and Mary, 1993; Compeau et al., 1991; Anderson, 1989). Modifications of classic plate methods include direct growth on vapors in a chamber, spray plate techniques, and agarose overlay (Bogardt and Hemmingsen, 1992; Marshall and Devinny, 1986). In response to the growing need to screen large numbers of samples rapidly, or for the use of potentially hazardous growth substrates (PAHs), traditional MPN methods have been reduced to a format using microtiter plates (Stieber et al., 1994). A summary of microbial plate methods for enumeration of hydrocarbon-degrading organisms is included in a review by Atlas (1992).

Methods for the *direct* enumeration of microorganisms circumvent some of the limitations inherent to culture techniques. Whereas culture methods require *ex situ* growth for detection, direct methods preclude growth and thereby enable direct observation within the soil matrix. Traditional methods for direct counting use vital or fluorescent dyes and microscopy to visualize microorganisms within an abiotic matrix (Byrd and Colwell, 1992; Song and Bartha, 1990). Although most microorganisms can be observed in this manner, many matrix interferences can occur. In particular, humic material, organic debris, and metals are mitigating compounds in the soil. More sophisticated methods have recently evolved, using immunological (Wright, 1992) and genetic (Atlas and Sayler, 1989) techniques. The latter include gene probe and hybridization procedures. Originally developed to track specific bacterial populations within the environment, the methodologies have since been broadened by application of the polymerase chain reaction (PCR) to enumerate larger, more diverse communities (Atlas and Bej, 1990). Quantitation in these advanced hybridization technologies is still poor and needs continued development and evaluation (Sayler and Layton, 1990; Bej et al., 1991; Steffan and Atlas, 1991).

Regardless of which enumeration methods are used to quantitate microbial populations, at best they will serve as an indicator for "microbial vitality" and shifts in populations, while providing only an estimate of the *in situ* microbiota. It must further be recognized that numbers do not always reflect actual microbial activity in the soil. For these reasons, activity measurements generally provide more useful information on biodegradation activities in the field, albeit sometimes by inference.

3.2.2. Activity

Direct, quantitative hydrocarbon analysis of soils is most appropriate for defined and engineered land treatment systems. However, such comprehensive techniques are not always feasible for some applications, especially for *in situ* operations where the costs of sampling and analysis are high or for laboratory studies that require nondestructive, long-term monitoring. Under these circumstances, measurement of microbial activity can effectively serve as an indirect indicator of remediation processes. Activity can be measured using many different parameters, depending on the information desired. For example, specific enzyme activities, e.g., dehydrogenases, phosphatases, amylases, ureases, esterases, and proteases, can be used to evaluate the activity of relatively narrow population groups, whereas energy charge, i.e., adenylate concentration, or respirometric methods, e.g., carbon dioxide production and oxygen consumption, serve as good measures for overall aerobic activity (Atlas and Bartha, 1993; Blackburn and Hafker, 1993; Ectors et al., 1992; Jorgenson et al., 1992; Hoeppel et al., 1991; Hinchee and Arthur, 1991; Dupont et al., 1991). The latter two strategies are most prevalently used for evaluating soils and will receive more emphasis here.

3.2.3. Respirometric Measurements

Biological conversion of contaminant carbon to carbon dioxide during soil incubation is a standard means to assess biodegradability in the laboratory. In the field, carbon dioxide (and oxygen) measurements can similarly be used to assess biodegradation activity; this is especially applicable to *in situ* field technologies such as bioventing.

Carbon dioxide production is a widely used measure of "mineralization" or the complete oxidation of the parent compound. Under ideal conditions, mineralization is proportional to the amount of contaminant present and can be used to establish a mass balance for biodegradation. In the laboratory, procedures and instrumentation to measure respiratory processes include biometer flasks (Bartha and Pramer, 1965; Anderson, 1989), electrolytic respirometry (Akgerman et al., 1992; Dooley et al., 1990; Andrews et al., 1990), and measurement of $[^{14}C]\text{-}CO_2$ from labeled compounds (Marinucci and Bartha, 1979).

Most soils readily respond to substrate addition, although the rate of response may vary considerably, depending on substrate type, concentration, and environmental factors. For example, in a recent study using agricultural soil incubations in biometer flasks to compare total mineralization to $[^{14}C]\text{-}CO_2$ release from a labeled substrate, Sharabi and Bartha (1993) showed that the substrate loading affected mineralization rates; higher amounts of $[^{14}C]$-glucose resulted in greater overall (nonlabeled) rates. In addition, it was demonstrated that mineralization of the background soil organic matter increased with substrate addition. They attributed the latter to a more robust soil microbiota, at a higher metabolic energy level. The results from these studies demonstrate the need for cautious interpretation of respirometric measurements; the physiological state of the test matrix, e.g., soil, will affect mineralization rates. Nevertheless, with appropriate controls, stoichiometric conversions provide a mass balance for the process; this can be used to assess both the rate and the extent of biodegradation (Bossert et al., 1991).

As with carbon dioxide production, oxygen consumption, or the oxygen uptake rate (OUR), also provides a rapid, simple evaluation of biodegradation as a measure of respirometry. Manometric methods, often carried out in a Warburg apparatus, have been traditionally used to conduct oxygen uptake measurements. More recently, oxygen consumption was measured electrolytically (Dang et al., 1989; Grady et al., 1989). Whereas carbon dioxide measures the ultimate endproduct of biodegradation, OUR reflects overall activity and includes transient uptake due to only partial oxidation of the substrate. As with the carbon dioxide data, metabolic stoichiometries can be established. For example, an estimated figure for theoretical oxygen demand during hydrocarbon degradation is approximately 3.5 mg molecular oxygen per mg hydrocarbon (Gibbs and Davis, 1976).

In summary, methods that allow for the direct determination of numbers or that evaluate specific metabolic activities, such as carbon dioxide production or oxygen consumption, are the preferred methods for evaluating biological activity in contaminated soils. Ultimately, the method(s) chosen to evaluate the progress of a bioremediation requires the direct measure of contaminant concentration, with additional information gained from the indirect measure of biodegradation activity, e.g., carbon dioxide evolution, OUR, total and substrate-specific microbial numbers, and nutrient utilization patterns. More advanced monitoring based on unique genetic attributes of degrading strains identified by DNA and mRNA probes and metabolic markers are under development but are not routinely used during full-scale remediation.

4. BIOREMEDIATION APPLICATIONS

Treatment of hydrocarbons in soils through the application of bioremediation technology is a proven and effective method for clean-up in many instances. Successful technology application depends on technology-specific factors, which include the type and concentration of hydrocarbons present, soil type, and microbial activity. In addition, site-specific factors such as negotiated clean-up levels, space to undertake the process, and the allowed time interval also affect the feasibility and selection of bioremediation technologies from among other remediation technologies.

The United States Environmental Protection Agency (U.S. EPA) Bioremediation In the Field program lists over 150 full-scale bioremediation projects that are currently being conducted. This does not reflect the enormous number of storage tank (UST and AST) closures and replacements requiring soil remediation, which are regulated at the state level.

Bioremediation technology applications to reduce hydrocarbon concentrations in soil are developed around strategies for delivering moisture, aeration, and nutrients to optimize microbial activity. The cost of remediation is typically driven by costs incurred during construction, for example, soil excavation or transport. While technologies such as land treatment, aeration piles, and composting are commonly practiced, research and development efforts are increasingly focused on *in situ* methods, such as bioventing, that allow soil to be left in place during treatment. In addition to advancing the technologies of bioremediation, progress in regulation and development of well-defined and attainable clean-up levels are also important.

The purpose of the following section is to discuss regulatory issues, treatability studies, and current approaches for performance and economics as they relate to bioremediation of hydrocarbon-contaminated soils.

4.1. Regulations

Federal government regulations enacted to control, reduce, or eliminate hazardous and problem wastes are the principal force driving the development and use of remediation technologies, including bioremediation. The Resource Conservation and Recovery Act (RCRA) regulates the management of hazardous waste from production through disposal, and the Comprehensive Environmental Response, Compensation, and Liability Act (CERCLA), also known as Superfund, provides the impetus for investigation, management, and use of clean-up technologies for remediation of existing hazardous waste sites. CERCLA exempts petroleum from the definition of a hazardous substance. In 1984, the EPA made a determination that this petroleum exclusion applies to refined products, including gasoline and unused cutting oils. Used cutting oils and crankcase oils are not excluded because of their potential to contain significant concentrations of heavy metals (Bakst and Devine, 1993). In 1984, Congress authorized RCRA Subtitle I, which provided an impetus for the EPA to develop guidelines for (among other issues) corrective actions for releases from underground storage tanks (USTs) (53 Fed. Reg. 37082 and 43322). Useful reviews of the regulatory context for bioremediation appear in *Bioremediation* (Baker and Herson, 1994) and *Bioremediation Field Experience* (Flathman et al., 1994).

Regulation and guidance for UST issues and hydrocarbons released in soil and groundwater have been delegated to the states by the federal government, with the expectation of meeting or exceeding federal guidelines established by RCRA Subtitle I. Each state is required to prepare comprehensive regulations regarding discovery, investigation, reporting, and remedial measures for hydrocarbon-affected environmental media. The promulgation of regulations and guidance for investigation, reporting, and clean up of petroleum releases varies widely from state to state. The Association for the Environmental Health of Soils (AEHS) compiles state-by-state clean-up levels for hydrocarbons in soils and water (Oliver et al., 1993). The considerable variation in clean-up levels and analytical methodologies for petroleum quantification adopted by the state regulatory agencies reflects a remarkable degree of interpretation concerning the risks inherent from hydrocarbons in soil and groundwater as well as lack of standardization in hydrocarbon analytical protocols.

Regulated hydrocarbon categories include gasoline, diesel, and waste oil. State soil clean-up levels for gasoline vary from <10 mg/kg to >500 mg/kg; diesel from <100 mg/kg to 5,000 mg/kg; and waste oil from 100 mg/kg to 5,000 mg/kg. In many states, standards are set at rigid values while in other states standards are open to risk assessment evaluation or site-specific considerations. Additionally, many states provide requirements for analysis of hydrocarbon substituents such as BTEX or PAHs, while other states have no such provisions. Oliver et al. (1993) also note that the clean-up standards are revised on a frequent basis as new information becomes available.

In part, the variation in clean-up levels is due to the nebulous goal of "effectiveness." Effectiveness in some instances is a quantifiable reduction in mobility, toxicity, or concentration. In other cases, effectiveness must also be measured by the ability for a technology to meet certain numeric standards that may be based on aesthetic criteria such as smell and taste of drinking water rather than health-based risk standards. Health-based hydrocarbon clean-up standards are based on exposure. Routes of exposure may include inhaling of contaminated dusts, ingesting soil, dermal contact, drinking contaminated water, or eating fish that swim in receiving waters of contaminated groundwater. Each of these potential exposures is considered during clean-up level development.

4.2. Treatability Studies

Variability in the source of hydrocarbon contamination, the nature of contaminated soils, and other concerns underlie the importance of treatability studies to demonstrate the efficacy of a potential remedial alternative. Also, treatability testing provides a mechanism for establishing proper treatment conditions and identifying potential problems to be encountered during a full-scale implementation.

Treatability studies are typically conducted to support a study of feasible remedial alternatives. The feasibility study is developed from site investigation, which, within some level of confidence, determines the extent of affected matrix, i.e., the volume of soil affected by hydrocarbons and their approximate average concentration. Together with an understanding of clean-up levels to be achieved, the results of site investigation analyses are used to develop a list of feasible options.

The feasibility study typically involves initial screening of parameters, followed by detailed analysis of options. A typical screening level analysis for a site contaminated by hydrocarbons is presented in Table 3.2. Remedial alternatives for petroleum-contaminated soils may include off-site disposal, incineration, asphalt incorporation, soil washing, as well as bioremediation. In addition to technical feasibility, a final determination of an appropriate alternative includes evaluation of criteria, such as protection of human health and the environment, permanence of the solution, time frame for restoration, and consideration of public concern. Bioremediation often meets the aforementioned criteria quite well and is also cost effective.

Screening remedial alternatives may require conducting of "treatability studies" to develop a quantitative basis for technology choice. In the case of hydrocarbons, there is little disagreement on "treatability" by biological technologies. The issue is

TABLE 3.2. General Treatment Alternatives for Hydrocarbon-Contaminated Soils

General Response Actions	Remedial Technologies	Process Options	Description	Determination of Applicability
Removal	Excavation	Mechanical excavation	Use of mechanical excavation equipment to remove and load contaminated soils for disposal or treatment	Potentially applicable for contaminated soils. May release VOCs to the atmosphere
		Hand excavation	Use of manual labor to excavate soil in areas with limited access	Potentially applicable. May be necessary in upper tank farm area
		Consolidation	Refers to consolidation under a landfill cap of excavated material from contaminated areas	Not applicable. Site is not suitable for landfill siting
Excavated soil disposal	Disposal on site	RCRA Subtitle C Landfill	Permanent storage facility on site, double lined with clay and a synthetic membrane liner and containing a leachate collection/detection system	Not applicable. Site is not suitable for landfill siting
	Disposal off site	RCRA Subtitle C Landfill	Transport of excavated soil to a RCRA Subtitle C permitted landfill	Potentially applicable for metals-contaminated soils. RCRA Land Disposal Restrictions may require treatment of waste prior to landfilling
		RCRA Subtitle D Landfill	Transport of excavated soils to a RCRA Subtitle D permitted landfill	Potentially applicable for excavated materials with immobile contaminants and/or low levels of contamination or TPH-impacted soils

Excavated soil treatment	Thermal treatment	Incineration	Excavated wastes are thermally destroyed in a controlled oxygen-sufficient environment	Not applicable for TPH-contaminated soils due to permitting, availability, and cost
		Low temperature thermal desorption	VOCs removed from soil in a drying unit, and off gases are incinerated	Potentially applicable. Primary soil contaminants TPH, which are often treated by this method
	Biological treatment	Aerobic biodegradation	Soil treated with nutrients to promote biological degradation in the presence of oxygen and proper moisture content	Potentially applicable. Pilot testing may be required to identify optimum conditions and degradation rates
	Chemical treatment	Chemical oxidation	Hydrocarbons are degraded by addition of chemical oxidants	Not applicable. Many other technologies are readily applied and more cost effective
	Physical treatment	Soil washing	Mixing solvent or water with soil in a controlled system to extract contaminants from soil	Potentially applicable. Potential solvents include water, surfactants, and chemical solvents
		Solidification/stabilization	Soil mixed with a pozzolanic/cement material that can solidify and reduce mobility of contaminants	Potentially applicable to soils containing elevated metals and CPAHs
		Asphalt incorporation	Soil used to make asphalt pavement and contaminants are immobilized	Potentially applicable
		Aeration	Soil is aerated in soil screen or tilledto volatilize organics	Not applicable. Some of the contaminants are heavy hydrocarbons, which do not readily volatilize

whether a specific process will achieve the necessary clean-up levels for the affected matrix. In this sense, the term *process evaluation* is more descriptive of laboratory-scale studies.

As noted, a key purpose of a process evaluation (treatability study) is to identify issues that may adversely affect the degradation process and to evaluate the rate and extent of hydrocarbon reduction that can be achieved. The evaluation provides an initial determination of the applicability of a process and whether or not proceeding to full-scale implementation is appropriate. Laboratory-scale process evaluations for hydrocarbon degradation typically mimic the full-scale treatment anticipated for the field. The laboratory set-up may involve placing soil and amendments in trays (land treatment or composting), flasks or small reactors (soil slurries), or columns (bioventing). The EPA, American Petroleum Institute (API), and National Environmental Technology Application Corporation (NETAC) are in the process of evaluating and publishing uniform guidelines for the conduct of treatability studies and process evaluations. The EPA has published a remedy screening for bioremediation (U.S. EPA, 1991). Recent reviews of bioremediation treatability testing are available (Nelson et. al., 1994; Skladany and Baker, 1994). A summary of hydrocarbon treatability testing is provided in the following sections.

Hydrocarbon treatability studies can be resolved into three elements: *material characterization, study conduct,* and *data evaluation*. The first and most important element is material characterization. Material characterization establishes the type and concentration of hydrocarbons present and also provides an indication of existing conditions that will impact hydrocarbon biodegradation. These conditions may include excessive hydrocarbon concentrations, the presence of heavy, condensed residues that are slow to degrade, low numbers of microorganisms, or soil conditions that make the process difficult to implement. The material characterization establishes the baseline status of process parameters such as nutrients, inorganic ions, hydrogen ion concentration (pH), total organic carbon, and moisture content.

The second treatability element, study conduct, evaluates effectiveness of various treatments on the rate and extent of contaminant degradation. Treatments evaluate and provide data for optimizing nutrient formulations, pH adjustment, organic amendments, surfactants, inoculation with specialized strains, and other factors. Proper selection and application of analytical and statistical methods are elemental to obtaining data that are both reproducible and correct. Laboratory demonstrations should include appropriate controls to account for biological as well as physical and abiotic losses such as volatilization, adsorption, leaching, or photo-oxidation.

The most important technical element of a treatability study, as well as subsequent pilot and full-scale remediation programs, is the analytical techniques chosen and the quality of data resulting from these methods. Following selection and execution of analytical protocols, proper statistical analysis and extrapolation of the data are second in importance. Without attention to these two criteria, poor or meaningless information will result from a treatability study. The sampling and analysis plan outlines appropriate sampling and determination of compounds, microbial numbers, and activity before, during, and after the process has been completed. All methods used to assess these parameters require that samples be taken for analysis and that proper sampling and statistical protocols be followed to obtain representative and reproducible results.

Samples used for treatability studies should reflect the distribution of contaminants expected in a full-scale application; they should be representative of the biological conditions to be encountered. Also, a critical concern for evaluation of treatability is the status of the samples at the time of collection. For example, samples taken when soil is dry will have dramatically different activity than moist soil samples. Fluctuations can also be expected from temperature change and nutrient additions. To avoid these complications, sampling events should be coordinated with site or laboratory operations such as tilling, nutrient addition, water adjustment, and other activities to ensure a measure of uniformity.

The final treatability element, data evaluation, should place experimental information into the specific project context and address whether contaminants can be degraded, whether clean-up levels can be attained, which types of nutrient or buffering agents are required, the time required for treatment, and other issues relating to implementation. Treatability studies may also provide information on the routes of compound loss. Shannon and Unterman (1993) examine interpretation of treatability study data and other issues related to effective design and evaluation of biodegradation testing. The conduct and evaluation of biodegradation testing and treatability testing is also reviewed by Grady (1985). Regardless of the specific treatment technology evaluated, the treatability test workplan should include a sampling and analysis plan, a quality assurance plan, and experimental design. In addition, site-specific details should be delineated to ensure that clean-up levels or action levels are clear and information gaps may be identified prior to commencing the study.

4.3. Full-Scale Bioremediation

Full-scale implementation of bioremediation for treatment of hydrocarbon contaminated soils combines microbiology, chemistry, engineering, and construction management. The following sections summarize selected reports in the literature concerning land treatment (land farming, solid-phase treatment), engineered biopiles, composting, bioventing, and rhizosphere remediation.

4.3.1. Land Treatment

Land treatment of petroleum hydrocarbons has been practiced in the United States since the 1950s primarily for reduction of the volume and toxicity of refinery sludges (Bosssert and Bartha, 1984; Loehr, 1984). Traditionally, these practices required incorporation of hydrocarbon wastes into soils. Recent legislation, however, e.g., Appendix Nine and Land Ban, have curtailed and eliminated this practice. Land treatment no longer includes direct incorporation into non-impacted soils.

Federal RCRA and CERCLA regulations, and state hazardous waste and underground storage tank regulations,typically promote technologies that provide perma-

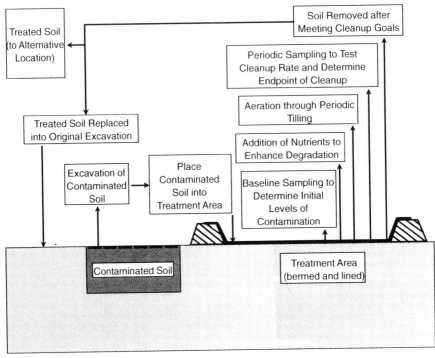

Figure 3.5. Solid-phase remediation.

nent waste solutions through degradation or recycling over waste transfer to land-fills. Soil treatment in engineered land treatment units is a commonly employed remedy for a variety of organic wastes, particularly hydrocarbon-contaminated soil from petroleum storage facilities and leaking underground storage tanks. Biotreatment of soil is carried out under controlled conditions that may include engineered liners to prevent contaminant transport in leachate, retention berms or auxiliary tanks to provide for stormwater retention, and post-process monitoring to confirm absence of residual effects. The schematic diagram presented in Figure 3.5 summarizes the process of land treatment.

Land treatment is a solid-phase technology that relies on long-standing principles on the biocycling of natural compounds as applied in agriculture and waste treatment. The conditions for biodegradation are optimized by regular tilling of the soil and by the addition of nutrients and water. The natural indigenous microbial populations of soil are diverse, and many of the appropriate microorganisms that degrade contaminants are found in the contaminated soil. Rates of bioremediation of contaminated soils are enhanced by optimizing the conditions of the site for oxygen levels, moisture content, available nutrients such as nitrogen and phosphorus, contaminant concentration, pH, and contact between appropriate microorganisms and the contaminants. Some of these factors are more readily manipulated in the field than others. For example, fertilizer supplementation is often the first measure used to stimulate the biodegradation of hydrocarbons in soil.

In general, nutrient addition stimulates hydrocarbon biodegradation in soils; however, the extent of stimulation reported in the literature is very inconsistent. In addition to the use of inorganic nutrients such as nitrogen and phosphorus (already discussed), organic supplements may also be added to soils, often as bulking agents to improve texture and mass transfer. As with inorganic amendments, results can vary widely. For example, Dibble and Bartha (1979) found no stimulation of oil biodegradation in soil treated with organic materials, such as yeast extract or activated sewage sludge, even though they were applied in conjunction with inorganic fertilizers. Whereas inorganic NPK fertilization did stimulate biodegradation at the intermediate loading rate (C:N:P:K = 1:60:800:400) in these studies, the organic supplements actually suppressed degradation. This may have been due to competition for oxygen or actual repression of hydrocarbon-degrading enzymes in the presence of more readily utilized substrates.

In other studies, manure has been used as a soil amendment in optimizing land treatment operations for the disposal of hazardous wastes (Sims, 1986; McGinnis, 1987). No single justification for the manure amendments is apparent from the literature. In some cases manure is used as an inorganic nutrient amendment (Linkenheil and Patnode, 1987) and/or as a source of organic matter. The organic matter may function as an additional carbon source to sustain an active microbial population (McGinnis, 1987; Sims, 1986) or to facilitate absorption of the waste components (Linkenheil and Patnode, 1987). McGill and Rowell (1977) found that manure, along with nitrate supplements, stimulated carbon dioxide production in hydrocarbon-contaminated soils more than nitrate addition alone. However, the excess carbon dioxide may have been derived more from the manure than the hydrocarbon contaminants.

Rates of hydrocarbon degradation depend on the type and age of the substrate in soil, microbial activity, and soil type. An estimation of biodegradation rates and remediation times is often extrapolated from laboratory studies. Biodegradation rates are often described by first-order kinetic equations, resulting in half-life calculations. However, these calculated estimates do not necessarily reflect biological kinetics and must be used with caution. Typically the reported half-life data describe the biodegradation kinetics of light, degradable hydrocarbons, e.g., less than C-20, and infer that biodegradation will continue on the remaining hydrocarbons at a similar rate.

According to the half-life concept, the time required for a 50% reduction of the contaminant concentration is constant. While first-order constants are often generated to describe the decrease in hydrocarbon concentrations, these calculations should be qualified because hydrocarbon substrates are complex and become proportionately enriched with high-molecular-weight material over time. Hydrocarbon weathering affects solubility, transport, and biodegradation rate. Ultimately, hydrocarbon residuals remain that are not amenable to biodegradation processes or, at best, degrade only very slowly.

TABLE 3.3. Summary of Laboratory and Field Treatment Data

Hydrocarbon Type	Initial Concentration (mg/kg)	Final Concentration (mg/kg)	Average Rate (mg/kg/day)	First-Order K (Day -1)	$T_{1/2}$ (days)	References
Diesel fuel	4,500	270	87	0.058	12	Jerger et al. (1993)
Diesel fuel	1,200	100	40	0.089[a]	8	Frankenberger (1993)
No. 6 diesel	60,000	24,000	400	0.010	68	Fogel (1993)
Diesel fuel	1,350	100	10	0.010[a]	70	Troy et al. (1993)
				0.018[b]	39	
Diesel fuel	100,000	42,000	518	0.015[a]	50	Song et al. (1990)
Crude oil	15,000	6,750	56	0.006	122	Huesemann and Moore (1993)[c]
Heavy oil	7,900	3,000	58	0.011	60	Ying et al. (1990)
Oils (refinery)	12,980	1,273	50	0.009	71	Ellis et al. (1990)
Crude residuals	6,000	1,000	65	0.020	38	Compeau et al. (1991)

[a]Laboratory value.
[b]Field value.
[c]Authors focus on HC compositional changes during treatment and do not report data as half-lives.

The biodegradation time frame may be best established by evaluating rates of decay, and careful evaluation of hydrocarbon residuals (typically by gas chromatography), to determine the potential for further degradation in order to meet regulatory clean-up standards or another endpoint determination. Too great an emphasis on theoretical kinetic data, especially those generated during treatability studies, can result in an unrealistic approximation of time and efficacy needed to achieve the desired endpoint of a bioremediation.

Table 3.3 summarizes rate data from selected petroleum hydrocarbon bioremediation projects; the calculated half-lives are presented for the purpose of comparison. From these, and several additional sources (Dibble and Bartha, 1979; Bossert et al., 1984; Ryan et al., 1987; Linkenheil and Patnode, 1987; Simms, 1986), it is concluded that hydrocarbon removal rates are proportional to initial concentration. Mass balance information is not available for the full-scale studies cited, so reported reductions are assumed to comprise volatile losses and biodegradation. Degradation half-lives are suprisingly similar and indicate a certain degree of predictability for the remediation of diesel fuel and heavy oils. For example, an average half-life of 54 days with a 90% confidence interval from 35 to 72 days is demonstrated for the tabulated studies, irrespective of hydrocarbon type. For those sites with diesel fuel contamination only, degradation half-lives averaged 41 days, with a 90% confidence interval from 22 to 57 days.

4.3.2. Engineered Biopiles (Aboveground)

Soil heap bioremediation is a modification of land treatment that is used when available space (area) is limited or when regulatory restrictions require the capture of air emmissions to treat volatile organic compounds. In soil pile bioreclamation, the contaminated soil is excavated and stockpiled into a heap within a lined treatment area to prevent further contamination. Figure 3.6 shows a schematic representation of a biopile.

For operation, the soil piles are covered with polyethylene. Water and nutrients are applied to the surface of the stockpile through perforated pipe and allowed to percolate down through the soil. Although usually not required, a microbial inoculum may be added to the soil, especially if the contamination is recent and extensive. Under such conditions, an initial, drastic decline in the number of hydrocarbon-degrading microorganisms may retard degradation rates, and bioaugmentation reduces the resulting lag periods. Aeration is provided for by a vacuum pump, and an air emissions recovery system may be installed. Leachate collection is used to collect the fluid, which is recycled. An internal piping system may also be installed to blow air upward through the soil and thus accelerate the biodegradation process through the addition of oxygen. During operation, pH and moisture content are maintained within ranges conducive to microbial activity. Typical costs are similar to conventional solid-phase treatment.

Biopile treatment design lends itself to recovery of volatile organic compounds and therefore provides benefits over land treatment in areas where regulations are sensitive to volatile compound release.

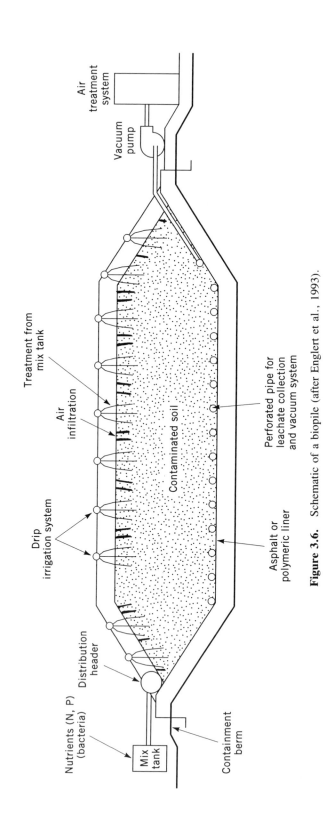

Figure 3.6. Schematic of a biopile (after Englert et al., 1993).

4.3.3. Composting Biotreatment

Composting processes are yet another modification of solid-phase treatment in which the system is operated at a higher temperature due to increased biological activity. The process relies on obligately aerobic, thermophilic microorganisms to biodegrade the hydrocarbon contaminants. This technology is best used for highly contaminated soils (such as refinery sludges) and in areas where temperature is critical to the sustained treatment process.

The contaminated soils are mixed with suitable bulking agents, such as straw, bark, or wood chips, and piled in mounds. Bulking agents improve soil texture for aeration and drainage. The system is optimized for pH, moisture, and nutrients via irrigation techniques; it can further be enclosed to contain volatile emissions. For a successful application, it is important that the bulking agent does interfere with the biodegradation of the contaminants. In addition, the accumulation of heavy metals and other potentially toxic components must be monitored, especially with repeated applications.

The costs of maintaining adequate oxygen levels and the increased bulk of the contaminated soil mass limit the use of composting for hydrocarbon-contaminated soils. Typical applications include heavy oils, PAHs, and munitions wastes. For example, McMillen et al. (1993) determined that composting showed promise for the treatment of production pit sludges. Concentrations of extractable hydrocarbon were reduced by over 90% from initial concentrations of nearly 11%. Renewed attention has also been directed toward the composting of munitions wastes, such as TNT, RDX, and HMX, which has been studied over the past 20 years (Kaplan and Kaplan, 1982; Osmon and Andrews, 1978). The issues surrounding efficacy for these compounds are beyond the scope of this review.

4.3.4. Bioventilation (In situ)

Bioventilation (bioventing) is a hybrid *in situ* technology that embodies principles of biodegradation and soil vapor extraction. Soil vapor extraction (SVE) is practiced routinely for the physical removal of volatile and semivolatile contaminants (Johnson et al., 1990). During SVE, a vacuum is used to induce negative pressure in the vadose zone, accelerating the volatilization of hydrocarbon compounds. As air is pulled through the soil, volatile organics are swept with the flow to the surface for aboveground treatment. The success of the technology is ultimately limited by the vapor pressure of organic compounds present. In general, organic compounds with vapor pressures greater than 0.1 atm are amenable to SVE (Dupont, 1992). The key advantages of SVE (and bioventilation) include limited disruption to ongoing site operations, potential for treatment of soils located below the practical limits of excavation, and potential cost savings related to decreased excavation costs. The technology is particularly promising for landfills, fueling operations, refineries and certain chemical production facilities. Considerable pilot-scale work has been performed at service stations and military installations.

Bioventing extends the advantages of SVE by also treating residual hydrocarbons in place and, because of the lower air flow rates required (as compared with vapor

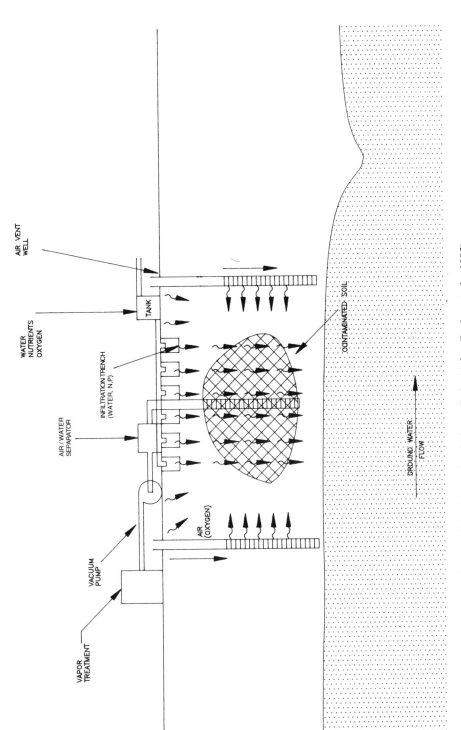

Figure 3.7. Schematic of bioventilation (after Englert et al., 1993).

extraction), secondary costs of vapor treatment are significantly reduced. Conceptually, bioventing builds on SVE, attempting to maximize the activity of indigenous microbial populations to bring about the biodegradation of organic compounds in place. As is shown in Figure 3.7, bioventilation requires airflow through unsaturated, contaminated zones; nutrients and/or moisture may be added to facilitate microbial activity and biodegradation. The zone of treatment may be increased by lowering the groundwater table through a dewatering system of wells. While the concept is simple, success depends on understanding complex physical, chemical, and biological processes. In effect, the technology turns the subsurface, unsaturated zone into a bioreactor.

As with most *in situ* technologies, bioventing is dependent on soil permeability, organic compound type, hydrocarbon concentration, and environmental constraints. Moreover, a successful operation requires an accurate assessment of the subsurface. Significant assumptions are sometimes necessary when interpreting data from subsurface investigations. The subsurface is often interspersed with lenses of materials of varying density and porosity. For example, it is not uncommon to find high hydrocarbon concentrations ($>1,000$ mg/kg) within centimeters of nondetectable concentrations in the same matrix. While this is not significant for excavation and mixing during *ex situ* treatment, these concentration extremes and the potential for channeling air flow are of importance to the implementation of subsurface remediation.

Evidence for the biological removal of contaminants during soil venting was first noted as a shift in the O_2/CO_2 ratios between inflow (atmospheric) and effluent gases. After passage through soil, oxygen levels are depleted and carbon dioxide concentrations are elevated, often within hours to days of operation. Field oxygen uptake rates have been recorded between 36 and 216 g O_2/ft^3 of soil/day (Dupont, 1992). The change in gas composition is primarily due to bacterial activity; molecular oxygen serves as the electron acceptor during biological oxidation of hydrocarbons by the soil microbiota.

Oxygen delivery in the gas phase is a critical aspect of bioventing. For example, approximately 3 pounds of molecular oxygen are required for each pound of benzene mineralized to carbon dioxide and water (Dupont, 1992). Air is a good medium for transferring oxygen to bulk soils in unsaturated zones. For example, assuming a 1,000 gallon (7,000 pound) fuel spill, the amount of oxygen required to biodegrade the contaminant is 21,000 pounds, or 200 hours of venting ambient air (20% O_2) at 100 cfm. Delivery of this same amount of oxygen in air-saturated water would require 2,300,000 gallons and that of a saturated peroxide solution would require 42,000 gallons.

Bioventing systems are constructed with the same materials as SVE. The philosophy of operation and design differs significantly. The primary purpose is to deliver the electron acceptor to the contaminant. As such, extraction wells may be placed on the periphery, and air is drawn through the contamination. The air flow rates are low (50–500 cfm) to maximize residence time of air (Dupont, 1992). Vapor extraction wells are advanced primarily for the purpose of inducing air flow through the contaminated zone collecting contaminated vapors.

Laboratory and field studies underscore the potential effectiveness of bioventing. Early reports by the Texas Research Institute (1984) suggested that approximately 30% of the gasoline removed from vapor-extracted soils may have been accounted for by biodegradation. These observations were further supported by Wilson and Ward (1988), who noted that vapor extraction systems are very effective in transferring oxygen to the subsurface and thereby stimulating biological activity and degradation of fuel hydrocarbons. Chevron Research received a patent for a bioventing system in 1988 (Ely and Heffner, 1988). Coho (1990) evaluated and reported ratios of physical to biological losses of hydrocarbon products as follows: gasoline 65%:35%; jet fuel 7%:93%; and fuel oil 100% biological degradation. Ely and Heffner (1988) reported approximately 50%:50% biological to physical reduction of mixed gasoline and diesel fuel hydrocarbons.

Recently, numerous authors have presented field-scale data demonstrating the effectiveness of bioventing for the remediation of fuel-contaminated soils. These reports include tests performed at military installations in the United States (Hinchee and Arthur, 1991; Dupont et al., 1991) and locations in the Netherlands. The results of these tests are presented by Hoeppel and coworkers (1991). Biodegradation rates were approximated by respirometric techniques involving injecting air into the subsurface and measuring oxygen utilization. Hydrocarbon degradation rates of 1–20 mg/kg/day based on a 3:1 oxygen to hydrocarbon ratio were reported for diesel and jet fuels. A full-scale system operating for over 2 years at Hill Air Force Base in Utah averaged 10 mg/kg/day. Authors note that the relationship between respirometric data and actual bioventing treatment rates (measured through reduction in contaminant concentrations) have not been established.

In discussion of data from widely diverse pilot-scale bioventing sites, Hoeppel et al. (1991) indicate that degradation rates (based on carbon dioxide evolution and oxygen uptake) differ from theoretical expectations. For example, degradation rates in sub-Artic regions are not significantly different from those measured in temperate regions; addition of nitrogen fertilizer to nitrogen-deficient soils did not accelerate degradation in all cases; and fine-grained clayish soils demonstrated excellent respiration rates.

Management of other bioprocess concerns such as moisture addition, pH balance, and nutrient addition are difficult since solutions may move along preferred pathways distinct from contaminant pathways. The additional issue of miscibility of aqueous nutrient solutions with hydrophobic hydrocarbons at depth is also difficult to manage or control. These issues remain significant challenges to technology development.

4.3.5. Rhizosphere Remediation

The rhizosphere is defined as that part of the soil adhering to a plant root system after shaking to remove loose soil. Rhizosphere size is determined by the size and complexity of a plant root system and may represent a significant contact area (Atlas and Bartha, 1993). Microbial density is high in rhizosphere soils, typically ranging from 5 to 20 times greater than soil without roots. Higher microbial metabolic activity results from plant root exudates, including organic acids, sugars, and other organic materials. Plants benefit from increased solubilization of minerals, synthesis of vitamins, and other growth-stimulating materials mediated by microorganisms.

The effects of rhizosphere activity on the fate of organic chemicals has been actively researched for the past 30 years. A principal focus of these studies has been to elucidate ecological mechanisms for persistence and degradation of herbicides and pesticides used in agriculture. Hazardous waste mitigation issues and regulations have focused new light on the beneficial activities of microorganisms in the rhizosphere. Interest in the capabilities of plant-associated microbes to stimulate remediation of soils containing high levels of a variety of organic compounds ranging from chlorinated solvents to pesticides and hydrocarbons are being examined. Anderson et al. (1993) provide an excellent recent review of rhizosphere bioremediation. Of particular note are studies performed by Sims and coworkers (Sorensen et al., 1994; Aprill and Sims, 1990), indicating increased mineralization of PAHs in root zone soils resulting from a mixture of prairie grasses as compared with unplanted controls. Walton and Anderson (1990) reported accelerated increased mineralization of [^{14}C]TCE in rhizosphere samples from four plant species and developed these data further through demonstration of similar results in whole plant studies.

Rhizosphere bioremediation is an emerging technology. Technical issues include the demonstration of mineralization versus humification or the translocation of compounds into plant tissues. A potential role of this technology and its associated phenomena may be for the clean-up of petroleum exploration sites in remote locations, as a means for accelerating natural degradation, or as a final phase of land-treatment operations to continue reduction of recalcitrant hydrocarbon residues cost-effectively.

5. SUMMARY

Petroleum hydrocarbons comprise naturally occurring materials that, by their nature, are biodegradable. Under the proper environmental conditions, most hydrocarbons are biodegraded by microorganisms already present in the environment. Those hydrocarbons that persist are generally unavailable to the microbiota. In addition to the bioavailability issue, a number of environmental factors can affect both the rate and the extent to which a compound or contaminant is biodegraded. The ability to manipulate better these parameters, particularly for *in situ* applications, will improve current bioremediation efforts, by increasing their efficacy while decreasing the treatment period.

A number of avenues of research need to be followed in order to develop the most expedient and cost-effective bioremediation technologies. These emerging biotechnologies need to be competitive with established technologies such as incineration, landfill, and injection. The latter often do not provide a permanent solution to the problem. Areas for further research include the development of better qualitative and quantitative methods for sampling and analysis of the contamination; improved and standardized techniques for assessment of biodegradation and the microbial mediators; development of methods for increasing the bioavailability of persistent materials; and the standardization of rational clean-up levels along with the development of methods for their assessment.

Bioremediation technologies offer a cost-effective, permanent solution to the clean-up of soils contaminated with hydrocarbons and other biodegradable organics. Through a better understanding of interactions between the environment, the contaminants, and the microbial mediators of biodegradation, improvements in the technology will ensue. Bioremediation is already a proven and effective option, but its full potential is yet to be realized for the clean-up of contaminated environments.

REFERENCES

Aeckersberg F, Bak F, Widdel F. (1991): Anaerobic oxidation of saturated hydrocarbons to CO_2 by a new type of sulfate-reducing *Bacterium*. Arch Microbiol 156:5–14.

Akgerman A, Erkey C, Ghoreishi SM. (1992): Supercritical extraction of hexachlorobenzene from soil. Ind Eng Chem Res 31:333–339.

Alexander M (1965): Biodegradation: Problems of molecular recalcitrance and microbial fallibilitiy. Adv Appl Microbiol 7:67–79.

Anderson TA, Guthrie EA, Walton BT (1993): Bioremediation in the rhizosphere. Environ Sci Technol 27:2360–2366.

Anderson JPE (1990): Principles of and assay systems for biodegradation. In Kamely D, Chakrabarty A, Omenn G (eds): Advances in Applied Biotechnology Series, vol 4. Houston: The Gulf Publishing Co, pp 129–145.

Anderson JPE, Domsch KH (1975): Measurement of bacterial and fungal contributions to respiration of selected agricultural and forest soils. Can J Microbiol 21:314–322.

Andrews AT, Ahlert RC, Kosson DS (1990): Supercritical fluid extraction of aromatic contaminants from a sandy loam soil. Environ Prog 9:204–210.

Aprill W, Sims RC (1990): Evaluation of the use of prairie grasses for stimulating polycyclic aromatic hydrocarbon treatment in soil. Chemosphere 20:253–265.

Aronstein BN, Alexander, M (1993): Effect of a non-ionic surfactant added to the soil surface on the biodegradation of aromatic hydrocarbons within the soil. Appl Microbiol Biotechnol 39:386–390.

Atlas R (1992): Detection and enumeration of microorganisms based upon phenotype. In Levin M, Seidler R, Rogul M (eds): Microbial Ecology: Principles, Methods, and Applications. New York: McGraw-Hill, pp 29–41.

Atlas RM (1981): Microbiological biodegradation of petroleum hydrocarbons: An environmental perspective. Microbiol Rev 45:180–209.

Atlas RM, Bartha R (1993): Microbial Ecology: Fundamentals and Applications, 3rd ed. New York: Benjamin-Cummings.

Atlas R, Bartha R (1973): Simulated biodegradation of oil slicks using oleophilic fertilizers. Environ Sci Technol 7:538–541.

Atlas RM, Bej AK (1990): Detecting bacterial pathogens in environmental water samples by using PCR and gene probes. In Innis MA, Gelfand DH, Sninsky JJ, White TJ (eds): A Guide to Methods and Applications. New York: Academic Press, pp 399–406.

Atlas RM, Sayler GS (1989): Tracking microorganisms and genes in the environment. In Omenn GS (ed): Environmental Biotechnology: Reducing Risks from Environmental Chemicals through Biotechnology. New York: Plenum, pp 31–50.

Bailey NJ, Jobson AM, Rogers MA (1973): Bacterial degradation of crude oil: Comparison of field and experimental data. Chem Geol 11:203–221.

Baker KH, Herson DS (1994): Bioremediation. New York: McGraw Hill.

Bakst J, Devine K (1993): Bioremediation: Environmental regulations and resulting market oppportunities. In Flathman P et al (eds): Bioremediation Field Experience. Boca Raton: Lewis Publishers, pp 11–48.

Baldwill DL (1989): Numbers, diversity, and morphological characteristics of aerobic chemoheterotrophic bacteria in deep subsurface sediments from a site in South Carolina. Geomicrobiol J 7:33–52.

Barbaro JR, Barker JF, Lemon LA, Mayfield CI (1992): Biotransformation of BTEX under anaerobic, denitrifying conditions: Field and laboratory observations, J Cont Hydrol 11:245–272.

Bartha R, Bossert I (1984): Treatment and disposal of petroleum refinery wastes. In Atlas R (ed): Petroleum Microbiology. New York: Macmillan, pp 553–578.

Bartha R, Pramer D (1965): Features of a flask and method for measuring the persistence and biological effects of pesticides in soil. Soil Sci 100:68–70.

Bauer JE, Capone DG (1988): Effects of co-occurring aromatic hydrocarbons on degradation of individual polycyclic aromatic hydrocarbons in marine sediment slurries. Appl Environ Microbiol 54:1649–1655.

Beerstecher E Jr (1954): "Petroleum Microbiology." New York: Elsevier Press.

Bej AK, Mahbubani MH, Atlas RM (1991): Amplification of nucleic acids by polymerase chain reaction (PCR) and other methods and their applications. Crit Rev Biochem Mol Biol 26:301–334.

Beliaeff B, Mary JY (1993): The most probable number estimate and its confidence limits. Wat Res 27:799–805.

Berger WD, Anderson KE (1981): Modern Petroleum, 2nd ed. Tulsa, OK: Penn Well.

Berry DF, Francis AJ, Bollag JM (1987): Microbial metabolism of homocyclic and heterocyclic aromatic compounds under anaerobic conditions. Microbiol Rev 51:43–59.

Blackburn JW, Hafker WR (1993): The impact of biochemistry, bioavailability and bioactivity on the selection of bioremediation techniques. Trends Biotechnol 11:328–333.

Bogardt AH, Hemmingsen BB (1992): Enumeration of phenanthrene-degrading bacteria by an overlay technique and its use in evaluation of petroleum-contaminated sites. Appl Environ Microbiol 58:2579–2582.

Bossert I, Bartha R (1984): The fate of petroleum in soil ecosystems. In Atlas R (ed): Petroleum Microbiology. New York: Macmillan, pp 435–474.

Bossert I, Bartha R (1986): Structure-biodegradability relationships of polycyclic aromatic hydrocarbons in soil. Bull Environ Contam Toxicol 37:490–495.

Bossert I, Robison P, Nelson E (1991): Implications of Nitrogen Fertilization for In Situ Bioremediation of Petroleum-Contaminated Soils. Presentation at the Spring National ACS Meeting, Atlanta, GA.

Bossert I, Kachel WM, Bartha R (1984): Fate of hydrocarbons during oily sludge disposal in soil. Appl Environ Microbiol 47:763–767.

Britton LH (1984): Microbial degradation of aliphatic hydrocarbons. In Gibson DT (ed): Microbial Degradation of Organic Compounds. New York: Marcel Dekker, pp 89–129.

Buchman HO, Brady NC (1969): The Nature and Properties of Soils. New York: MacMillan.

Byrd JJ, Colwell RR (1992): Microscopy applications for analysis of environmental samples. In Levin MA, Seidler RJ, Rogul M (eds): Microbial Ecology: Principles, Methods, and Applications. New York: McGraw-Hill, pp 93–112.

Carmichel LM, Pfaender FK (1994): Fractionation of phenanthrene biodegradation in soil ecosystems. Presentation at the National ASM meeting, Las Vegas, NV. Abstract Q137.

Cerniglia CE (1993): Biodegradation of polycyclic aromatic hydrocarbons. Curr Opin Biotechnol 4:331–338.

Cerniglia C (1984a): Microbial metabolism of polycyclic aromatic hydrocarbons. Adv Appl Microbiol 30:30–71.

Cerniglia CE (1984b): Microbial transformations of aromatic hydrocarbons. In Atlas R (ed): Petroleum Microbiology. New York: MacMillan, pp 99–152.

Coho JW (1990): Biodegradation of Jet Fuel in Vented Columns of Water-Unsaturated Sandy Soil. Thesis. University of Florida.

Compeau GC, Mahaffey WD, Patras L (1991): Full-scale bioremediation of contaminated soil and water. In Sayler GS (ed): Environmental Biotechnology for Waste Treatment. New York: Plenum Press, pp 91–109.

Connaughton DF, Stedinger JR, Lion LW, Shuler ML (1993): Description of time-varying desorption kinetics: Release of naphthalene from contaminated soils. Environ Sci Technol 27:2397–2403.

Dang JS, Harvey DM, Jobbagy S, Grady CPL Jr (1989): Evaluation of biodegradation kinetics with respirometric data. Res J WPCF 61:1711–1721.

Dibble JT, Bartha R (1979): Effect of environmental parameters on the degradation of oil sludge. Appl Environ Microbiol 37:729–739.

Dooley KM, Ghonasgi D, Knopf FC (1990): Supercritical CO_2-cosolvent extraction of contaminated soils and sediments. Environ Prog 9:197–202.

Dupont RR, Doucette WJ, Hinchee RE (1991): Assessment of In Situ bioremediation potential and the application of bioventing at a fuel-contaminated site. In Proceedings of the In Situ and On-Site Bioreclamation International Symposium, San Diego, CA, March 19–21, 1991.

Dupont RR (1992): Application of Bioremediation Fundamentals to the Design and Evaluation of In Situ Soil Bioventing Systems. 85th Annual Meeting & Exhibition Air and Waste Management Association, 1992.

Ectors A, Wagemakers E, Ide G (1992): Gravimetric measuring of carbon dioxide production as alternative bench-scale method for sapromat tests in soil bioremediation projects. Med Fac Landbouww Univ Gent 57:1709–1711.

Efroymson RA, Alexander M (1991): Biodegradation by and *Arthrobacter* Species of hydrocarbons partitioned into an organic solvent. Appl Environ Microbiol 57:1441–1447.

Ellis B, Balba MT, Theile P (1990): Bioremediation of oil contaminated land. Environ Technol 11:443–455.

Ely DL, Heffner DA (1988): Process for In Situ Biodegradation of Hydrocarbon Contaminated Soil. U.S. Patent No. 4,765,902.

Englert CJ, Kenzie E, Dragun J (1993): Bioremediation of petroleum products in soils. In Calabrese EJ, Kostecki PT (eds): Principles and Practices for Petroleum Hydrocarbon-Contaminated Soils. Lewis Publishers.

Evans PJ, Ling W, Goldschmidt B, Ritter ER, Young LY (1992): Metabolites formed during anaerobic transformation of toluene and *o*-xylene and their proposed relationship to the initial steps of toluene mineralization. Appl Environ Microbiol 58:496–501.

Fedorak PM (1990): Microbial metabolism of organosulfur compounds in petroleum. In Oar WL, White CM (eds): Geochemistry of Sulfur in Fossil Fuels. Washington, DC: ACS Books, pp 93–112.

Fedorak PM, Westlake DWS (1984): Microbial degradation of alkyl carbazoles in Norman Wells crude oil. Appl Environ Microbiol 47:858–862.

Fetter CW (1993): Contaminant Hydrogeology. New York: MacMillan.

Finnerty WR, Shockley K, Attaway H (1983): Microbial desulfurization and denitrogenation of hydrocarbons. In Zajic JE, Cooper DG, Jack TR, Kosaric N (eds): Microbial Enhanced Oil Recovery. Tulsa OK: PennWell, pp 83–91.

Flathman PE, Jurger D, Exner J (eds) (1994): Bioremediation: Field Experience. Lewis Publishers.

Flyvbjerg J, Arvin E, Jensen BK, Olsen SK (1993): Microbial degradation of phenols and aromatic hydrocarbons in creosote-contaminated groundwater under nitrate-reducing conditions. J Cont Hydrol 12:133–150.

Fogel S (1993): Full-scale bioremediation of No. 6 fuel oil-contaminated soil: 6 Months of active and 3 years of passive treatment. In Flathman P, Jurger D, Exner J (eds): Bioremediation Field Experience. Lewis Publishers.

Frankenberger WT (1993): Microbial degradation of petroleum hydrocarbons. In Calabrese EJ, Kostecki PT (eds): Principles and Practices for Petroleum Hydrocarbon-Contaminated Soils. Lewis Publishers.

Fredrickson JK, Balkwill DM, Zachara JM, Li SW, Brockman FJ, Simmons MA (1991): Physiological diversity and distributions of heterotrophic bacteria in deep cretaceous sediments of the Atlantic Coastal Plain. Appl Environ Microbiol 57:402–411.

Gallagher JR, Olson ES, Stanley DC (1993): Microbial desulfurization of dibenzothiophene: A sulfur-specific pathway. FEMS Microbiol Lett 107:31–36.

Gardner WS, Lee RF, Tenore KR, Smith LW (1979): Degradation of selected polycyclic aromatic hydrocarbons in coastal sediments: Importance of microbes and polychaete worms. Water Air Soil Pollut 11:339–347.

Gary JH, Handwerk GE (1984): Petroleum Refining: Technology and Economics, 2nd ed. New York; Marcel Dekker.

Gerhardt P, Murray R, Costilow R, Nester E, Wood W, Krieg N, Phillips G (eds) Manual of Methods for General Bacteriology. Washington, DC: American Society for Microbiology.

Gerson DF (1985): The biophysics of microbial growth on hydrocarbons: Insoluble substrates. In Zajic JE, Donaldson SG (eds): International Bioresources J, vol I, Microbes and Oil Recovery. pp 39–53.

Gibbs CF, Davis SJ (1976): The rate of microbial degradation of oil in a beach gravel column. Microbial Ecol 3:55–64.

Grady CPL Jr, Dang JS, Harvey DM, Jobbagy A, Wang XL (1989): Determination of biodegradation kinetics through use of electrolytic respirometry. Wat Sci Technol 21:957–968.

Grady CPL Jr (1985): Biodegradation: Its measurement and microbiological basis. Biotechnol Bioeng 27:660–674.

Grbic-Galic D, Vogel TM (1987): Transformation of toluene and benzene by mixed methanogenic cultures. Appl Environ Microbiol 53:254–260.

Greenberg AE, Clesceri LS, Eaton AD (eds) (1992): Standard Methods. Washington DC: American Public Health Association.

Hamilton WA (1985): Sulphate-reducing bacteria and anaerobic corrosion. Annu Rev Microbiol 39:195–217.

Harder H, Kurzel-Seidel B, Hopner T (1991): Hydrocarbon biodegradation in sediments and soils. Erdoel Kohle-Erdgas 44:59–62.

Häusler A, Müller-Hurtig R, Wagner F (1992): Influence of Chemical- and Biosurfactants on the Microbial Degradation of a Fuel Oil Spill. DECHEMA Biotechnology Conference 5, VCH Verlagsgesellschaft, 1992, pp 1037–1041.

Herbert BN (1985): Sulphate-reducing bacteria in oil-bearing reservoirs. Conference Proceedings, World Biotechnology Report, vol 1. Pinner, UK: Online Publishers, pp 417–427.

Hill EC (1984): Biodegradation of petroleum products. In Atlas R (ed): Petroleum Microbiology. New York: MacMillan, pp 579–617.

Hinchee RE, Arthur M (1991): Bench-scale studies of the soil aeration process for bioremediation of petroleum hydrocarbons. Appl Biochem Biotechnol 28/29:901–906.

Hoeppel RE, Hinchee RE, Arthur MF (1991): Bioventing soils contaminated with petroleum hydrocarbons. J Ind Microbiol 8:141–146.

Hosler KR, Bulman TL, Booth RM (1992): The persistence and fate of aromatic constituents of heavy oil production waste during landfarming. Hydrocarbon Contam Soils 2:591–609.

Huesemann MH, Moore KO (1993): Compositional changes during landfarming of weathered Michigan crude oil-contaminated soil. J Soil Contam 2:245–264.

Hutchins SR, Sewell GW, Kovacs DA, Smith GA (1991): Biodegradation of aromatic hydrocarbons by aquifer microorganisms under denitrifying conditions. Environ Sci Technol 25:68–76.

Hutchins SR, Moolenaar SW, Rhodes DE (1992): Column studies on BTEX biodegradation under microaerophilic and denitrifying conditions. J Hazard Mater 32:195–214.

Iverson WP, Olson GJ (1984): Problems related to sulfate-reducing bacteria in the petroleum industry. In Atlas R (ed): Petroleum Microbiology. New York: MacMillan, pp 619–642.

Jerger DE, Woodhull PM, Flathman PE, Exner JH (1993): Solid-phase bioremediation of petroleum-contaminated soil: Laboratory treatability study through site closure. In Flathman PE, Jerger DE, Exner JH (eds): Bioremediation Field Experience. Lewis Publishing, pp 177–193.

Jobson AM, Cook FD, Westlake DWS (1979): Interaction of aerobic and anaerobic bacteria in petroleum biodegradation. Chemical Geol 24:355–365.

Johnson PC, Kemblowski MW, Colthart JD (1990): Quantitative analysis for the cleanup of hydrocarbon-contaminated soils by in situ soil venting. Groundwater 28:413–429.

Jones SH, Alexander M (1988): Phosphorus enhancement of mineralization of low concentrations of p-nitrophenol by *Flavobacterium* sp. in lake water. FEMS Microbiol Lett 52:121–126.

Jorgensen PE, Eriksen T, Jensen BK (1992): Estimation of viable biomass in wastewater and activated sludge by determination of ATP, oxygen utilization rate and FDA hydrolysis. Wat Res 26:1495–1501.

Kaplan AM, Kaplan DL (1982): Thermophilic biotransformations of 2,4,6-trinitrotoluene under simulated composting conditions. Appl Environ Microbiol 44:757–760.

Kim TS, Kim HY, Kim BH (1990): Petroleum desulfurization by *Desulfovibrio desulfuricans* M6 using electrochemically supplied reducing equivalent. Biotechnol Lett 12: 757–760.

Klevens HB (1950): Solubilization of polycyclic hydrocarbons. J Phys Chem 54:283–298.

Laha S, Luthy RG (1992): Effects of nonionic surfactants on the solubilization and mineralization of phenanthrene in soil–water systems. Biotechnol Bioeng 40:1367–1380.

Leahy JG, Colwell RR (1990): Microbial degradation of hydrocarbons in the environment. Microbiol Rev 54:305–315.

Linkenheil R, Patnode T (1987): Bioremediation of Contamination by Heavy Organics at Wood Preserving Plant Site. Brainard Final Report. Remediation Technologies and Glacier Park Company.

Liu S, Suflita J (1993): Ecology and evolution of microbial populations for bioremediation. TIB Technol 11:344–352.

Loehr RC (1984): Land treatment as a waste management technology. In Loehr RC, Malina JF (eds): Land Treatment—A Hazardous Waste Management Alternative. Water Resources Symposium No. 13. Austin, TX: University of Austin, pp 41–61.

Madsen EL (1991): Determining In situ biodegradation. ES&T 25:1663–1673.

Madsen EL, Sinclair JL, Ghiorse WC (1991): In situ biodegradation: microbiological patterns in a contaminated aquifer. Science 252:830–833.

Manilal VB, Alexander M (1991): Factors affecting the microbial degradation of phenanthrene in soil. Appl Microbiol Biotechnol 35:401–405.

Magor AM, Warburton J, Trower MK, Griffen M (1986): Comparative study of the ability of three *Xanthobacter* species to metabolize cycloalkanes. Appl Environ Microbiol 52:665–671.

Marinucci AC, Bartha R (1979): Apparatus for monitoring the mineralization of volatile ^{14}C-labeled compounds. Appl Environ Microbiol 38:1020–1022.

Marshall TR, Devinny JS (1986): Methods for enumeration of microorganisms in petroleum land treatment soils. Hazard Waste Hazard Mater 3:175–181.

McGill WB, Rowell MJ, Westlake DWS (1981): Biochemistry, ecology, and microbiology of petroleum components in soil. In Paul EA, Ladd JN (eds): Soil Biochemistry. New York: Marcel Dekker, pp 229–295.

McGill W, Rowell M (1977): In Toogood JA (ed): The Reclamation of Agricultural Soils after Oil Spills. I. Research. Publication No. M-77-11. Alberta: Alberta Institute of Pedology, pp 69–132.

McGinnis GD (1987): Potential for Migration of Hazardous Wood Treating Chemicals During Land Treatment Operations. Technical Completion Report. Project G1234-03. U.S. Department of the Interior.

McMillen SJ, Kerr JM, Gray NR (1993): Microcosm studies of factors that influence bioremediation of crude oils in soils. In Proceedings of EPA Exploration and Production Environmental Conference, March 3–7, 1993, pp 389–401.

Mihelcic JR, Lueking DR, Mitzell RJ, Stapleton JM (1993): Bioavailability of sorbed- and separate-phase chemicals. Biodegradation 4:141–153.

Miller R (1994): Surfactant-enhanced bioavailability of slightly soluble organic compounds. In Skipper H (ed): Bioremediation—Science and Applications. Madison WI: Soil Science Society of American Publications.

Miller RM, Singer GM, Rosen JD, Bartha R (1988): Photolysis primes biodegradation of benzo[a]pyrene. Appl Environ Microbiol 54:1724–1730.

Milner CW, Rogers MA, Evans CR (1977): Petroleum transformation in reservoirs. J Geochem Explor 7:101–153.

Monticello DJ, Finnerty WR (1985): Microbial desulfurization of fossil fuels. Annu Rev Microbiol 39:371–389.

Morgan P, Watkinson RJ (1992): Factors limiting the supply and efficiency of nutrient and oxygen supplements for the in situ biotreatment of contaminated soil and groundwater. Wat Res 26:73–78.

Morgan P, Watkinson RJ (1989): Hydrocarbon degradation in soils and methods for soil biotreatment. Crit Rev Biotechnol 8:305–333.

Mulkin-Phillips G, Stewart JE (1975): Effects of environmental parameters on bacterial degradation of Bunker C oil, crude oils, and hydrocarbons. Appl Microbiol 28:915–922.

Nelson MJ, Compeau CG, Mariarz T, Mahaffey WR (1994): Laboratory treatability testing for assessment of field applicability. In Flathman PE, Jurger DE, Exner JE (eds): Bioremediation Field Experience. Lewis Publishers, pp 59–78.

Obuekwe CO, Westlake DWS, Cook FD (1983): Corrosion of Pembina Crude Oil Pipeline: The origin and mode of formation of hydrogen sulphide. Eur J Appl Biotechnol 17: 173–177.

Oliver T, Kostecki P, Calabrese E (1993): State summary of soil and groundwater cleanup standards. Soils. December, pp 12–63.

Osmon JL, Andrews CC (1978): The biodegradation of TNT in enhanced soil and compost systems. Large Caliber Weapon Syst Lab Army Armament Res Dev Command. Dover, NJ: NTIS Publ AD-A054375.

Pennell KD, Abriola LM, Weber WJ Jr (1993): Surfactant-enhanced solubilization of residual dodecane in soil columns: I. Experimental investigation ES&T 27:2332–2340.

Price LC (1980): Crude oil biodegradation as an explanation of the depth rule. Chem Geol 28:1–30.

Puttman W (1988): Microbial degradation of petroleum in contaminated soil-analytical aspects. In Wolf K, van den Brink WJ, Colon FJ (eds): Contaminated Soil. Dordrecht, Germany: Kluwer Academic Publishers, pp 189–199.

Rasiah V, Voroney RP, Kachanoski RG (1992a): Biodegradation of an oily waste as influenced by nitrogen forms and sources. Water, Air, Soil Pollut 65:143–151.

Rasiah V, Voroney RP, Kachanoski RG (1992b): Bioavilability of stabilized oily waste organics in ultrasonified soil aggregates. Water, Air, Soil Pollut 63:179–186.

Robertson BK, Alexander M (1992): Influence of calcium, iron, and pH on phosphate availability for microbial mineralization of organic chemicals. Appl Environ Microbiol 58:38–41.

Rogerson A, Berger J (1981): Effect of crude oil and petroleum-degrading microorganisms on the growth of freshwater and soil protozoa. J Gen Microbiol 124:53–59.

Rogerson A, Berger J (1983): Enhancement of the microbial degradation of crude oil by the ciliate Colpidium colpoda. J Gen Appl Microbiol 29:41–50.

Rogers JA, Tedaldi DJ, Kavanaugh MC (1993): A screening protocol for bioremediation of contaminated soil. Environ Prog 12:146–156.

Rosenberg M, Gutnick DL, Rosenberg E (1983): Bacterial adherence to hydrocarbons. In

Zajic JE, Cooper DG, Jack TR, Kosaric N (eds): Microbial Enhanced Oil Recovery. Tulsa, OK: PennWell, pp 114–123.

Ryan J, Loehr R, Sims R (1987): The Land Treatability of Appendix VIII Constituents Present in Petroleum Refinery Wastes: Laboratory and Modeling Studies. Report to the American Petroleum Institute.

Sayler GS, Layton AC (1990): Environmental application of nucleic acid hybridization— DNA probe for detection of genetically engineered microorganisms by hybridization. Annu Rev Microbiol 44:625–648.

Scheibenbogen K, Zytner RG, Lee H, Trevors JT (1994): Enhanced removal of selected hydrocarbons from soil by *Pseudomonas aeruginosa UG2* biosurfactants and some chemical surfactants. J Chem Tech Biotechnol 59:53–59.

Schink B (1985): Degradation of unsaturated hydrocarbons by methanogenic enrichment cultures. FEMS Microbiol Ecol 31:69–77.

Schlautman MA, Morgan JJ (1993): Binding of a fluorescent hydrophobic organic probe by dissolved humic substances and organically-coated aluminum oxide surfaces. ES&T 27:2523–2532.

Schwarzenbach RP, Gschwend PM, Imboden DM (1993): Environmental Organic Chemistry. New York: John Wiley and Sons.

Shannon M, Unterman R (1993): Evaluating bioremediation—Distinguishing fact from fiction. Annu Rev Microbiol 47:715–738.

Sharabi NED, Bartha R (1993): Testing of some assumptions about biodegradability in soil as measured by carbon dioxide evolution. Appl Environ Microbiol 59:1201–1205.

Sims R (1986): Waste/soil treatability studies for four complex wastes: Methodologies and results, Volume 2: Waste Loading Impacts on Soil Degradation, Transformation and Immobilization. EPA/600/6-86/003b.

Sinclair JL, Ghiorse WC (1989): Distribution of aerobic bacteria, protozoa, algae, and fungi in deep subsurface sediments. Geomicrobiol J 7:15–31.

Sinclair JL, Kampbell DH, Cook ML, Wilson JT (1993): Protozoa in subsurface sediments from sites contaminated with aviation gasoline or jet fuel. Appl Environ Microbiol 59:467–472.

Song HG, Bartha R (1990): Effects of jet fuel spills on the microbial community of soil. Appl Environ Microbiol 56:646–651.

Song H, Wang X, Bartha R (1990): Bioremediation potential of terrestrial fuel spills. Appl Environ Microbiol 56:652–656.

Song HG, Pedersen TA, Bartha R (1986): Hydrocarbon mineralization in soil: Relative bacterial and fungal contributuion. Soil Biol Biochem 18:109–111.

Sorensen DL, Sims RC, Qiu X (1994): Field scale evaluation of grass-enhanced bioremediation of PAH contaminated soils. EPA Risk Reduction Eng Lab 20th Annu Res Symp, Cincinnati, OH.

Steffan RJ, Atlas RM (1991): Polymerase chain reaction: Applications in environmental microbiology. Annu Rev Microbiol 45:137–161.

Stieber M, Haeseler F, Werner P, Frimmel FH (1994): A rapid screening method for microorganisms degrading polycyclic aromatic hydrocarbons in microplates. Appl Microbiol Biotechnol 40:753–755.

Sullivan B, Johnson S (1993): "Oil" you need to know about crude. Soils May, 8.

Sutherland JB (1992): Detoxification of polycyclic aromatic hydrocarbons by fungi. J Ind Microbiol 9:53–62.

Tan K (1982): Principles of Soil Chemistry. New York: Marcel Dekker.

Tate RL III (1987): Soil Organic Matter: Biological and Ecological Effects. New York: John Wiley and Sons.

Tehrani J (1993): Successful Supercritical Fluid Extraction Strategies. American Laboratory, February, pp 40HH–40MM.

Texas Research Institute (1984): Forced Venting to Remove Gasoline Vapor from a Large-Scale Model Aquifer. American Petroleum Institute, Report No. 82101-F-TAV.

Thai LT, Maier WJ (1992): Solubilization and Biodegradation of Octadecane in the Presence of Two Commercial Surfactants. In 47th Purdue Industrial Waste Conference Proceedings. Lewis Publishers, pp 167–175.

Thomas JM, Ward CH (1992): Subsurface microbial ecology and bioremediation. J Hazard Mater 32:179–194.

Tiehm A, Zumft WG (1992): Bioavailability of Polycyclic Aromatic hydrocarbons—Solubilizing Potential and Biological Efficiency of Technical Surfactants. In DECHEMA Biotechnology Conference 5. VCH Verlagsgesellschaft, pp 1029–1032.

Travis C, Macinnis JM (1992): Vapor extraction of organics from subsurfce soils: Is it effective? ES&T 26:1885–1887.

Troy MA, Berry SW, Jerger DE (1993): Biological land treatment of diesel fuel-contaminated soil: Emergency response through closure. In Flathman PE, Jerger DE, Exner JH (eds): Bioremediation Field Experience. Lewis Publishing, pp 145–160.

U.S. EPA (1991): Guide for Conducting Treatability Studies Under CERCLA: Aerobic Biodegradation Remedy Screening. Interim Guidance.

Vance I, Brink DE (1994): Propionate-driven sulphate-reduction by oil-field bacteria in a pressurized porous rock bioreactor. Appl Microbiol Biotechnol 40:920–925.

Van Dyke MI, Couture P, Brauer M, Lee H, Trevors JT (1993a): *Pseudomonas aeruginosa* UG2 rhamnolipid biosurfactants: Structural characterization and their use in removing hydrophobic compounds from oil. Can J Microbiol 39:1071–1078.

Van Dyke MI, Gulley SL, Lee H, Trevors JT (1993b): Evaluation of microbial surfactants for recovery of hydrophobic pollutants from soil. J Ind Microbiol 11:163–170.

Vogel TM, Grbic-Galic D (1986): Incorporation of oxygen from water into toluene and benzene during anaerobic fermentative transformation. Appl Environ Microbiol 52:200–202.

Volkering F, Breure AM, van Andel JG (1993): Effect of microorganisms on the bioavailability and biodegradation of crystalline naphthalene. Appl Microbiol Biotechnol 40:535–540.

Walton BT, Anderson TA (1990): Microbial degradation of trichlorethylene in the rhizosphere: Potential application to biological remediation of waste sites—soil decontamination. Appl Environ Microbiol 56:1012–1016.

Ward DM, Brock TD (1978): Anaerobic metabolism of hexadecane in sediments. Geomicrobiol J 1:1–9.

Weissenfels WD, Klewer HJ, Langhoff J (1992): Adsorption of polycyclic aromatic hydrocarbons by soil particles: Influence on biodegradability and biotoxicity. Appl Microbiol Biotechnol 36:689–696.

Wodzinski RS, Coyle JE (1974): Physical state of phenanthrene for utilization by bacteria. Appl Microbiol 27:1081–1084.

Wright S (1992): Immunological techniques for detection, identifiction, and enumeration of microorganisms in the environment. In Levin MA, Seidler RJ, Rogul M (eds): Microbial Ecology: Principles, Methods, and Applications. New York: McGraw-Hill, pp 45–64.

Ying A, Duffy J, Shepard G, Wright D. (1990): Bioremediation of heavy petroleum oil in soil at a railroad maintenance yard. In Kostecki PT, Calabrese EJ (eds): Petroleum Contaminated Soils, vol 3. Lewis Publishing, pp 227–238.

Yoshiba T, Chantal PD, Sawatzky H (1988): Characterization of heteroatomic compounds in various synthetic crude naphthas. In Proceedings of the American Chemical Society, Division of Petrolelum Chemistry, Toronto Meeting, June 5–11, pp 341–346.

Zeyer J, Kuhn EP, Schwarzenbach RP (1986): Rapid microbial mineralization of toluene and 1,3-dimethylbenzene in the absence of molecular oxygen. Appl Environ Microbiol 52:944–947.

Zhang Y, Miller RM (1992): Enhanced octadecane dispersion and biodegradation by a *Pseudomonas* rhamnolipid surfactant. Appl Environ Microbiol 58:3276–3282.

4

MICROBIAL REDUCTIVE DECHLORINATION OF POLYCHLORINATED BIPHENYLS

DONNA L. BEDARD

Bioremediation Research Program, Environmental Laboratory, GE Corporate Research and Development, Schenectady, New York 12301

JOHN F. QUENSEN III

Department of Crop and Soil Sciences, Michigan State University, East Lansing, Michigan 48824

1. INTRODUCTION

conclusion

Polychlorinated biphenyls (PCBs) were widely used for a variety of industrial purposes between 1929 and 1978. Because of their low chemical reactivity, heat stability, nonflammability, and high electrical resistance, they were widely used as dielectric fluids in capacitors and transformers, hydraulic fluids, solvent extenders, and flame retardants. They were also used in carbonless copy paper. Some 1.4 billion pounds were manufactured, and several hundred million pounds were released into the environment (Hutzinger and Veerkamp, 1981), where they have persisted and become ubiquitous environmental contaminants. Because they are hydrophobic, PCBs adsorb to organic soils and sediments and tend to accumulate in biota. PCBs are stringently regulated because of concerns about their potential toxicity.

The reductive dechlorination of PCBs in anaerobic environments is a newly discovered and potentially important environmental fate of PCBs. It has significant

Microbial Transformation and Degradation of Toxic Organic Chemicals, pages 127–216
© *1995 Wiley-Liss, Inc.*

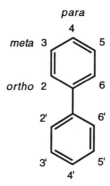

Figure 4.1. Structure of a PCB congener showing the numbering system for chlorine positions. Positions 2, 6, 2′, and 6′ are oriented *ortho* to the opposite phenyl ring, positions 3, 5, 3′, and 5′ are oriented *meta* to the opposite phenyl ring, and positions 4 and 4′ are oriented *para* to the opposite phenyl ring.

implications for both risk assessment and remediation strategies. Dechlorination is expected to reduce the toxicity of PCB mixtures. It should also make them more aerobically degradable; hence sequential anaerobic/aerobic microbial processes have the potential to destroy PCBs biologically.

PCBs were manufactured by catalytic chlorination of biphenyl to produce complex mixtures containing specified weight percents of chlorine. There are 209 theoretically possible PCB molecules, referred to as *congeners,* that differ in the number (from 1 to 10) and position of chlorines (Fig. 4.1) (Hutzinger et al., 1974). Sets of PCBs grouped according to the number of chlorines, e.g., trichlorophenyls and tetrachlorophenyls, are referred to as *homologs,* while PCBs containing the same number of chlorines but differing in the positions substituted are referred to as *isomers*. Most commercial mixtures contained from 60 to 90 different congeners (Schulz et al., 1989). In the United States and Great Britain, nearly all PCBs were manufactured by Monsanto under the trade name Aroclor. Most formulations were distinguished by a four digit number; the first two digits were usually 12 for 12 carbon atoms, and the last two digits usually indicated the percent chlorine by weight. Thus Aroclor 1242 contains 42% chlorine by weight. Trade names for PCB formulations manufactured in other countries include Fenclor (Caffaro, Italy), Phenoclor and Pyralene (Prodelec, France), Clophen (Bayer, Germany), and Kanechlor (Kanegafuchi, Japan) (Hutzinger et al., 1974).

In this review, we refer to the individual PCB congeners by listing the substituted positions on each ring, separated by a hyphen. Thus 2-4-CB is the congener substituted at positions 2 and 4′ in Figure 4.1. This is a nonconventional means of naming congeners, but we believe it is more easily understood by the reader and conserves space in tables and figures.

1.1. Discovery of PCB Dechlorination and Designation of Patterns

The first report of microbial dechlorination of PCBs was not based on a laboratory study, but rather on observed differences between the congener distribution patterns of the PCBs in sediments of the upper Hudson River and those of the commercial PCB mixtures discharged into the river (Brown et al., 1984). General Electric discharged PCBs into the upper Hudson from a pair of capacitor-manufacturing plants at Hudson Falls and Fort Edward, and the major contaminant was Aroclor 1242, a mixture of PCBs with an average of 3.2 chlorines per molecule. Comparative analyses of the sediment PCBs and the original input mixture by high-resolution gas chromatography (GC) revealed three major differences. Relative to the original Aroclor composition, the congener distribution patterns of sediment PCBs were characterized by 1) a much higher proportion of mono- and dichlorinated congeners, 2) a higher proportion of *ortho* chlorines, and 3) selective depletion of tri- through pentachlorobiphenyls. These compositional alterations of sediment PCBs in the upper Hudson River were confirmed by Bopp et al. (1984). No previously identified biological or physical alteration processes could account for such differences, and Brown et al. (1984) attributed them to reductive *meta* and *para* dechlorination mediated by microorganisms in the anaerobic sediments.

The specific PCB congeners that were decreased or increased relative to the original Aroclor composition varied among samples. A few repeatedly observed congener distribution profiles, each exhibiting a distinct pattern of congener selectivity and chlorophenyl reactivity, were identified as different dechlorination patterns and designated by letters (Brown et al., 1984). In 1987, Brown and colleagues presented evidence of PCB dechlorination in aquatic sediments for several other locations: Silver Lake (Pittsfield, MA), the Sheboygan River (Sheboygan, WI), Waukegan Harbor (IL), the Acushnet Estuary (New Bedford, MA), and the Hoosic River (North Adams, MA). At the same time the PCB dechlorination patterns in the upper Hudson (B, B',C,E) and several new dechlorination patterns (Silver Lake F and G) were described in greater detail (Brown et al., 1987a,b). Brown and colleagues (1984, 1987a,b) proposed that the different patterns of dechlorination resulted from the action of enzymes that had distinct congener selectivities and resided in anaerobic bacteria and that these enzymes were analogous to those responsible for position-specific dechlorination of chlorobenzoates and chlorophenols (Suflita et al., 1982; Boyd et al., 1983). Brown et al. (1987b) also proposed that anaerobic bacteria having such enzymes could use PCBs as electron acceptors and could derive energy from the dechlorination of PCBs. Thus such bacteria would have a competitive advantage in sediments that were contaminated with PCBs but otherwise limited in available electron acceptors. Finally, Brown et al. (1984, 1987a,b) pointed out that the partially dechlorinated mixtures remaining in aquatic sediments were sharply depleted in those congeners that are of concern with regard to persistence, P_{448} cytochrome induction, or toxicity in higher animals.

1.2. Laboratory Confirmation to Microbial Dechlorination of PCBs

Because of the importance of these findings, Brown's publications stimulated a great deal of research. In 1988, Quensen et al. demonstrated that microorganisms from Hudson River sediment could reductively dechlorinate most of the congeners in Aroclor 1242. The PCB congener distribution pattern obtained after a 16 week incubation was very similar, although not identical, to the pattern seen in the

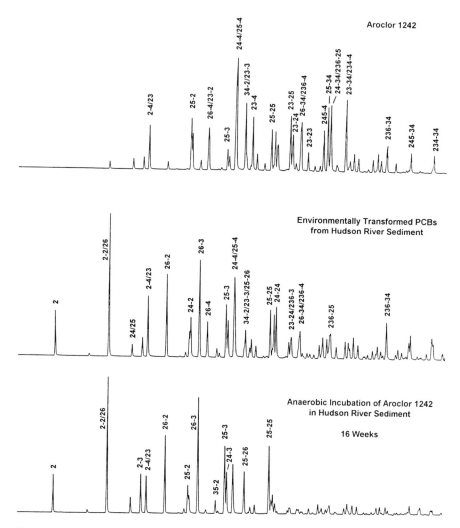

Figure 4.2. Comparison of Aroclor 1242, environmentally dechlorinated Aroclor 1242, and Aroclor 1242 dechlorinated in a laboratory study. GC/ECD DB-1 capillary chromatograms. **Top:** Aroclor 1242. **Center:** PCBs extracted from a sediment sample collected in the upper Hudson River south of Hudson Falls. The dechlorination is typical of an early pattern C (i.e., it shows evidence of processes M and Q in near-equal proportions). **Bottom:** Heavily dechlorinated Aroclor 1242 obtained after a 16 week anaerobic incubation with live Hudson River sediment. In this instance the dechlorination shows evidence of a combination of processes Q (primarily) and M. See sections on dechlorination processes M, Q, and C and on the Hudson River for more details. (Adapted from Quensen et al., 1988, with permission of the publisher.)

environment (Fig. 4.2). At the highest concentration tested (700 μg/g sediment), 53% of the total chlorine was removed, but only from the *meta* and *para* positions, and the proportion of mono- and dichlorobiphenyls increased from 9% to 88% (Quensen et al., 1988).

In 1990, Quensen and colleagues reported the dechlorination of Aroclors 1242, 1248, 1254, and 1260 by anaerobic microorganisms eluted from Hudson River sediment and of Aroclors 1242 and 1260 by microorganisms eluted from Silver Lake sediment. The Hudson River microorganisms removed virtually all of the *meta* and *para* chlorines from Aroclors 1242, 1248, and 1254, a type of dechlorination known as *process C* (Brown et al., 1984, 1987a,b). In contrast, the dechlorination of Aroclor 1260 mediated by Hudson River microorganisms (Quensen et al., 1990), identified as *process H* (Brown and Wagner, 1990; Brown, 1990), was much more selective. The dechlorination of Aroclors 1242 and 1260 by microorganisms from Silver Lake sediments yielded congener distribution profiles that were distinctly different from those seen with the Hudson River organisms (Quensen et al., 1990). Thus Brown's proposal that microorganisms with different characteristic specificities for PCB dechlorination existed in PCB-contaminated sediments was confirmed for two different locations. Since that time many other laboratories have also reported the dehalogenation of PCBs and polybrominated biphenyls by anaerobic microorganisms from sediments (Nies and Vogel, 1990; Van Dort and Bedard, 1991; May et al., 1992; Morris et al., 1992a,b, 1993; Ye et al., 1992; Alder et al., 1993; Abramowicz et al., 1993; Bedard et al., 1993; Boyle et al., 1993; Fava et al., 1993; Rhee et al., 1993 a–e; Williams, 1994).

1.3. Scope of This Review

The objective of this review is to elucidate and integrate what is known from reports of environmental dechlorination of PCBs and from laboratory investigations. Because there are different types of PCB dechlorination, recognizable by congener specificities, it is necessary to define these clearly before attempting to evaluate the effects of various treatments. As a first step, we provide, within a common framework, detailed descriptions of the various types of dechlorination that have been observed and compile all of the examples of each type of dechlorination that have been reported.

A second objective is to establish guidelines for interpreting and reporting the dechlorination of Aroclors in a way that will permit direct comparison of results from various laboratories. Aroclors are dechlorinated by the loss of *meta* and/or *para* chlorines, but this is a gross oversimplification. There are multiple factors that govern exactly which *meta* and/or *para* chlorines are lost. As will become apparent, there are at least six distinct microbial dechlorination processes that can be identified through careful analysis of the patterns of congener loss and product formation. Several of these processes have been observed using sediment microorganisms from multiple locations; hence they may represent broad classes of dechlorination. The six individual dechlorination processes may occur separately, or in combination. Therefore, a basic understanding of each is essential for accurate interpretation of Aroclor dechlorination.

Whenever possible, the particular type or pattern of dechlorination observed in the experiments reviewed is taken into account in evaluating treatment effects. This has often not been done in the original references. Thus we discuss the importance of various environmental factors and culture conditions for the individual dechlorination processes. Other material to be reviewed includes attempts to enhance dechlorination through the addition of carbon sources, surfactants, and selected halogenated biphenyls and attempts to characterize and isolate the dechlorinating microorganisms present in various sediments. We review the published literature on PCB dechlorination in a critical way, comparing the results of similar studies by different investigators, giving likely reasons for observed differences, and indicating where further work is particularly needed. Finally, we discuss the implications of this newly discovered environmental fate of PCBs with regard to detoxication, natural restoration, and bioremediation.

2. BIOLOGICAL BASIS OF PCB DECHLORINATION PATTERNS

Numerous investigators have now demonstrated that the anaerobic dechlorination of PCBs in aquatic sediments is a biological process (Quensen et al., 1988, 1990; Nies and Vogel, 1990; Ye et al., 1992; Morris et al., 1992a; Alder et al., 1993; Rhee et al., 1993a–c). However, little is known about the microorganisms responsible for aryl dehalogenation because until very recently only one such organism, *Desulfomonile tiedjei* (which dehalogenates halobenzoates, not PCBs), had been isolated and studied in any detail (Mohn and Tiedje, 1992). The fundamental characteristics of aryl dehalogenation exhibited by *D. tiedjei* may also be valid for other aryl dehalogenators. These characteristics are 1) aryl reductive dehalogenation is catalyzed by inducible enzymes, 2) these enzymes exhibit distinct substrate specificities, 3) aryl dehalogenators function in syntrophic communities and may be dependent on such communities, and 4) aryl dehalogenators derive metabolic energy from reductive dehalogenation (Mohn and Tiedje, 1992). We hypothesize that these characteristics apply to the organisms responsible for reductive dechlorination of PCBs and that a variety of sediment microorganisms with distinct dehalogenating enzymes, each exhibiting a unique pattern of congener selectivity, are responsible for the various patterns of PCB dechlorination observed both in the laboratory and in the environment. In addition, it appears that some of the dechlorination patterns that have been observed may result from the action of a single population of dechlorinating microorganisms, whereas others may reflect either the concerted or the sequential activity of several distinct dechlorinating populations.

Of necessity, all of the laboratory investigations of microbial PCB dechlorination have been carried out using sediment slurries both as the source of the microorganisms and as the growth matrix. Attempts at isolating the microorganisms responsible for PCB dechlorination have not yet been successful, and dechlorination activity is dependent on the presence of sediment. The possibility exists that the dechlorination observed in such undefined systems results from the action of multiple microbial populations dechlorinating PCBs either simultaneously or in succession. For this reason, we must be cautious in ascribing dechlorination activity to a single microbial population. Nevertheless, close examination of the congener distribution profiles reported by various investigators does allow us to identify several distinct patterns of dechlorination that are quite selective and may represent the action of individual dechlorinating populations.

3. DECHLORINATION TERMINOLOGY

A more complete understanding of the microorganisms and enzymes that mediate PCB dechlorination will be possible only when these microorganisms have been isolated and characterized or at least established in defined consortia. For the present, we refer to a set of dechlorination reactions exhibiting a particular pattern of congener selectivity and chlorophenyl reactivity as a *dechlorination process* or a *dechlorination activity*. Hence, each dechlorination process can be described by a set or series of reactions that indicates which congeners are substrates, which chlorines will be removed from these congeners, and the order in which they will be removed. The resulting congener distribution profile is referred to as the corresponding *dechlorination pattern*. For example, the set of reactions that defines one specific type of *meta* dechlorination is referred to as *process* or *activity M,* and the resulting PCB congener distribution profile as *pattern M*. Because the congener distribution profile of the dechlorinated PCBs depends on that of the initial PCB mixture and on the extent to which the dechlorination has progressed, the term *dechlorination pattern* (e.g., *pattern M*) refers to any one of a continuum of congener distribution profiles that are consistent with a particular dechlorination process (e.g., process M). In some cases the dechlorination pattern may actually reflect the combined activity of two or more distinct dechlorination processes.

4. RECOGNITION AND CHARACTERIZATION OF ENVIRONMENTAL DECHLORINATION OF PCBs

Commercial PCBs, such as the Aroclors, are complex mixtures of about 60–90 congeners that were produced in fixed and essentially invariant proportions during the manufacturing process (Hutzinger et al., 1974). However, the individual PCB congeners vary greatly in their susceptibility to evaporation, solubilization, microbial dechlorination, and other alteration processes that may occur in specific environmental compartments. Because of this, the PCB congener distribution pattern in an environmental sample will present a record of all of the alteration processes to which that sample has been exposed (Brown and Wagner, 1990). Accurate analysis and interpretation of PCB alterations in an environmental sample requires quantitative congener-specific analysis, such as that afforded by high resolution capillary gas chromatography (GC), to determine the range and relative proportions of PCB congeners present. The GC analyses can be done using either an electron capture detector (ECD), which is highly sensitive to halogenated compounds, or a mass spectrometric detector. Recently, complete congener assignments and quantitation for the Aroclors have become available for fused silica columns coated with poly(dimethylsiloxane) (DB-1, SE-30, OV-1, HP-1, Ultra-1, SPB-1) (Brown et al., 1987a; Bedard et al., 1987; R.E. Wagner, J.C. Carnahan, and R.J. May, unpublished data) or with the closely related copolymer containing 5%-phenyl-95%-methyl-siloxane units (DB-5, SE-54, OV-5, HP-5, Ultra-2, SPB-5) (Mulin et al., 1984; Schulz et al., 1989). As a result, these are the columns of choice for quantitative congener-specific PCB analysis.

Many congener pairs that co-elute on the two stationary phases listed above can be resolved by use of phases having more hydrocarbon character such as pure hydrocarbon phases (C-87, Apiezon L) (Bush et al., 1985) or a poly(dimethylsiloxane) phase containing 50% n-octyl groups (SB-Octyl 50) (Fischer and Ballschmiter, 1988, 1989). Unfortunately, complete PCB congener assignments and quantitation for the Aroclors are not yet available for columns with the latter phases.

If chromatograms of the dechlorinated PCBs contain PCB congeners not present in the Aroclor(s) or present in much higher proportions in the environmental sample than in the Aroclor(s), then it is necessary to identify those congeners and to add them to the appropriate Aroclor(s) in order to construct a standard suitable for accurate quantitative analysis. Such congeners might include 2-CB and 2-2-CB; 24-3-CB, 25-3-CB, and 26-3-CB; or 25-26-CB, 24-26-CB, and 24-24-CB, depending on the original Aroclor and the nature of the dechlorination.

To interpret correctly the alteration processes that have occurred, it is helpful to know which of the commercial Aroclors were originally present. If there is a primary source of the PCB contamination, records of the purchase or usage of Aroclors may be available. Clearly the delineation of an environmental alteration process is much easier if a single Aroclor was responsible for all or at least most of the contamination at any particular location. However, even when this is not the case it may be possible to calculate the original proportions of the Aroclors in the samples under study by using an indicator peak procedure (Brown and Wagner, 1990; Lake et al., 1992). This approach is based on the assumption that the ratio of two specific PCB peaks in the chromatogram can be used to calculate the original proportions of the different Aroclors. The approach will be of value only if the PCB congeners in the chosen peaks have not been affected by dechlorination or any other chemical transformations. One such procedure was used by two different laboratories to establish the relative proportions of Aroclors 1242 and 1254 in sediments from New Bedford Harbor (Brown and Wagner, 1990; Lake et al., 1992). In that case it was found that the relative proportions of the two Aroclors varied widely among samples collected. Hence, it was necessary to redetermine the relative contributions of the two Aroclors in each individual sample.

PCB dechlorination may be extensive, resulting in the removal of virtually all *meta* and *para* chlorines as seen in the Hudson River (Brown et al., 1987a,b), or it may be highly selective, resulting in very limited decreases of particular PCB congeners and corresponding increases in others (Bedard and May, submitted; Bedard et al., submitted). Clearly, limited dechlorination is more difficult to identify than extensive dechlorination, and the ability to detect and correctly interpret dechlorination will be directly related to the sensitivity, accuracy, and thoroughness of the congener-specific analysis. Until very recently analysts have routinely reported PCBs in environmental samples in terms of whichever commercial Aroclor had about the same average chlorine number. This explains why environmental dechlorination was often not recognized in the past and may still be overlooked. Such misreporting also underscores the importance of a complete and quantitative congener-specific analysis.

The most sensitive indicator of dechlorination is the appearance of typical PCB dechlorination products in proportions not found in the Aroclors. These will differ depending on the Aroclor(s) present and on the specificity of the dechlorination. Some examples are 2-CB, 2-2-CB, 2-3-CB, 2-4-CB, 24-3-CB, 25-3-CB, 26-2-CB, 26-3-CB, 26-4-CB, 24-24-CB, 24-25-CB, 24-26-CB, 25-25-CB, and 25-26-CB. Selective decreases in congeners sharing the same chlorine configuration on one ring, or sharing a common feature such as a flanked *para* chlorine, is the second indicator of dechlorination. But unambiguous confirmation of dechlorination depends on the correlation of the disappearance of specific congeners with increases in others and on the ability to achieve a reasonable mass balance. "Mother–daughter" correlations of this kind are actually far easier to make when the pattern of dechlorination is very restricted. For example, when the dechlorination is restricted to loss of the *para* chlorine from 34-, 234-, and 245-chlorophenyl groups, decreases in congeners carrying these chlorophenyl groups and increases in congeners with 3-, 23-, and 25-chlorophenyl groups are fairly easy to recognize.

5. DECHLORINATION PROCESSES OBSERVED IN THE LABORATORY

As mentioned earlier, Brown was the first to recognize that there are specific patterns of PCB dechlorination. He described various patterns of environmental PCB dechlorination in terms of congener selectivity and chlorophenyl reactivity and gave each a letter designation (Brown et al., 1984, 1987a,b; Brown and Wagner, 1990). In addition, he compared these environmental PCB dechlorination patterns with those obtained by other investigators in laboratory studies (several investigators sent him chromatograms for this purpose prior to publishing their own results) and assigned designations to the dechlorination processes observed in the laboratory. Brown (1990) summarized the dechlorination pattern assignments for both environmental and laboratory samples and the key characteristics of each in a table, relying in many cases on unpublished chromatograms or congener distribution profiles.

To bring clarity to a complex field, we believe that it is necessary to establish a common frame of reference that all investigators can use to describe their own results and to compare them with the results of others. Dechlorination process designations based on clearly defined rules of congener selectivity can serve this purpose. Therefore, we provide detailed descriptions and interpretations of PCB dechlorination processes that have been observed in the laboratory, including time sequences, when available, and references to published congener distributions and/or chromatograms.

In some cases we include new information from our own evaluation of the time sequence of dechlorination gleaned from unpublished chromatograms, histograms, or tables of congener distributions provided to us by others. These instances are clearly identified in the text, and in all cases the unpublished data were drawn from studies that have themselves been published.

We have kept the pattern designations given by Brown (1990) in most cases and have assigned them to dechlorination processes observed in the laboratory when none were previously reported. But, in the light of data not available in 1990, and with the benefit of hindsight, we have reinterpreted the pattern of dechlorination in several instances and have assigned different pattern designations when appropriate.

We present detailed descriptions of PCB dechlorination observed in laboratory studies before attempting to describe the results of environmental studies. This approach allows us to begin with quantitative analyses of a known PCB mixture and to follow the successive changes that occur with time. As a result, it is much easier to correlate the disappearance of PCB congeners with the appearance of daughter products and, from that information, to deduce the specificity of the dechlorination. In addition, because laboratory experiments provide control over incubation conditions that would change in the environment due to seasonal variations, tidal flushing, carbon input, and so forth, it is possible to begin to sort out the importance of these variables. Finally, the ability to follow the time course of dechlorination in the laboratory permits observation of changes in dechlorination specificity that result from microbial succession.

As we show in the ensuing examples, the specificity of microbial dechlorination varies widely, even within the same sediment. There are at least five major factors that determine whether a chlorine will be removed from any particular congener in each sediment: 1) the microbial populations present, 2) the position (*ortho, meta,* or *para*) of the chlorine relative to the opposite phenyl ring, 3) the surrounding chlorine configuration, 4) the chlorine configuration on the *opposite* ring, and 5) the incubation conditions (the Aroclor added, temperature, carbon nature and availability, electron acceptors present, salinity, oil, other contaminants, and so forth).

Most studies have used sediments from the upper Hudson River because that is where environmental dechlorination was discovered, and most have focused on Aroclor 1242 because that was the primary contaminant. Unless otherwise specified, all experiments were carried out under methanogenic conditions. The first laboratory reports of dechlorination of Aroclor 1242 by Hudson River microorganisms described both *meta* and *para* dechlorination to yield primarily *ortho*-substituted PCBs (Quensen et al., 1988, 1990). In the 1990 experiments, 87 mole% of the total PCBs were converted to 2-CB (39 mole%), 2-2-CB/26-CB (44 mole%), and 26-2-CB (4 mole%) after 16 weeks of incubation (Quensen et al., 1990). This congener distribution profile, pattern C, results from loss of virtually all *meta* and *para* chlorines, leaving primarily *ortho*-substituted PCBs. As previously proposed (Brown, 1990; Quensen et al., 1990), it appears that pattern C results from two separate and partially complementary dechlorination activities, M and Q, which will be described below.

5.1. Individual Dechlorination Processes

5.1.1. Process M

The clearest example available of dechlorination process M is that published by Ye et al. (1992). Untreated inocula of sediment microorganisms from the Hudson River removed *meta* and *para* chlorines from Aroclor 1242 to yield pattern C, but inocula heated at 85°C removed only *meta* chlorines. The characteristic dechlorination products from the heated inocula included 2-CB, 2-2-CB/26-CB, and 26-2-CB, but in lower amounts than in pattern C, plus large amounts of 2-4-CB, 24-2-CB, and 24-4-CB. Most of the *meta* chlorines but none of the *para* chlorines in Aroclor 1242 were removed (Fig. 4.3, Table 4.1). Histograms of the progress of process M over an 8 week incubation period (Ye, Quensen, and colleagues, unpublished results) revealed that the first conversions were those of 25-2-CB and 23-25-CB to 2-2-CB; of 34-2-CB, 23-4-CB, and 23-34-CB to 2-4-CB; and of 26-34-CB (and possibly 236-4-CB) to 26-4-CB. Next, 25-25-CB and 24-25-CB were dechlorinated to 2-2-CB and 24-2-CB, respectively, and then 25-4-CB was dechlorinated to 2-4-CB. Finally, 25-34-CB and 24-34-CB were dechlorinated to 2-4-CB, 24-4-CB, respectively. There was never any accumulation of 2-3-CB; 24-3-CB, 25-3-CB, or 26-3-CB, although these are prominent products of other PCB dechlorination processes. It appears from the time sequence that 23-chlorophenyl groups were more reactive than 25-chlorophenyl groups and that the removal of the *meta* chlorine from a 25-chlorophenyl group was facilitated by the presence of an *ortho* chlorine in the 2′ position. Hence 25-2-CB, 25-25-CB, and 24-25-CB were dechlorinated earlier than 25-34-CB. There was no observed increase or decrease of 2-3-CB, but the modest increase in 2-CB (from 3 to 7 mole%) may have come from the dechlorination of transiently produced 2-3-CB in addition to 23-CB and 25-CB. This was probably the dechlorination process observed by Nies and Vogel (1990) in their experiments with Hudson River sediment, although it is difficult to be sure from the way their data were presented.

Process M has also been observed in sediment inoculated with microorganisms derived from Silver Lake (Pittsfield, MA) sediment and incubated with Aroclor 1242 (Quensen et al., 1990). This dechlorination was originally assigned to N because it was assumed to be an extension of the *meta*-dechlorinating activity of Woods Pond and Silver Lake microorganisms on Aroclor 1260 that had previously been designated process N (Quensen et al., 1990). However, re-examination of the data indicates that the *meta*-dechlorinating activity exhibited by Silver Lake microorganisms on Aroclor 1242 is nearly identical to Hudson River dechlorination process M (Fig. 4.3) and differs in several important respects from process N. Like M, the dechlorination activity seen with Silver Lake microorganisms on Aroclor 1242 removes *meta*-chlorines from 23-, 25-, and possibly 3-chlorophenyl groups (Table 4.1)—an ability that distinguishes this activity from N. Two minor differences were noted between the Hudson River pattern M and the Silver Lake M. In the Hudson River M, 2-3-CB was more persistent than in the Silver Lake M. Conversely, 25-4-CB was less persistent in the Hudson River M.

DECHLORINATION PROCESS M

		23	$\longrightarrow$	2	
		25	$\longrightarrow$	2	
		2-3	$\longrightarrow$	2 ?	

	23-2	$\longrightarrow$	2-2	
23-25 $\longrightarrow$	25-2	$\longrightarrow$	2-2	
25-25 $\longrightarrow$	25-2	$\longrightarrow$	2-2	

23-34 $\longrightarrow$	34-2	$\longrightarrow$	2-4	
	23-4	$\longrightarrow$	2-4	
25-34 $\longrightarrow$	25-4	$\longrightarrow$	2-4	

24-25 $\longrightarrow$	24-2	
24-34 $\longrightarrow$	24-4	
26-34 $\longrightarrow$	26-4	
236-4 $\longrightarrow$	26-4 ?	

Specificity:	Flanked and unflanked meta Cl
Reactive Groups:	3?, 23, 25, 34, 234, 236
Relative Reactivity:	23 > 25 > 34 > 3
Reactivity Unknown:	245, 2345, 2346, 2356, 23456
Comments:	2' Cl facilitates dechlorination of 25-CP group
	No increase in 2-3, 24-3, 25-3, 26-3
	Less effective on more highly chlorinated PCBs

Aroclor	Reported as	Source of Microorganisms	Reference
1242	M	Hudson River	Ye et al., 1992, Fig 3
1242	--	Hudson River	Quensen et al., unpublished data
1242	N	Silver Lake	Quensen et al., 1990, Tab 3, Fig 7
1242	--	Hudson River	Nies and Vogel, 1990
* 1242	M	Hudson River	Abramowicz and Brennan, 1991, Fig 4.5.2B
* 1242	M	Hudson River	Morris et al., 1992, Fig 2, Pyruvate
* 1242	--	Hudson River	Rhee et al., 1993b, Fig 1

* These are examples of M + H or H' which do show increases in 2-3, 24-3, 25-3, and 26-3

Figure 4.3. Key characteristics of dechlorination process M as observed in the references indicated. There have been no studies of dechlorination process M with Aroclors 1248, 1254, or 1260, so it is not clear which penta- through octachlorobiphenyls are also substrates for dechlorination process M.

A key distinguishing feature between the Hudson River M and the dechlorination of Aroclor 1242 exhibited by Silver Lake microorganisms (Quensen et al., 1990) is that 24-24-CB was also dechlorinated in the latter case, presumably by *para*-dechlorination. A likely explanation is that the *para* dechlorination of 24-24-CB was due to the activity of a second microbial population that followed process M. This explanation is supported by the observation that although nearly all of the

TABLE 4.1. Chlorophenyl Reactivity Patterns of Various PCB Dechlorination Processes

Dechlorination Reaction			Dechlorination Process						
		M	Q	C	H	H′	P	N
3[a]	→ 0	?		+				
4	→ 0		+	+				
23	→ 2	+	+	+		+		
24	→ 2		+	+				
25	→ 2	+		+				
34	→ 3		+	+	+	+	+	
34	→ 4	+		+				+
234	→ 2		+	+				
234[b]	→ 23		?				+	
234[b]	→ 24	+	?		+	+		+
236	→ 26	+	?	+		?		+
245	→ 2			+				
245	→ 24	?						+
245	→ 25		+	+	+	+	+	
2345	→ 24	NA[c]	NA	NA				+
2345	→ 235	NA	NA	NA	+	+	+	

[a]It is not clear whether the ability ro remove this chlorine is due to process M or to a separate activity that sometimes occurs with process M.

[b]For process Q it is not clear which chlorine is removed first, but the ultimate product is the 2-chlorophenyl group.

[c]Data not available.

chlorine loss occurred in the first 8 weeks of incubation, 24-24-CB persisted during that time and decreased only after 8 weeks (Quensen et al., 1990).

5.1.2. Process Q

Dechlorination process Q was also observed with microorganisms from Hudson River sediment. This process is characterized primarily by *para* dechlorination, although *meta* dechlorination of 23- and possibly 236-chlorophenyl groups (but not 3-, 34-, or 25-chlorophenyl groups) also occurred. Dechlorination process Q was first observed by Quensen et al. (1988). Figure 4.2 (bottom panel) shows a chromatogram of Aroclor 1242 transformed by process Q (primarily) in combination with process M. A chromatogram exhibiting an Aroclor mixture dechlorinated by process Q alone was published by Abramowicz and Brennan (1991), and we have examined additional chromatograms (Quensen, unpublished data) that show the same activity, but at a more advanced stage. Dechlorination process Q removed virtually all *para* chlorines regardless of the surrounding chlorine configuration; hence 4-, 24-, 34-, and 245-chlophenyl groups were all dechlorinated by loss of the *para* chlorine (Fig. 4.4, Table 4.1). The primary dechlorination products were 2-CB, 2-2-CB/26-CB, and 2-3-CB, with lesser amounts of 25-2-CB, 26-2-CB, 25-3-CB, and 26-3-CB (Fig. 4.4).

DECHLORINATION PROCESS Q

234-4	$\longrightarrow$	24-4?	$\longrightarrow$ 2-4	$\longrightarrow$	2
234-4	$\longrightarrow$	23-4?	$\longrightarrow$ 2-4	$\longrightarrow$	2
245-4	$\longrightarrow$	25-4	$\longrightarrow$ 25		
		26-4	$\longrightarrow$ 26		

23-23	$\longrightarrow$	23-2	$\longrightarrow$ 2-2
23-24	$\longrightarrow$	24-2	$\longrightarrow$ 2-2
24-24	$\longrightarrow$	24-2	$\longrightarrow$ 2-2

23-25	$\longrightarrow$	25-2
24-25	$\longrightarrow$	25-2

24-34	$\longrightarrow$	24-3	$\longrightarrow$ 2-3
		34-2	$\longrightarrow$ 2-3
		23-3	$\longrightarrow$ 2-3

236-2	$\longrightarrow$	26-2 ?
23-26	$\longrightarrow$	26-2

25-34	$\longrightarrow$	25-3
26-34	$\longrightarrow$	26-3

Specificity:	Flanked and unflanked para Cl
	Meta Cl of 23- and maybe 234-CP
Reactive Groups (para Cl):	4, 24, 34, 245, (234?)
Reactive Groups (meta Cl):	23, (234?), (236?)
Relative Reactivity:	34 > 24 = 23? > 4
Reactivity Unknown:	2345, 2346, 23456
Comments:	2-2 substitution hinders reactivity of 24 groups
	Limited data available

Aroclor	Reported as	Source of Microorganisms	Reference
1242	Q	Hudson River	Abramowicz and Brennan, 1991, Fig 4.5.2C
1242	--	Hudson River	Quensen et al., 1988, unpublished chromatograms

Figure 4.4. Key characteristics of dechlorination process Q as observed in the references indicated. It is not clear which chlorine is removed first from 234-chlorophenyl (234-CP) groups. Ultimately the chlorines are removed from both the *meta* and *para* positions. There have been no quantitative analyses of process Q dechlorination and no studies of this dechlorination process with Aroclors 1248, 1254, or 1260. Therefore, it is not clear which penta- through octachlorobiphenyls are also substrates for dechlorination process Q.

Examination of chromatograms over the time course of the incubation indicates that the earliest decreases occurred in 24-4-CB, 24-2-CB, 34-2-CB, 24-34-CB, and 25-34-CB and that dechlorination of 24-25-CB and 24-24-CB occurred later. Examination of congener depletion and appearance patterns indicates that 2-4-CB and 24-4-CB were dechlorinated to 2-CB and that 23-2-CB, 24-2-CB, 23-23-CB, 23-24-CB, and 24-24-CB were each dechlorinated to 2-2-CB (Fig. 4.4). In addition, 26-4-CB was dechlorinated to 26-CB; 34-2-CB and 23-3-CB to 2-3-CB; and 236-2-CB and 23-26-CB to 26-2-CB. Finally, 26-34-CB and 236-34-CB were dechlorinated to 26-3-CB; 25-34-CB to 25-3-CB; and 24-34-CB to 24-3-CB and then 2-3-CB.

5.1.3. Process H′

A more limited type of dechlorination, process H′, was obtained by Alder, Häggblom, Oppenheimer, and Young (1993) using sediment from New Bedford Harbor. In this case they were examining dechlorination of the pre-existing PCBs in the sediment, a mixture of Aroclor 1242 and 1254 that had undergone limited environmental dechlorination. They saw essentially the same activity when Aroclor 1242 was added to the sediment. It is notable that they did not observe accumulation of 2-CB and observed only limited accumulation of 2-2-CB. The largest increases were seen for 25-2-CB, 2-3-CB, 24-3-CB, 25-3-CB, 26-3-CB, and 25-4/24-4-CB. Additional dechlorination products and the congeners that were dechlorinated are shown in Figure 4.5. It appears that this type of dechlorination is characterized by the dechlorination of 23-, 234-, and possibly 236-chlorophenyl groups from the *meta* position and 34- and 245-chlorophenyl groups from the *para* position.

Analysis of the depletion and appearance of PCB congeners indicates that 23-2-CB and 23-23-CB were dechlorinated to 2-2-CB; 23-25-CB to 25-2-CB; and 23-34-CB, 23-3-CB, and 34-2-CB to 2-3-CB. In addition, 23-24-CB was dechlorinated to 24-2-CB; 24-34-CB to 24-3-CB; 25-34-CB to 25-3-CB; and 234-4-CB and 245-4-CB to 24-4-CB/25-4-CB.

A very similar pattern of dechlorination has also been observed in incubations of Hudson River sediment with a PCB mixture that was predominantly Aroclor 1242 (unpublished chromatograms provided by Dan Abramowicz) and in incubations of Hudson River microorganisms in sediment slurries with added Aroclor 1242 (Rhee et al., 1993a), indicating that a microbial population with H′-like activity also exists in Hudson River sediment.

Dechlorination activities M and Q were both able to remove isolated chlorines, i.e., chlorines that were not flanked by other chlorines—for example, the *meta* chlorine from a 3- or 25-chlorophenyl ring (process M) or the *para* chlorine from a 4- or 24-chlorophenyl ring (process Q). In contrast, dechlorination process H′ did not remove isolated chlorines, but only *meta* or *para* chlorines that were positioned adjacent to other chlorines (Fig. 4.5). Hence PCBs containing 3-, 4-, 24-, and 25-chorophenyl rings accumulated in addition to PCBs substituted only at *ortho* positions.

5.1.4. Process H

Dechlorination process H is very much like H′ except that there is no activity against 23-chlorophenyl groups (Fig. 4.6). Process H dechlorination of Aroclor 1260, seen with microorganisms from Hudson River sediment (Quensen et al., 1990), was characterized by the accumulation of 25-25-CB (8 mole%), 235-25-CB (9 mole%), and other dechlorination products containing a 25-, 235-, or 236-

DECHLORINATION PROCESS H'

23-23 $\longrightarrow$ 23-2 $\longrightarrow$ 2-2

23-34 $\longrightarrow$ 23-3 $\longrightarrow$ 2-3
            34-2 $\longrightarrow$ 2-3
            23-4 $\longrightarrow$ 2-4

23-24 $\longrightarrow$ 24-2
23-25 $\longrightarrow$ 25-2
23-26 $\longrightarrow$ 26-2

234-4 $\longrightarrow$ 24-4
245-4 $\longrightarrow$ 25-4
236-4 $\longrightarrow$ 26-4?

* 234-34 $\longrightarrow$ 24-34 $\longrightarrow$ 24-3
* 245-34 $\longrightarrow$ 25-34 $\longrightarrow$ 25-3
* 236-34? $\longrightarrow$ 26-34 $\longrightarrow$ 26-3

245-24 $\longrightarrow$ 24-25
245-25 $\longrightarrow$ 25-25

Specificity: Flanked and doubly flanked para chlorines
                                Meta Cl on 23, 234, and possibly 236-CP
Reactive Groups (para Cl):   34, 245, (2345?), (234567?)
Reactive Groups (meta Cl):   23, 234, (236?) (2346?)
Reactivity Unknown: 235
Unreactive Groups: 3, 4, 24, 25, 246

Aroclor	Reported as	Source of Microorganisms	Reference
1242	--	New Bedford	Alder et al., 1993, Fig 4
1242	--	Hudson	Rhee et al., 1993a, Fig 1B

* Reactivity for these pentachlorobiphenyls is inferred from chlorophenyl reactivity pattern

Figure 4.5. Key characteristics of dechlorination process H' as observed in the references indicated. There have been no studies of dechlorination process H' with Aroclors 1248, 1254, or 1260. However, it is expected that flanked *para* chlorines on hexa- through octa-chlorobiphenyls would also be removed as in dechlorination process H (Fig. 4.6).

chlorophenyl group. Hence the major dechlorination activity was the *para* dechlorination of 245-, 2345-, and 2346-chlorophenyl groups (Fig. 4.6). Small increases of 24-3-CB and 25-3-CB from the dechlorination of 24-34-CB and 25-34-CB were also seen, indicating *para* dechlorination of 34-chlorophenyl groups as well. However, there was no accumulation of 23-25-CB or 235-23-CB, the products expected from *para* dechlorination of 234-25-CB and 2345-234-CB, respectively. This suggests that the 234-chlorophenyl group in these congeners was dechlorinated by loss of the *meta* chlorine to generate 24-25-CB and 235-24-CB, which did accumulate. Hence, for process H the position from which the chlorine

DECHLORINATION PROCESS H

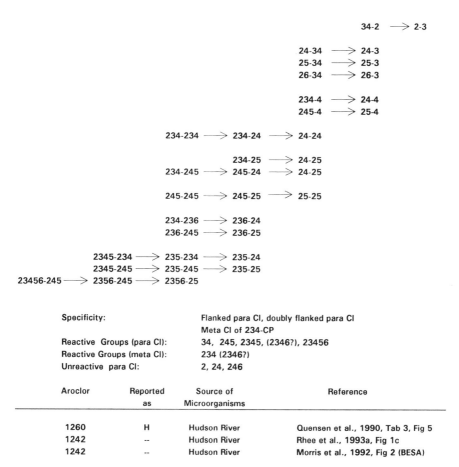

Figure 4.6. Key characteristics of dechlorination process H as observed in the references indicated. The actual products will depend on the congener composition of the PCB mixture undergoing dechlorination.

was lost was determined by the chlorine configuration rather than the absolute location.

Process H dechlorination of Aroclor 1242 was observed by Morris et al. (1992a) in Hudson River sediment slurries that had been enriched on Aroclor 1242–pyruvate and then treated with 10 mM bromoethane sulfonic acid (BESA). The most marked changes were decreases in 34-2-CB, 24-34-CB, 25-34-CB, 26-34-CB, and 245-4-CB and corresponding increases in 2-3-CB, 24-3-CB, 25-3-CB, 26-3-CB, and 25-4-CB. All of these changes indicate *para* dechlorination.

We mentioned above that the dechlorination of Aroclor 1242 observed by Rhee et al. (1993a) in methanogenic sediment slurries inoculated with microorganisms eluted from Hudson River sediment was process H' (our interpretation). In parallel cultures amended once with sulfate (10 mM), process H dechlorination occurred (our interpretation). The degree of *para* dechlorination, and the specific congener losses and increases due to *para* dechlorination of 34- and 245-chlorophenyl groups, were identical in both sets of cultures. However, the *meta* dechlorination of congeners bearing 23- and possibly 236-chlorophenyl groups that occurred in the methanogenic cultures did not occur in samples amended with sulfate. This suggests that process H' may in fact be a composite of process H and a second process that is limited to *meta* dechlorination of 23- and possibly 236-chlorophenyl groups. (see section on electron acceptors for further discussion).

5.1.5. Process P

Process P dechlorination of Aroclor 1260 is one of two major environmental de-chlorination activities occurring in the sediments of Woods Pond (Lenox, MA) and has also been observed in the laboratory (Bedard et al., 1993; submitted; Bedard and May, submitted). This dechlorination process is quite similar to H, but is restricted to *para* dechlorination, primarily of tetra- through heptachlorobiphenyls containing 34-, 234-, 245-, and 2345-chlorophenyl groups (Fig. 4.7). The largest decreases were seen in the major penta- and hexachlorobiphenyl peaks (234-25-CB, 234-24-CB, 245-245-CB, 234-245-CB, and 236-245-CB). A large increase of 25-25-CB occurred, as well as moderate increases of 23-25-CB, 24-25-CB, 235-25-CB, and other congeners containing 23-, 25-, and 235-chlorophenyl rings. Only *para* chlorines with at least one adjacent chlorine were removed by this dechlorination activity; hence *para* chlorines were removed from 34-, 234-, 245-, 2345-, and 2346-chlorophenyl groups, but not from 4-, 24-, or 246-chlorophenyl groups.

5.1.6. Process N

Dechlorination process N has been observed with sediment microorganisms derived from Silver Lake (Quensen et al., 1990; Alder et al., 1993), Woods Pond (Bedard et al., 1993; Bedard and Van Dort, in preparation), and the Hudson River (Rhee et al., 1993c). Process N removes all *meta* chlorines from 34-, 234-, 236-, 245-, 2345-, 2346-, and 23456-chlorophenyl groups, but not from 3-, 23-, or 25-chlorophenyl groups (Fig. 4.8, Table 4.1). Hence only *meta* chlorines flanked by at least one other chlorine are removed by this dechlorination activity. Nearly all hexa- and heptachlorobiphenyls and many pentachlorobiphenyls in Aroclor 1260 are de-creased by process N, and the products are primarily tri-, tetra- and penta-chlorobiphenyls. The most striking characteristic of this dechlorination is the high accumulation of 24-24-CB resulting from the removal of all *meta* chlorines from 234-245-CB, 245-245-CB, 234-2345-CB, and 245-2345-CB, which are all major components of Aroclor 1260. In addition, many of the other dechlorination prod-ucts have 24-chlorophenyl groups, for example, 24-4-CB, 24-25-CB, 24-26-CB, 246-24-CB, and 2356-24-CB.

Process N dechlorination of Aroclor 1254 (our interpretation) was observed in a

DECHLORINATION PROCESS P

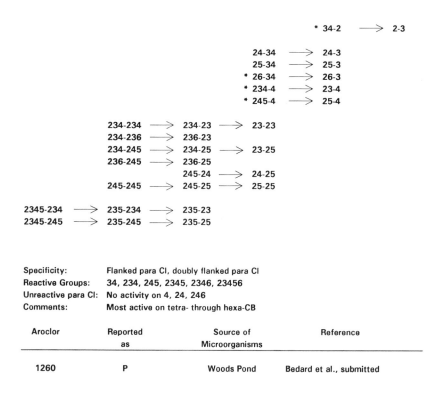

Specificity:	Flanked para Cl, doubly flanked para Cl
Reactive Groups:	34, 234, 245, 2345, 2346, 23456
Unreactive para Cl:	No activity on 4, 24, 246
Comments:	Most active on tetra- through hexa-CB

Aroclor	Reported as	Source of Microorganisms	Reference
1260	P	Woods Pond	Bedard et al., submitted

*These congeners are not present in Aroclor 1260. The activities are inferred from the chlorophenyl reactivity pattern.

Figure 4.7. Key characteristics of dechlorination process P as observed in the reference indicated. The actual products will depend on the congener composition of the PCB mixture undergoing dechlorination. The reactivity of several key congeners in Aroclor 1242 was inferred from the chlorophenyl reactivity pattern.

study with microorganisms from the Hudson River (Rhee et al., 1993c). Although many of the congener peaks were not reported, it is clear that all tetra-, penta-, and hexachlorobiphenyls containing 34-, 234-, 245-, or 236-chlorophenyl groups were dechlorinated. The principal products were 24-4-CB and 24-24-CB. According to our analysis of their Figure 1B, 24-4-CB was derived from 24-34-CB, 245-34-CB, and 234-34-CB and probably also 245-4-CB and 234-4-CB (the latter congeners were not reported but are significant peaks in Aroclor 1254); 24-24-CB was derived from 245-24-CB, 234-24-CB, 245-245-CB, 234-245-CB, and 234-234-CB. Other prominent products included 26-4-CB (from 236-34-CB), 25-4-CB, 24-25-CB, 24-26-CB, and 25-26-CB. (Congeners 24-26-CB and 25-26-CB co-elute in the peaks identified by Rhee et al. [1993c] as 23-4-CB and 34-2-CB, respectively.)

DECHLORINATION PROCESS N

				234-4	$\longrightarrow$	24-4
				236-4	$\longrightarrow$	26-4
				245-4	$\longrightarrow$	24-4

234-4 $\longrightarrow$ 24-4
236-4 $\longrightarrow$ 26-4
245-4 $\longrightarrow$ 24-4

25-34 $\longrightarrow$ 25-4
234-34 $\longrightarrow$ 24-34 $\longrightarrow$ 24-4
245-34 $\longrightarrow$ 24-34 $\longrightarrow$ 24-4
236-34 $\longrightarrow$ 26-34 $\longrightarrow$ 26-4

234-25 $\longrightarrow$ 24-25
245-25 $\longrightarrow$ 24-25
236-25 $\longrightarrow$ 25-26

234-236 $\longrightarrow$ 236-24 $\longrightarrow$ 24-26
236-245 $\longrightarrow$ 236-24 $\longrightarrow$ 24-26

234-234 $\longrightarrow$ 234-24 $\longrightarrow$ 24-24
2345-234 $\longrightarrow$ 234-245 $\longrightarrow$ 245-24 $\longrightarrow$ 24-24
2356-234 $\longrightarrow$ 2356-24 $\longrightarrow$ 236-24 $\longrightarrow$ 24-26

2345-2345 $\longrightarrow$ 2345-245 $\longrightarrow$ 245-245 $\longrightarrow$ 245-24 $\longrightarrow$ 24-24
2345-2346 $\longrightarrow$ 2345-246 $\longrightarrow$ 245-246 $\longrightarrow$ 246-25
2345-2356 $\longrightarrow$ 2356-245 $\longrightarrow$ 2356-24 $\longrightarrow$ 236-24 $\longrightarrow$ 24-26

Specificity:	Flanked meta Cl, doubly flanked meta Cl
Reactive Groups:	34, 234, 236, 245, 2345, 2346, 23456
	Weak activity against 235, 2356
Unreactive meta Cl:	3, 23, 25
Comments:	Very reactive on all tetra- through octa-CB with susceptible CP groups

Aroclor	Reported As	Source	Reference
1260	N	Silver Lake	Quensen et al., 1990, Tab 3, Fig 8
1260	--	Silver Lake	Alder et al., 1993, Fig 6
1260	N	Woods Pond	Bedard et al., 1993, Bedard & Van Dort, in prep.
1254	H	Hudson River	Rhee et al., 1993e, Fig 1B (biphenyl)

Figure 4.8. Key characteristics of dechlorination process N as observed in the references indicated. The actual products will depend on the congener composition of the PCB mixture undergoing dechlorination.

5.2. Combined Dechlorination Processes

5.2.1. Process C

As we stated earlier, dechlorination pattern C, which is characterized by extensive loss of *meta* and *para* chlorines and accumulation of primarily *ortho*-substituted PCBs, appears to result from the combination of two distinct dechlorination activities, processes M and Q, which are presumably mediated by different microorganisms. There are several lines of evidence to support this conclusion. The first comes from re-examination of histograms of the PCB congener distribution profile throughout the time course of an incubation that culminated in dechlorination pattern C (Quensen et al., 1990). The earliest time point (4 weeks) showed strong decreases in 25-2-CB, 34-2-CB, 23-4-CB, and 23-34-CB and strong increases in 2-CB, 2-2-CB, 2-4-CB, and 24-2-CB. Decreases were also seen in most other *meta*-substituted congeners, including 23-25-CB, 24-25-CB, 25-25-CB, 23-24-CB, 25-34-CB, and 26-34-CB, but activity was notably weaker on 25-4-CB. These changes all result from *meta*-dechlorination and are identical to the earliest changes seen in dechlorination process M. At 8 weeks the same congeners had decreased further; 2-4-CB and 24-2-CB had increased by a total of 13 and 1.5 mole%, respectively; and 24-4-CB, 25-4-CB, 26-4-CB, and 24-24-CB were unchanged. Subsequent *para* dechlorination of these congeners, consistent with dechlorination process Q, produced the congener distribution profile characteristic of dechlorination pattern C. This sequential *meta* and then *para* dechlorination has been observed frequently (Quensen et al., unpublished data; William Williams, personal communication) and is consistent with the interpretation that different microorganisms are responsible for the two stages of dechlorination.

The second line of evidence comes from the pasteurization and ethanol treatment studies done by Ye et al. (1992). Untreated inocula from fresh Hudson River sediment exhibited process C dechlorination of Aroclor 1242, but pasteurized or ethanol-treated inocula exhibited only process M. In these experiments the PCB congener distribution profile and homolog distribution profile for both the untreated and the treated inocula were the same during the first 4 weeks. After this time, no additional dechlorination occurred in sediments that had received treated inocula, but repeated transfers to sediment spiked with Aroclor 1242 established that the microorganisms responsible for the *meta*-dechlorinating activity were still present. In contrast, in sediment slurries that had received untreated inocula, *para* dechlorination commenced and continued for the next 8 weeks, culminating in the removal of virtually all *meta* and *para* chlorines and the PCB congener distribution profile typical of dechlorination pattern C. In additional sets of experiments by the same investigators, untreated inocula from sediment that had been stored in a cold room for extended periods of time lost the *para*-dechlorinating component of the activity and also exhibited only process M activity. These experiments indicate that dechlorination process M (exclusively *meta*-dechlorinating activity) and dechlorination process Q (primarily *para*-dechlorinating activity) represent the activities of two distinct microbial populations. The microorganisms responsible for dechlorination process M are apparently spore-formers and appear to be more viable in long-term storage. The microorganisms responsible for process Q appear to be less hardy, at least under laboratory conditions.

Other evidence indicates that dechlorination pattern C can also result from the simultaneous activity of dechlorination processes M and Q. Alder et al. (1993) incubated Hudson River sediment slurries with Aroclor 1242 and with or without a supplement of fatty acids. Although the dechlorination progressed more rapidly in fatty acid–supplemented cultures, the ultimate dechlorination achieved, pattern C (albeit less extensively developed than that reported by Quensen et al. [1990]), was the same in both cases after 11 months of incubation at 30°C. (Max Häggblom kindly provided us with the congener distribution profiles obtained in these experiments over the time course of the incubation.) In contrast to the results described above, in these incubations it appears that processes M and Q were operating at the same time rather than in a sequential fashion. Hence modest amounts of 2-3-CB (a product of process Q dechlorination) and considerable amounts of 2-4-CB (a product of process M) accumulated over the first several months of incubation. At the same time 25-2-CB decreased sharply (M activity), as did 24-4-CB (Q activity). The mono-*ortho*-tetrachlorobiphenyls, 24-34-CB, 25-34-CB, 245-4-CB, and 23-34-CB, all decreased in the first few months (process Q), whereas decreases in 24-2-CB (process Q) and the di-*ortho*-tetrachlorobiphenyl peaks began later: 25-25-CB (process M), 24-25-CB (both M and Q), and 24-24-CB (process Q). Subsequently both 2-3-CB and 2-4-CB decreased to generate the high levels of 2-CB typical of process C. A second interpretation of these results is that they indicate a new type of dechlorination. However, because these experiments were conducted using undefined communities in sediment slurries that are known to harbor microorganisms responsible for both activities M and Q, we believe that the best interpretation of these results is that they reflect the concerted activity of two distinct microbial populations, namely, those separately responsible for dechlorination processes M and Q.

Dechlorination pattern C has also been obtained in experiments using Hudson River sediment to dechlorinate Aroclors 1248 and 1254, which contain high proportions of tetra-, pentra-, and hexachlorobiphenyls (Quensen et al., 1990). The data show that even these higher congeners are dechlorinated to mono-, di-, and trichlorobiphenyls. Thus it would appear 1) that either process M or process Q or both are capable of attacking the trichlorophenyl rings that are constituents of the PCB congeners in these Aroclors, and 2) that their combined activity is relatively insensitive to the trichlorophenyl substitution pattern on the unattacked ring.

5.2.2. *Processes M and H or H'*

Rhee et al. (1993b) studied the dechlorination of Aroclor 1242 in slurries of autoclaved sediment from Owasco Lake, NY, inoculated with microorganisms eluted from Hudson River sediment. They observed that the first PCB congeners to undergo dechlorination were 34-2-CB, 24-34-CB, and 25-34-CB, each of which was dechlorinated from the *para* position to produce 2-3-CB, 24-3-CB, and 25-3-CB, respectively. Subsequently, *meta* dechlorination of 25-2-CB, 23-25-CB, 25-25-CB, and 23-4-CB was observed with consequent accumulation of 2-2-CB and 2-4-CB. *Meta* dechlorination continued until late in the incubation, when indications of *para* dechlorination of 24-4-CB and 24-2-CB, and then later 26-4-CB and 24-24-CB, were seen. These observations are consistent with the interpretation that several microbial populations with different dechlorinating capabilities were active at different times during the incubation. The earliest dechlorination was typical of process H or H' (*para* dechlorination of 34-chlorophenyl groups), the subsequent *meta* dechlorination is typical of process M, and the early indications of *para* dechlorination of 24-2-CB and 24-4-CB late in the incubation are consistent with incipient process Q. Processes H and H' are distinguished only by the ability of H to remove *meta* chlorines from 23-chlorophenyl groups. Since process M also dechlorinates 23-chlorophenyl groups, it is not possible to distinguish H from H' in the presence of M.

A particularly interesting example of M and H or H' comes from pyruvate–Aroclor 1242 enrichments of microorganisms from Hudson River sediment under methanogenic conditions (Morris et al., 1992a). The second transfer of this enrichment resulted in 45% removal of the *meta* plus *para* chlorines in 4 weeks. The dechlorination pattern looked very much like M (Morris et al., 1992a) and was characterized by high accumulations of 2-CB, 2-2/26-CB, 2-4-CB, and 24-4-CB. However, increased levels of 2-3-CB, 24-3-CB, and 25-3-CB indicated that process H was also active (our interpretation). More support for this interpretation comes from the observation that the pattern of dechlorination changed when the same enrichment was used to inoculate a sediment slurry that was then incubated with sulfate (10 mM) or BESA (10 mM). The dechlorination was much slower and much less extensive with sulfate or BESA. In addition, in the cultures with sulfate the increases of 2-CB, 2-2-CB/26-CB, 2-4-CB, 24-2-CB, and 24-4-CB were at most half as large as in cultures without sulfate, but there were larger increases in 2-3-CB, 24-3-CB, and 25-3-CB. Thus, in the sulfate-amended cultures process M was much less active, and the effects of process H or H' were more apparent. In the cultures treated with BESA there was no evidence of M; the dechlorination pattern was that of process H or H' (see above).

Additional examples of H (or H') activity followed immediately by process M were seen in incubations of an Aroclor mixture with Hudson River sediment (Abramowicz and Brennan, 1991; also PCB concentration experiments, Abramowicz et al., 1993).

5.3. Conclusions

The previous examples support the conclusion that there are at least six separable activities that dechlorinate Aroclors: M, Q, H, H', N, and P. These can be distinguished by their congener selectivity patterns (Fig. 4.3–4.8) and also by their chlorophenyl group reactivity patterns (Table 4.1). None of these processes removes *ortho* chlorines. Processes M and N remove exclusively *meta* chlorines and process P exclusively *para* chlorines. The other three dechlorination processes remove both *meta* and *para* chlorines, but with distinct specificities (Tables 4.1, 4.2). Most of these processes are apparently influenced by the chlorine configuration on both the attacked ring and the opposite ring.

Only processes Q and M are capable of removing unflanked chlorines such as those on a 3- or 25-chlorophenyl ring (M) or a 4- or 24-chlorophenyl ring (Q). The ability to remove *meta* or *para* chlorines from flanked positions, such as those on 23-, 34-, 234-, 236-, and 245-chlorophenyl rings, is apparently more common, but individual preferences and limitations are still seen among the six dechlorination processes (Tables 4.1, and 4.2).

Clearly, processes M and Q, which attack the broadest range of mono- and

TABLE 4.2. Characteristics of PCB Dechlorination Processes

Dechlorination Process	Characteristic Dechlorination Products[a]	Susceptible Chlorines	Susceptible Aroclors	Source of Microorganisms
M	2 2-2/26 2-4 24-2 24-4 26-2	Flanked and unflanked *meta*	1242 1248? 1254?	Upper Hudson Silver Lake
Q	2 2-2/26 2-3 25-2 26-2 26-3	Flanked and unflanked *para* *Meta* of 23	1242 1248 1254	Upper Hudson
C	2 2-2/26 26-2 26-3	Flanked and unflanked *meta* and *para*	1242 1248 1254	Upper Hudson
H′	2-3 2-4 24-2 25-2 24-3 25-3 26-3 24-4/25-4 24-24[b] 24-25 25-25 235-24[b] 235-25[b] 236-24[b] 236-25[b]	Flanked *para* *Meta* of 23, 234	1242 1248 1254 1260	Upper Hudson Lower Hudson? New Bedford
H	2-3 24-3 25-3 26-3 24-4/25-4 24-24 24-25 25-25 235-24	Flanked *para* Doubly flanked *meta*	1242 1248 1254 1260	Upper Hudson Lower Hudson New Bedford Silver Lake?

(continued)

TABLE 4.2. (*Continued*)

Dechlorination Process	Characteristic Dechlorination Products[a]	Susceptible Chlorines	Susceptible Aroclors	Source of Microorganisms
H	235-25 236-24 236-25			
P	23-25 24-25 25-25 235-23 235-25	Flanked *para*	1254? 1260	Woods Pond Silver Lake?
N	24-4 24-24 24-25 24-26 246-24 2356-24	Flanked *meta*	1254 1260	Upper Hudson Silver Lake Woods Pond

[a]Products will vary depending on the congener composition of the PCB mixture being dechlorinated.
[b]Proposed products from Aroclors 1254 and 1260.

dichlorophenyl groups, will have the greatest impact on Aroclors 1242, 1248, and, to some extent, 1254, because these Aroclors are composed largely of congeners with mono- and dichlorophenyl rings. Processes Q and M may, however, be less effective on more highly chlorinated PCBs due to steric hindrance from the unattacked ring. Conversely, processes H, N, and P preferentially dechlorinate more heavily substituted chlorophenyl rings. They will therefore have little effect on Aroclors 1242 and 1248, but a greater impact on Aroclors 1254 and 1260 because these Aroclors contain a higher proportion of tri- and tetrachlorophenyl groups. For these reasons the effects of process H, P, or N might not be recognized at sites contaminated with Aroclor 1242 or 1248, and the effects of process Q or M might not be evident at sites contaminated with Aroclor 1260.

It is apparent that some dechlorination activities, like M and Q, complement each other and work together to effect virtually complete removal of *meta* and *para* chlorines (Process C, Tables 4.1, 4.2). In theory, process Q, or a dechlorination process with similar specificity, would also complement process N. Thus the sequential activity of process N and process Q acting on Aroclor 1260 would result in far more effective chlorine removal than either process alone. In the same way, process M would complement process H, H', or P. In other cases two dechlorination processes would not complement each other. For example, processes P and N compete for the same chlorophenyl substrates and generate di- and trichlorophenyl groups such as 24- and 246-chlorophenyls (N) or 23-, 25-, and 235-chlorophenyls (P) that are not substrates for either process.

It is evident that a variety of dechlorination activities have been observed within a single sediment sample and even within the same laboratory using the same culturing procedures. For example, Quensen and colleagues (1988, 1990) and Ye et al. (1992) have observed processes Q, M, C, and H with Hudson River microorganisms and processes M and N with Silver Lake microorganisms. Rhee and coworkers (1993a,b,e) have observed processes H, H', M, and N (our interpretation of their congener distribution profiles) with Hudson River microorganisms from a different site, and Bedard et al. (1993; Bedard and May, submitted) have observed processes N and P in sediments from Woods Pond. In addition, dechlorination processes H', M, and N have each been observed in sediments from two or more locations (see Figs. 4.3, 4.5, and 4.8). These observations raise questions about the nature of the enzymes responsible for PCB dechlorination. It is difficult to imagine that enzymes with nearly identical substrate specificities on complex Aroclor mixtures evolved separately at diverse locations in response to a chemical that has only existed for about 60 years. More likely the enzymes responsible for PCB dechlorination also serve some other important function and are therefore fairly common. It is not clear whether the specificities for the dechlorination processes described above reside at the level of the enzyme, the microbial species, or a broad physiological group of microorganisms. The answers to such questions await the isolation of the dehalogenating microorganisms and their enzymes.

6. ENVIRONMENTAL DECHLORINATION OF PCBs

6.1. New Bedford Harbor

There have been several reports of environmental dechlorination of the PCBs in upper New Bedford Harbor (Brown and Wagner, 1990; Lake et al., 1991, 1992). This area is a small, highly protected salt marsh estuary that received inputs of both Aroclors 1242 and 1254 from a capacitor plant. The sediments vary widely in PCB concentration and in the relative proportions of the two Aroclors and show varying degrees of dechlorination. Accordingly, both laboratories used an indicator peak procedure to estimate the original proportions of Aroclors 1242 and 1254 (see previous section on recognition and characterization of environmental dechlorination of PCBs; the indicator peaks used by both groups of investigators were 26-34-CB/236-4-CB and 236-34-CB). The highest PCB concentrations (as high as 3,800 ppm) were found closest to the plant and at a depth of 5–17.5 cm (Brown and Wagner, 1990; Lake et al., 1991,1992). Brown and Wagner reported that chromatograms from all of the tidal flat cores that they examined showed evidence of either process H or H' dechlorination. These chromatograms were characterized by heavy losses of PCB congeners with 34-, 234-, and 2345-chlorophenyl groups and, to a lesser extent, 245-chlorophenyl groups. They also showed marked increases of 24-3-CB, 25-3-CB, 25-25-CB, 24-25-CB, and 24-24-CB, and some 2-3-CB as terminal dechlorination products. Brown and Wagner concluded that the *meta* chlorine of 234-chlorophenyl groups had been removed to produce 24-chlorophenyl groups, but the *para* chlorine had been lost from 34-, 245-, and 2345-chlorophenyl groups. In all cases, dechlorination was impaired by juxtaposition of a 26-, 236-, or 2356-

chlorophenyl group on the opposite ring. Some chromatograms showed limited *meta* dechlorination of 23-chlorophenyl groups. The latter were identified as dechlorination pattern H′.

Lake and colleagues (1991, 1992) reported that the seven midchannel cores they sampled showed evidence of dechlorination, but to varying degrees and in at least two distinct patterns. Cores taken closest to the plant showed very extensive dechlorination characterized by marked increases of 24-3-CB, 25-3-CB, 24-25-CB, and 25-25-CB. The published chromatogram also showed increases in 24-24-CB and 236-24-CB (our interpretation). No information was given on the dechlorination of 23-chlorophenyl groups, and none could be inferred from the chromatogram presented. This alteration pattern is consistent with the H/H′ dechlorination described by Brown and Wagner (1990) with the exception that the published chromatogram (Fig. 3d in Lake et al., 1992) also showed extensive decreases of 24-4-CB and 25-4-CB. This is significant because these congeners account for about 15% of the PCBs in Aroclor 1242. Lake and colleagues did not comment on these losses, nor did they present the results from other chromatograms, so it is not possible to assess how common this phenomenon was.

Lake and colleagues also reported that the PCBs in one core showed a very selective pattern of *meta* dechlorination limited to losses of congeners containing 234-chlorophenyl groups. The heaviest losses were in 234-25-CB and 234-245-CB, which were dechlorinated to 24-25-CB and 245-24-CB, respectively. This particular core showed no increase of 24-3-CB or 25-3-CB, which are the most sensitive indicators of process H. Conversely, other cores located further south of the plant showed increases of 24-3-CB and 25-3-CB, but no losses of congeners with 234-chlorophenyl groups. From these data it appears that the *meta* dechlorination of 234-chlorophenyl groups represents a distinct and very limited type of dechlorination in New Bedford Harbor that can occur alone. Therefore, it is possible that the New Bedford process H dechlorination is actually the result of at least *two separate* activities that usually occur together, namely, 1) *para* dechlorination of 34, 245-, and 2345-chlorophenyl groups and 2) *meta* dechlorination of 234-chlorophenyl groups.

Lake et al. (1991, 1992) further observed that dechlorination of congeners containing 245-chlorophenyl groups occurred only in the most most heavily dechlorinated samples. This may indicate that 245-chlorophenyl groups are dechlorinated only after more susceptible chlorophenyl groups (34-, 234-, 2345-chlorophenyls) have first been removed, or it could be an indication that a separate dechlorination activity is responsible for the dechlorination of 245-chlorophenyl groups.

The process H′ dechlorination of pre-existing PCBs in sediment from New Bedford Harbor that was observed in laboratory studies (Alder et al., 1993) resulted in essentially the same changes as those described by Brown and Wagner (1990) in the environmental samples. However, in addition to the changes seen in environmental samples, the laboratory incubation revealed 1) more extensive losses of congeners containing 23-chlorophenyl rings and 2) substantial decreases of congeners containing 34-chlorophenyl groups juxtaposed to 26- and 236-chlorophenyl groups. (In the environmental samples these di-*ortho*-chlorophenyl groups inhibited the dechlorination of the attached ring.) These differences resulted in a decrease in

23-2-CB, a higher increase in 25-2-CB (from dechlorination of 23-25-CB), and an increase in 26-3-CB (from dechlorination of 26-34-CB). These changes are consistent with what might be expected from more advanced dechlorination within the parameters of the previously defined specificity of dechlorination process H' (Brown and Wagner, 1990). Alder et al. suggested that dechlorination *in situ* might be inhibited by the continuous input of sulfate from tidal flushing. Although sulfate concentration was not measured in the sediments, no dechlorination occurred in sediments from New Bedford Harbor incubated in the laboratory under sulfidogenic conditions (Alder et al., 1993). Lower temperatures *in situ* might also limit dechlorination either directly, through effects on the dechlorinating microorganisms themselves, or indirectly, through effects on other microorganisms that interact with the dechlorinating microorganisms.

6.2. Hudson River

The original PCB contamination in upper Hudson River sediment was primarily Aroclor 1242 with small amounts of Aroclor 1254 (typically 5%–10%) (Brown et al., 1987b). The PCBs in nearly all sediment samples collected between Fort Edward and Troy in the last few years show fairly extensive losses of *meta* and *para* chlorines. Thus the PCBs in most of these sediment samples are characterized by high proportions of *ortho*-substituted mono- and dichlorinated PCBs and fit one of three distinct environmental dechlorination patterns: B, B', or C (Brown et al., 1984, 1987a,b; Abramowicz et al., 1995). In the tidal Hudson, below Troy, where the PCB concentrations are considerably lower, PCB sediment samples occasionally show dechlorination pattern B, but dechlorination patterns H and H' are more typical. These two patterns are characterized by strong decreases in 245-4-CB, 25-34-CB, 24-34-CB, and 23-34-CB, and increases in 2-3-CB, and certain trichlorobiphenyls, especially 24-3-CB and 25-3-CB (Brown et al., 1987b; Brown, 1990).

Brown et al. (1987a,b) proposed that dechlorination patterns B, B', and C resulted from alternative, rather than successive, transformation processes. However, we propose that all three patterns can be reasonably explained as the result of various combinations of dechlorination activities M, Q, and H or H'. Brown et al. noted that 23- and 34-chlorophenyl groups were the most reactive groups in environmental dechlorination patterns B, B', and C, leading to marked declines in 23-4-CB, 34-2-CB, 23-23-CB, 23-25-CB, 245-4-CB, (note that a 245-chlorophenyl group is the same as a 34-chlorophenyl group with an additional chlorine in the 6-position), 25-34-CB, and 24-34-CB. This is consistent with our hypothesis that these environmental patterns result from various combinations of M, Q, and H', because each of the latter dechlorination processes exhibits activity against 23- and 34-chlorophenyl groups (Table 4.1). Alteration patterns B, B', and C also share a number of other features. Each is characterized by markedly decreased levels of 25-2-CB (process M) and 24-4-CB (process Q) and by increased levels of 2-CB, 2-2-CB/26-CB, 26-2-CB, and 26-3-CB. However, patterns B, B', and C differ in the relative proportions of dechlorination products and in the extent of dechlorination of 3-, 4-, 24-, and 25-chlorophenyl groups. We shall discuss the distinguishing features of each of these environmental alteration patterns in turn.

6.2.1. Pattern B

Pattern B exhibits less complete removal of *meta* and *para* chlorines than either pattern B' or C; hence 2-3-CB and 2-4-CB are major dechlorination products (Table 4.3). The amounts of 2-CB, 2-2-CB/26-CB, and 26-2-CB are all elevated relative to Aroclor 1242, but are significantly lower than in dechlorination pattern C (Table 4.3). In addition, 25-3-CB and 26-3-CB are elevated, and only small decreases are seen in 24-2-CB, 26-4-CB, 25-25-CB, and 24-25-CB. Taken together, this congener distribution is what would be expected from process H or H' dechlorination followed by M and then Q. The increases of 2-3-CB, 25-3-CB, and 26-3-CB most likely result from *para* dechlorination of 34-2-CB, 25-34-CB, and 26-34-CB (process H or H'), the increase in 2-2-CB from *meta* dechlorination of 25-2-CB (process M), and the increase in 2-4-CB from *para* dechlorination of 24-4-CB (process Q). The minimal decreases in 24-2-CB, 26-4-CB, 25-25-CB, and 24-25-CB indicate that neither the M nor the Q activity has been fully developed; hence only those losses that usually occur first (the losses of 25-2-CB due to M, and of 24-4-CB due to Q), are evident.

6.2.2. Pattern B'

Pattern B' is distinguished from B most clearly by a substantial increase in 2-CB and the virtual absence of 2-3-CB (Table 4.3). In addition, congeners carrying 3- and 25-chlorophenyl rings show greater decreases than in B (indicative of more advanced M activity), and more of the congeners carrying 24-chlorophenyl groups are decreased (indicative of moderately advanced process Q). The low value of

TABLE 4.3. Comparison of Weight Percent Contributions of Key Congeners in Hudson River Dechlorination Patterns B, B', and C With Those of Aroclor 1242[a]

	Observed Range of PCB Congener (% by Weight)			
PCB Congener	Aroclor 1242	Pattern B	Pattern B'	Pattern C
2	0.7	10–17	28–52	13[b]–43
2-2 (+ 26)	2.6	12–19	14–27	30–41
2-3	1.3	4–9	0.3–0.9	0.7–1.6
2-4 (+ 23)	6.2	15–18	6–16	2.5–4.6
26-2	0.9	2.8–4.1	2.9–3.7	5.1–5.4
26-3	0.8	2.8–4.3	2.6–3.5	2.6–3.1
25-4/24-4	14.4	3.1–8.4	1.9–4.2	1.6–3.1
25-34	3.3	0.1–0.4	0.0–0.5	0.1–0.8

[a]Adapted from Brown et al. (1987a).
[b]This value was seen in a single surface-grab sample taken from a midchannel area that was subject to scouring.

2-3-CB suggests that this congener never accumulated to any extent and, therefore, that dechlorination process H (or H′) probably was not active. The levels of 2-4-CB are lower than in pattern B, but higher than in C. This congener profile is what would be expected from a fairly advanced process M dechlorination followed by less developed process Q dechlorination.

6.2.3. Pattern C

Pattern C, as we have described above, is the result when both dechlorination activities M and Q have been more fully developed. It is distinguished from pattern B′ primarily by higher levels of 2-CB and 2-2-CB/26-CB (Table 4.3, Fig. 4.2, center panel) resulting from *para* dechlorination of 2-4-CB, 24-2-CB, and 26-4-CB (process Q). We propose that patterns B and B′ are not endpoints and that both could eventually arrive at pattern C by further M and Q activity. In addition, as suggested by the activity seen by Alder et al. (1993) (see Section 5.2.1.), pattern C dechlorination can result from the action of dechlorination processes M and Q working in concert rather than in sequence. Hence, we predict that ultimately the extent of dechlorination in all sediments of the upper Hudson will reach that of pattern C.

6.2.4. Pattern E

Dechlorination pattern E was seen only in combination with pattern B or C and is distinguished primarily by marked decreases in most *para*-substituted PCBs with five or more chlorines. (In the Hudson River the primary source of higher chlorinated congeners is low level contamination with Aroclors 1254 and 1260.) Brown et al. (1987b) reported approximately 90% decreases of 245-245-CB, 2345-245-CB, and 2345-234-CB and substantial decreases in all other higher congeners carrying a *para* chlorine on either ring. The most prominent dechlorination products were 2356-25-CB and 2356-235-CB, with lesser amounts of 236-35-CB, 236-236-CB, 235-236-CB, 235-235-CB, 2356-24-CB, and possibly 2356-2356-CB. These products result from *para* dechlorination of PCB congeners containing 245-, 2345-, 2346-, and 23456-chlorophenyl rings and are consistent with those of process H dechlorination of Aroclor 1260 (see Fig. 4.6) mediated by Hudson River microorganisms in a laboratory study (Quensen et al., 1990). The primary products reported by Quensen and colleagues, 25-25-CB and 235-25-CB, would have been further dechlorinated by dechlorination systems B and C that occur with dechlorination pattern E in the sediment and, hence, would not have accumulated. Therefore, it is quite possible that dechlorination process E is the same as process H.

6.3. Woods Pond

Woods Pond (near Lenox, MA) is a shallow impoundment on the Housatonic River. For many years PCBs accumulated in this pond from storm sewer discharges and drainage from the areas in and around transformer manufacturing operations in Pittsfield, 11 miles upstream. Aroclor 1260 was used at the transformer manufacturing operation from 1934 to 1970. Aroclor 1254 was used from 1970 to 1978, but by this time procedures for handling PCBs had been improved because of increased environmental awareness. Bedard and May (submitted) used high-resolution GC to analyze the PCBs in sediment samples collected from more than 90 locations in the pond. The sediments contain 30–150 µg/g (sediment dry weight) PCBs and an unidentified hydrocarbon oil (0.5%–2%, dry weight basis). Congener distribution profiles of the sediment PCBs collected from all locations indicated that the original contaminant was virtually all Aroclor 1260 and that the PCBs had been partially dechlorinated by processes P and N. Relative to Aroclor 1260, the major hexa- and heptachlorobiphenyls were significantly decreased in the sediment PCBs, and specific tri-, tetra-, and pentachlorobiphenyls were increased. On a mole percent basis, the largest losses were those of 236-245-CB, 245-245-CB, 234-245-CB, 2345-245-CB, and 2345-234-CB. Smaller losses of 2345-25-CB, 2345-236-CB, 2356-25-CB, 235-245-CB, and 2345-246-CB/2356-245-CB were also evident. The strongest indication of dechlorination was the presence of substantial amounts (4–11 mole%) of three tetrachlorobiphenyls, 24-24-CB, 24-25-CB, and 25-25-CB, which are virtually absent in Aroclor 1260. These three congeners and other congeners that were increased relative to their abundance in Aroclor 1260 correspond to the products expected from dechlorination by processes P and N. Elevated amounts of 23-25-CB, 25-25-CB, 235-25-CB, and, in more dechlorinated samples, 24-3-CB and 25-3-CB provided evidence of process P dechlorination (Fig. 4.7), whereas elevated amounts of 24-4-CB, 25-4-CB, 26-4-CB, 24-24-CB, 24-26-CB, 25-26-CB, 236-24-CB, 245-24-CB, and 2356-24-CB provided evidence of process N dechlorination (Fig. 4.8). The increases in 24-25-CB were consistent with both dechlorination processes. Finally, a mother–daughter analysis of key congeners in Aroclor 1260 and their putative dechlorination products yielded an excellent mass balance (Bedard and May, submitted).

All sediment samples showed evidence of both dechlorination processes P and N, but the extent of dechlorination and the relative contributions of the two processes varied considerably among samples. Along the eastern edge of the pond process P was more dominant. Along the western edge there was more variability; in many samples process N was more prominent, in some N and P were equally evident, and in some process P was dominant. The extent of dechlorination was modest compared with that seen in nearby Silver Lake or in the Hudson River. Even the most dechlorinated samples exhibited losses of only 13% of the *meta* plus *para* chlorines.

Laboratory studies with Woods Pond sediment have shown that more extensive dechlorination of the sediment PCBs by process P and by process N could be selectively stimulated by the addition of single congeners that act as preferred dechlorination substrates for one of the processes (Bedard et al., 1993; submitted). These studies will be discussed at greater length in the section on enhancing dechlorination. They are mentioned here because they support the conclusion that the pattern of dechlorination observed in Woods Pond sediments is a composite of patterns P and N. Furthermore, they demonstrate that the microorganisms responsible for both dechlorination processes are still present in the sediment and that it is possible to stimulate substantial dechlorination of the Aroclor 1260 that has persisted in these sediments for decades.

6.4. Silver Lake

Silver Lake is a 26 acre urban pond in Pittsfield, MA, that is polluted from past discharges of municipal sanitary sewage, storm sewage, and industrial wastes. The sediments are composed of a highly methanogenic black, oily, organic material overlying sand and silt. The most likely source of the PCBs found in Silver Lake sediments is an electrical equipment manufacturing plant that included capacitor and transformer manufacturing operations. The capacitor manufacturing operation, which was located close to Silver Lake, used Aroclor 1254 from the mid-1930s until it ceased production around 1950. A transformer manufacturing operation, which was located somewhat more distant from Silver Lake, used Aroclor 1260 from 1934 to 1970 and Aroclor 1254 from 1970 to 1978. The PCBs probably entered Silver Lake through storm sewer discharges and drainage from the areas in and around these manufacturing operations.

In a survey of Silver Lake sediments conducted in 1980 and 1982, 78 sediment samples were analyzed by packed column GC/ECD (Stewart Laboratories, 1982). All samples were reported as containing Aroclors 1254 and 1260, but Aroclor 1254 was reported as the predominant Aroclor in 80% of the sediments examined. The study also reported "pattern alterations" of the Aroclors in 40 of 72 samples from the perimeter of Silver Lake and in all six samples from the center of the pond. Three of the samples showing pattern alterations were subjected to extensive clean-up procedures and reanalyzed by packed column GC/MS. These analyses confirmed that the observed pattern alterations were due to many di-, tri- and tetra-chlorobiphenyls that did not match the patterns of Aroclor 1242, 1248, or 1254 (Stewart Laboratories, 1982).

Based on a re-evaluation of the three packed column GC/MS analyses performed by Stewart Laboratories and a single capillary GC/ECD analysis of a new sediment sample performed at General Electric, Brown et al. (1987a,b) described two patterns of PCB dechlorination, F and G, observed in Silver Lake sediments. Brown and colleagues were not aware of the location of storm sewer lines that drained the areas around the capacitor and transformer operations, and because transformers were manufactured over a much longer time, they assumed that the original contaminant of Silver Lake was virtually all Aroclor 1260. In support of this assumption they noted that the distribution of residual hexachlorobiphenyls in the three GC/MS samples matched that of Aroclor 1260 and that dechlorinated Aroclor 1260 could easily have been mistaken for 1254 in the packed column GC/ECD analyses performed by Stewart Laboratories. Based on the assumption that the original contaminant was all Aroclor 1260, Brown et al. (1987a,b) concluded that the PCBs in Silver Lake had undergone *ortho* dechlorination as well as *meta* and *para* dechlorination. However, the storm sewers that empty into Silver Lake primarily drain the area around the capacitor manufacturing area, not the transformer manufacturing area

TABLE 4.4. Comparison of PCB Homolog Distributions[a] in Aroclors 1254 and 1260 and in Silver Lake Sediment Samples[b] Exhibiting Dechlorination Patterns F or G

Homolog Group	Mole Percent[a] in Sample of Type			
	1254	1260[b]	F(+G)[b,c]	G(+F)[b,c]
Monochlorobiphenyl	0.0	0.0	0.1	17[d]
Dichlorobiphenyl	0.2	0.0	0.1	17
Trichlorobiphenyl	3.2	0.0	32	26
Tetrachlorobiphenyl	25.9	0.6	50	31
Pentachlorobiphenyl	41.0	16	12	6
Hexachlorobiphenyl	26.7	53	5	3.4
Heptachlorobiphenyl	3.0	25	0.5	0.6
Octachlorobiphenyl	0.1	5	0.1	0.1
ortho-Cl per biphenyl	1.9	2.5	1.8	1.8
m,p-Cl per biphenyl	3.1	3.6	2.1	1.2
Total Cl per biphenyl	5.0	6.1	3.9	3.0

[a]Approximate distribution in each case except for Aroclor 1254 estimated from relative sizes of parent ion peaks in packed column GC/MS. Distribution for Aroclor 1254 was determined by capillary GC/ECD.
[b]Brown et al. (1987b).
[c]Sample F(+G), Silver Lake 1982 core C1, 48–64 cm section; specimen G(+F) 1982 core C2, 80–96 cm section.
[d]Estimated from a single GC/MS of another sample showing mono- and dichlorobiphenyls at equal levels.

(T. Rouse, personal communication); therefore, the most likely contaminant of Silver Lake was Aroclor 1254 with only small amounts of Aroclor 1260. Further investigations would be required to determine the actual proportions of Aroclors 1254 and 1260 in Silver Lake sediments. Table 4.4 compares the homolog distribution and chlorine distribution of Aroclors 1254 and 1260 with those of sediment PCBs displaying dechlorination patterns F and G and shows that, relative to Aroclor 1254, the sediment PCBs show losses in the *meta* and *para* chlorines, but not in *ortho* chlorines. However, since we do not know the actual proportions of Aroclor 1260 in the sediment, we cannot rule out the possibility that some *ortho* dechlorination occurred as well. We should note that *ortho* dechlorination of 246-CB was observed in cultures of Silver Lake sediment (Williams, 1994).

In addition to the data reported previously by Brown et al. (1987a,b), we have studied congener distributions from more recent capillary analyses of sediments displaying patterns F and G (some provided by John F. Brown, Jr., and others from our own analyses). Relative to Aroclor 1254, pattern F is characterized by large increases of 24-3-CB and 25-3-CB (6–15 mole% each); moderate increases 25-25-CB, 24-25-CB, 24-24-CB, and 236-3-CB/23-24-CB; and lesser increases of 26-3-CB, 235-3-CB, 235-23-CB, and 235-25-CB. Most of these correspond to the products expected from process H dechlorination of Aroclor 1254. Alternatively, a combination of process P and, to a lesser extent, N could also give this pattern (See Figs. 4.6–4.8).

Pattern G appears to be the result of further dechlorination of the products of pattern F, which culminates in the removal of about 61% of the *meta* plus *para* chlorines in Aroclor 1254. The most prominent congeners in pattern G are 2-CB, 2-2-CB/26-CB, 24-2-CB, 26-2-CB, and 26-3-CB. Some of the congeners characteristic of pattern F remain, but in significantly lower amounts. Thus it would appear that the products of pattern F have been further dechlorinated by removal of *meta* chlorines from 3-, 23-, 25-, and 235-chlorophenyl groups and, to a lesser extent, isolated *para* chlorines from 24-chlorophenyl groups. This *meta* dechlorination activity is probably M, which has been seen in laboratory incubations with microorganisms from Silver Lake sediments (Quensen et al., 1990, as reinterpreted in this review). Little is known about the *para* dechlorination activity except that, like process Q, it can dechlorinate 24-chlorophenyl groups.

Laboratory studies with microorganisms from Silver Lake have demonstrated dechlorination processes M and N. Both of these processes are limited to *meta* dechlorination. Yet the large amounts of 24-3-CB, 25-3-CB, and other *meta*-enriched congeners seen in Silver Lake pattern F provide clear evidence of *para* dechlorination of the abundant congeners in Aroclor 1254 that have 34-, 245-, and 2345-chlorophenyl groups. This kind of activity could be due to process P, which has been demonstrated in nearby Woods Pond, or to process H.

In summary, the congener distribution of the dechlorinated PCBs in Silver Lake sediments can be explained by *meta* and *para* dechlorination of Aroclor 1254, and the original report that it was due to *ortho, meta,* and *para* dechlorination of Aroclor 1260 (Brown et al., 1987a,b) is probably erroneous. Pattern F most likely results from process H dechlorination, or from a combination of processes P (primarily) and N. Pattern G is most likely a composite resulting from processes H and M with some Q-like activity, or possibly from processes P, N and M with some Q-like activity.

6.5. Other Locations

Limited evidence for *in situ* dechlorination at several additional locations has been reported. Complete quantitative congener-specific analyses of sediment PCBs have not been published for any of these locations, but the data that have been published suggest that PCB dechlorination has occurred.

6.5.1. *Escambia Bay, Hudson Estuary and River, and Hoosic River*

Brown and colleagues (1987a) reported that capillary chromatograms of sediment PCBs from the Hoosic River (North Adams, MA) showed enhanced levels of the unusual 24-3-CB and 25-3-CB and in 1990 that chromatograms of sediment samples from Escambia Bay (near the mouth of the Pensacola River, FL) and from the tidal Hudson River near Troy, Mechanicville, Albany, and Kingston, NY, showed well-developed pattern H/H′ alterations. In 1990 they also reported less extensively developed pattern H/H′ alterations in chromatograms of sediment samples from the Hudson Estuary near Catskill and Poughkeepsie. No chromatograms, congener distributions, or further details have been published for these locations.

6.5.2. Waukegan Harbor

Brown et al. (1987a) interpreted previously published (Stalling, 1982) capillary GC patterns and congener distributions for five sediment samples from Waukegan Harbor, IL, a site contaminated with Aroclor 1248. They reported that the patterns showed various degrees of loss of most of the original tri-, tetra-, and penta-chlorobiphenyls, including losses of 47%–98.5% for 34-34-CB and 15%–89% for 234-34-CB and 245-34-CB. Corresponding increases were noted for lower chlorinated biphenyls, especially 2-2-CB, 2-3-CB, 2-4-CB, 4-4-CB, 24-2-CB, 26-4-CB, 24-3-CB, and 24-24-CB. This pattern, which they called W, is similar to Hudson River alteration pattern B.

6.5.3. Sheboygan River and Harbor

The Sheboygan River is a Wisconsin tributary of Lake Michigan. It is polluted with waste hydraulic fluids containing Aroclors 1248 and 1254 (David, 1990) from its mouth to about 14 miles upstream (Sonzogni et al., 1991). Brown and colleagues (1987a, 1990) reported alterations "similar, but not identical, to pattern H" in the sediments of Sheboygan Harbor and the Sheboygan River, but did not publish details. Sonzogni et al. (1991) used multidimensional capillary chromatography to determine total PCB concentrations and the weight percents of four coplanar PCBs and four mono-*ortho* analogs of the coplanar PCBs in Sheboygan River sediments. Five of these, 245-345-CB, 2345-4-CB, 345-34-CB, 345-345-CB, and 345-4-CB, were either undetectable or present at less than 0.1 weight percent in all eight sediment samples. The remaining three, 34-34-CB, 234-34-CB, and 245-34-CB, were 10-fold lower than in the original contaminant. Sonzogni and colleagues attributed these decreases to anaerobic dechlorination.

6.5.4. Lake Ketelmeer

Lake Ketelmeer is a sedimentation area of the Rhine River located in the central part of the Netherlands. The Rhine, a major European river, drains a large industrialized area that includes many countries in Europe. Hence the sediments of Lake Ketelmeer are contaminated with a wide variety of pollutants, including heavy metals, PCBs, polychlorinated dioxins, polychlorinated dibenzofurans, and polyaromatic hydrocarbons. Through the use of radionuclide time tracers and area-specific time markers Beurskens et al. (1993) determined present day and historical concentrations of six coplanar and mono-*ortho* PCB congeners in five sediment cores from this lake. The data showed 70%–88% decreases in the concentrations of 234-34-CB, 2345-34-CB, 345-34-CB, and 345-345-CB deposited in 1970 as compared with the levels of the same congeners in sediment samples from the same location that had been collected and dried in 1972 and stored in the dark. Buerskens and colleagues attributed these decreases to microbial dechlorination. Decreases of 34-34-CB (52%) and 245-34-CB (30%) were also noted, but were not significant at the 0.05 confidence level.

6.5.5. Lake Shinji

The sediments of Lake Shinji, Japan, are contaminated with low levels (0.03–1 ppm) of Kanechlor 500 (Sugiura, 1992), a commercial PCB mixture similar to Aroclor 1254. A nearby capacitor assembly plant that used Kanechlor 500 from 1966 to 1972 is assumed to be the source of the pollution. Capillary chromatograms of the sediment PCBs and Kanechlor 500 with only some of the peaks identified were published without comment (Fig. 1 in Sugiura, 1992). The remaining peaks were identified by comparison with Aroclor 1254 and reveal clear evidence of dechlorination (John F. Brown, Jr., personal communication).

According to our own analysis of these chromatograms (aided by John F. Brown, Jr.), several penta- and hexachlorobiphenyls containing 234-, 236-, 2345-, and 2346-chlorophenyl groups were substantially decreased, and several tetra- and penta-chlorobiphenyls containing 24-, 26-, and 245-groups were increased. The most obvious changes were decreases in 25-34-CB, 234-23-CB, 234-25-CB, 234-34-CB, 236-34-CB, 2345-25-CB, and 2346-34-CB and increases in 24-24-CB, 24-25-CB, 245-24-CB, and possibly 25-26-CB. These data suggest incipient *meta* dechlorination of 34-, 234-, 236-, 2345-, and 2346-chlorophenyl groups, but not 245-chlorophenyl groups. The data also suggest that dechlorination can occur at very low PCB concentrations.

6.5.6. Otonabee River–Rice Lake

Ferguson and Metcalf (1989) analyzed the distribution of 19 PCB congeners in sediments at five locations along the Otonabee River–Rice Lake system in Peterborough, Canada. The contamination was attributed to "accidental discharges from various industries in Peterborough" that are "located within the drainage area or have storm sewer access to Little Lake." The specific Aroclor contaminant was not identified. Ferguson and Metcalf concluded that "microbial degradation is not a major fate process in this system."

Ferguson and Metcalf did not publish chromatograms, and the 19 peaks reported in histograms did not include major congeners that are most susceptible to dechlorination (24-34-CB, 25-34-CB, 234-4-CB, 23-34-CB, and 245-4-CB) or the most likely dechlorination products (2-3-CB, 24-3-CB, and 25-3-CB). However, the relative proportions of 25-25-CB, 24-25-CB, and 24-24-CB in these sediments were reported (identified as IUPAC numbers 52, 49, and 47, respectively, in Fig. 3 of Ferguson and Metcalf, 1989). In all five sediments these three congeners accounted for 30%–40% of the 19 congeners reported. In the Aroclors the relative concentrations of these tetrachlorobiphenyls are 25-25-CB > 24-25-CB >> 24-24-CB (Schulz et al., 1989). But in the data presented for all locations along the Otonabee River–Rice Lake system, 24-25-CB was the highest of these three congeners and was at least 50% higher than 25-25-CB, and 24-24-CB was equal to or higher than 25-25-CB. This distribution cannot be explained on the basis of physical or chemical processes. Because 24-25-CB and 24-24-CB are common dechlorination products of penta- and hexachlorobiphenyls that are abundant in the Aroclors (Figs. 4.6–4.8), the unusual distribution of these three congeners is most likely the result of *in situ* dechlorination. A complete congener-specific analysis of the sediment PCBs as described earlier would be necessary to confirm dechlorination.

6.6. Conclusions

Conclusive identification of dechlorination of PCBs requires a complete quantitative congener-specific analysis and the correlation of losses in specific congeners with increases in daughter products. This type of analysis has been done for sediments of the upper Hudson River, the Acushnet Estuary, Woods Pond, and Silver Lake and has provided compelling proof of *in situ* dechlorination at each of these locations. Moreover, laboratory studies have confirmed the existence of PCB dechlorinating microorganisms at each site and have revealed that environmental dechlorination patterns are often composites resulting from several distinct dechlorination processes.

Less complete evidence of dechlorination has been summarized for eight additional locations. In some cases, such as Waukegan Harbon and Lake Shinji, chromatograms showing specific congener losses and increases have been published. In other cases, the interpretation of the data has been given, but the data have not been published. In yet other cases, such as Lake Ketelmeer and the Otonabee River–Rice Lake system, only data for selected congeners have been reported. Such data are the least conclusive, especially if no evidence is provided for the appearance of dechlorination products, because they might easily be misinterpreted as evidence either for or against dechlorination, depending on which congeners are selected.

Despite the complexities of a conclusive analysis, we hope that we have not discouraged investigators from embarking on studies of environmental dechlorination of commercial PCBs, but have instead provided them with knowledge and tools that will aid them in their analyses. The eight locations for which only limited evidence of dechlorination is available are logical places to begin more detailed studies. In addition, there are many locations that have been totally ignored to date. The demonstration of dechlorination in New Bedford Harbor and the reports of dechlorination in Escambia Bay and the Hudson Estuary indicate that marine sites are reasonable candidates for study. Nearly all of the sites previously studied are in cold northern climates. It is likely that conditions in warmer climates would be more favorable for dechlorination, but contaminated sites in such areas have not yet been investigated. It will be interesting to see whether further investigations reveal a commonality of dechlorination processes at all sites.

7. DECHLORINATION OF PURE CONGENERS

Experiments involving the addition of specific PCB congeners to anaerobic sediment slurries have been undertaken primarily to determine 1) the routes by which congeners are dechlorinated, 2) the effects of chlorine substitution patterns on reactivity,

3) the dechlorination capabilities of microbial populations, and 4) the terminal products of PCB dechlorination. Sediments from the Hudson River, Silver Lake, and Woods Pond have been used in most of these studies. The Aroclor studies described earlier in this review have revealed that each of these sediments exhibits several distinct dechlorination processes that are presumably mediated by different microbial populations and that may act alone or in combination. We must bear this in mind in interpreting the results of the experiments with pure congeners.

7.1. Pathways and Microbial Succession

Abramowicz and colleagues (1990, 1993) studied the stepwise dechlorination of 234-34-CB added in a high concentration (500 μg/ml, ~2,875 μg/g sediment dry weight) to a slurry of Hudson River sediment. Over a 12 week period they observed stepwise dechlorination of this pentachlorobiphenyl according to the following sequence: 234-34-CB → 24-34-CB → 24-3-CB → 2-3-CB → 2-CB. According to the congener selectivities we have described for dechlorination processes in the Hudson River (Figs. 4.3–4.6, 4.8), it appears that the dechlorination of this congener was mediated by at least two, and possibly three or four, different dechlorination processes. Removal of the doubly flanked *meta* chlorine of the 234-chlorophenyl group is consistent with processes H, H′, Q, M, and N; removal of the *para* chlorine from the 34-chlorophenyl group is consistent with processes H, H′, and Q; and removal of the *para* chlorine from 24-3-CB is consistent with process Q. Hence the first three steps could all be mediated by process Q, but the final step, removal of the isolated *meta* chlorine on the 3-chlorophenyl ring, seems to be a unique activity of process M. These findings are in agreement with those of laboratory studies with Aroclors and analyses of environmental samples that have demonstrated that in Hudson River sediment several dechlorination processes can act together (see Sections 5.2–6.6).

Rhee et al. (1993d) studied the dechlorination of 234-245-CB, 245-245-CB, and 23456-CB by microorganisms from Hudson River sediment transferred to autoclaved sediments from Owasco Lake. The dechlorination products they observed (described in detail in the section below on mass balance) also indicated that both processes M and Q were active and suggested that the same congener was dechlorinated by several different routes. Because these cultures were sampled infrequently, it is not clear whether dechlorination by various routes occurred simultaneously or sequentially. In the case of 245-245-CB the dechlorination removed only two chlorines and the major products were 25-25-CB, 24-25-CB, and 24-24-CB, which are the expected products of the combined activity of dechlorination processes H or H′ and M or N; or, alternatively, Q and M or N. Very small amounts of 2-2-CB were also detected, demonstrating again that complete removal of *meta* and *para* chlorines was possible. These studies used a much lower proportion of sediment than those of most other investigators: about 2.5% by volume versus 35%–40% for most others. This may explain, in part, why the dechlorination observed was somewhat more limited than that observed by others.

Van Dort and Bedard (1991) studied the dechlorination of 2356-CB (100 μg/ml,

~ 700 μg/g sediment dry weight) in a slurry of Woods Pond sediment. After a 21 week lag the 2356-CB was dechlorinated in two separate stages. In the first stage, between 21 and 28 weeks of incubation, 92% of the parent congener was dechlorinated by *meta* dechlorination to 236-CB (79%) and by sequential *ortho* and then *meta* dechlorination to 25-CB (21%). After this, no further dechlorination of 2356-CB was observed, but in the second stage, from 29 weeks to 37 weeks of incubation, 236-CB was dechlorinated to 26-CB. Van Dort and Bedard attributed these two stages to different dechlorinating populations and proposed that a microbial succession had occurred. They supported this proposal by demonstrating that inocula collected at 37 weeks and transferred to autoclaved sediment could dechlorinate 236-CB, but not 2356-CB or 235-CB.

The study by Van Dort and Bedard also provided the first conclusive demonstration of microbially mediated *ortho* dechlorination of PCBs and showed that microorganisms capable of this activity exist in Woods Pond sediment. Yet *ortho* dechlorination does not appear to play a major role in the dechlorination of the Aroclor 1260 in Woods Pond sediment. Instead, *meta* and *para* dechlorination mediated by processes N and P, respectively, appear to be the predominant dechlorination activities occurring *in situ* (see Sections 6.3 and 9.3).

7.2. Reactivity Preference

Williams (1994) compared the dechlorination of 234-CB, 235-CB, 236-CB, 245-CB, 246-CB, and 345-CB (90 μg/ml) in slurries of sediment from the Hudson River, Silver Lake, and Woods Pond. These congeners represent all possible trichlorobiphenyls with all chlorines on a single ring and were intentionally selected to avoid any influence of the non-target ring on the dechlorination process. For five of the congeners and in all three sediments the chlorine located between the other two chlorines was removed first, regardless of its position (*meta* or *para*) relative to the opposite phenyl ring. Hence 234-CB was dechlorinated to 24-CB, 235-CB to 25-CB, 236-CB to 26-CB, 245-CB to 25-CB, and 345-CB to 35-CB. No known single dechlorination process in any of these sediments can account for all of these transformations. For example, processes P and N are present in Woods Pond sediment, but N removes only *meta* chlorines and does not remove the *para* chlorine from 245-chlorophenyl rings (Fig. 4.8). Likewise, P removes only *para* chlorines (Fig. 4.7) and cannot account for the *meta* dechlorination of any of these congeners. This suggests that the chlorine configuration of the congener added to the culture can determine which microbial population will become active. This might be the result of a specific induction or simply a matter of competition between microbial populations that can dechlorinate the same substrate.

The fate of the sixth congener, 246-CB, was quite different for the three sediments. Sediment microorganisms from the Hudson River removed only the *para* chlorine of 246-CB (probably by process Q) after a 3–4 week lag. In contrast, sediment microorganisms from Woods Pond and Silver Lake removed both *ortho* chlorines from 246-CB to generate 4-CB, but only after extremely long lag times (24 to >52 weeks, respectively). This confirms the *ortho* dechlorination activity of

Woods Pond microorganisms reported by Van Dort and Bedard (1991) and demonstrates that microorganisms with a similar capability also exist in Silver Lake. It also demonstrates that a congener that is recalcitrant to dechlorination by the predominant dechlorinating populations may elicit an unexpected microbial activity.

The second and third *meta* and *para* chlorines, if present, were removed only in the Hudson River cultures. This is consistent with the results of Aroclor studies using Hudson River and Woods Pond sediments, but it was surprising that the Silver Lake cultures did not remove all of the *meta* chlorines because dechlorination process M has been demonstrated in Silver Lake sediments (see Sections 5.1.1 and 6.4).

The study by Williams also demonstrated the stoichiometric conversion of 345-CB to biphenyl in Hudson River sediment slurries. This suggests that the "dioxin-like" coplanar PCBs such as 34-34-CB, 345-34-CB, and 345-345-CB would also be dechlorinated to biphenyl in this sediment. Indeed, Rhee et al. (1993c) have reported the dechlorination of 34-34-CB to biphenyl by microorganisms from Hudson River sediment.

Williams noted that the acclimation time preceding dechlorination in Hudson River sediments varied only slightly (18–28 days) for all six trichlorobiphenyls. This was not the case for Woods Pond and Silver Lake sediments. In the latter two sediments the acclimation time was only 21 days for congeners with adjacent *meta* and *para* chlorines (234-CB, 245-CB, and 345-CB), but was longer for 236-CB (35 and 56 days, respectively), and much longer for 235-CB and 246-CB (14 weeks to more than 1 year). Williams ranked the relative reactivities of various chlorophenyl groups based on the acclimation times he observed and on dechlorination patterns reported by others, and he concluded, as we have, that the reactivity of a *meta* or *para* chlorine is partly determined by the number and position of the chlorines on the same ring. Specifically, doubly flanked chlorines, such as the center chlorine in 234- and 345-chlorophenyl rings, are more reactive than singly flanked chlorines such as the *meta* and *para* chlorines in 245- and 34-chlorophenyl rings, which are in turn more reactive than unflanked chlorines, such as the *para* chlorine of 246- and 24-chlorophenyl rings and the *meta* chlorine of 25-chlorophenyl rings. Isolated chlorines on monochlorophenyl rings are least reactive. However, whether a particular type of dechlorination actually occurs in a sediment will ultimately be determined by the capabilities of the dechlorinating microorganisms that exist in the sediment.

Although the studies described above demonstrated *ortho* dechlorination of 2356-CB, 246-CB, and 24-CB, *ortho* dechlorination of 26-CB has not been demonstrated in any of these sediments despite prolonged incubation. Furthermore, because the *ortho*-substituted congeners 2-CB, 2-2-CB, and 26-CB were the dominant dechlorination products of Aroclors 1242, 1248, and 1254 produced by Hudson River microorganisms (Quensen et al., 1990), Tiedje and colleagues (1992) investigated the fate of these three congeners and biphenyl in anaerobic cultures of Hudson River sediment microorganisms. No dechlorination or degradation of any of these compounds was observed even after more than 1 year of incubation. These studies suggest that 2-CB, 2-2-CB, 26-CB, and, to a lesser extent, biphenyl are terminal dechlorination products in Hudson River sediment.

Rhee et al. (1993d) addressed the issue of dechlorination specificity in a study involving 234-CB and 245-CB. In a series of six transfers at 2 month intervals, Hudson River sediment microorganisms consistently removed only the *meta* chlorine from 234-CB (300 μg/g dry sediment weight) to form 24-CB. But, when subsequently transferred to slurries of autoclaved sediment spiked with 245-CB, both 24-CB and 25-CB were produced in a ratio of 33:1. Thus, a small amount of *para* dechlorination occurred using an inoculum from a culture that had removed only *meta* chlorines from 234-CB. Over the next three transfers the ratio of products from 245-CB gradually shifted, and by the fifth transfer the sole product was 25-CB, a product of *para* dechlorination. But when an inoculum from this *para*-dechlorinating culture was transferred to autoclaved sediment slurries spiked with 234-CB, the only product produced (96% conversion) was 24-CB, a product of *meta* dechlorination.

Rhee and colleagues commented that the product shift of 245-CB dechlorination from 24-CB to 25-CB suggested that the repeated transfers might have enriched populations that could remove the *para* chlorine from 245-CB, but not from 234-CB. We concur with this interpretation and offer the following explanation of these results. As we discussed earlier, dechlorination processes M, N, H, H′, and probably Q all remove the *meta* chlorine from a 234-chlorophenyl group. Therefore, any combination of these microorganisms could have been enriched by the first six transfers on 234-CB. However, of the dechlorination processes known to be indigenous to the Hudson River, only processes M and N remove the *meta* chlorine from a 245-chlorophenyl group, while processes Q, H, and H′ remove the *para* chlorine. By the process of deduction, it appears that the population enriched on 234-CB must have been predominantly M or N, which remove only *meta* chlorines, and that during the transfers on 245-CB the population shifted to predominantly Q, H, or H′. Each of the latter processes would remove the *para* chlorine from 245-CB, but the *meta* chlorine from 234-CB, hence explaining Rhee's observations.

It is clear from this example and from Figures 4.3–4.8 that the dechlorination of individual PCB congeners often does not provide sufficient information to distinguish dechlorination processes. Therefore, until pure PCB-dechlorinating cultures are available, conclusions about which dechlorination process is active should be based on Aroclor dechlorination studies, which do permit distinction of the various dechlorination processes.

7.3. Mass Balance and Quantitation

All Aroclor dechlorination studies have shown decreases in highly chlorinated PCBs and corresponding increases in less chlorinated congeners. The apparent terminal products have varied depending on the Aroclor used and on the pattern of dechlorination (Figs. 4.3–4.8), but in all cases the increases of less chlorinated congeners appeared to correspond on a molar basis to the decreases of more chlorinated congeners. Alder et al. (1993) measured PCB concentration in dechlorination experiments using sediments from New Bedford Harbor, Silver Lake, and the Hudson River and determined that the total molar PCB concentration in each sediment remained constant for the entire 17 months of incubation, thus confirming that dechlorination was the only significant transformation that occurred. Rhee and colleagues (1993b,e) reported the same for their Aroclor dechlorination experiments using microorganisms from the Hudson River and stated: "A mass balance of the transformation showed that the total molar concentration of PCBs remained unchanged (P > 0.05) during the seven months. Therefore, all transformation during this period involved dechlorination without any loss of PCB molecules."

In contrast to the mass balance they observed in their Aroclor studies, Rhee et al. (1993c,d) reported that total molar PCB concentrations decreased 35%–90% in their studies of dechlorination of 23456-CB, 234-245-CB, 34-34-CB, and 245-CB by microorganisms from the same sediment. They explained their failure to obtain mass balance in these experiments with single congeners by proposing that "anaerobic PCB biotransformation may include mechanisms other than dechlorination and that the mechanisms are congener dependent" (1993c). We find this explanation unlikely and propose instead that their inability to obtain mass balance with these PCB congeners may have been due to analytical problems and sensitivity issues.

Rhee et al. (1993c) could account for the fate of only 50% of 23456-CB after 15 months of incubation. During the first 3 months about two-thirds of the 23456-CB was dechlorinated to 2356-CB and 246-CB (ratio 2:1) with no decrease in total PCB concentration (Figs. 3 and 4 in Rhee et al., 1993c). In the next 3 months another 30% of the 23456-CB was dechlorinated, and a corresponding increase of 2356-CB was observed. In the same time period the concentration of 246-CB decreased by 75%, but the expected product, 26-CB, was not detected. As a consequence the total concentration of PCBs reported decreased. When next measured at 12 months, the 2356-CB and the remaining 23456-CB and 246-CB had declined further, but the expected products, 236-CB and 26-CB, were not observed; hence total recovery was only 63% of expected. At 15 months, when the experiment was ended, a significant amount of 26-CB was detected for the first time, but not in sufficient amounts to account for the decreases in the higher chlorinated congeners, and the final concentration of PCBs was only 50% of the starting value. The lack of correspondence in time between the disappearance of 2356-CB and 246-CB and the appearance of the expected terminal product, 26-CB, suggests to us that the 26-CB was not detected until high levels had accumulated and that the 50% decrease in total PCB concentration reported by Rhee and colleagues (1993c) was not real (see below).

In a second experiment Rhee et al. (1993c) reported that only 25% of the starting concentration of 234-245-CB could be accounted for after 15 months of incubation. About half of the 234-245-CB was dechlorinated to 245-24-CB, 24-24-CB, and 24-25-CB in the first 3 months, with a minimal change in the total PCB concentration. By 9 months, 90% of the 234-245-CB was gone, 245-24-CB and 24-24-CB showed net decreases, and the total molar PCB concentration had decreased to 78% of that expected. A single product, 24-2-CB, was detected, but not in sufficient amounts to account for the losses of the more chlorinated congeners.

The further decrease in total molar PCB concentration to 25% at 15 months occurred concomitantly with net losses of the intermediates 24-2-CB and 24-25-CB. The expected terminal product, 2-2-CB, was *not* observed, probably because it has a very low response factor. According to published values (Mullin et al., 1984), its response is only 4.5%–9% that of 24-2-CB, 24-25-CB, and 234-245-CB, and therefore 2-2-CB would be much harder to detect than its precursors.

Rhee and colleagues (1993c,d) also reported decreases in total PCB concentration for the conversion of 245-CB to 25-CB (a 35% decrease) and of 34-34-CB to biphenyl (>90% decrease). No time courses were presented for these experiments. In the case of 34-34-CB, the only intermediate detected was 3-3-CB. The subsequent dechlorination product, 3-CB, has an extremely low response factor and would be difficult to detect by ECD. The biphenyl concentration was determined by FID. When ^{14}C-labelled 34-34-CB was used, all ^{14}C activity was recovered in the hexane extract, indicating that no transformation to polar products occurred.

In contrast to the above results, mass balance was obtained for the dechlorination of 234-CB to 24-CB; of 245-CB to 24-CB; and of 245-245-CB to 24-24-CB, 24-25-CB, and 25-25-CB (Rhee et al., 1993c,d). In the latter experiment the tetrachlorobiphenyl products were not further dechlorinated. Hence they accounted for essentially all of the 245-245-CB, even after 20 months, and the detection of products with low response factors was not an issue.

We discount Rhee and colleagues' proposal (1993c) that "anaerobic PCB biotransformation may include mechanisms other than dechlorination and that the mechanisms are congener dependent" in part because no products other than dechlorination products were demonstrated and because aromatic compounds that lack aliphatic or polar substituents are extremely difficult to metabolize anaerobically (Schink, 1988). (For example, many studies of groundwater contaminated with benzene, toluene, and xylene have demonstrated anaerobic metabolism of toluene and xylene, but not benzene.) But the main reason we question such an explanation of their results is that the information provided about how PCB concentrations were determined was inadequate (see below) to support their conclusions. In particular, we question whether their analytical conditions were sensitive enough to detect and accurately measure small amounts of daughter congeners with low response factors.

Rhee and colleagues conducted appropriate controls to confirm full recovery of products in their experiments and used authentic pure congener standards to identify products by matching GC retention times. But the only information given about how congener concentrations were determined was the statement that "all PCB concentrations were determined in the linear range of the ECD response" (Rhee et al., 1993c) and a reference to calibration using a single concentration of a mixture of Aroclors 1221, 1016, 1254, and 1260 (Rhee et al., 1993b). In our experience Aroclor standards are not well suited for quantitative analysis of pure congeners and are no longer necessary because all PCB congeners are now commercially available.

The detection levels for individual PCB congeners will differ because their ECD response factors vary over several orders of magnitude (Erickson, 1986). Hence it is possible to be in the linear response range of the ECD for certain congeners and below the level of detection for other congeners that are present at or near the same molar concentrations. The problem is compounded when the congeners with lower response factors (mono-, di-, and trichlorobiphenyls) are present in lower molar quantities than the parent congener as is the case when a higher chlorinated congener is dechlorinated to multiple daughter products. This is far less of an issue in experiments with Aroclors, because then several parent congeners are dechlorinated to the same daughter product (e.g., 2-CB, 2-2-CB). It is also unreliable to use published response factors or relative response factors for individual congeners, because these will vary for each instrument depending on the detector design, the manufacturer, the individual detector, the chromatography conditions (flow, temperature, volume, makeup gas, and so forth), and even the daily instrumental response (Rust, 1984). In fact, the relative response factors published by Mullin et al. (1981, 1984) for the same set of PCB isomers on two different occasions varied considerably both in value and in trend.

Therefore, for reliable quantitation of dechlorination in single congener experiments, it is necessary to use an authentic standard for each parent compound and each daughter product. When using ECD it is preferable to use a multipoint calibration over the entire concentration range being examined. This was done by Van Dort and Bedard (1991), who were able to obtain reliable quantitation for 2356-CB and its daughter products (236-CB, 25-CB, and 26-CB) by use of an 18 point third-order calibration curve. Alternatively, samples can be analyzed in the linear range of the ECD by using appropriate dilutions and concentrations to bring the parent congener and each daughter congener on scale in the linear range. Analysis by GC/MS is simpler because the response factors for parent and daughter congeners do not differ as greatly as for ECD and because detector response is linear over a broader concentration range. Williams (1994) was able to obtain reliable quantitation of 345-CB and its daughter products, including biphenyl, by using GC/MS and a 3 point linear calibration curve.

In summary, the decrease in total molar PCB concentration reported by Rhee and colleagues should be considered questionable until 1) their findings are confirmed with appropriate calibration standards and quantitation methods as described above and 2) products other than PCBs are demonstrated. In any event, an alternative anaerobic biotransformation mechanism such as that proposed by Rhee and colleagues would appear irrelevant to the dechlorination of Aroclors because no losses in total PCB concentration were observed in Aroclor studies exhibiting five different dechlorination processes in three different sediments (Alder et al., 1993; Rhee et al., 1993b,e).

7.4. Unusual Dechlorination Activities

Mavoungou and colleagues (1991) screened 20 environmental samples for their ability to dechlorinate 4-4-CB under anaerobic conditions. Anaerobic medium (30 ml) containing 500 μg/ml 4-4-CB was inoculated with 10 ml or 10 g of water, soil, aquatic sediments, or sludges, and incubated at 29°C for 60 days. Three of the samples tested, each an activated sludge from a different sewage treatment plant, dechlorinated 5%–10% of the 4-4-CB to 4-CB and biphenyl. Although the conversion was low, this result is exciting because it provides evidence that sewage treatment plants may be another source of PCB-dechlorinating organisms. It is also an unexpected result because activated sludge is an aerobic environment. None of the inocula from sediment, soil, or water dechlorinated 4-4-CB. However, of the six dechlorination processes that have been identified, only one, Q, can remove the isolated chlorine of a 4-chlorophenyl group. This suggests that the results might have been different if a congener containing a chlorophenyl group more susceptible to dechlorination had been used.

Montgomery and Vogel (1992) reported dechlorination of 2356-CB (20 μg/ml) by an anaerobic phototrophic enrichment culture. During 14 months of incubation, 58% of the 2356-CB was dechlorinated to di- and trichlorobiphenyls, and 68% of the dechlorination was from the *ortho* position. Photochemical dechlorination of PCBs also preferentially targets *ortho* chlorines (Ruzo et al., 1974; Bunce et al., 1978), but Montgomery and Vogel discounted photochemical dechlorination because their experiments were conducted in glass vessels, which would eliminate most ultraviolet light, and because they had observed only trace amounts of dechlorination in abiotic controls exposed to higher levels of fluorescent lighting than used in the experiment with the phototrophic enrichment culture. It is unfortunate that they did not conduct a killed-cell control that was otherwise identical to the live culture, since this would have made their conclusions more definitive. In any event, the experimental results are intriguing and merit further study even though a light-dependent process is unlikely to impact the fate of PCBs associated with sediment.

7.5. Conclusions

The studies described above demonstrate both the usefulness and the limitations of dechlorination studies using pure congeners. Pure congeners have been useful for screening environmental samples for dechlorination activity and for identifying unusual PCB dechlorination activities such as *ortho* dechlorination. In some instances studies with pure congeners have permitted observation of microbial succession, but success in this area depends on careful choice of congeners, careful quantitation, and frequent sampling to observe changes in specificity and to identify all products. Quantitation requires the use of standards that span the entire concentration range examined for the parent and all daughter congeners, and it can still be difficult when the detector response for some products is much lower than for the parent congener. Pure congener studies have limited value for determining the effect of chlorine substitution pattern on dechlorination when more than one dechlorination activity is present, as is generally the case in sediments, because there is usually insufficient information to determine which process(es) is responsible for the observed dechlorination. In addition, the congener used may select for a certain (even rare) dechlorination process, or there may be a succession of processes during the experiment. The congener distribution profiles generated by Aroclor studies provide a better means of discriminating between dechlorination processes that share some congener specificity.

8. FACTORS AFFECTING RATE AND EXTENT OF PCB DECHLORINATION

8.1. General Ecological Parameters

Environmental factors that are important to microbial ecology in general may be expected to influence PCB dechlorination. These factors include oxygen tension, redox level, temperature, pH, salinity, available carbon, and trace metals (Atlas and Bartha, 1981). No detailed studies have been conducted to determine how these factors affect PCB dechlorination processes, so very little is known.

PCB dechlorination is believed to occur only at low redox levels in the absence of oxygen. In the environment, more extensively dechlorinated PCB mixtures are found in deeper, potentially anaerobic sediments than in oxygenated surficial sediments (Brown et al., 1984, 1987a,b). In laboratory experiments, dechlorination is concomitant with methanogenesis, except in methanogen-free enrichments (May et al., 1992; Ye et al., 1992). This suggests that PCB dechlorination requires a low redox potential similar to that required for methanogenesis (Eh less than -400 mV, Oremland, 1988), although methanogenesis itself is not required for all dechlorination processes. On the other hand, Mavoungou et al. (1991) found that microorganisms from activated sludge, an aerobic environment, were capable of limited dechlorination of 4-4-CB when incubated anaerobically.

The effect of temperature is largely unknown. It is notable that extensive *in situ* dechlorination occurred in the upper Hudson River, where sediment temperatures are above 12°C for only 5–6 months of the year. In experiments with Hudson River microorganisms, dechlorination was observed at 12°C, but was roughly twice as fast at 25°C and did not occur at 37°C or above (Tiedje et al., 1993). In these experiments, dechlorination processes M and H occurred at 12°C, while process C dechlorination occurred at 25°C (Quensen and Schimel, unpublished data).

Complex interrelationships between temperature, pH, and salinity are known to affect the ecology of microorganisms (Atlas and Bartha, 1981). Most of the sites where PCB dechlorination has been investigated are freshwater sites, but dechlorination has been documented in New Bedford Harbor where the salinity was $26^0/_{00}$ (Alder et al., 1993) and is suspected in Escambia Bay, FL (Brown et al., 1987a,b). (The salinity of sea water is $35^0/_{00}$). Abramowicz et al. (1993) reported that sea salts completely inhibited dechlorination of Aroclor 1242 added to Hudson River sediment slurries. Like seawater (Weyl, 1970), this sea salt solution contained 28 mM sulfate (Abramowicz, personal communication), and this potential electron acceptor probably had a greater adverse impact on the dechlorination process than salinity (see below). Nothing is known about the effect of pH on the dechlorination process; all laboratory experiments have been conducted at a near-neutral pH.

PCB-dechlorinating microorganisms do not metabolize the biphenyl skeleton of PCBs; therefore they must require a source of carbon other than PCBs. This is presumably provided by the organic matter present in sediments. The carbon content of sediments used in slurries in laboratory experiments has usually not been characterized. The Lake Owasco sediments used by Rhee et al. (1993b) contained 9% organic carbon (by dry weight), while Morris et al. (1993) reported that the Hudson River sediments they used contained 7.66% organic carbon. Alder et al. (1993) found that additions of volatile fatty acid stimulated the onset of PCB dechlorination in slurries of Hudson River sediment amended with Aroclor 1242, but after 11 months the extent of dechlorination was the same in slurries that had not been so supplemented. The addition of fatty acids had no influence on the rate or extent of PCB dechlorination in sediment slurries with higher organic content.

Trace metals are required for microbial activity, but specific requirements of the dechlorination process are unknown. Omission of the trace metals normally included in the media resulted in a slight negative impact on the rate and extent of Aroclor 1242 dechlorination by Hudson River microorganisms (Abramowicz et al., 1993).

8.2. Electron Acceptors

It has been proposed that dechlorinating microorganisms might use PCBs as electron acceptors and that in sediments where concentrations of electron acceptors are limiting, the ability to dechlorinate PCBs might provide a selective growth advantage (Brown et al., 1987a,b; Quensen et al., 1988). Most investigations of PCB dechlorination have explored only methanogenic conditions. However, there have been a few attempts to examine how bromoethane sulfonic acid (BESA), an inhibitor of methanogenesis, or alternate electron acceptors, such as sulfate, nitrate, and iron, affect dechlorination.

Alder and colleagues (1993) investigated PCB dechlorination in slurries of sediments from the Hudson River, Silver Lake, and New Bedford Harbor incubated under both methanogenic and sulfidogenic conditions. As described above (see Sections 5.2.1, 5.1.6, and 5.1.3), they observed dechlorination processes C, N, and H', respectively, in these three sediments under methanogenic conditions. For the sulfidogenic experiments, wet sediment and medium containing sulfate (20 mM) were combined in a ratio of 35:65 (by volume). After 1 month, small amounts of methane (about 10% of that observed in methanogenic cultures) were detected in several of the sulfate-amended cultures that had been fed fatty acids, including some from the Hudson River, Silver Lake, and New Bedford Harbor, suggesting that localized depletion of the sulfate had occurred (Häggblom and Young, personal communication). Excess sulfate (30 mM) was then added to all cultures in order to maintain sulfidogenic conditions. After 3–4 months of incubation small amounts of methane were again detected in cultures that were fed fatty acids, but no methane was ever detected in any of the cultures that were not supplemented with fatty acids. No PCB dechlorination was observed in any of the latter cultures that remained truly sulfidogenic throughout the 17 month incubation.

May et al. (1992) also studied the effect of sulfate on dechlorination. This study compared the dechlorination of a mixture of Aroclor 1242, 236-CB, 246-CB, and 26-26-CB by Hudson River sediment microorganisms in methanogenic cultures and in cultures amended once with sulfate (20 mM). May and colleagues noted that the pattern of dechlorination in sulfate-amended cultures was different than in the methanogenic cultures and proposed that the sulfate treatment had enriched different dechlorinating organisms. Our analysis of their congener distribution profiles (their Fig. 1) indicates that the predominant dechlorination process in the methanogenic sediment slurries was M (as evidenced by depletion of many *meta*-substituted congeners such as 236-CB, 25-2-CB, and 25-25-CB and accumulation of *para*-substituted products such as 2-4-CB and 24-2-CB), but that processes Q (evidenced by partial depletion of 246-CB and 24-4-CB) and possibly H (evidenced by strong depletion of 25-34-CB, 24-34-CB, 234-4-CB, and 245-4-CB) also occurred to a lesser extent. In contrast, process H alone occurred in sediment slurries incubated with sulfate, but only after the sulfate was depleted. Process H in these cultures was evidenced by the dechlorination of 34-2-CB to 2-3-CB, 25-34-CB to 25-3-CB, 24-34-CB to 24-3-CB, and 234-4-CB and 245-4-CB to 24-4-CB/25-4-CB and by the lack of dechlorination of 236-CB and 246-CB.

Morris et al. (1992a) investigated the effects of sulfate (10 mM), nitrate (10 mM), ferric oxyhydroxide (50 mM), and BESA (10 mM) on the dechlorination of Aroclor 1242 added to sediment slurries. The source of microorganisms was an inoculum from a second-generation methanogenic enrichment of Hudson River sediment with pyruvate and Aroclor 1242. All cultures were incubated in an atmosphere of N_2 and amended with pyruvate. Some CO_2 was produced in all cultures from metabolism of the pyruvate. The most extensive dechlorination was seen in cultures amended with pyruvate alone (methanogenic) and those amended with nitrate. In these cultures 26% of the *meta* plus *para* chlorines were removed in 4 weeks and 39% in 12 weeks. The pattern of dechlorination, shown in Figure 4.9B, was predominantly M, but extensive losses of 245-4-CB, 24-34-CB, and 25-34-CB and elevated levels of 2-3-CB and 24-3-CB indicate that process H was also active. Dechlorination in cultures incubated with ferric oxyhydroxide was somewhat slower (34% loss of *meta* plus *para* chlorines in 12 weeks).

The dechlorination was much slower and far less extensive in cultures amended with sulfate. Only 11% of the *meta* plus *para* chlorines were removed in 12 weeks. Of greater significance is the change we noted in the pattern of dechlorination. The congener distribution (Fig. 4.9C, Table 4.5) reveals that both processes M and H were active, but in different proportions than in the methanogenic cultures. The concentrations of principal products of process M dechlorination, 2-CB, 2-2-CB, 2-4-CB, and 24-2-CB, were much lower than in the methanogenic cultures, and the concentrations of products of process H dechlorination, 2-3-CB, 24-3-CB, 25-3-CB, and 25-4-CB, were much higher.

It is unknown if either dechlorination process M or H occurred in the presence of sulfate in the experiments described above because Morris and colleagues did not monitor sulfate concentrations. Dechlorination was already evident in their experiments at 4 weeks (Fig. 4A in Morris et al., 1992a). May et al. (1992) reported that

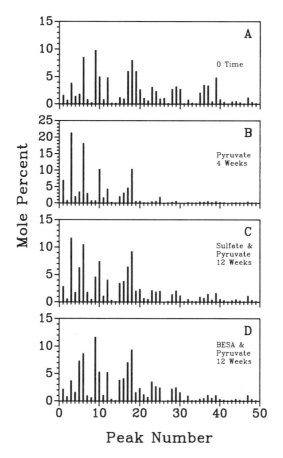

Figure 4.9. Congener profiles for Aroclor 1242 before (**A**) and after (**B–D**) dechlorination by Hudson River microorganisms enriched on pyruvate. Both dechlorination processes M and H are evident in B. Addition of sulfate or BESA resulted in partial (C) or complete (D) suppression of process M activity. See Table 4.5 for identification of congeners in peaks that represent at least 1 mole% of the total PCBs and Quensen et al. (1990) for complete congener identifications. (Adapted from Morris et al., 1992, with permission of the publisher.)

the sulfate (20 mM) in their cultures was depleted by 9 weeks. However, because Morris et al. inoculated with an actively dechlorinating enriched culture, included pyruvate in the medium, and used only 10 mM sulfate, the sulfate might have been depleted sooner in their cultures. Regardless of when sulfate depletion occurred, the data suggest that process M was partially inhibited by the sulfate addition and that process H was relatively unaffected.

In parallel cultures treated with BESA (10 mM), which inhibits methanogenesis, no dechlorination was observed in the first 4 weeks, but after this, slow dechlorination occurred, resulting in a 10% reduction of *meta* plus *para* chlorines in 12 weeks

TABLE 4.5. Congener Identification for Peaks
Representing More Than One Mole% of Total PCBs
in Samples Depicted in Figure 4.9

Peak No.	Congener ID[a]
1	<u>2</u>
3	<u>2-2</u> 26
4	24 25
5	<u>2-3</u>
6	<u>2-4</u> 23
7	26-2
9	<u>25-2</u> 4-4
10	<u>24-2</u>
11	26-3
12	<u>26-4</u> 23-2
15	<u>25-3</u>
16	<u>24-3</u>
17	<u>25-4</u>
18	<u>24-4</u> (246-2)
19	<u>34-2</u> 23-3 (25-26)
20	23-4 (24-26)
21	236-2
23	<u>25-25</u> (26-35)
24	24-25
25	24-24
26	245-2 246-4
28	<u>23-25</u>
29	23-24 236-3 34-4
30	26-34 234-2 236-4 25-35
35	245-4 (235-26)
36	<u>25-34</u> 345-2
37	<u>24-34</u> (236-25 245-26)
38	234-3 236-24
39	<u>23-34</u> <u>234-4</u>
47	34-34 236-34

[a]Congeners most important for identification of dechlorinatin processes M and H are underlined. Congeners present in only trace
amounts are in parentheses.

(Morris et al., 1992a). The investigators noted that 2-CB and 2-2-CB did not increase in these cultures, and our analysis of the congener distribution profile (Fig.
4.9d) indicates that the dechlorination pattern obtained was H, with no evidence of
M. This suggests that process M was specifically inhibited by BESA, while process
H was unaffected. This result is surprising because Ye et al. (1992) have shown
process M dechlorination of Aroclor 1242 in experiments using pasteurized inocula
of organisms from the same sediment. In the Ye et al. experiments, the microor-

ganisms exhibiting process M dechlorination appeared to be spore-formers, and the dechlorination occurred in the absence of methanogenesis. This could mean that the PCB dechlorinating population exhibiting process M in the Morris et al. experiments was a different population than that observed in the Ye et al. experiments, but there are other possible explanations for the observed inhibition by BESA: 1) Debromination of the BESA could have caused competitive inhibition. 2) The sulfonic acid group of BESA might have acted as a preferred electron acceptor (Oremland and Capone, 1988) for the microorganisms exhibiting process M. 3) The BESA might have changed the community structure in a way that was unfavorable for the dechlorinating population exhibiting process M. 4) The BESA might have been toxic to the microorganisms exhibiting process M. The BESA was not monitored to determine whether it was dehalogenated or desulfonated, so no firm conclusions can be drawn about the nature of the inhibition observed (see Section 10.2 for more information on the effects of BESA).

Over a 6 month period, Rhee et al. (1993a) compared the dechlorination of added Aroclor 1242 in methanogenic cultures with that observed in cultures amended once with sulfate (10 mM) or nitrate (16 mM). Their experiments were done with autoclaved sediment slurries inoculated with microorganisms eluted from Hudson River sediment. They observed no dechlorination in cultures incubated with nitrate. Rhee and colleagues did observe dechlorination in methanogenic and sulfate-amended cultures and noted that the total extent of dechlorination in sulfate-amended cultures was lower than that in methanogenic cultures, 7% versus 12% of total chlorines removed. They further stated that "the congeners involved in the dechlorination under sulfidogenic conditions were not qualitatively different from those under methanogenic conditions." But our analysis of their congener distributions (their Fig. 1) reveals that process H' dechlorination occurred in the methanogenic cultures, whereas process H occurred in cultures amended with sulfate. The degree of *para* dechlorination and the specific congener losses and increases due to *para* dechlorination of 34- and 245-chlorophenyl groups were identical in the methanogenic cultures and in those amended with sulfate. However, the *meta* dechlorination of 23- and possibly 236-chlorophenyl groups that was seen in the methanogenic cultures did not occur in the cultures amended with sulfate. Hence, decreases of 23-4-CB, 23-23-CB, and 23-25-CB and concomitant increases in 2-4-CB, 2-2-CB, and 25-2-CB that occurred under methanogenic conditions did not occur in sulfate-amended cultures. Dechlorination of 23-2-CB, which co-elutes with 26-4-CB, can also be inferred from the increase in 2-2-CB.

These findings are particularly important because they suggest that process H' may actually be a combination of process H and a still unidentified process that is restricted to *meta* dechlorination of 23- and possibly 236-chlorophenyl groups, and that the sulfate amendment completely inhibited the latter activity. The total dechlorination due to process H was the same in both methanogenic and sulfate-amended cultures, but it is not possible to determine if process H dechlorination occurred in the presence of sulfate because the sulfate concentration was not monitored. It is also not possible to determine if there was an increased lag time before dechlorination in the sulfate-amended cultures because they were only sampled once, after 6 months of incubation.

Rhee and colleagues (1993a) studied the effect of 10 mM BESA on the dechlorination of a single congener, 234-CB, and found that BESA had no effect on its dechlorination to 24-CB. Based on this observation, they concluded in their abstract that "methane production per se was not essential for dechlorination." The context of their statement implies that this conclusion extends to the dechlorination of Aroclors. We disagree with such a generalization. Although their finding shows that dechlorination of this single congener can occur in the absence of methanogenesis, it does nothing to elucidate the dechlorination of Aroclors because 1) this congener is not a key component of any Aroclor, 2) the data described in this review indicate that there are at least six different processes by which Aroclors are dechlorinated, and 3) according to the chlorophenyl reactivity preferences that we have described (Table 4.1), 234-CB could have been dechlorinated to 24-CB by any one of five of these processes: H, H′, M, N, or Q. The effect of a particular condition on the activity of one dechlorination process cannot necessarily be extrapolated to other dechlorination processes. This is clearly illustrated by the differential effects of sulfate and BESA on different dechlorination processes in the examples discussed above.

The various electron acceptors investigated, the dechlorination pattern observed, and the percent chlorine removal normalized to that seen in methanogenic conditions are summarized in Table 4.6. The study by May et al. (1992) showed that a single addition of sulfate inhibited process M dechlorination, and that of Morris et al. (1992a) indicated that sulfate partially inhibited process M. BESA also inhibited process M, even though pasteurization studies with the same sediment have shown that methanogenesis is not required for process M dechlorination. The similar effects of BESA and sulfate on process M dechlorination (Morris et al., 1992a) suggest that sulfate and the sulfonate group of BESA might act as preferred electron acceptors for population M and that population M might belong to the physiological group of sulfate reducers. The first aryl dehalogenator to be isolated, *Desulfomonile tiedjei*, is classed as a sulfate reducer, but when sulfoxy anions are not available it can use halobenzoates as electron acceptors (Mohn and Tiedje, 1992). Further evidence suggesting that process M is mediated by sulfate reducers is discussed below (see Section 10.2). However, since it is likely that dehalogenating microorganisms live in syntrophic communities, it is also possible that the inhibition of process M was due to an unfavorable change in the community structure.

In three studies process H dechlorination occurred in sulfate-amended cultures (Table 4.6), but this does not mean that it occurred in the presence of sulfate. No dechlorination occurred in any of the cultures that were amended with 30–40 mM sulfate and carefully monitored (Alder et al., 1993) to ensure that sulfidogenic conditions were maintained (including slurries of New Bedford Harbor sediment, even though process H dechlorination has occurred *in situ* in New Bedford Harbor). When added only once, sulfate may become depleted in micro-niches, allowing methanogenesis and sulfidogenesis to occur simultaneously in different micro-niches

TABLE 4.6. Effect of Electron Acceptors on Dechlorination Pattern and on Extent of Dechlorination

Source of Microorganisms	Carbon Dioxide		Sulfate		BESA		Nitrate		Ferric Oxyhydroxide		Reference
	Decl Pat'n	% Decl[a]	Decl Pat'n	% Decl	Decl Pat'n	% Decl	Decl Pat'n	% Decl	Decl Pat'n	% Decl	
Hudson River	H'	100	H	58			None	0			Rhee et al. (1993a)
Hudson River	M(+H)	100	H(+M)	30	H	28	M(+H)	100	M(+H?)[b]	87	Morris et al. (1992a)
Hudson River	M	100	H	NA[c]							May et al. (1992)
Hudson River	C	100	None	0							Alder et al. (1993)
New Bedford Harbor	H'	100	None	0							Alder et al. (1993)
Silver Lake	N	100	None	0							Alder et al. (1993)

[a]Percent dechlorination normalized to dechlorination observed under methanogenic conditions.
[b]Pattern not available.
[c]Data not available.

of the same sediment (Boyer, 1986), or it may be depleted throughout the sediment over time. Temperature, carbon and electron donor availability, bioturbation, and microbial activity probably all play a role in determining the rate of sulfate depletion. Indeed, when May et al. (1992) monitored sulfate concentrations during incubation, they determined that dechlorination did not begin until after sulfate (initially 20 mM) had been depleted. Because Rhee et al. (1993a) and Morris et al. (1992a) used a lower sulfate concentration (10 mM), it is possible that sulfate was also depleted in their cultures. In addition, because the sulfonic acid moiety of BESA may also serve as a terminal electron acceptor for sulfate reducers (Oremland and Capone, 1988), it may also have eventually depleted. Thus, in all cases process H dechlorination might have occurred only after sulfate or BESA depletion.

Process H dechlorination has also been observed in the tidal Hudson and in New Bedford Harbor, also an estuary. Although seawater and the mixed water in estuaries contain sulfate, the fact that PCB dechlorination occurs in marine or estuarine sediments does not prove that it occurs in a sulfate-reducing environment. The subsurface sediments may have much lower sulfate levels and may even be sulfate depleted (Capone and Kiene, 1988). Further investigations that carefully monitor sulfate concentrations, methane production, and PCB dechlorination over time are necessary to determine whether process H dechlorination can occur in the presence of sulfate or only after sulfate has been depleted.

8.3. PCB Concentration

The first laboratory report of PCB dechlorination indicated that the rate and extent of dechlorination were dependent on total PCB concentration. When Aroclor 1242 was added to anaerobic sediment slurries inoculated with upper Hudson River microorganisms, dechlorination was greatest in both rate and extent at 700 μg/g (sediment dry weight), less at 140 μg/g, and undetectable at 14 μg/g (Quensen et al., 1988). Environmental evidence for a concentration effect is less compelling because PCBs may be dechlorinated in an area of high concentration, then scoured and redeposited elsewhere at low concentrations. Still, 93% of sediment samples containing more than 100 μg/g PCBs were extensively dechlorinated compared with only 63% of samples containing 5–10 μg/g (Abramowicz et al., 1995). These observations suggest that PCB concentrations might have to be above some critical concentration for dechlorination to occur. The issue is particularly relevant to the environmental fate of PCBs, because low level PCB contamination of sediments is widespread.

The relationship between PCB concentration and the rate and extent of dechlorination can be more thoroughly evaluated in laboratory experiments where other factors can be controlled. The only appropriate way of comparing rates for samples with different PCB concentrations is in terms of the amount of Cl⁻ removed per unit time. Rate comparisons based on changes in the homolog distribution or average number of chlorines per biphenyl are valid only for comparing treatments at the same PCB concentration because changes in these expressions are themselves functions of the total PCB concentration. Because we have no measure directly related to the number of dechlorinating microorganisms, rates cannot be standardized. Perhaps the best we can do is to normalize rates in terms of grams of sediment dry weight. Because this is such a crude means of normalizing rates, it may not be appropriate to compare rates between different experiments.

Two laboratories have investigated the effect of PCB concentration on the rate of dechlorination (Rhee et al., 1993b; Abramowicz et al., 1993). In each case the dechlorination was a combination of processes M and H. To facilitate comparison we have recalculated the data so that concentrations and rates are expressed in the same terms. Rhee et al. expressed PCB concentrations in terms of $\mu g/g$ sediment dry weight, while Abramowicz et al. used $\mu g/ml$ of slurry. We have converted $\mu g/ml$ slurry to $\mu g/g$ sediment dry weight using a factor of 5.75 (Dan Abramowicz, personal communication). For both studies we have calculated maximum observed dechlorination rates in terms of nanogram-atoms Cl^- removed/g sediment dry weight/week. (A nanogram-atom, ng-at, is equivalent to a nanomole). For the study by Abramowicz et al. (1993) we used the data (shown in their Fig. 1) from the interval of 4–23 weeks to calculate maximal observed dechlorination rates; rates were nearly constant over this interval at all concentrations. To calculate maximal rates for the study by Rhee et al. (1993b) we used data (shown in their Fig. 3) from the interval of 3–4.5 months for PCB concentrations of 120 and 800 $\mu g/g$, and the interval of 0–3 months for PCB concentrations of 300 and 500 $\mu g/g$. The results of our calculations are plotted in Figure 4.10.

It should be noted that there are several discrepancies in the units in which the data are reported by Rhee et al. (1993b). We have used the data from their Figure 3 for our calculations. These data, plotted as Cl/biphenyl versus time, are consistent with their descriptions in their Materials and Methods, with published values for the average number of chlorines per biphenyl of Aroclor 1242, and with the changes in the congener distribution profile shown in their Figure 1. However, the text, p 1027, incorrectly describes the units for Cl/biphenyl as milligram-atoms of Cl per mole of biphenyl. (The correct unit in this case would be gram-atoms/mole or, more simply, the average number of chlorines per biphenyl). Furthermore, the calculations of the cumulative amounts of Cl^- removed per gram reported by Rhee et al. as 1,800 μg-at Cl/g sediment (1993b, Figure 5) and as 1.8 mol per gram (p 1027) are inconsistent with each other and with our calculation for the same data of a value of 2 μg-at Cl^- removed per gram of sediment.

The data from both studies show approximately linear relationships between maximal observed dechlorination rates and total PCB concentrations from 115 to 800 $\mu g/g$ (Fig. 4.10a). Rhee and colleagues (1993b) observed no dechlorination at concentrations of 1,000 and 1,500 $\mu g/g$ and on the basis of those results proposed that there is "an upper threshold concentration for the inhibition of dechlorination between 800 and 1,000 $\mu g/g$." However, Abramowicz et al. (1993) observed PCB dechlorination at all concentrations tested, including 8,600 $\mu g/g$, although the dependence of dechlorination rate on concentration began to decrease between 575 and 1,450 $\mu g/g$, and the rate of dechlorination leveled off at PCB concentrations above 4,000 $\mu g/g$ sediment (Fig. 4.10a). They explained their observations on the basis of partitioning of PCBs between the organic matter in the sediment and the

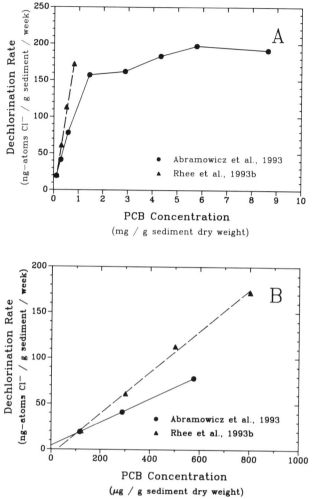

Figure 4.10.   Rate of dechlorination in cultures of Hudson River microorganisms at various total PCB concentrations. Based on data from Abramowicz et al. (1993) and Rhee et al. (1993b). Rhee et al. used Aroclor 1242 and Abramowicz et al. used a mixture of Aroclors 1242, 1254, and 1260. **A:** Data for all concentrations examined; **B:** The linear ranges observed. See text for details.

aqueous phase. The assumptions underlying their explanation are that 1) only aqueous phase PCBs were available to the dechlorinating microorganisms, 2) the rate of PCB dechlorination was directly proportional to the concentration of PCBs in the aqueous phase, and 3) the pore water became saturated with PCBs at 300 μg/ml of slurry, or 1,725 μg/g sediment. Below this critical total concentration the dechlorination rate would be directly proportional to the total concentration. Above this critical concentration both aqueous phase PCB concentration and dechlorination rate would be independent of total PCB concentration. Unfortunately, the available information is insufficient to permit conclusions. No measurements were reported for the aqueous concentration of PCBs. A reliable estimate of the total PCB concentration required to saturate the pore water is not possible, in part because the total organic content of the sediments was not given. Furthermore, the sorption and partitioning of Aroclors are poorly understood complex processes that are not well predicted by simple partitioning coefficients. Finally, in unstirred sediment slurries with added PCBs, such as those described, it is not clear when, if ever, equilibrium was achieved.

Because both of the studies reported above examined the effect of freshly added PCBs, their relevance to PCBs that have been associated with the sediment for decades is questionable. PCBs in contaminated sediments may exist as microdroplets, may be dissolved in other contaminants such as oil or grease, may be dissolved in the organic phase of the sediment, and/or may be strongly sorbed to or dissolved in the polymeric humic matrix of aged organic matter. In addition, the effects of PCB concentrations below 115 $\mu g/g$ were not examined. However, a detail of the linear range for rate versus concentration (Fig. 4.10b) shows that regression lines for both studies would pass near the origin if extended and suggests that extrapolation of the data to lower PCB concentrations might be defensible.

Indeed, despite all of the considerations mentioned, we find that such an extrapolation accurately predicts the rate of dechlorination of pre-existing PCBs in slurries of contaminated Hudson River sediment in a 24 week study (Abramowicz et al., 1993). The total PCB concentration in this sediment was 20 $\mu g/g$ sediment dry weight, but, because the PCBs had been dechlorinated *in situ,* half of the PCBs present (9.8 $\mu g/g$ by our calculations) were substituted only in the *ortho* position and hence could not be further dechlorinated. Taking this into account, the dechlorination rate was 79% of that expected based on the linear relationship between dechlorination rate and PCB concentration observed in the sediment slurries spiked with the Aroclor mixture. Our calculations thus support the conclusions of Abramowicz et al. (1993) even though these investigators mistakenly considered PCB concentrations of 20 $\mu g/g$ sediment and 20 $\mu g/ml$ slurry to be equivalent in their original calculations.

Relating dechlorination rates to total PCB concentration is an obvious simplification. As in the example above with PCB-contaminated sediments, rates should instead be related to the concentration of congeners that can be dechlorinated or, even more specifically, to the concentrations of congeners that induce the activity, if such exist. These concentrations would also depend on the congener specificity of the dechlorinating microorganisms, and biological thresholds (such as for enzyme induction) might exist. Thus it is not surprising that experiments with single congeners reveal different relationships between concentration and rate and sometimes the existence of thresholds or residual levels for dechlorination activity. Rhee et al. (1993d) noted a "residual" concentration of 245-CB that was relatively constant in successive transfers (30–40 nmol or 15–20 $\mu g/g$ sediment) and suggested that PCB concentrations need to be above this concentration for dechlorination to occur.

However, they did not observe this phenomenon with 234-245-CB or 23456-CB. It should also be noted that the dechlorination of many PCB congeners in Aroclor mixtures has been observed even when their individual concentrations were below 1 μg/g (Schulz et al., 1989; Quensen et al., 1990).

8.4. Bioavailability

The microbial degradation of poorly water soluble nonionic organic contaminants (NOCs) like PCBs may be slow because of their limited bioavailability to the degrading microorganisms. NOCs tend to sorb to soil and sediment particles. Their degradation rates may then be limited by the rate at which they desorb from the soil or sediment into the aqueous phase. Their desorption rates are often found to be biphasic; a "labile" portion of the NOC desorbs rapidly, but another "nonlabile" portion desorbs much more slowly (DiToro et al., 1982; DiToro and Horzempa, 1982; Karickoff and Morris, 1985). Frequently the nonlabile fraction appears to increase with the length of time the soil or sediment is exposed to the NOC (Karickoff, 1980).

One physical model used to explain the biphasic desorption of NOC is that the non-labile portion is NOC that has diffused into the interior of organo-mineral aggregates. Movement back out is retarded by sorption–desorption interactions along tortuous diffusion paths (Wu and Gschwend, 1986). Another physical model considers that the humic organic matter in sediments into which PCBs partition exists in both swollen and condensed polymeric phases. PCBs partition out of the swollen phase relatively rapidly, but partition out of the condensed phase three orders of magnitude more slowly (Carroll et al., 1994).

The concern from a bioremediation standpoint is that, for soils or sediments that have been contaminated with PCBs for years, the desorption rate for this nonlabile portion may be so slow that it precludes achieving target levels in a reasonable length of time. For example, the results of an *in situ* study of aerobic degradation of PCBs conducted in the Hudson River suggested that short-term biodegradation was limited by the desorption kinetics of PCBs from the sediments (Harkness et al., 1993). Desorption studies of the same sediment revealed that about half of the PCBs desorbed within 170 hours but that only 50% of the remaining PCBs desorbed over the next 6 months (Carroll et al., 1994). On the other hand, three laboratory studies of dechlorination of pre-existing PCBs in sediment suggest that bioavailability may not be a major issue for long-term processes such as PCB dechlorination, where the limiting step may be the rate of dechlorination (Abramowicz et al., 1993; Alder et al., 1993; Bedard et al., 1993; Bedard and Van Dort, in preparation). The entire issue of bioavailability and its impact on PCB dechlorination warrants additional study, preferably using environmentally contaminated sediment rather than freshly added PCBs.

8.5. Co-Contaminants

Many PCB-contaminated sediments also contain other contaminants that may affect PCB dechlorination. Especially common are heavy metals and oil. The tolerance of the dechlorination process for oil and some metals can be inferred from the fact that substantial *in situ* dechlorination has occurred at Silver Lake and New Bedford Harbor despite co-contamination of the sediments at these sites with high levels of oil and heavy metals. Silver Lake sediments contain as much as 6%–10% oil by weight (Quensen, unpublished data; Bedard, unpublished data), and oil is visible in New Bedford Harbor sediments as well. Oil could serve as a sorptive phase for PCBs, thereby significantly decreasing their aqueous concentrations (Boyd and Sun, 1989). This could slow dechlorination if the dechlorinating microorganisms only have access to PCBs in aqueous solution and if the rate of dechlorination depends on the aqueous concentration of PCBs (see Section 8.3). However, as individual PCB congeners in the aqueous phase are depleted by either biodegradation or dechlorination, additional PCB substrates should partition out of an oil phase relatively rapidly and completely because there should be no nonlabile fraction of PCBs (see Section 8.4) in an oil phase. Thus dechlorination could continue at a steady rate as long as the rate at which PCBs partition out of oil is at least as fast as the potential rate of dechlorination. It is also possible that some or all dechlorinating microorganisms have direct access to PCBs in an oil phase, in which case oil might have no adverse effect or could even aid dechlorination. Further research is needed to determine the actual impact of oils on PCB dechlorination. Measured concentrations (μg/g) of some heavy metals in New Bedford Harbor sediments were Cr, 620 $\pm$ 20; Cd, 15 $\pm$ 0.5; Pb, 550 $\pm$ 30; and Zn, 1,440 $\pm$ 40. In Silver Lake sediments heavy metal concentrations were Cr, 940 $\pm$ 30; Cd, 170 $\pm$ 5; Pb, 1,680 $\pm$ 290; and Zn, 5,490 $\pm$ 40 (Alder et al., 1993). Under the highly reduced conditions that seem to be required for dechlorination, most metals probably occur as insoluble sulfides and would appear to have little impact on the dechlorination process. The significance of other organics including PAHs and halogenated compounds is completely unknown but conceivably important as they could influence the microbial community present.

8.6. Culture Conditions

There are several aspects of laboratory methodology that probably have no environmental relevance, but may impact the results of laboratory experiments. These factors include the methods by which sediment slurries were prepared, Aroclors were added, and cultures were incubated (stationary vs. shaking), as well as the composition of head-space gases and aspects related to scale. While it is difficult to judge the importance of these factors because they have not been investigated in any systematic way, a few tentative inferences may be made about some of them. It is probably more significant that anaerobic PCB dechlorination has been observed under such diverse experimental conditions.

The sediment slurries used in dechlorination studies have been prepared either from wet PCB-contaminated sediments or from air-dried uncontaminated sediments. Wet sediments from a PCB-contaminated site have usually been slurried with mineral medium and either uninoculated (Abramowicz et al., 1993; Alder et al., 1993; Williams, 1994) or inoculated with an enrichment culture (Nies and

Vogel, 1990). Other studies have used air-dried noncontaminated sediments slurried with mineral medium, autoclaved, and subsequently inoculated with microorganisms eluted from sediments (Quensen et al., 1988, 1990; Rhee et al., 1993a–c, e; Ye et al., 1992) or enrichment cultures (Morris et al., 1992a; Rhee et al., 1993d). Because of the diversity of inocula and the imposition of other treatments on all of these methods, it is difficult to determine if the different methods of preparing sediment slurries influenced experimental results in any significant way. Abramowicz et al. (1993) reported that the dechlorination rate doubled and the lag time before dechlorination decreased by half when a reduced mineral salts medium (Shelton and Tiedje, 1984), instead of water, was used to prepare slurries from wet sediments. There was no discernible effect due to the various reductants (sodium sulfide, cysteine, titanium citrate) that have been used, and in fact Rhee et al. (1993a) noted no difference when the reductant was omitted. There does appear to be a trend for more extensive dechlorination in cases where highly reduced slurries prepared from air-dried sediments were inoculated with microorganisms eluted from sediments (Quensen et al., 1988, 1990). This is possibly because highly reduced conditions were established in the sediment slurries by preincubating them prior to autoclaving and because autoclaving increased the availability of carbon solubilized from the sediments.

Most studies have been performed with added Aroclors or PCB congeners, but two studies (Alder et al., 1993; Abramowicz et al., 1993) reported further dechlorination of PCBs already present in sediments from New Bedford Harbor and the upper Hudson River, respectively. In addition, Bedard et al. (1993; submitted; Bedard and Van Dort, in preparation) stimulated the dechlorination of the pre-existing Aroclor 1260 in Woods Pond sediment by the addition of small excesses of pure PCB or brominated biphenyl congeners. Since it is most important to understand the dechlorination of PCBs already present in sediments, further attention should be given to studies of dechlorination of pre-existing PCBs.

Some concern has been expressed over the use of a carrier solvent, usually acetone, to add PCBs to cultures because it could serve as a carbon source. Nies and Vogel (1990) reported that dechlorination was dependent on the amount of acetone added, but their results were based on sediment slurries inoculated with microorganisms that had been previously enriched on acetone. Rhee et al. (1993a–e) observed dechlorination of PCBs added without any carrier solvent. (They used hexane to add PCBs to dried sediments and then evaporated the hexane before making slurries with the sediment.)

The initial composition of head-space gases used in dechlorination studies appears to have little if any effect on PCB dechlorination. In most cases, N_2 or N_2/CO_2 mixtures have been used, but even in cases where CO_2 is omitted it is probably present from metabolism or from bicarbonate added to the medium. Hydrogen has sometimes been included also. Morris et al. (1992a) stated that hydrogen could be substituted for pyruvate as an electron donor with equal effect, but in fact their experiment lacked the proper control without pyruvate or hydrogen. It is possible that natural organic matter in their sediment slurry served as the actual electron donor.

Most incubations have been carried out under static conditions in the dark at room temperature or at 30°C. Rhee et al. (1993a) incubated with continuous shaking, but in other studies (1993c,e) used static conditions, justifying this change by citing unpublished data showing no effect due to shaking. Tiedje et al. (1991) reported no difference in the extent of dechlorination among three shaking regimens (static, shaking once a week, and shaking continuously). Most investigators have incubated their cultures in the dark to minimize the growth of photosynthetic or phototropic organisms and to minimize the possibility of photodechlorination.

We have noted two potential problems related to scale that may limit dechlorination in culture. Tiedje et al. (1991) reported that the extent of dechlorination observed was dependent on inoculum volume. Balch tubes (28 ml capacity) containing 1 g air-dried sediment were inoculated with 1, 2, or 5 ml of an inoculum consisting of eluted sediment microorganisms suspended in reduced mineral medium. Hence all of the liquid was provided by the inoculum. Regardless of volume, all inocula contained the same number of microorganisms as they were prepared by dilution with reduced mineral medium. Twice as much dechlorination occurred in treatments receiving the 5 ml inoculum. These results were interpreted as possibly being related to the greater capacity of 5 ml of reduced medium to reduce traces of oxygen that might have diffused into the cultures during the incubation period. This problem is probably minimized when much larger volumes of sediments are used because any oxygen that might diffuse into the incubation vessel would be consumed in the top few millimeters of sediment. In fact, dechlorination has been observed in Winogradsky columns and bottles containing 20 g or more of sediments covered with liquid medium even though they were open to the atmosphere (Quensen, Schimel, and Griffith, unpublished data). Another problem of scale is suggested by the results of Rhee et al. (1993a,b,e). The proportion of sediment to media (0.5 g to 20 ml) is much lower in their experiments than in those of others, and they generally report more limited dechlorination than others using microorganisms from Hudson River sediments.

8.7. Conclusions

Despite 6 years of research on PCB dechlorination, little can be said definitively about what factors control PCB dechlorination in the environment. Temperature, pH, salinity, and trace metals are all known to be important to microbial ecology, but their influence on PCB dechlorination is virtually unknown. While PCB dechlorination processes apparently require a low redox level and some carbon source other than PCBs, the exact requirements for these factors are also unknown. This situation has arisen in part because most experiments have used culture conditions quite different from the environmental situation and in part because evaluation of PCB dechlorination has often been simplistic, focusing on the overall rate or extent of dechlorination without recognizing that there are distinct dechlorination processes, each of which may be affected differently by the same environmental variable.

The importance of recognizing distinct dechlorination processes is evident from our review of experiments involving the addition of the potential electron acceptors sulfate, nitrate, ferric oxyhydroxide, and BESA (which might act as an electron acceptor for sulfate reducers). By recognizing M and H as distinct dechlorination processes, we noted that sulfate additions favor process H over process M. Further studies are needed to delineate clearly the effects of electron acceptors (and BESA) on PCB dechlorination processes. Such studies will be most conclusive if the concentrations of the electron acceptors and the congener specificity of dechlorination are frequently monitored throughout the course of the experiments.

From the earliest observations it was evident that higher PCB concentrations increased the rate of dechlorination in culture and possibly the extent of dechlorination in the environment. A likely explanation is that concentrations above some level allow the growth of dechlorinating populations so that higher rates occur with higher concentrations of dechlorinating microorganisms (which can be viewed as catalysts for the reaction).

There are two important aspects to the question of how PCB concentration affects dechlorination. The first experiments and some early environmental data suggested that there might be a critical concentration below which dechlorination would not occur. Recently, however, Abramowicz et al. (1993) demonstrated that dechlorination can occur at total PCB concentrations as low as 20 μg/g sediment dry weight. This example is especially significant because it involves dechlorination of PCBs already present in the sediment for decades rather than added PCBs, but it is a single incidence and more work on the dechlorination of low concentrations of PCBs in contaminated sediments is needed.

The second aspect concerns the effect of PCB concentration on the rate of dechlorination. In investigations of the fate of freshly added PCBs, the dechlorination rate increased with increasing PCB concentration up to a point and then leveled off. Abramowicz et al. (1993) proposed that the rate of dechlorination is dependent on the aqueous concentration of PCBs in sediment pore water and that, below saturation level, the aqueous PCB concentration depends on the total PCB concentration. The strength of this hypothesis cannot be evaluated with the data at hand, and alternative explanations are possible. Thus more rigorous research in this area is also needed.

The effect of PCB concentration is closely tied to the concept of bioavailability. Most thinking to date has considered that only aqueous phase PCBs are available to microorganisms. PCBs tend to sorb to organic matter in sediment so that, in cases where total PCB concentrations are not high enough to form a bulk phase, the rate of desorption from sediment potentially controls the rate of dechlorination. Desorption of the labile PCB component associated with sediment is potentially fast enough not to limit PCB dechlorination. Desorption of the nonlabile PCB component occurs much more slowly, but even so this may not limit PCB dechlorination, which is itself a long-term process.

Co-contaminants might present a practical problem because most PCB-contaminated sediments also contain other contaminants. Oil might limit the rate of dechlorination if dechlorination rate is a function of the aqueous concentration of PCBs, but oil seems less likely to be important in determining the eventual extent of dechlorination.

PCB dechlorination processes appear to have a high tolerance for most heavy metals, perhaps because heavy metals probably exist as insoluble sulfides under the highly reduced conditions required for dechlorination. The effects of other halogenated or aromatic pollutants are unknown, but potentially important because such compounds could influence the nature of the microbial community.

It is clear that PCB dechlorination can occur under a variety of culture conditions. Scaling down the size of cultures can apparently impose a limit on the extent of dechlorination and lead to less reproducible results.

To understand better what is important in controlling environmental dechlorination, investigations using more environmentally relevant conditions are needed. These experiments should focus on the ability of indigenous microorganisms to dechlorinate PCBs already present in the sediments and should include incubations in site-water, rather than medium, and at environmentally relevant temperatures.

9. ATTEMPTS TO ENHANCE PCB DECHLORINATION

9.1. Carbon Additions

Most experiments involving the addition of carbon sources have been performed in an attempt to enhance the rate of PCB dechlorination. If the anaerobic biotransformation of PCBs is limited to reductive dechlorination, then the responsible microorganisms must be using compounds other than PCBs as their source of carbon for energy and growth. By providing an appropriate carbon source, it should be possible to increase the reducing equivalents available for PCB reduction and/or to support growth of the dechlorinating population(s), thereby increasing PCB dechlorination activity.

Attempts to enhance PCB dechlorination through carbon additions have been performed in four different kinds of systems: 1) autoclaved slurries prepared from air-dried sediments inoculated with microorganisms eluted from sediments (Rhee et al., 1993a,e; Tiedje et al., 1991), 2) unautoclaved sediment slurries prepared from air-dried sediments and inoculated with microorganisms from an enrichment (Morris et al., 1992a), 3) uninoculated slurries prepared from wet sediments (Alder et al., 1993; Abramowicz et al., 1993), and 4) slurries prepared from wet sediments and inoculated with microorganisms from enrichments (Nies and Vogel, 1990). In addition, PCBs were usually added in a solvent that could have served as a carbon source. All of these variations may be expected to influence the outcome of making carbon additions.

The interpretation of carbon addition experiments is complicated by the fact that the carbon additions may affect dechlorination through more than one mechanism. They may enhance dechlorination by serving as substrates for facultative anaerobic microorganisms so that oxygen or other potential electron acceptors are more rapidly depleted in the system. Or, they may provide a means of selection for a change in the community structure that may or may not be advantageous to the dechlorination process. Changes in the relative activity or in the relative abundance of the PCB-dechlorinating strains present may cause a change in the observed pattern of dechlorination. Large additions of rich carbon sources are more likely to inhibit dechlorination by selecting for nondechlorinating microorganisms (Quensen, unpublished data).

9.1.1. Primary Cultures

A major complication in interpreting the results of experiments involving carbon additions is determining if carbon is a limiting factor in unamended sediments. Sediments themselves provide relatively large amounts of utilizable carbon, and dechlorination has not been obtained in the absence of sediments. The influence of natural organic matter was demonstrated in the experiments of Alder et al. (1993), who found that small repeated additions of volatile fatty acids (acetate, propionate, butyrate, and hexanoic acid) stimulated dechlorination of added PCBs in carbon-limited Hudson River sediment slurries, but not in Silver Lake or New Bedford Harbor sediment slurries, which had higher organic carbon content. The availability of degradable organic matter was assessed by the relative amounts of methane production with and without the fatty acids addition.

The addition of fluid thioglycolate medium with beef extract (FTMBE) to slurries prepared from wet Hudson River sediment shortened or eliminated the lag time before dechlorination of added PCBs, but did not increase the observed rate of dechlorination (Abramowicz et al., 1993). In contrast, small additions of acetate, but not sludge supernatant or yeast extract, slightly increased the rate of dechlorination of 23456-CB (Tiedje et al., 1991) in autoclaved sediment slurries inoculated with microorganisms eluted from Hudson River sediment. The amount of natural organic matter available should have been relatively high in the latter cultures as a result of the autoclaving.

The "biphenyl enrichment" experiments of Rhee et al. (1993a,e) are more accurately termed *carbon addition* or *carbon amendment* experiments. The sediment slurries were inoculated with microorganisms eluted from Hudson River sediments rather than from enrichment cultures, and no evidence was presented for the metabolism of the biphenyl. In one set of experiments (Rhee et al., 1993e), added Aroclor 1254 was predominantly dechlorinated by process N (our interpretation, see Section 5.1.6), and the principal products were 24-4-CB and 24-24-CB. In unamended cultures, the 24-4-CB was subsequently dechlorinated to 2-4-CB, but the addition of biphenyl (1 mg/g sediment dry weight) selectively inhibited the *para* dechlorination (Rhee et al., 1993e). It would appear that two different dechlorination activities were involved and that biphenyl had no effect on process N but did inhibit the dechlorination of 24-4-CB. Process H' dechlorination (our interpretation, see Section 5.1.3) of Aroclor 1242 was more severely inhibited, or at least delayed, in cultures amended with biphenyl (Rhee et al., 1993a). It is not clear why biphenyl had any adverse effect on dechlorination.

9.1.2. Enriched Cultures

Perhaps not unexpectedly, the dechlorination activities of enriched cultures exhibit some dependence on the carbon source used for the enrichment. This is most evident in the experiments of Nies and Vogel (1990), who compared the dechlorination activity of second-generation cultures enriched on acetone, acetate, glucose, or methanol. No nonenriched cultures were included for comparison, and no results were reported for the primary enrichment cultures. The slurries in their experiments (15 g wet sediment and 50 ml mineral medium plus Aroclor 1242, 300 μg/g sediment dry weight) were inoculated from slurries of Hudson River sediment that had been previously enriched on each of the carbon sources. During the dechlorination experiments the second-generation cultures were fed the same carbon source on which they had previously been enriched. The acetone-fed cultures received 200 mg of acetone once, at the time of inoculation. The acetate-fed and methanol-fed cultures received 1,945 mg acetate or 988 mg methanol over the first 10 weeks, and the glucose-fed cultures received 1,680 mg glucose over the first 12 weeks. Comparisons were made to autoclaved controls and to a set of unfed cultures inoculated from the acetone enrichment. In these unfed cultures, no dechlorination of added Aroclor 1242 was observed in 22 weeks. However, the parallel cultures that were fed acetone removed 39% of the *meta* plus *para* chlorines in 22 weeks. This suggests that the dechlorinating microorganisms in the inoculum from the acetone enrichment had become dependent on acetone.

Nies and Vogel (1990) reported that dechlorination patterns were similar for all enrichments but that there were significant differences in both the rate and extent of dechlorination among enrichments. They stated that PCBs were dechlorinated earlier and more extensively in the glucose-fed batches than in the others, but that, although dechlorination was initially the slowest in acetone-fed batches, it eventually progressed to the same extent as in the glucose-fed batches. Because their means of presenting data does not allow comparisons of rates, we used the data in Table 3 of Nies and Vogel (1990) to plot decreases in the average number of chlorines per biphenyl over time (Fig. 4.11). From this plot it is evident that the initial rates were in fact similar for the acetate, glucose, and methanol enrichments, and slightly slower for the acetone enrichment. The differences in the extent of dechlorination among these treatments is due to the fact that dechlorination ceased in the acetate enrichment after 6 weeks and in the methanol treatment after 10 weeks, while the glucose and acetone enrichments continued to

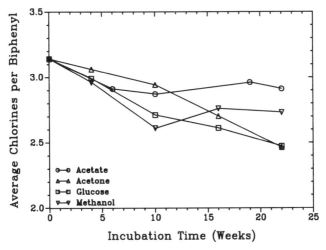

Figure 4.11. Dechlorination of Aroclor 1242 by Hudson River microorganisms enriched on four carbon sources. During the 22 week dechlorination study, each enrichment was fed with the carbon source on which it was enriched. Based on data from Nies and Vogel (1990).

dechlorinate until the end of the experiment at 22 weeks. It would also appear that acetone was the best substrate, because PCB dechlorination in slurries that were fed 200 mg of acetone only once ultimately progressed to the same extent as that in slurries that received multiple feedings (totalling 1,680 mg) of glucose. All of these conclusions are tentative, however, because apparently only a single sample was taken from each enrichment at each time point, and sampling or quantitation problems are evident for the methanol enrichment.

Because their enrichment cultures exhibited 70%–80% removal of tetra- through hexachlorobiphenyls in 22 weeks, Nies and Vogel (1990) proposed that "*in situ* organic substrate addition to these PCB-contaminated sites might accelerate the anaerobic dechlorination of the more-toxic highly chlorinated congeners." Unfortunately, their data do not provide a basis for evaluating whether carbon additions would accelerate *in situ* dechlorination of decades-old PCBs because 1) their experiments actually investigated the effect of adding both a carbon substrate and an actively dechlorinating inoculum from an enrichment culture and 2) their experiments investigated the dechlorination of freshly added PCBs. It is possible that in sediments where carbon is limiting, organic amendments might enhance dechlorination, as the results of Alder et al. (1993) suggest. But even in the latter case organic amendments did not stimulate further dechlorination of pre-existing PCBs in carbon-limited Hudson River sediment, possibly because the PCBs had already been extensively dechlorinated *in situ*. Therefore, further research is required to determine whether organic substrate additions hold promise for accelerating PCB dechlorination *in situ*.

That carbon additions can affect PCB dechlorination in complex ways is evident from the results of Morris et al. (1992a). Their initial, primary cultures consisted of nonsterile slurries prepared from air-dried uncontaminated sediments and inoculated with microorganisms eluted from Hudson River sediments. In these primary cultures, pyruvate (16 mM final concentration) had no effect on the dechlorination of added Aroclor 1242; after 12 weeks of incubation 48% of the *meta* plus *para* chlorines had been removed both with and without the pyruvate amendment, and dechlorination pattern C was observed (Morris et al., 1992a). Since pattern C results from the combined action of dechlorination processes M and Q (and sometimes H), the naturally occurring organic matter in the sediments was apparently adequate to support several PCB-dechlorinating populations. A slight inhibition of dechlorination occurred in cultures amended with formate (8 mM final concentration). When transfers of the pyruvate-fed and unamended cultures were made, however, dechlorination was more rapid and extensive in the pyruvate-enriched cultures. Then, 55% of *meta* plus *para* chlorines were removed in 4 weeks in the pyruvate-enriched cultures, but only 9% were removed in transfers of the unamended cultures. This suggests that a greater increase in the number of dechlorinating microorganisms had occurred in the primary culture amended with pyruvate. Dechlorination activity was maintained through nine serial transfers, indicating that growth of dechlorinating microorganisms did occur, but there was no increase in the rate or extent of dechlorination after the second transfer. Only activities M (Morris et al., 1992a) and H (our interpretation of their Fig. 2; see Fig. 4.9 and Section 8.2) were evident in these subsequent transfers; process Q activity was apparently lost during the repeated transfers on pyruvate.

Pyruvate was not the immediate electron donor for PCB dechlorination because dechlorination did not occur until after all pyruvate was depleted, and hydrogen could be substituted for pyruvate with no loss in dechlorination activity (Morris et al., 1992a). Acetate accumulated transiently as an intermediate, and its disappearance coincided with the onset of dechlorination and methanogenesis. However, substituting acetate (16 mM) for pyruvate caused a delay before dechlorination (Morris et al., 1992a) and the loss of process H dechlorination (our interpretation of their Fig. 2).

Perhaps the losses of dechlorination activities described above can best be explained as due to changes in community structure caused initially by enrichment on pyruvate and secondly by substitution of acetate for pyruvate.

9.2. Surfactant Additions

It is generally believed that the biodegradation of poorly water-soluble compounds like PCBs in soil or sediment systems is limited by low bioavailability due to strong sorption of the compounds to natural organic matter. The use of surfactants to increase aqueous concentrations of such compounds has often been suggested as a way of overcoming this problem, but little research with regard to PCB dechlorination has been reported.

Three nonionic surfactants from the Triton series and the ionic surfactant sodium dodecyl benzenesulfonate (SDBS), all at 600 ppm, were compared for their effects on dechlorination of a freshly added Aroclor mixture in Hudson River sediment slurries (Abramowicz et al., 1993). SDBS completely inhibited dechlorination. Dechlorination was detected earlier in cultures amended with Triton X-405 or Triton X-705 than in cultures with no surfactant (at 8 weeks rather than 12), but Triton X-705 had no effect on dechlorination rate, and Triton X-100 and Triton X-405 each reduced the rate of dechlorination about eightfold. Thus, because of the earlier onset, the most extensive dechlorination achieved within the 22 weeks of incubation was with the Triton X-705.

The above results do not seem very encouraging and are probably not relevant to PCB-contaminated sediments because the PCBs in sediments are sorbed to organic matter and probably exist in both labile and nonlabile fractions (see Section 8.4). It is unlikely that surfactants will enhance the dechlorination of the nonlabile fraction of PCBs if these PCBs are dissolved in a condensed polymeric humic matrix (Carroll et al., 1994).

9.3. Priming With Halogenated Biphenyls

Brown and others (Brown et al., 1987a,b; Quensen et al., 1988) have proposed that PCB-dechlorinating microorganisms use PCBs as electron acceptors and that they may even derive energy from dechlorination. If this hypothesis is correct, then it should be possible to selectively enrich PCB-dechlorinating microorganisms by supplying excess PCBs, provided more favorable electron acceptors are limiting. Furthermore, it should be possible to enrich microorganisms that carry out a specific dechlorination process by adding a PCB congener that acts as a preferred dechlorination substrate and electron acceptor for that particular process. We have called this procedure *priming* for lack of a better term.

In their studies with Woods Pond sediment, Bedard and colleagues (1993; submitted) found that they could stimulate process P dechlorination of the sediment PCBs by adding excess 25-34-CB. The 25-34-CB was added in acetone to a sediment slurry at a final concentration of 350 µM (equivalent to 100 µg/ml slurry or approximately 700 µg/g sediment dry weight). On a weight basis, this is roughly 7–14 times the concentration of the Aroclor 1260 in the sediment. The 25-34-CB began to dechlorinate to 25-3-CB within 3 weeks, and by 7 weeks, when the 25-34-CB had decreased by 45%–60%, it was apparent that process P dechlorination of the Aroclor 1260 in the sediment was also occurring. After 12 weeks, decreases of 17%–83% were seen in most of the major penta-, hexa-, and heptachlorobiphenyl components of the sediment PCBs, especially 245-25-CB, 245-24-CB, 245-245-CB, 2345-25-CB, 234-245-CB, and 2345-236-CB. The products were tetra- and pentachlorobiphenyls containing 23-, 25-, and 235-chlorophenyl groups. No dechlorination occurred in identical controls to which no 25-34-CB was added. Various carbon sources added without 25-34-CB stimulated methanogenesis, but not dechlorination.

In parallel studies Bedard and colleagues (1993; Bedard and Van Dort, in preparation) have also been able to stimulate process N dechlorination by adding excess 236-CB, 2346-CB, or 23456-CB to slurries of Woods Pond sediment. In each case predominantly *meta* dechlorination of the added congener occurred; hence 236-CB was dechlorinated to 26-CB, and 2346-CB to 246-CB, and 23456-CB was dechlorinated to 246-CB (74%) by the loss of both *meta* chlorines and to 2356-CB (24%) by the loss of the *para* chlorine. Small amounts of 236-CB and 26-CB were also formed. After most of the added congener had been dechlorinated, the Aroclor 1260 in the sediment began to dechlorinate, undergoing extensive and almost indiscriminate decreases of the hexa-, hepta-, and octachlorobiphenyls and large increases in tetra- and pentachlorobiphenyls. After 19 weeks 20%–30% of the total sediment PCBs had been dechlorinated to a single product, 24-24-CB.

Based on their results with individual PCB congeners, Bedard et al. (1993) proposed that bromobiphenyls (BB), like PCBs, might also be used to stimulate microbial dechlorination of the sediment PCBs in Woods Pond. A variety of mono-through tetrabromobiphenyls were tested. All were dehalogenated to biphenyl by the indigenous microbial population, and most stimulated process N dechlorination of the sediment PCBs. The dechlorination was stimulated most strongly by 25-BB, 26-BB, and 25-3-BB. A single addition of 26-BB (350 µM, equivalent to about 750 µg/g sediment dry weight) and disodium malate (10 mM) to sediment slurried in water and incubated at 25°C resulted in 99% dehalogenation of the 26-BB to biphenyl (97%) and 2-BB (2%) in 35 days. After 93 days the hexa- through nonachlorobiphenyls (initially 73 mole% of the total PCBs) had been decreased by 79%. The major dechlorination products were tri- to pentachlorobiphenyls, especially 24-24-CB and 24-26-CB. No dechlorination occurred in parallel samples incubated without bromobiphenyl.

These findings indicate that microorganisms exhibiting at least two different PCB dechlorination activities are present in Woods Pond sediment and that these activities can be selectively stimulated by the addition of a relatively small excess of an appropriate halogenated biphenyl congener. This suggests that the excess halogenated biphenyl confers an advantage on those microorganisms that are capable of dehalogenating it. Presumably the halogenated biphenyls act as electron acceptors and selectively promote the growth or activity of the individual dechlorinating populations.

9.4. Conclusions

Before attempting to enhance PCB dechlorination through the use of amendments, attention should be given to what dechlorination process(es) have occurred naturally. Dechlorination may have already proceeded as far as possible given the dechlorinating microorganisms present, in which case attempts to enhance further dechlorination through amendments are likely to prove futile. In such cases, inoculation with dechlorinating microorganisms with complementary specificities may be a future option.

Attempts to enhance PCB dechlorination by adding carbon substrates to primary cultures have not been successful. At best they have shortened the lag time before dechlorination was observed in cultures, or increased initial rates, but have not increased the extent of subsequent dechlorination. Continued carbon additions may eliminate some (as happened in the pyruvate enrichments of Morris et al., 1992a) or all dechlorinating microorganisms initially present and so have a detrimental effect on dechlorination. In special situations (when carbon is actually limiting) carbon additions could be advantageous, but do not appear to be generally advisable.

Surfactants have the potential to enhance dechlorination of the labile fraction of PCBs, but no relevant experimental data exist to demonstrate their effectiveness, and it is unlikely that surfactants would enhance the bioavailability of the nonlabile component of PCBs if this fraction is dissolved in a condensed polymeric humic matrix (Carroll et al., 1994). Future experiments should compare the effectiveness of surfactants at concentrations related to their critical micelle concentration rather than their concentration as percent by weight and should investigate their effectiveness at enhancing the dechlorination of pre-existing sediment PCBs instead of freshly added PCBs.

Priming, the addition of a single PCB congener or other halogenated aromatic compound to a sediment, is the most intriguing and effective means of enhancing PCB dechlorination yet demonstrated. The added agent is presumed to serve as an electron acceptor and promote the growth or activity of the dechlorinating microorganisms to a population density where they are effective. More research is needed in this area to determine under what environmental conditions priming is effective and what environmentally acceptable agents may be effective.

10. MICROBIOLOGY OF PCB DECHLORINATION

10.1. Isolation Attempts

Microorganisms from a variety of PCB-contaminated sediments have now been shown to dechlorinate PCBs, but no PCB-dechlorinating isolates have been obtained. Isolation is inherently difficult because PCB dechlorination has not been obtained in defined medium. Most PCB dechlorination experiments have of necessity been conducted using sediment slurries, and carbon for the PCB-dechlorinating microorganisms was presumably supplied by the sediment. Efforts to identify appropriate simple carbon sources that would sustain PCB-dechlorinating populations in the absence of sediment have not been successful. Furthermore, it is characteristic of anaerobic microbial communities that member species are strongly dependent on interactions with each other, and this is probably also true for PCB-dechlorinating microorganisms. However, the results of several studies provide some insight into methods that might be used to enrich PCB-dechlorinating populations and might eventually lead to isolation of PCB-dechlorinating strains.

Boyle et al. (1993) described a method of enrichment that resulted in a 100-fold increase in the volumetric rate of dechlorination of 236-CB to 26-CB in a sediment slurry. The microbial population in the slurry consisted of microorganisms from Hudson River sediment that had initially been enriched on a mixture of Aroclor 1242, 236-CB, and 246-CB and had subsequently been enriched on 236-CB alone. The original enrichment culture exhibited *meta* and *para* dechlorination of Aroclor 1242, *meta* dechlorination of 236-CB, and *para* dechlorination of 246-CB. This enrichment was transferred twice to slurries of sterile sediment in medium supplemented with 236-CB, and was subsequently maintained at constant volume for more than a year by periodic resupplementation with sterile sediment and medium (to replace amounts removed during sampling) and with 236-CB (to replenish amounts removed by dechlorination). The authors demonstrated that the rate of dechlorination of 236-CB to 26-CB could be increased by raising the concentration of 236-CB and increasing the frequency of supplementation with sterile sediment medium. The enrichment, designated C-146, dechlorinated 236-CB at a rate of 2.1 μmol L^{-1} day $^{-1}$ when provided with 84 μmol/L of 236-CB and not resupplemented. Weekly supplementation with 84 μmol/L of 236-CB and sediment medium increased the rate of dechlorination fourfold, to 8.2 μmol L^{-1} day $^{-1}$. But daily supplementation with 1 to 2 mmol/L of 236-CB and sediment medium increased the rate of dechlorination more than 100-fold, to 346 μmol L^{-1} day^{-1}. The results also demonstrated that daily supplementation with high levels (1 mmol/L) of 236-CB decreased the rate of methanogenesis from 628 μmol L^{-1} day^{-1} (observed in parallel cultures supplemented with sediment medium but not PCBs) to 126 μmol L^{-1} day^{-1}. No such decrease in methanogenesis occurred in cultures that were supplemented weekly with low levels (84 μmol/L) of 236-CB. Thus the proportion of electrons channelled to dechlorination increased relative to the proportion channelled to methanogensis. These results suggest that the frequent supplementation with high levels of 236-CB changed the microbial community and most likely increased the number of PCB-dechlorinating microorganisms.

Boyle et al. (1993) estimated the most probable number (MPN) of PCB-dechlorinating microorganisms in their cultures by assaying dechlorination of 236-CB in sterile sediment slurries inoculated with an actively dechlorinating culture in dilutions ranging from 1:10 to 1:10^{12}. From these experiments they estimated that cultures dechlorinating 236-CB at a rate of 339 μmol L^{-1} day^{-1} contained ~3 × 10^5 236-CB–dechlorinating cells per liter. Based on this, they calculated that the rate of dechlorination of 236-CB to 26-CB was 1.13 pmol day^{-1} per dechlorinating cell, a rate 10-fold higher than that reported for the dechlorination of 3-chlorobenzoic acid by *D. tiedjei* (Tiedje et al., 1987).

Continued resupplementation with a specific PCB congener, as described by Boyle et al. (1993), might provide a means of enriching a specific PCB-dechlorinating population. This would be consistent with the hypothesis that PCBs can act as electron acceptors (Brown et al., 1987b; Quensen et al., 1988) and can therefore act as a means of selective enrichment. Most likely the accelerated volumetric rate of dechlorination observed in enrichments that were frequently supplemented with high concentrations of 236-CB and fresh sediment medium involved an increase in a PCB-dechlorinating population. Unfortunately, no estimate of the number of PCB-dechlorinating cells was given for the original sediment slurry, so this was not demonstrated. In addition, the enrichment cultures were not challenged with Aroclor 1242 to determine the range of congener specificity after prolonged enrichment on 236-CB. It will be important to learn whether this type of enrichment selects for a PCB-dechlorinating population that can dechlorinate a broad range of PCBs (as do the six dechlorination processes we have described) or a population with a very narrow dechlorination specificity, perhaps even limited to congeners containing 236-chlorophenyl groups.

Using conditions known to be favorable to *D. tiedjei,* an anaerobic microorganism that obtains energy from the dechlorination of chlorobenzoates (Dolfing and Tiedje, 1987; Dolfing, 1990; Mohn and Tiedje, 1991), Morris et al. (1992a) obtained a pyruvate-degrading enrichment capable of removing primarily *meta* chlorines from PCBs. Nine serial transfers from the initial enrichment were carried out. These transfers exhibited a decrease in the lag time preceding dechlorination and a change from process C dechlorination to a combination of processes M and H (see Sections 9.1 to 9.1.2). The results clearly indicate that growth of PCB-dechlorinating microorganisms occurred because dechlorination activity was maintained through nine serial transfers. Apparently pyruvate was not the immediate electron donor in the enriched culture because dechlorination did not begin until after all pyruvate had been consumed and because the substitution of hydrogen for pyruvate had no effect on dechlorination. Six hydrogen-utilizing bacteria (including a presumed methanogen and five sulfate reducers) were isolated from this enrichment and tested for the ability to dechlorinate PCBs. Unfortunately, all of these isolates failed to dechlorinate PCBs even when tested in non-PCB–contaminated Hudson River sediments with the indigenous microbial community intact. *D. tiedjei* and three pure cultures of methanogens (*Methanobacter* sp. strain DG1, *Methanospirillum* sp. strain PM, and *Methanosarcina* sp. strain MS) also failed to dechlorinate PCBs when tested under the same conditions.

Despite these failed attempts, the demonstration that PCB-dechlorinating microorganisms can be grown successfully on solid media in the absence of PCBs (May et al., 1992) has raised hopes that the isolation of PCB-dechlorinating microorganisms will eventually be successful. May et al. (1992) noted that the dechlorination of Aroclor 1242 in Hudson River sediment slurries amended with sulfate differed from that which occurred in unamended slurries. They further noted that the dechlorination in the sulfate-amended cultures occurred only after depletion of the sulfate and appeared to be limited to the removal of *para* chlorines from congeners that had adjacent *meta* and *para* chlorines. We have identified this dechlorination as process H (see Section 8.2). In an attempt to enrich the microorganisms responsible for this *para*-dechlorinating activity (process H), May and colleagues transferred the microorganisms from the sulfate-amended culture into sterile sediment slurries containing a mixture of four PCB congeners with adjacent *meta* and *para* chlorines (234-CB, 245-CB, 34-2-CB, and 25-34-CB).

After three such transfers, the microorganisms were transferred to anaerobic slants prepared with a mineral medium and river sediments without PCBs. A colony picked from one of these slants produced methane and dechlorinated the four congeners when transferred back into a sterile anaerobic sediment slurry. It is not possible to interpret clearly the dechlorination process that occurred from the data presented because 234-CB and 34-2-CB co-elute, and because several of the congeners could be dechlorinated to the same products or to products that co-elute. Hence 234-CB could be dechlorinated to either 24-CB or 23-CB, 245-CB could be dechlorinated to either 24-CB or 25-CB, and 34-2-CB could be dechlorinated to either 2-3-CB or 2-4-CB. The products shown were primarily 24-CB/25-CB (co-eluting congeners), lesser amounts of 23-CB/2-4-CB and 25-3-CB, and a trace of 2-3-CB. May et al. did not note the appearance of 25-3-CB, but it can be seen in their Figure 2B as a new peak that is only partially resolved from 245-CB. The 25-3-CB and 2-3-CB result from *para* dechlorination of 25-34-CB and 34-2-CB, respectively, which is characteristic of dechlorination process H. The dechlorinating culture was plated on solid media a second and third time. Colonies from the second and third platings did not produce methane and apparently lost the ability to dechlorinate 25-34-CB and 34-2-CB. Hence 24-CB/25-CB was the only product peak detected from the mixture of four congeners, and 2-3-CB and 25-3-CB were no longer produced. Thus the colony simultaneously lost its methanogenic activity and a dechlorination activity characteristic of process H.

10.2. Characterization of Microorganisms

Because of difficulties inherent in isolating PCB-dechlorinating microorganisms, Ye and colleagues have tried to characterize the responsible species based on their physiology. Their approach has included the use of both selective culture conditions and inhibitors.

10.2.1. Dechlorination Population M

Ye et al. (1992) demonstrated that pasteurized and ethanol-treated inocula prepared from Hudson River sediments retained their ability to effect PCB dechlorination by process M and failed to produce any methane. Thus the dechlorinating microorganisms in these cultures are most likely spore-formers. Ye (1994) also used molybdate, an analog of sulfate that inhibits sulfate-reducing bacteria by depleting ATP (Taylor and Oremland, 1979), to help characterize these dechlorinating spore-formers. Molybdate prevented PCB dechlorination in the pasteurized cultures when added at a concentration of 16 mM and partially inhibited dechlorination when added at lower concentrations (2–8 mM). This suggested that sulfate-reducing microorganisms were responsible for dechlorination in the pasteurized cultures. Further, Ye found that 2 mM sulfate added once at the beginning of the experiment also prevented dechlorination, presumably by serving as the preferred electron acceptor for sulfate reducers. From these data Ye proposed that the microorganisms responsible for process M dechlorination in the pasteurized cultures were sulfate-reducing spore-formers, perhaps *Desulfotomaculum,* which is the only known genus of sulfate-reducing spore-formers.

Unexpectedly, Ye (1994) found that BESA (1 mM), an analog of coenzyme M that inhibits methanogens (Gunsalus et al., 1978), also inhibited PCB dechlorination in 4 week incubations of his nonmethanogenic pasteurized cultures. This suggested some other mode of action for BESA in these cultures. Ye proposed that the sulfonic acid group of the BESA was used in preference to the PCBs as an electron acceptor (Oremland and Copeland, 1988) by sulfate-reducing spore-formers in his cultures. His hypothesis was supported by the fact that the pasteurized cultures amended with 16 mM BESA turned black (characteristic of sulfides) and produced H_2S (Ye, 1994). Actual measurements to determine whether the BESA was depleted or transformed would further strengthen this hypothesis.

If sulfate reducers are in fact responsible for process M dechlorination and if sulfate acts as a preferred electron acceptor for these organisms, then it is logical to expect them to increase in number when sulfate is present and to resume dechlorination of PCBs when sulfate is depleted. However, the data (see Section 8.2) suggest that process M dechlorination does not resume following sulfate depletion. There are a number of possible explanations: 1) Sulfate might cause an unfavorable change in the community structure. 2) Hydrogen sulfide could be toxic to population M. 3) Other more robust sulfate reducers might out-compete population M when sulfate is present. 4) Population M could lose its ability to dechlorinate when grown in the presence of sulfate. Although the first two possibilities do not rule out the interpretation that microorganisms other than sulfate reducers might be responsible for dechlorination process M, all four are consistent with the hypothesis that population M might belong to the physiological group of sulfate reducers. We believe that the hypothesis that best fits all of the data is that the microorganisms that mediate dechlorination process M are sulfate-reducing spore-formers.

10.2.2. Dechlorination Population H

Evidence suggesting that methanogens are responsible for process H dechlorination comes from dechlorination assays conducted under culture conditions that selected for methanogenic bacteria (Ye, 1994). The Hudson River microorganisms Ye used in these experiments normally dechlorinate Aroclor 1242 by processes M, H, and Q to produce pattern C. When added together, penicillin G and D-cycloserine totally inhibited methanogenesis and dechlorination of Aroclor 1242 in anaerobic sediment slurries inoculated with the Hudson River microorganisms. These antibiotics directly inhibit eubacteria and indirectly inhibit methanogens that are dependent on substrates produced by eubacteria. When the antibiotics and methanogenic substrates (a mixture of methanol, formate, and acetate) were added in combination, methanogenesis was not inhibited and dechlorination of Aroclor 1242 was observed. The dechlorination was much more selective than in cultures not treated with antibiotics and was characterized by decreases in 34-2-CB/23-3-CB (co-eluting congeners), 25-34-CB, 24-34-CB, 23-34-CB/234-4-CB, and 245-4-CB; corresponding increases in 2-3-CB, 25-3-CB, 24-3-CB, 24-4-CB, and 25-4-CB; and an unexplained increase in 25-2-CB (Ye, 1994). Except for the increase in 25-2-CB, these changes match those we have described for process H.

The finding of May et al. (1992) that cultures repeatedly plated on solid media simultaneously lost the ability to produce methane and to dechlorinate 25-34-CB and 34-2-CB to 25-3-CB and 2-3-CB, respectively (see above), is also consistent with Ye's hypothesis that methanogens are responsible for process H dechlorination.

Process H dechlorination has been observed in cultures derived from Hudson River sediments and amended with sulfate or higher concentrations of BESA in several studies (see Section 8.3). BESA is known to inhibit methanogens (Gunsalus et al., 1978), but it is not clear whether process H dechlorination actually occurred in the presence of sulfate or BESA or only after these compounds had been depleted. The microorganisms in unamended Hudson River sediments normally have the potential to dechlorinate PCBs by processes M, H, and sometimes Q (Figs. 4.3, 4.4 and 4.6) The interesting result of these experiments is that process H dechlorination was much more prominent following sulfate or BESA amendments. This suggests that the sulfate or BESA altered the community structure in a way that was more detrimental to process M than H.

If the population responsible for dechlorination process H is methanogenic, it might be able to dechlorinate PCBs in sulfidogenic cultures within micro-niches where sulfate (or BESA) has been locally depleted. Alternatively, population H might not dechlorinate PCBs when sulfate is present, but might resume dechlorination activity when sulfate has been depleted and the redox level becomes low enough. This might explain why H is prevalent in estuarine New Bedford Harbor sediments. There the sulfate in the seawater may select for faster growing, nondechlorinating sulfate reducers, while dechlorinating methanogens could still carry out their activity in microzones or subsurface sediments where sulfate is depleted.

10.3. Conclusions

We believe that the proposal (Ye, 1994) that sulfate-reducing spore-formers and methanogens, respectively, mediate dechlorination processes M and H is a good working hypothesis that can provide direction to future attempts to isolate these dechlorinating microorganisms. Several lines of evidence suggest that growth of PCB-dechlorinating microorganisms in sediment slurries occurs during PCB dechlorination and can be enhanced by the addition of high concentrations of PCBs or brominated biphenyls (Morris et al., 1992a; Boyle et al., 1993; Bedard et al., 1993). In addition, the demonstration by May et al. (1992) that PCB-dechlorinating microorganisms can be plated in the absence of PCBs suggests that isolation of some PCB-dechlorinating microorganisms may be possible in the near future. Success may still be difficult, however, because even if some dechlorinating microorganisms are sulfate reducers or methanogens, their numbers might be low relative to the total number of sulfate reducers or methanogens present in sediments. It should also be kept in mind that dechlorination activity might be dependent on metabolic activity of nondechlorinating members of the microbial community. For example, nondechlorinating microorganisms might cross-feed dechlorinating organisms or might maintain a favorable H_2 concentration by producing or consuming H_2. Thus pure cultures of the dechlorinating microorganisms might not dechlorinate PCBs. Growth and chlorobenzoate dechlorination by pure cultures of *D. tiedjei* were extremely slow and could be achieved only in undefined media for years until all nutritive conditions could be identified (Apajalahti et al., 1989; DeWeerd et al., 1990), but reasonable growth and dechlorination were obtained earlier when *D. tiedjei* was grown in a consortium (Dolfing and Tiedje, 1986). Thus we advise that isolates be screened for dechlorination activity both in pure culture and in nondechlorinating mixed cultures obtained from sediments.

11. PCB DECHLORINATION: ENZYMATIC OR NONENZYMATIC, SELECTED OR FORTUITOUS?

It is now generally accepted that PCB dechlorination in anaerobic sediments is biologically mediated. The activity can be serially transferred from culture to culture (Morris et al., 1992a) and even from bacterial colonies picked from agar plates to sediment slurries (May et al., 1992), but is not observed in sterile control cultures. It is our interpretation that the distinct congener selectivities of the various dechlorination processes we have documented here most likely result from enzymatic reactions carried out by several different strains or species of PCB-dechlorinating bacteria. This argument is strengthened by the fact that one can select for a given dechlorination activity by manipulating the culture conditions to favor a certain physiological group of microorganisms (Ye et al., 1992; Ye, 1994).

But our conclusion that biologically mediated PCB dechlorination is most likely enzymatic is not based solely on the fact that distinct dechlorination patterns can be recognized. Nonenzymatic processes also exhibit regiospecificities, as demonstrated by the work of Roth et al. (1994). Roth and colleagues studied the dechlorination of PCBs by nickel boride and the tetrakis (triphenylphosphine) nickel(0) complex, $Ni(0)(PPh_3)_4$. Distinct regiospecific dechlorination was obtained in both reactions. For the catalytic dechlorination by nickel boride, the order of reactivity was *meta* > *ortho* > *para*. This is in contrast to the biologically mediated processes we have described in which *ortho* positions are generally not dechlorinated. Dechlorination of mono- and dichlorobiphenyls by $Ni(0)(PPh_3)_4$ was more similar to biological processes in that the *ortho* position was less reactive than the *meta* and *para* positions, but the intramolecular reactivities of the trichlorobiphenyls examined were quite different from those reported by Williams (1994). Roth and colleagues found that removal of the least-crowded chlorine was favored by a ratio of more than 100:1 (specifically, 234-CB was dechlorinated to 23-CB, 235-CB to 23-CB, 245-CB to 24-CB, and 345-CB to 34-CB), while in each case Williams reported that biologically the central chlorine was preferentially removed. In addition, Roth et al. noted that the $Ni(0)(PPh_3)_4$ acted as a stoichiometric reagent, not a catalyst, because it was consumed in the reaction.

Thus the question remains whether the observed selectivity results from enzyme and coenzyme variations, enzyme active-site structural differences, intrinsic PCB structural differences, or some combination of these (Roth et al., 1994). Clearly PCB structural differences cannot account for all of the differences because the same congeners are dechlorinated to different products by different dechlorination processes. On the other hand, none of the six dechlorination processes described in this review can remove *ortho* chlorines, and five out of six remove the doubly flanked *meta* chlorine from 234-chlorophenyl groups. These facts indicate that PCB structural differences do play some role in determining specificity.

Aryl dehalogenation has been demonstrated for compounds other than PCBs, including chlorobenzoates, chlorophenols, and chlorobenzenes (see Mohn and Tiedje, 1992, for a review). In all cases, the actual mechanism of dechlorination is unknown, but it has been demonstrated that cofactors such as vitamin B_{12} and hematin might be involved. Gantzer and Wackett (1991) reported the dechlorination of hexa- and pentachlorobenzenes using both Vitamin B_{12} and hematin as electron carriers and titanium (III) citrate as a reductant. The rate of dechlorination of the pentachlorobenzene was about two orders of magnitude slower than the dechlorination of the hexachlorobenzene, and the products of pentachlorobenzene were 1,2,4,5-tetrachlorobenzene (80%) and 1,2,3,5-tetrachlorobenzene (20%). This type of work has been extended to include the dechlorination of a PCB congener. Vitamin B_{12} transferred electrons from dithiothreitol to 23456-CB in a 1:1 mixture of aqueous buffer and 1,4-dioxane (used to enhance PCB solubility) with the result that the 23456-CB was dechlorinated to 2356-CB (68%) and 2346-CB (31%) (Assaf-Anid et al., 1992). The dechlorination rate was slow; only 3.5% of the initial 1 mM 23456-CB was dechlorinated in 4 weeks, and no tri- or lesser chlorinated products were reported. Both in this experiment and in culture, the proton added to the biphenyl ring in place of the leaving Cl^- was abstracted from water (Nies and Vogel, 1992; Assaf-Anid et al., 1992).

Several aspects of these results are inconsistent with the results of microbially mediated dechlorination experiments. Gantzer and Wackett (1991) found that the rate of chlorobenzene dechlorination decreased drastically as chlorines were removed from the ring. Pentachlorobenzene was dechlorinated 100 times more slowly than hexachlorobenzene, and no dechlorination of the tetrachlorobenzenes was detected. In contrast, Fathepure et al. (1988) found that sewage sludge dechlorinated hexachlorobenzene primarily to 1,3,5-trichlorobenzene, but also detected all possible dichlorobenzenes. The high proportion of 1,3,5-trichlorobenzene formed indicates that the regiospecificity of the second dechlorination step was different from that in Gantzer and Wackett's experiments. In addition, Bosma et al. (1988) demonstrated the microbial dechlorination of all three trichlorobenzene isomers to monochlorobenzene in columns of anaerobic Rhine River sediment. The dechlorination was specific: 1,2,3- and 1,3,5-trichlorobenzene were initially dechlorinated to 1,3-dichlorobenzene, and 1,2,4-trichlorobenzene was first dechlorinated to 1,4-dichlorobenzene.

The studies of Gantzer and Wackett (1991) and Assaf-Anid et al. (1992) both suggest that the aryl dechlorination reactions mediated by vitamin B_{12} and hematin are limited to dechlorination of aromatic rings containing five or six chlorines. The fact that microbially mediated dechlorination does not have this limitation suggests that if vitamin B_{12} and hematin play a significant role in biological dechlorination, their activity must be greatly enhanced by association with an enzyme.

Assaf-Anid et al. (1992) concluded that their results demonstrated that nonspecific PCB dechlorination can occur and suggested that microbially mediated reductive dechlorination might use "available electrons fortuitously, without providing any significant energetic benefit for the microorganisms that synthesize the coenzyme derivative of Vitamin B_{12} for use in normal metabolic functiors." However, their observations do not exclude the hypothesis that PCB dechlorination is advantageous to the responsible microorganisms, and their argument ignores several important pieces of information that suggest that selection for PCB dechlorination activity does occur. Mohn and Tiedje (1992) concluded their review of microbial reductive dehalogenation with statements that "aryl dehalogenations appear more likely [than alkyl] to be catalyzed by native enzymes" and that "particularly with aryl halides, inducible enzymes with substrate specificity appear to be required." They arrived at these conclusions primarily because it is possible to enrich for substrate-specific dehalogenation activities. This is apparently also the case with PCB dechlorination.

Enrichment for PCB dechlorination activity appears to have occurred both in the environment and in the laboratory. PCB-dechlorinating microorganisms appear to be more abundant in PCB-contaminated sediments than in uncontaminated sediments. Several investigators have noticed that the lag time before the onset of dechlorination in experiments is much shorter when the microorganisms are derived from contaminated sediments rather than uncontaminated sediments, and have attributed this shorter lag time to the greater abundance of dechlorinating microorganisms at the contaminated sites (Quensen et al., 1988; Tiedje et al., 1991; Abramowicz et al., 1993). The experiments of Morris et al. (1992a) demonstrate enrichment because, while the primary cultures exhibited a 4-week lag time, there was no observable lag time in transfers of those cultures. The experiments of Rhee et al. (1993d) demonstrate that the regiospecificity of a culture can be altered by enrichment on a single PCB congener (see Section 7.2). And, the ability to selectively prime two different dechlorination activities in Woods Pond sediment by the addition of specific halogenated biphenyl congeners (Bedard et al., 1993; submitted; Bedard and Van Dort, in preparation) implies enrichment for PCB-dechlorinating microorganisms, induction of PCB-dechlorinating enzymes, or both.

The ability to enrich for an activity implies that there is a selective advantage to that activity. One possibility is that in highly reduced sediments where terminal electron acceptors are limiting, microorganisms that are able to use PCBs as terminal electron acceptors have a selective advantage (Quensen et al., 1988; Brown et al., 1987a,b). Alternatively, energy for growth could be derived from the dechlorination process. The capture of energy from the dechlorination of chlorobenzoate by *D. tiedjei* has been demonstrated (Dolfing, 1990; Mohn and Tiedje, 1992) and is theoretically possible for PCBs also (Holmes et al., 1993).

In conclusion, we believe the weight of available evidence suggests that microbially mediated dechlorination of PCBs is enzymatic and that it confers a selective advantage on the microorganisms that exhibit the activity. We discount the hypothesis that the abiotic aqueous model system of vitamin B_{12} is "biomimetic" (Assaf-Anid et al., 1992) because the reactions observed do not agree with the results of biological experiments in terms of relative rates or relative proportions of products formed. In addition, none of the transition metal complexes that have been tested exhibits a pattern of regiospecificity corresponding to any of the biological dechlorination systems that have been described. Therefore, although we concur with Gantzer and Wackett's suggestion that transition-metal coenzymes may play a role in the reductive dechlorination of polychlorinated aromatic compounds, we would qualify this by proposing that these coenzymes do not act alone, but together with enzymes that greatly enhance the reaction rates and modify the regiospecificity.

12. IMPLICATIONS

12.1. Detoxication

A variety of toxic effects have been ascribed to PCBs. This is not surprising since they occur as complex mixtures, and congener-specific differences in mode of toxicity and potency might be expected. Because anaerobic dechlorination alters the relative proportions of congeners present in a PCB mixture, it could affect the toxicity of the mixture as a whole; we believe its general effect is to reduce toxicity.

While there is no conclusive evidence for PCB-related carcinogenicity in humans, even in occupationally exposed groups (ATSDR, 1993), PCBs are considered potential carcinogens because Aroclor 1260 has been shown to increase the incidence of hepatic tumors in rats (Kimbrough et al., 1975; Norback and Weltman, 1985). Lesser chlorinated PCB mixtures were not as potent as Aroclor 1260 (ATSDR, 1993) and did not cause statistically significant elevations of liver tumors (Moore, 1991). Since carcinogenic potential appears directly related to the degree of chlorination, reductive dechlorination of PCB mixtures may be expected to reduce the overall risk of cancer even though the total molar quantity of PCBs is unchanged by the process.

The most sensitive eco-toxicological endpoint for PCBs is related to reproductive success and deformities in some fish and fish-eating birds (Ludwig et al., 1993). This is believed to be related to "dioxin-like" toxicity mediated through the Ah receptor and hence is the form of PCB toxicity for which structure–activity relationships are best understood. The most potent congeners are approximate stereoisomers of 2,3,7,8-tetrachloro-dibenzo-*p*-dioxin (TCDD). These "coplanar" congeners lack *ortho* chlorines, have two *para* chlorines, and at least two *meta* chlorines (34-34-CB, 345-34-CB, and 345-345-CB) (Safe, 1990, 1993) and are present in only minute quantities (0.015 to 0.3 weight %; Schwartz et al., 1993) in Aroclors. The mono-*ortho*-derivatives of the coplanar PCBs, such as 234-34-CB and 245-34-CB, exhibit similar toxicity but are orders of magnitude less potent (Safe, 1990). The adjacent *meta* and *para* chlorines present in all of these congeners make them susceptible to at least partial dechlorination by all of the microbial processes described in this review (Fig. 4.3–4.8). In fact, Tiedje et al. (1993) have demonstrated that microorganisms from Hudson River sediment dechlorinate 34-34-CB in the presence of Aroclor 1242. In addition, Sonzogni et al. (1991) have reported that the weight percents of 34-34-CB, 245-34-CB, and 234-34-CB in Sheboygan River sediments are 10-fold less than in the original Aroclors and have attributed these decreases to *in situ* microbial dechlorination. Similar decreases in congeners responsible for dioxin-like toxicity have been reported in Waukegan Harbor sediments (Stalling, 1982, cited by Brown et al., 1987a) and Lake Ketel-meer sediments (Beurskens et al., 1993). These decreases have also been attributed to *in situ* dechlorination.

While many toxic effects of PCBs seem to be mediated through the Ah receptor, the possibility of other modes of toxicity cannot be ruled out. High dose levels of *ortho*-substituted congeners have been implicated in causing reduction in dopamine content of macaque brain tissue *in vivo* and PC12 cells *in vitro* through an unidentified mechanism not involving the Ah receptor (Seegal et al., 1991). But, according to a review of PCB toxicity studies recently published by the Agency for Toxic Substances and Disease Registry (ATSDR, 1993), data linking PCB exposure to physiologically relevant neurotoxic effects are inconclusive. In particular, widely cited studies purporting to correlate certain behavioral deficits in infants and young children with *in utero* PCB exposure (Jacobson et al., 1985, 1990a,b; Fein et al., 1984) have been criticized for not taking important cofactors into account (ATSDR, 1993).

Whatever the toxicological importance of PCBs, because human exposure is primarily through food, only congeners that show bioaccumulation potential are likely to be important in causing adverse human health effects. Most congeners that accumulate in human tissue are characterized by adjacent *meta* and *para* chlorines (Safe et al., 1985), a feature that makes them particularly susceptible to at least partial dechlorination. Even those that lack this feature, such as 24-4-CB and 235-3-CB (Jones, 1988), have been shown to dechlorinate in the environment. Thus the overall effect of microbial dechlorination of PCBs is to reduce the toxicity and exposure potential resulting from bioaccumulation.

12.2. Biodegradation

Microbially mediated dechlorination of Aroclors typically removes *meta* and/or *para* chlorines to generate primarily *ortho*-enriched mono- through tetrachlorobiphenyls. The actual products of dechlorination will depend on the original Aroclor(s), the specificity of the dechlorination process(es), and the extent to which the dechlorination has progressed. But in all cases the dechlorination products will be more biodegradable than the original PCBs. Numerous bacterial strains with the ability to degrade mono- through tetrachlorobiphenyls have been isolated from soils and aquatic sediments (Furukawa, 1982; Bedard et al., 1986; Abramowicz, 1990; Harkness et al., 1993). Like the dechlorination processes we have described, aerobic bacteria also exhibit congener specificity, but the majority readily degrade the most frequently observed dechlorination products: 2-CB, 2-2-CB, 2-3-CB, 2-4-CB, 24-2-CB, 24-4-CB, 25-2-CB, 25-3-CB, and 25-4-CB (Brown et al., 1984; Bedard et al., 1986; Bedard and Harberl, 1990; Harkness et al., 1993). The PCB degradation pathway is a four-step process beginning with a dioxygenase attack at carbons 2 and 3 and subsequent metabolism through the *meta*-fission pathway to chlorobenzoic acid. The mono- and dichlorobenzoic acids that are generated from the oxidation of dechlorinated PCBs are in turn mineralized by aerobic bacteria (Focht, 1993) and by anaerobic bacteria (Mohn and Tiedje, 1992).

Bedard and colleagues (1987) proposed that PCBs could be totally degraded by anaerobic dechlorination followed by aerobic oxidation. Furthermore, they demonstrated that environmentally dechlorinated Aroclor 1242 extracted from Hudson River sediment could be extensively degraded by *Alcaligenes eutrophus* H850, a bacterium isolated from dredge spoils of the Hudson River. Their experiments demonstrated that the constituents of every congener peak in Aroclor 1242 were transformed by dechlorination, oxidation, or a combination of both. The net effect was that 81% of the dechlorinated PCBs, including all of the 2-CB, were degraded in 48 hours (Bedard et al., 1987) and that the combination of dechlorination and oxidation resulted in decreases of 76%, 78%, 93%, and 80%, respectively, for the total di-, tri-, tetra-, and pentachlorobiphenyls originally present in Aroclor 1242 (Bedard et al., 1987; Bedard, 1990).

The isolation of PCB-degrading bacteria from aquatic sediments does not prove that oxidation of PCBs actually occurs in these sediments, but several recent publications provide evidence that aerobic degradation of dechlorinated PCBs can and does occur in aquatic sediments. Anid et al. (1991) studied the sequential anaerobic–aerobic degradation of PCBs in sealed batch incubations of Hudson River sediment that had been spiked with Aroclor 1242 (300 μg/g wet sediment). The dechlorination resulted in a net decrease of tri- through pentachlorobiphenyls and an increase in mono- and dichlorobiphenyls (from 33% in Aroclor 1242 to 78% in the dechlorinated PCBs). Following dechlorination, the head-space was flushed with oxygen and the sediment was inoculated with a bacterial isolate from the Hudson River. The total PCB concentration decreased by 43% in 96 hours, but only mono- and dichlorobiphenyls were degraded (85% and 30%, respectively). Oxygen delivery to the sediment was restricted to passive diffusion from the head-space through the water into the sediment. Because such diffusion is very inefficient in static systems, it is surprising that such extensive short-term degradation occurred. This experiment does not model conditions in the PCB-contaminated regions of the upper Hudson River, field measurements have shown that the water in these regions remains saturated with oxygen (Kenneth Fish, personal communication), presumably because of natural turbulence.

A recent field study in the Hudson River was conducted in caisson reactors that had been driven into the river bottom to isolate the natural biota and sediment from the river environment (Harkness et al., 1993). The experiment was conducted at ambient river temperatures over a 73 day period from August through October, 1991. Dissolved oxygen concentrations in the water column were maintained at 6–6.5 mg/L through automatically controlled addition of hydrogen peroxide. Ammonia-nitrogen, phosphate, and biphenyl were added periodically in a successful effort to stimulate growth of the indigenous PCB-degrading bacteria. The sediments were stirred to enhance aerobic conditions within the sediments. The concentration of biphenyl-metabolizing bacteria increased by six orders of magnitude, and congener-specific degradation of the dechlorinated sediment PCBs occurred in all experimental caissons. Mono- and dichlorobenzoic acids appeared transiently, concomitantly with the observed increase in biphenyl-degrading bacteria. Furthermore, the specific chlorobenzoic acids produced were those expected from the observed degradation of specific mono- through trichlorobiphenyls. The chlorobenzoic acids did not persist and were presumably degraded by indigenous bacteria. Only 37%–55% of the PCBs in Hudson River sediments were degraded in this field test (Harkness et al., 1993), although more than 90% should be biodegradable on the basis of past laboratory experiments in the absence of sediment (Bedard et al., 1986, 1987; Bedard and Haberl, 1990). Subsequent investigations of the desorption kinetics of PCBs in the same sediments demonstrated that 55% of the PCBs desorbed within 1 day, but only half of the remaining resistant fraction desorbed over the next year (Carroll et al., 1994). Both groups of investigators (Harkness et al., 1993; Carroll et al., 1994) proposed that the resistant fraction consists of PCBs that are dissolved in a condensed phase of the polymeric humic matrix and that diffusion out of this matrix must occur before desorption can occur. Hence, the resistant fraction is likely not bioavailable and represents a primary limitation to short-term biodegradation (Harkness et al., 1993).

The field test described above clearly demonstrated that the indigenous bacteria can degrade the dechlorinated Aroclor 1242 in the sediments when supplements are added and oxygen levels are maintained in the sediment by agitation. But more convincing evidence of biodegradation under natural conditions comes from the recent discovery of chlorobenzoic acids and other PCB metabolites in cores of PCB-contaminated Hudson River sediments. Using a highly sensitive GC/MS procedure, Flanagan and May (1993) detected 2-chlorobenzoic acid (2-CBA), 4-CBA, 24-CBA, 25-CBA, and hydroxylated intermediates of 2-CB corresponding to the expected products of the first two steps of PCB oxidation. The specific chloro-benzoic acids detected were those that would be produced from the oxidation of the dechlorinated PCBs in the sediment. The PCBs in these sediments consisted of 60%–70% mono- and dichlorobiphenyls, primarily 2-CB, 2-2-CB, 2-3-CB, and 2-4-CB. All of these can be degraded to 2-CBA (Bedard and Haberl, 1990), which was the most abundant chlorobenzoic acid detected. Likewise, the 4-CBA is an expected product of 4-CB and 2-4-CB, and 24-CBA and 25-CBA are the expected degradation products of 24-2-CB, 24-3-CB, 24-4-CB, and 25-2-CB, 25-3-CB, and 25-4-CB, respectively. No chlorobenzoic acids were detected in samples of river sediments that were not contaminated with PCBs. Thus the presence of these PCB metabolites is a clear indication of naturally occurring aerobic biodegradation of PCBs and suggests that sequential anaerobic–aerobic PCB degradation is occurring *in situ* in Hudson River sediments (Flanagan and May, 1993). Experiments that model river conditions will be necessary to determine the rates at which such biodegradation occurs.

13. GENERAL CONCLUSIONS

Research has demonstrated that PCB dechlorination is occurring *in situ* in many aquatic sediments and is biologically mediated. Several distinct dechlorination processes can be recognized on the basis of the congener specificity observed at each site and in laboratory experiments. The available evidence is consistent with the hypotheses that the dechlorinating microorganisms derive a benefit from the dechlorination of PCBs and that enzymes account for the observed congener specificity.

The process of anaerobic dechlorination of PCBs has important implications for deciding remediation issues involving anaerobic sediments. Virtually all PCBs with *meta* and *para* chlorines can be dechlorinated. The less chlorinated congeners produced from dechlorination are expected to be less toxic and more readily degraded by aerobic bacteria. The aerobic metabolism of dechlorinated PCBs *in situ* has been demonstrated. Thus the potential exists for the complete biodegradation of PCBs. The challenge that remains is to find ways to facilitate and accelerate these processes.

In designing and evaluating experiments related to enhancement of dechlorination it is imperative that the dechlorination processes potentially present are taken into account. We have provided descriptions of the various known processes so that they may be readily identified. We encourage the evaluation of treatment effects with regard to the various dechlorination processes initially or potentially present. This is best done if Aroclor mixtures, rather than pure congeners, are used to assess the effects on dechlorination, because only the former provide sufficient information to allow discrimination among dechlorination processes.

Enhancement of dechlorination will likely involve identifying and overcoming site-specific limiting factors. Much more basic research is needed to allow the determination of potential site-specific limiting factors. In conducting such experiments, greater emphasis should be placed on using environmentally relevant conditions. Investigations of the fate of PCBs in additional aquatic sediments is needed. Ideally such investigations should include freshwater, estuarine, and marine sites subject to various climates. The significance of basic environmental factors such as temperature, pH, salinity, available carbon, and co-contaminants are important topics for future investigation. Experiments with electron acceptors in which the congener selectivity (dechlorination process) and concentration of electron acceptor(s) are monitored throughout the experiment are also needed. And it is important to determine if dechlorinating microorganisms can take up PCBs only from aqueous solution or if they can also access bulk phase PCBs or PCBs sorbed to surfaces or in an oil phase.

Many of the most definitive experiments cannot be undertaken until microorganisms that mediate PCB dechlorination are isolated. The hypotheses that sulfate-reducing spore-formers are responsible process M and methanogens are responsible for process H can furnish guidance for isolation strategies for these microorganisms, but the involvement of other physiological groups cannot be excluded.

ACKNOWLEDGMENTS

We thank Pam Morris, Dingyi Ye, Max Häggblom and Lily Young, Dan Abramowicz and Michael Brennan, and John F. Brown, Jr., for providing access to unpublished chromatograms, histograms, and tables of congener distributions. We also thank Sherry Mueller for cross-checking references, John Bergeron for many helpful comments on the manuscript, and our families and colleagues for their support and encouragement.

NOTE ADDED IN PROOF

After receiving a pre-print of this review, G.-Y. Rhee wrote us to clarify the analytical techniques that he and his colleagues used in their mass balance experiments with pure congeners. Their clarifications are presented verbatim: "after reviewing our own paper (Rhee et al., 1993c) we realize that the description of how congener concentrations were determined was rather vague and could be left open to misinterpretation. The pure congener standards mentioned in our materials and methods section were used both for product identification as well as for calibration. . . . The calibration standard we used contained a pure congener for each parent compound and each probable daughter product. A 4-point calibration curve was constructed for each individual congener to determine the linearity of the ECD response. All samples were subsequently diluted or concentrated to bring them within the linear range of the calibration mixture. This calibration mixture was created with different concentrations of individual congeners that fell within each congener's linear response range. Therefore we are confident that [un]reliable quantitation was not the reason for the apparent loss of molar mass. . . . the molar loss was not due to the inability to detect daughter products such as 2/2 and 2,6 at low levels. In those cases where we did not detect these daughter [products] we concentrated and re-ran the samples."

Despite these clarifications it is still not clear that each daughter congener in the samples was within the linear range of that congener in the calibration mixture. To bring the parent congener and each daughter congener into the linear range of the calibration standard it would probably be necessary to analyze several concentrations and dilutions of each sample. Failure to do so could lead to serious underestimates, especially of mono- and dichlorobiphenyls. Concentration can also result in physical loss of these volatile congeners. The incomplete mass balance that Rhee and colleagues obtained with some congeners remains unexplained. Our cautions about the quantitation difficulties inherent in experiments using pure congeners are based on our own experiences and remain valid.

REFERENCES

Abramowicz DA (1990): Aerobic and anaerobic biodegradation of PCBs—A review. Crit Rev Biotechnol 10:241–249.

Abramowicz DA, Brennan MJ (1991): Anaerobic and aerobic biodegradation of endogenous PCBs. In Jafvert CT, Rogers JE (eds): Biological Remediation of Contaminated Sediment With Emphasis on Great Lakes. Environmental Research Laboratory, Office of Research and Development, U.S. Environmental Protection Agency, Athens, GA. EPA/600/9-91/001, pp 78–86.

Abramowicz DA, Brennan MJ, Van Dort HM, Gallagher EL (1993): Factors influencing the rate of polychlorinated biphenyl dechlorination in Hudson River sediments. Environ Sci Technol 27:1125–1131.

Abramowicz DA, Brown JF Jr, Harkness MR, O'Donnell MK (1995): *In situ* anaerobic PCB dechlorination and aerobic PCB biodegradation in Hudson River sediments. In Hickey RF (ed): The Implementation of Biotechnology in Industrial Waste Treatment and Bioremediation. Lewis Publishers, pp 57–92.

Alder AC, Häggblom MM, Oppenheimer SR, Young LY (1993): Reductive dechlorination of polychlorinated biphenyls in anaerobic sediments. Environ Sci Technol 27:530–538.

Anid PJ, Nies L, Vogel TM (1991): Sequential anaerobic–aerobic biodegradation of PCBs in the river model. In Hinchee RE, Olfenbuttel RF (eds): On-Site Bioreclamation. Stoneham, MA: Butterworth Meinemann, pp 428–436.

Apajalahti J, Cole J, Tiedje J (1989): Characterization of a dechlorination cofactor: An essential activator for 3-chlorobenzoate dechlorination by the bacterium DCB-1. Abstracts of the 89th Annual Meeting of the American Society for Microbiology, Washington, DC: Abstract Q-36, p. 336.

Assaf-Anid N, Nies L, Vogel TM (1992): Reductive dechlorination of a polychlorinated biphenyl congener and hexachlorobenzene by vitamin B_{12}. Appl Environ Microbiol 58:1057–1060.

Atlas RM, Bartha R (1981): Microbial Ecology: Fundamentals and Applications. Reading, MA: Addison-Wesley Publishing Company.

ATSDR, (1993): Update: Toxicological Profile for Selected PCBs (Aroclor-1260, -1254, -1248, -1242, -1232, -1221, and -1016). Atlanta, GA: U.S. Department of Human Health & Human Services, Public Health Service, Agency for Toxic Substances and Disease Registry, TP-92/16.

Bedard DL (1990): Bacterial transformation of polychlorinated biphenyls. In Kamely D., Chakrabarty A, Omenn GS (eds): Biotechnology and Biodegradation. The Woodlands, Houston, TX: Portfolio Publishing Co., pp 369–388.

Bedard DL, Bunnell SC, Smullen LA: Stimulation of *para* dechlorination of pre-existing Aroclor 1260 in sediment microcosms. Environ Sci Technol (submitted).

Bedard DL, Haberl M (1990): Influence of chlorine substitution pattern on the degradation of polychlorinated biphenyls by eight bacterial strains. Microb Ecol 20:87–102.

Bedard DL, May RJ: Characterization of the polychlorinated biphenyls (PCBs) in the sediments of Woods Pond: Evidence for *in situ* dechlorination. Environ Sci Technol (submitted).

Bedard DL, Unterman R, Bopp LH, Brennan MJ, Haberl ML, Johnson C (1986): Rapid assay for screening and characterizing microorganisms for the ability to degrade polychlorinated biphenyls. Appl Environ Microbiol 51:761–768.

Bedard DL, Wagner RE, Brennan MJ, Haberl ML, Brown JF Jr (1987): Extensive degradation of Aroclors and environmentally transformed polychlorinated biphenyls by *Alcaligenes eutrophus* H850. Appl Environ Microbiol 53:1094–1102.

Bedard DL, Van Dort HM: Stimulation and enrichment of extensive microbial dechlorination of pre-existing Aroclor 1260 in sediment. In preparation.

Bedard DL, Van Dort HM, Bunnell SC, Principe JM, DeWeerd KA, May RJ, Smullen LA: (1993): Stimulation of reductive dechlorination of Aroclor 1260 contaminant in anaerobic slurries of Woods Pond sediment. In Rogers JE, Abramowicz DA (eds): Anaerobic Dehalogenation and its Environmental Implications. Abstracts of the 1992 American Society for Microbiology Conference. Athens, GA: Office of Research and Development, U.S. Environmental Protection Agency, pp 19–21.

Beurskens JEM, Mol GAJ, Barreveld, HL, van Munster B, Winkels HJ (1993): Geochronology of priority pollutants in a sedimentation area of the Rhine River. Environ Toxicol Chem 12:1549–1566.

Bopp RF, Simpson HJ, Deck BI, Kostyk N (1984): The persistence of PCB components in sediments of the lower Hudson. Northeast Environ Sci 3:180–184.

Bosma TNP, van der Meer JR, Schraa G, Tros ME, Zehnder AJB. (1988): Reductive dechlorination of all trichloro- and dichlorobenzene isomers. FEMS Microbiol Ecol 53:223–229.

Boyd SA, Shelton DR, Berry D, Tiedje JM (1983): Anaerobic biodegradation of phenolic compounds in digested sludge. Appl Environ Microbiol 46:50–54.

Boyd SA, Sun S (1989): Residual petroleum and polychlorobiphenyl oils as sorptive phases for organic contaminants in soils. Environ Sci Technol 24:142–144.

Boyer JN (1986): End products of anaerobic chitin degradation by salt march bacteria as substrates for dissimilatory sulfate reduction and methanogenesis. Appl Environ Microbiol 52:1415–1418.

Boyle AW, Blake CK, Price WA II, May HD (1993): Effects of polychlorinated biphenyl congener concentration and sediment supplementation on rates of methanogenesis and 2,3,6-trichlorobiphenyl concentration in an anaerobic enrichment. Appl Environ Microbiol 59:3027–3031.

Brown JF Jr (1990): Identification of environmental PCB transformation processes. In Hutzinger O, Fiedler H (eds): Dioxin '90—EPRI Seminar Short Papers, Organohalogen Compounds. Bayreuth, Germany: Ecoinforma Press, vol 2, pp 21–24.

Brown JF, Bedard DL, Brennan MJ, Carnahan JC, Feng H, Wagner RE (1987a): Polychlorinated biphenyl dechlorination in aquatic sediments. Science 236:709–712.

Brown JF, Wagner RE (1990): PCB movement, dechlorination, and detoxication in the Auschnet estuary. Environ Toxicol Chem 9:1215–1233.

Brown JF, Wagner RE, Bedard DL, Brennan MJ, Carnahan JC, May RJ, Tofflemire TJ (1984): PCB transformations in upper Hudson sediments. Northeast Environ Sci 3:167–179.

Brown JF, Wagner RE, Feng H, Bedard DL, Brennan MJ, Carnahan JC, May RJ (1987b): Environmental dechlorination of PCBs. Environ Toxicol Chem 6:579–593.

Bunce NJ, Kumar Y, Ravanal L, Safe S (1978): Photochemistry of chlorinated biphenyls in iso-octane solution. J Chem Soc Perkin II:880–884.

Bush B, Murphy MJ, Connor S, Snow J, Barnard E (1985): Improvements in glass capillary chromatographic polychlorobiphenyl analysis. J Chromatog Sci 23:509–515.

Capone DC, Kiene RP (1988): Comparison of microbial dynamics in marine and freshwater sediments: Contrasts in anaerobic carbon catabolism. Limnol Ocean 33:725–749.

Carroll KM, Harkness MR, Bracco AA, Balcarcel RR (1994): Application of a permeant/polymer diffusional model to the desorption of polychlorinated biphenyls from Hudson River sediments. Environ Sci Technol 28:253–258.

David M (1990): PCB Congener Distribution in Sheboygan River Sediment, Fish, and Water. MS Thesis. Water Chemistry Program, University of Wisconsin, Madison.

DeWeerd KA, Mandelco L, Tanner RS, Woese CR, Suflita JM (1990): *Desulfomonile tiedjei* gen. nov. and sp. nov., a novel anaerobic, dehalogenating, sulfate-reducing bacterium. Arch Microbiol 154:23–30.

DiToro DM, Horzempa LM (1982): Reversible and resistant components of PCB adsorption-desorption: Isotherms. Environ Sci Technol 16:594–602.

DiToro DM, Horzempa LM, Casey MM, Richardson W (1982): Reversible and resistant components of PCB adsorption-desorption: Adsorbent concentration effects. J Great Lakes Res 8:336–349.

Dolfing J (1990): Reductive dechlorination of 3-chlorobenzoate is coupled to ATP production and growth in an anaerobic bacterium, strain DCB-1. Arch Microbiol 153:264–266.

Dolfing J, Tiedje JM (1986): Hydrogen cycling in a three-tiered food web growing on the methanogenic conversion of 3-chlorobenzoate. FEMS Microbiol Ecol 38:293–298.

Dolfing J, Tiedje JM (1987): Growth yield increase linked to reductive dechlorination in a defined 3-chlorobenzoate degrading methanogenic co-culture. Arch Microbiol 149:102–105.

Erickson MD (1986): Analytical Chemistry of PCBs. Boston: Butterworth Publishers.

Fathepure BZ, Tiedje JM, Boyd SA (1988): Reductive dechlorination of hexachlorobenzene to tri- and dichlorobenzenes in anaerobic sewage sludge. Appl Environ Microbiol 54:327–330.

Fava F, Cinti S, Marchetti L (1993): Dechlorination of Fenclor-54 and of a synthetic mixture of polychlorinated biphenyls by anaerobic microorganisms. Appl Microbiol Biotechnol 38:808–814.

Fein GG, Jacobson JL, Jacobson SW, Schwartz PM, Dowler JK (1984): Prenatal exposure to polychlorinated biphenyls: Effects on birth size and gestation age. J Pediatr 105:315–320.

Ferguson ML, Metcalf CD (1989): Distribution of PCB congeners in sediments of the Otonabee River–Rice Lake system, Petersborough, Canada. Chemosphere 19:1321–1328.

Fischer R, Ballschmiter K (1988): Ortho-substituent correlated retention of polychlorinated biphenyls on a 50% N-octyl-methylpolysiloxane stationary phase by HRGC/MSD. Fresenius Z Anal Chem 332:441–446.

Fischer R, Ballschmiter K (1989): Congener-specific identification of technical PCB mixtures by capillary gas chromatography on a N-octyl-methyl silicone phase (SB-octyl-50) with electron capture and mass selective detection. Fresenius Z Anal Chem 335:457–463.

Flanagan WP, May RJ (1993): Metabolite detection as evidence for naturally occurring aerobic PCB biodegradation in Hudson River sediments. Environ Sci Technol 27:2207–2212.

Focht DD (1993): Microbial degradation of chlorinated biphenyls. In Bollag J-M, Stotzky G (eds): Soil Biochemistry. New York: Marcel Dekker, Inc., vol 8, pp 341–407.

Furukawa K (1982): Microbial degradation of polychlorinated biphenyls. In Chakrabarty AM (ed): Biodegradation and Detoxification of Environmental Pollutants. Boca Raton, FL: CRC Press, Inc., pp 33–57.

Gantzer CJ, Wackett LP (1991): Reductive dechlorination catalyzed by bacterial transition-metal coenzymes. Environ Sci Technol 25:715–722.

Gunsalus RP, Romesser JA, Wolfe RS (1978): Preparation of coenzyme M analogues and their activity in the methyl coenzyme M reductase system of *Methanobacterium thermotrophicum*. Biochemistry 17:2374–2376.

Harkness MR, McDermott JB, Abramowicz DA, Salvo JJ, Flanagan WP, Stephens ML, Mondello FJ, May RJ, Lobos JH, Carroll KM, Brennan MJ, Bracco AA, Fish KM, Warner GL, Wilson PR, Dietrich DK, Lin DT, Morgan CB, Gately WL (1993): *In situ* stimulation of aerobic PCB biodegradation in Hudson River sediments. Science 259:503–507.

Holmes DA, Harrison BK, Dolfing J (1993): Estimation of Gibbs free energies of formation for polychlorinated biphenyls. Environ Sci Technol 27:725–731.

Hutzinger OH, Safe S, Zitko V (1974): The Chemistry of the PCBs. Cleveland, OH: CRC Press, Inc.

Hutzinger O, Veerkamp W (1981): Xenobiotic chemicals with pollution potential. In Leisinger T, Hutter R, Cook AM, Nuesch J (eds): Microbial Degradation of Xenobiotics and Recalcitrant Compounds. New York: Academic Press, pp 3–45.

Jacobson JL, Jacobson SW, Humphrey HEB (1990a): Effects of in utero exposure to polychlorinated-biphenyls and related contaminants on cognitive-functioning in young children. J Pediatr 116:38–45.

Jacobson JL, Jacobson SW, Humphrey HEB (1990b): Effects of exposure to PCBs and related compounds on growth and activity in children. Neurotoxicol Teratol 12:319–326.

Jacobson SW, Fein GG, Jacobson JL, Schwartz PM, Dowler JK (1985): The effect of intrauterine PCB exposure on visual recognition memory. Child Dev 56:853–860.

Jones KC (1988): Determination of polychlorinated biphenyls in human foodstuffs and tissues: Suggestions for a selective congener analytical approach. Sci Total Environ 68:141–159.

Karickoff SW (1980): Sorption kinetics of hydrophobic pollutants in natural sediments. In Baker RA (ed): Contaminants and Sediments. Ann Arbor, MI: Ann Arbor Science Publisher, vol II, pp 193–205.

Karickoff SW, Morris KR (1985): Sorption dynamics of hydrophobic pollutants in sediment suspensions. Environ Toxicol Chem 4:469–479.

Kimbrough RD, Squire RA, Linder RE, Strandberg JD, Montali RJ, Burse VW (1975): Induction of liver tumors in Sherman strain female rats by polychlorinated biphenyl Aroclor 1260. J Natl Cancer Inst 55:1453–1459.

Lake JL, Pruell RJ, Osterman FA (1991): Dechlorination of polychlorinated biphenyls in sediments of New Bedford Harbor. In Baker RA (ed): Organic Substances and Sediments in Water. Chelsea, MI: Lewis Publishers Inc., pp 173–197.

Lake JL, Pruell RJ, Osterman FA (1992): An examination of dechlorination processes and pathways in New Bedford Harbor sediments. Marine Environ Res 33:31–47.

Ludwig JP, Giesy JP, Summer CL, Bowerman W, Aulerich R, Bursian S, Auman HJ, Jones

PD, Williams LL, Tillitt DE, Gibertson M (1993): A comparison of water quality criteria for the Great Lakes based on human and wildlife health. J Great Lakes Res 19:789–807.

Mavoungou R, Massé R, Sylvestre M (1991): Microbial dehalogenation of 4,4'-dichlorobiphenyl under anaerobic conditions. Sci Total Environ 101:263–268.

May HD, Boyle AW, Price WA II, Blake CK (1992): Subculturing of a polychlorinated biphenyl-dechlorinating anaerobic enrichment on solid media. Appl Environ Microbiol 58:4051–4054.

Mohn WW, Tiedje JM (1991): Evidence for chemiosmotic coupling of reductive dechlorination and ATP synthesis in *Desulfomonile tiedjei*. Arch Microbiol 157:1–6.

Mohn WW, Tiedje JM (1992): Microbial reductive dehalogenation. Microbiol Rev 56:482–507.

Montgomery L, Vogel TM (1992): Dechlorination of 2,3,5,6-tetrachlorobiphenyl by a phototrophic enrichment culture. FEMS Microbiol Lett 94:247–250.

Moore JA (1991): Reassessment of Liver Findings in PCB Studies for Rats. Washington, DC: Institute for Evaluating Health Risks, July 1.

Morris PJ, Mohn WW, Quensen JF III, Tiedje JM, Boyd SA (1992a): Establishment of a polychlorinated biphenyl-degrading enrichment culture with predominantly *meta* dechlorination. Appl Environ Microbiol 58:3088–3094.

Morris PJ, Quensen JF III, Tiedje JM, Boyd SA (1992b): Reductive debromination of the commercial polybrominated biphenyl mixture firemaster BP6 by anaerobic microorganisms from sediments. Appl Environ Microbiol 58:3249–3256.

Morris PJ, Quensen JF III, Tiedje JM, Boyd SA (1993): An assessment of the reductive debromination of polybrominated biphenyls in the Pine River reservoir. Environ Sci Technol 27:1580–1586.

Mullin M, Sawka G, Safe L, McCrindle S, Safe S (1981): Synthesis of the octa- and nonachlorobiphenyl isomers and congeners and their quantitation in commercial polychlorinated biphenyls and identification in human breast milk. J Anal Toxicol 5:138–142.

Mullin MD, Pochini CM, McCrindle S, Romkes M, Safe SH, Safe LM (1984): High-resolution PCB analysis: Synthesis and chromatographic properties of all 209 PCB congeners. Environ Sci Technol 18:468–476.

Nies L, Vogel TM (1990): Effects of organic substrates on dechlorination of Aroclor 1242 in anaerobic sediments. Appl Environ Microbiol 56:2612–2617.

Nies L, Vogel TM (1991): Identification of the proton source for the microbial reductive dechlorination of 2,3,4,5,6-pentachlorobiphenyl. Appl Environ Microbiol 57:2771–2774.

Norback DH, Weltman RH (1985): Polychlorinated biphenyl induction of hepatocellular carcinoma in the Sprague-Dawley rat. Environ Health Perspect 60:97–105.

Oremland RS (1988): Biogeochemistry of methane production. In Zehnder AJB (ed): Biology of Anaerobic Microorganisms. New York: John Wiley & Sons, pp 641–705.

Oremland RS, Capone DG (1988): Use of "specific" inhibitors in biogeochemistry and microbial ecology. In Marshall KC (ed): Advances in Microbial Ecology. New York: Plenum, vol 10, pp 285–383.

Quensen JF III, Boyd SA, Tiedje JM (1990): Dechlorination of four commercial polychlorinated biphenyl mixtures (Aroclors) by anaerobic microorganisms from sediments. Appl Environ Microbiol 56:2360–2369.

Quensen JF III, Tiedje JM, Boyd SA (1988): Reductive dechlorination of polychlorinated biphenyls by anaerobic microorganisms from sediments. Science 242:752–754.

Rhee G-Y, Bush B, Bethoney CM, DeNucci A, Oh H-M, Sokol RC (1993a): Anaerobic dechlorination of Aroclor 1242 as affected by some environmental conditions. Environ Toxicol Chem 12:1033–1039.

Rhee G-Y, Bush B, Bethoney CM, DeNucci A, Oh H-M, Sokol RC (1993b): Reductive dechlorination of Aroclor 1242 in anaerobic sediments: Pattern, rate, and concentration dependence. Environ Toxicol Chem 12:1025–1032.

Rhee G-Y, Sokol RC, Bethoney CM, Bush B (1993c): A long-term study of anaerobic dechlorination of PCB congeners by sediment microorganisms: Pathways and mass balance. Environ Toxicol Chem 12:1829–1834.

Rhee G-Y, Sokol RC, Bethoney CM, Bush B (1993d): Dechlorination of polychlorinated biphenyls by Hudson River sediment organisms: Specificity to the chlorination pattern of congeners. Environ Sci Technol 27:1190–1192.

Rhee G-Y, Sokol RC, Bush B, Bethoney CM (1993e): Long-term study of the anaerobic dechlorination of Aroclor 1254 with and without biphenyl enrichment. Environ Sci Technol 27:714–719.

Roth JA, Dakoji SR, Hughes RC, Carmody RE (1994): Hydrogenolysis of polychlorinated biphenyls by sodium borohydride with homogeneous and heterogeneous nickel catalysts. Environ Sci Technol 28:80–87.

Rust SW (1984): Study of Errors Generated in the Chemical Analysis of Environmental Samples. Final Report, Task 67, Contract No. 68-01-6721. Washington, DC: Office of Toxic Substances, U.S. Environmental Protection Agency.

Ruzo LO, Zabik MJ, Schuetz RD (1974): Photochemistry of bioactive compounds. Photochemical processes of polychlorinated biphenyls. J Am Chem Soc 96:3809–3813.

Safe S (1990): Polychlorinated biphenyls (PCBs) and polybrominated biphenyls (PBBs): Biochemistry, toxicology, and mechanism of action. Crit Rev Toxicol 21:51–88.

Safe S (1993): Toxicology, structure–function relationship, and human and environmental health impacts of polychlorinated biphenyls: Progress and problems. Environ Health Perspect 100:259–268.

Safe S, Safe L, Mullin M (1985): Polychlorinated biphenyls: Congener-specific analysis of a commercial mixture and a human milk extract. J Agric Food Chem 33:24–29.

Schink B (1988): Principles and limits of anaerobic degradation: Environmental and technological aspects. In Zehnder AJB (ed): Biology of Anaerobic Microorganisms. New York: John Wiley & Sons, pp 771–846.

Schulz DE, Petrick G, Duinker JC (1989): Complete characterization of polychlorinated biphenyl congeners in commercial Aroclor and Clophen mixtures by multidimensional gas chromatography–electron capture detection. Environ Sci Technol 23:852–859.

Schwartz TR, Tillitt DE, Feltz KP, Peterman PH (1993): Determination of mono- and non-O,O'-chlorine substituted polychlorinated biphenyls in Aroclors and environmental samples. Chemosphere 26:1443–1460.

Seegal RF, Bush B, Shain W (1991): Neurotoxicology of ortho-substituted polychlorinated biphenyls. Chemosphere 23:1941–1949.

Shelton DR, Tiedje JM (1984): General method for determining general anaerobic biodegradation potential. Appl Environ Microbiol 47:850–857.

Sonzogni W, Maack L, Gibson T, Lawrence J (1991): Toxic polychlorinated biphenyl congeners in Sheboygan River (USA) sediments. Bull Environ Contam Toxicol 47:398–405.

Stalling DL (1982): Isomer Specific Composition of PCB Residues in Fish and Sediment from Waukegan Harbor and Other Great Lakes Fish. Columbia, MO: Columbia National Fisheries Research Laboratory.

Stewart Laboratories, Inc. (1982): Housatonic River Study 1980 and 1982 Investigations, Final Report. Knoxville, TN: Stewart Laboratories, Inc.

Suflita JM, Horowitz A, Shelton DR, Tiedje JM (1982): Dehalogenation: A novel pathway for the anaerobic biodegradation of haloaromatic compounds. Science 218:1115–1117.

Sugiura K (1992): Congener-specific PCB analyses of a sediment core of Lake Shinji, Japan. Chemosphere 24:427–432.

Taylor BF, Oremland RS (1979): Depletion of adenosine triphosphate in *Desulfovibrio* by oxyanions of group VI elements. Curr Microbiol 3:101–103.

Tiedje JM, Boyd SA, Fathepure BZ (1987): Anaerobic degradation of chlorinated aromatic hydrocarbons. Dev Ind Microbiol 27:117–127.

Tiedje JM, Quensen JF III, Chee-Sanford J, Schimel JP, Boyd SA (1993): Microbial reductive dechlorination of PCBs. Bioremediation 4:231–240.

Tiedje JM, Quensen JF III, Mohn WW, Schimel JP, Cole JA, Boyd SA (1991): Reductive dechlorination of chlorinated aromatic pollutants. In Rossmore RW (ed): Biodeterioration and Biodegradation 8. New York: Elsevier Applied Science, pp 292–307.

Van Dort HM, Bedard DL (1991): Reductive *ortho*-dechlorination and *meta*-dechlorination of a polychlorinated biphenyl congener by anaerobic microorganisms. Appl Environ Microbiol 57:1576–1578.

Weyl PK (1970): Oceanography: An Introduction to the Marine Environment. New York: John Wiley & Sons, p 324.

Williams WA (1994): Microbial reductive dechlorination of trichlorobiphenyls in anaerobic sediment slurries. Environ Sci Technol 28:630–635.

Wu SC, Gschwend PM (1986): Sorption kinetics of hydrophobic organic compounds in natural sediments and soils. Environ Sci Technol 20:717–725.

Ye DY (1994): Characterization of PCB (Polychlorobiphenyls) Dechlorination by Anaerobic Microorganisms From Hudson River Sediments. PhD Dissertation, Department of Crop & Soil Sciences, Michigan State University, East Lansing.

Ye DY, Quensen JF, Tiedje JM, Boyd SA (1992): Anaerobic dechlorination of polychlorobiphenyls (Aroclor 1242) by pasteurized and ethanol-treated microorganisms from sediments. Appl Environ Microbiol 58:1110–1114.

5

BACTERIAL CO-METABOLISM OF HALOGENATED ORGANIC COMPOUNDS

LAWRENCE P. WACKETT

Department of Biochemistry and Institute for Advanced Studies in Biological Process Technology, University of Minnesota, St. Paul, Minnesota 55108

1. INTRODUCTION

1.1. Chemistry and Distribution of Organohalides

The halogens occupy the penultimate column on the right edge of the Periodic Table of the elements (Fig. 5.1). These group 17 elements are fluorine, chlorine, bromine, iodine, and astatine. The total amount of astatine on the earth at any given moment is estimated to be 50 mg, so this element will not be considered here (Kirk, 1991). As a class, the halogens are strongly electronegative and often exist as anions in aqueous solution. All of the halogens form stable bonds with carbon atoms. Bond strengths decrease markedly with increasing molecular weight to give the order: C–F > C–Cl > C–Br > C–I. Indeed, carbon–fluorine bonds are remarkably strong, with bond dissociation energies in the range of 106–115 Kcal/mol (Reineke, 1984). By contrast, carbon–iodine bonds are relatively weak ($\sim$ 70 Kcal/mol); for example, they are labile to photocatalytic cleavage. The reactivity of a particular carbon–halogen bond is strongly influenced by the type of carbon atom to which it is attached. For example, nucleophilic displacement of fluoride from a perfluoroalkane is difficult to accomplish. In nucleophilic aromatic substitution, by contrast, aryl fluorides are more reactive than corresponding aryl chlorides, bromides, and iodides (March, 1985).

Microbial Transformation and Degradation of Toxic Organic Chemicals, pages 217–241
© *1995 Wiley-Liss, Inc.*

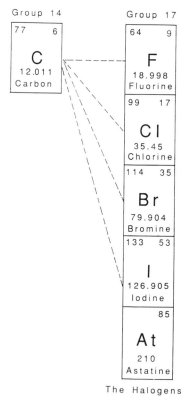

Group 14   Group 17

77	6
C	
12.011	
Carbon	

64	9
F	
18.998	
Fluorine	

99	17
Cl	
35.45	
Chlorine	

114	35
Br	
79.904	
Bromine	

133	53
I	
126.905	
Iodine	

	85
At	
210	
Astatine	

The Halogens

Figure 5.1. Carbon and group 17 elements in the periodic table. The numbers in the upper left, upper right, and lower middle of the boxes refer to covalent radius (in pm), atomic number, and relative atomic mass (^{12}C = 12.000), respectively.

The great variation in ionic radii for the halogen series is also relevant to organohalide metabolism. Fluorine is not much larger than a hydrogen substituent; this may "fool" enzymes into sterically mistaking a C–F bond for a C–H bond (Walsh, 1983). A chlorine substituent is on the same order of size as a hydroxyl group. Organic compounds containing bulky bromine and iodine substituents are sometimes poorly metabolized because of steric crowding.

Thousands of halogenated organic compounds are listed in the Chemical Abstracts registry (Hutzinger and Veerkamp, 1981). This includes a significant number of natural products and important industrial materials. Most of the latter are chlorinated compounds and include important pesticides, solvents, and chemical feedstocks. Their usage expanded greatly in the early to middle twentieth century, as indicated by the explosion in patents issued for new organochloride formulations during this period (Hutzinger and Veerkamp, 1981). There is also a vast literature describing halogenated natural products (Neidleman and Geigert, 1986). 3,5-Diiodotyrosine was first detected in 1896 (by Drechsel) as an algal biosynthetic endproduct. Since then, over 700 halogenated natural products have been discovered.

1.2. Toxicology of Organohalides

The rush to produce new organohalide compounds in the period 1940–1960 was predicated on the importance of their applications. DDT, 1,1-*bis*(chlorophenyl)-2,2,2-trichloroethane, was responsible for controlling pest populations and thereby saved millions of people from malaria and other insect-borne diseases. Organohalide insecticides and herbicides contributed significantly to the large increase in crop yields attained by farmers in industrialized countries. Moreover, organohalide pesticides were persistent in the field, allowing their infrequent application. Although the acute toxicity of most of these formulations is low, negative impacts on human health have been uncovered since the midcentury. The mammalian metabolism of organohalides has been extensively investigated. The metabolic reactions known to transform xenobiotic compounds entering the body usually serve to protect the organism from toxic effects. With organohalides, these same reactions sometimes render the compound more reactive and hence more toxic (Anders and Pohl, 1985). General examples are shown in Figure 5.2.

Mammalian liver glutathione transferases catalyze the nucleophilic attack of the thiol group of glutathione on electrophilic halogen-bearing carbon centers (Jakoby and Keen, 1977; Keen et al., 1976). The resultant glutathione conjugate is transported to the kidney, metabolically processed, and excreted in the urine. This series of responses can protect critical macromolecules from being damaged by their reaction with electrophilic organohalides. Mammalian liver cytochrome P_{450} monooxygenases also serve to functionalize hydrophobic xenobiotic compounds for fur-

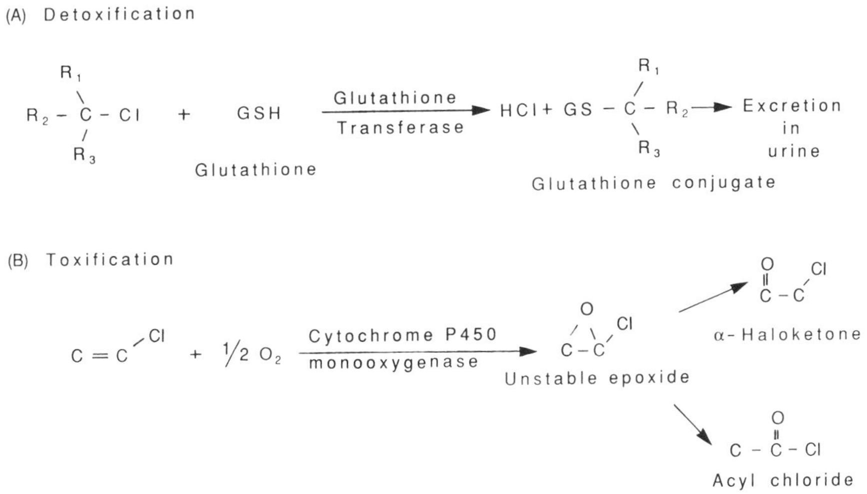

Figure 5.2. General routes for metabolism of organochlorides in mammalian liver.

ther processing and excretion from the body (Porter and Coon, 1991). In some instances, oxygen insertion reactions give rise to intermediates significantly more electrophilic than the starting compound. Figure 5.2B shows oxygen addition to a chloroalkene yielding an unstable epoxide. Subsequent spontaneous oxirane opening produces an α-haloketone or an acyl chloride, and both are highly reactive chemical groups. They are known to alkylate critical intracellular molecules.

The modification of DNA can lead to birth defects or cancer (Miller and Miller, 1985). In general, the toxic and/or carcinogenic effects of organohalides are due to metabolic activation to reactive intermediates. A notable exception is 2,3,7,8-tetrachlorodibenzodioxin (dioxin), which is largely inert to metabolism but can still be exceedingly toxic to mammals (Hutzinger and Veerkamp, 1981).

1.3. Challenges in the Microbial Metabolism of Organohalides

The diversity of structures and the chemical inertness of organohalides pose particular problems for microorganisms. The general mechanisms of carbon–halogen bond cleavage catalyzed by bacteria are shown in Figure 5.3. For an organohalogen to enter intermediary metabolism, all halogens must be removed and the carbon skeleton must be transformed into common intermediate(s). Furthermore, the metabolic route should not generate highly reactive chemical groups such as those shown in Figure 5.2B. These constraints serve to limit the possibilities for microbial assimilation of organohalides. Additionally, one must consider whether microorganisms have had sufficient time to evolve the series of enzymes necessary to construct new metabolic pathways for handling highly chlorinated compounds.

Despite the apparent limitations, bacteria succeed in transforming organohalides in three general ways:

1. Metabolism into trunk pathways
2. Use of halocarbons as final electron acceptors for ATP-generating electron transport
3. Via co-metabolic, or gratuitous, reactions

Funneling into trunk pathways usually involves overlaying a new metabolic reaction onto an existing metabolic pathway. As an example, the C_1 pathway used by methanotrophic and methylotrophic bacteria is shown in Figure 5.4. This pathway generates energy and assimilates carbon at the oxidation level of formaldehyde (Anthony, 1982). Some methylotrophs contain an additional gene encoding for an inducible dichloromethane dehalogenase (LaRoche and Leisinger, 1990). This enzyme catalyzes a net hydrolysis of dichloromethane yielding formaldehyde (Kohler-Staub and Leisinger, 1985). Bacteria containing dichloromethane dehalogenase and the basic C_1 pathway are capable of growing on dihalomethanes as their sole source of carbon and energy.

Currently, two pure cultures are indicated to link organohalide reduction with the generation of ATP. One example is the reduction of 3-chlorobenzoate to benzoate by *Desulfomonile tiedjei* (Dolfing, 1990). The other is the reductive dechlorination

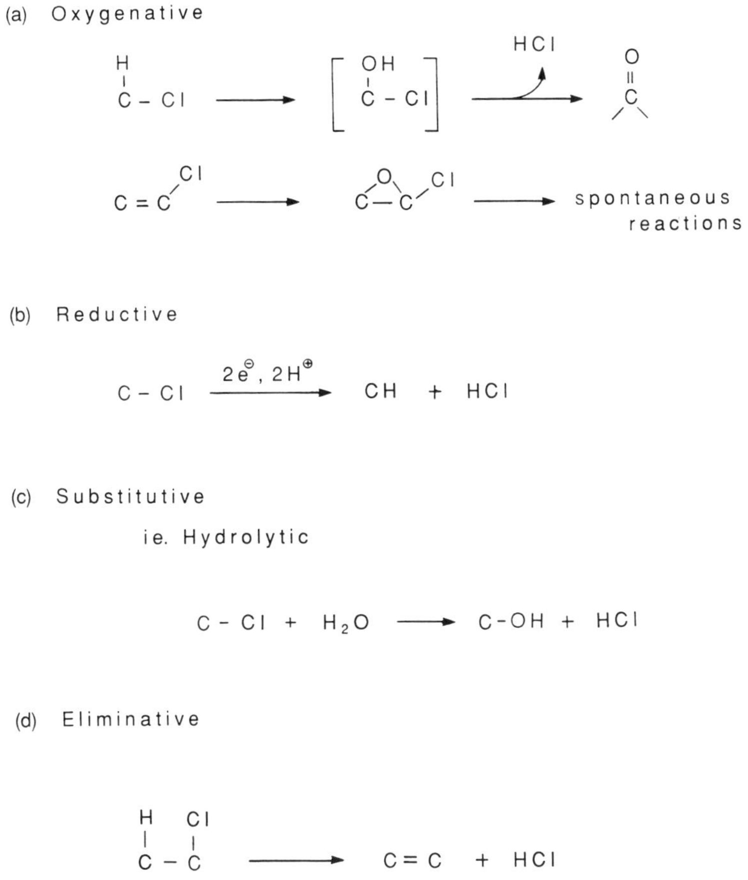

(a) Oxygenative

$$
\begin{array}{ccc}
\underset{\underset{\displaystyle C - Cl}{|}}{\overset{\displaystyle H}{|}} & \longrightarrow & \left[\underset{\underset{\displaystyle C - Cl}{|}}{\overset{\displaystyle OH}{|}} \right] \xrightarrow{\ HCl\ } \overset{O}{\underset{\diagup \diagdown}{\overset{\|}{C}}}
\end{array}
$$

$$
C = C \diagup \overset{\displaystyle Cl}{} \longrightarrow \overset{O}{\underset{C - C}{\diagup \diagdown}}\diagdown Cl \longrightarrow \begin{array}{l} \text{spontaneous} \\ \text{reactions} \end{array}
$$

(b) Reductive

$$
C - Cl \xrightarrow{2e^{\ominus},\ 2H^{\oplus}} CH + HCl
$$

(c) Substitutive

ie. Hydrolytic

$$
C - Cl + H_2O \longrightarrow C-OH + HCl
$$

(d) Eliminative

$$
\underset{\underset{\displaystyle C}{|} - \underset{\displaystyle C}{|}}{\overset{\displaystyle H \quad Cl}{|\quad |}} \longrightarrow C = C + HCl
$$

Figure 5.3. General mechanisms of bacterial catalyzed dehalogenation.

of tetrachloroethylene to *cis*-1,2-dichloroethylene by *Dehalobacter restrictus* (Holliger, 1991). As efforts increase to understand anaerobic bacteria, more examples of halocarbons serving as bacterial final electron acceptors are likely to emerge.

Lastly, bacteria expressing broad-specifity catabolic enzymes may transform organohalides without any direct linkage to their carbon and energy metabolism. This falls under the commonly used description of *co-metabolism*. In those instances where carbon–halogen bonds are cleaved, the gratuitous or co-metabolic reaction(s) can be important for environmental detoxification. The evolutionary raw material for co-metabolism is the diverse biocatalysts assembled by bacteria. Nature recycles many chemically inert molecules, for example, methane and diatomic nitrogen (Fig. 5.5). These reactions are exclusively mediated by bacteria and are dependent on complex metalloenzyme catalysts (Wackett et al., 1989b). There has been great interest in the metal clusters present in nitrogenase and methane mono-oxygenases, respectively. These and other bacterial enzymes are responsible for the

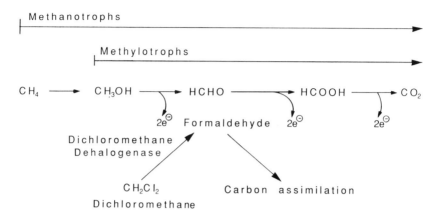

Figure 5.4. C_1 metabolism by methanotrophs and methylotrophs.

global cycling of carbon, nitrogen, oxygen, and sulfur on a megaton scale. Large quantities of environmental organohalide pollutants could conceivably be transformed by some of these same catalysts acting co-metabolically. For example, many broad-specificity bacterial oxygenases function physiologically in the oxygenation of hydrophobic substrates. Oxygenases must accomplish at least two functions to effect catalysis: 1) substrate binding and 2) activation of either or both oxygen and organic substrate. In most cases, oxygenases generate a highly reactive monoatomic or diatomic active oxygen species. The juxtaposition of the organic substrate next to the activated oxygen species governs reactivity. The binding of hydrophilic substrates can be tightly controlled by enzymes. Hydrophobic interactions offer less potential for discrimination. Thus, many hydrophobic compounds, including organohalides, readily partition into the active site of oxygenases such as toluene dioxygenase or methane monooxygenase and are subject to electrophilic attack by the respective reactive oxygenating species. Reductive dechlorination,

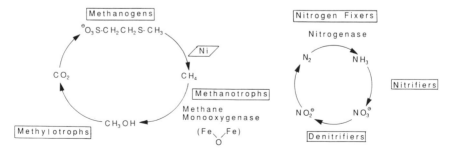

Figure 5.5. Elemental cycles of single carbon compounds and nitrogen showing key reactions in which bacteria figure prominently.

likewise, proceeds in two steps: 1) substrate binding and 2) electron transfer. These oxidative and reductive reactions with organohalide substrates are discussed below.

2. COOXIDATIVE METABOLISM OF ORGANOHALIDES

Microbial oxygenases are of fundamental importance in the recycling of organic matter on earth (Dagley, 1972). Oxygenases functionalize unreactive compounds such as hydrocarbons as a prelude to further oxidation by other oxidoreductases. Relatively unreactive halogenated organic compounds are also functionalized by oxygenase enzymes, and this often leads to unstable products that decompose spontaneously (see Fig. 5.3). Some prominent examples of halocarbon co-oxidation are known with trichloroethylene (TCE) (Table 5.1). TCE is an unreactive, widely used industrial solvent. It is rather nontoxic to mammals, and it is questionably a weak carcinogen. TCE is the most common organic groundwater pollutant in the United States (Storck, 1987). The wide distribution of TCE is of concern because of recent observations that anaerobic bacteria can reductively dehalogenate TCE to form vinylchloride (Vogel and McCarty, 1985). Vinylchloride is strongly mutagenic and carcinogenic in experimental animals, and it is a known human carcinogen (Maltoni and Lefemine, 1974). This knowledge has provided impetus to efforts in obtaining TCE-degrading bacteria. In over a decade, however, no one has succeeded in obtaining a well-documented example of a bacterium able to grow on TCE as a carbon and energy source. This has driven recent studies on the cometabolism of TCE by bacteria and their catabolic oxygenases.

Theoretically, bacterial oxygenases can generate unstable TCE oxidation products that unravel via spontaneous and predictable routes to give nonchlorinated, metabolizable endproducts. A monooxygenase might yield TCE-epoxide; a dioxygenase might produce 1,2-dihydroxy-TCE (Fig. 5.6). TCE-epoxide readily undergoes hydration in aqueous media to form 1,2-dihydroxy-TCE. This latter compound rapidly *gem*-eliminates two equivalents of HCl to form glyoxylate and rearranges via a carbon–carbon bond scission reaction to yield formate and carbon monoxide (Miller and Guengerich, 1982). Carbon monoxide is readily oxidizable to CO_2 by carboxydotrophic and other bacteria. It was difficult to predict beforehand whether monooxygenase or dioxygenase catalysts would perform very differently in TCE oxidation. Furthermore, not all oxygenases will bind TCE, so only a subset will catalyze the reactions shown in Figure 5.6 (Wackett et al., 1989a). To date, a minimum of nine oxygenases have been implicated in TCE oxidation (Table 5.1). We have investigated TCE oxidation by soluble methane monooxygenase and toluene dioxygenase in detail, and these data are presented below.

2.1. Methanotrophs and Soluble Methane Monooxygenase

Methanotrophs were shown to oxidize TCE, and this ability was attributed to methane monooxygenase (MMO) (Table 5.1). MMO oxidizes methane to methanol in the first step of C_1 oxidative metabolism by methanotrophs (Anthony, 1982) (Fig.

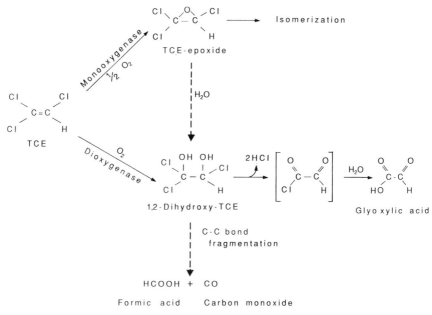

Figure 5.6. Theoretical pathways of TCE oxidation by monooxygenases and dioxygenases.

TABLE 5.1. Oxygenases and Organisms Implicated in TCE Oxidation and Their Relative *In Vivo* Rates

Organism	Enzyme	Rate (nmol/min/mg)	Reference
Methylosinus trichosporium OB3b	Soluble methane monooxygenase	20–150	Oldenhuis et al. (1989), Tsien et al. (1989)
Pseudomonas cepacia G4	Toluene 2-monooxygenase	8	Folsom et al. (1990)
P. mendocina	Toluene 4-monooxygenase	2	Winter et al. (1989)
P. putida	Toluene dioxygenase	2	Wackett and Gibson (1988)
Nitrosomonas europaea	Ammonia monooxygenase	1	Arciero et al. (1989)
Methylocystis parvus OBBP	Particulate methane monooxygenase	0.7	DiSpirito et al. (1992)
Mycobacterium sp.	Propane monooxygenase	0.5	Wackett et al. (1989a)
Alcaligenes eutrophus JMP 134	Phenol hydroxylase	0.2	Harker and Kim (1990)
Rhodococcus erythropolis	Isoprene oxygenase	0.2	Ewers et al. (1990)

TABLE 5.2. Steady-Rate Kinetic Data for Oxidation of Chlorinated Ethylenes by Soluble Methane Monooxygenase of *Methylosinus trichosporium* OB3b[a]

Compound	V_{max} (nmol/min/mg)	k_{cat} (s^{-1})	K_m (μM)	k_{cat}/K_m (s^{-1} M^{-1})
Vinyl chloride	748	3.0	33	9×10^4
cis-Dichloroethylene	935	3.8	28	14×10^4
trans-Dichloroethylene	888	3.6	38	10×10^4
1,1-Dichloroethylene	648	2.7	18	15×10^4
Trichloroethylene	682	2.8	35	8×10^4

[a]Data are from Fox et al. (1990).

5.4). There are, however, two distinct forms of MMO, and the role of each in TCE oxidation was undefined in early studies. A membrane MMO, which may be a copper protein, is expressed under growth conditions of copper sufficiency. At low copper concentrations, a soluble MMO is biosynthesized by some methanotrophs. The soluble MMO has been purified from *Methylococcus capsulatus* (Green and Dalton, 1985; Woodland and Dalton, 1984) and *Methylosinus trichosporium* OB3b (Fox et al., 1989). Both enzyme systems are comprised of three protein components, and the hydroxylase component contains a binuclear iron center thought to be the site of oxygen binding and activation (Ericson et al., 1988; Fox et al., 1988). Originally, it was proposed that the membrane MMO was most responsible for *in vivo* TCE oxidation by methanotrophs (Henry et al., 1989). Subsequently, two independent investigations clearly established that soluble MMO expression in *Methylosinus trichosporium* OB3b is required for high rates of TCE oxidation (Oldenhuis et al., 1989; Tsien et al., 1989). Other methanotrophs also require soluble MMO expression under low copper growth conditions to show substantial ability to oxidize TCE (Brusseau et al., 1990). Most recently, studies with purified soluble MMO confirmed the conclusions of the *in vivo* studies (Fox et al., 1990). The three soluble MMO component proteins from *Methylosinus trichosporium* OB3b oxidize TCE, dichloroethylenes, and vinylchloride at rates approaching those observed with the physiological substrate methane (Table 5.2). The K_m values were, likewise, similar to the K_m for the physiological substrate. Soluble MMO is among the most nonspecific enzymes known; over 100 compounds are transformed by this enzyme (Colby et al., 1977; Green and Dalton, 1989).

Complete product stoichiometries for TCE oxidation by soluble MMO were determined. Bioremediation schemes that are efficient and safe must take into account the fate of the enzymatic reaction products. Due to the potential for generating reactive intermediates, the experiments were conducted using purified enzyme components in a well-defined *in vitro* system. Multiple methods were used to determine quantitatively the full range of expected products; organic acids were determined by HPLC, 2,2,2-trichloroacetaldehyde by gas chromatography, and carbon monoxide by spectrophotometry (Fox et al., 1990). The products and the complete reaction stoichiometry are shown in Figure 5.7. TCE-epoxide was synthe-

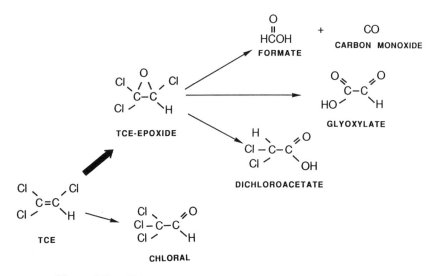

Figure 5.7. TCE oxidation by soluble methane monooxygenase.

sized by published methods (Miller and Guengerich, 1982) and allowed to decompose under the same conditions as the enzyme incubations. The product ratios were similar, suggesting that TCE epoxide is the major product of TCE oxidation by soluble MMO. Enzyme incubation mixtures uniquely contained 2,2,2-trichloroacetaldehyde or chloral. Incubation of TCE-epoxide with soluble MMO did not yield chloral, indicating that chloral did not arise from enzyme-mediated rearrangement of the epoxide. We have proposed that chloral is derived from an enzyme active-site polar transition state intermediate that can decompose to TCE-epoxide or rearrange via chloride migration to yield chloral (Fox et al., 1990).

Regardless of its mechanistic origins, chloral formation is a cause for some concern with respect to using methanotrophs in TCE bioremediation. Chloral is also known as *chloral hydrate* or *knockout drops* in detective novels (Christie, 1963). In admixture with alcoholic beverages, it is known as a *Mickey Finn*; the combination is stupifying and sometimes fatal to the consumer. Certainly, bioremediation schemes designed to remove TCE must not accumulate chloral in the process. This potential problem has been investigated in the laboratory (Newman and Wackett, 1991). *In vivo,* methanotrophs expressing soluble MMO produce chloral as a minor product (3%–6% yield). However, chloral does not accumulate in methanotrophic cultures but undergoes either oxidation to trichloroacetic acid or reduction to trichloroethanol. Further work is required to determine if these same transformations occur with methanotrophs in the field.

A critical parameter in organohalide co-oxidation is the potential for enzyme inactivation by reactive intermediates generated during oxidation. For example, TCE oxidation by soluble MMO generates TCE-epoxide, glyoxyl chloride, dichloroacetyl chloride, and formyl chloride (Fig. 5.7). The reactive compounds can react with water or other nucleophiles in solution, for example, amino acid side chains on proteins. The latter reactions can inactivate soluble MMO by a brute force modification of the protein structure or by modifying one highly reactive critical amino acid. In our experiments, the former was indicated by kinetic studies and covalent attachment of multiple [^{14}C]TCE molecules to each enzyme component (Fox et al., 1990). The enzyme system is more robust *in vivo,* suggesting that other cellular nucleophiles trap out reactive intermediates and protect soluble MMO (Brusseau and Wackett, unpublished data). The inactivation kinetics are important to consider in any sustained biotreatment regime that must allow for enzyme regeneration (Alvarez-Cohen and McCaarty, 1991a,b).

2.2. Toluene Dioxygenase from *Pseudomonas putida* F1

Toluene dioxygenase is a three-component enzyme system catalyzing the enantiospecific dioxygenation of toluene to (+)-*cis*-2,3-dihydroxy-1-methylcyclohexa-4,6-diene (Gibson et al., 1970). The component proteins have been purified to homogeneity (Subramanian et al., 1979, 1981, 1985). They are a flavoprotein reductase, a ferredoxin, and a large iron sulfur protein, ISP$_{TOL}$, which is indicated to be the oxygenase component. Toluene dioxygenase has a broad substrate specificity; over 40 substrates are now known (Gibson et al., 1990). It was first implicated in TCE oxidation by Nelson et al. (1988), and this was supported by the work of Wackett and Gibson (1988). The latter study showed that *cis*-dichloroethylene, *trans*-dichloroethylene, 1,1-dichloroethylene, but not vinylchloride and tetrachloroethylene, are also substrates. Toluene dioxygenase alone is required for TCE oxidation. This was demonstrated by cloning the *tod ABC$_1$C$_2$* genes from *P. putida* F1 into *Escherichia coli* (Fig. 5.8), thus conferring TCE-oxidizing activity onto the recipient strain (Zylstra et al., 1989). The DNA transferred had been mapped and sequenced, and it contained only the toluene dioxygenase structural genes (Zylstra et al., 1988; Zylstra and Gibson, 1989). These data further indicated that a specific transport system for TCE was not required. The hydrophic substrate can likely partition across the *E. coli* membrane to gain access to toluene dioxygenase in the cytoplasm.

TCE co-metabolism by *P. putida* F1 is marked by a time-dependent inactivation of toluene dioxygenase activity (Wackett and Gibson, 1988). Further study indicated that cell growth was inhibited by TCE only in TCE-metabolizing cells (Wackett and Householder, 1989). Mutants of *P. putida* F defective in toluene dioxygenase grew at an uninhibited rate, arguing against toxicity mediated by solvent effects on the membranes. Furthermore, cells incubated with [^{14}C]TCE for 1 hour showed significant incorporation of radiolabel into protein, nucleic acid, and lipid cell fractions. Acid hydrolysis of the cell fractions released radiolabel as formate and glyoxylate. These data suggested that metabolic activation of TCE by toluene dioxygenase led to alkylation of cellular macromolecules and cytotoxicity. This is reminiscent of metabolic activation of chlorinated compounds by cytochrome P$_{450}$ monooxygenase in mammalian liver (Fig. 5.2B).

In light of the observed enzyme inactivation, it was important to understand the

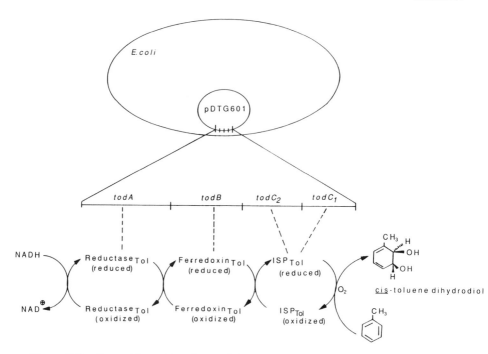

Figure 5.8. Recombinant *E. coli* containing plasmid DTG601 expressing toluene dioxygenase.

mechanism of TCE oxidation by toluene dioxygenase. This required a full stoichiometric accounting of all reaction products in an *in vitro* system using purified enzyme components. Toluene dioxygenase purification was facilitated by the use of recombinant *E. coli* strains each containing high levels of one or more enzyme components (Zylstra and Gibson, 1991). [^{14}C]TCE was used to follow and quantitatively determine the products. Figure 5.9 shows the elution profile of an HPLC organic acids column injected with a reaction mixture containing TCE and toluene dioxygenase (Li and Wackett, 1992). The major TCE oxidation products were formic acid and glyoxylic acid. Relatively small amounts of [^{14}C]label became covalently attached to NADPH (Fig. 5.9) and to the protein components (not shown). The latter observation may underlie the observed *in vivo* enzyme inactivation.

The TCE oxidation product studies provide insights into the mechanism of dioxygen insertion catalyzed by toluene dioxygenase (Li and Wackett, 1992). The formation of significant concentrations of formate and the absence of detectable dichloroacetate and carbon monoxide (<1% of TCE oxidized) argues against TCE-epoxide or free 1,2-dihydroxy-TCE intermediates in the reaction (Fig. 5.9) (Miller and Guengerich, 1982). Deuterated-TCE yielded 50% D-formate, indicating both carbon atoms give rise to formic acid. These data, and previous observations with toluene dioxygenase, are consistent with a symmetrical iron-bridged dioxygen spe-

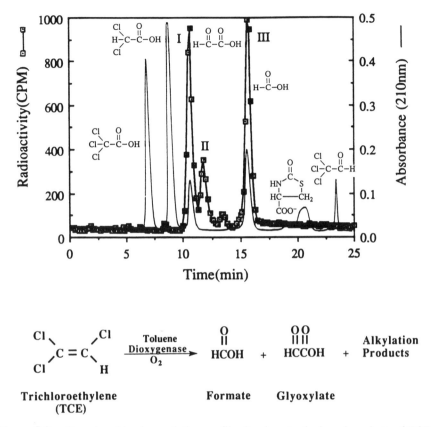

Figure 5.9. Organic acid column elution profile showing standards and products of TCE oxidation by toluene dioxygenase.

cies that can partition to yield formate and glyoxylate (Fig. 5.10). A similar intermediate has been proposed for pthallate dioxygenase (Ballou and Batie, 1988). The results obtained here with toluene dioxygenase should stimulate further study on bacterial mechanisms of dioxygen fixation.

3. COREDUCTIVE METABOLISM OF ORGANOHALIDES

Broad-specificity bacterial oxygenases, despite their considerable potential, fail to oxidize some highly chlorinated molecules. For example, toluene dioxygenase and soluble MMO do not oxidize tetrachloroethylene (perchloroethylene, PCE). Other bacterial oxygenases also fail to oxidize PCE (Ewers et al., 1990; Arciero et al., 1989). Highly chlorinated biphenyls likewise resist attack by aerobic organisms that can dioxygenate less chlorinated PCBs (Abramowicz, 1990). This highlights the importance of reductive dechlorination. Anaerobes have been shown to reductively

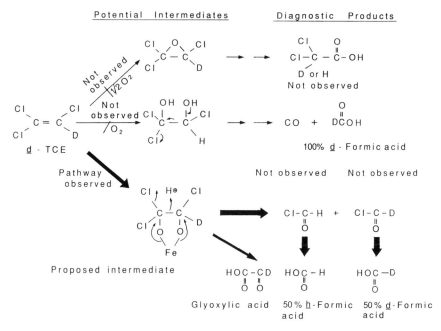

Figure 5.10. Diagnostic products and proposed intermediate in TCE dioxygenation by toluene dioxygenase.

dechlorinate PCE to TCE (Fathepure and Boyd, 1988) and highly chlorinated PCBs to less chlorinated PCBs (Quensen et al., 1988). Similar observations have been made with other highly chlorinated compounds, and these materials are often toxic and environmentally persistent. This give impetus to efforts in obtaining efficient biological systems catalyzing reductive dechlorination.

A molecular understanding of microbial reductive dehalogenation has lagged behind that of aerobic mechanisms. Bacterial pure cultures catalyzing reductive dehalogenation are rare. This emanates from the complex ecological interactions in anaerobic ecosystems and the attendant difficulties in cultivating anaerobic pure cultures. The known organisms are *Desulfomonile tiedjei* (DeWeerd et al., 1990), *Dehalobacter restrictus* (Holliger, 1991), and the aerobes *Rhodococcus chlorophenolicus* (Apajalahti and Salkinoja-Salonen, 1987) and *Flavobacterium* sp. (Xun et al., 1992). Each organism carries out one or more reductive dechlorination reactions with certain substrates as part of its carbon or energy metabolism. Enzyme structures and mechanisms relevant to these reductive dechlorination reactions are just beginning to be delineated.

In this context a co-metabolic approach to reductive dechlorination could prove valuable. This could yield a bacterial system with a broader substrate range and enhanced rates and lead to new insights into molecular mechanisms derived from studying a more well-defined system. To begin with, we investigated the *in vitro* chemical reactivity of bacterial transition metal cofactors (Fig. 5.11) with chloro-

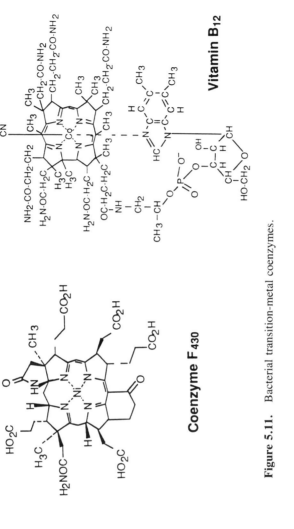

Figure 5.11. Bacterial transition-metal coenzymes.

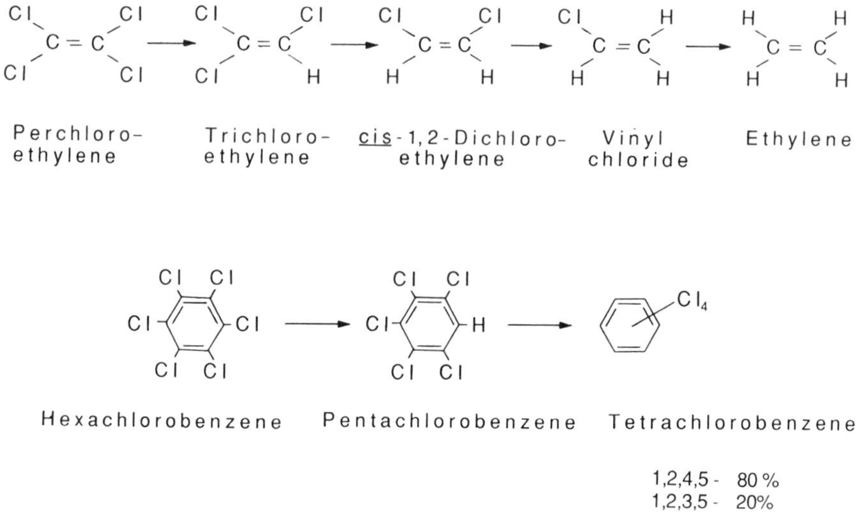

Perchloro-   Trichloro-   cis-1,2-Dichloro-   Vinyl   Ethylene
ethylene   ethylene   ethylene   chloride

Hexachlorobenzene   Pentachlorobenzene   Tetrachlorobenzene

1,2,4,5 - 80 %
1,2,3,5 - 20%

Figure 5.12.   Reductive dechlorination of tetrachloroethylene and hexachlorobenzene by bacterial transition metal coenzymes. The isomeric yield of tetrachlorobenzene is given.

alkanes, alkenes, and aromatic compounds. The reactions of heme, vitamin B_{12}, and coenzyme F_{430} with alkyl halides were well established, but other reactions were ill-defined. The coenzymes were reduced with titanium (III) citrate and incubated with the organohalide substrate under an argon atmosphere (Gantzer and Wackett, 1991). Under these conditions, chloroethylenes underwent two electron reduction to their respective product bearing one less chlorine atom (Fig. 5.12). Hexachlorobenzene also underwent coenzyme-dependent reductive dechlorination to pentachlorobenzene and this to a mixture of tetrachlorobenzene isomers. The rates of dechlorination decreased with decreasing chlorination of the substrate. The data indicate that these bacterial transition metal coenzymes contain the requisite chemical reactivity to catalyze a broad range of reductive dechlorination reactions. They may be involved in some of the widely reported reductive dechlorination reactions observed in anaerobic sediments. Furthermore, these coenzymes, when inside enzymes *in vivo,* are potential biocatalysts for effecting novel co-metabolic reductive transformations.

In the search for a broadly active transition-metal reductive biocatalyst, several features were sought: 1) the ability to bind hydrophobic organohalides at the active site, 2) the ability to transfer electrons repeatedly to effect multiple turnovers, 3) the ability to receive electrons from intermediary metabolism, and 4) the ability of a reductive dechlorinating catalyst to function in an aerobic organism. Ideally, there should be extensive structural and mechanistic information available on the biocatalyst to allow potential protein engineering to improve specificity and catalytic turnover (Ornstein, 1991). Cytochrome $P_{450_{CAM}}$ from *P. putida* G786 best fits the features above, and so it was chosen for study.

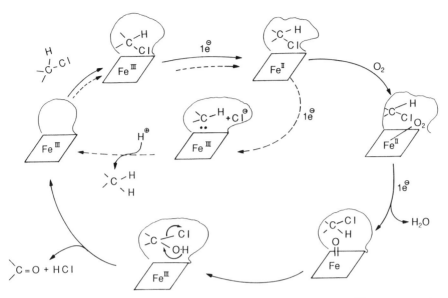

Figure 5.13. Reaction cycle of cytochrome $P_{450_{CAM}}$ monooxygenase. Shown are established steps in the oxygenative cycle (solid arrows) and possible steps in the reductive cycle (dashed arrows).

Bacterial growth on camphor is mediated by a series of largely oxidative reactions initiated by a hydroxylation reaction (Trudgill, 1990). In *P. putida* G786, cytochrome $P_{450_{CAM}}$ monooxygenase catalyzes a regio- and enantiospecific hydroxylation of camphor, yielding 5-*exo*-hydroxycamphor (Bradshaw et al., 1959). Thirty years of study have made cytochrome $P_{450_{CAM}}$ the paradigm for understanding oxygenase structure and mechanisms (Gunsalus et al., 1974; Gunsalus and Wagner, 1978). The *in vivo* hydroxylation activity requires NADH and three protein components, putidaredoxin reductase, putidaredoxin, and cytochrome $P_{450_{CAM}}$, which comprise an electron transfer assembly analogous to that of toluene dioxygenase (Fig. 5.8). The functional hydroxylase component is the cytochrome P_{450}. It is a heme protein of 45,000 molecular weight with an axial cysteinyl thiolate ligand modulating the reactivity of the heme iron in oxygen activation. A reactive monoatomic oxygen species is generated at the iron center to carry out camphor hydroxylation (Fig. 5.13). There is still active investigation of the precise mechanisms of oxygen activation and substrate hydroxylation (Fisher and Sligar, 1985; Sligar et al., 1991; Poulos and Raag, 1992).

Under anaerobic conditions, the heme iron of cytochrome P_{450} monooxygenases can act as a reductive dehalogenation catalyst (Fig. 5.13). The reductive reactions can be envisioned as a short circuit in which electrons are transferred directly to the organic substrate rather than to molecular oxygen as in the physiological reaction. Although the oxygenative cycle is well investigated, the kinetic order of events and the mechanism of electron transfer in the reductive cycle are ill-defined by compari-

TABLE 5.3. Chlorinated Substrate Reduction and Binding by *P. putida* G786 and Cytochrome P450$_{CAM}$, Respectively

Substrate	Product (Yield, %)	*In vivo* Rate (pmol/min/mg protein)	*In vitro* Binding	
			K$_D$(μM)	% High Spin
Hexachloroethane	Tetrachloroethane (92%)	1,380	0.7	>95
Pentachloroethane	Trichloroethylene (90%)	890	7	70
1,1,1,2-Tetrachloroethane	1,1-Dichloroethylene (85%)	138	150	50
1,1,2,2-Tetrachloroethane	None detected	—	—	—
1,1,1-Trichloroethane	None detected	—	—	—

son. Studies are being conducted to define these reactions further. This knowledge may prove useful for the goal of developing a broad-specity, high-turnover, aerotolerant reductive dehalogenation biocatalyst system.

P. putida G786 and/or purified cytochrome P$_{450_{CAM}}$ have been shown to catalyze reductive dehalogenation (Castro et al., 1985; Castro and Belser, 1990). The substrates include carbon tetrachloride, bromotrichloromethane, trichloronitromethane, 1,1,2-trichloroethane, and 1,2-dibromo-3-chloropropane. Most recently, we have begun to screen cytochrome P$_{450_{CAM}}$ systematically for reductive dehalogenation activity with low-molecular-weight chlorinated solvents under anaerobic conditions. The polychlorinated ethane series was examined in detail. *In vivo* reactions were confirmed as being cytochrome P$_{450_{CAM}}$ dependent by parallel *in vitro* studies with purified enzyme. Camphor-induced *P. putida* G786, under an argon atmosphere, transformed hexachloroethane, pentachloroethane, 1,1,1,2-tetrachloroethane but not 1,1,2,2-tetrachloroethane and 1,1,1-trichloroethane (Table 5.3) over a time course of minutes to hours (Logan et al., 1992). A representative reaction course is shown in Figure 5.14. The disappearance of substrate was accompanied by the appearance of an alkene product bearing two less chlorine substituents. The general course of the reaction is a two electron reduction with a concomitant β-elimination to yield the observed products (Fig. 5.15). Similar reaction products have been observed under anaerobic conditions with reduced iron porphyrins (hematin) in solution (Wade and Castro, 1973; Schanke and Wackett, 1992).

Previous studies with *P. putida* G786 have not systematically investigated the effects of oxygen on the reductive reactions, so this was considered further. Such studies are particularly relevant to applications as aerotolerance in reductive dechlorination would be greatly advantageous. In fact, earlier studies implied that reductive dechlorination reactions are unaffected by oxygen (Castro and Belser, 1990). These experiments used sealed incubation vessels with heavy cell densities (100 mg dry weight cells per milliliter) for time scales as long as 24 hours. It is likely that oxygen tensions decreased significantly within several hours. Our experi-

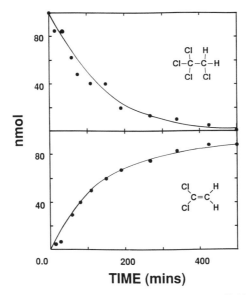

Figure 5.14. 1,1,1,2-Tetrachloroethane disappearance and 1,1-dichloroethylene appearance in an incubation with *P. putida* G786 expressing cytochrome $P_{450_{CAM}}$ monooxygenase.

ments avoided this ambiguity by adding known oxygen concentrations to sealed reaction vials just prior to substrate addition and analyzing the reaction mixture after 5 *minutes* (Logan et al., 1992).

Experiments with indicator dyes suggested that significant oxygen depletion did not occur within 5 minutes under these conditions. Rates of reductive dehalogenation were compared with a control 5 minute incubation conducted under an atmosphere of 100% argon. The data indicated that reductive dechlorination is strongly inhibited by oxygen. This is not surprising since O_2 binding and reduction by the Fe^{2+} center of reduced cytochrome $P_{450_{CAM}}$ will compete with organic substrate reduction. An ambient oxygen concentration (20%) inhibited *in vivo* reductive dechlorination activity >95%. Significant reductive dechlorination did occur at lower O_2 tensions. For example, at a 5% (v/v) O_2 atmosphere in argon, the reduc-

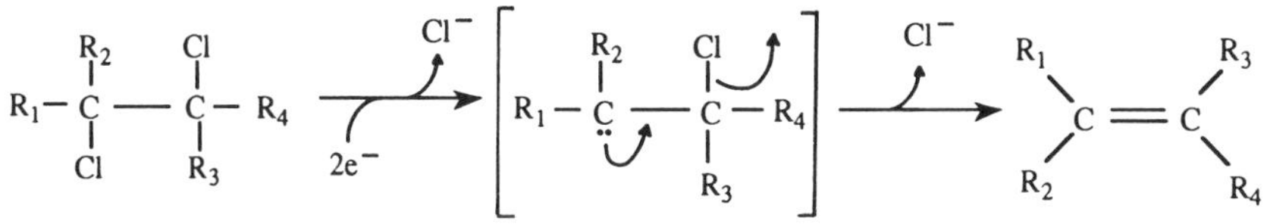

Figure 5.15. Reductive elimination reaction of a polychlorinated ethane yielding the respective ethylene, catalyzed by bacterial transition-metal coenzymes and cytochrome $P_{450_{CAM}}$. The hypothetical intermediate in brackets is a two electron reduced species that may be coordinated to the coenzyme metal.

tive elimination of hexachloroethane and pentachloroethane was inhibited by 70% and 30%, respectively. This suggests that significant reductive dehalogenation would occur in soil and water environments at moderate oxygen tensions. Strict anaerobes such as methanogens would be unable to grow and metabolize at this degree of aerobicity (Kiener and Leisinger, 1983).

The data suggest that reductive dehalogenation can occur under conditions that could support concomitant cooxidative metabolism. There is currently great interest in combining anaerobic and aerobic treatment regimes to biodegrade highly chlorinated compounds (Tiedje et al., 1987). To date, the efforts have focused on engineering sequential anaerobic and aerobic transformations, each occurring in separate reactors (Fathepure and Vogel, 1991). The use of an aerotolerant *Pseudomonas* containing an aerotolerant enzyme operative in a reductive mode would greatly simplify the engineering of sequential reductive–oxidative biodegradation schemes.

ACKNOWLEDGMENTS

I thank the talented coworkers in my laboratory who have contributed to our current understanding of co-metabolic dehalogenation processes. The assistance of Mary Jo Keefe in the preparation of the manuscript is gratefully appreciated. This research was supported by National Institutes of Health grant GM 41235 and Environmental Protection Agency cooperative agreement EPA/CR820771-01-0.

REFERENCES

Abramowicz DA (1990): Aerobic and anaerobic biodegradation of PCBs: A review. Crit Rev Biotechnol 10:241–251.

Alvarez-Cohen L, McCarty PL (1991a): Two-stage dispersed-growth treatment of halogenated aliphatic compounds by cometabolism. Environ Sci Technol 25:1387–1393.

Alvarez-Cohen L, McCarty PL (1991b): Product toxicity and cometabolic competitive inhibition modeling of chloroform and trichloroethylene transformation by methanotrophic resting cells. Appl Environ Microbiol 57:1031–1037.

Anders MW, Pohl LR (1985): Halogenated alkanes. In Anders MW (ed): Bioactivation of Foreign Compounds. New York: Academic Press.

Anthony C (1982): The Biochemistry of Methylotrophs. London: Academic Press.

Apajalahti JHA, Salkinoja-Salonen MS (1987): Complete dechlorination of tetrachlorohydroquinone by cell extracts of pentachlorophenol-induced *Rhodococcus chlorophenolieus*. J Bacteriol 169:5125–5130.

Arciero D, Vannelli T, Logan M, Hoper AB (1989): Degradation of trichloroethylene by the ammonia-oxidizing bacterium *Nitrosomonas europaea*. Biochem Biophys Res Commun 159:640–643.

Ballou D, Batie C (1988): Oxidases and Related Redox Systems. New York: Alan R. Liss, pp 211–216.

Bradshaw WH, Conrad HE, Corey EJ, Gunsalus, IC, Lednicer D (1959): Microbiological degradation of (+)-camphor. J Am Chem Soc 81:5507.

Brusseau GA, Tsien H-C, Hanson RS, Wackett LP (1990): Optimization of trichloroethylene oxidation by methanotrophs and the use of a colorimetric assay to detect soluble methane monooxygenase activity. Biodegradation 1:19–29.

Castro CE, Belser NO (1990): Biodehalogenation: Oxidative and reductive metabolism of 1,1,2-trichloroethane by *Pseudomonas putida*—Biogeneration of vinyl chloride. Environ Toxicol Chem 9:707–714.

Castor CE, Wade RS, Belser NO (1985): Biodehalogenation: Reactions of cytochrome P-450 with polyhalomethanes. Biochemistry 24:204–210.

Christie A (1963): The Clocks. Anstey, United Kingdom: F.A. Thorpe, p 230.

Colby J, Stirling DI, Dalton H (1977): The soluble methane monooxygenase of *Methylococcus capsulatus* (Bath): Its ability to oxygenate *n*-alkanes, *n*-alkenes, ethers and alicyclic, aromatic and heterocyclic compounds. Biochem J 165:395–402.

Dagley S (1972): Microbial degradation of stable chemical structures: General features of metabolic pathways. In Degradation of Synthetic Organic Molecules in the Biosphere. Washington, DC: National Academy of Sciences, pp 1–16.

DeWeerd KA, Mandelco L, Tanner RS, Woese CR, Suflita JM (1990): *Desulfomonile tiedgei* gen. nov. and sp. nov., a novel anaerobic, dehalogenating, sulfate-reducing bacterium. Arch Microbiol 154:23–30.

DiSpirito AA, Gulledge J, Schiemke AK, Murrell JC, Lidstrom ME, Drema CL (1992): Trichloroethylene oxidation by the membrane-associated methane monooxygenase in type I, type II and type X methanotrophs. Biodegradation 2:151–164.

Dolfing J (1990): Reductive dechlorination of 3-chlorobenzoate is coupled to ATP production and growth in an anaerobic bacterium, strain DCB-1. Arch Microbiol 153:264–266.

Drechsel E (1896): Contribution to the chemistry of a sea animal. Z Biol 33:8–107.

Ericson A, Hedman B, Hodgson KO, Green J, Dalton H, Bentsen JG, Beer RH, Lippard SJ (1988): Structural characterization by EXAFS spectroscopy of the binuclear iron center in protein A of methane monooxygenase from *Methylococcus capsulatus* (Bath). J Am Chem Soc 110:2330–2332.

Ewers J, Freier-Schröder D, Knackmuss H-J (1990): Selection of trichloroethene (TCE) degrading bacteria that resist inactivation by TCE. Arch Microbiol 154:410–413.

Fathepure BZ, Boyd SA (1988): Reductive dechlorination of perchloroethylene and the role of methanogens. FEMS Microbiol Lett 49:149–156.

Fathepure BZ, Vogel TM (1991): Complete degradation of polychlorinated hydrocarbons by a two-stage biofilm reactor. Appl Environ Microbiol 57:3418–3422.

Fisher MT, Sligar SG (1985): Control of heme protein potential and reduction rate: Linear free energy relation between potential and ferric spin state equilibrium. J Am Chem Soc 107:5018–5019.

Folsom BR, Chapman PJ, Pritchard PH (1990): Phenol and trichloroethylene degradation by *Pseudomonas cepacia* G4: Kinetics and interaction between substrates. Appl Environ Microbiol 56:1279–1285.

Fox BG, Borneman JG, Wackett LP, Lipscomb JD (1990): Haloalkene oxidation by the soluble methane monooxygenase from *Methylosinus trichosporium* OB3b: Mechanistic and environmental implications. Biochemistry 29:6419–6427.

Fox BG, Froland WA, Dege JE, Lipscomb JD (1989): Methane monooxygenase from *Methylosinus trichosporium* OB3b: Purification and properties of a three-component system with high specific activity from a type II methanotroph. J Biol Chem 264:10023–10033.

Fox BG, Surerus KK, Münck E, Lipscomb JD (1988): Evidence for a μ-oxo-bridged binuclear iron cluster in the hydroxylase component of methane monooxygenase. J Biol Chem 253:10553–10556.

Gantzer GJ, Wackett LP (1991): Reductive dechlorination catalyzed by bacterial transition-metal coenzymes. Environ Sci Tech 715–722.

Gibson DT, Hensley M, Yoshioka H, Mabry TJ (1970): Formation of (+)-cis-2,3-dihydroxy-1-methylcyclohexa-4,6-diene from toluene by *Pseudomonas putida*. Biochemistry 7:3795–3802.

Gibson DT, Zylstra GJ, Chauhan S (1990): Biotransformations catalyzed by toluene dioxygenase from *Pseudomonas putida* F1. In Silver S, Chakrabarty AM, Iglewski B, Kaplan S (eds): *Pseudomonas:* Biogransformations, Pathogenesis and Emerging Biotechnology. Washington, DC: American Society for Microbiology Press, pp 121–132.

Green J, Dalton H (1985): Protein B of soluble methane monoxygenase from *Methylococcus capsulatus* (Bath). J Biol Chem 260:15795–15801.

Green J, Dalton H (1989): Substrate specificity of soluble methane monooxygenase: Mechanistic implications. J Biol Chem 264:17698–17703.

Gunsalus IC, Meeks JR, Lipscomb JD, Debrunner P, Münck E (1974): Bacterial monooxygenases—The P-450 cytochrome system. In Hayaishi O (ed): Molecular Mechanisms of Oxygen Activation. New York: Academic Press, pp 559–613.

Gunsalus IC, Wagner GC (1978): Bacterial P-450$_{CAM}$ methylene monooxygenase components: Cytochrome *m,* putidaredoxin, and putidaredoxin reductase. Methods Enzymol 52:166–188.

Harker AR, Kim Y (1990): Trichloroethylene degradation by two independent aromatic-degrading pathways in *Alcaligenes eutrophus* JMP134. Appl Environ Microbiol 56:1179–1181.

Henry SM, DiSpirito AA, Lidstrom ME, Grbic-Galic D (1989): Effects of mineral medium on trichloroethylene oxidation and involvement of a particulate methane monooxygenase. In Abstracts of the 89th Annual Meeting of the American Society for Microbiology. Washington, DC: American Society for Microbiology, abstract K69, p 91.

Holliger C (1992): Reductive dehalogenation by anaerobic bacteria. Ph.D. dissertation, the Netherlands: Wageningen Agricultural University.

Hutzinger O, Veerkamp W (1981): Xenobiotic compounds with pollution potential. In Leisinger T, Cook A, Hutter R, Nuesch J (eds): Microbial Degradation of Xenobiotic and Recalcitrant Compounds. London: Academic Press, p 3.

Jakoby WB, Keen JH (1977): A triple treat in detoxification: The glutathione S-transferases. Trends Biochem Sci 2:229–231.

Keen JH, Habig WH, Jakoby WB (1976): Mechanism for the several activities of the glutathione S-transferases. J Biol Chem 251:6183–6188.

Kiener A, Leisinger T (1983): Oxygen sensitivity of methanogenic bacteria. Sys Appl Microbiol 4:305–312.

Kirk KL (1991): Biochemistry of the Elemental Halogens and Inorganic Halides. New York: Plenum Press, vol 9A.

Kohler-Staub D, Leisinger T (1985): Dichloromethane dehalogenase of *Hyphomicrobium* sp. strain DM2. J Bacteriol 162:676–681.

La Roche S, Leisinger T (1990): Sequence analysis and expression of the bacterial dichloromethane dehalogenase structural gene, a member of the glutathione S-transferase supergene family. J Bacteriol 172:164–171.

Li S, Wackett LP (1992): Trichloroethylene oxidation by toluene dioxygenase. Biochem Biophys Res Commun 185:443–451.

Logan MSP, Newman LM, Schanke CA, Wackett LP (1993): Co-substrate effects in reductive dehalogenation by *Pseudomonas putida* G786 expressing cytochrome P450$_{CAM}$. Biodegradation 4:39–50.

Maltoni C, Lefemine G (1974): Carcinogenicity bioassays of vinylchloride research plan and early results. Environ Res 7:387–396.

March J (1985): Advanced Organic Chemistry: Reactions, Mechanisms and Structure. New York: John Wiley & Sons.

Miller EC, Miller JA (1985): Some historical perspectives on the metabolism of xenobiotic chemicals to reactive electrophiles. In Anders MW (ed): Bioactivation of Foreign Compounds. London: Academic Press, pp 3–28.

Miller RE, Guengerich FP (1982): Oxidation of trichloroethylene by liver microsomal cytochrome P450: Evidence for chlorine migration in a transition state not involving trichloroethylene oxide. Biochemistry 21:1090–1097.

Neidleman SL, Geigert J (1986): Biohalogenation: Principles, Basic Roles and Applications. Chichester, England: Ellis Horwood Ltd.

Nelson MJ, Montgomery SO, Pritchard PH (1988): Trichloroethylene metabolisms by microorganisms that degrade aromatic compounds. Appl Environ Microbiol 54: 604–606.

Newman LM, Wackett LP (1991): Fate of 2,2,2-trichloroacetaldehyde (chloral hydrate) produced during trichloroethylene oxidation by methanotrophs. Appl Environ Microbiol 57:2399–2402.

Oldenhuis R, Vink RL, Vink JM, Janssen DB, Witholt B (1989): Degradation of chlorinated aliphatic hydrocarbons by *Methylosinus trichosporium* OB3b expressing soluble methane monooxygenase. Appl Environ Microbiol 55:2819–2826.

Ornstein RL (1991): Why timely bioremediatio of synthetics may require rational enzyme redesign: Preliminary report on redesigning cytochrome P450$_{CAM}$ for trichloroethylene dehalogenation. In Hinchee RE, Olfenbuttel RF (eds): On Site Bioreclamation: Processes for Xenobiotic and Hydrocarbon Treatment. Boston: Butterworth-Heinemann, pp 509–514.

Porter TD, Coon MJ (1991): Cytrochrome P450: Multiplicity of isoforms, substrates, and catalytic and regulatory mechanisms. J Biol Chem 266:13469–13472.

Poulos TL, Finzel BC, Gunsalus IC, Wagner GC, Kraut J (1985): The 2.6 Å crystal structure of *Pseudomonas putida* cytochrome P-450. J Biol Chem 260:16122–16130.

Poulos TL, Raag R (1992): Cytochrome P450$_{CAM}$: Crystallography, oxygen activation, and electron transfer. FASEB J 6:674–679.

Quensen JF III, Tiedje JM, Boyd SA (1988): Reductive dechlorination of polychlorinated biphenyls by anaerobic microorganisms from sediment. Science 242:752–754.

Reineke W (1984): Microbial degradation of halogenated aromatic compounds. In Gibson DT (ed): Microbial Degradation of Organic Compounds. New York: Marcel Dekker, pp 319–360.

Schanke CA, Wackett LP (1992): Environmental reductive elimination reactions of polychlorinated ethanes mimicked by transition-metal coenzymes. Environ Sci Technol 26:830–833.

Sligar SG, Fillipovic D, Stayton PS (1991): Mutagenesis of cytochrones P450$_{CAM}$ and b$_5$. Methods Enzymol 206:31–49.

Storck W (1987): Chlorinated solvent use hurt by federal rules. Chem Eng News 65:11.

Subramanian V, Liu T-N, Yeh W-K, Gibson DT (1979): Toluene dioxygenase: Purification of an iron sulfur protein by affinity chromatography. Biochem Biophys Res Commun 91:1131–1139.

Subramanian V, Liu T-N, Yeh W-K, Narro M, Gibson DT (1981): Purification and properties of NADH-ferredoxin$_{TOL}$ reductase. J Biol Chem 256:2723–2730.

Subramanian V, Liu T-N, Yeh W-K, Serdar CM, Wackett LP, Gibson DT (1985): Purification and properties of ferredoxin$_{TOL}$: A component of toluene dioxygenase from *Pseudomonas putida* F1. J Biol Chem 260:2355–2363.

Tiedje JM, Boyd SA, Fathepure BZ (1987): Anaerobic degradation of chlorinated aromatic hydrocarbons. Dev Ind Microbiol 27:117–127.

Trudgill PW (1990): Microbial metabolism of monoterpenes—Recent developments. Biodegradation 1:93–106.

Tsien H-C, Brusseau GA, Hanson RS, Wackett LP (1989): Biodegradation of trichloroethylene by *Methylosinus trichosporium* OB3b. Appl Environ Microbiol 55:3155–3161.

Vogel TM, McCarty P (1985): Biotransformation of tetrachloroethylene to trichloroethylene, dichloroethylene, vinyl chloride, and carbon dioxide under methanogenic conditions. Appl Environ Microbiol 49:1080–1083.

Wackett LP, Brusseau GA, Hanson RS (1989a): Survey of microbial oxygenases: Trichloroethylene degradation by propane-oxidizing bacteria. Appl Environ Microbiol 55:2960–2964.

Wackett LP, Gibson DT (1988): Degradation of trichloroethylene by toluene dioxygenase in whole cell studies with *Pseudomonas putida* F1. Appl Environ Microbiol 54:1703–1708.

Wackett LP, Householder SR (1989): Toxicity of trichloroethylene to *Pseudomonas putida* F1 is mediated by toluene dioxygenase. Appl Environ Microbiol 55:2723–2725.

Wackett LP, Orme-Johnson WH, Walsh CT (1989b): Transition metal enzymes in bacterial metabolism In Beveridge TJ, Doyle RJ (eds): Metal Ions and Bacteria. New York: John Wiley & Sons, pp 165–206.

Wade RS, Castro CE (1973): Oxidation of iron(II) porphyrins by alkyl halides. J Am Chem Soc 95:226–230.

Walsh DT (1983): Fluorinated substrate analogs: Routes of metabolism and selective toxicity. Adv Enzymol Rel Areas Mol Biol 55:197–289.

Winter RB, Yen K-M, Ensley BD (1989): Efficient degradation of trichloroethylene by a recombinant *Escherichia coli*. Biotechnology 7:282–285.

Woodland MP, Dalton H (1984): Purification and characterization of component A of the methane monooxygenase from *Methylococcus capsulatus* (Bath). J Biol Chem 259:53–59.

Xun L, Topp E, Orser CS (1992): Glutathione is the reducing agent for the reductive dehalogenation of tetrachloro-*p*-hydroquinone by extracts from a *Flavobacterium* sp. Biochem Biophys Res Commun 182:361–366.

Zylstra GJ, Gibson DT (1989): Toluene degradation by *Pseudomonas putida* F1: Nucleotide sequence of the *todC1C2BADE* genes and their expression in *Escherichia coli*. J Biol Chem 264:14940–14946.

Zylstra GJ, Gibson DT (1991): Aromatic hydrocarbon degradation: A molecular approach. Genetic Engineering 13:183–203.

Zylstra GJ, McCombie WR, Gibson DT, Finette BA (1988): Toluene degradation by *Pseudomonas putida* F1: Genetic organization of the *tod* operon. Appl Environ Microbiol 54:1498–1503.

Zylstra GJ, Wackett LP, Gibson DT (1989): Trichloroethylene degradation by *Escherichia coli* containing the cloned *Pseudomonas putida* F1 toluene dioxygenase genes. Appl Environ Microbiol 55:3162–3166.

6

THE MICROBIAL ECOLOGY AND PHYSIOLOGY OF ARYL DEHALOGENATION REACTIONS AND IMPLICATIONS FOR BIOREMEDIATION

JOSEPH M. SUFLITA
G. TODD TOWNSEND

Department of Botany and Microbiology, University of Oklahoma, Norman, Oklahoma 73019

1. INTRODUCTION

Synthetic halogenated aromatic compounds have been used extensively as biocides, disinfectants, solvents, hydraulic fluids, degreasers, fire retardants, and in the manufacture of rubbers, plastics, dyes, lubricants, and electrical goods (Gerhartz, 1986). Halogenation of the aromatic nucleus generally leads to increased chemical stability and toxicity, characteristics deemed desirable when these compounds are exploited for commercial purposes. However, upon release to the environment, their increased resistance to biotic and abiotic decay, coupled with their toxicity, carcinogenicity, and tendency to accumulate in food chains, renders haloaromatic compounds as some of the most pervasive and troubling contaminants in existence.

Microbial Transformation and Degradation of Toxic Organic Chemicals, pages 243–268
© *1995 Wiley-Liss, Inc.*

Anthropogenic production of halogenated organic compounds does not represent the only source of these materials to the environment. Chloro-, bromo-, and iodo-substituted chemicals of biological origin are widely distributed in the biosphere, particularly in marine environments. There are hundreds of halo-organic chemicals isolated from natural sources, including some homocyclic and heterocyclic aromatic compounds (Fenical, 1982; Neidleman, 1975; Neidleman and Geigert, 1987). Therefore, it should not be surprising that microorganisms have evolved mechanisms to transform haloaromatic materials.

The aerobic microbial metabolism of haloaromatic compounds has been extensively studied, and the reader is referred to a recent review (Rochkind-Dubinsky et al., 1987) and to other chapters in this book for detailed summaries of this information. However, many haloaromatic pollutants have been found to resist aerobic decay. Furthermore, once released in the environment, halogenated pollutants often partition to and reside in habitats where oxygen is unavailable. Anaerobic processes must be relied upon for the biotransformation of haloaromatic materials in habitats such as aquatic sediments, digester sludges, flooded soils, polluted aquifers, landfills, as well as the gastrointestinal tracts of many organisms.

Under anaerobic conditions, some microorganisms are able to remove halogens from an aromatic ring and replace them with a proton. This process, known as *aryl reductive dehalogenation,* results in the formation of lesser halogenated congeners. Unlike aerobic processes, an increasing degree of halogenation does not hinder this transformation, and, in fact, highly halogenated pollutants appear to be more susceptible to dehalogenation (see below). Often the dehalogenation process proceeds in a sequential manner until most or all of the aryl halides are removed from the parent molecule. The dehalogenated metabolites are generally more amenable to further anaerobic or aerobic biodegradation processes. In many cases, the decrease in aryl halide content is also associated with decreased toxicity. For these reasons, there has been a great deal of interest in aryl reductive dehalogenation as a mechanism for the bioremediation of sites contaminated with haloaromatic pollutants (for recent reviews on reductive dehalogenation, see Kuhn and Suflita, 1989a; Mohn and Tiedje, 1992). Success in such endeavors requires information on both the physiological basis for such bioconversions as well as the ecological factors that influence dehalogenation.

Since the recognition that aryl reductive dehalogenation represents a major anaerobic fate process for haloaromatic compounds (Suflita et al., 1982), the number of classes of homocyclic aromatic compounds transformed by this mechanism has steadily grown (Fig. 6.1). There are also several reports of the reductive dehalogenation of halogenated heterocyclic molecules (Adrian and Suflita, 1990, 1992; Liu and Rogers, 1991; Ramanand et al., 1993). Aryl reductive dehalogenation reactions have been observed in diverse anaerobic habitats, including marine and freshwater sediments, sewage sludges, aquifer materials, and a variety of soil samples. Evidence to date indicates that aryl reductive dehalogenation activity is widespread in most anaerobic environments, and it is likely an obligatory transformation for haloaromatic pollutants prior to the anaerobic cleavage of the aromatic nucleus.

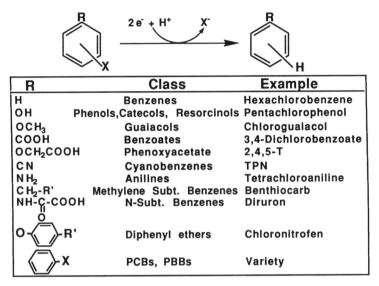

R	Class	Example
H	Benzenes	Hexachlorobenzene
OH	Phenols, Catecols, Resorcinols	Pentachlorophenol
OCH₃	Guaiacols	Chloroguaiacol
COOH	Benzoates	3,4-Dichlorobenzoate
OCH₂COOH	Phenoxyacetate	2,4,5-T
CN	Cyanobenzenes	TPN
NH₂	Anilines	Tetrachloroaniline
CH₂-R′	Methylene Subt. Benzenes	Benthiocarb
NH-C-COOH	N-Subt. Benzenes	Diruron
O-⬡-R′	Diphenyl ethers	Chloronitrofen
⬡-X	PCBs, PBBs	Variety

Figure 6.1. Classes of homocyclic aromatic compounds known to undergo reductive dehalogenation and representative examples.

Although many haloaromatic pollutants undergo aryl reductive dehalogenation in anoxic habitats, knowledge of the microorganisms catalyzing this reaction is limited. Dehalogenation can be demonstrated in anoxic environmental samples with relative ease, and increased dehalogenation rates in these incubations suggest that the requisite microorganisms are being enriched. However, isolation of microorganisms capable of catalyzing this reaction has proven extremely difficult. This difficulty has confined the majority of research on aryl dehalogenation to acclimated enrichments and consortia. Accordingly, the development of remedial strategies that seek to exploit this process is hampered by the lack of information on the factors controlling the metabolism and ultimately the growth of dehalogenating microorganisms.

Table 6.1 lists the aerobic and anaerobic isolates presently known to catalyze aryl reductive dehalogenation reactions. Although this transformation occurs in some aerobic degradative pathways, it must be emphasized that aryl reductive dehalogenation reactions occur most frequently and with a wider range of substrates under anaerobic conditions. At this time, only one anaerobic bacterium capable of catalyzing an aryl reductive dehalogenation has been well studied. *Desulfomonile tiedjei,* formerly strain DCB-1, reductively dehalogenates 3-chlorobenzoate to benzoate and presently serves as a model microorganism for probing the physiological basis of aryl reductive dehalogenation reactions (DeWeerd et al., 1990).

Successful bioremedial efforts depend on identifying and overcoming the factors that limit biodegradation. This would appear to be true regardless of the particular environment, pollutant, or transformation of interest. A consideration of the factors

TABLE 6.1. Isolates Known to Catalyze Aryl Reductive Dehalogenation Reactions

Organisms	Substrate	Reference
Aerobic		
Staphylococcus epidermidis	1,2,4-Trichlorobenzene	Tsuchiya and Yamaha (1984)
44 aerobic isolates	Tetrachloroisophthalonitrile	Sato and Tanaka (1987)
Alcaligenes denitrificans	2,4-Dichlorobenzoate	van den Tweel et al. (1987)
Rhodococcus chlorophenolicus	Tetrachlorohydroquinone	Apajalahti and Salkinoja-Salonen (1987) Haggblom et al. (1989)
Flavobacterium sp.	Tetrachlorohydroquinone	Steiert and Crawford (1986)
PCP-degrading bacterium	Tetrachlorohydroquinone	Reiner et al. (1978)
Anaerobic		
Desulfomonile tiedjei	3-Chlorobenzoate	Shelton and Tiedje (1984) DeWeerd et al. (1990)
Desulfitobacterium dehalogens	*Ortho*-Chlorophenols	Utkin et al. (1994)
Strain DCB-2	Tri- and Dichlorophenols	Madsen and Licht (1992)
Undesignated strain	2-Chlorophenol	Foxworthy et al. (1992)

that limit the microbial metabolism of haloaromatic pollutants logically leads to bioremedial strategies that attempt to overcome these barriers and stimulate such dehalogenation reactions. In this chapter, we examine some of the physiological, environmental, and chemical factors influencing aryl reductive dehalogenation of haloaromatic compounds in anaerobic environments. When applicable, we use information on dehalogenation by the bacterium *D. tiedjei* to help explain the fate of haloaromatic compounds in more complex environments.

One particularly troubling characteristic of aryl dehalogenation reactions in laboratory investigations is an extended lag or acclimation period of varying length prior to the onset of biodegradation followed by rapid transformation. Subsequent substrate amendments are rapidly degraded without lag. These acclimation periods are frequently on the order of weeks, months, and sometimes even years. Such time frames are difficult to ignore and are likely more important in governing the exposure of recipient ecosystems to a haloaromatic pollutant than the rate of transformation once it does occur. Thus, remedial efforts must be directed toward understanding and overcoming this phase of degradation.

2. PHYSIOLOGICAL BARRIERS TO DEHALOGENATION

Although microorganisms are quite efficient in the biogeochemical cycling of naturally occurring organic compounds, society has learned through difficult experience that there are physiological limits in the ability of microbes to degrade unwanted pollutants at rates that prevent undesirable environmental impact. In order for biodegradation to play a substantial role in the destruction of a xenobiotic organic compound, the organic contaminant should be able to enter a suitable microorganism, be converted to central metabolic intermediates by existing pathways, and supply the carbon and energy needs of the cell. Conversely, a contaminant will be recalcitrant if there are no microorganisms present in the environment that have the ability to degrade it. While studies have indicated that reductive dehalogenation activity is found in many anaerobic habitats, it is premature to suggest that the requisite microorganisms are ubiquitously distributed in all of them. Sites chronically impacted by haloaromatic pollutants may contain significant numbers of dehalogenating microorganisms because the pollutants provide some positive selection pressure favoring the enrichment of these organisms (see below).

Conversely, without such environmental selection pressure, dehalogenating anaerobes may be present only in low numbers or may even be absent. Surveys of the dehalogenation potential of marine and freshwater sediments have found sites impacted by industrial effluents have greater dehalogenation capabilities than samples taken from nearby but unimpacted areas (Genthner et al., 1989; Gerritse et al., 1992). Also, deep regions of the terrestrial subsurface may not possess the metabolic diversity that surface or shallow subsurface microbial communities possess. Nevertheless, the absence of degradation should not be equated with the absence of suitable microorganisms, and, conversely, the presence of suitable microorganisms does not ensure that degradation of the pollutant will proceed. This is evidenced by the inability to predict dehalogenation in lake sediments by the number of dehalogenating microorganisms found in those samples (Hale et al., 1991). Indeed, several known physiological factors have been identified that may partially inhibit or completely preclude the indigenous microbiota's degradative capabilities.

2.1. Failure To Derepress the Catabolic Enzymes

Degradation of a pollutant may be inhibited because the compound may only poorly derepress the enzymes necessary for its transformation. Often the enzymes responsible for such transformations are inducible, and their derepression is believed to play a role in the acclimation period prior to the onset of aryl reductive dehalogenation activity (Linkfield et al., 1989; Madsen and Aamand, 1992). Studies with *D. tiedjei* have shown that aryl reductive dehalogenation in this bacterium is indeed an inducible activity (Cole and Tiedje, 1990; DeWeerd and Suflita, 1990). A variety of *meta*-substituted aromatic compounds, including those that are substrates for reductive dehalogenation (*m*-chloro-, bromo-, and iodobenzoate) as well as those that are not (*m*-fluorobenzoate and *m*-cresol) are capable of derepressing dehalogenation activity in *D. tiedjei*. Other compounds such as *ortho*- and *para*-substituted bromo- and iodobenzenoids as well as *meta*-substituted polychlorophenols can be transformed by *D. tiedjei,* but dehalogenation activity is not induced upon exposure to these chemicals.

There is evidence to suggest that information on the derepression of dehalogena-

TABLE 6.2. Selected Examples of Studies That Have Shown Aryl Dehalogenation Activity in Response to the Amendment of a Cross-Acclimating Substrate

Chemical Used to Accelerate Dehalogenation	Substrates Dehalogenated	System[a]	Reference
3-Bromobenzoate	Bromo- and chlorobenzoates	Lake sed	Horowitz et al. (1983)
4-Amino-3,5-dichlorobenzoate	Bromo- and chlorobenzoates	Lake sed	Horowitz et al. (1983)
Dichloroanilines	Dichlorophenols	Pond sed	Struijs and Rogers (1989)
Dichlorophenols	Dichloroanilines	Pond sed	Struijs and Rogers (1989)
Dichlorophenols	Dichlorophenoxyacetate	Pond sed	Bryant (1992)
PCB congeners	Aroclor 1242, 1260	River sed	Bedard et al. (1991)
Bromobiphenyls	Aroclor 1260	River sed	Bedard and Van Dort (1992)
3-Chlorobenzoate	3-Bromobenzoate and 3-iodobenzoate	Bacterial consortium	DeWeerd et al. (1986)
3-Chlorobenzoate	Perchloroethylene	*D. tiedjei*	Fathepure et al. (1987)
3-Fluorobenzoate	3-Chlorobenzoate	*D. tiedjei*	Cole and Tiedje (1990)
3-Chlorobenzoate	Pentachlorophenol	*D. tiedjei*	Mohn and Kennedy (1992)

[a]sed, Sediment.

tion activity might be used to stimulate the degradation of recalcitrant haloaromatic substances (Table 6.2). An interesting remedial approach can be suggested based on the amendment of suitable substrates to ellicit a more rapid dehalogenation response. In the process of cross-acclimation, a chemical analog that can be relatively easily degraded is added to an environment to facilitate the decomposition of more recalcitrant substances.

For instance, 4-amino-3,5-dichlorobenzoic acid is a relatively recalcitrant chemical that exhibits approximately an 8 week acclimation period before dehalogenation starts to occur in freshwater sediment samples (Horowitz et al., 1983). However, once the resident microflora has been acclimated to the degradation of this chemical, subsequent amendments of this compound were completely degraded in only 2–3 weeks. Essentially, the same effect could also be obtained with 3-bromobenzoic acid, except that this substrate normally exhibited an acclimation period of only a few weeks. However, when the microflora was acclimated to the destruction of the bromobenzoate and then exposed to the 4-amino-3,5-dichlorobenzoate, degradation of the latter compound occurred without a measurable acclimation period. Presumably, the brominated derivative was better at derepressing the requisite activity. The sediment microflora acclimated to either substrate exhibited the same range and rate of substrate dehalogenation. However, when the microflora was acclimated to the degradation of 3-iodobenzoate, a different pattern of dehalogenation was observed and no cross-acclimation to other halobenzoates could be demonstrated.

A similar strategy has been used to stimulate the anaerobic biodegradation of polychlorinated biphenyls (PCBs) (Table 6.2). No significant reductive dehalogenation in river sediments contaminated with a commercial PCB mixture was noted until bromobiphenyls were added as cross-acclimating substrates (Bedard and Van Dort, 1992). Concomitant with the loss of bromines from the amendments was the loss of chlorines from the endogenous PCB mixture present in the samples. These findings may help provide a basis for stimulating PCB metabolism *in situ*.

2.2. Multiple Substrate Effects

Another factor that may influence the acclimation period prior to dehalogenation is the presence of structurally related chemicals in the same environment. These additional substrates may either inhibit or stimulate dehalogenation activity.

The inhibition of reductive dehalogenation due to a structurally related substrate was observed with an enrichment that was capable of dehalogenating and completely mineralizing 3-chlorobenzoate (Suflita et al., 1983). When the enrichment was amended with the structural analog 3,5-dichlorobenzoate, 3-chlorobenzoate accumulated to about 70% of that theoretically possible. Despite being the substrate on which the enrichment was grown, 3-chlorobenzoate was not degraded when it was produced as an intermediate until the parent substrate was reduced to low levels (<5 μM). An examination of the kinetics of dehalogenation revealed that the enrichment had a higher apparent affinity for 3,5-dichlorobenzoate but a slower maximum rate of decay relative to 3-chlorobenzoate. The relative differences in

Competing Substrate		Dehalogenation Rate ± SD of Substrate 1 (μmol/liter/min)
1	2	
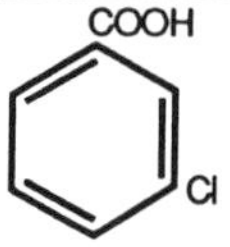 COOH, Cl (chlorobenzoate)	None	3.19 ± 0.23
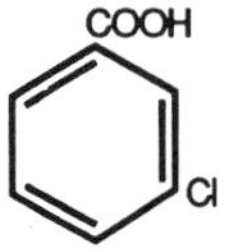 COOH, Cl	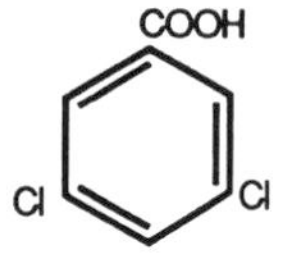 COOH, Cl, Cl	0.90 ± 0.11
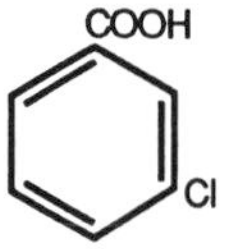 COOH, Cl	None	2.70 ± 0.09
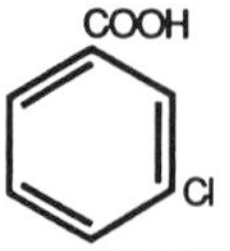 COOH, Cl	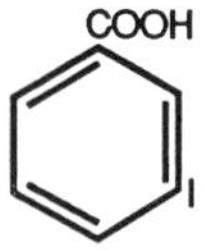 COOH, I	1.49 ± 0.10
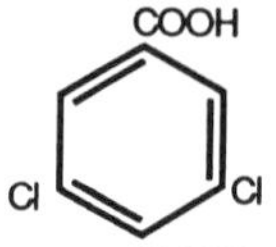 COOH, Cl, Cl	None	2.70 ± 0.29
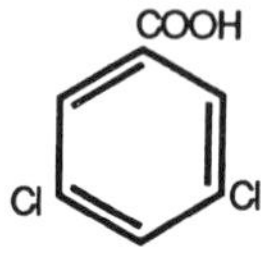 COOH, Cl, Cl	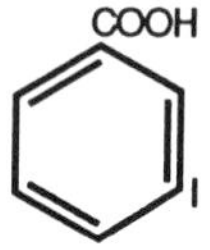 COOH, I	1.85 ± 0.13

Figure 6.2. The effect of competing substrates on dehalogenation activity in cell extracts of *D. tiedjei* (adapted from DeWeerd and Suflita, 1990).

observed kinetic parameters were not enough to account for the preferential dehalogenation of the dichlorosubstrate and the accumulation of the monochlorinated derivative. However, a variety of experiments indicated that the dichlorobenzoate substrate was acting as a competitive inhibitor of the dehalogenation of the mono-haloaromatic substrate (Suflita et al., 1983). Presumably, this interaction occurred at either a substrate transport site or an active site of the requisite enzyme. Since similar competitive inhibition effects were seen in cell-free extracts of *D. tiedjei*, the latter speculation is more likely (DeWeerd and Suflita, 1990) (Fig. 6.2).

    Therefore, the presence of structurally related chemicals can effectively decrease or even preclude the degradation of another substrate by a combination of relative kinetic differences and competitive inhibition effects. This may lead to prolonged acclimation periods that can only be overcome by the removal of preferred substrates. Alternately, mixtures of structurally related haloaromatic compounds may result in increased dehalogenation. A single polybrominated biphenyl congener resisted dehalogenation when amended to anaerobic incubations. However, when added as part of a polybrominated biphenyl mixture in parallel incubations, the congener was readily dehalogenated (Morris et al., 1992b). Whether this increase in dehalogenation is due to the derepression of the catabolic enzymes or some other interaction is unknown.

2.3. Substrate Concentration

Substrate concentration also influences the susceptibility of chemicals to anaerobic decomposition. Figure 6.3 compares the length of the acclimation time for several haloaromatic chemicals incubated in freshwater sediment slurries. Not surprisingly, 3-chloro- and 3-fluorobenzoate are degraded relatively rapidly at low concentrations, but the same two chemicals are persistent at higher concentrations. The halobenzoates are known bacteriostatic agents, and their effect on microbial growth is concentration dependent. Similar inhibition patterns have been noted with chlorophenols (Mohn and Kennedy, 1992; Madsen and Aamand, 1992), and the literature is replete with examples of substrates that effectively prolong the acclimation period by resisting decomposition when they are present in high concentrations.

Arguably a more interesting case is observed with 4-amino-3,5-dichlorobenzoate (Fig. 6.3). At higher concentrations (up to 800 μM), this chemical degrades by reductive dehalogenation following a typical acclimation period. However, at low concentrations ($\leq$20 μM), the chemical persists. This experiment was followed for

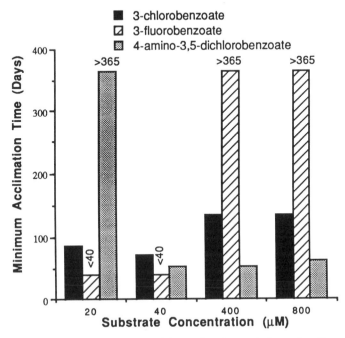

Figure 6.3. Comparison of the length of the acclimation period as a function of substrate concentration prior to the dehalogenation of several substrates in sediment incubations.

over a year, and the normally degradable substrate persisted throughout the incubation period. However, when the concentration of 4-amino-3,5-dichlorobenzoate was merely doubled, dehalogenation proceeded in typical fashion. For some chemicals, there may be a concentration threshold below which degradation is limited in anaerobic environments. Similar observations have been made with the metabolism of xenobiotic chemicals by aerobic microorganisms. (Boethling and Alexander, 1979; Rubin et al., 1982).

This may also be the case for some problem contaminants that commonly are found in the environment at low concentrations. The dehalogenation of both PCBs and polybrominated biphenyls in laboratory experiments has been shown to occur at 500–700 ppm but not at 14–50 ppm (Quensen et al., 1988; Morris et al., 1992b). While this is an interesting finding, it logically leads to a remediation scenario that is not likely to be warmly embraced by the regulatory community in the United States.

2.4. Lack of Required Growth Substrates or Growth Factors

The amendment of environmental samples and enrichment cultures with complex nutrients has been found to stimulate aryl reductive dehalogenation in some cases. Whether these nutrients are used for growth or as growth factors required in only catalytic amounts is unknown. Fathepure et al. (1987) found that amendments of yeast extract, trypticase, or rumen fluid greatly enhanced the rate of dehalogenation by a 4-chlororesorcinol enrichment without an increase in total biomass. Similar amendments stimulated the dehalogenation of 3,4-dichloroaniline (Kuhn et al., 1990) and trichlorophenol (Madsen and Aamand, 1992). Others have found that small amounts of such complex nutrients were required for dehalogenation but apparently did not constitute a primary source of carbon for the dehalogenating microorganisms (Dietrich and Winter, 1990). Similarly, *D. tiedjei* initially required medium supplemented with 20% rumen fluid to obtain only modest growth (Linkfield and Tiedje, 1990). Subsequently, both growth rate and yield were greatly enhanced by culturing *D. tiedjei* in a defined medium supplemented with niacin, nicotinamide, and 1,4-naphthoquinone (DeWeerd et al., 1990). Anaerobic microbial communities are well known for syntrophic or mutualistic relationships between component members, and the cross-feeding of micronutrients is believed to be one of the reasons for the difficulty in isolating dehalogenating microorganisms in pure culture.

2.5. Benefits of Dehalogenation

One question may be asked regarding any biotransformation: What is the physiological benefit to microorganisms catalyzing this transformation? As already noted, an acclimation period is often observed prior to the onset of rapid degradation, and in the case of aryl reductive dehalogenation these acclimation periods can be quite lengthy. Acclimation periods are often attributed to the growth of an initially small number of microorganisms into a large enough biomass to effect a quantifiable substrate depletion. In the simplest case, the cells possess the metabolic ability to use the amended substrate as a source of carbon and energy. However, in the case of aryl reductive dehalogenation reactions, acclimation periods are observed regardless of whether the substrates are mineralized. Polychlorinated biphenyls, benzenes, and anilines have been demonstrated to undergo initial dehalogenation reactions to yield lesser halogenated congeners (Quensen et al., 1988; Fathepure et al., 1988; Kuhn and Suflita, 1989b). The dehalogenated metabolites from these starting materials often accumulate to stoichiometric levels, and the aromatic ring remains intact. Since the haloaromatic compounds do not provide a source of carbon for the dehalogenating cells, what is the potential advantage of aryl dehalogenation reactions to these organisms?

While reductive dehalogenation does not provide a source of carbon to the requisite microorganisms, it may provide energy to the cells. Dehalogenating microorganisms transfer electrons, perhaps by a series of intermediate carriers, to the haloaromatic compound. This process may be coupled with energy conservation, with the haloaromatic compound serving as the terminal electron acceptor.

Thermodynamic calculations suggest that the standard free energy change for the oxidation of hydrogen using a haloaromatic compound as an electron acceptor can yield more energy than some of the more common electron acceptors available in anaerobic environments (Dolfing and Harrison, 1992). The opportunity to use an alternate electron acceptor, particularly one for which there may be only a limited degree of competition in energy-limited environments, may provide dehalogenating cells with some selective advantage. This advantage could ultimately manifest itself in increased growth rates and in the selective enrichment of dehalogenation communities in environments that receive substantial amounts of haloaromatic compounds.

Evidence supporting the hypothesis that reductive dehalogenation may be a form of anaerobic respiration resulting in some energetic benefit for dehalogenating microorganisms has been shown in studies of resting cell suspensions of *D. tiedjei* (Mohn and Tiedje, 1991). Upon the addition of 3-chlorobenzoate to cell suspensions, an increase in ATP pool size concomitant with dehalogenation of the substrate was observed. Additionally, when 3-chlorobenzoate was coupled with a suitable electron donor, proton translocation was observed as evidenced by a decrease in pH of the medium, while in separate experiments an artificially imposed pH gradient led to increased ATP pools. This evidence is consistent with the chemiosmotic coupling of reductive dehalogenation of 3-chlorobenzoate with ATP synthesis by *D. tiedjei*.

2.6. Need For Electron Donors

If aryl reductive dehalogenation serves as a novel form of anaerobic respiration in dehalogenating microorganisms, there is an obvious need to couple this reaction with suitable electron donors. Conceivably, the lack of appropriate donors can limit the transformation of haloaromatic chemicals in the environment. Some credence to this suggestion has been gained in studies on the anaerobic fate of the pesticide 2,4,5-trichlorophenoxyacetate (2,4,5-T) (Gibson and Suflita, 1990). This pesticide is degraded in anaerobic aquifer slurries through a variety of intermediate metabolites. The primary conversion results in the formation of either 2,4- or 2,5-

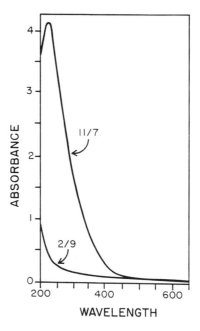

Figure 6.4.   Absorbance characteristics of ground water sampled at different times of the year from the anoxic aquifer underlying the Norman municipal landfill.

dichlorophenoxyacetate as initial reductive dehalogenation products. Subsequent degradation of intermediates also involves a number of other aryl dehalogenation reactions.

By repeatedly examining the decomposition of 2,4,5-T in samples taken at different times of the year from an anoxic aquifer impacted by landfill leachate, it was noted that the onset of degradation varied from several weeks to several months. Shorter acclimation times were correlated with periods when the groundwater was highly colored relative to other times of the year when the groundwater was essentially colorless (Fig. 6.4). The color was believed to be due to the presence of leachate constituents and/or humic materials originating from the landfill. Field evidence suggested that anaerobic biodegradation processes in the aquifer slowed dramatically or ceased during the winter months (Adrian et al., 1991). The correlation of shorter acclimation periods with highly colored water sampled during the warmer parts of the year (summer and fall) suggested that reductive dehalogenation reactions might be limited at certain times of the year by the availability of suitable electron donors.

In a test of this hypothesis, aquifer samples were taken during the winter and amended with a variety of low-molecular-weight fatty acids or alcohols, typical intermediates of fermentation processes, as potential electron donors for reductive dehalogenation (Gibson and Suflita, 1990). Most of the supplements stimulated both the onset and the rate of 2,4,5-T dehalogenation, resulting in an overall decrease in halogenated residues. This was the first report to suggest that the addition of suitable electron donors could shorten the acclimation period and stimu-late reductive dehalogenation processes. A similar approach was found successful in stimulating the dehalogenation of an Arochlor 1242 PCB mixture (Nies and Vogel, 1990). Other investigators have found that the addition of an electron donor was essential for efficient pentachlorophenol dehalogenation in sludge blanket reac-tors (Hendriksen et al., 1992) and for the maintenance of 1,2,3-trichlorobenzene–(Holliger et al., 1992) and PCB-dehalogenating enrichments (Morris et al., 1992a).

Therefore, the addition of electron donors to complex environmental samples has been shown to stimulate dehalogenation. However, the mechanism by which these donors act is unclear, and no single electron donor has proven to be universally successful in promoting reductive dehalogenation activity. The amended electron donors may provide reducing equivalents for reductive dehalogenation either direct-ly or indirectly. They also may provide a source of carbon for the dehalogenating microorganisms or for other microorganisms that may live in synergistic, mutualis-tic, or even commensalistic association with the dehalogenators. Alternatively, the addition of exogenous electron donors may stimulate the consumption of preferred or more energetically favorable electron acceptors. The latter process may eventu-ally result in a haloaromatic compound becoming a preferred electron acceptor in environments in which alternate electron acceptors reach short supply. Inter-estingly, a recent report has presented evidence that enrichment of anaerobes on naturally occurring nonhalogenated aromatic compounds led to an increase in the dehalogenation of seemingly unrelated chlorocatechols (Allard et al., 1992). The precise ways in which microbial communities and/or habitats are altered by amend-ments intended to favor dehalogenation reactions remains to be determined and awaits the isolation and study of the participating microorganisms.

In this respect, *D. tiedjei* serves as an illustrative example. The number of electron donors that support dehalogenation in *D. tiedjei* appears to be quite limited. Resting cell suspensions were used to assess directly the effect of potential electron donors on dehalogenation by *D. tiedjei* (DeWeerd et al., 1991). The addition of hydrogen, formate, and pyruvate to the cells resulted in a 500%–600% increase in the dehalogenation rate over the endogenous rate of dehalogenation, whereas acetate-amended suspensions exhibited less than a 20% increase. In experiments with a defined triculture capable of mineralizing 3-chlorobenzoate, Dolfing and Tiedje (1991) presented limited evidence that the oxidation of acetate provides reducing equivalents for the dehalogenation of 3-chlorobenzoate by *D. tiedjei*, but hydrogen was the preferred source of reducing equivalents. Additionally, *D. tiedjei* is unable to grow on acetate with or without either 3-chlorobenzoate or sulfate as electron acceptors. Therefore, it is not likely that acetate serves as a major source of reducing equivalents for the reductive dehalogenation activity in this microor-ganism.

3. ENVIRONMENTAL BARRIERS TO BIODEGRADATION

One of the attractive features of bioremediation is the possibility of destruction of pollutants *in situ* without costly excavation or other removal processes. However, *in*

TABLE 6.3. Environmental Factors Potentially Limiting
Biodegradation Activity

pH	Free water limitations
Salinity	Radiation
Other synthetic chemicals	Nutrient availability
Heavy metals	Permeability/porosity constraints
Osmotic pressures	Electron acceptor availability
Hydrostatic pressures	Redox conditions
	Others

situ biodegradation attempts must contend with a number of environmental parameters that may impact the rate or extent of biodegradation. Not surprisingly, extremes of environmental parameters such as those listed in Table 6.3 may alone or in combination adversely influence the rate of biodegradation and render even very labile chemicals quite recalcitrant. If aryl reductive dehalogenation is to be exploited as part of a bioremedial strategy, environmental parameters that pose barriers to dehalogenation must be identified and overcome.

The effect of some environmental parameters on aryl dehalogenation reactions has been documented. For instance, temperature is known to influence both the acclimation period and the reaction rate of 2,4-dichlorophenol dehalogenation in anaerobic lake sediments (Kohring et al., 1989a). Dehalogenation could be measured at temperatures ranging from 5° to 50°C with optima at 30° and 43°C. The dehalogenation of a PCB congener has also been shown to have two similar temperature optima, likely due to separate mesophilic and thermophilic dehalogenating populations (Wu et al., 1992). A separate study demonstrated PCB dehalogenation at 12° and 25°C, but no activity occurred at temperatures ≥37°C (Tiedje et al., 1992). Similarly, Holliger et al. (1992) measured a temperature optimum of 30°C for 1,2,3-trichlorobenzene dehalogenation by a mixed culture, but no dehalogenation at or above 37°C *D. tiedjei,* which was isolated from an anaerobic sludge, exhibits dehalogenation activity at 37°C but not 20° or 46°C (Linkfield and Tiedje, 1990); cell-free extracts demonstrated a similar temperature optimum (DeWeerd and Suflita, 1990). Although the ability to alter the temperature of a polluted site substantially may be limited, the wide range of temperature conditions under which dehalogenation has been observed argues that remedial approaches may be slower but not necessarily limited by seasonal temperature variations.

Laboratory studies largely address the fate of haloaromatic compounds that have been freshly amended to environmental samples. With few exceptions, there has been little examination of the dehalogenation of contaminants already present in polluted sites that have had a much greater time of exposure to their biotic and abiotic surroundings. Many haloaromatic contaminants are hydrophobic, and the effect that sorption has on limiting reductive dehalogenation in environmentally mature samples is unknown. Interestingly, the addition of an individual PCB congener to contaminated sediment samples resulted in the dehalogenation of both the amended congener and the PCBs already in place (Bedard et al., 1991). Presumably, abiotic effects such as sorption could have limited the bioavailability of the PCBs and the metabolic ability of the resident microbiota. The bioavailability of hydrophobic contaminants and its influence on dehalogenation will likely be a fruitful area for future investigations.

Another environmental parameter influencing the onset of aryl dehalogenation activity is the presence of alternative electron acceptors. As mentioned above, it is currently believed that haloaromatic compounds serve as alternate electron acceptors for dehalogenating anaerobes. In complex anaerobic communities, electron acceptors are frequently a limiting resource. Moreover, it is quite likely that there is substantial competition for reducing equivalents in any climax microbial community. It is not surprising then that the availability of other, perhaps more favored, electron acceptors can alter electron flow and disrupt aryl dehalogenation.

Many of the observations on the influence of electron acceptors on dehalogenation come from experiments designed to examine the fate of haloaromatic compounds in samples from an anoxic aquifer polluted by municipal landfill leachate. A variety of field and laboratory evidence collected over several years indicates that two spatially distinct sites exist within the aquifer (Suflita et al., 1988; Beeman and Suflita, 1987). At one site the predominant flow of carbon and energy is through methanogenesis, while sulfate reduction predominates at the other. These two spatially proximate sites provided an opportunity to examine the biodegradation of contaminant chemicals under two different anaerobic conditions.

To date, over 30 haloaromatic compounds have been examined with such samples, including a variety of halogenated benzoates, phenols, phenoxyacetates, anilines, and several heterocyclic molecules (Gibson and Suflita, 1986; Kuhn and Suflita, 1989b; Suflita et al., 1988). A rather consistent pattern emerged when the biodegradation of these material was compared with such samples. In all cases, the reductive dehalogenation of the substrates was detected as the primary degradative event under mathanogenic conditions. However, the same compounds proved extremely recalcitrant when assayed in sulfate-reducing aquifer slurries or when sulfate was added to the previously methanogenic incubations. Such findings do not specifically implicate the involvement of methanogens in the dehalogenation of these substrates. The inhibition or delay of aryl reductive dehalogenation by the presence of alternate electron acceptors has been confirmed by other investigators (Madsen and Aamand, 1991; Genthner et al., 1989; Allard et al., 1992; Hale et al., 1991; Kohring et al., 1989b; May et al., 1992a,b; Morris et al., 1992a; Alder et al., 1993).

From an environmental perspective, an increase in length of the acclimation period prior to the onset of dehalogenation has been noted in the presence of alternate electron acceptors. This is shown in Table 6.4, which compares 2,4,5-T acclimation times in aquifer slurries that received sulfate in addition to an electron donor.

The adaptation time increased in all cases relative to the sample that did not receive the sulfate treatment. There seems to be a relatively consistent pattern of increased acclimation periods in environmental incubations that have either an endogenous or exogenous supply of potential terminal electron acceptors. This

TABLE 6.4. Effect of Sulfate and an Organic Amendment on the Acclimation Time Prior to the Onset of 2,4,5-T Dehalogenation in Aquifer Slurries

Organic Amendment	Sulfate Added	Acclimation Period (Months) in Replicates	
		A	B
None	No	3–4	>4
None	Yes	>4	>4
Formate	No	0–1	1–2
Formate	Yes	2–3	2–3
Acetate	No	1–2	2–3
Acetate	Yes	3–4	>4
Propionate	No	0–1	0–1
Propionate	Yes	2–3	2–3
Butyrate	No	0–1	0–1
Butyrate	Yes	2–3	2–3
Methanol	No	1–2	2–33
Methanol	Yes	3–4	>4
Ethanol	No	0–1	0–1
Ethanol	Yes	2–3	>4

Adapted from Gibson and Suflita (1990).

point notwithstanding, recent reports indicate that once acclimation periods are overcome, the amendments of environmental samples with alternate electron acceptors may or may not influence dehalogenation activity (Kohring et al., 1989b; May et al., 1992a; Allard et al., 1992). In other cases, the microbial metabolism of a haloaromatic compound in unacclimated environmental samples may be either unaffected or even dependent on another electron acceptor (Haggblom and Young, 1990; Haggblom et al., 1992; King, 1988). However, the relationship of several of these studies to aryl dehalogenation *per se* remains to be clarified.

Some insight into the apparent extension of the adaptation period can be gained by looking at the effect of sulfur oxyanions and other potential electron acceptors on aryl dehalogenation reactions catalyzed by *D. tiedjei*. An examination of the consumption of hydrogen by resting cell suspensions of *D. tiedjei* in response to various electron acceptors is illustrated in Figure 6.5.

The rate of hydrogen consumption is about six times faster when sulfite is provided as an electron acceptor compared with the rate observed with 3-chlorobenzoate as the electron acceptor. Therefore, a competition for reducing equivalents when both electron acceptors were present could be envisioned. In similar fashion, when both sulfite and 3-chlorobenzoate are available as potential electron acceptors, an intermediate rate of hydrogen consumption is observed. The same effect on

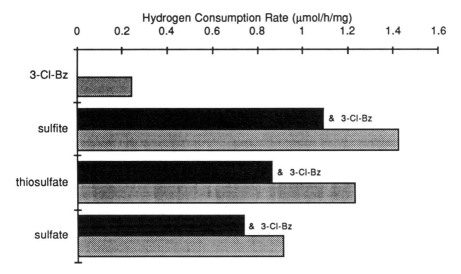

Figure 6.5. The effect of various electron acceptor combinations on the consumption of hydrogen by resting cell suspensions of *D. tiedjei*.

hydrogen consumption can be seen with the other sulfur oxyanions; that is, the rate of hydrogen consumption in the presence of both electron acceptors is less than that observed with the sulfur oxyanions as the only acceptor. It is therefore tempting to speculate that a mutual electron carrier for dehalogenation and sulfur oxyanion reduction could possibly account for the intermediate hydrogen consumption rates. More credence is lent to this speculation when the effect of various potential inhibitors are examined for their influence on both dehalogenation and hydrogen consumption (Table 6.5). The inhibitors range from ferredoxin and hy-

TABLE 6.5. **The Effect of Various Inhibitors on Hydrogen Uptake by *D. tiedjei* Cells Exposed to 3-Chlorobenzoate or One of Several Sulfur Oxyanions as Electron Acceptors**

Substance	Concentration (≥50% Inhib.)	H$_2$ Uptake Inhibition With	
		3-Chlorobenzoate	Sulfur Oxyanion
Acetylene	100%	Yes	Yes
Molybdate	2 mM	Yes	Yes
Selenate	2 mM	Yes	Yes
CCCP	10 μM	Yes	Yes
CuCl$_2$	1 mM	Yes	Yes
Diphenylamine	100 μM	Yes	Yes
8-Hydroxyquinoline	2 mM	Yes	Yes
Dimercaprol	10 mM	Yes	Yes
Metronidazole	1 mM	Yes	Yes
NaCl	4 mM	No	No

drogenase inhibitors to others that would disrupt electron flow in the cell. In all
cases they acted similarly, disrupting dehalogenation as well as hydrogen consumption.

Such findings argue that sulfur oxyanions should inhibit aryl dehalogenation
reactions. When 3-chlorobenzoate metabolism was examined in these experiments,
it was found that dehalogenation was inhibited in the presence of sulfite and thiosulfate, but not by sulfate. Comparable results were obtained with aryl dehalogenation
in cell-free extracts as well (DeWeerd et al., 1990). Consequently, the intracellular
competition for reducing equivalents helps explain some, but not all, of the effects
associated with the presence of alternate electron acceptors on dehalogenation.

The environmental implications of such findings are worth noting. The kinetics
of hydrogen metabolism in the presence of sulfur oxyanions and halobenzoates
suggest that, in environments predominated by sulfate reduction, the levels of
hydrogen may be too low to support significant amounts of aryl dehalogenation.
This suggestion is consistent with the bulk of the literature but contrasts with recent
findings of bromophenol and chlorophenol dehalogenation in sulfate-rich environments (Haggblom et al., 1992; Haggblom and Young, 1990; King, 1988).

The ability of dehalogenating bacteria to scavenge reducing equivalents in anaerobic environments may be dependent on other microbial electron-accepting processes present in the particular habitats. Thus, the regulation of dehalogenation in
anaerobic habitats may ultimately depend on the availability of alternate electron
acceptors and the ability of the dehalogenating cells inhabiting such environments to
compete for electron donors with other bacteria.

5. CHEMICAL BARRIERS TO DEHALOGENATION

It is well known that the chemical structure of a particular pollutant influences the
rate and extent of its degradation (e.g., Alexander, 1981; Howard et al., 1987;
Niemi et al., 1987). While broad generalizations are neither useful nor possible,
several trends can be gleaned from the literature regarding the influence of chemical
structure on susceptibility of haloaromatic compounds to reductive dehalogenation
reactions.

The majority of aryl reductive dehalogenation studies have been conducted specifically with chlorinated aromatic compounds. However, brominated and iodinated
aromatic compounds are also known to be environmentally significant pollutants.
Molecules possessing the latter halogens are also susceptible to reductive dehalogenation reactions. The strength of aryl halide bond decreases with increasing
halogen radius. Not surprisingly, then, the strength of this bond can be an important
factor in the rate or extent of dehalogenation. For example, there is no conclusive
evidence for the reductive cleavage of the small and tightly bound fluoride atom
from the aromatic nucleus. Based on bond strength alone, one might predict that the
larger halogens would be the most labile, and studies with environmental samples
seem to support this contention (Linkfield et al., 1989). However, whole cells of *D.
tiedjei* dehalogenated halobenzoates with the smallest halogen radius most rapidly

(exclusive of the fluorobenzoates), while in cell-free extracts the opposite proved true (Linkfield and Tiedje, 1990; DeWeerd and Suflita, 1990). It is likely that difficulties in the transport of haloaromatic substrates by whole cells may account for some of the differences observed when dehalogenation rates were compared in whole cells and cell-free preparations of *D. tiedjei.*

The degree of halogenation also influences the susceptibility of aromatic contaminants to reductive dehalogenation. The more highly halogenated aromatic compounds (and thus more oxidized) appear generally more amenable to this type of bioconversion. Often, the polyhalogenated substrates undergo sequential reactions to yield lesser halogenated congeners that may sometimes resist subsequent dehalogenation (e.g., Fathepure et al., 1988; Bosma et al., 1988; Kuhn and Suflita, 1989b; Gerritse et al., 1992; Holliger et al., 1992). Several factors may help to explain the preferential release of halogens from multihalogenated substrates.

Thermodynamically, the free energy available from the reductive dehalogenation of polyhalogenated substrates coupled to a given electron donor is more negative (i.e., more exergonic) than the comparable reaction involving a monohalogenated substrate. Dolfing and Harrison (1992) found that the reduction of hexa-, penta-, and tetrachlorobenzene with hydrogen as an electron donor releases $\sim$20 kJ/mol more free energy than the dechlorination of tri-, di-, or monochlorobenzene. The higher free energy associated with these substrates might preferentially favor aryl halide release from a multihalogenated aromatic nucleus. Bacteria capable of benefiting from this energy release would also likely have a selective advantage in the environment and be more easily enriched. However, as Dolfing and Harrison have noted, a cause and effect relationship cannot be assumed, as free energy calculations provide no information on the ease of a reaction or the ease with which a reaction proceeds. Mechanistically, the presence of multiple electrophilic halogens on an aromatic ring causes a decreased electron density around the ring. An electron-poor aromatic ring would be more susceptible to a reductive (nucleophilic) attack such as aryl reductive dehalogenation reactions.

The presence and nature of other aryl substituents also impacts the susceptibility of substrates to dehalogenation. Aryl dehalogenation reactions seem to proceed easier when the ring is substituted with electron-destabilizing groups such as carboxy, hydroxy, or cyano groups (Kuhn and Suflita, 1989a). Still other substituents bonded to the aromatic ring may have additional effects, but broad generalizations in this regard are difficult to make (Dolfing and Tiedje, 1991b). Steric and electronic effects of the presence of aryl substituents may alter the fit of the substrate in the enzyme active site. Additionally, the conformation of the enzyme may be altered with some substrates, and thus its ability to catalyze dehalogenation reactions may be changed. Dolfing and Tiedje (1991b) have suggested that the high-energy-state complex that is believed to be formed during reductive dehalogenation reactions may be affected by other aryl substituents. These authors found that *m*-chlorobenzoate metabolism by *D. tiedjei* was inhibited by *para*-hydroxy and -amino groups, thus implicating a nucleophilic attack on the π-electron cloud of the benzene nucleus as the rate-limiting step in dehalogenation.

As already pointed out, the structure of a chemical also influences the acclimation period prior to aryl dehalogenation reactions. Linkfield et al. (1989) found that the acclimation period prior to the anaerobic dehalogenation of six *meta*-substituted halobenzoates were reproducible over time and among freshwater sediment sampling locations. Moreover, the acclimation times were characteristic of the particular chemical under investigation. The acclimation periods appeared to reflect the time necessary for the derepression of dehalogenation activity rather than other phenomena.

Aryl dehalogenation reactions are often characterized by their specificity for molecules within a particular chemical class. This specificity often extends to the position of the halogens within a particular chemical class. For instance, aryl dehalogenation reactions are generally favored when benzoates possess a *meta*-substituted halogen (Kuhn and Suflita, 1989a). Furthermore, acclimated mixtures of microorganisms exhibit a high degree of substrate specificity favoring the removal of *meta*-substituted aryl halides from benzoate molecules (Tiedje et al., 1987; Horowitz et al., 1983). While capable of catalyzing the dehalogenation of 3-chlorobenzoate, *D. tiedjei* cannot remove a chlorine substituent from the corresponding *ortho*- or *para*-substituted isomers. As noted above, other substituents can have drastic effects on the specificity of aryl dehalogenation reactions. Dicamba is a pesticide possessing a halogen *meta* to a carboxy group. Dehalogenation of dicamba is known following its initial bioconversion to 3,6-dichlorosolicylic acid (Taraban et al., 1993). Yet dicamba was not transformed when it was tested in a bacterial enrichment containing *D. tiedjei* as the dehalogenating organism (unpublished observation).

Cautious generalizations can be made regarding the susceptibility of other classes of haloaromatic compounds to dehalogenation. For instance, the tendency of the amino group of haloanilines to release electrons to aromatic resonance structures makes the ring more susceptible to attack, particularly at the *ortho* and *para* positions. This resonance effect might contribute to the preferential release of halides attached at these positions. Halogenated phenols possess analogous resonance structures, and the preferential removal of *ortho* and *para* halogens from these substrates has been observed on many occasions. In consistent fashion, halogenated pesticides containing an aryl ether or a nitrogen-substituted aromatic nucleus were often dehalogenated faster if they possessed an *ortho* or *para* halogen (Kuhn and Suflita, 1989a). In this regard the findings of Struijs and Rogers (1989) and Bryant (1992) are notable. Both studies found that preexposure of sediments to dichlorophenols resulted in the transformation of analogous dichloroaniline or dichlorophenoxyacetate isomers without a measurable acclimation period. However, acclimation of sediments to 2,4- or 3,4-dichlorophenol did not shorten the acclimation period for 2,4,5-T degradation. Such findings implicate the importance of mesomeric effects of ring substituents. However, there is no doubt that all ring positions are potentially amenable to reductive dehalogenation reactions among widely divergent classes of haloaromatic compounds.

As noted earlier, substrate concentration also influences the susceptibility of haloaromatic compounds to dehalogenation reactions (Linkfield et al., 1989). Relatively high concentrations of halobenzoates were associated with increased acclimation periods. A similar toxicity of high concentrations of pentachlorophenol to *D*.

tiedjei has been cited as the reason why the dehalogenation potential of this organism for chlorophenols was not recognized previously (Mohn and Kennedy, 1992). However, one can also observe the extension of the acclimation period prior to reductive dehalogenation when substrate concentrations are too low (Linkfield et al., 1989). That is, with some halobenzoates there appears to be a concentration threshold below which dehalogenation does not occur to any appreciable degree (see above). Increased recalcitrance as a function of low substrate concentrations has also been observed with other classes of haloaromatic materials such as the PCBs (Quensen et al., 1988).

Still another important factor that influences the transformation of a contaminant by microorganisms is the physical state of the contaminant. Dehalogenation rates may be positively or negatively influenced, depending on whether the contaminant is soluble, sorbed, volatile, conjugated, and so forth. As noted earlier, the reductive dehalogenation of extremely hydrophobic haloorganic molecules may conceivably be precluded by the partitioning of contaminants to environmental compartments in which microorganisms (or their enzymes) have only limited accessibility. On the other hand, the partitioning of haloaromatic substrates to nonaqueous phases can also have a beneficial effect on biodegradation. For instance, the partitioning of 1,2,3-trichlorobenzene to a solvent phase (Holliger et al., 1992) and pentachlorophenol to bark chips (Apajalahti and Salkinoja-Salonen, 1984) was believed to reduce the aqueous concentrations of these substances below levels that were normally toxic to the requisite microorganisms. Strategies that are designed to modify the bioavailability of pollutants to the catalyzing microorganisms are likely to be very important tools for the bioremediation of haloaromatic contaminants in the future.

6. CONCLUSIONS

Haloaromatic compounds have been manufactured in enormous quantities and are an inexorable part of our industrialized society. Once these compounds are introduced to the environment, they often persist and may pose a public health hazard. Laboratory and field studies have demonstrated that microbially mediated aryl dehalogenation reactions can transform even the most highly halogenated compounds into intermediates that are more amenable to subsequent aerobic or anaerobic decay. Often such reactions lead to the eventual mineralization of the problem contaminant.

Bioremediation is an emerging technology that seeks to exploit such microbial processes for the transformation of environmental contaminants into innocuous endproducts. There is a great deal of interest in taking advantage of the nutritional versatility of anaerobic microorganisms to catalyze dehalogenation reactions *in situ*. If microbial processes are to be used to this end, there is little doubt that the rates of biocatalysis must be greatly accelerated. It would appear that the best chance for stimulating such bioconversions is a thorough consideration of the major environmental, physiological, and ecological constraints of the process. Through such an understanding, rational steps can conceivably be taken to overcome such barriers. We believe this approach can be extrapolated well beyond aryl reductive dehalogenation reactions and that similar approaches will prove useful for stimulating the *in situ* microbial destruction of a variety of organic contaminants.

REFERENCES

Adrian NR, Robinson JA, Suflita JM (1994): Spatial variability in biodegradation rates as evidenced by methane production from an aquifer. Appl Environ Microbiol 60:3632–3639.

Adrian NR, Suflita JM (1993): Anaerobic biodegradation of halogenated and nonhalogenated *N*-, *S*-, and *O*-heterocyclic compounds in aquifer slurries. Environ Toxicol Chem 13:1551–1557.

Adrian NR, Suflita JM (1990): Reductive dehalogenation of a nitrogen heterocyclic herbicide in anoxic aquifer slurries. Appl Environ Microbiol 56:292–294.

Alder AC, Häggblom MM, Oppenheimer SR, Young LY (1993): Reductive dehalogenation of polychlorinated biphenyls in anaerobic sediments. Environ Sci Technol 27:530–538.

Alexander M (1981): Biodegradation of chemicals of environmental concern. Science 211: 132–138.

Allard A-S, Hynning P-A, Remberger M, Neilson AH (1992): Role of sulfate concentration in dechlorination of 3,4,5-trichlorocatechol by stable enrichment cultures grown with coumarin and flavanone glycones and aglycones. Appl Environ Microbiol 58:961–968.

Apajalahti JH, Salkinoja-Salonen MS (1984): Absorption of pentachlorophenol (PCP) by bark chips and its role in microbial PCP degradation. Microb Ecol 10:359–367.

Apajalahti JHA, Salkinoja-Salonen MS (1987): Dechlorination and *para*-hydroxylation of polychlorinated phenols by *Rhodococcus chlorophenolicus*. J Bacteriol 169:675–681.

Bedard DL, Bunnell SC, Van Dort HM (1991): Reductive dechlorination of PCBs (Araclor 1260) in methanogenic sediment slurries. In Abstracts of the 91st General Meeting of the American Society for Microbiology. Washington, DC: American Society for Microbiology, abstract Q-36, p 282.

Bedard DL, Van Dort HM (1992): Brominated biphenyls can stimulate dechlorination of endogenous Araclor 1260 in methanogenic sediment slurries. In Abstracts of the 92nd Annual Meeting of the American Society for Microbiology. Washington, DC: American Society for Microbiology, abstract Q-26, p 339.

Beeman RE, Suflita JM (1987): Microbial ecology of a shallow unconfined ground water aquifer polluted by municipal landfill leachate. Microb Ecol 14:39–54.

Boethling RS, Alexander M (1979): Effect of concentration of organic chemicals on their biodegradation by natural microbial communities. Appl Environ Microbiol 37:1211–1216.

Bosma TNP, van der Meer JR, Schraa G, Tros ME, Zehnder AJB (1988): Reductive dechlorination of all trichloro- and dichlorobenzene isomers. FEMS Microbiol Ecol 53:223–229.

Bryant FO (1992): Biodegradation of 2,4-dichlorophenoxyacetic acid and 2,4,5-trichlorophenoxyacetic acid by dichlorophenol-adapted microorganisms from freshwater, anaerobic sediments. Appl Microbiol Biotechnol 38: 276–281.

Cole JR, Tiedje JM (1990): Induction of anaerobic dechlorination of chlorobenzoate in strain DCB-1. In Abstracts of the 90th Annual Meeting of the American Society for Microbiology. Washington, DC: American Society for Microbiology, abstract Q-43, p 295.

DeWeerd KA, Concannon F, Suflita JM (1991): Relationship between hydrogen consumption, dehalogenation, and the reduction of sulfur oxyanions by *Desulfomonile tiedjei.* Appl Environ Microbiol 57:1929–1934.

DeWeerd KA, Mandelco L, Tanner RS, Woese CR, Suflita JM (1990): *Desulfomonile tiedjei* gen. nov. and sp. nov., a novel anaerobic dehalogenating, sulfate-reducing bacterium. Arch Microbiol 154:23–30.

DeWeerd KA, Suflita JM (1990): Anaerobic aryl reductive dehalogenation of halobenzoates by cell extracts of "*Desulfomonile tiedjei.*" Appl Environ Microbiol 56:2999–3005.

DeWeerd KA, Suflita JM, Linkfield T, Tiedje JM, Pritchard PH (1986): The relationship between reductive dehalogenation and other aryl substituent removal reactions catalyzed by anaerobes. FEMS Microbiol Ecol 38:331–339.

Dietrich G, Winter J (1990): Anaerobic degradation of chlorophenol by an enrichment culture. Appl Microbiol Biotechnol 34:253–258.

Dolfing J, Harrison BK (1992): The Gibbs free energy of formation of halogenated aromatic compounds and their potential role as electron acceptors in anaerobic environments. Environ Sci Technol 26:2213–2218.

Dolfing J, Tiedje JM (1991a): Acetate as a source of reducing equivalents in the reductive dehalogenation of 2,5-dichlorobenzoate. Arch Microbiol 156:356–361.

Dolfing J, Tiedje JM (1991b): Influence of substituents on reductive dehalogenation of 3-chlorobenzoate analogs. Appl Environ Microbiol 57:820–824.

Fathepure BZ, Tiedje JM, Boyd SA (1987): Reductive dechlorination of 4-chlororesorcinol by anaerobic microorganisms. Environ Toxicol Chem 6:929–934.

Fathepure BZ, Tiedje JM, Boyd SA (1988): Reductive dechlorination of hexachlorobenzene to tri- and dichlorobenzenes in anaerobic sewage sludge. Appl Environ Microbiol 54:327–330.

Fenical W (1982): Natural products chemistry in the marine environment. Science 215:923–928.

Foxworthy AL, Mohn WW, Cole JR (1992): Enrichment and isolation of a novel bacterium growing by anaerobic reductive dehalogenation. In Abstracts of the 92nd General Meeting of the American Society for Microbiology. Washington, DC: American Society for Microbiology, abstract Q-197, p 368.

Genthner BRS, Price WA II, Pritchard PH (1989): Anaerobic degradation of chloroaromatic compounds in aquatic sediments under a variety of enrichment conditions. Appl Environ Microbiol 55:1466–1471.

Gerhartz W (1986): Ullmann's Encyclopedia of Industrial Chemistry. New York: Weinheim, vol A6, pp 233–398.

Gerritse J, van der Woude BJ, Gottschal JC (1992): Specific removal of chlorine from the *ortho*-position of halogenated benzoic acids by reductive dechlorination in anaerobic enrichment cultures. FEMS Microbiol Lett 100:273–280.

Gibson SA, Suflita JM (1986): Extrapolation of biodegradation results to groundwater aquifers: Reductive dehalogenation of aromatic compounds. Appl Environ Microbiol 52:681–688.

Gibson SA, Suflita JM (1990): Anaerobic biodegradation of 2,4,5-trichlorophenoxyacetic acid in samples from a methanogenic aquifer: Stimulation by short-chain organic acids and alcohols. Appl Environ Microbiol 56:1825–1832.

Häggblom MM, Janke D, Salkinoja-Salonen MS (1989): Hydroxylation and dechlorination of tetrachlorohydroquinone by *Rhodococcus* sp. strain CP-2 cell extracts. Appl Environ Microbiol 55:516–519.

Häggblom MM, Rivera MD, Oliver D, Young LY (1992): Anaerobic degradation of halogenated phenols by a sulfate reducing consortium. In Abstracts of the 92nd General Meeting of the American Society for Microbiology. Washington, DC: American Society for Microbiology, abstract Q-198, p 368.

Häggblom MM, Young LY (1990): Chlorophenol degradation coupled to sulfate reduction. Appl Environ Microbiol 56:3255–3260.

Hale DD, Rogers JE, Wiegel J (1991): Environmental factors correlated to dichlorophenol dechlorination in anoxic freshwater sediments. Environ Toxicol Chem 10: 1255–1265.

Hendriksen HV, Larsen S, Ahring BK (1992): Influence of a supplemental carbon source on anaerobic dechlorination of pentachlorophenol in granular sludge. Appl Environ Microbiol 58:365–370.

Holliger C, Schraa G, Stams AJM, Zehnder AJB (1992): Enrichment and properties of an anaerobic mixed culture reductively dechlorinating 1,2,3-trichlorobenzene to 1,3-dichlorobenzene. Appl Environ Microbiol 58:1636–1644.

Horowitz A, Suflita JM, Tiedje JM (1983): Reductive dehalogenations of halobenzoates by anaerobic lake sedimentmicroorganisms. Appl Environ Microbiol 45:1459–1465.

Howard PH, Hueber AE, Boethling RS (1987): Biodegradation data evaluation for structure/biodegradability relations. Environ Toxicol Chem 6:1–10.

King GM (1988): Dehalogenation in marine sediments containing natural sources of halophenols. Appl Environ Microbiol 54:3079–3085.

Kohring G-W, Rogers JE, Wiegel J (1989a): Anaerobic biodegradation of 2,4-dichlorphenol in freshwater lake sediments at different temperatures. Appl Environ Microbiol 55:348–353.

Kohring G-W, Zhang X, Wiegel J (1989b): Anaerobic dechlorination of 2,4-dichlorophenol in freshwater sediments in the presence of sulfate. Appl Environ Microbiol 55:2735–2737.

Kuhn EP, Suflita JM (1989a): Dehalogenation of pesticides by anaerobic microorganisms in soils and groundwater—A review. In Sawhney BL, Brown K (eds): Reactions and Movements of Organic Chemicals in Soils. Madison, WI: Soil Science Society of America and American Society of Agronomy, pp 111–180.

Kuhn EP, Suflita JM (1989b): Sequential reductive dehalogenation of chloroanilines by microorganisms from a methanogenic aquifer. Environ Sci Technol 23:848–852.

Kuhn EP, Townsend GT, Suflita JM (1990): Effect of sulfate and organic carbon supplements of reductive dehalogenation of chloroanilines in anaerobic aquifer slurries. Appl Environ Microbiol 56:2630–2637.

Linkfield TG, Suflita JM, Tiedje JM (1989): Characterization of the acclimation period before anaerobic dehalogenation of halobenzoates. Appl Environ Microbiol 55:2773–2778.

Linkfield TG, Tiedje JM (1990): Characterization of the requirements and substrates for reductive dehalogenation by strain DCB-1. J Ind Microbiol 5:9–16.

Liu S-M, Rogers JE (1991): Anaerobic biodegradation of chlorinated pyridines. In Abstracts of the 91st General Meeting of the American Society for Microbiology. Washington, DC: American Society for Microbiology, abstract K-3, p 215.

Madsen T, Aamand J (1991): Effects of sulfuroxy anions on degradation of pentachlorophenol by a methanogenic enrichment culture. Appl Environ Microbiol 57:2453–2458.

Madsen T, Aamand J (1992): Anaerobic transformation and toxicity of trichlorophenols in a stable enrichment culture. Appl Environ Microbiol 58:557–561.

Madsen T, Licht D (1992): Isolation and characterization of an anaerobic chlorophenol-transforming bacterium. Appl Environ Microbiol 58:2874–2878.

May HD, Boyle AW, Price WA II, Blake CK (1992a): Dechlorination of PCBs with Hudson River bacterial cultures under anaerobic conditions: Effect of sulfate, sulfite and nitrate. In Abstracts of the 92nd General Meeting of the American Society for Microbiology. Washington, DC: American Society for Microbiology, abstract Q-28, p 340.

May HD, Boyle AW, Price WA II, Blake CK (1992b): Subculturing of a polychlorinated biphenyl-dechlorinating anaerobic enrichment on solid media. Appl Environ Microbiol 58:4051–4054.

Mohn WW, Kennedy KJ (1992): Reductive dehalogenation of chlorophenols by *Desulfomonile tiedjei* DCB-1. Appl Environ Microbiol 58:1367–1370.

Mohn WW, Tiedje JM (1991): Evidence for chemisomotic coupling of reductive dechlorination and ATP synthesis in *Desulfomonile tiedjei*. Arch Microbiol 157:1–6.

Mohn WW, Tiedje JM (1992): Microbial reductive dehalogenation. Microbiol Rev 56:482–507.

Morris PJ, Mohn WW, Quensen JF III, Tiedje JM, Boyd SA (1992a): Establishment of a polychlorinated biphenyl-degrading enrichment culture with predominantly *meta* dechlorination. Appl Environ Microbiol 58:3088–3094.

Morris PJ, Quensen JF III, Tiedje JM, Boyd SA (1992b): Reductive debromination of the commercial polybrominated biphenyl mixture Firemaster BP6 by anaerobic microorganisms from sediments. Appl Environ Microbiol 58:3249–3256.

Neidleman SL (1975): Microbial halogenation. CRC Crit Rev Microbiol 3:333–358.

Neidleman SL, Geigert J (1987): Biological halogenation: Roles in nature, potential in industry. Endeavor, New Series 11:5–15.

Niemi GJ, Veith GD, Regal RR, Vaishnav DD (1987): Structural features associated with degradable and persistent chemicals. Environ Toxicol Chem 6:515–527.

Nies L, Vogel TM (1990): Effects of organic substrates on dechlorination of Aroclor 1242 in anaerobic sediments. Appl Environ Microbiol 56:2612–2617.

Quensen JF III, Tiedje JM, Boyd S (1988): Reductive dechlorination of polychlorinated biphenyls by anaerobic microorganisms from sediments. Science 242:752–754.

Ramanand K, Nagarajan A, Suflita JM (1993): Reductive dechlorination of the nitrogen heterocyclic herbicide picloram. Appl Environ Microbiol 59:2251–2256.

Reiner EA, Chu J, Kirsch EJ (1978): Microbial metabolism of pentachlorophenol. In Rao KR (ed): Environmental Science Research, Vol 12: Pentachlorophenol: Chemistry, Pharmacology, and Environmental Toxicology. New York: Plenum, pp 67–81.

Rochkind-Dubinsky ML, Sayler GS, Blackburn JW (1987): Microbial Decomposition of Chlorinated Aromatic Compounds. New York: Marcel Dekker.

Rubin HE, Subba-Rao RV, Alexander M (1982): Rates of mineralization of trace concentrations of aromatic compounds in lake water and sewage sludge. Appl Environ Microbiol 48:840–848.

Sato K, Tanaka H (1987): Degradation and metabolism of a fungicide, 2,4,5,6-tetrachlroisophthalonitrile (TPN) in soil. Biol Fertil Soils 3:205–209.

Shelton DR, Tiedje JM (1984): Isolation and partial characterization of bacteria in an anaerobic consortium that mineralizes 3-chlorobenzoic acid. Appl Environ Microbiol 48:840–848.

Steiert JG, Crawford RL (1986): Catabolism of pentachlorophenol by a *Flavobacterium* sp. Biochem Biophys Res Commun 141:825–830.

Struijs J, Rogers JE (1989): Reductive dehalogenation of dichloroanilines by anaerobic microorganisms in fresh and dichlorophenol-acclimated pond sediment. Appl Environ Microbiol 55:2527–2531.

Suflita JM, Gibson SA, Beeman RE (1988): Anaerobic biotransformations of pollutant chemicals in aquifers. J Ind Microbiol 3:179–194.

Suflita JM, Horowitz A, Shelton DR, Tiedje JM (1982): Dehalogenation: A novel pathway for the anaerobic biodegradation of haloaromatic compounds. Science 218:1115–1117.

Suflita JM, Robinson JA, Tiedje JM (1983): Kinetics of microbial dehalogenation of haloaromatic substrates in methanogenic environments. Appl Environ Microbiol 45:1466–1473.

Taraban RH, Berry DF, Berry DA, Walker HL Jr. (1993): Degradation of Dicamba by an anaerobic consortium enriched from wetland soil. Appl Environ Microbiol 59:2332–2334.

Tiedje JM, Quensen JF III, Chee-Sanford J, Schimel JP, Boyd SA (1992): Microbial Reductive Dechlorination of PCBs. Session 15. Pacific Basin Conference on Hazardous Wastes, 1992 Proceedings, Bangkok, Thailand. Honolulu, HI: East–West Center.

Tsuchiya T, Yamaha T (1984): Reductive dechlorination of 1,2,4-trichlorobenzene by *Staphylococcus epidermidis* isolated from intestional contents of rats. Agric Biol Chem 48:1545–1550.

Utkin I, Woese C, Wiegel J (1994): Isolation and characterization of *Desulfitobacterium dehalogens* gen. nov., sp. nov., an anaerobic bacterium which reductively dechlorinates chlorophenolic compounds. Int J Syst Bacteriol 44:612–619.

van den Tweel WJJ, Kok JB, De Bont JAM (1987): Reductive dechlorination of 2,4-dichlorobenzoate to 4-chlorobenzoate and hydrolytic dehalogenation of 4-chloro, 4-bromo, and 4-iodobenzoate by *Alcaligenes denitrificans* NTB-1. Appl Environ Microbiol 53:810–815.

Wu Q, Bedard DL, Wiegel J (1992): Effect of temperatures on the anaerobic transformation of 2,3,4,6-tetrachlorobiphenyl in methanogenic pond sediment slurries. In Abstracts of the 92nd General Meeting of the American Society for Microbiology. Washington, DC: American Society for Microbiology, abstract Q-25, p 339.

MECHANISMS OF POLYCYCLIC AROMATIC HYDROCARBON DEGRADATION

JOHN B. SUTHERLAND
FATEMEH RAFII
ASHRAF A. KHAN
CARL E. CERNIGLIA

Division of Microbiology, National Center for Toxicological Research, U.S. Food and Drug Administration, Jefferson, Arkansas 72079

1. INTRODUCTION

Polycyclic aromatic hydrocarbons (PAHs) (Fig. 7.1) are ubiquitous environmental pollutants that have been found to have toxic, mutagenic, and carcinogenic properties (International Agency for Research on Cancer [IARC] 1983). Consequently, the U.S. Environmental Protection Agency (EPA) has listed PAHs among the priority pollutants to be monitored in aquatic and terrestrial ecosystems (Keith and Telliard, 1979).

At present, many microorganisms are known to metabolize the lower molecular weight PAHs, such as naphthalene, phenanthrene, and anthracene (Cerniglia, 1992). Less is known about the potential for biodegradation of higher molecular weight PAHs, such as chrysene and benzo[a]pyrene. Since many PAHs and substituted PAHs have been implicated as probable human carcinogens (IARC, 1983), innovative bioremediation techniques are being developed to remove them from contaminated soils and waters (Davis et al., 1993; Mueller et al., 1993).

Microbial Transformation and Degradation of Toxic Organic Chemicals, pages 269–306
© 1995 Wiley-Liss, Inc.

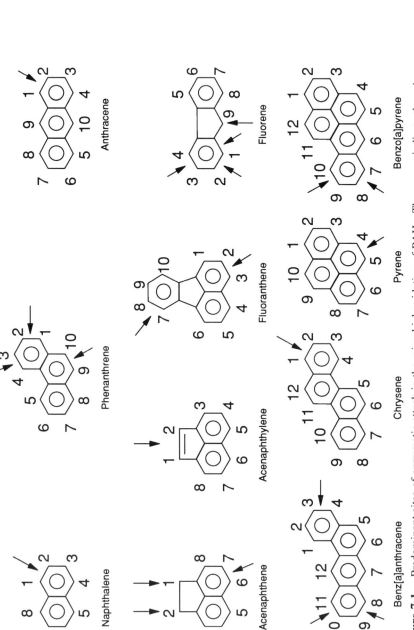

Figure 7.1. Predominant sites of enzymatic attack in the microbial oxidation of PAHs. The arrows indicate the primary sites of metabolism. Benzo[α]pyrene, benz[α]anthracene, and chrysene are biologically active as carcinogens.

The biodegradability of specific PAHs in the environment depends on their physical and chemical properties, their concentrations and rates of diffusion in the soil or water, and their bioavailability to microorganisms. It also depends on soil type, availability of water and oxygen, presence of nutrients, activity of microbial populations, previous exposure to chemicals, sediment toxicity, pH, temperature, and other seasonal effects (Cerniglia and Heitkamp, 1989; Weissenfels et al., 1992; Cerniglia, 1993).

The metabolism of PAHs by aerobic microorganisms has been summarized in several recent reviews (Cerniglia, 1984a,b, 1992, 1993; Gibson and Subramanian, 1984; Pothuluri and Cerniglia, 1994). In some reviews the pathways used by bacteria for metabolism of PAHs have been emphasized (Smith, 1990), and in others the pathways used by fungi have been emphasized (Cerniglia et al., 1992; Sutherland, 1992). Ecological aspects of the microbial metabolism of PAHs in aquatic ecosystems, including eutrophic lakes and polluted estuaries, have also been summarized (Cerniglia and Heitkamp, 1989; Cerniglia, 1991). The intent of this chapter is to discuss the toxicology of PAHs briefly and then describe the enzymatic mechanisms used by microorganisms to metabolize and detoxify various PAHs.

Human exposure to PAHs can occur by inhalation of tobacco smoke and polluted air (Lewtas, 1993), ingestion of contaminated foods (Perfetti et al., 1992; Gomaa et al., 1993; Saxton et al., 1993), or exposure to environmental contaminants in soil and water (Blumer and Youngblood, 1975; Huntley et al., 1993). Workers in several industries may also be exposed through contact with creosote, oils, tars, or asphalt (dell'Omo and Lauwerys, 1993; Grimmer et al., 1993; Ny et al., 1993).

The carcinogenicity of PAHs has been known since the 1930s. Hieger (1933) and Cook et al. (1933) purified the yellow crystals of benzo[a]pyrene from coal tar and found that they induced tumors in experimental animals. Although PAHs were the first pure chemicals to be identified as causes of cancer, it was not until the 1970s that an understanding of the mechanisms of carcinogenesis by PAHs emerged (Miller and Miller, 1985).

Many PAHs, including benzo[a]pyrene, are known to function as precarcinogens that require metabolic activation (Fig. 7.2) before they are able to bind to DNA, RNA, or proteins (Kadlubar and Hammons, 1987; Hall and Grover, 1990). The initial products of the oxidation of PAHs are arene oxides, whose formation is catalyzed by multiple forms of cytochrome P_{450} monooxygenase enzymes (Guengerich, 1992, 1993a). The regioselectivity and stereoselectivity of cytochrome P_{450} isozymes are important determinants of whether a PAH will be detoxified or metabolically activated to a proximate or ultimate carcinogen (Koreeda et al., 1978; Guengerich, 1993b). Arene oxide metabolites of PAHs may be hydrated by epoxide hydrolase to *trans*-dihydrodiols or rearranged nonenzymatically to phenols. Glucuronide, glutathione, and sulfate conjugates of PAH dihydrodiols and phenols are usually considered detoxification products (Guengerich, 1992).

The pioneering research of Miller and Miller (1985), as well as research in many other laboratories over the last 20 years, has shown that cancer induction by PAHs is a complex, multistep process that depends on many factors. These factors include size of the PAH molecules, polarity constraints, stereochemistry and chemical

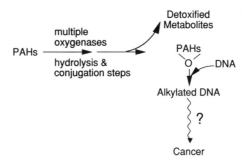

Figure 7.2. Schematic representation of the mammalian metabolism of PAHs.

reactivity of the metabolites, and electronic factors that affect the binding of metabolites to macromolecules (Hall and Grover, 1990).

Benzo[*a*]pyrene is one of the most potent chemical carcinogens known (Miller and Miller, 1985). Because of its genotoxicity, benzo[*a*]pyrene was one of the first PAHs studied to determine the mechanism of biological activity. The activation of benzo[*a*]pyrene to an ultimate carcinogen requires the oxidation of the terminal benzo ring to form benzo[*a*]pyrene 7,8-diol-9,10-epoxide (Sims et al., 1974). Four stereoisomers of this 7,8-diol-9,10-epoxide have been found (Fig. 7.3).

Studies of DNA binding, bacterial mutagenesis, and mammalian tumorigenesis

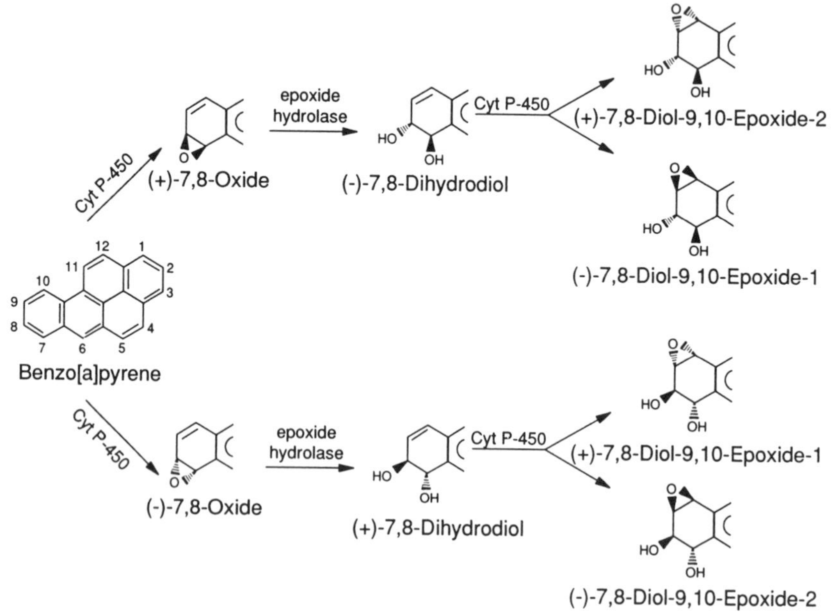

epoxide
hydrolase

Cyt P-450

(+)-7,8-Oxide

(-)-7,8-Dihydrodiol

(+)-7,8-Diol-9,10-Epoxide-2

(-)-7,8-Diol-9,10-Epoxide-1

Cyt P-450

Benzo[a]pyrene

Cyt P-450

epoxide
hydrolase

Cyt P-450

(-)-7,8-Oxide

(+)-7,8-Dihydrodiol

(+)-7,8-Diol-9,10-Epoxide-1

(-)-7,8-Diol-9,10-Epoxide-2

Figure 7.3. Pathways of metabolic activation in the mammalian metabolism of benzo[*a*]pyrene. (After Hall and Grover, 1990.)

have shown that the isomer benzo[*a*]pyrene (+)-7,8-diol-9,10-epoxide-2, which has the 7-hydroxyl group on the opposite side of the molecular plane from the epoxide oxygen and the 8-hydroxyl group on the same side (Fig. 7.3), is the most potent ultimate carcinogen produced from benzo[*a*]pyrene in mammalian metabolism (Hall and Grover, 1990). In the pathway leading to this isomer, benzo[*a*]pyrene is metabolized stereoselectively to the (+)-7,8-epoxide and then to the (−)-7,8-dihydrodiol. The latter compound then is metabolized stereoselectively to the (+)-7,8-diol-9,10-epoxide-2 (Koreeda et al., 1978; Kadlubar and Hammons, 1987).

2. PATHWAYS USED BY MICROORGANISMS FOR THE METABOLISM OF PAHs

2.1. Metabolism of PAHs to *cis*-Dihydrodiols, Phenols, and Ring-Fission Products by Bacteria and Green Algae

2.1.1. Formation of cis-Dihydrodiols

In most bacteria and some green algae, a principal mechanism for the aerobic metabolism of PAHs involves oxidation of the rings by dioxygenases to form *cis*-dihydrodiols (Cerniglia, 1992) (Table 7.1). These dihydrodiols are transformed further to diphenols, which are cleaved by other dioxygenases. The initial sites of enzymatic attack have been determined for several PAHs (Fig. 7.1).

2.1.2. Naphthalene

The metabolism of naphthalene has been studied more extensively than that of any other PAH. *Pseudomonas* spp. metabolize naphthalene via naphthalene *cis*-1,2-dihydrodiol, 1,2-dihydroxynaphthalene, 2-hydroxychromene-2-carboxylic acid (HCCA), *trans-o*-hydroxybenzylidenepyruvic acid (tHBPA), salicylaldehyde, salicylic acid, and either catechol or gentisic acid (Davies and Evans, 1964; Utkin et al., 1990; Eaton and Chapman, 1992) (Fig. 7.4). *Acinetobacter calcoaceticus* also metabolizes naphthalene via salicylate (Ryu et al., 1989). *Mycobacterium* sp. converts naphthalene initially to naphthalene *cis*-1,2-dihydrodiol and then to salicylate and catechol, although it also can form a *trans*-dihydrodiol (Heitkamp et al., 1988a; Kelley et al., 1990). *Rhodococcus* sp. metabolizes naphthalene via salicylaldehyde, salicylate, and gentisate (Walter et al., 1991; Grund et al., 1992).

2.1.3. Acenaphthene

Acenaphthene is oxidized on the five-membered ring by a biphenyl-degrading *Beijerinckia* sp. The process involves monooxygenation to form metabolites that include 1-acenaphthenol, 1-acenaphthenone, acenaphthene *cis*-1,2-dihydrodiol, 1,2-acenaphthenedione, and possibly 1,2-dihydroxyacenaphthylene (Schocken and Gibson, 1984). Acenaphthene is catabolized by a *Pseudomonas* sp., which uses it as a sole carbon and energy source, via 1-acenaphthenol and 1-acenaphthenone

TABLE 7.1. Polycyclic Aromatic Hydrocarbons Oxidized to Dihydrodiols by Bacteria, Cyanobacteria, and Green Algae[a]

Substrate	Dihydrodiol	Organism	Reference
Naphthalene	1,2[b]	Gram-negative rod	Walker and Wiltshire (1953)
	cis-1,2 and trans-1,2	Mycobacterium sp.	Kelley et al. (1990)
	1,2[b]	Nocardia sp.	Treccani et al. (1954)
	1,2[b]	Pseudomonas sp.	Treccani et al. (1954)
	cis-1,2	Pseudomonas sp.	Catterall et al. (1971), Jerina et al. (1971), Jeffrey et al. (1975), Ensley et al. (1982)
	1,2[b]	P. acidovorans	Treccani et al. (1954)
	cis-1,2	P. fluorescens	Jeffrey et al. (1975)
	cis-1,2	P. putida	Jeffrey et al. (1975)
	cis-1,2	Oscillatoria sp.	Cerniglia et al. (1980c)
Acenaphthylene	cis-1,2	Beijerinckia sp.	Schocken and Gibson (1984)
Fluorene	cis-1,1a[c]	Pseudomonas sp.	Selifonov et al. (1993)
Anthracene	cis-1,2	Beijerinckia sp.	Akhtar et al. (1975), Jerina et al. (1976)
	1,2[b]	Flavobacterium sp.	Colla et al. (1959)
Phenanthrene	cis-1,2 and cis-3,4	Beijerinckia sp.	Jerina et al. (1976)
	3,4[b]	Flavobacterium sp.	Colla et al. (1959)
	3,4	Mycobacterium sp.	Boldrin et al. (1993)
	cis-1,2 and cis-3,4	Pseudomonas putida	Jerina et al. (1976)
	trans-9,10	Streptomyces flavovirens	Sutherland et al. (1990)
	trans-9,10	Agmenellum quadruplicatum	Narro et al. (1992b)
Pyrene	cis-4,5 and trans-4,5	Mycobacterium sp.	Heitkamp et al. (1988b)
Benz[a]anthracene	cis-1,2, cis-8,9, and cis-10,11	Beijerinckia sp.	Gibson et al. (1975), Jerina et al. (1984)
Benzo[a]pyrene	cis-7,8 and cis-9,10	Beijerinckia sp.	Gibson et al. (1975)
	cis-4,5, cis-7,8, and cis-11,12	Selenastrum capricornutum	Lindquist and Warshawsky (1985a,b), Warshawsky et al. (1988)

[a]*Oscillatoria* and *Agmenellum* spp. are cyanobacteria; *Selenastrum* spp. are green algae.
[b]Methods for differentiating *cis* and *trans* isomers were not available.
[c]*cis*-1,1a-Dihydroxy-1-hydrofluoren-9-one.

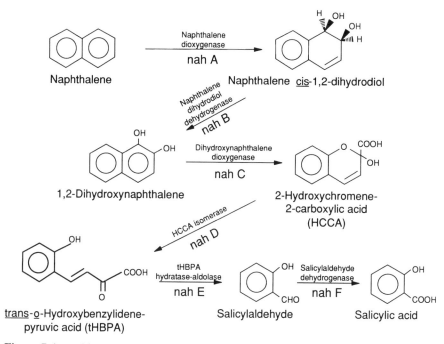

Figure 7.4. Initial steps in the metabolism of naphthalene to salicylic acid by *Pseudomonas putida*. (After Davies and Evans, 1964; Eaton and Chapman, 1992.) The genes coding for enzymes involved in the metabolism of naphthalene are designated by *nah*.

(Komatsu et al., 1993). *Alcaligenes eutrophus* and *A. paradoxus* use acenaphthene as a sole carbon and energy source, metabolizing it via 7,8-diketonaphthyl-1-acetic acid, 1,8-naphthalenedicarboxylic acid, and 3-hydroxyphthalic acid (Selifonov et al., 1993b).

2.1.4. Acenaphthylene

Beijerinckia sp. dioxygenates the five-membered ring of acenaphthylene to form acenaphthene *cis*-1,2-dihydrodiol, 1,2-acenaphthenedione, and possibly 1,2-dihydroxyacenaphthylene (Schocken and Gibson, 1984). Acenaphthylene is catabolized by a *Pseudomonas* sp., which uses it as a sole carbon and energy source, via 1,8-naphthalenedicarboxylic acid (Komatsu et al., 1993).

2.1.5. Fluorene

Whereas *Pseudomonas vesicularis* mineralizes fluorene by an unknown pathway (Weissenfels et al., 1990), another *Pseudomonas* sp. metabolizes fluorene to 9-fluorenoné and then, by an unusual angular-carbon dioxygenation, to another intermediate, 1,1a-dihydroxy-1-hydrofluoren-9-one (Selifonov et al., 1993a). A dibenzofuran-utilizing *Brevibacterium* sp. uses similar modes of enzymatic attack (Engesser et al., 1989). *Rhodococcus* sp. and *Mycobacterium* sp. also metabolize fluorene; the metabolites produced by the latter include 9-fluorenol, 9-fluorenone, and 1-indanone (Walter et al., 1991; Boldrin et al., 1993). *Arthrobacter* sp. metabolizes fluorene by two different pathways; one involves monooxygenation at C-9, leading to 9-fluorenol and 9-fluorenone, and the other involves dioxygenation and *meta*-cleavage to form 3,4-dihydrocoumarin as an intermediate (Grifoll et al., 1992). Similarly, *Staphylococcus auriculans* metabolizes fluorene to 9-fluorenol and 9-fluorenone, but it oxidizes the latter product further to 4- and 1-hydroxy-9-fluorenone as dead-end products (Monna et al., 1993).

2.1.6. Anthracene

Anthracene is metabolized by *Pseudomonas aeruginosa* via anthracene *cis*-1,2-dihydrodiol, 1,2-dihydroxyanthracene, *cis*-4-(2-hydroxynaphth-3-yl)-2-oxobut-3-enoic acid, 2-hydroxy-3-naphthaldehyde, 2-hydroxy-3-naphthoic acid, and 2,3-dihydroxynaphthalene (Evans et al., 1965). The latter compound is mineralized via salicylic acid and catechol (Dagley and Gibson, 1965; Yamamoto et al., 1965) (Fig. 7.5). 2-Hydroxy-3-naphthoic acid has also been found in cultures of *P. fluorescens* grown with anthracene (Menn et al., 1993). A *Rhodococcus* sp. also utilizes anthracene for growth but the pathway is unknown (Walter et al., 1991).

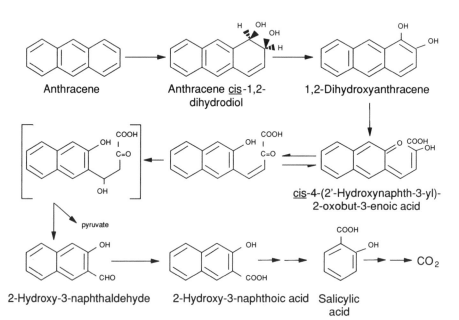

Figure 7.5. Pathway for the bacterial metabolism of anthracene by *Pseudomonas aeruginosa*. (After Evans et al., 1965.)

2.1.7. Phenanthrene

Pseudomonas sp. metabolizes phenanthrene via phenanthrene *cis*-3,4-dihydrodiol, 3,4-dihydroxyphenanthrene, *cis*-4-(1-hydroxynaphth-2-yl)-2-oxobut-3-enoic acid, 1-hydroxy-2-naphthaldehyde, 1-hydroxy-2-naphthoic acid, and 1,2-dihydroxynaphthalene (Evans et al., 1965; Jerina et al., 1976). The latter compound is mineralized via the naphthalene pathway (Davies and Evans, 1964) (Fig. 7.6). *Pseudomonas paucimobilis* and *P. fluorescens* both mineralize phenanthrene (Weissenfels et al., 1990; Menn et al., 1993). *Aeromonas* sp. metabolizes phenanthrene to 1-hydroxy-2-naphthaldehyde, 2-carboxybenzaldehyde, *o*-phthalic acid, and protocatechuic acid (Kiyohara et al., 1976). *Beijerinckia* sp. also metabolizes phenanthrene in the 1,2- and 3,4-positions (Jerina et al., 1976; Strandberg et al., 1986) to form *cis*-dihydrodiols. *Mycobacterium* spp. mineralize phenanthrene via 1-hydroxy-2-naphthoic acid, *o*-phthalic acid, and protocatechuic acid (Guerin and Jones, 1988; Heitkamp et al., 1988a; Boldrin et al., 1993). *Rhodococcus* sp. and *Arthrobacter polychromogenes* utilize phenanthrene for growth; the latter probably uses a pathway with *cis*-4-(1-hydroxynaphth-2-yl)-2-oxobut-3-enoic acid (Keuth and Rehm, 1991; Walter et al., 1991).

2.1.8. Fluoranthene

Stationary-phase cultures of *Pseudomonas putida* and another *Pseudomonas* sp. that can grow on naphthalene are able to co-metabolize fluoranthene (Barnsley, 1975a). *Pseudomonas paucimobilis* and *Alcaligenes denitrificans* utilize fluoranthene for growth; the metabolic intermediates produced by *A. denitrificans* include 7-acenaphthenone, 7-hydroxyacenaphthylene, and 3-hydroxymethyl-4,5-benzocoumarin (Mueller et al., 1990; Weissenfels et al., 1990, 1991). *Mycobacterium* spp. produce CO_2 from fluoranthene but are not known to use this PAH as a sole carbon and energy source (Heitkamp et al., 1988a; Kelley and Cerniglia, 1991; Boldrin et al., 1993). The intermediates include 9-fluorenone-1-carboxylic acid, 8-hydroxy-7-methoxyfluoranthene, 9-hydroxyfluorene, 9-fluorenone, 1-acenaphthenone, 9-hydroxy-1-fluorenecarboxylic acid, phthalic acid, 2-carboxybenzaldehyde, benzoic acid, phenylacetic acid, and adipic acid (Kelley and Cerniglia, 1991; Kelley et al., 1991, 1993). *Rhodococcus* sp. also utilizes fluoranthene as a sole carbon and energy source (Walter et al., 1991).

2.1.9. Pyrene

Pyrene is catabolized by *Mycobacterium* spp. to produce CO_2; the intermediates include pyrene *cis*-4,5-dihydrodiol, 4-hydroxyperinaphthenone, 4-phenanthroic acid, phthalic acid, and cinnamic acid, as well as a *trans*-dihydrodiol (Heitkamp et al., 1988a,b; Boldrin et al., 1993). Several pathways have been proposed by Cerniglia (1992) (Fig. 7.7). In contrast, *Rhodococcus* sp. produces a different set of metabolites from pyrene, including 1,2- and 4,5-dihydroxypyrene, *cis*-2-hydroxy-3-(perinaphthenone-9-yl)propenic acid, and 2-hydroxy-2-(phenanthren-5-one-4-enyl)acetic acid (Walter et al., 1991).

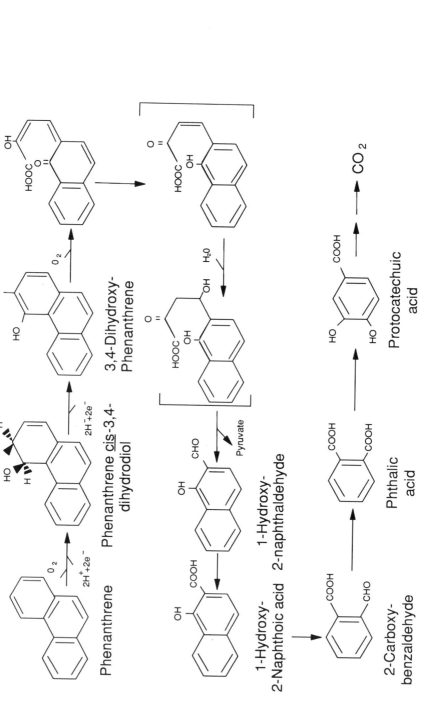

Figure 7.6. Pathway for the bacterial metabolism of phenanthrene. (After Evans et al., 1965; Kiyohara et al., 1976.)

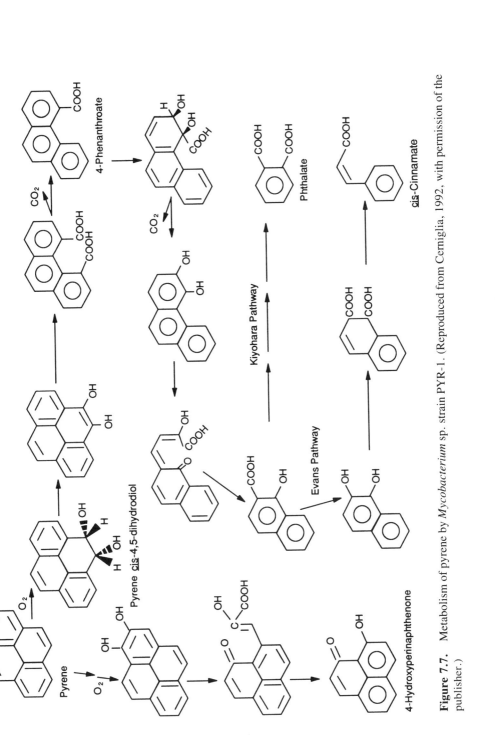

Figure 7.7. Metabolism of pyrene by *Mycobacterium* sp. strain PYR-1. (Reproduced from Cerniglia, 1992, with permission of the publisher.)

2.1.10. Benz[a]anthracene

A *Beijerinckia* sp. oxidizes benz[*a*]anthracene to the *cis*-1,2-, *cis*-8,9-, and *cis*-10,11-dihydrodiols (Gibson et al., Jerina et al., 1984). Further metabolism yields 1-hydroxy-2-anthranoic acid, 2-hydroxy-3-phenanthroic acid, 3-hydroxy-2-phenanthroic acid, and CO_2 (Mahaffey et al., 1988).

2.1.11. Chrysene

Chrysene has been reported to be utilized as a carbon source by *Rhodococcus* sp. (Walter et al., 1991), although the pathways and products are unknown.

2.1.12. Benzo[a]pyrene

Beijerinckia sp. oxidizes benzo[*a*]pyrene to the *cis*-7,8- and *cis*-9,10-dihydrodiols Gibson et al., 1975). The green alga *Selenastrum capricornutum* produces benzo[*a*]pyrene *cis*-4,5-, *cis*-7,8-, and *cis*-11, 12-dihydrodiols (Lindquist and Warshawsky, 1985a,b; Warshawsky et al., 1988) as well as sulfate and glucoside conjugates of the *cis*-4,5-dihydrodiol (Warshawsky et al., 1990). Benzo[*a*]pyrene is also metabolized by *Pseudomonas* spp. but the products are unknown (Barnsley, 1975a).

2.2. Metabolism of PAHs to Phenols by Methylotrophic Bacteria

The methane monooxygenase system of *Methylococcus capsulatus,* in the presence of NADH, oxidizes benzene to phenol and naphthalene to 1- and 2-naphthol (Dalton et al., 1981) (Fig. 7.8).

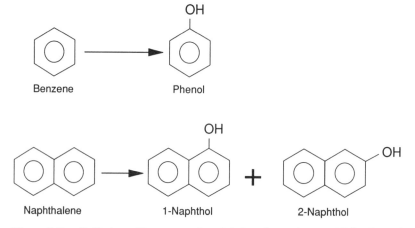

Figure 7.8. Oxidation of benzene and naphthalene by methane-oxidizing bacteria.

TABLE 7.2. **Polycyclic Aromatic Hydrocarbons Oxidized to Dihydrodiols by Fungi**

Substrate	Dihydrodiol	Organism	Reference
Naphthalene	*trans*-1,2	Many fungi	Cerniglia and Gibson (1977), Cerniglia et al. (1978)
Fluoranthene	*trans*-2,3	*Cunninghamella elegans*	Pothuluri et al. (1990)
Anthracene	*trans*-1,2	*Cunninghamella elegans*	Cerniglia (1982), Cerniglia and Yang (1984)
	trans-1,2	*Rhizoctonia solani*	Sutherland et al. (1992)
Phenanthrene	*trans*-1,2, *trans*-3,4 and *trans*-9,10	*Cunninghamella elegans*	Cerniglia and Yang (1984), Cerniglia et al. (1989)
	trans-3,4 and *trans*-9,10	*Phanerochaete chrysosporium*	Sutherland et al. (1991)
	trans-3,4 and *trans*-9,10	*Syncephalastrum racemosum*	Sutherland et al. (1993)
Benz[*a*]anthracene	*trans*-3,4, *trans*-8,9, and *trans*-10,11	*Cunninghamella elegans*	Cerniglia et al. (1980a, 1985)
Benzo[*a*]pyrene	*trans*-7,8 and *trans*-9,10	*Cunninghamella elegans*	Cerniglia and Gibson (1979), Cerniglia et al. (1980b)
	trans-7,8	*Saccharomyces cerevisiae*	Wiseman and Woods (1979)
	trans-4,5, *trans*-7,8, and *trans*-9,10	*Aspergillus ochraceus*	Datta and Samanta (1988)

2.3. Metabolism of PAHs to *trans*-Dihydrodiols by Fungi, Bacteria, and Cyanobacteria

2.3.1. Formation of *trans*-Dihydrodiols

Many species of fungi, a few bacteria, and some cyanobacteria produce cytochrome P_{450} monooxygenases. These enzymes transform PAHs to arene oxides, which are then either hydrated by epoxide hydrolase to form *trans*-dihydrodiols or rearranged nonenzymatically to form phenols (Table 7.2). Microorganisms that have only these pathways cannot utilize PAHs as carbon sources but may be able to detoxify them (Cerniglia et al., 1992; Sutherland, 1992).

2.3.2. Naphthalene

Naphthalene is metabolized by many fungi to naphthalene *trans*-1,2-dihydrodiol, 1- and 2-naphthol, 4-hydroxy-1-tetralone, and glucuronide and sulfate conjugates

(Cerniglia et al., 1978; Hofmann, 1986). Some gram-positive bacteria, including *Bacillus cereus, Streptomyces griseus,* and *Mycobacterium* sp., metabolize naphthalene to 1-naphthol (Cerniglia et al., 1984; Trower et al., 1988; Kelley et al., 1990). *Mycobacterium* sp. produces naphthalene *trans*-1,2-dihydrodiol in addition to the *cis*-dihydrodiol noted previously (Kelley et al., 1990). The marine cyanobacterium *Oscillatoria* sp. also metabolizes naphthalene to 1-naphthol (Narro et al., 1992a).

2.3.3. Acenaphthene

The fungus *Cunninghamella elegans* metabolizes acenaphthene to 1-acenaphthenol, 1,5-dihydroxyacenaphthene, *cis*- and *trans*-1,2-dihydroxyacenaphthene, 1-acenaphthenone, 1,2-acenaphthenedione, and 6-hydroxyacenaphthenone (Pothuluri et al., 1992a).

2.3.4. Fluorene

Fluorene is metabolized by *Cunninghamella elegans* to 9-fluorenol, 9-fluorenone, and 2-hydroxy-9-fluorenone (Pothuluri et al., 1993). *Laetiporus sulphureus* and other fungi metabolize fluorene to unknown products (Sack and Günther, 1993).

2.3.5. Anthracene

Anthracene is metabolized by *Cunninghamella elegans* and *Rhizoctonia solani* to anthracene *trans*-1,2-dihydrodiol (Cerniglia, 1982; Cerniglia and Yang, 1984; Sutherland et al., 1992). It then is conjugated by *Cunninghamella elegans* to form 1-anthryl sulfate (Cerniglia, 1982) and by *Rhizoctonia solani* to three different xyloside conjugates (Sutherland et al., 1992). *Penicillium* sp. metabolizes anthracene to unknown products (Sack and Günther, 1993).

2.3.6. Phenanthrene

Cunninghamella elegans metabolizes phenanthrene to phenanthrene *trans*-1,2-, *trans*-3,4-, and *trans*-9,10-dihydrodiols and a glucoside conjugate (Cerniglia and Yang, 1984; Cerniglia et al., 1989; Sutherland et al., 1993) (Fig. 7.9). A related fungus, *Syncephalastrum racemosum,* also produces the *trans*-3,4- and *trans*-9,10-dihydrodiols (Sutherland et al., 1993). The white-rot fungus *Phanerochaete chrysosporium* metabolizes phenanthrene to the *trans*-3,4- and *trans*-9,10-dihydrodiols, the 3-, 4-, and 9-phenanthrols, and a glucoside conjugate (Sutherland et al., 1991; 1993). Cytochrome P_{450} has not been demonstrated in this species, and the mechanism is unknown. *Laetiporus sulphureus, Flammulina velutipes, Marasmiellus* sp., *Penicillium* sp., *Trichosporon penicillatum,* and other fungi metabolize phenanthrene to unknown products (MacGillivray and Shiaris, 1993; Sack and Günther, 1993). The filamentous bacterium *Streptomyces flavovirens* produces phenanthrene *trans*-9,10-dihydrodiol (Sutherland et al., 1990), and a *Mycobacterium* sp. probably produces the *trans*-3,4-dihydrodiol (Boldrin et al., 1993). The marine cyanobacterium *Agmenellum quadruplicatum* metabolizes phenanthrene to phe-

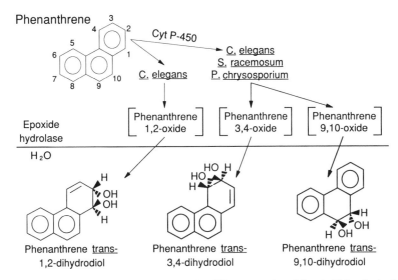

Figure 7.9. Metabolism of phenanthrene by different species of fungi. (After Sutherland et al., 1993.)

nanthrene *trans*-9,10-dihydrodiol and 1-methoxyphenanthrene (Narro et al., 1992a).

2.3.7. Fluoranthene

Fluoranthene is metabolized by *Cunninghamella elegans* to fluoranthene *trans*-2,3-dihydrodiol, 8- and 9-hydroxyfluoranthene *trans*-2,3-dihydrodiol, and two glucoside conjugates (Pothuluri et al., 1990, 1992b). *Penicillium* sp., *Laetiporus sulphureus*, and other fungi also metabolize fluoranthene, but the products are unknown (Sack and Günther, 1993).

2.3.8. Pyrene

Cunninghamella elegans metabolizes pyrene to 1-hydroxypyrene, 1,6- and 1,8-pyrenequinones, and three glucoside conjugates (Cerniglia et al., 1986). *Penicillium* sp. and other fungi metabolize pyrene to unknown products (Sack and Günther, 1993). *Mycobacterium* sp. produces pyrene *trans*-4,5-dihydrodiol and a pyrenol as intermediates in addition to the *cis*-dihydrodiol and other products mentioned above (Heitkamp et al., 1988b).

2.3.9. Benz[a]anthracene

Benz[a]anthracene is metabolized by *Cunninghamella elegans* to the *trans*-3,4-, *trans*-8,9-, and *trans*-10,11-dihydrodiols and to several other derivatives and conjugates (Cerniglia et al., 1980a; Cerniglia et al., 1985). The yeasts *Candida krusei* and *Rhodotorula minuta* metabolize benz[*a*]anthracene to unknown products (Mac-Gillivray and Shiaris, 1993).

2.3.10. Benzo[a]pyrene

Several metabolites are produced from benzo[*a*]pyrene by *Cunninghamella elegans,* including *trans*-7,8- and *trans*-9,10-dihydrodiols, 3- and 9-hydroxybenzo[*a*]pyrene, two quinones, various conjugates, and trace amounts of the highly carcinogenic derivative benzo[*a*]pyrene (+)-7,8-diol-9,10-epoxide-2 (Cerniglia and Gibson, 1979; Cerniglia et al., 1992). *Aspergillus ochraceus* also metabolizes benzo[*a*]pyrene, but the major product is the *trans*-4,5-dihydrodiol (Datta and Samanta, 1988).

2.4. Metabolism of PAHs to Quinones by White-Rot Fungi

2.4.1. Lignin Peroxidases

Some white-rot fungi, which decay lignin and cellulose in wood, metabolize PAHs to quinones and other metabolites by mechanisms that do not appear to involve *cis*- or *trans*-dihydrodiols (Fig. 7.10). In some cases, but not all, lignin peroxidases are involved.

2.4.2. Fluorene

Phanerochaete chrysosporium, Trametes versicolor, and *Pleurotus ostreatus* metabolize fluorene to unknown products (George and Neufeld, 1989; Sack and Günther, 1993).

2.4.3. Anthracene

Anthracene is metabolized by *Phanerochaete chrysosporium,* via 9,10-anthraquinone and phthalic acid, to CO_2 (Hammel et al., 1991). *Bjerkandera* sp. and several other white-rot fungi metabolize anthracene to 9,10-anthraquinone as a final product; *Trametes* sp. metabolizes anthracene but does not accumulate the anthraquinone (Field et al., 1992; Sack and Günther, 1993).

2.4.4. Phenanthrene

Although phenanthrene is not a substrate for lignin peroxidase H8 of *Phanerochaete chrysosporium* (Hammel et al., 1986), it is nevertheless metabolized by *Phanerochaete chrysosporium, Trametes versicolor,* and *Pleurotus ostreatus* to CO_2 (Bumpus, 1989; Morgan et al., 1991; Dhawale et al., 1992; Sack and Günther, 1993) as well as to the *trans*-dihydrodiols, phenols, and conjugates noted above (Sutherland et al., 1991). Lignin peroxidase may, however, oxidize 9-phenanthrol to phenanthrene 9,10-quinone (Tatarko and Bumpus, 1993). The intermediate metabolites in the catabolic pathway of *Phanerochaete chrysosporium* include phenanthrene 9,10-quinone and the ring cleavage product 2,2'-diphenic acid (Hammel et al., 1992).

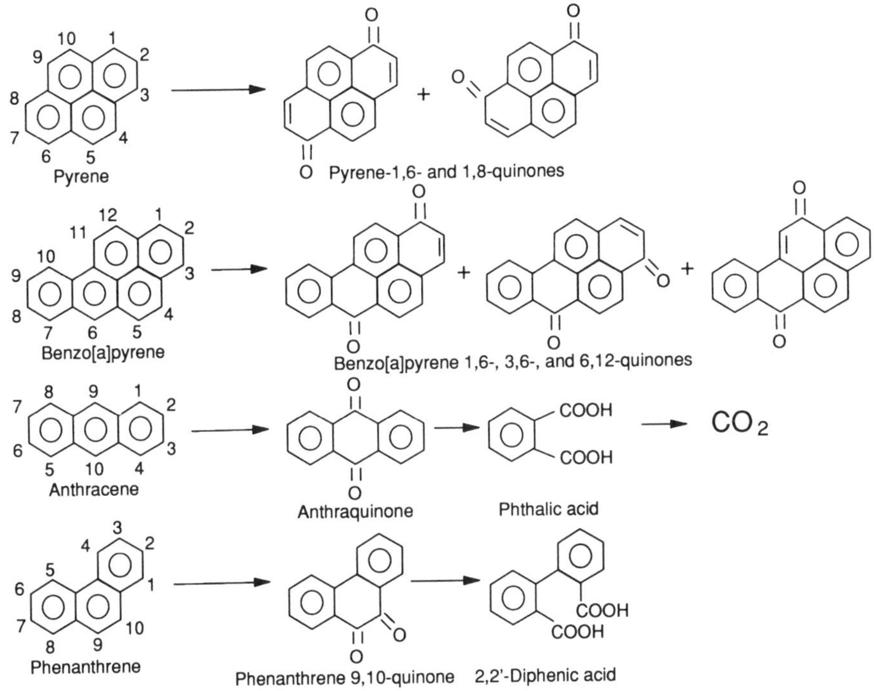

Figure 7.10. Oxidation of PAHs by *Phanerochaete chrysosporium.*

2.4.5. Fluoranthene

The white-rot fungi *Trametes versicolor* and *Pleurotus ostreatus* metabolize fluoranthene to unknown products (Sack and Günther, 1993).

2.4.6. Pyrene

Pyrene is oxidized, either by cultures of *Phanerochaete chrysosporium* or by purified lignin peroxidase, to pyrene 1,6- and 1,8-quinones (Hammel et al., 1986).

2.4.7. Benz[a]anthracene and Perylene

Benz[a]anthracene and perylene show spectral changes, when incubated with lignin peroxidase and H_2O_2, that indicate that they are substrates for this enzyme (Hammel et al., 1986).

2.4.8. Benzo[a]pyrene

Phanerochaete chrysosporium metabolizes benzo[a]pyrene to CO_2 (Bumpus et al., 1985; Sanglard et al., 1986). The first intermediate produced is apparently 6-hydroxybenzo[a]pyrene, which is quickly oxidized to the 1,6-, 3-6-, and 6,12-quinones (Haemmerli et al., 1986). *Bjerkandera* sp., *Trametes versicolor,* and *Trametes* sp.

metabolize benzo[*a*]pyrene to unknown products (Field et al., 1992; Morgan et al., 1993).

2.5. Metabolism of PAHs by Anaerobic Bacteria

Little is known about the potential of PAHs for anaerobic metabolism. Although naphthalene and acenaphthene are removed to nondetectable levels by mixed cultures of denitrifying bacteria (Mihelcic and Luthy, 1988), the metabolic pathways that they use are unknown.

3. ENZYMES INVOLVED IN MICROBIAL PAH METABOLISM

3.1. Characteristics of the Enzymes

Of all the enzymes involved in the microbial metabolism of PAHs (Fig. 7.11), the enzymes for naphthalene metabolism produced by bacteria in the genus *Pseudomonas* have been studied most extensively. The characteristics of several of these enzymes, of the cytochrome P_{450} monooxygenases, and of the lignin peroxidases are discussed here.

3.2. Principal Pathways for Naphthalene Metabolism

3.2.1. Naphthalene Dioxygenase

The first reaction in the metabolism of naphthalene by a *Pseudomonas* sp. is the incorporation of two atoms of oxygen to form naphthalene *cis*-1,2-dihydrodiol (Ensley et al., 1982). This reaction is catalyzed by naphthalene dioxygenase, a multicomponent enzyme system containing three proteins: a flavoprotein (ferredoxin$_{NAP}$ reductase), a ferredoxin (ferredoxinl$_{NAP}$), and a terminal oxidase (ISP$_{NAP}$) (Ensley et al., 1982; Ensley and Gibson, 1983; Haigler and Gibson, 1990a,b).

Ferredoxin$_{NAP}$ reductase is an NADH oxidoreductase with NAD(P)H-cytochrome c reductase activity (Ensley et al., 1982; Haigler and Gibson, 1990a). This enzyme is a red iron-sulfur flavoprotein with 1 mol of FAD, which is loosely

Enzymes for Microbial Transformation
of Aromatic Hydrocarbons

- Dioxygenases
- Monooxygenases
  - Methane monoxygenase
  - Cytochrome P-450 monooxygenase
- Lignin Peroxidases
- Laccase

Figure 7.11. Types of enzymes involved in the microbial metabolism of aromatic compounds.

bound and readily lost during purification, per 1 mol of protein. Ferredoxin$_{NAP}$ reductase consists of a single polypeptide with a molecular weight of 36,300 and approximately 1.8 g-atoms of iron and 2.0 g-atoms of acid-labile sulfur per mole. As the initial electron acceptor, ferredoxin$_{NAP}$ reductase shuttles electrons from NADH to ferredoxin$_{NAP}$ (Haigler and Gibson, 1990a,b). NADPH may be substituted for NADH as a cofactor, but the resulting enzyme activity is lower (Haigler and Gibson, 1990b).

Ferredoxin$_{NAP}$ functions as an intermediate electron transfer protein in the naphthalene dioxygenase system of *Pseudomonas* sp. (Haigler and Gibson, 1990b). It is a red-brown iron-sulfur protein with a molecular weight of 13,600 and an isoelectric point of 4.6. The UV/visible absorption spectrum has maxima at 280, 325, and 460 nm, with a broad shoulder at 550 nm. The extinction coefficients of ferredoxin$_{NAP}$ at 325 and 460 nm are 9.85 mM^{-1}cm^{-1} and 4.92 mM^{-1}cm^{-1}, respectively. It contains 2 g-atoms each of iron and acid-labile sulfur per mole. Ferredoxin$_{NAP}$ is specifically required for the functioning of the naphthalene dioxygenase system of *Pseudomonas* sp.; substitution with ferredoxin from spinach or other sources does not restore activity (Haigler and Gibson, 1990a,b).

ISP$_{NAP}$, the terminal oxygenase that binds to naphthalene, requires both ferredoxin$_{NAP}$ reductase and ferredoxin$_{NAP}$ for activity (Ensley and Gibson, 1983). In the presence of NADH and the other two components, ISP$_{NAP}$ catalyzes the addition of two oxygen atoms to one of the aromatic rings of naphthalene (Ensley and Gibson, 1983). The ISP$_{NAP}$ of *Pseudomonas* sp. has a molecular weight of 158,000, including two subunits with molecular weights of 55,000 and two subunits with molecular weights of 20,000. It also has 6 g-atoms of iron and 5 g-atoms of acid-labile sulfur per mole of enzyme. Since the spectral characteristics of ferredoxin$_{NAP}$ and ISP$_{NAP}$ are similar, the iron-sulfur chromophores of these two proteins are probably located in similar chemical environments (Haigler and Gibson, 1990b).

3.2.2. *Naphthalene Dihydrodiol Dehydrogenase*

The enzyme naphthalene dihydrodiol dehydrogenase oxidizes the (+)-isomer of naphthalene *cis*-1,2-dihydrodiol to 1,2-dihydroxynaphthalene in the presence of NAD$^+$ (Patel and Gibson, 1974). The (+)-*cis*-naphthalene dihydrodiol dehydrogenase of *P. putida* strain NP, which is induced by growth on naphthalene as the sole source of carbon and energy, has been purified (Patel and Gibson, 1974, 1976). The same enzyme has been found in other *Pseudomonas* and *Nocardia* spp. (Patel and Gibson, 1974, 1976). The molecular weight of the purified dehydrogenase from *P. putida* strain NP is 102,000. It is a tetramer of four identical subunits, each with a molecular weight of 25,500, and it has an absolute requirement for NAD$^+$ as an electron acceptor. Although the dihydrodiol dehydrogenase of *P. putida* oxidizes only the (+)-isomer of naphthalene *cis*-1,2-dihydrodiol, it is also able to oxidize *cis*-dihydrodiols of anthracene, phenanthrene, biphenyl, toluene, ethylbenzene, and benzene. It is inhibited by 2,2'-dipyridyl, 8-hydroxyquinoline, EDTA, and *p*-chloromercuribenzoate (Patel and Gibson, 1974, 1976).

Other forms of naphthalene dihydrodiol dehydrogenase have been purified from *Pseudomonas* sp. NCIB 9816, *Nocardia* sp. grown on naphthalene, and *P. putida* biotype A grown on toluene or benzene (Patel and Gibson, 1976). The molecular weights of the naphthalene dihydrodiol dehydrogenases from *Nocardia* sp., *P. putida* biotype A, and *Pseudomonas* sp. NCIB 9816 are 92,000, 160,000, and 112,000, respectively (Patel and Gibson, 1976). These enzymes also oxidize the *cis*-dihydrodiols of PAHs and require NAD^+ for activity. The dihydrodiol dehydrogenases from *Pseudomonas* spp. have pH optima of 8.8–9.0, whereas the enzyme from *Nocardia* sp. shows maximum activity at pH 8.4 (Patel and Gibson, 1976).

Cell extracts from *P. putida* NCIB 9816 and *P. putida* biotype A cross-react with IgG antibodies raised against the naphthalene dihydrodiol dehydrogenase of *P. putida* NP (Patel and Gibson, 1976). The dehydrogenase from *P. putida* NP also has some antigenic determinants not found in the other strains, as shown by spur formation in gel diffusion tests (Patel and Gibson, 1976). The antibody raised against the enzyme from *P. putida* NP did not inhibit dehydrogenase activity in cell extracts of *P. putida* biotype A, showing that the antigenic determinants of biotype A do not include regions of the active site (Patel and Gibson, 1976).

Although the strain of *P. putida* biotype A used by Patel and Gibson (1976) does not utilize naphthalene as a growth substrate, the dehydrogenase from cultures of this strain (grown on toluene or benzene) oxidizes naphthalene *cis*-1,2-dihydrodiol and other *cis*-dihydrodiols at higher rates than the enzymes of other strains that can be grown with naphthalene. Cell extracts of both *Nocardia* sp., grown on naphthalene, and *P. putida* biotype A, grown on toluene, have higher affinity for naphthalene *cis*-1,2-dihydrodiol, as shown by the apparent K_m, than do cell extracts of other bacterial strains grown on naphthalene (Patel and Gibson, 1976).

3.2.3. *Dihydroxynaphthalene Dioxygenase*

The enzyme dihydroxynaphthalene dioxygenase has been purified from *P. putida* NCIB 9816 (Patel and Barnsley, 1980) and from an unidentified soil bacterium, strain BN6, that grows on naphthalenesulfonic acids (Kuhm et al., 1991b). Both forms of the enzyme require Fe^{2+} for maximum activity.

The 1,2-dihydroxynaphthalene dioxygenase from *P. putida* is induced by growth on naphthalene as the sole source of carbon and energy. The enzyme is colorless with an absorption maximum at 280 nm, has a molecular weight of at least 275,000, and contains several subunits with molecular weights of 19,000 (Patel and Barnsley, 1980). This dioxygenase oxidizes 1,2-dihydroxynaphthalene to 2-hydroxychromene-2-carboxylate (HCCA), which is apparently produced by cyclization of a ring-fission product even before it is released from the enzyme (Barnsley, 1967a; Eaton and Chapman, 1992). Dihydroxynaphthalene dioxygenase also can use 3- and 4-methylcatechol as substrates, although these reactions are slower. 3-Methylcatechol is oxidized to 2-hydroxy-6-oxoheptadienoic acid by this enzyme.

The corresponding dioxygenase from the unidentified bacterial strain BN6 has a molecular weight of 290,000 and is composed of eight identical subunits with molecular weights of about 33,000 (Kuhm et al., 1991b). In molecular weight and subunit structure, this dioxygenase resembles the 2,3-dihydroxybiphenyl dioxygenases of *P. pseudoalcaligenes* and *P. paucimobilis* (Kuhm et al., 1991b). The purified dioxygenase from strain BN6 cleaves 1,2-dihydroxynaphthalene by *meta*-cleavage (extradiol cleavage or α-keto acid pathway) to a compound at pH 5.5 with the spectral characteristics of HCCA (Kuhm et al., 1991b). The enzymatic product can be converted at pH 11 to *cis*-2'-hydroxybenzalpyruvate (*cis-o*-hydroxybenzylidenepyruvate or cHBPA). The dioxygenase cleaves 1,2,6-trihydroxynaphthalene to a different product with spectral characteristics at pH 5.5 similar to those of HCCA; under either acidic or alkaline conditions, the spectrum of this enzymatic product also undergoes changes that are analogous to those of HCCA. The same dioxygenase also cleaves catechol, 3- and 4-methylcatechol, 2,3- and 3,4-dihydroxybiphenyl, and 1,2,5- and 1,2,7-trihydroxynaphthalene; the compounds produced from catechol, 3-methylcatechol, and 4-methylcatechol have been shown to arise by *meta*-cleavage (Kuhm et al., 1991b).

The 1,2-dihydroxynaphthalene dioxygenase from strain BN6 is even able to produce a *meta*-cleavage product from 2,3-dihydroxybiphenyl, namely, 2-hydroxy--6-oxo-phenylhexa-2,4-dienoate. Since the NH_2-terminal amino acid sequence of the dihydroxynaphthalene dioxygenase shows homology with that of the 2,3-dihydroxybiphenyl dioxygenase of *P. paucimobilis,* Kuhm et al. (1991a,b) suggest that the pathways for metabolism of naphthalene and biphenyl in pseudomonads may be closely related.

3.2.4. HCCA Isomerase

In *Pseudomonas* sp. NCIB 9816, an inducible isomerase catalyzes the conversion of HCCA to an isomer of *o*-hydroxybenzylidenepyruvic acid. Barnsley (1976a) believed this product to be the *cis*-isomer (cHBPA). The HCCA isomerase from *Pseudomonas* sp. NCIB 9816 has a K_m of 0.2 mM, an optimum pH of 10, and an uninduced specific activity of 0.13 μmol/min/mg protein. The specific activity increases to 0.39 μmol/min/mg protein if cells are induced by salicylate (Barnsley, 1976a).

Eaton and Chapman (1992) have recently proposed that cHBPA is a transient intermediate and that the HCCA isomerase actually converts HCCA to an equilibrium mixture also containing 45% *trans-o*-hydroxybenzylidenepyruvic acid (tHBPA). Their conclusions were based on studies of a strain of *Escherichia coli* containing plasmid pRE718, which has a 1.95 kb insert from plasmid NAH7 of *P. putida* PpG1064.

3.2.5. Hydratase-Aldolase

An aldolase converts the *o*-hydroxybenzylidenepyruvate produced from HCCA to salicylaldehyde with the loss of pyruvate (Barnsley, 1967a; Eaton and Chapman, 1992). The latter authors, after cloning a 1 kb DNA fragment from plasmid NAH7 into *E. coli,* suggested that the enzyme encoded by this fragment was responsible for both hydration of tHBPA and aldol cleavage and that the enzyme should be considered a tHBPA hydratase-aldolase. In succinate-grown *Pseudomonas* sp. NCIB 9816, the specific activity of this enzyme is 0.27 μmol/min/mg protein, but after induction by salicylate the specific activity increases to 0.93 μmol/min/mg protein (Barnsley, 1976a).

3.2.6. Salicylaldehyde Dehydrogenase

Salicylaldehyde dehydrogenase converts salicylaldehyde to salicylate. It requires NAD$^+$ for activity, and NADP cannot be substituted. It uses one oxygen atom for each molecule of salicylaldehyde (Davies and Evans, 1964).

3.2.7. Salicylate Hydroxylase

Yamamoto et al. (1965) purified salicylate hydroxylase from a *Pseudomonas* sp. grown on sodium salicylate as the sole source of carbon and energy. It has a molecular weight of 57,200 and is loosely bound to 1 mol of FAD per 1 mol of enzyme. In the presence of NADH, it catalyzes the decarboxylation and hydroxylation of salicylate to catechol.

3.2.8. Catechol 2,3-Dioxygenase

Catechol 2,3-dioxygenase transforms catechol by *meta*-cleavage to 2-hydroxymuconic semialdehyde (Dagley and Gibson, 1965). In *Pseudomonas* sp. NCIB 9816, *P. testosteroni,* and *P. stutzeri,* catechol 2,3-dioxygenase is induced by growth on naphthalene (Catterall et al., 1971; García-Valdés et al., 1988) and in the latter two species also by growth on salicylate or salicylaldehyde (García-Valdés et al., 1988).

3.2.9. 2-Hydroxymuconic Semialdehyde Dehydrogenase and Hydrolase

Growth on naphthalene apparently induces two other enzymes in *Pseudomonas* sp. that are involved in the further metabolism of 2-hydroxymuconic semialdehyde (Catterall et al., 1971). The NAD$^+$-dependent enzyme 2-hydroxymuconic semialdehyde dehydrogenase converts 2-hydroxymuconic semialdehyde and NAD$^+$ into γ-oxalocrotonate and NADH. The non-NAD$^+$-dependent enzyme 2-hydroxymuconic semialdehyde hydrolase causes the hydrolytic fission of 2-hydroxymuconic semialdehyde to formate and 2-oxopent-4-enoic acid (Dagley and Gibson, 1965; Farr and Cain, 1968; Bayly and Dagley, 1969; Catterall et al., 1971). Since the activity of 2-hydroxymuconic semialdehyde dehydrogenase is generally higher than that of 2-hydroxymuconic semialdehyde hydrolase, the most important reaction *in vivo* may be the NAD$^+$-dependent dehydrogenation of 2-hydroxymuconic semialdehyde to form γ-oxalocrotonate (Catterall et al., 1971; Yen and Serdar, 1988).

3.2.10. Other Enzymes

The remaining enzymes of the naphthalene pathway that convert γ-oxalocrotonate to further metabolites are 4-oxalocrotonate tautomerase, 4-oxalocrotonate decarboxylase, 4-hydroxy-2-oxovalerate aldolase, and oxo-4-pentenoate hydratase (Sala-Trepat et al., 1972; Yen and Serdar, 1988). These enzymes have not been fully characterized.

3.3. Alternative Pathways for Naphthalene Metabolism

3.3.1. Gentisate Pathway

In contrast to other species of *Pseudomonas*, cells of *P. fluorescens* grown on naphthalene use a different pathway for the metabolism of salicylate. These cultures convert salicylate to gentisic acid instead of catechol (Starovoitov, 1975). The gentisate is transformed by gentisate 1,2-dioxygenase to maleylpyruvic acid, which is transformed by other enzymes to fumaric and pyruvic acids (Skryabin and Starovoitov, 1975; Utkin et al., 1990).

3.3.2. β-Ketoadipate Pathway

Growth of *Pseudomonas* sp. strain PG on salicylate, catechol, or benzoate results in induction of the enzymes of the *ortho* or β-ketoadipate pathway instead of the *meta* pathway (Williams et al., 1975). Either *cis,cis*-muconate or one of its metabolites may induce the enzymes of the *ortho* pathway. In *Pseudomonas* sp. PG, salicylate, catechol, and benzoate are metabolized by the *ortho* pathway without the induction of either the early enzymes of naphthalene catabolism or the later enzymes of the *meta* pathway, such as catechol 2,3-dioxygenase. The *ortho* pathway enzymes are produced in strain PG only when the *meta* pathway enzymes have not been induced and catechol has accumulated. In the *ortho* pathway, catechol 1,2-dioxygenase metabolizes catechol to *cis,cis*-muconate. The *cis,cis*-muconate lactonizing enzyme then converts *cis,cis*-muconate to a muconolactone, which is then metabolized to succinate and acetyl CoA by other enzymes of the *ortho* pathway (Williams et al., 1975).

3.4. Enzyme Regulation: Induction of Enzymes for Naphthalene Metabolism

Several *Pseudomonas* species produce naphthalene dioxygenase when grown either on salicylate or on a mixture of salicylate and succinate (Shamsuzzaman and Barnsley, 1974a,b). Acetylnaphthalene, 2-hydroxybenzyl alcohol, or 2-aminobenzoate may also function as a gratuitous inducer of naphthalene dioxygenase in cultures grown on succinate. In *Pseudomonas* sp. NCIB 9816, *P. putida* ATCC 17483, and *P. putida* PpG7, either 2-aminobenzoate or salicylate is able to induce naphthalene dioxygenase, 1,2-dihydroxynaphthalene dioxygenase, salicylaldehyde dehydrogenase, and salicylate hydroxylase (Shamsuzzaman and Barnsley, 1974a,b). Salicylate hydroxylase activity, however, is not detected in *Pseudomonas* sp. NCIB 9816 unless cell extracts have been fortified with FAD. In *P. putida* ATCC 17483 and PpG7, catechol 2,3-dioxygenase is also inducible.

The induction of all of the enzymes in the pathway from naphthalene to salicylate by the same gratuitous inducer suggests coordinate regulation. Shamsuzzaman and Barnsley (1974a; 1974b) proposed that all of these enzymes may be specified by genes in a single operon, and Barnsley (1975b) suggested that all of the enzymes to convert naphthalene to 2-hydroxymuconic semialdehyde may be regulated coordinately.

3.5. Other Enzymes Involved in Microbial PAH Metabolism

3.5.1. Cytochrome P_{450}

The yeast *Saccharomyces cerevisiae* produces a cytochrome P_{450} aryl hydrocarbon hydroxylase (Woods and Wiseman, 1979). Although the apparent K_m of the enzyme in noninduced cells is 0.60 mM, yeast cells grown in the presence of benzo[a]pyrene produce an isozyme with a higher affinity for benzo[a]pyrene so that the apparent K_m becomes 0.16 mM. The major metabolites produced are benzo[a]pyrene, 7,8-dihydrodiol, 9-hydroxybenzo[a]pyrene, and 3-hydroxybenzo[a]pyrene (Wiseman and Woods, 1979). The inducible isozyme is formed preferentially in 20% glucose-grown cells, in which mitochondrial enzymes are repressed and cyclic AMP levels are low (Wiseman, 1980).

Cytochrome $P_{450_{soy}}$, produced by *Streptomyces griseus*, hydroxylates naphthalene to 1-naphthol but is not induced by naphthalene. Cytochrome $P_{450_{soy}}$ is a single polypeptide with a molecular weight of 47,500 and ferriprotoporphyrin IX as a prosthetic group (Trower et al., 1988). An NAD(P)H-dependent flavoprotein reductase transfers electrons to a ferredoxin and then to the cytochrome P_{450} monooxygenase component, which uses molecular oxygen to oxidize naphthalene. When reconstituted with spinach ferredoxin:NADP$^+$ oxidoreductase and spinach ferredoxin, the cytochrome $P_{450_{soy}}$ from *S. griseus* catalyzes the NADPH-dependent oxidation of naphthalene *in vitro* (Trower et al., 1988).

3.5.2. Lignin Peroxidase

In the presence of H_2O_2, purified lignin peroxidase preparations from the white-rot fungus *Phanerochaete chrysosporium* oxidize benzo[a]pyrene to the 1,6-, 3,6-, and 6,12-quinones and oxidize pyrene to the 1,6- and 1,8-quinones (Haemmerli et al., 1986; Hammel et al., 1986). The addition of veratryl (3,4-dimethoxybenzyl) alcohol enhances benzo[a]pyrene oxidation by preventing the inactivation of lignin peroxidase by H_2O_2 (Haemmerli et al., 1986).

Haemmerli et al. (1986) proposed that lignin peroxidase oxidizes benzo[a]pyrene initially to 6-hydroxybenzo[a]pyrene and that the subsequent reactions to form quinones occur rapidly and spontaneously. An isomeric ratio of 46.7% of the 1,6-quinone, 33.3% of the 3,6-quinone, and 20.0% of the 6,12-quinone is produced by chemical or electrochemical oxidation of benzo[a]pyrene and is similar to the isomeric ratio produced by lignin peroxidase (Haemmerli et al., 1986). According to these workers, H_2O_2 oxidizes lignin peroxidase, which then reacts with veratryl alcohol to produce a cation radical, which in turn produces a benzo[a]pyrene cation radical and regenerates the veratryl alcohol. The quinones then are formed by nonenzymatic oxidation of the benzo[a]pyrene radical.

4. GENETICS OF PAH METABOLISM

4.1. Genetics of Naphthalene Metabolism

4.1.1. The NAH Plasmids

The biochemistry and genetics of bacteria capable of utilizing naphthalene or other polycyclic aromatic hydrocarbons as sole growth substrates have been reviewed extensively (Yen and Serdar, 1988). The genes for naphthalene degradation are located on self-transmissible NAH plasmids in most bacteria (Dunn and Gunsalus, 1973; Yen and Gunsalus, 1982; Yen and Serdar, 1988).

Dunn and Gunsalus (1973) suggested that NAH plasmids can be transferred to other strains by conjugation and that the genes can be expressed in the recipient strains. The NAH7 plasmid has two operons involved in naphthalene metabolism: a *nah* operon that encodes an upper metabolic pathway, for the metabolism of naphthalene to salicylate, and a *sal* operon that encodes a lower pathway, for the metabolism of salicylate via catechol to pyruvate and acetaldehyde (Yen and Gunsalus, 1982). Eleven genes encoded by plasmid NAH7 are responsible for the upper and lower pathways of naphthalene metabolism (Yen and Gunsalus, 1982). The organization and regulation of this plasmid have been investigated with Tn5 polar mutants (Yen and Gunsalus, 1982).

4.1.2. Gene Organization

The gene order of the *nah* operon has recently been reinvestigated (Eaton and Chapman, 1992) by a combined genetic–biochemical approach. The construction of a recombinant bacterium that contained a plasmid encoding only the first three enzymes allowed them to study ring cleavage with 1,2-dihydroxynaphthalene as a substrate. The gene for HCCA isomerase, encoded by *nah*D, was localized on a 1.95-kb *Kpn*I–*Bgl*II fragment. This was confirmed by using an enzyme extract from an *E. coli* harboring a subclone that only catalyzed the rapid equilibration of HCCA and tHBPA. The tHBPA was subsequently metabolized by an enzyme extract from a 1-kb *Mlu*I–*Stu*I subclone that encodes the *nah*E enzyme tHBPA hydratase-aldolase, which transforms tHBPA to salicylaldehyde and pyruvate. On the basis of these experiments, Eaton and Chapman (1992) concluded that the gene order for the *nah* operon is *nahABFCED* (Fig. 7.12). Molecular cloning and restriction map analysis of plasmid NAH7 also indicate that there are two regulatory genes, *nah*N and *nah*L, in the *sal* operon between the *nah*I and *nah*J genes (Schell, 1986) (Fig. 7.12).

The genes for naphthalene catabolism located on several plasmids of *P. putida* have been analyzed for DNA sequence homology. A 15-kb *Eco*RI fragment of plasmid pDTG1 that includes the *nah* operon was nick-translated and hybridized with *Eco*RI fragments of NAH7 and other plasmids (Serdar and Gibson, 1989). The extensive DNA sequence homology found between plasmids pDTG1 and NAH7 shows that the enzymes encoded by each of the two plasmids are responsible for

Naphthalene Catabolic Gene Organization and Regulation

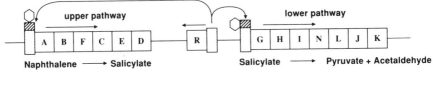

Figure 7.12. Naphthalene catabolic gene organization and regulation. (After Sayler and Layton, 1990.)

naphthalene degradation (Serdar and Gibson, 1989). Similar observations have been reported for other catabolic pathways; DNA–DNA hybridization studies show homology between the 3-chlorobenzoate plasmids pJP4 and pAC27 (Ghosal et al., 1985) as well as between the *nah*H gene of plasmid NAH7 and the *xyl*E gene of the TOL (toluene catabolism) plasmid pWWO (Ghosal et al., 1987). Recently the naphthalene dioxygenase genes from two *P. putida* strains have been sequenced; the nucleotide sequences of both the *nah*Aa (ferredoxin$_{NAP}$ reductase) and the *nah*Ab (ferredoxin$_{NAP}$) genes of these strains show 93% homology (Simon et al., 1993).

4.1.3. Regulatory Genes

The activation of the *nah* operon needs both an inducer, such as salicylate, and the product of the regulatory gene *nah*R (Yen and Gunsalus, 1982; Schell, 1983) (Fig. 7.12). Salicylate also serves as an inducer for the *sal* operon enzymes. Induction experiments by Yen and Gunsalus (1988) on the *nah* operon indicate an increased level of mRNA in salicylate-grown cells, confirming that induction of these enzymes by salicylate is at the transcription level. Naphthalene is not an inducer for enzymes of the *nah* operon, as evidenced by experiments with Tn5 transposon insertion mutants (Yen and Gunsalus, 1982).

The regulatory gene *nah*R, which activates transcription of both the *nah* and *sal* operons, was cloned and mapped by S1 nuclease to determine the approximate start site for transcription of the *nah* operon (Schell, 1986). The direction of the transcription of *nah*R is the opposite of the *nah*G gene (Schell, 1986). Since the amounts of mRNA encoded by the *nah*R region are not significantly different in cells grown with and without salicylate, it appears that the *nah*R gene is transcribed constitutively and that the *nah*R protein can exist in two forms, an active form (*nah*Ra) and an inactive form (*nah*Ri), which are in equilibrium (Schell, 1985). In the presence of a suitable inducer, *nah*Ra predominates; in its absence, the equilibrium shifts towards *nah*Ri.

4.1.4. Comparison With Toluene Plasmids

Schell (1986) also compared the nucleotide sequences of the promoter regions of the *nah* and *sal* operons of the NAH7 plasmid with those of the *xylABC* and *xylDEFG* operons of the TOL plasmid. There was little homology between the 5′ regions, although both plasmids contain similar sets of positively regulated operons for hydrocarbon degradation (Schell, 1986). The nucleotide sequences upstream from the *nah*Aa genes of two *P. putida* strains have 77% homology (Simon et al., 1993).

4.1.5. Location of Genes

Evidence of the chromosomal localization of naphthalene or salicylate catabolic genes (Zuniga et al., 1981) is not convincing for most strains. Indirect evidence, however, indicates that genes for naphthalene catabolism in some pseudomonads may be located on the chromosome or on plasmid-like transposable elements (Dunn and Gunsalus, 1973; Kleckner, 1981; García-Valdés et al., 1988).

4.2. Genetics of Phenanthrene and Anthracene Metabolism

4.2.1. Phenanthrene Metabolism

In contrast to naphthalene, much less is known about the genetics of phenanthrene metabolism. Kim et al. (1986), studying an *Acinetobacter* sp. that utilizes phenanthrene as a sole source of carbon, found that the enzymes for phenanthrene metabolism are encoded by genes on two plasmids. These plasmids, of 4 and 40 kb, are lost when the cells are cured with mitomycin (Kim et al., 1986).

Guerin and Jones (1988) isolated a *Mycobacterium* sp. that utilizes phenanthrene as a sole source of carbon and harbors three plasmids with molecular weights of 21, 58, and 77 megadaltons (32, 88, and 117 kb). In a medium containing nutrients other than phenanthrene, the *Mycobacterium* sp. lost all three plasmids as well as the ability to metabolize phenanthrene, suggesting that one or more of these plasmids may be involved in the mineralization of phenanthrene (Guerin and Jones, 1988).

Kiyohara et al. (1990) isolated a phenanthrene-degrading *Alcaligenes faecalis* strain that harbors two plasmids, pHK1 and pHK2; the molecular weights of the plasmids are 66.9 and 33.4 kb. Cells that were cured with mitomycin C lost the ability to metabolize phenanthrene. After the purified plasmid pHK2 had been used to transform several bacterial strains, the expression of phenanthrene dioxygenase in transformants was shown by measuring oxygen consumption in the presence of phenanthrene and the production of indigo by cells incubated with indole (Kiyohara et al., 1990).

4.2.2. NAH7 Plasmid–Encoded Enzymes and Metabolism of Anthracene and Phenanthrene

Recently, direct evidence for the mineralization of phenanthrene and anthracene by enzymes encoded by NAH7 and similar plasmids in *P. putida* was reported by

Sanseverino et al. (1993). Using transposon Tn5, King et al. (1990) constructed a defective plasmid, pUTK21, that lacks a functional *nah*G gene for salicylate hydroxylase. A strain with this mutant plasmid accumulated 1-hydroxy-2-naphthoic acid from phenanthrene (Sanseverino et al., 1993).

Sayler and Layton (1990) reported that DNA probes constructed from the NAH7 plasmid can be successfully used to identify and isolate novel microorganisms from the microbial community.

4.2.3. Gene Organization

Several reports (Yen and Gunsalus, 1982; Schell, 1986; Ghosal et al., 1987; Menn et al., 1993) indicate that the organization, regulation, structural diversity, and instability of the NAH7 plasmid may have evolved by the assembly of distinct gene clusters. Structural changes in this plasmid may be mediated by transposable elements that promote nonhomologous rearrangement of DNA, including insertions, inversions, deletions, and other complex changes (Kleckner, 1981; Haas and Riess, 1983).

4.3. Genetic Engineering for Bioremediation

Understanding the genetic basis of the biodegradation of PAHs and applying genetic engineering techniques may allow the construction of more efficient degradative strains of bacteria. As an example, when the *xyl*D (toluate 1,2-dioxygenase) and *xyl*L (dihydro-dihydroxybenzoic acid dehydrogenase) genes of the TOL plasmid and the *nah*G gene of the NAH plasmid were recruited into a 3-chlorobenzoate-degrading *Pseudomonas* sp. strain, it was then able to metabolize salicylate and chlorosalicylates (Lehrbach et al., 1984). The use of DNA fragments containing only the essential genes for metabolism of xenobiotic compounds avoids the introduction of genes for counterproductive and unproductive enzymes. The *cbpABCD* genes for the degradation of polychlorinated biphenyls, like the genes for naphthalene metabolism, are arranged in clusters that can be cloned to construct microorganisms with whole new metabolic pathways (Khan and Walia, 1991; Eaton and Chapman, 1992). A detailed knowledge of the biochemistry and genetics of the existing pathways and the stability of the vectors in hosts will be essential before making attempts to construct new pathways.

5. CONCLUSIONS

Microorganisms use several different mechanisms to metabolize PAHs. These mechanisms usually involve enzymatic oxidation to arene oxides, *cis*- and *trans*-dihydrodiols, phenols, quinones, and conjugates. The enzymology and genetics of naphthalene metabolism in bacteria now are reasonably well understood, and the mechanisms involved in the microbial metabolism of phenanthrene, anthracene, benzo[*a*]pyrene, and other PAHs are beginning to yield to investigation.

ACKNOWLEDGMENTS

We thank Dr. Kay L. Shuttleworth for helpful comments on the manuscript, Ms. Kat Wheeler for computer graphics, and Ms. Patricia Fleischer and Mr. Robert Couston for the preparation of the manuscript.

REFERENCES

Akhtar MN, Boyd DR, Thomas NJ, Koreeda M, Gibson DT, Mahadevan V, Jerina DM (1975): Absolute stereochemistry of the dihydroanthracene-*cis*- and *trans*-1,2-diols produced from anthracene by mammals and bacteria. J Chem Soc Perkin Trans I 1975:2506–2511.

Barnsley EA (1975a): The bacterial degradation of fluoranthene and benzo[*a*]pyrene. Can J Microbiol 21:1004–1008.

Barnsley EA (1975b): The induction of the enzymes of naphthalene metabolism in pseudomonads by salicylate and 2-aminobenzoate. J Gen Microbiol 88:193–196.

Barnsley EA (1976a): Naphthalene metabolism by pseudomonads: The oxidation of 1,2-dihydroxynaphthalene to 2-hydroxychromene-2-carboxylic acid and the formation of 2′-hydroxybenzalpyruvate. Biochem Biophys Res Commun 72:1116–1121.

Barnsley EA (1976b): Role and regulation of the *ortho* and *meta* pathways of catechol metabolism in pseudomonads metabolizing naphthalene and salicylate. J Bacteriol 125:404–408.

Bayly RC, Dagley S (1969): Oxoenoic acids as metabolites in the bacterial degradation of catechols. Biochem J 111:303–307.

Blumer M, Youngblood WW (1975): Polycyclic aromatic hydrocarbons in soils and recent sediments. Science 188:53–55.

Boldrin B, Tiehm A, Fritzsche C (1993): Degradation of phenanthrene, fluorene, fluoranthene, and pyrene by a *Mycobacterium* sp. Appl Environ Microbiol 59:1927–1930.

Bumpus JA (1989): Biodegradation of polycyclic aromatic hydrocarbons by *Phanerochaete chrysosporium*. Appl Environ Microbiol 55:154–158.

Bumpus JA, Tien M, Wright D, Aust SD (1985): Oxidation of persistent environmental pollutants by a white rot fungus. Science 228:1434–1436.

Catterall FA, Sala-Trepat JM, Williams PA (1971): The coexistence of two pathways for the metabolism of 2-hydroxymuconic semialdehyde in a naphthalene-grown pseudomonad. Biochem Biophys Res Commun 43:463–469.

Cerniglia CE (1982): Initial reactions in the oxidation of anthracene by *Cunninghamella elegans*. J Gen Microbiol 128:2055–2061.

Cerniglia CE (1984a): Microbial metabolism of polycyclic aromatic hydrocarbons. Adv Appl Microbiol 30:31–71.

Cerniglia CE (1984b): Microbial transformation of aromatic hydrocarbons. In Atlas RM (ed): Petroleum Microbiology. New York: Macmillan, pp 99–128.

Cerniglia CE (1991): Biodegradation of organic contaminants in sediments: Overview and examples with polycyclic aromatic hydrocarbons. In Baker RA (ed): Organic Substances and Sediments in Water. Chelsea, MI: Lewis Publishers, vol 3, pp 267–281.

Cerniglia CE (1992): Biodegradation of polycyclic aromatic hydrocarbons. Biodegradation 3:351–368.

Cerniglia CE (1993): Biodegradation of polycyclic aromatic hydrocarbons. Curr Opin Biotechnol 4:331–338.

Cerniglia CE, Campbell WL, Freeman JP, Evans FE (1989): Identification of a novel metabolite in phenanthrene metabolism by the fungus *Cunninghamella elegans*. Appl Environ Microbiol 55:2275–2279.

Cerniglia CE, Dodge RH, Gibson DT (1980a): Studies on the fungal oxidation of polycyclic aromatic hydrocarbons. Bot Mar 23:121–124.

Cerniglia CE, Freeman JP, Evans FE (1984): Evidence for an arene oxide-NIH shift pathway in the transformation of naphthalene to 1-naphthol by *Bacillus cereus*. Arch Microbiol 138:283–286.

Cerniglia CE, Gibson DT (1977): Metabolism of naphthalene by *Cunninghamella elegans*. Appl Environ Microbiol 34:363–370.

Cerniglia CE, Gibson DT (1979): Oxidation of benzo[*a*]pyrene by the filamentous fungus *Cunninghamella elegans*. J Biol Chem 254:12174–12180.

Cerniglia CE, Hebert RL, Szaniszlo PJ, Gibson DT (1978): Fungal transformation of naphthalene. Arch Microbiol 117:135–143.

Cerniglia CE, Heitkamp MA (1989): Microbial degradation of polycyclic aromatic hydrocarbons (PAH) in the aquatic environment. In Varanasi U (ed): Metabolism of Polycyclic Aromatic Hydrocarbons in the Aquatic Environment. Boca Raton, FL: CRC Press, pp 41–68.

Cerniglia CE, Kelly DW, Freeman JP, Miller DW (1986): Microbial metabolism of pyrene. Chem Biol Interact 57:203–216.

Cerniglia CE, Mahaffey W, Gibson DT (1980b): Fungal oxidation of benzo[*a*]pyrene: Formation of (−)-*trans*-7,8-dihydroxy-7,8-dihydrobenzo[*a*]pyrene by *Cunninghamella elegans*. Biochem Biophys Res Commun 94:226–232.

Cerniglia CE, Sutherland JB, Crow SA (1992): Fungal metabolism of aromatic hydrocarbons. In Winkelmann G (ed): Microbial Degradation of Natural Products. Weinheim, Germany: VCH Verlagsgesellschaft, pp 193–217.

Cerniglia CE, Van Baalen C, Gibson DT (1980c): Metabolism of naphthalene by the cyanobacterium *Oscillatoria* sp., strain JCM. J Gen Microbiol 116:485–494.

Cerniglia CE, White GL, Heflich RH (1985): Fungal metabolism and detoxification of polycyclic aromatic hydrocarbons. Arch Microbiol 143:105–110.

Cerniglia CE, Yang SK (1984): Stereoselective metabolism of anthracene and phenanthrene by the fungus *Cunninghamella elegans*. Appl Environ Microbiol 47:119–124.

Colla C, Fiecchi A, Treccani V (1959): Ricerche sul metabolismo ossidativo microbico dell'antracene e del fenantrene. Nota II. Isolamento e caratterizzazione del 3,4-diidro-3,4-diossifenantrene. Ann Microbiol Enzimol 9:87–91.

Cook JW, Hewett CL, Hieger I (1933): Isolation of a cancer-producing hydrocarbon from coal tar. II. Isolation of 1,2- and 4,5-benzopyrenes, perylene and 1,2-benzanthracene. J Chem Soc 1933:396–398.

Dagley S, Gibson DT (1965): The bacterial degradation of catechol. Biochem J 95:466–474.

Dalton H, Golding BT, Waters BW, Higgins R, Taylor JA (1981): Oxidations of cyclopropane, methylcyclopropane, and arenes with the mono-oxygenase system from *Methylococcus capsulatus*. J Chem Soc Chem Commun 1981:482–483.

Datta D, Samanta TB (1988): Effect of inducers on metabolism of benzo[a]pyrene *in vivo* and *in vitro:* Analysis by high pressure liquid chromatography. Biochem Biophys Res Commun 155:493–502.

Davies JI, Evans WC (1964): Oxidative metabolism of naphthalene by soil pseudomonads: The ring-fission mechanism. Biochem J 91:251–261.

Davis MW, Glaser JA, Evans JW, Lamar RT (1993): Field evaluation of the lignin-degrading fungus *Phanerochaete sordida* to treat creosote-contaminated soil. Environ Sci Technol 27:2572–2576.

dell'Omo M, Lauwerys RR (1993): Adducts to macromolecules in the biological monitoring of workers exposed to polycyclic aromatic hydrocarbons. Crit Rev Toxicol 23:111–126.

Dhawale SW, Dhawale SS, Dean-Ross D (1992): Degradation of phenanthrene by *Phanerochaete chrysosporium* occurs under ligninolytic as well as nonligninolytic conditions. Appl Environ Microbiol 58:3000–3006.

Dunn NW, Gunsalus IC (1973): Transmissible plasmid coding early enzymes of naphthalene oxidation in *Pseudomonas putida*. J Bacteriol 114:974–979.

Eaton RW, Chapman PJ (1992): Bacterial metabolism of naphthalene: Construction and use of recombinant bacteria to study ring cleavage of 1,2-dihydroxynaphthalene and subsequent reactions. J Bacteriol 174:7542–7554.

Engesser KH, Strubel V, Christoglou K, Fischer P, Rast HG (1989): Dioxygenolytic cleavage of aryl ether bonds: 1,10-Dihydro-1,10-dihydroxyfluoren-9-one, a novel arene dihydrodiol as evidence for angular dioxygenation of dibenzofuran. FEMS Microbiol Lett 65:205–209.

Ensley BD, Gibson DT (1983): Naphthalene dioxygenase: Purification and properties of a terminal oxygenase component. J Bacteriol 155:505–511.

Ensley BD, Gibson DT, Laborde AL (1982): Oxidation of naphthalene by a multicomponent enzyme system from *Pseudomonas* sp. strain NCIB 9816. J Bacteriol 149:948–954.

Evans WC, Fernley HN, Griffiths E (1965): Oxidative metabolism of phenanthrene and anthracene by soil pseudomonads: The ring-fission mechanism. Biochem J 95:819–831.

Farr DR, Cain RB (1968): Catechol oxygenase induction in *Pseudomonas aeruginosa*. Biochem J 106:879–885.

Field JA, de Jong E, Feijoo Costa G, de Bont JAM (1992): Biodegradation of polycyclic aromatic hydrocarbons by new isolates of white rot fungi. Appl Environ Microbiol 58:2219–2226.

García-Valdés E, Cozar E, Rotger R, Lalucat J, Ursing J (1988): New naphthalene-degrading marine *Pseudomonas* strains. Appl Environ Microbiol 54:2478–2485.

George EJ, Neufeld RD (1989): Degradation of fluorene in soil by fungus *Phanerochaete chrysosporium*. Biotechnol Bioeng 33:1306–1310.

Ghosal D, You IS, Chatterjee DK, Chakrabarty AM (1985): Genes specifying degradation of 3-chlorobenzoic acid in plasmids pAC27 and pJP4. Proc Natl Acad Sci USA 82:1638–1642.

Ghosal D, You IS, Gunsalus IC (1987): Nucleotide sequence and expression of gene *nahH* of plasmid NAH7 and homology with gene *xylE* of TOL pWWO. Gene 55:19–28.

Gibson DT, Mahadevan V, Jerina DM, Yagi H, Yeh HJC (1975): Oxidation of the carcinogens benzo[a]pyrene and benz[a]anthracene to dihydrodiols by a bacterium. Science 189:295–297.

Gibson DT, Subramanian V (1984): Microbial degradation of aromatic hydrocarbons. In Gibson DT (ed): Microbial Degradation of Organic Compounds. New York: Marcel Dekker, pp 181–252.

Gomaa EA, Gray JI, Rabie S, Lopez-Bote C, Booren AM (1993): Polycyclic aromatic hydrocarbons in smoked food products and commercial liquid smoke flavourings. Food Addit Contam 10:503–521.

Grifoll M, Casellas M, Bayona JM, Solanas AM (1992): Isolation and characterization of a fluorene-degrading bacterium: Identification of ring oxidation and ring fission products. Appl Environ Microbiol 58:2910–2917.

Grimmer G, Dettbarn G, Jacob J (1993): Biomonitoring of polycyclic aromatic hydrocarbons in highly exposed coke plant workers by measurement of urinary phenanthrene and pyrene metabolites (phenols and dihydrodiols). Int Arch Occup Environ Health 65:189–199.

Grund E, Denecke B, Eichenlaub R (1992): Naphthalene degradation via salicylate and gentisate by *Rhodococcus* sp. strain B4. Appl Environ Microbiol 58:1874–1877.

Guengerich FP (1992): Metabolic activation of carcinogens. Pharmacol Ther 54:17–61.

Guengerich FP (1993a): Cytochrome P450 enzymes. Am Sci 81:440–447.

Guengerich FP (1993b): Bioactivation and detoxication of toxic and carcinogenic chemicals. Drug Metab Dispos 21:1–6.

Guerin WF, Jones GE (1988): Mineralization of phenanthrene by a *Mycobacterium* sp. Appl Environ Microbiol 54:937–944.

Haas D, Riess G (1983): Spontaneous deletions of the chromosome-mobilizing plasmid R68.45 in *Pseudomonas aeruginosa* PAO. Plasmid 9:42–52.

Haemmerli SD, Leisola MSA, Sanglard D, Fiechter A (1986): Oxidation of benzo[*a*]pyrene by extracellular ligninases of *Phanerochaete chrysosporium*. J Biol Chem 261:6900–6903.

Haigler BE, Gibson DT (1990a): Purification and properties of NADH-ferredoxin$_{NAP}$ reductase, a component of naphthalene dioxygenase from *Pseudomonas* sp. strain NCIB 9816. J Bacteriol 172:457–464.

Haigler BE, Gibson DT (1990b): Purification and properties of ferredoxin$_{NAP}$, a component of naphthalene dioxygenase from *Pseudomonas* sp. strain NCIB 9816. J Bacteriol 172:465–468.

Hall M, Grover PL (1990): Polycyclic aromatic hydrocarbons: Metabolism, activation and tumour initiation. In Cooper CS, Grover PL (eds): Chemical Carcinogenesis and Mutagenesis. Berlin: Springer-Verlag, vol 1, pp 327–372.

Hammel KE, Gai WZ, Green B, Moen MA (1992): Oxidative degradation of phenanthrene by the ligninolytic fungus *Phanerochaete chrysosporium*. Appl Environ Microbiol 58:1832–1838.

Hammel KE, Green B, Gai WZ (1991): Ring fission of anthracene by a eukaryote. Proc Natl Acad Sci USA 88:10605–10608.

Hammel KE, Kalyanaraman B, Kirk TK (1986): Oxidation of polycyclic aromatic hydrocarbons and dibenzo[*p*]dioxins by *Phanerochaete chrysosporium* ligninase. J Biol Chem 261:16948–16952.

Heitkamp MA, Cerniglia CE (1988): Mineralization of polycyclic aromatic hydrocarbons by a bacterium isolated from sediment below an oil field. Appl Environ Microbiol 54:1612–1614.

Heitkamp MA, Franklin W, Cerniglia CE (1988a): Microbial metabolism of polycyclic aromatic hydrocarbons: Isolation and characterization of a pyrene-degrading bacterium. Appl Environ Microbiol 54:2549–2555.

Heitkamp MA, Freeman JP, Miller DW, Cerniglia CE (1988b): Pyrene degradation by a *Mycobacterium* sp.: Identification of ring oxidation and ring fission products. Appl Environ Microbiol 54:2556–2565.

Hieger I (1993): Isolation of a cancer-producing hydrocarbon from coal tar. I. Concentration of the active substance. J Chem Soc 1933:395–396.

Hofmann KH (1986): Oxidation of naphthalene by *Saccharomyces cerevisiae* and *Candida utilis*. J Basic Microbiol 26:109–111.

Huntley SL, Bonnevie NL, Wenning RJ, Bedbury H (1993): Distribution of polycyclic aromatic hydrocarbons (PAHs) in three northern New Jersey waterways. Bull Environ Contam Toxicol 51:865–872.

International Agency for Research on Cancer (1983): IARC Monographs on the Evaluation of the Carcinogenic Risk of Chemicals to Humans. Polynuclear Aromatic Compounds, Part 1, Chemical, Environmental and Experimental Data, vol 32. Geneva: World Health Organization.

Jeffrey AM, Yeh HJC, Jerina DM, Patel TR, Davey JF, Gibson DT (1975): Initial reactions in the oxidation of naphthalene by *Pseudomonas putida*. Biochemistry 14:575–584.

Jerina DM, Daly JW, Jeffrey AM, Gibson DT (1971): *cis*-1,2-Dihydroxy-1,2-dihydronaphthalene: A bacterial metabolite from naphthalene. Arch Biochem Biophys 142:394–396.

Jerina DM, Selander H, Yagi H, Wells MC, Davey JF, Mahadevan V, Gibson DT (1976): Dihydrodiols from anthracene and phenanthrene. J Am Chem Soc 98:5988–5996.

Jerina DM, van Bladeren PJ, Yagi H, Gibson DT, Mahadevan V, Neese AS, Koreeda M, Sharma ND, Boyd DR (1984): Synthesis and absolute configuration of the bacterial *cis*-1,2-, *cis*-8,9-, and *cis*-10,11-dihydrodiol metabolites of benz[*a*]anthracene formed by a strain of *Beijerinckia*. J Org Chem 49:3621–3628.

Kadlubar FF, Hammons GJ (1987): The role of cytochrome P-450 in the metabolism of chemical carcinogens. In Guengerich FP (ed): Mammalian Cytochromes P-450. Boca Raton, FL: CRC Press, vol 2, pp 81–130.

Keith LH, Telliard WA (1979): Priority pollutants I—A perspective view. Environ Sci Technol 13:416–423.

Kelley I, Cerniglia CE (1991): The metabolism of fluoranthene by a species of *Mycobacterium*. J Ind Microbiol 7:19–26.

Kelley I, Freeman, JP, Cerniglia CE (1990): Identification of metabolites from degradation of naphthalene by a *Mycobacterium* sp. Biodegradation 1:283–290.

Kelley I, Freeman JP, Evans FE, Cerniglia CE (1991): Identification of a carboxylic acid metabolite from the catabolism of fluoranthene by a *Mycobacterium* sp. Appl Environ Microbiol 57:636–641.

Kelley I, Freeman JP, Evans FE, Cerniglia CE (1993): Identification of metabolites from the degradation of fluoranthene by *Mycobacterium* sp. strain PYR-1. Appl Environ Microbiol 59:800–806.

Keuth S, Rehm HJ (1991): Biodegradation of phenanthrene by *Arthrobacter polychromogenes* isolated from a contaminated soil. Appl Microbiol Biotechnol 34:804–808.

Khan AA, Walia SK (1991): Expression, localization, and functional analysis of polychlorinated biphenyl degradation genes *cbpABCD of Pseudomonas putida*. Appl Environ Microbiol 57:1325–1332.

Kim CK, Kim JW, Kim YC, Mheen TI (1986): Isolation of aromatic hydrocarbon-degrading bacteria and genetic characterization of their plasmid genes [in Korean]. Kor J Microbiol 24:67–72.

King JMH, DiGrazia PM, Applegate B, Burlage R, Sanseverino J, Dunbar P, Larimer F, Sayler GS (1990): Rapid, sensitive bioluminescent reporter technology for naphthalene exposure and biodegradation. Science 249:778–781.

Kiyohara H, Nagao K, Nomi R (1976): Degradation of phenanthrene through *o*-phthalate by an *Aeromonas* sp. Agric Biol Chem 40:1075–1082.

Kiyohara H, Takizawa N, Date H, Torigoe S, Yano K (1990): Characterization of a phenanthrene degradation plasmid from *Alcaligenes faecalis* AFK2. J Ferment Bioeng 69:54–56.

Kleckner N (1981): Transposable elements in prokaryotes. Annu Rev Genet 15:341–404.

Komatsu T, Omori T, Kodama T (1993): Microbial degradation of the polycyclic aromatic hydrocarbons acenaphthene and acenaphthylene by a pure bacterial culture. Biosci Biotechnol Biochem 57:864–865.

Koreeda M, Moore PD, Wislocki PG, Levin W, Conney AH, Yagi H, Jerina DM (1978): Binding of benzo[*a*]pyrene 7,8-diol-9,10-epoxides to DNA, RNA, and protein of mouse skin occurs with high stereoselectivity. Science 199:778–781.

Kuhm AE, Stolz A, Knackmuss HJ (1991a): Metabolism of naphthalene by the biphenyl-degrading bacterium *Pseudomonas paucimobilis* Q1. Biodegradation 2:115–120.

Kuhm AE, Stolz A, Ngai KL, Knackmuss HJ (1991b): Purification and characterization of a 1,2-dihydroxynaphthalene dioxygenase from a bacterium that degrades naphthalenesulfonic acids. J Bacteriol 173:3795–3802.

Lehrbach PR, Zeyer J, Reineke W, Knackmuss HJ, Timmis KN (1984): Enzyme recruitment in vitro: Use of cloned genes to extend the range of haloaromatics degraded by *Pseudomonas* sp. strain B13. J Bacteriol 158:1025–1032.

Lewtas J (1993): Complex mixtures of air pollutants: Characterizing the cancer risk of polycyclic organic matter. Environ Health Perspect 100:211–218.

Lindquist B, Warshawsky D (1985a): Identification of the 11,12-dihydroxybenzo[*a*]pyrene as a major metabolite produced by the green alga, *Selenastrum capricornutum*. Biochem Biophys Res Commun 130:71–75.

Lindquist B, Warshawsky D (1985b): Stereospecificity in algal oxidation of the carcinogen benzo[*a*]pyrene. Experientia 41:767–769.

MacGillivray AR, Shiaris MP (1993): Biotransformation of polycyclic aromatic hydrocarbons by yeasts isolated from coastal sediments. Appl Environ Microbiol 59:1613–1618.

Mahaffey WR, Gibson DT, Cerniglia CE (1988): Bacterial oxidation of chemical carcinogens: Formation of polycyclic aromatic acids from benz[*a*]anthracene. Appl Environ Microbiol 54:2415–2423.

Menn FM, Applegate BM, Sayler GS (1993): NAH plasmid-mediated catabolism of anthracene and phenanthrene to naphthoic acids. Appl Environ Microbiol 59:1938–1942.

Mihelcic JR, Luthy RG (1988): Microbial degradation of acenaphthene and naphthalene under denitrification conditions in soil–water systems. Appl Environ Microbiol 54:1188–1198.

Miller EC, Miller JA (1985): Some historical perspectives on the metabolism of xenobiotic chemicals to reactive electrophiles. In Anders MW (ed): Bioactivation of Foreign Compounds. Orlando, FL: Academic Press, pp 3–28.

Monna L, Omori T, Kodama T (1993): Microbial degradation of dibenzofuran, fluorene, and dibenzo-*p*-dioxin by *Staphylococcus auriculans* DBF63. Appl Environ Microbiol 59:285–289.

Morgan P, Lee SA, Lewis ST, Sheppard AN, Watkinson RJ (1993): Growth and biodegradation by white-rot fungi inoculated into soil. Soil Biol Biochem 25:279–287.

Morgan P, Lewis ST, Watkinson RJ (1991): Comparison of abilities of white-rot fungi to mineralize selected xenobiotic compounds. Appl Microbiol Biotechnol 34:693–696.

Mueller JG, Chapman PJ, Blattmann BO, Pritchard PH (1990): Isolation and characterization of a fluoranthene-utilizing strain of *Pseudomonas paucimobilis*. Appl Environ Microbiol 56:1079–1086.

Mueller JG, Lantz SE, Ross D, Colvin RJ, Middaugh DP, and Pritchard PH (1993): Strategy using bioreactors and specially selected microorganisms for bioremediation of groundwater contaminated with creosote and pentachlorophenol. Environ Sci Technol 27:691–698.

Narro ML, Cerniglia CE, Van Baalen C, Gibson DT (1992a): Metabolism of phenanthrene by the marine cyanobacterium *Agmenellum quadruplicatum* PR-6. Appl Environ Microbiol 58:1351–1359.

Narro ML, Cerniglia CE, Van Baalen C, Gibson DT (1992b): Evidence for an NIH shift in oxidation of naphthalene by the marine cyanobacterium *Oscillatoria* sp. strain JCM. Appl Environ Microbiol 58:1360–1363.

Ny ET, Heederik D, Kromhout H, Jongeneelen F (1993): The relationship between polycyclic aromatic hydrocarbons in air and in urine of workers in a Söderberg potroom. Am Ind Hyg Assoc J 54:277–284.

Patel TR, Barnsley EA (1980): Naphthalene metabolism by pseudomonads: Purification and properties of 1,2-dihydroxynaphthalene oxygenase. J Bacteriol 143:668–673.

Patel TR, Gibson DT (1974): Purification and properties of (+)-*cis*-naphthalene dihydrodiol dehydrogenase of *Pseudomonas putida*. J Bacteriol 119:879–888.

Patel TR, Gibson DT (1976): Bacterial *cis*-dihydrodiol dehydrogenases: Comparison of physicochemical and immunological properties. J Bacteriol 128:842–850.

Perfetti GA, Nyman PJ, Fisher S, Joe FL, Diachenko GW (1992): Determination of polynuclear aromatic hydrocarbons in seafood by liquid chromatography with fluorescence detection. J AOAC Int 75:872–877.

Pothuluri JV, Cerniglia CE (1994): Microbial metabolism of polycyclic aromatic hydrocarbons. In Chaudhry GR (ed): Biological Degradation and Bioremediation of Toxic Chemicals. Portland, OR: Dioscorides Press.

Pothuluri JV, Freeman JP, Evans FE, Cerniglia CE (1990): Fungal transformation of fluoranthene. Appl Environ Microbiol 56:2974–2983.

Pothuluri JV, Freeman JP, Evans FE, Cerniglia CE (1992a): Fungal metabolism of acenaphthene by *Cunninghamella elegans*. Appl Environ Microbiol 58:3654–3659.

Pothuluri JV, Freeman JP, Evans FE, Cerniglia CE (1993): Biotransformation of fluorene by the fungus *Cunninghamella elegans*. Appl Environ Microbiol 59:1977–1980.

Pothuluri JV, Heflich RH, Fu PP, Cerniglia CE (1992b): Fungal metabolism and detoxification of fluoranthene. Appl Environ Microbiol 58:937–941.

Ryu BH, Oh YK, Bae KC, Bin JH (1989): Biodegradation of naphthalene by *Acinetobacter calcoaceticus* R-88 [in Korean]. Han'guk Nonghwa Hakhoechi 32:315–320. (Chem Abstr 112:154982q.)

Sack U, Günther T (1993): Metabolism of PAH by fungi and correlation with extracellular enzymatic activities. J Basic Microbiol 33:269–277.

Sala-Trepat JM, Murray K, Williams PA (1972): The metabolic divergence in the *meta* cleavage of catechols by *Pseudomonas putida* NCIB 10015: Physiological significance and evolutionary implications. Eur J Biochem 28:347–356.

Sanglard D, Leisola MSA, Fiechter A (1986): Role of extracellular ligninases in biodegradation of benzo[*a*]pyrene by *Phanerochaete chrysosporium*. Enzyme Microb Technol 8:209–212.

Sanseverino J, Applegate BM, King JMH, Sayler GS (1993): Plasmid-mediated mineralization of naphthalene, phenanthrene, and anthracene. Appl Environ Microbiol 59:1931–1937.

Saxton WL, Newton RT, Rorberg J, Sutton J, Johnson LE (1993): Polycyclic aromatic hydrocarbons in seafood from the Gulf of Alaska following a major crude oil spill. Bull Environ Contam Toxicol 51:515–522.

Sayler GS, Layton AC (1990): Environmental application of nucleic acid hybridization. Annu Rev Microbiol 44:625–648.

Schell MA (1983): Cloning and expression in *Escherichia coli* of the naphthalene degradation genes from plasmid NAH7. J Bacteriol 153:822–829.

Schell MA (1985): Transcriptional control of the *nah* and *sal* hydrocarbon-degradation operons by the *nahR* gene product. Gene 36:301–309.

Schell MA (1986): Homology between nucleotide sequences of promoter regions of *nah* and *sal* operons of NAH7 plasmid of *Pseudomonas putida*. Proc Natl Acad Sci USA 83:369–373.

Schocken MJ, Gibson DT (1984): Bacterial oxidation of the polycyclic aromatic hydrocarbons acenaphthene and acenaphthylene. Appl Environ Microbiol 48:10–16.

Selifonov SA, Grifoll M, Gurst JE, Chapman PJ (1993a): Isolation and characterization of (+)-1,1a-dihydroxy-1-hydrofluoren-9-one formed by angular dioxygenation in the bacterial catabolism of fluorene. Biochem Biophys Res Commun 193:67–76.

Selifonov SA, Slepenkin AV, Adanin VM, Grechkina GM, Starovoitov II (1993b): Acenaphthene catabolism by strains of *Alcaligenes eutrophus* and *Alcaligenes paradoxus*. Microbiology 62:85–92.

Serdar CM, Gibson DT (1989): Studies of nucleotide sequence homology between naphthalene-utilizing strains of bacteria. Biochem Biophys Res Commun 164:772–779.

Shamsuzzaman KM, Barnsley EA (1974a): The regulation of naphthalene metabolism in pseudomonads. Biochem Biophys Res Commun 60:582–589.

Shamsuzzaman KM, Barnsley EA (1974b): The regulation of naphthalene oxygenase in pseudomonads. J Gen Microbiol 83:165–170.

Simon MJ, Osslund TD, Saunders R, Ensley BD, Suggs S, Harcourt A, Suen WC, Cruden DL, Gibson DT, Zylstra GJ (1993): Sequences of genes encoding naphthalene dioxygenase in *Pseudomonas putida* strains G7 and NCIB 9816-4. Gene 127:31–37.

Sims P, Grover PL, Swaisland A, Pal K, Hewer A (1974): Metabolic activation of benzo[*a*]pyrene proceeds by a diol-epoxide. Nature 252:326–328.

Skryabin GK, Starovoitov II (1975): Alternative path of naphthalene catabolism by *Pseu-*

domonas fluorescens [in Russian]. Dokl Akad Nauk SSSR 221:493–495. (Chem Abstr 83:4691n.)

Smith MR (1990): The biodegradation of aromatic hydrocarbons by bacteria. Biodegradation 1:191–206.

Starovoitov II, Nefedova MY, Yakovlev GI, Zyakun AM, Adanin VM (1975): Gentisic acid, product of the microbiological oxidation of naphthalene [in Russian]. Izv Akad Nauk SSSR Ser Khim 1975:2091–2092. (Chem Abstr 83:203594h.)

Strandberg GW, Abraham TJ, Frazier GC (1986): Phenanthrene degradation by *Beijerinckia* sp. B8/36. Biotechnol Bioeng 28:142–145.

Sutherland JB (1992): Detoxification of polycyclic aromatic hydrocarbons by fungi. J Ind Microbiol 9:53–62.

Sutherland JB, Freeman JP, Selby AL, Fu PP, Miller DW, Cerniglia CE (1990): Stereoselective formation of a K-region dihydrodiol from phenanthrene by *Streptomyces flavovirens*. Arch Microbiol 154:260–266.

Sutherland JB, Fu PP, Yang SK, Von Tungeln LS, Casillas RP, Crow SA, Cerniglia CE (1993): Enantiomeric composition of the *trans*-dihydrodiols produced from phenanthrene by fungi. Appl Environ Microbiol 59:2145–2149.

Sutherland JB, Selby AL, Freeman JP, Evans FE, Cerniglia CE (1991): Metabolism of phenanthrene by *Phanerochaete chrysosporium*. Appl Environ Microbiol 57:3310–3316.

Sutherland JB, Selby AL, Freeman JP, Fu PP, Miller DW, Cerniglia CE (1992): Identification of xyloside conjugates formed from anthracene by *Rhizoctonia solani*. Mycol Res 96:509–517.

Tatarko M, Bumpus JA (1993): Biodegradation of phenanthrene by *Phanerochaete chrysosporium:* On the role of lignin peroxidase. Lett Appl Microbiol 17:20–24.

Treccani V, Walker N, Wiltshire GH (1954): The metabolism of naphthalene by soil bacteria. J Gen Microbiol 11:341–348.

Trower MK, Sariaslani FS, Kitson FG (1988): Xenobiotic oxidation by cytochrome P-450—enriched extracts of *Streptomyces griseus*. Biochem Biophys Res Commun 157:1417–1422.

Utkin IB, Yakimov MM, Matveeva LN, Kozlyak EI, Rogozhin IS, Solomon ZG, Bezborodov AM (1990): Catabolism of naphthalene and salicylate by *Pseudomonas fluorescens*. Folia Microbiol 35:557–560.

Walker N, Wiltshire GH (1953): The breakdown of naphthalene by a soil bacterium. J Gen Microbiol 8:273–276.

Walter U, Beyer M, Klein J, Rehm HJ (1991): Degradation of pyrene by *Rhodococcus* sp. UW1. Appl Microbiol Biotechnol 34:671–676.

Warshawsky D, Keenan TH, Reilman R, Cody TE, Radike MJ (1990): Conjugation of benzo[*a*]pyrene metabolites by freshwater green alga *Selenastrum capricornutum*. Chem Biol Interact 74:93–105.

Warshawsky D, Radike M, Jayasimhulu K, Cody T (1988): Metabolism of benzo[*a*]pyrene by a dioxygenase system of the freshwater green alga *Selenastrum capricornutum*. Biochem Biophys Res Commun 152:540–544.

Weissenfels WD, Beyer M, Klein J (1990): Degradation of phenanthrene, fluorene and fluoranthene by pure bacterial cultures. Appl Microbiol Biotechnol 32:479–484.

Weissenfels WD, Beyer M, Klein J, Rehm HJ (1991): Microbial metabolism of fluoranthene:

Isolation and identification of ring fission products. Appl Microbiol Biotechnol 34:528–535.

Weissenfels WD, Klewer JH, Langhoff J (1992): Adsorption of polycyclic aromatic hydrocarbons (PAHs) by soil particles: Influence on biodegradability and biotoxicity. Appl Microbiol Biotechnol 36:689–696.

Williams PA, Catterall FA, Murray K (1975): Metabolism of naphthalene, 2-methylnaphthalene, salicylate, and benzoate by *Pseudomonas* P_G: Regulation of tangential pathways. J Bacteriol 124:679–685.

Wiseman A (1980): Xenobiotic-metabolising cytochromes P-450 from micro-organisms. Trends Biochem Sci 5:102–104.

Wiseman A, Woods LFJ (1979): Benzo[*a*]pyrene metabolites formed by the action of yeast cytochrome P-450/P-448. J Chem Technol Biotechnol 29:320–324.

Woods LFJ, Wiseman A (1979): Metabolism of benzo[*a*]pyrene by the cytochrome P-450/P-448 of *Saccharomyces cerevisiae*. Biochem Soc Trans 7:124–127.

Yamamoto S, Katagiri M, Maeno H, Hayaishi O (1965): Salicylate hydroxylase, a monooxygenase requiring flavin adenine dinucleotide. I. Purification and general properties. J Biol Chem 240:3408–3413.

Yen KM, Gunsalus IC (1982): Plasmid gene organization: Naphthalene/salicylate oxidation. Proc Natl Acad Sci USA 79:874–878.

Yen KM, Serdar CM (1988): Genetics of naphthalene catabolism in pseudomonads. CRC Crit Rev Microbiol 15:247–268.

Zuniga MC, Durham DR, Welch RA (1981): Plasmid- and chromosome-mediated dissimilation of naphthalene and salicylate in *Pseudomonas putida* PMD-1. J Bacteriol 147:836–843.

8

ANAEROBIC DEGRADATION OF NONHALOGENATED HOMOCYCLIC AROMATIC COMPOUNDS COUPLED WITH NITRATE, IRON, OR SULFATE REDUCTION

PATRICIA J. S. COLBERG

Department of Zoology and Physiology,
University of Wyoming,
Laramie, Wyoming 82071

LILY Y. YOUNG

Center for Agricultural Molecular Biology,
Cook College, Rutgers University,
New Brunswick, New Jersey 08903

1. INTRODUCTION

Interest in the microbial degradation of aromatic compounds was first generated by a 1934 report claiming that the benzene nucleus was amenable to attack by bacteria in the absence of molecular oxygen (Tarvin and Buswell, 1934). This realization, coupled with the urgency to remediate soil and subsurface sites that have been

Microbial Transformation and Degradation of Toxic Organic Chemicals, pages 307–330
© *1995 Wiley-Liss, Inc.*

contaminated with a plethora of anthropogenic organic compounds, has contributed to current efforts to elucidate the mechanisms of aromatic degradation as mediated by microbial communities indigenous to low redox environments.

This chapter focuses on what is currently known about the anaerobic biodegradation of several homocyclic monoaromatic hydrocarbons (benzene, toluene, xylenes, ethylbenzene) and some of their oxygen-substituted counterparts (benzoate, phthalates, phenol, cresols). Our discussion will exclude environmental contaminants that possess halogen substituent groups (see Chapter 7, this volume) and that are classified as heterocyclic (see recent review by Bollag and Kaiser, 1991).

We center our discussion on aromatic metabolism that occurs under three redox regimes: nitrate reduction (denitrification), iron reduction, and sulfate reduction. The reader is referred to several recent papers for consideration of aromatic transformations that occur during anoxygenic photometabolism (see, e.g., Harwood and Gibson, 1986, 1988; Berry et al., 1987; Evans and Fuchs, 1988; Khanna et al., 1992; Rahalkar et al., 1993). Aromatic metabolism under fermentative/methanogenic conditions is also reviewed elsewhere in the recent literature (see, e.g., Grbić-Galić 1989, 1990a,b; Schink et al., 1992).

2. BENZENE

Despite significant progress made in documenting the susceptibility of aromatic compounds to microbial degradation under anaerobic conditions, benzene itself generally eludes biotransformation in the absence of molecular oxygen. This is primarily due to its chemical stability, conferred by the even distribution of n-electrons over a symmetrical planar ring structure. Aerobic bacteria overcome this barrier to attack by using molecular oxygen as a cosubstrate in an oxygenase-mediated reaction that results in the formation of catechol (structure 3 in Fig. 8.1). Additional oxygenase reactions subsequently cleave the catechol intermediate, either between the carbon atoms that carry the hydroxyl groups (*ortho* cleavage) or between two other carbon atoms (*meta* cleavage), only one of which has a hydroxyl substituent (see Fig. 8.1).

Except for the well-documented report of partial benzene mineralization by a fermentative/methanogenic consortium (Vogel and Grbić-Galić, 1986; Grbić-Galić and Vogel, 1987), to our knowledge there is only one other study that maintains that benzene transformation is possible under anoxic conditions. In a paper published by Major and his colleagues in 1988, microcosms were established with subsurface core material from a shallow groundwater aquifer that had a history of BTEX contamination. After 62 days of incubation at 10°C, only 5% of the original benzene (C_o = 3 mg/liter) was detected in microcosms which were amended with nitrate (NO_3^-), while autoclaved controls exhibited about a 20% loss in benzene over the same period. Acetylene block administered to the nitrate-reducing microcosms halted benzene transformation, which the authors inferred as evidence that benzene mineralization was coupled with nitrate reduction. It is important to remember, however, that acetylene only inhibits the final step in dentrification (i.e., nitrous

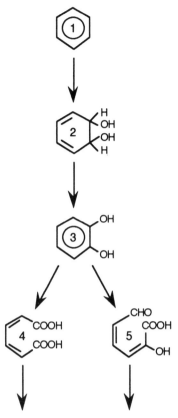

Figure 8.1. Initial steps in the aerobic microbial oxidation of benzene (1) to catechol (3) via *cis*-benzene dihydrodiol (2). Subsequent ring cleavage follows one of two routes: the *ortho* pathway (4) or the *meta* pathway (5).

oxide reductase). It does not block denitrification up to that point, meaning that neither nitrate reduction nor nitrite reduction activities should be ruled out.

Results of other field and laboratory studies, in which attempts have been made to document benzene transformation under denitrifying conditions, lend support to the more usual consensus of benzene's recalcitrance (see, e.g., Kuhn et al., 1988; Hutchins, 1991; Hutchins et al., 1991; Ball et al., 1991; Reinhard et al., 1991; Barbaro et al., 1992; Flyvbjerg et al., 1993). At this writing, there are no published reports of benzene biotransformation occurring under iron-reducing conditions.

The only unequivocal report of benzene degradation under strictly anaerobic conditions is that of Edwards and Grbić-Galić (1992) in which aquifer-derived microorganisms grown in a sulfide-reduced mineral medium supplemented with 20 mM sulfate were shown to mineralize benzene (40–200 μM). Using ^{14}C-labeled benzene, more than 90% of the label was recovered as $^{14}CO_2$. The terminal electron acceptor used during this transformation was not established, although sulfate is the most probable candidate given the experimental conditions.

3. TOLUENE

Unlike the enigmatic benzene molecule, reports of the microbial degradation of toluene under low redox conditions have been repeatedly corroborated by numerous investigators since about 1986. Evidence of toluene degradation under nitrate-reducing, iron-reducing, and sulfate-reducing conditions has been unequivocal. Table 8.1 includes a list of investigations that report anaerobic toluene mineralization under these three redox states. Several citations refer to the results of batch microcosm studies (e.g., Lovley et al., 1989; Evans et al., 1991b; Hutchins, 1991; Jørgensen et al., 1991; Beller et al., 1992a; Edwards et al., 1992; Flyvbjerg et al., 1993). Others refer to transformations observed in laboratory columns (e.g., Zeyer et al., 1986; Kuhn et al., 1988; Haag et al., 1991). One study involved a combination of laboratory work and field experiments (Barbaro et al., 1992).

Since 1990, at least seven pure cultures of bacteria that can use toluene as a sole source of carbon have been described. Five of these organisms use NO_3^- as their terminal electron acceptor (Dolfing et al., 1990; Schocher et al., 1991; Evans et al., 1991a). Another of the toluene degraders is a dissimilatory iron reducer (Lovley and Lonergan, 1990; Lovley et al., 1993). The most recently described isolate is a sulfidogen, the first sulfate-reducing pure culture to be shown to degrade an aromatic hydrocarbon that does not contain an oxygen substituent group (Rabus et al., 1993).

Several mechanisms involved in the initial steps of anaerobic toluene degradation have been suggested and include 1) oxidation of the methyl group to form benzoate; 2) carboxylation of the aromatic ring to form toluate; 3) hydroxylation of the methyl group to form benzyl alcohol; and 4) *para*-hydroxylation of the aromatic ring resulting in *p*-cresol. Four general pathways proposed for anaerobic toluene degradation that include these postulated initial reactions are summarized in Figure 8.2.

Kuhn et al. (1988) and Lovley and Lonergan (1990) based their suspected routes of transformation on growth studies rather than on identification of metabolic intermediates. And while Tschech and Fuchs (1989) and Schnell and Schink (1991) have demonstrated carboxylation of the aromatic ring of phenol and aniline, neither they nor Schocher et al. (1991) have observed toluate formation from toluene. On the basis of degradation studies with fluoroacetate, Schocher et al. (1991) concluded that the initial catabolic steps taken by their *Pseudomonas* strains involve neither carboxylation nor hydroxylation reactions, but are due to oxidation of the methyl group of toluene to benzoate. They hypothesize further that the formation of benzoate is probably due to dehydroxylation of *p*-hydroxybenzoyl-CoA and hydrolysis of the thioester bond. Recent work by Altenschmidt and Fuchs (1992) has implicated benzyl alcohol as the initial oxidation product during ^{14}C-toluene transformation by their denitrifying *Pseudomonas* strain K172.

A novel pathway of anaerobic toluene transformation by a single organism was

TABLE 8.1. Some Homocyclic Monoaromatic Hydrocarbons That Are Amenable to Microbial Degradation Under Nitrate-, Iron-, and Sulfate-Reducing Conditions

Compound	Conditions	References
 benzene	Nitrate reduction	Major et al. (1988)
	Sulfate reduction	Edwards and Grbić-Galić (1992)
 toluene	Nitrate reduction	Zeyer et al. (1986), Kuhn et al. (1988), Major et al. (1988), Dolfing et al. (1990), Altenschmidt and Fuchs (1991, 1992), Ball et al. (1991), Evans et al. (1991a,b), Hutchins (1991), Hutchins et al. (1991), Schocher et al. (1991), Jørgensen et al. (1991), Barbaro et al. (1992), Flyvbjerg et al. (1993), Frazer et al. (1993)
	Iron reduction	Lovley et al. (1989), Lovley and Lonergan (1990)
	Sulfate reduction	Haag et al. (1991), Beller et al. (1992a,b), Edwards et al. (1992), Flyvbjerg et al. (1993), Rabus et al. (1993)
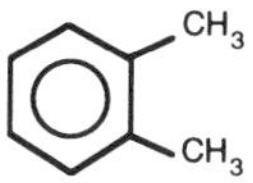 o-xylene	Nitrate reduction	Kuhn et al. (1985), Major et al. (1988), Evans et al. (1991b), Hutchins (1991), Hutchins et al. (1991), Hutchins and Wilson (1991), Barbaro et al. (1992), Evans et al. (1992)
m-xylene	Sulfate reduction	Edwards et al. (1992)
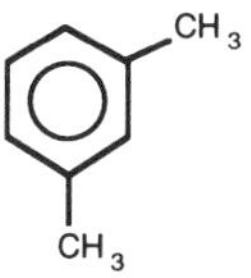	Nitrate reduction	Kuhn et al. (1985), Zeyer et al. (1986), Kuhn et al. (1988), Major et al. (1988), Dolfing et al. (1990), Ball et al. (1991), Evans et al. (1991a,b), Hutchins (1991), Hutchins et al. (1991), Hutchins and Wilson (1991), Barbaro et al. (1992)
p-xylene	Sulfate reduction	Edwards et al. (1992)
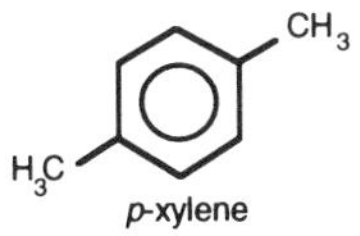	Nitrate reduction	Kuhn et al. (1985), Major et al. (1988), Hutchins et al. (1991), Barbaro et al. (1992)
	Sulfate reduction	Haag et al. (1991), Edwards et al. (1992)
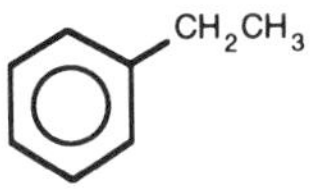 ethylbenzene	Nitrate reduction	Ball et al. (1991), Hutchins (1991), Hutchins et al. (1991), Barbaro et al. (1992)

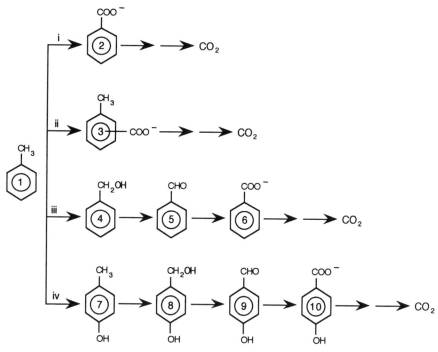

Figure 8.2. Postulated pathways of anaerobic toluene degradation based on results from several laboratories. Initial steps in the pathways shown are (**i**) oxidation of the methyl group; (**ii**) carboxylation of the aromatic ring; (**iii**) hydroxylation of the methyl group; (**iv**) *para*-hydroxylation of the ring. 1, Toluene; 2, benzoate; 3, toluate; 4, benzyl alcohol; 5, benzaldehyde; 6, benzoate; 7, *p*-cresol; 8, *p*-hydroxybenzylalcohol; 9, *p*-hydroxybenzaldehyde; 10, *p*-hydroxybenzoate.

recently proposed by Evans et al. (1992) for a denitrifying bacterium (strain T1) and is shown in Figure 8.3. This pathway is a significant departure from those proposed by other groups in that it involves none of the initiation mechanisms included in Figure 8.2.

The first step in the mineralization pathway (lower pathway in Fig. 8.3) involves the nucleophilic attack of the methyl group (structure 1 in Fig. 8.3) by acetyl-CoA to form phenylpropionyl-CoA (structure 5 in Fig. 8.3). Phenylpropionyl-CoA then undergoes β-oxidation to form benzoyl-CoA (structure 6 in Fig. 8.3), which is known to be an intermediate in the degradation of benzoate during anaerobic photometabolism (Harwood and Gibson, 1986; Geissler et al., 1988). Several investigators have presented convincing cases for benzoyl-CoA serving as a central intermediate in the metabolism of a number of aromatic compounds (e.g., Dangel et al., 1991; Rudolphi et al., 1991; Schnell and Schink, 1991) and toluene in particular (Altenschmidt and Fuchs, 1991). Since CoA esters would remain intracellular because of their large size, Evans et al. (1992) believe this may explain why inter-

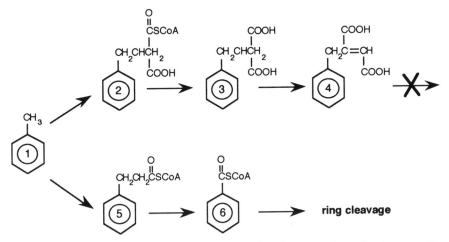

Figure 8.3. Proposed pathways of the initial steps in toluene transformation (**upper pathway**) and toluene mineralization (**lower pathway**) under denitrifying conditions by strain T1. 1, Toluene; 2, benzylsuccinyl-CoA; 3, benzylsuccinic acid; 4, benzylfumaric acid; 5, phenylpropionyl-CoA; 6, benzoyl-CoA. (Based on an original drawing by Evans et al., 1992.)

mediates of anaerobic toluene oxidation have generally not been observed in denitrifying and iron-reducing systems (Dolfing et al., 1990; Evans et al., 1991b; Kuhn et al., 1988; Lovley and Lonergan, 1990).

The toluene transformation pathway (upper pathway depicted in Fig. 8.3) results in the formation of two dead-end metabolites, benzylsuccinic acid and benzylfumaric acid (structures 3 and 4, respectively, in Fig. 8.3). Apparently, this novel yet nonproductive pathway is not restricted to Evans' denitrifying strain T1. Beller and his colleagues (1992b) recently reported accumulation of the same two metabolites in sulfate-reducing enrichment cultures that were fed toluene as a sole carbon source. Subsequent work with strain T1 by Frazer et al. (1993) has shown anaerobic toluene degradation to be inducible. They have also documented benzoate accumulation during toluene transformation; metabolites such as benzyl alcohol and benzaldehyde were not seen. Their results also suggest that the dead-end products benzylfumarate and benzylsuccinate formed only from toluene, since neither phenylpropionate nor benzaldehyde could serve as precursors.

4. XYLENES

Results of a field study tracing the fate of alkyl-substituted benzenes during infiltration of river water to groundwater provided the first indication that xylenes are transformed by microbiological processes (Schwarzenbach et al., 1983). Subsequent studies utilizing saturated-flow laboratory columns filled with subsurface sediments from the field site concluded that *o-, m-,* and *p-*xylenes were used as sole sources of carbon by sediment microbial communities under denitrifying conditions (Kuhn et al., 1985). Even though the investigators clearly demonstrated that xylene removal was correlated with NO_3^- removal, the concentrations of the xylene isomers were too low (<5.5 μM) to allow reliable calculation of carbon and electron mass balances. Additional column studies in the same laboratory demonstrated that [*ring-*UL-^{14}C]*m*-xylene was oxidized to $^{14}CO_2$ with a concomitant reduction in NO_3^- (Zeyer et al., 1986). *m*-Xylene degradation continued when NO_3^- was replaced by nitrous oxide, but was inhibited in the presence of O_2 or after substitution of NO_3^- with nitrite (NO_2^-). The *m*-xylene–adapted microbial community in the aquifer column also transformed toluene, benzaldehyde, benzoate, *m*-toluylaldehyde, *m*-toluate, *m*-cresol, and *p*-hydroxybenzoate (Kuhn et al., 1988).

Field and field-related microcosm studies performed by various investigators have confirmed that xylene degradation occurs under nitrate-reducing conditions, although the rates and extent of transformation are highly variable (see complete list of studies in Table 8.1). Hutchins et al. (1991) established denitrifying microcosms with core materials from a shallow water table aquifer contaminated with JP-4 jet fuel. A 30 day lag period was required before removal of *o-, m-,* and *p-*xylenes was observed in uncontaminated aquifer material, while microcosms that contained contaminated core materials from the same site exhibited no significant xylene removal during the first 60 days of incubation. In addition to the longer lag period, biodegradation rates were three to seven times lower than in microcosms established with the uncontaminated material, results the authors suggest are consistent with either preferential metabolism of indigenous carbon or the potential toxicity of components of JP-4 jet fuel.

Subsequent work with materials from this site yielded somewhat different results. For example, when added as a single compound, *m*-xylene was deemed recalcitrant in uncontaminated core microcosms and credited with inhibiting basal rates of denitrification (Hutchins et al., 1991). In another report of work with the contaminated aquifer materials, Hutchins (1991) demonstrated that *m-,* and *o-*xylenes were biotransformed under both nitrate-reducing and nitrous oxide-reducing conditions with only a 7–14 day lag period.

Despite the consistent reports of *m*-xylene degradation under anaerobic conditions (see Table 8.1), only two studies have attempted to characterize the microorganism(s) responsible for this seemingly ubiquitous transformation. Dolfing et al. (1990) were the first to describe a nitrate-reducing, *m*-xylene–degrading pure culture, a *Pseudomonas* sp. designated strain T, which was isolated from the *m*-xylene–degrading aquifer columns discussed earlier (Zeyer et al., 1986; Kuhn et al., 1988). Although the paper describing this organism focused mainly on its ability to mineralize toluene under denitrifying conditions, experiments with strain T were also described in which more than 50% of [*ring-*UL-^{14}C]*m*-xylene was recovered as $^{14}CO_2$ when NO_3^- was provided as the sole electron acceptor.

Evans et al. (1991b) reported interesting results from their studies with nitrate-reducing enrichment cultures in which *o*-xylene transformation was dependent on toluene mineralization. A pure culture, designated strain T1, was eventually isolated (Evans et al., 1991a) and, as discussed previously, is able to mineralize toluene under nitrate-reducing conditions. *o*-Xylene does not serve as a source of carbon for T1 and is not mineralized; instead, it is transformed during growth on toluene to two dead-end metabolites, 2-methylbenzyl succinic acid and 2-methylbenzyl fumaric acid. *m*-Xylene degradation was not observed in any of their initial enrichment cultures, but was mineralized to CO_2 in subcultures provided *m*-xylene as a sole carbon source (Evans et al., 1991b).

To date, there are no published reports of xylene biotransformation occurring under iron-reducing conditions; however, two recent papers report that xylene degradation occurs during sulfidogenesis. Haag and her colleagues (1991) examined the transformation of several aromatic compounds, including *o*- and *p*-xylenes, in columns filled with aquifer material from a gasoline-contaminated site. Although their results suggest that xylenes were degraded under anaerobic conditions, the relationship between xylene disappearance and sulfate reduction was not established. Microcosm-based studies using sediments from the same site were much more conclusive about the fate of the three xylene isomers during bacterial sulfate reduction (Edwards et al., 1992). Xylene degradation in the microcosms ceased when sulfate was depleted and resumed upon its addition. Addition of molybdate, an inhibitor of sulfate reduction, stopped xylene degradation, while BESA, an inhibitor of methanogenic activity, had no effect. The complete oxidation of *o*-xylene in actively sulfidogenic microcosms was confirmed by recovery of $^{14}CO_2$ from [*methyl*-^{14}C]*o*-xylene. A degradation sequence (*m*-xylene > *p*-xylene >*o*-xylene) was also demonstrated in sulfate-reducing mixed cultures, which included toluene as one of four electron donors.

5. ETHYLBENZENE

Despite the fact that ethylbenzene is a significant component of many refined petroleum products (e.g., gasoline, kerosene, diesel fuel, fuel oil) and is relatively water soluble, reports of its biotransformation under nitrate-reducing, iron-reducing, or sulfate-reducing conditions are rare. One report by Ball et al. (1991) clearly demonstrates, however, that ethylbenzene may be biotransformed during nitrate reduction. In consortia enriched from sediment obtained from an oil refinery treatment pond, ethylbenzene degradation was shown to occur as long as NO_3^- was present. Barbaro et al. (1992) observed some evidence of partial ethylbenzene transformation in the presence of NO_3^- in natural-gradient injection experiments in an aquifer, although its transformation in laboratory microcosms was less apparent.

Results from denitrifying microcosms amended with ethylbenzene that contained the jet fuel–contaminated sediments discussed earlier (Hutchins et al., 1991) were similar to those reported for the xylene isomers: A 30-day lag period before ethylbenzene removal was observed in uncontaminated aquifer material; a 60-day lag period before ethylbenzene removal was observed in contaminated core material; and biodegradation rates that were several times lower in microcosms established with contaminated core material than in microcosms containing uncontaminated materials. Rate constants (mg/liter/day) calculated for ethylbenzene removal from microcosms incubated under nitrate- and nitrous oxide–reducing conditions were also similar to those determined for the xylene isomers (Hutchins, 1991) and were remarkably close to values calculated for *in situ* ethylbenzene removal in a field demonstration project (Hutchins and Wilson, 1991).

6. BENZOATE

Taylor and his colleagues were the first to describe a benzoate-degrading bacterium that required NO_3^- as an electron acceptor (Taylor et al., 1970; Taylor and Heeb, 1972). Originally designated *Pseudomonas* strain PN-1, their isolate aerobically metabolized *p*-hydroxybenzoate using the *meta* pathway of ring cleavage (see Fig. 8.1). Using [*ring*-UL-^{14}C]benzoate, they confirmed benzoate mineralization to CO_2 under denitrifying conditions. A nonreductive mechanism was proposed that required the addition of three molecules of H_2O to the benzene nucleus prior to ring fission; direct evidence in support of their pathway was, however, not provided.

Using the strain originally isolated by Taylor, which has since been reclassified as *Alcaligenes xylosoxidans* subsp. *denitrificans* PN-1, Blake and Hegeman (see Evans and Fuchs, 1988) have demonstrated that extracts of cells provided NO_3^- as an electron acceptor convert radiolabeled benzoate to benzoyl-CoA. Using Tn*5* transposon mutagenesis, they recently obtained mutants of PN-1 that can no longer metabolize benzoate anaerobically. These mutants were simultaneously cured of a 17.4 kb resident plasmid (pCB1) that can be transferred by conjugation to *P. aeruginosa* and *P. stutzeri,* neither strain of which is able to use benzoate aerobically (Blake and Hegeman, 1987). After acquisition of pCB1, however, both organisms are able to metabolize benzoate anaerobically.

Williams and Evans (1973, 1975) also isolated a bacterium able to mineralize benzoate during nitrate reduction. It was initially identified as *P. stutzeri* but later classified as a *Moraxella* sp. It is now believed to be a strain of *Paracoccus denitrificans* (see Evans and Fuchs, 1988). When grown aerobically, this organism decarboxylates benzoate to catechol and subsequently metabolizes it by the *ortho* pathway (see Fig. 8.1). When cells are grown anaerobically on benzoate, however, they are devoid of any oxygenase activity. In early experiments, *P. stutzeri* transformed both [*ring*-^{14}C]benzoate and [*carboxy*-^{14}C]benzoate to *trans*-2-hydroxycyclohexanecarboxylate, which suggested that degradation proceeded via a reductive pathway. Evidence supporting this novel route of aromatic metabolism was subsequently obtained by isolation and identification of additional metabolic intermediates (Williams and Evans, 1975). Their proposed pathway is summarized in Figure 8.4. It is noteworthy, however, that *Paracoccus denitrificans* (NCIB No. 11085) has apparently lost its ability to grow on benzoate (Evans and Young, unpublished data).

Lovley et al. (1989) were the first to document the microbial oxidation of aromatic compounds under iron-reducing conditions. When anaerobic subsurface sediments located downstream from a crude oil pipeline break were amended with

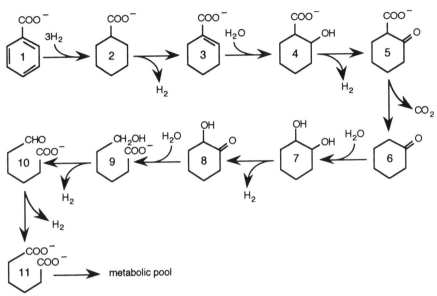

Figure 8.4. Proposed pathway for the degradation of benzoate during nitrate reduction. 1, Benzoate; 2, cyclohexanecarboxylate; 3, cyclohex-1-enecarboxylate; 4, 2-hydroxycyclo-hexane-carboxylate; 5, 2-oxocyclohexane; 6, cyclohexanone; 7, 1,2-dihydroxycyclohexane; 8, 2-hydroxycyclohexanone; 9, 6-hydroxyhexonate; 10, adipate semialdehyde; 11, adipate. (Excerpted from an original drawing by Williams and Evans, 1975.)

radiolabeled benzoate, $^{14}CO_2$ was recovered, but only when biological activity in the sediment samples was uninhibited, suggesting that the observed transformation was due to a microbiological mechanism.

Lovley and his colleagues described the isolation of a dissimilatory iron-reducing bacterium in 1988, originally designated GS-15 (Lovley and Phillips, 1988) and recently named *Geobacter metallireducens* (Lovley et al., 1993), that was subsequently found able to use benzoate as a carbon source with Fe(III) serving as the sole electron acceptor. Growth was associated with metabolism of benzoate and the concomitant reduction of Fe(III) to Fe(II) (Lovley et al., 1989). Benzoate mineralization agreed with the following stoichiometry:

$$C_6H_5COO^- + 30Fe^{3+} + 19H_2O \rightarrow 30Fe^{2+} + 7HCO_3^- + 36H^+$$

G. metallireducens was also shown to use the native Fe(III) in the crude oil–contaminated sediments as an electron acceptor during benzoate metabolism. The pathway of benzoate degradation under iron-reducing conditions has not yet been elucidated.

Widdel (1980) provided the first conclusive evidence that sulfate-reducing bacteria are able to use aromatic compounds as sole sources of carbon and electrons. At

TABLE 8.2.   Some Substituted Monoaromatic Hydrocarbons That Are Amenable to Microbial Degradation Under Nitrate-, Iron-, and Sulfate-Reducing Conditions

Compound	Conditions	References
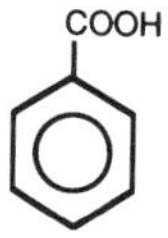 benzoic acid	Nitrate reduction	Taylor et al. (1970), Taylor and Heeb (1972), Williams and Evans (1973, 1975), Aftring and Taylor (1981), Schennen et al. (1985), Dolfing et al. (1990)
	Iron reduction	Lovley et al. (1989)
	Sulfate reduction	Balba and Evans (1980), Widdel et al. (1983), Cord-Ruwisch and Garcia (1985), Gibson and Suflita (1986), Szewzyk and Pfennig (1987), Schnell et al. (1989), Sharak Genthner et al. (1989)
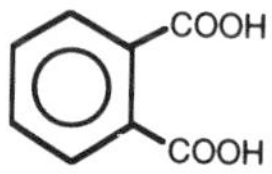 o-phthalic acid	Nitrate reduction	Taylor and Ribbons (1983), Aftring and Taylor (1981), Aftring et al. (1981), Nozawa and Maruyama (1988a,b)
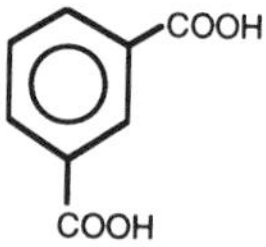 m-phthalic acid	Nitrate reduction	Aftring and Taylor (1981), Aftring et al. (1981), Nozawa and Maruyama (1988a,b)
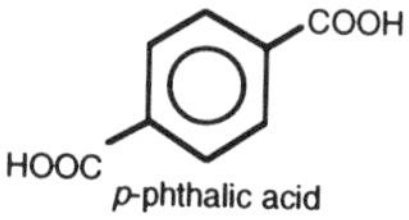 p-phthalic acid	Nitrate reduction	Aftring and Taylor (1981), Aftring et al. (1981), Nozawa and Maruyama (1988a,b)
 phenol	Nitrate reduction	Bakker (1977), Ehrlich et al. (1983), Bossert et al. (1986), Hu and Shieh (1987), Tschech and Fuchs (1987), Glöckler et al. (1989), Rudolphi et al. (1991), Khoury et al. (1992a), Flyvbjerg et al. (1993)
	Iron reduction	Lovley and Lonergan (1990)
	Sulfate reduction	Bak and Widdel (1986), Gibson and Suflita (1986), King (1988), Schnell et al. (1989), Sharak Genthner et al. (1989), Flyvbjerg et al. (1993)
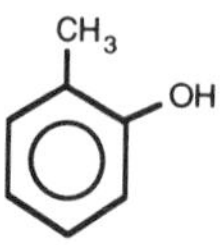 o-cresol	Nitrate reduction	Rudolphi et al. (1991), Flyvbjerg et al. (1993)
	Sulfate reduction	Suflita et al. (1989), Flyvbjerg et al. (1993)

(continued)

TABLE 8.2. *(Continued)*

Compound	Conditions	References
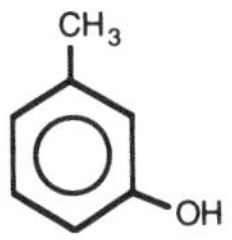 m-cresol (CH₃, OH)	Nitrate reduction	Rudolphi et al. (1991), Flyvbjerg et al. (1993)
	Sulfate reduction	Suflita et al. (1989), Ramanand and Suflita (1991), Flyvbjerg et al. (1993)
 p-cresol (CH₃, OH)	Nitrate reduction	Bakker (1977), Bossert and Young (1986), Bossert et al. (1986, 1989), Tschech and Fuchs (1987), Dolfing et al. (1990), Häggblom et al. (1990), Rudolphi et al. (1991), Khoury et al. (1992b), Flyvbjerg et al. (1993)
	Iron reduction	Lovley and Lonergan (1990)
	Sulfate reduction	Bak and Widdel (1986), Smolenski and Suflita (1987), Schnell et al. (1989), Suflita et al. (1989), Häggblom et al. (1990)

least six pure cultures of sulfidogens isolated from a variety of natural habitats have been described that are able to metabolize benzoate for growth (Widdel et al., 1983; Cord-Ruwisch and Garcia, 1985; Szewzyk and Pfennig, 1987; Schnell et al., 1989). Mixed consortia that mineralize benzoate under sulfate-reducing conditions have also been reported (refer to studies listed in Table 8.2). To our knowledge, pathways for the sulfate-mediated metabolism of benzoate have not yet been described.

7. PHTHALIC ACIDS

Mineralization of phthalate might be viewed as analogous to that of benzoate once the second carboxylic acid substituent is removed. During aerobic phthalate degradation, decarboxylation occurs only after at least one hydroxyl group has been introduced in the position *para* to the –COOH group undergoing elimination (Keyser et al., 1976). Investigations of anaerobic phthalate degradation by mixed microbial consortia (Aftring et al., 1981) and by pure cultures of nitrate-reducing bacteria (Aftring and Taylor, 1981) indicate, however, that hydroxylation reactions are not involved and that benzoate is a metabolic intermediate. Taylor and Ribbons (1983) have suggested that decarboxylation of *o*-phthalic acid involves initial reduction to 1,2-dihydrophthalic acid (3,5-cyclohexadiene-1,2-dicarboxylic acid) followed by oxidative decarboxylation to benzoic acid as shown in Figure 8.5.

Nozawa and Maruyama (1988a,b) have proposed a more comprehensive pathway for phthalate degradation under denitrifying conditions based on their studies with a *Pseudomonas* sp. that was isolated from garden soil. Benzoate, cyclohex-1-enecarboxylate, 2-hydroxycyclohexanecarboxylate, and pimelate were found to be the predominant metabolic intermediates during growth on *o-*, *m-*, and *p*-phthalic

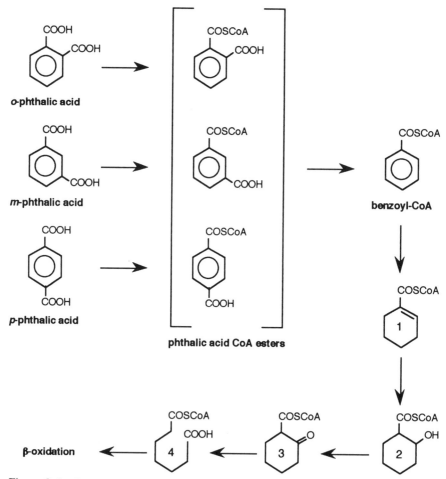

Figure 8.5. Suggested mechanism for the transformation of *o*-phthalic acid (1) to benzoic acid (3) via an initial reduction to 1,2-dihydrophthalic acid (2). (Based on an original drawing by Taylor and Ribbons, 1983.)

Figure 8.6. Pathway for the metabolism of phthalate isomers under denitrifying conditions by *Pseudomonas* sp. strain P136 as proposed by Nozawa and Maruyama (1988b). The numbered compounds are CoA esters of cyclohex-1-enecarboxylate (1), 2-hydroxycyclohex-anecarboxylate (2), 2-oxocyclohexane-carboxylate (3), and pimelic acid (4).

acids. Inducible acyl-coenzyme A synthetase activities were also detected in cells grown on the phthalate isomers benzoate, cyclohex-1-enecarboxylate, and cyclohex-3-enecarboxylate. On the basis of these findings, they hypothesized that the initial step in phthalate catabolism under nitrate-reducing conditions is the formation of CoA esters, which are subsequently decarboxylated to benzoyl-CoA. Their proposed pathway is shown in Figure 8.6.

8. PHENOL

One of the earliest reports of anaerobic phenol degradation was by a mixed culture that mineralized phenol to CO_2 under denitrifying conditions (Bakker, 1977). Even though little experimental evidence was provided, Bakker (1977) proposed that phenol was reduced to cyclohexanol, then dehydrogenated to cyclohexanone with subsequent hydrolytic cleavage of the aromatic ring. Tschech and Fuchs (1987, 1989) and Glöckler et al. (1989) have studied several pure cultures of pseudomonads able to grow on phenol during nitrate reduction, but their isolates neither form cyclohexanol during anaerobic growth on phenol nor degrade cyclohexanol. They have demonstrated instead that phenol is first carboxylated to 4-hydroxybenzoate, followed by coenzyme A activation and reductive elimination of the hydroxyl group to form benzoyl-CoA (see Fig. 8.7). Other reports of phenol degradation coupled with the reduction of NO_3^- are listed in Table 8.2.

Lovley and Lonergan (1990) have described the oxidation of phenol to CO_2 by *G. metallireducens,* the dissimilatory iron reducer discussed previously. *p*-Hydroxybenzoate is the only metabolic intermediate detected during growth on phenol, suggesting that the first step in phenol mineralization is carboxylation of the aromatic ring, which is consistent with the mechanism proposed for nitrate-respiring bacteria (Tschech and Fuchs, 1987, 1989).

Although there are several reports of phenol transformation under sulfate-reducing conditions (see Table 8.2), only Bak and Widdel (1986) have described a pure culture—*Desulfobacterium phenolicum*—that oxidizes phenol to CO_2 using SO_4^{2-} as terminal electron acceptor. Pathways for the anaerobic metabolism of phenol via sulfate respiration have not yet been published.

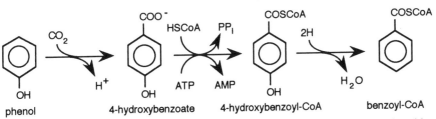

Figure 8.7. Degradation of phenol via initial carboxylation to 4-hydroxybenzoic acid as proposed by Tschech and Fuchs (1987). Subsequent steps in the pathway are based on additional studies with denitrifying *Pseudomonas* sp. strain K 172 (Glöckler et al., 1989).

9. CRESOLS

Bakker (1977) was the first to report cresol degradation under anaerobic conditions when he observed that his phenol-utilizing, nitrate-reducing mixed culture was also able to mineralize p-cresol (4-methylphenol). Since then a number of studies have focused on the anaerobic degradation of o-, m-, and p-cresols during denitrification (see, e.g., Häggblom et al., 1990; Khoury et al., 1992b, Flyvbjerg et al., 1993), during Fe(III) reduction (Lovley and Lonergan, 1990) and under sulfate-reducing conditions (see, e.g., Smolenski and Suflita, 1987; Suflita et al., 1989; Häggblom et al., 1990; Ramanand and Suflita, 1991). Significant progress has also been made in elucidating metabolic pathways and in measuring *in vitro* enzyme activities, especially during nitrate reduction (Bossert et al., 1986, 1989; Rudolphi et al., 1991).

Using polluted river sediment for primary enrichment, Bossert et al. (1986) isolated two bacterial species that used p-cresol as a sole carbon source when grown in coculture under nitrate-reducing conditions. One species transformed the p-cresol to p-hydroxybenzoate (pOHB); the second species metabolized the pOHB. Nitrate was required for transformation of both substrates and was reduced to N_2 via NO_2^- and N_2O.

Subsequent studies with their p-cresol–transforming pure culture (Bossert and Young, 1986) suggested that it oxidizes p-cresol to p-hydroxybenzylalcohol (pHB-zalc) by the same oxygen-independent reaction used by bacteria growing under aerobic conditions (Hopper, 1978). (The oxygen has been shown to originate from water that is added to a quinone methide intermediate [see Fig. 8.8]). The pHBzalc then undergoes hydrogenation resulting in p-hydroxybenzaldehyde (pHBzald). A p-cresol methylhydroxylase has been partially purified that catalyzes oxidation of p-cresol to pHBzald, while conversion of pHBzald to pOHB has been shown to be mediated by an NAD^+-dependent dehydrogenase (Bossert et al., 1989). The pathway proposed by Bossert and Young (1986) for p-cresol transformation under denitrifying conditions is shown in Figure 8.8.

Rudolphi and her colleagues (1991) have studied two bacteria able to mineralize cresols during nitrate reduction: a *Paracoccus* sp. that degrades both o- and p-cresols to CO_2 and a *Pseudomonas*-like strain that completely oxidizes the m- and p-isomers. Two of their proposed pathways are summarized in Figure 8.9.

The o-cresol pathway proceeds via 3-methyl-benzoyl-CoA as a central intermediate. o-Cresol initially undergoes a carboxylation reaction mediated by an o-cresol carboxylase that results in the formation of 4-hydroxy-3-methylbenzoic acid, which is subsequently activated to its coenzyme A thioester by 4-hydroxy-3-methylbenzoyl-CoA synthetase. 4-Hydroxy-3-methylbenzoyl-CoA reductase then catalyzes the reductive dehydroxylation of the p-hydroxyl group to yield 3-methylbenzoyl-CoA.

Oxidation of p-cresol by both denitrifying strains proceeds via benzoyl-CoA as a central intermediate. Oxidation of the methyl substituent to pOHB is catalyzed by p-cresol methylhydroxylase. Following oxidation of the aldehyde to pOHB, p-hydroxybenzoyl-CoA is formed by p-hydroxybenzyl-CoA synthetase. Reductive

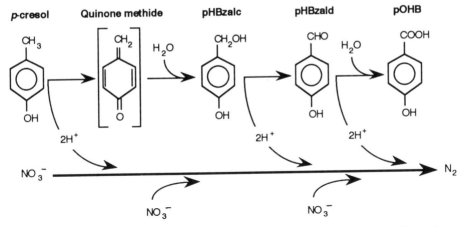

Figure 8.8. Proposed pathway for the anaerobic transformation of p-cresol to p-hydroxybenzoic acid by a denitrifying *Achromobacter* sp. as mediated by p-cresol methylhydroxylase (p-cresol → pHBzald) and benzaldehyde dehydrogenase (pHBzald → pOHB). pHBzalc, p-hydroxybenzylalcohol; pHBzald, p-hydroxybenzaldehyde; pOHB, p-hydroxybenzoic acid. (Adapted from an original drawing by Bossert and Young, 1986.)

dehydroxylation of p-hydroxylbenzoyl-CoA to benzoyl-CoA is mediated by p-hydroxybenzoyl-CoA reductase.

Rudolphi et al. (1991) were unable to elucidate the initial reactions involved in the nitrate-mediated oxidation of the m-isomer of cresol. They did not observe any evidence of oxidation or carboxylation of the methyl group, even though 2,4- and

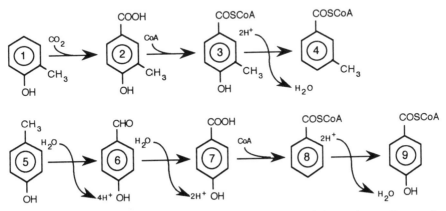

Figure 8.9. Initial steps (1–4) in the proposed anaerobic degradation pathway of o-cresol as mediated by a denitrifying *Paracoccus* sp. Proposed initial steps (5–9) during mineralization of p-cresol by denitrifying *Pseudomonas* and *Paracoccus* spp. 1, o-Cresol; 2, 4-hydroxy-3-methylbenzoic acid; 3, 4-hydroxy-3-methylbenzoyl-CoA; 4, 3-methylbenzoyl-CoA; 5, p-cresol; 6, p-hydroxybenzaldehyde; 7, p-hydroxybenzoic acid; 8, p-hydroxybenzoyl-CoA; 9, benzoyl-CoA. (Adapted from Rudolphi et al., 1991.)

3,4- dimethylphenols were apparently hydroxylated to corresponding hydroxyme-thylbenzoates.

Lovley and Lonergan (1990) have published the only report of methylphenol degradation during Fe(III) reduction. p-Cresol was completely oxidized to CO_2 by *G. metallireducens,* and p-hydroxybenzoate was detected in the culture fluid during growth. In addition, two potential metabolic intermediates (pHBzalc and pHBzald) were oxidized by this organism.

Except for the testing of a few pure cultures for their ability to utilize meth-ylphenols as substrates (e.g., Bak and Widdel, 1986; Schnell et al., 1989), most of the work documenting cresol degradation under sulfate-reducing conditions has involved laboratory microcosms of one type or another using materials obtained from various environmental sources (e.g., sediment from a shallow anoxic aquifer underlying a municipal landfill [Smolenski and Suflita, 1987; Suflita et al., 1989], freshwater pond sediments [(Häggblom et al., 1990], and anaerobic ground water from a creosote-contaminated site [Flyvbjerg et al., 1993]).

General agreement has been reached that under sulfate-reducing conditions p-cresol degradation proceeds via initial oxidation of the methyl group to pHBzald and pOHB (see Häggblom et al., 1990). Suflita et al. (1989) have shown that o-cresol is anaerobically oxidized to o-hydroxybenzoate by an analogous reaction. It is clear, however, that m-cresol oxidation is initiated by a very different mecha-nism, one that Ramanand and Suflita (1991) suggest involves carboxylation of the parent substrate. They enriched for a bacterial consortium from anoxic aquifer slurries that was able to metabolize m-cresol under sulfate-reducing conditions. 4-Hydroxy-2-methylbenzoic acid and acetate were detected as transient intermedi-ates, suggesting that m-cresol had undergone carboxylation. The presence of pOHB in the culture medium temporarily inhibited m-cresol metabolism such that pOHB was oxidized preferentially. Their findings are in agreement with those of Roberts et al. (1987, 1990), who have proposed the only anaerobic (methanogenic) pathway of m-cresol degradation.

10. CONCLUDING REMARKS

The microbially mediated degradation of toluene under denitrifying conditions was first documented in aquifer column studies published in 1986 (Zeyer et al., 1986). Degradation of aromatic contaminants coupled with Fe(III) reduction was unknown until a few years ago (Lovley et al., 1989). The first description of an aromatic compound being used by a sulfate reducer was the subject of a 1980 doctoral dissertation (Widdel, 1980). In fact, most of the papers cited in this review have been published within the last decade.

Since many soils and subsurface sites that are contaminated with hazardous compounds are anaerobic, have become anoxic, or have the potential to become oxygen-limited, microbial transformations that occur under low redox conditions are of great interest. Much of the work discussed here has no doubt been initiated, at least in part, with an eye toward bioremediation.

Despite significant progress made in documenting biotransformations of numerous nonhalogenated homocyclic aromatic compounds, careful physiological and enzymatic studies remain to be done on most of the organisms described in this chapter, while investigations involving mixed consortia are perhaps our best models of indigenous microbial life. The list of organic compounds that may be amenable to bioremediation is limited by what is known about their microbial transformation. The key to developing successful remedial strategies clearly depends on improving our understanding of microbial processes operating in anaerobic regimes.

ACKNOWLEDGMENTS

The library assistance of Stephanie Schroeder and expert technical editing by Linda D. Mickley are greatly appreciated.

REFERENCES

Aftring RP, Chalker BE, Taylor BF (1981): Degradation of phthalic acids by denitrifying, mixed cultures of bacteria. Appl Environ Microbiol 41:1177–1183.

Aftring RP, Taylor BF (1981): Aerobic and anaerobic catabolism of phthalic acid by a nitrate-respiring bacterium. Arch Microbiol 130:101–104.

Altenschmidt U, Fuchs G (1991): Anaerobic degradation of toluene in denitrifying *Pseudomonas* sp.: Indication for toluene methylhydroxylation and benzoyl-CoA as central intermediate. Arch Microbiol 156:152–158.

Altenschmidt U, Fuchs G (1992): Anaerobic toluene oxidation to benzyl alcohol and benzaldehyde in a denitrifying *Pseudomonas* strain. J Bacteriol 174:4860–4862.

Bak F, Widdel F (1986): Anaerobic degradation of phenol and phenol derivatives by *Desulfobacterium phenolicum* sp. nov. Arch Microbiol 146:177–180.

Bakker G (1977): Anaerobic degradation of aromatic compounds in the presence of nitrate. FEMS Microbiol Lett 1:103–108.

Balba MT, Evans WC (1980): The anaerobic dissimilation of benzoate by *Pseudomonas aeruginosa* coupled with *Desulfovibrio vulgaris* with sulphate as terminal electron acceptor. Biochem Soc Trans 8:624–627.

Ball HA, Reinhard M, McCarty PL (1991): Biotransformation of monoaromatic hydrocarbons under anoxic conditions. In Hinchee RE, Olfenbuttel RF (eds): *In Situ* Bioreclamation. Stoneham, MA: Butterworth-Heinemann, pp 458–463.

Barbaro JR, Barker JF, Lemon LA, Mayfield CI (1992): Biotransformation of BTEX under anaerobic, denitrifying conditions: Field and laboratory observations. J Contam Hydrol 11:245–272.

Beller HR, Grbić-Galić D, Reinhard M (1992a): Microbial degradation of toluene under sulfate-reducing conditions and the influence of iron on the process. Appl Environ Microbiol 58:786–793.

Beller HR, Reinhard M, Grbić-Galić D (1992b): Metabolic by-products of anaerobic toluene degradation by sulfate-reducing enrichment cultures. Appl Environ Microbiol 58:3192–3195.

Berry DF, Francis AJ, Bollag J-M (1987): Microbial metabolism of homocyclic and heterocyclic aromatic compounds under anaerobic conditions. Microbiol Rev 51:43–59.

Blake CK, Hegeman GC (1987): Plasmid pCB1 carries genes for anaerobic benzoate catabolism in *Alcaligenes xylosoxidans* subsp. *denitrificans* PN-1. J Bacteriol 169:4878–4883.

Bollag J-M, Kaiser J-P (1991): The transformation of heterocyclic aromatic compounds and their derivatives under anaerobic conditions. Crit Rev Environ Control 21:297–329.

Bossert ID, Rivera MD, Young LY (1986): *p*-Cresol biodegradation under denitrifying conditions: Isolation of a bacterial coculture. FEMS Microbiol Ecol 38:313–319.

Bossert ID, Young LY (1986): Anaerobic oxidation of *p*-cresol by a denitrifying bacterium. Appl Environ Microbiol 52:1117–1122.

Bossert ID, Whited G, Gibson DT, Young LY (1989): Anaerobic oxidation of *p*-cresol mediated by a partially purified methylhydroxylase from a denitrifying bacterium. J Bacteriol 171:2956–2962.

Cord-Ruwisch R, Garcia JL (1985): Isolation and characterization of an anaerobic benzoate-degrading spore-forming sulfate-reducing bacterium, *Desulfotomaculum sapomandes* sp. nov. FEMS Microbiol Lett 29:325–330.

Dangel W, Brackman R, Lack A, Mohamed M, Koch J, Oswald B, Seyfried B, Tschech A, Fuchs G (1991): Differential expression of enzyme activities initiating anoxic metabolism of various aromatic compounds via benzoyl-CoA. Arch Microbiol 155:256–262.

Dolfing J, Zeyer J, Binder-Eicher P, Schwarzenbach RP (1990): Isolation and characterization of a bacterium that mineralizes toluene in the absence of oxygen. Arch Microbiol 154:336–341.

Edwards EA, Grbić-Galić D (1992): Complete mineralization of benzene by aquifer microorganisms under strictly anaerobic conditions. Appl Environ Microbiol 58:2663–2666.

Edwards EA, Wills LE, Reinhard M, Grbić-Galić D (1992): Anaerobic degradation of toluene and xylene by aquifer microorganisms under sulfate-reducing conditions. Appl Environ Microbiol 58:794–800.

Ehrlich GG, Godsy EM, Goerlitz DF, Hult MF (1983): Microbial ecology of a creosote-contaminated aquifer at St. Louis Park, Minnesota. Dev Ind Microbiol 24:235–245.

Evans WC, Fuchs G (1988): Anaerobic degradation of aromatic compounds. Annu Rev Microbiol 42:289–317.

Evans PJ, Ling W, Goldschmidt B, Ritter ER, Young LY (1992): Metabolites formed during anaerobic transformation of toluene and *o*-xylene and their proposed relationship to the initial steps of toluene mineralization. Appl Environ Microbiol 58:496–501.

Evans PJ, Mang DT, Kim KS, Young LY (1991a): Anaerobic degradation of toluene by a denitrifying bacterium. Appl Environ Microbiol 57:1139–1145.

Evans PJ, Mang DT, Young LY (1991b): Anaerobic degradation of toluene and *m*-xylene and transformation of *o*-xylene by denitrifying enrichment cultures. Appl Environ Microbiol 57:450–454.

Flyvbjerg J, Arvin E, Jensen BK, Olsen SK (1993): Microbial degradation of phenols and aromatic hydrocarbons in creosote-contaminated groundwater under nitrate-reducing conditions. J Contam Hydrol 12:133–150.

Frazer AC, Ling W, Young LY (1993): Substrate induction and metabolic accumulation during anaerobic toluene utilization by the denitrifying strain T1. Appl Environ Microbiol 59:3157–3160.

Geissler JF, Harwood CS, Gibson J (1988): Purification and properties of benzoate-coenzyme A ligase, a *Rhodopseudomonas palustris* enzyme involved in the anaerobic degradation of benzoate. J Bacteriol 170:1709–1714.

Gibson SA, Suflita JM (1986): Extrapolation of biodegradation results to groundwater aquifers: Reductive dehalogenation of aromatic compounds. Appl Environ Microbiol 52:681–688.

Glöckler R, Tschech A, Fuchs G (1989): Reductive dehydroxylation of 4-hydroxybenzoyl-CoA to benzoyl-CoA in a denitrifying, phenol degrading *Pseudomonas* species. FEBS Microbiol Lett 251:237–240.

Grbić-Galić D (1989): Microbial degradation of homocyclic and heterocyclic aromatic hydrocarbons under anaerobic conditions. Dev Ind Microbiol 30:237–253.

Grbić-Galić D (1990a): Anaerobic microbial transformation of nonoxygenated aromatic and acyclic compounds in soil, subsurface, and freshwater sediments. In Bollag J-M, Stotzky G (eds): Soil Biochemistry. New York: Marcel Dekker, vol 6, pp 117–189.

Grbić-Galić D (1990b): Methanogenic transformation of aromatic hydrocarbons and phenols in groundwater aquifers. Geomicrobiol J 8:167–200.

Grbić-Galić D, Vogel TM (1987): Transformation of toluene and benzene by mixed methanogenic cultures. Appl Environ Microbiol 53:254–260.

Haag F, Reinhard M, McCarty PL (1991): Degradation of toluene and *p*-xylene in anaerobic microcosms: Evidence for sulfate as a terminal electron acceptor. Environ Tox Chem 10:1379–1389.

Häggblom MM, Rivera MD, Bossert ID, Rogers JE, Young LY (1990): Anaerobic biodegradation of *p*-cresol under three reducing conditions. Microbial Ecol 20:141–150.

Harwood CS, Gibson J (1986): Uptake of benzoate by *Rhodopseudomonas palustris* grown in light. J Bacteriol 165:504–509.

Harwood CS, Gibson J (1988): Anaerobic and aerobic metabolism of diverse aromatic compounds by the photosynthetic bacterium *Rhodopseudomonas palustris*. Appl Environ Microbiol 54:712–717.

Hopper DJ (1978): Incorporation of [^{18}O]water in the formation of *p*-hydroxybenzyl alcohol by the *p*-cresol methylhydroxylase from *Pseudomonas putida*. Biochem J 175:345–347.

Hu LZ, Shieh WK (1987): Anoxic biofilm degradation of monocyclic aromatic compounds. Biotechnol Bioeng 30:1077–1083.

Hutchins SR (1991): Biodegradation of monoaromatic hydrocarbons by aquifer microorganisms using oxygen, nitrate, or nitrous oxide as the terminal electron acceptor. Appl Environ Microbiol 57:2403–2407.

Hutchins SR, Sewell GW, Kovacs DA, Smith GA (1991): Biodegradation of aromatic hydrocarbons by aquifer microorganisms under denitrifying conditions. Environ Sci Technol 25:68–76.

Hutchins SR, Wilson JT (1991): Laboratory and field studies on BTEX biodegradation in a fuel-contaminated aquifer under denitrifying conditions. In Hinchee RE, Olfenbuttel RF (eds): *In Situ* Bioreclamation. Stoneham, MA: Butterworth-Heinemann, pp 157–172.

Jørgensen C, Mortensen E, Jensen BK, Arvin E (1991): Biodegradation of toluene by a denitrifying enrichment culture. In Hinchee RE, Olfenbuttel RF (eds): *In Situ* Bioreclamation. Stoneham, MA: Butterworth-Heinemann, pp 480–487.

Keyser P, Pujar BG, Eaton RW, Ribbons DW (1976): Biodegradation of the phthalates and their esters by bacteria. Environ Health Perspect 18:159–166.

Khanna P, Rajkumar B, Jothikumar N (1992): Anoxygenic degradation of aromatic substances by *Rhodopseudomonas palustris*. Curr Microbiol 25:63–67.

King GM (1988): Dehalogenation in marine sediments containing natural sources of halophenols. Appl Environ Microbiol 54:3079–3085.

Khoury N, Dott W, Kampfer P (1992a): Anaerobic degradation of phenol in batch and continuous cultures by a denitrifying bacterial consortium. Appl Microbiol Biotechnol 37:524–528.

Khoury N, Dott W, Kampfer P (1992b): Anaerobic degradation of *p*-cresol in batch and continuous cultures by a denitrifying bacterial consortium. Appl Microbiol Biotechnol 37:529–534.

Kuhn EP, Colberg PJ, Schnoor JL, Wanner O, Zehnder AJB, Schwarzenbach RP (1985): Microbial transformation of substituted benzenes during infiltration of river water to groundwater: Laboratory column studies. Environ Sci Technol 19:961–968.

Kuhn EP, Zeyer J, Eicher P, Schwarzenbach RP (1988): Anaerobic degradation of alkylated benzenes in denitrifying laboratory aquifer columns. Appl Environ Microbiol 54:490–496.

Lovley DR, Baedecker MJ, Lonergan DJ, Cozzarelli IM, Phillips EJP, Siegel DI (1989): Oxidation of aromatic contaminants coupled to microbial iron reduction. Nature 329:297–299.

Lovley DR, Giovannoni SJ, White DC, Champine JE, Phillips EJP, Gorby YA, Goodwin S (1993): *Geobacter metallireducens* gen. nov. sp. nov., a microorganism capable of coupling the complete oxidation of organic compounds to the reduction of iron and other metals. Arch Microbiol 159:336–344.

Lovley DR, Lonergan DJ (1990): Anaerobic oxidation of toluene, phenol, and *p*-cresol by the dissimilatory iron-reducing organism, GS-15. Appl Environ Microbiol 56:1858–1864.

Lovley DR, Phillips EJP (1988): Novel mode of microbial energy metabolism: Organic carbon oxidation coupled to dissimilatory reduction of iron or manganese. Appl Environ Microbiol 54:1472–1480.

Major DW, Mayfield CI, Barker JF (1988): Biotransformation of benzene by denitrification in aquifer sand. Ground Water 26:8–14.

Nozawa T, Maruyama Y (1988a): Denitrification by a soil bacterium with phthalate and other aromatic compounds as substrates. J Bacteriol 170:2501–2505.

Nozawa T, Maruyama Y (1988b): Anaerobic metabolism of phthalate and other aromatic compounds by a denitrifying bacterium. J Bacteriol 170:5778–5784.

Rabus R, Nordhaus R, Ludwig W, Widdel F (1993): Complete oxidation of toluene under strictly anoxic conditions by a new sulfate-reducing bacterium. Appl Environ Microbiol 59:1444–1451.

Rahalkar SB, Joshi SR, Shivaraman N (1993): Photometabolism of aromatic compounds by *Rhodopseudomonas palustris*. Curr Microbiol 26:1–9.

Ramanand K, Suflita JM (1991): Anaerobic degradation of *m*-cresol in anoxic aquifer slurries: Carboxylation reactions in a sulfate-reducing bacterial enrichment. Appl Environ Microbiol 57:1689–1695.

Reinhard M, Wills LE, Ball HA, Harmon T, Phipps DW, Ridgeway HF, Eisman MP (1991): A field experiment for the anaerobic biotransformation of aromatic hydrocarbon compounds at Seal Beach, California. In Hinchee RE, Olfenbuttel RF (eds): *In Situ* Bioreclamation. Stoneham, MA: Butterworth-Heinemann, pp 487–496.

Roberts DJ, Fedorak PM, Hrudey SE (1987): Comparison of the fates of the methyl carbons of *m*-cresol and *p*-cresol in methanogenic consortia. Can J Microbiol 33:335–338.

Roberts DJ, Fedorak PM, Hrudey SE (1990): CO_2 incorporation and 4-hydroxy-2-methylbenzoic acid formation during anaerobic metabolism of *m*-cresol by a methanogenic consortium. Appl Environ Microbiol 56:472–478.

Rudolphi A, Tschech A, Fuchs G (1991): Anaerobic degradation of cresols by denitrifying bacteria. Arch Microbiol 155:238–248.

Sharak Genthner BR, Price WA II, Pritchard PH (1989): Anaerobic degradation of chloroaromatic compounds under a variety of enrichment conditions. Appl Environ Microbiol 55:1466–1471.

Schennen U, Braun K, Knackmuss H-J (1985): Anaerobic degradation of 2-fluorobenzoate by benzoate-degrading, denitrifying bacteria. J Bact 161:321–325.

Schink B, Brune A, Schnell S (1992): Anaerobic degradation of aromatic compounds. In Winkelman G (ed): Microbial Degradation of Natural Products. Weinheim: Verlag Chemie, pp 219–242.

Schnell S, Bak F, Pfennig N (1989): Anaerobic degradation of aniline and dihydroxybenzenes by newly isolated sulfate-reducing bacteria and description of *Desulfobacterium anilini*. Arch Microbiol 152:556–563.

Schnell S, Schink B (1991): Anaerobic aniline degradation via reductive deamination of 4-aminobenzoyl-CoA in *Desulfobacterium anilini*. Arch Microbiol 155:183–190.

Schocher RJ, Seyfried B, Vazquez F, Zeyer J (1991): Anaerobic degradation of toluene by pure cultures of denitrifying bacteria. Arch Microbiol 157:7–12.

Schwarzenbach RP, Giger W, Hoehn E, Schneider JK (1983): Behavior of organic compounds during infiltration of river water to groundwater. Field studies. Environ Sci Technol 17:472–479.

Smolenski WJ, Suflita JM (1987): Biodegradation of cresol isomers in anoxic aquifers. Appl Environ Microbiol 53:710–716.

Suflita JM, Liang L, Saxena A (1989): The anaerobic biodegradation of *o*-, *m*-, and *p*-cresol by sulfate-reducing enrichment cultures obtained from a shallow anoxic aquifer. J Ind Microbiol 4:255–266.

Szewzyk R, Pfennig N (1987): Complete oxidation of catechol by the strictly anaerobic sulfate-reducing *Desulfobacterium catecholicum* sp. nov. Arch Microbiol 147:163–168.

Tarvin D, Buswell AM (1934): The methane fermentation of organic acids and carbohydrates. J Am Chem Soc 56:1751–1755.

Taylor BF, Campbell WL, Chinoy I (1970): Anaerobic degradation of the benzene nucleus by a facultatively anaerobic microorganism. J Bacteriol 102:430–437.

Taylor BF, Heeb MJ (1972): The anaerobic degradation of aromatic compounds by a denitrifying bacterium. Arch Microbiol 83:165–171.

Taylor BF, Ribbons DW (1983): Bacterial decarboxylation of *o*-phthalic acids. Appl Environ Microbiol 46:1276–1281.

Tschech A, Fuchs G (1987): Anaerobic degradation of phenol by pure cultures of newly isolated denitrifying pseudomonads. Arch Microbiol 148:213–217.

Tschech A, Fuchs G (1989): Anaerobic degradation of phenol via carboxylation to 4-hydroxybenzoate: *In vitro* study of isotope exchange between $^{14}CO_2$ and 4-hydroxybenzoate. Arch Microbiol 152:594–599.

Vogel TM, Grbić-Galić (1986): Incorporation of oxygen from water into toluene and benzene during anaerobic fermentative transformation. Appl Environ Microbiol 52:200–202.

Widdel F (1980): Anaerober Abbau von Fettsäuren and Benzosäure durch neu isolierte Arten sulfat-reduzierender Bakterien. [Anaerobic Degradation of Fatty Acids and Benzoic Acid by Newly Isolated Types of Sulfate-Reducing Bacteria]. Doctoral Dissertation, Georg-August-Universität, Göttingen.

Widdel F, Kohring GW, Mayer F (1983): Studies on dissimilatory sulfate-reducing bacteria that decompose fatty acids. II. Characterization of the filamentous gliding *Desulfonema limicola* gen. nov., sp. nov., and *Desulfonema magnum* sp. nov. Arch Microbiol 134:286–294.

Williams RJ, Evans WC (1973): Anaerobic metabolism of aromatic substrates by certain microorganisms. Biochem Soc Trans 1:186–187.

Williams RJ, Evans, WC (1975): The metabolism of benzoate by *Moraxella* sp. through anaerobic nitrate respiration. Biochem J 148:1–10.

Zeyer J, Kuhn EP, Schwarzenbach RP (1986): Rapid microbial mineralization of toluene and 1,3-dimethylbenzene in the absence of molecular oxygen. Appl Environ Microbiol 52:944–947.

9

ORGANOPOLLUTANT DEGRADATION BY LIGNINOLYTIC FUNGI

KENNETH E. HAMMEL

Forest Products Laboratory, U.S. Department of Agriculture, Madison, Wisconsin 53705

1. INTRODUCTION

The ligninolytic fungi that cause white rot of wood degrade an impressive array of organopollutants (for earlier reviews, see Bumpus and Aust, 1987b; Hammel, 1989, 1992). As this chapter shows, the xenobiotic metabolism of white-rot fungi is at least in part a consequence of the mechanisms these organisms have to degrade lignin. Although observations of aromatic pollutant breakdown by white-rot fungi date back 30 years (Lyr, 1963), the connection of this process with ligninolytic metabolism was noted more recently, when several groups observed that the basidiomycete *Phanerochaete chrysosporium* degrades benzo[*a*]pyrene, DDT, chlorinated anilines, various alkylhalide insecticides, and a polychlorinated biphenyl mixture under culture conditions that promote the expression of ligninolytic metabolism (Arjmand and Sandermann, 1985; Bumpus et al., 1985; Eaton, 1985; Huynh et al., 1985; Bumpus and Aust, 1987a; Kennedy et al., 1990). In some cases, significant mineralization of the pollutants was obtained.

The ability of *P. chrysosporium* to degrade most organopollutants correlates closely with ligninolytic activity in culture, both processes depending on secondary (idiophasic) metabolism, which is triggered by nutrient starvation and the cessation

Microbial Transformation and Degradation of Toxic Organic Chemicals, pages 331–346
© *1995 Wiley-Liss, Inc.*

of cell growth. There are some, as yet unexplained, exceptions to this observation (Köhler et al., 1988), as well as many instances in which idiophasic metabolism is required but no direct involvement of ligninolytic enzymes has been shown (Bumpus et al., 1985; Eaton, 1985; Bumpus and Aust, 1987a; Kennedy et al., 1990). Here I focus only on those xenobiotic oxidations that show clear mechanistic connections with ligninolytic metabolism.

2. BIOCHEMICAL FEATURES OF FUNGAL LIGNINOLYSIS

Lignins are complex phenylpropane wood polymers, formed in plant cell walls via the coupling of *p*-hydroxycinnamyl alcohols, that make up a significant proportion (10%–30%) of the biomass in terrestrial higher plants. The function of lignin is to give these plants structural rigidity and to protect their cellulose and hemicelluloses from microbial attack. Lignins are amorphous, stereoirregular, water insoluble, nonhydrolyzable, and highly resistant to degradation by most organisms (Adler, 1977; Kirk and Farrell, 1987).

Rapid lignin degradation appears to be limited to white-rot basidiomycetes, which are inhabitants of dead wood and soil litter, and depend for their growth on the polysaccharides in these substrates (Kirk and Farrell, 1987; Gold et al., 1989). Ligninolysis *per se* does not support growth, but is rather intended to open up the structure of woody materials so that fungal polysaccharide-degrading agents can penetrate. The large size, irregular structure, and nonhydrolyzable nature of lignin dictate key features of the ligninolytic process: It is extracellular, highly non-specific, and oxidative. One review refers to fungal ligninolysis as "enzymatic combustion" to reflect these properties (Kirk and Farrell, 1987), and it is this low specificity that has led many investigators to suggest that organopollutants might be fortuitous targets of fungal ligninolytic metabolism (Bumpus et al., 1985; Haem-merli et al., 1986; Hammel et al., 1986a; Sanglard et al., 1986; Bumpus and Aust, 1987b).

Considerable progress has been made in our understanding of fungal lig-ninolysis. One of the principal reactions in this process is oxidative C_α–C_β cleavage of the lignin propyl side chain. For example, the principal arylglycerol-β-aryl ether structures of lignin, which comprise 50%–60% of the total polymer, are cleaved *in vivo* to give C_α-benzylic aldehydes (Fig. 9.1) (Kirk and Farrell, 1987; Gold et al., 1989). This reaction is catalyzed, both in simple lignin model compounds and in lignin, by extracellular, highly oxidizing peroxidases known as lignin peroxidases (LiPs) (Glenn et al., 1983; Tien and Kirk, 1983; Gold et al., 1984; Tien and Kirk, 1984; Kirk and Farrell, 1987; Gold et al., 1989; Hammel and Moen, 1991; Hammel et al., 1993). LiPs ionize their aromatic substrates to give aryl cation radicals, acting in effect as minute mass spectrometers (Kersten et al., 1985; Dolphin et al., 1987). These cation radicals then undergo a variety of nonenzymatic reactions, including nucleophilic attack by water in the case of some simple aromatics (Ker-sten et al., 1985, 1990) or carbon–carbon bond fission in the case of phenylpropane models that imitate natural lignin substructures (Hammel et al., 1985, 1986b; Kirk

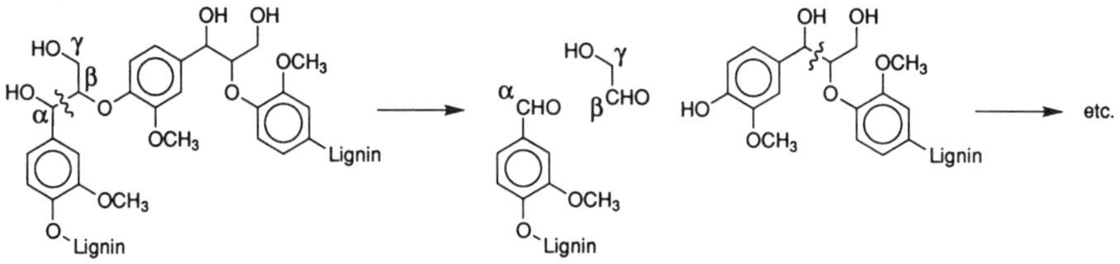

Figure 9.1. A linear region of softwood lignin, showing the major arylglycerol-β-aryl ether structure of the polymer and principal sites of fungal side-chain cleavage.

et al., 1986). The initial steps of fungal ligninolysis thus result from enzymatic one-electron oxidations of lignin.

White-rot fungi also produce other extracellular one-electron oxidants, in the form of laccases and manganese-dependent peroxidases (MnPs) (Glenn and Gold, 1985; Glenn et al., 1986; Kirk and Farrell, 1987; Gold et al., 1989). Laccases oxidize substrates directly, whereas MnPs operate by oxidizing manganese(II) in the presence of various organic chelators to give complexes of manganese(III), which act as diffusible oxidants at a distance from the enzyme active site (Glenn and Gold, 1985; Glenn et al., 1986). The low molecular weight and diffusibility of Mn(III) make MnPs particularly attractive candidates for ligninolytic catalysis in wood. However, chelated Mn(III) is a relatively weak oxidant that is limited to attack on easily oxidized substrates such as phenols. MnPs have been shown to depolymerize lignins *in vitro*, presumably via attack on phenolic structures in the polymer (Wariishi et al., 1991), and are now thought to constitute an important part of the fungal ligninolytic machinery. Both LiPs and MnPs, because of their low specificity, stand out as obvious candidates for the catalysis of xenobiotic oxidations by white-rot fungi.

3. FUNGAL ORGANOPOLLUTANT DEGRADATION

3.1. Polycyclic Aromatic Hydrocarbons

3.1.1. *Fungal Metabolism*

Eukaryotes in general do not degrade polycyclic aromatic hydrocarbons (PAHs). Instead, they metabolize these compounds, using well-described monooxygenase systems, to hydroxylated metabolites that can be excreted directly or as conjugates with more polar molecules (Gibson and Subramanian, 1984). Hydroxylation and excretion are standard eukaryotic strategies for xenobiotic detoxification, and most fungi fit this pattern. For example, members of the genus *Cunninghamella* utilize membrane-bound cytochrome P_{450} systems to oxidize naphthalene, phenanthrene, anthracene, benz[*a*]anthracene, 3-methylcholanthrene, and benzo[*a*]pyrene (Cerniglia and Gibson, 1979; Cerniglia and Yang, 1984; Gibson and Subramanian, 1984). Recently, Sutherland et al. (1991) showed that the white-rot basidiomycete

P. chrysosporium metabolizes phenanthrene to *trans*-dihydrodiols and phenanthrol conjugates when grown under nonligninolytic conditions in a rich nutrient medium. PAH metabolism by white-rot fungi under nonligninolytic conditions is accordingly unexceptional, resembling the process in nonligninolytic fungi and other eukaryotes.

It is therefore singular that idiophasic *P. chrysosporium,* grown under nutrient-limiting conditions that induce the ligninolytic system, is able to mineralize benzo[*a*]pyrene. The rate and extent of degradation initially reported were low—about 10% of initially added (0.125 μM) benzo[*a*]pyrene was oxidized to CO_2 in 30 days—but the finding was nonetheless unprecedented (Bumpus et al., 1985). Later work (Sanglard et al., 1986) demonstrated greater degradation of this PAH by *P. chrysosporium.* Low-nitrogen fungal cultures with biomass levels of about 2 g dry weight L^{-1} were able to metabolize 3 μM benzo[*a*]pyrene completely in 7 days. About 15% of the starting material was mineralized, about 60% was oxidized to water-soluble products, and about 20% was oxidized to organic-soluble metabolites, with the remainder (approx. 5%) present as mycelium-bound material. The maximum rate of benzo[*a*]pyrene mineralization observed with this PAH load was about 2 pmol h^{-1} mg^{-1} mycelial dry weight. When the initial benzo[*a*]pyrene concentration was increased to 30 μM, the maximal mineralization rate was about 6 pmol h^{-1} mg^{-1} dry weight.

Recent studies by Field et al. (1992) have shown that relatively high initial loads (10 mg L^{-1}) of anthracene or benzo[*a*]pyrene can be removed by a variety of white-rot fungi growing in liquid media. Most of the strains tested, including various species of *Trametes* and *Bjerkandera,* were more effective than the standard laboratory organism *P. chrysosporium.* In soils, the results are similar, but less clear-cut. Fluorene, added at 75 ppm to sterile soil, was effectively removed by *P. chrysosporium* in laboratory studies (George and Neufeld, 1989). 9-Fluorenone was reported to be an intermediary metabolite and was also depleted from soil by the fungus. In a study done on creosote-contaminated soil, fluorene and a variety of other PAHs with fewer than five rings were depleted by the related fungus *P. sordida,* but high variability in the data made a reliable assessment of the rates and extents of degradation difficult. PAHs with more than five rings, *e.g.,* benzo[*a*]pyrene, were not degraded by the fungus in soil, presumably because their greater hydrophobicity promoted tight binding to organic constituents in the soil, thereby preventing biological attack (Davis et al., 1993).

It is pertinent to compare fungal rates of PAH degradation with those exhibited by bacteria. To take one example, a *Mycobacterium* sp. has been described that co-oxidizes pyrene during growth on a complex organic medium (Heitkamp et al., 1988). This organism mineralized approximately 50% of initially added (2.5 μM) pyrene in 2 days, during which time it grew to a biomass level of about 0.1 g dry weight L^{-1}. Stationary-phase *P. chrysosporium,* by contrast, mineralized 16% of initially added (0.2 μM) pyrene in 2 weeks, at a biomass level of about 2 g dry weight L^{-1} (Hammel et al., 1991). We can conclude from this large difference that bacteria appear more promising for practical application in PAH bioremediation than white-rot fungi do. However, from an ecological standpoint, ligninolytic fungi probably play a significant role in the natural breakdown of PAH contaminants: Although these organisms are slow biodegraders, they are highly nonspecific, and they are ubiquitous inhabitants of woodland soils (Frankland, 1982).

3.1.2. Oxidation by Ligninolytic Enzymes

The possible involvement of the ligninolytic system in PAH degradation by white-rot fungi was shown some years ago in work with purified LiP. Benzo[a]pyrene was found to be a substrate for the enzyme, and it was proposed that LiPs might participate in the degradation of this PAH *in vivo* (Haemmerli et al., 1986). The products of enzymatic benzo[a]pyrene oxidation were shown to be the benzo[a]pyrene 1,6-, 3,6-, and 6,12-quinones, which is consistent with a reaction mechanism in which LiP oxidizes the PAH to its cation radical. However, the quinones were not identified as metabolic intermediates in fungal cultures, and it remained unclear whether LiP is actually involved in the degradative pathway.

Concurrent work showed that PAHs in general are LiP substrates, provided that they have ionization potentials lower than about 7.6 eV (Hammel et al., 1986a). Perylene (7.1 eV), benzo[a]pyrene (7.2 eV), anthracene (7.4 eV), pyrene (7.5 eV), and benz[a]anthracene are oxidized, whereas benzo[e]pyrene (7.7 eV), chrysene (7.8 eV), benzo[c]phenanthrene (7.9 eV), phenanthrene (8.1 eV), and naphthalene (8.1 eV) give no reaction. Classical peroxidases such as the horseradish enzyme have also been reported to oxidize PAHs, but only when the ionization potential is lower than about 7.35 eV (Cavalieri et al., 1983). These results suggest that the oxidized states (compounds I and II) of LiP are more oxidizing than those of typical peroxidases. Experiments on the relative ability of LiP and horseradish peroxidase to oxidize methoxybenzenes with known ionization potentials have confirmed this conclusion (Kersten et al., 1990), as have measurements of the Fe(II)–Fe(III) redox potentials of the two enzymes (Millis et al., 1989).

The hypothesis that LiP-catalyzed PAH oxidations proceed through a PAH cation radical intermediate received strong support in $H_2^{18}O$-labeling studies of pyrene oxidation (Hammel et al., 1986a). Pyrene was oxidized by the enzyme to its 1,6- and 1,8-quinones, and the newly introduced oxygens were derived from water. These results indicate that a cationic reaction intermediate undergoes nucleophilic attack by water at its positions of minimum charge density. Deprotonation and further oxidation then presumably lead to the formation of hydroxylated PAHs, which undergo further, similar reactions to yield the quinones (Fig. 9.2). It was also observed in this study that intact cultures of *P. chrysosporium* produce low levels of pyrene quinones as transient products of pyrene metabolism, but it was not possible to conclude whether these products are precursors for ring fission and mineralization.

Recent work has now shown that LiPs are actually involved in the oxidation by *P. chrysosporium* of one PAH, anthracene, that is a LiP substrate (Hammel et al., 1991). Both purified LiP and idiophasic (nitrogen-limited) cultures of *P. chrysosporium* oxidized anthracene to 9,10-anthraquinone. Chromatographic (Fig. 9.3) and isotope dilution experiments with [14C]anthracene showed that the fungal oxi-

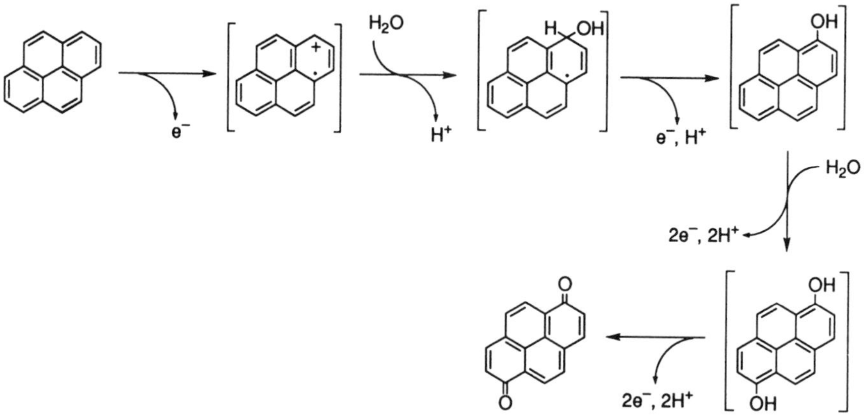

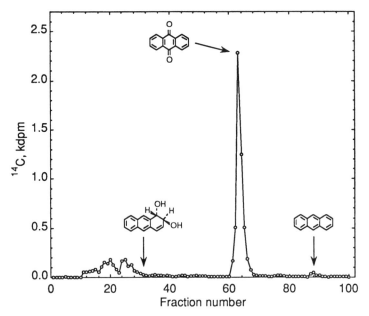

Figure 9.2. Mechanism proposed for the LiP-catalyzed oxidation of pyrene to pyrene-1,6-quinone. (Reproduced from Hammel, 1992, with permission of the publisher.)

Figure 9.3. High-performance liquid chromatography of neutral metabolites formed from [^{14}C]anthracene by nitrogen-limited *P. chrysosporium* in 2 days. The major anthracene metabolite formed is 9,10-anthraquinone. Anthracene *trans*-1,2-dihydrodiol, the major anthracene metabolite found in nonligninolytic fungi (Cerniglia and Yang, 1984), was not found. (Reproduced from Hammel et al., 1991, with permission of the publisher.)

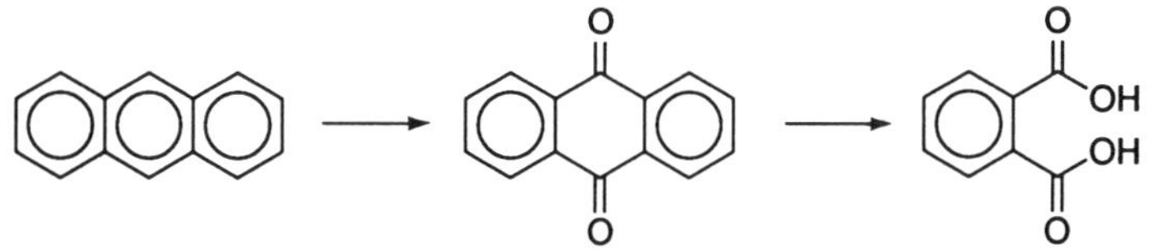

Figure 9.4.   Known reactions of anthracene metabolism in idiophasic *P. chrysosporium.*

dation of anthracene to anthraquinone was rapid, preceding the onset of anthracene mineralization, and accounted after 48 hours for about 40% of the ^{14}C initially added. Both anthracene and anthraquinone were mineralized to the same extent in culture (approx. 13% of 0.2 µM in 2 weeks), and both compounds were cleaved by the fungus to give equivalent yields (12%–13%) of the same diagnostic ring-fission metabolite, phthalic acid. Phthalate production from the quinone was demonstrated to occur only under ligninolytic culture conditions. These results show that the major pathway for anthracene degradation in *P. chrysosporium* proceeds via the 9,10-quinone to phthalate (Fig. 9.4) and that it is probably mediated by LiPs and other enzymes of ligninolytic metabolism. The mechanism that the fungus employs for the cleavage of anthraquinone to phthalate remains unknown.

The degradation of PAHs by ligninolytic fungi is not limited to those compounds that are LiP substrates. Phenanthrene, which repeated investigations have shown is not oxidized by the enzyme, is nevertheless rapidly metabolized by ligninolytic *P. chrysosporium.* The major metabolite found in fungal cultures is 2,2'-diphenic acid, which results from cleavage of the phenanthrene 9,10-bond (Fig. 9.5) (Hammel et al., 1992). Recent work suggests that the MnPs of *P. chrysosporium* are instrumental in this process (Moen and Hammel, 1994).

3.2.   Chlorinated Phenols

3.2.1.   Fungal Metabolism

Early work by Lyr (1963) in Germany showed that the white-rot fungus *Trametes versicolor* was able to dechlorinate polychlorinated phenols to unspecified products. Subsequent research has confirmed this finding and showed that white-rot fungi

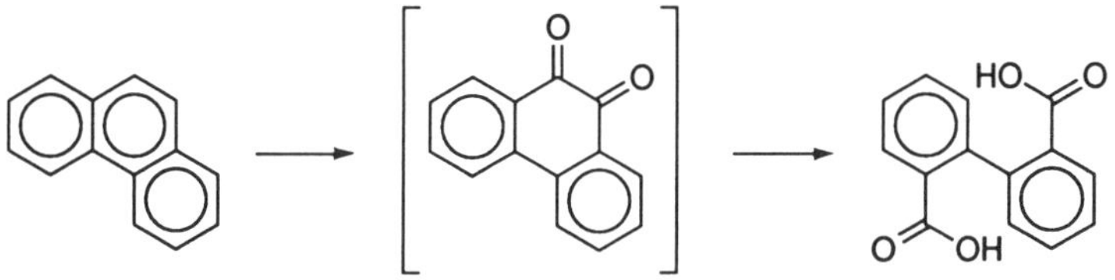

Figure 9.5.   Oxidation of phenanthrene by idiophasic *P. chrysosporium.* The intermediacy of the quinone is likely, but not firmly established.

degrade a wide variety of chlorinated phenols, some of which are significant pollutants formed during the chlorine bleaching of wood pulp. For example, *P. chrysosporium* degrades 2,4,6-trichlorophenol, various polychlorinated guaiacols, and several polychlorinated vanillins, all of which are toxic constituents of pulp bleachery effluents (Huynh et al., 1985). Products in these studies were identified by gas chromatography/mass spectrometry, and although direct precursor/product relationships were not determined, it could be concluded that the mechanisms of removal include oxidation, dechlorination, and methylation.

Recent work with individual chlorophenols has confirmed the ability of *P. chrysosporium* to degrade this class of compound. 2-Chlorophenol, added at initial concentrations as high as 0.5 g L^{-1}, was removed from the extracellular medium in packed-bed bioreactors (Lewandowski et al., 1990). Although no organic degradation products were identified, dehalogenation of the aromatic ring must have occurred, because inorganic chloride was produced. 2,4,6-Trichlorophenol and pentachlorophenol are also degraded in *P. chrysosporium* liquid cultures (Guo et al., 1990). In this study, both phenols were rapidly depleted from the extracellular medium, in part because they were methylated to give the corresponding anisoles, which were also removed by unspecified mechanisms.

Some polychlorinated phenols are mineralized by *P. chrysosporium* when the fungus is grown in nitrogen-limited liquid culture. It was recently reported that *ring*-^{14}C-labeled 2,4-dichlorophenol and 2,4,5-trichlorophenol, added at 180 μM initial concentration, were mineralized over 30 days in 50%–60% yield (Valli and Gold, 1991; Joshi and Gold, 1993). Pentachlorophenol is also mineralized by *P. chrysosporium*. In nitrogen-limited liquid cultures to which 0.5–36 μM [*ring*-^{14}C]pentachlorophenol was added, variable proportions (20%–50%) of the total were reportedly trapped as [14]CO_2. The times required to produce this much CO_2 varied considerably, from 4 to 30 days (Mileski et al., 1988). Subsequent studies have suggested that some of the volatile ^{14}C produced from pentachlorophenol by *P. chrysosporium* actually consists of organic products, thus leading to overestimates of mineralization (Lamar et al., 1990b). The ability of *P. chrysosporium* to degrade pentachlorophenol is markedly better in pregrown cultures with appreciable biomass levels than it is when the pollutant is added to freshly inoculated cultures (Mileski et al., 1988), a result attributable to the not unexpected finding that pentachlorophenol, a fungicidal wood preservative, inhibits the growth of *P. chrysosporium* and other white-rot fungi.

Experiments with liquid cultures of ligninolytic fungi provide useful information on what these organisms might do to chlorophenols during wastewater biotreatment. However, when ligninolytic fungi are grown in soil, the picture is evidently different. The mineralization of pentachlorophenol by *P. chrysosporium* or the related white-rot fungus *P. sordida* in this environment was found to be negligible. Nevertheless, both fungi did deplete pentachlorophenol from soil (Lamar and Dietrich, 1990; Lamar et al., 1990a). Some of the depletion was attributable to the formation of soil-bound metabolites, which have not thus far been characterized, but might have resulted from oxidative radical coupling between phenolic pentachlorophenol metabolites and phenolic constituents in the soil. The amount of soil-bound versus organic-extractable pentachlorophenol-derived material was influenced by soil type, with high clay content favoring binding. Another prominent fate of pentachlorophenol in soil was methylation to give pentachloroanisole, which was in turn slowly converted to unknown products by both of the fungi examined.

3.2.2. Oxidation by Ligninolytic Enzymes

Two frequently made observations suggest that the fungal degradation of polychlorinated phenols might involve ligninolytic metabolism: Degradation is highly non-specific, and it is stimulated under nutrient limitation. Moreover, polychlorinated phenols have been shown to be substrates for *P. chrysosporium* lignin peroxidases (Hammel and Tardone, 1988). The results showed that these enzymes catalyze the oxidative 4-dechlorination of polychlorinated phenols to give 1,4-benzoquinones. For example, 2,4,6-trichlorophenol is quantitatively oxidized to 2,6-dichloro-1,4-benzoquinone and inorganic chloride by LiP *in vitro*. Similarly, pentachlorophenol is oxidized to tetrachloro-1,4-benzoquinone. Reactions of this type presumably involve two sequential one-electron oxidations of the phenol to give a cyclohexadienone cation, which then eliminates halide upon nucleophilic attack by water (Fig. 9.6). Other studies on the LiP-catalyzed oxidation of pentachlorophenol by LiP have given results in agreement with this scheme (Mileski et al., 1988; Lin et al., 1990). Varying degrees of enzyme inactivation are observed during LiP-catalyzed phenol oxidations (Aitken et al., 1989), and MnP and laccase are now known to catalyze these oxidative dechlorinations more efficiently than does LiP (Roy-Arcand and Archibald, 1991).

The observations that LiP, MnP, and laccase can dehalogenate polychlorinated phenols were, in retrospect, not surprising: Peroxidases and laccases have long been known to oxidize a wide variety of phenols. Lyr demonstrated in 1963 that a crude laccase from *T. versicolor* oxidized chlorophenols to unknown products with the release of chloride, and subsequent studies with plant peroxidases showed that chlorophenols were enzymatically dechlorinated to give uncharacterized products (Saunders and Stark, 1967). However, it was not immediately evident how these (per)oxidative reactions might be integrated into metabolic pathways that lead to the extensive degradation of chlorinated phenols.

Gold and coworkers have recently provided the first evidence that fungal peroxidases are directly involved in polychlorophenol degradation by *P. chrysosporium* (Valli and Gold, 1991; Joshi and Gold, 1993). These biodegradations evidently result from the simultaneous operation of several extracellular and intracellular

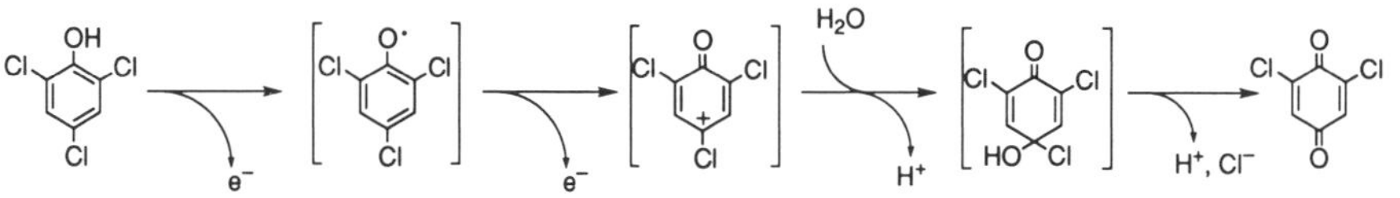

Figure 9.6. Mechanism proposed for the LiP-catalyzed oxidation of 2,4,6-trichlorophenol to 2,6-dichloro-1,4-benzoquinone. (Reproduced from Hammel, 1992, with permission of the publisher.)

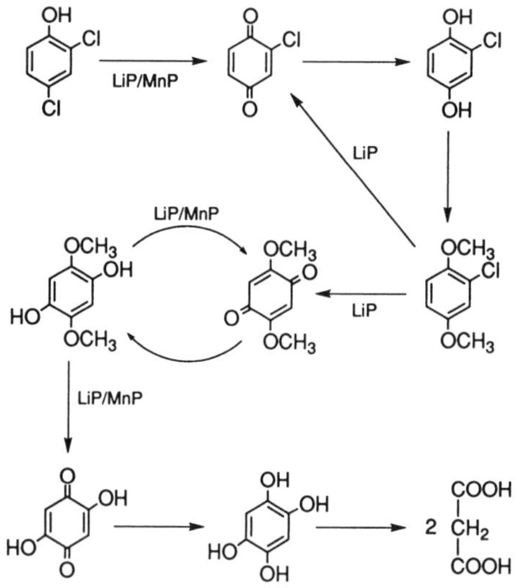

Figure 9.7. Route proposed by Valli and Gold (1991) for 2,4-dichlorophenol degradation in cultures of *P. chrysosporium*. (Reproduced from Hammel, 1992, with permission of the publisher.)

fungal processes: (1) peroxidative dechlorination of the phenols (by LiPs and MnPs) to give quinones; (2) reduction of these quinones to hydroquinones, possibly by a previously described intracellular oxidoreductase (Buswell and Eriksson, 1988); (3) methylation, in some cases, of the hydroquinones to give dimethoxybenzenes; and (4) oxidative dechlorination of these hydroquinones (by MnP and LiP) and dimethoxybenzenes (by LiP) to give hydroxyquinones (Fig. 9.7).

All of these dechlorination pathways contain futile cycles. In particular, the oxidation of chlorinated hydroquinones and methoxybenzenes by peroxidases leads predominantly to the formation of quinones without dechlorination. These chloroquinones can then be reduced to hydroquinones, which can be attacked again by peroxidases, and so on. Oxidative dechlorination occurs as a low-efficiency side reaction during these cycles, but iterations of the process evidently result in the complete dechlorination of some polychlorinated phenols.

A quantitative measure of these pathways is hard to obtain, because the crucial dehalogenation reactions occur in low yield when chlorinated hydroquinones or dimethoxybenzenes are oxidized *in vitro* with LiP or MnP, which makes it difficult to assess the importance of specific chloride-releasing reactions *in vivo*. It is noteworthy in this connection that certain polychlorinated phenols, e.g., the 2,4-dichloro-, 2,4,6-trichloro-, and pentachloro-isomers are also methylated to give polychlorinated anisoles in *P. chrysosporium* liquid cultures. These anisoles are not substrates for LiP, MnP, or laccase, yet they gradually disappear when added to fungal cultures (Guo et al., 1990), suggesting that mechanisms in addition to those proposed by Gold et al. play a role in polychlorophenol degradation.

3.3. Amino-, Nitro-, and Azoaromatics

The ability of *P. chrysosporium* to mineralize aminoaromatics was demonstrated by Arjmand and Sandermann (1985). A variety of chlorinated anilines and chloroaniline–lignin conjugates was shown to be mineralized by the fungus in idiophasic culture. Later, the aniline dye crystal violet (Bumpus and Brock, 1988) and the explosive 2,4,6-trinitrotoluene (Fernando et al., 1990) were shown to be mineralized by the fungus. However, high concentrations of trinitrotoluene inhibit fungal growth, raising doubts about the utility of the fungus for bioremediation of this pollutant (Spiker et al., 1992).

Work by Gold's group indicates that MnPs and LiPs are likely participants in many of these reactions (Valli et al., 1992). These investigators have recently shown that *P. chrysosporium* reduces 2,4-dinitrotoluene to 2-amino-4-nitrotoluene, which is oxidatively demethylated and deaminated by MnP to give 4-nitro-1,2-benzoquinone. 4-Nitro-1,2-benzoquinone is then reduced and methylated by the fungus to give 4-nitro-1,2-dimethoxybenzene, which in turn is oxidized by LiP, with the elimination of methanol and nitrite, to give 2-methoxy-1,4-benzoquinone. 2-Methoxy-1,4-benzoquinone is a central intermediate in the intracellular ligninolytic metabolism of white-rot fungi. It is degraded by *P. chrysosporium,* via reactions shown by Valli et al. (1992) and others (Buswell et al., 1979; Buswell and Eriksson, 1979, 1988) to give ring-fission products and CO_2.

Work in several other laboratories has implicated LiPs in the degradation of azo dyes by white-rot fungi (Cripps et al., 1990; Paszczynski and Crawford, 1991). Research shows that dyes with ring substituents that make them better LiP substrates, e.g., methoxyl groups, are more easily degraded. These findings have resulted in innovative attempts to synthesize new LiP-susceptible azo dyes that might easily be degraded in waste treatment schemes (Paszczynski et al., 1991).

3.4. Polymerized Organopollutants

The demonstration by Arjmand and Sandermann (1985, 1986) that *P. chrysosporium* degrades polymerized anilines is a particularly significant finding, because peroxidases readily catalyze free radical coupling reactions between aminoaromatics, hydroxyaromatics, or mixtures of the two. That is, the peroxidative metabolism of anilines or phenols by white-rot fungi leads not only to dechlorination, deamination, and other degradative reactions, but also to covalent cross-linking of the organopollutant with polymeric constituents of soil or wood (Bollag and Siu, 1985). It is therefore important that these basidiomycetes can degrade not only the original aromatic pollutants, but also polymers of them. Subsequent work has shown that idiophasic *P. chrysosporium* is able to degrade a variety of organopollutants when they are covalently linked to humic acids (Haider and Martin, 1989). Conjugates between chloroanilines, chlorophenols, and chlorocatechols were all significantly mineralized. Humic acids themselves are also degraded by the fungus (Bollag and Siu, 1985; Blondeau, 1989; Haider and Martin, 1989; Dehorter et al., 1992).

It remains to be seen whether the degradation of these xenobiotic polymers is catalyzed by ligninolytic enzymes, and much work obviously remains to be done. However, these results argue convincingly that the biodegradative abilities of ligninolytic fungi are not just curiosities of the microbiological laboratory, but rather play a significant ecological role in the natural breakdown of organopollutants.

REFERENCES

Adler E (1977): Lignin chemistry. Past, present and future. Wood Sci Technol 11:169–218.

Aitken MD, Venkatadri R, Irvine RL (1989): Oxidation of phenolic pollutants by a lignin degrading enzyme from the white-rot fungus *Phanerochaete chrysosporium*. Water Res 23:443–450.

Arjmand M, Sandermann H Jr (1985): Mineralization of chloroaniline/lignin conjugates and of free chloroanilines by the white rot fungus *Phanerochaete chrysosporium*. J Agric Food Chem 33:1055–1060.

Arjmand M, Sandermann H Jr (1986): Plant biochemistry of xenobiotics. Mineralization of chloroaniline/lignin metabolites from wheat by the white-rot fungus, *Phanerochaete chrysosporium*. Z Naturforsch 41c:206–214.

Blondeau R (1989): Biodegradation of natural and synthetic humic acids by the white rot fungus *Phanerochaete chrysosporium*. Appl Environ Microbiol 55:1282–1285.

Bollag J-M, Siu S-Y (1985): Formation of hybrid-oligomers between anthropogenic chemicals and humic acid derivatives. Org Geochem 8:131.

Bumpus JA, Aust SD (1987a): Biodegradation of DDT [1,1,1-trichloro-2,2-bis(4-chlorophenyl)ethane] by the white rot fungus *Phanerochaete chrysosporium*. Appl Environ Microbiol 53:2001–2008.

Bumpus JA, Aust SD (1987b): Biodegradation of environmental pollutants by the white rot fungus *Phanerochaete chrysosporium*. Bioessays 6:166–170.

Bumpus JA, Brock BJ (1988): Biodegradation of crystal violet by the white rot fungus *Phanerochaete chrysosporium*. Appl Environ Microbiol 54:1143–1150.

Bumpus JA, Tien M, Wright D, Aust SD (1985): Oxidation of persistent environmental pollutants by a white rot fungus. Science 228:1434–1436.

Buswell JA, Ander P, Pettersson B, Eriksson K-E (1979): Oxidative decarboxylation of vanillic acid by *Sporotrichum pulverulentum*. FEBS Lett 103:98–101.

Buswell JA, Eriksson K-E (1979): Aromatic ring cleavage by the white rot fungus *Sporotrichum pulverulentum*. FEBS Lett 104:258–260.

Buswell JA, Eriksson K-E (1988): NAD(P)H dehydrogenase (quinone) from *Sporotrichum pulverulentum*. Methods Enzymol 161:271–274.

Cavalieri EL, Rogan EG, Roth RW, Saugier RK, Hakam A (1983): The relationship between ionization potential and horseradish peroxidase/hydrogen peroxide-catalyzed binding of aromatic hydrocarbons to DNA. Chem Biol Interact 47:87–109.

Cerniglia CE, Gibson DT (1979): Oxidation of benzo[*a*]pyrene by the filamentous fungus *Cunninghamella elegans*. J Biol Chem 254:12174–12180.

Cerniglia CE, Yang SK (1984): Stereoselective metabolism of anthracene and phenanthrene by the fungus *Cunninghamella elegans*. Appl Environ Microbiol 47:119–124.

Cripps C, Bumpus JA, Aust SD (1990): Biodegradation of azo and heterocyclic dyes by *Phanerochaete chrysosporium*. Appl Environ Microbiol 56:1114–1118.

Davis MW, Glaser JA, Evans JW, Lamar RT (1993): Field evaluation of the lignin-degrading fungus *Phanerochaete sordida* to treat creosote-contaminated soil. Environ Sci Technol 27:2572–2576.

Dehorter B, Kontchou CY, Blondeau R (1992): ^{13}C NMR spectroscopic analysis of soil humic acids recovered after incubation with some white rot fungi and actinomycetes. Soil Biol Biochem 24:667–673.

Dolphin D, Nakano T, Maione TE, Kirk TK, Farrell R (1987): Synthetic model ligninases. In Odier E (ed): Lignin Enzymic and Microbial Degradation. Paris: Institute National de la Recherche Agronomique, pp 157–162.

Eaton DC (1985): Mineralization of polychlorinated biphenyls by *Phanerochaete chrysosporium*. Enzyme Microb Technol 7:194–196.

Fernando T, Bumpus JA, Aust SD (1990): Biodegradation of TNT (2,4,6-trinitrotoluene) by *Phanerochaete chrysosporium*. Appl Environ Microbiol 56:1666–1671.

Field JA, de Jong E, Costa GF, de Bont JAM (1992): Biodegradation of polycyclic aromatic hydrocarbons by new isolates of white rot fungi. Appl Environ Microbiol 58:2219–2226.

Frankland JC (1982): Biomass and nutrient cycling by decomposer basidiomycetes. In Frankland JC, Hedger JN, Swift MJ (eds): Decomposer Basidiomycetes. Cambridge: Cambridge University Press, pp 241–261.

George EJ, Neufeld RD (1989): Degradation of fluorene in soil by fungus *Phanerochaete chrysosporium*. Biotechnol Bioeng 33:1306–1310.

Gibson DT, Subramanian V (1984): Microbial degradation of aromatic hydrocarbons. In Gibson DT (ed): Microbial Degradation of Organic Compounds. New York: Marcel Dekker, Inc, pp 181–252.

Glenn JK, Akileswaran L, Gold MH (1986): Mn(II) oxidation is the principal function of the extracellular Mn-peroxidase from *Phanerochaete chrysosporium*. Arch Biochem Biophys 251:688–696.

Glenn JK, Gold MH (1985): Purification and characterization of an extracellular Mn(III)-dependent peroxidase from the lignin-degrading basidiomycete, *Phanerochaete chrysosporium*. Arch Biochem Biophys 242:329–341.

Glenn JK, Morgan MA, Mayfield MB, Kuwahara M, Gold MH (1983): An extracellular H_2O_2-requiring enzyme preparation involved in lignin biodegradation by the white-rot basidiomycete *Phanerochaete chrysosporium*. Biochem Biophys Res Commun 114:1077–1083.

Gold MH, Kuwahara M, Chiu AA, Glenn JK (1984): Purification and characterization of an extracellular H_2O_2-requiring diarylpropane oxygenase from the white rot basidiomycete *Phanerochaete chrysosporium*. Arch Biochem Biophys 234:353–362.

Gold MH, Wariishi H, Valli K (1989): Extracellular peroxidases involved in lignin degradation by the white rot basidiomycete *Phanerochaete chrysosporium*. ACS Symp Ser 389:127–140.

Guo H, Chang H-m, Joyce TW, Glaser J (1990): Degradation of chlorinated phenols and guaiacols by the white-rot fungus *Phanerochaete chrysosporium*. In Kirk TK, Chang H-m (eds): Biotechnology in Pulp and Paper Manufacture. Applications and Fundamental Investigations. Boston: Butterworth-Heinemann, pp 223–230.

Haemmerli SD, Leisola MSA, Sanglard D, Fiechter A (1986): Oxidation of benzo[*a*]pyrene by extracellular ligninase of *Phanerochaete chrysosporium:* Veratryl alcohol and stability of ligninase. J Biol Chem 261:6900–6903.

Haider KM, Martin JP (1989): Mineralization of ¹⁴C-labelled humic acids and of humic-acid bound ¹⁴C xenobiotics by *Phanerochaete chrysosporium.* Soil Biol Biochem 20:425–429.

Hammel KE (1989): Organopollutant degradation by ligninolytic fungi. Enzyme Microb Technol 11:776–777.

Hammel KE (1992): Oxidation of aromatic pollutants by lignin-degrading fungi and their extracellular peroxidases. In Sigel H, Sigel A (eds): Metal Ions in Biological Systems, vol 28, Degradation of Environmental Pollutants by Microorganisms and Their Metalloenzymes. New York: Marcel Dekker, Inc, pp 41–60.

Hammel KE, Gai WZ, Green B, Moen MA (1992): Oxidative degradation of phenanthrene by the ligninolytic fungus *Phanerochaete chrysosporium.* Appl Environ Microbiol 58:1832–1838.

Hammel KE, Green B, Gai WZ (1991): Ring fission of anthracene by a eukaryote. Proc Natl Acad Sci USA 88:10605–10608.

Hammel KE, Jensen KA Jr, Mozuch MD, Landucci LL, Tien M, Pease EA (1993): Ligninolysis by a purified lignin peroxidase. J Biol Chem 268:12274–12281.

Hammel KE, Kalyanaraman B, Kirk TK (1986a): Oxidation of polycyclic aromatic hydrocarbons and dibenzo[*p*]dioxins by *Phanerochaete chrysosporium* ligninase. J Biol Chem 261:16948–16952.

Hammel KE, Kalyanaraman B, Kirk TK (1986b): Substrate free radicals are intermediates in ligninase catalysis. Proc Natl Acad Sci USA 83:3708–3712.

Hammel KE, Moen MA (1991): Depolymerization of a synthetic lignin *in vitro* by lignin peroxidase. Enzyme Microb Technol 13:15–18.

Hammel KE, Tardone PJ (1988): The oxidative 4-dechlorination of polychlorinated phenols is catalyzed by extracellular fungal lignin peroxidases. Biochemistry 27:6563–6568.

Hammel KE, Tien M, Kalyanaraman B, Kirk TK (1985): Mechanism of oxidative $C_\alpha-C_\beta$ cleavage of a lignin model dimer by *Phanerochaete chrysosporium* ligninase. J Biol Chem 260:8348–8353.

Heitkamp MA, Franklin W, Cerniglia CE (1988): Microbial metabolism of polycyclic aromatic hydrocarbons: Isolation and characterization of a pyrene-degrading bacterium. Appl Environ Microbiol 54:2549–2555.

Huynh V-B, Chang H-m, Joyce TW, Kirk TK (1985): Dechlorination of chloro-organics by a white-rot fungus. Tappi J 68:98–102.

Joshi DK, Gold MH (1993): Degradation of 2,4,5-trichlorophenol by the lignin-degrading basidiomycete *Phanerochaete chrysosporium.* Appl Environ Microbiol 59:1779–1785.

Kennedy DW, Aust SD, Bumpus JA (1990): Comparative biodegradation of alkyl halide insecticides by the white rot fungus, *Phanerochaete chrysosporium* (BKM-F-1767). Appl Environ Microbiol 56:2347–2353.

Kersten PJ, Kalyanaraman B, Hammel KE, Reinhammar B, Kirk TK (1990): Comparison of lignin peroxidase, horseradish peroxidase and laccase in the oxidation of methoxybenzenes. Biochem J 268:475–480.

Kersten PJ, Tien M, Kalyanaraman B, Kirk TK (1985): The ligninase of *Phanerochaete chrysosporium* generates cation radicals from methoxybenzenes. J Biol Chem 260:2609–2612.

Kirk TK, Farrell RL (1987): Enzymatic "combustion": The microbial degradation of lignin. Annu Rev Microbiol 41:465–505.

Kirk TK, Tien M, Kersten PJ, Mozuch MD, Kalyanaraman B (1986): Ligninase of *Phanerochaete chrysosporium*. Mechanism of its degradation of the non-phenolic arylglycerol β-aryl ether substructure of lignin. Biochem J 236:279–287.

Köhler A, Jäger A, Willershausen H, Graf H (1988): Extracellular ligninase of *Phanerochaete chrysosporium* Burdsall has no role in the degradation of DDT. Appl Microbiol Biotechnol 29:618–620.

Lamar RT, Dietrich DM (1990): *In situ* depletion of pentachlorophenol from contaminated soil by *Phanerochaete* spp. Appl Environ Microbiol 56:3093–3100.

Lamar RT, Glaser JA, Kirk TK (1990a): Fate of pentachlorophenol (PCP) in sterile soils inoculated with *Phanerochaete chrysosporium:* Mineralization, volatilization and depletion of PCP. Soil Biol Biochem 22:433–440.

Lamar RT, Larsen MJ, Kirk TK (1990b): Sensitivity to and degradation of pentachlorophenol by *Phanerochaete* spp. Appl Environ Microbiol 56:3519–3526.

Lewandowski GA, Armenante PM, Pak D (1990): Reactor design for hazardous waste treatment using a white rot fungus. Water Res 24:75–82.

Lin J-E, Wang HY, Hickey RF (1990): Degradation kinetics of pentachlorophenol by *Phanerochaete chrysosporium*. Biotechnol Bioeng 35:1125–1134.

Lyr H (1963): Enzymatische Detoxifikation chlorierter Phenole. Phytopathol Z 47:73–83.

Mileski GJ, Bumpus JA, Jurek MA, Aust SD (1988): Biodegradation of pentachlorophenol by the white rot fungus *Phanerochaete chrysosporium*. Appl Environ Microbiol 54:2885–2889.

Millis CD, Cai D, Stankovich MT, Tien M (1989): Oxidation-reduction potentials and ionization states of extracellular peroxidases from the lignin-degrading fungus *Phanerochaete chrysosporium*. Biochemistry 28:8484–8489.

Moen MA, Hammel KE (1994): Lipid peroxidation by the manganese peroxidase of *Phanerochaete chrysosporium* is the basis for phenanthrene oxidation by the intact fungus. Appl Environ Microbiol 60:1956–1961.

Paszczynski A, Crawford RL (1991): Degradation of azo compounds by ligninase from *Phanerochaete chrysosporium:* Involvement of veratryl alcohol. Biochem Biophys Res Commun 178:1056–1063.

Paszczynski A, Pasti MB, Goszczynski S, Crawford DL, Crawford RL (1991): New approach to improve degradation of recalcitrant azo dyes by *Streptomyces* spp. and *Phanerochaete chrysosporium*. Enzyme Microb Technol 13:378–384.

Roy-Arcand L, Archibald FS (1991): Direct dechlorination of chlorophenolic compounds by laccases from *Trametes (Coriolus) versicolor*. Enzyme Microb Technol 13:194–203.

Sanglard D, Leisola MSA, Fiechter A (1986): Role of extracellular ligninases in biodegradation of benzo[*a*]pyrene by *Phanerochaete chrysosporium*. Enzyme Microb Technol 8:209–212.

Saunders BC, Stark BP (1967): Studies in peroxidase action—XVII. Some general observations. Tetrahedron 23:1867–1872.

Spiker JK, Crawford DL, Crawford RL (1992): Influence of 2,4,6-trinitrotoluene (TNT) concentration on the degradation of TNT in explosive-contaminated soils by the white rot fungus *Phanerochaete chrysosporium*. Appl Environ Microbiol 58:3199–3202.

Sutherland JB, Selby AL, Freeman JP, Evans FE, Cerniglia CE (1991): Metabolism of phenanthrene by *Phanerochaete chrysosporium*. Appl Environ Microbiol 57:3310–3316.

Tien M, Kirk TK (1983): Lignin-degrading enzyme from the hymenomycete *Phanerochaete chrysosporium* Burds. Science 221:661–663.

Tien M, Kirk TK (1984): Lignin-degrading enzyme from *Phanerochaete chrysosporium:* Purification, characterization, and catalytic properties of a unique H_2O_2-requiring oxygenase. Proc Natl Acad Sci USA 81:2280–2284.

Valli K, Brock BJ, Joshi DK, Gold MH (1992): Degradation of 2,4-dinitrotoluene by the lignin-degrading fungus *Phanerochaete chrysosporium*. Appl Environ Microbiol 58:221–228.

Valli K, Gold MH (1991): Degradation of 2,4-dichlorophenol by the lignin-degrading fungus *Phanerochaete chrysosporium*. J Bacteriol 173:345–352.

Wariishi H, Valli K, Gold MH (1991): *In vitro* depolymerization of lignin by manganese peroxidase of *Phanerochaete chrysosporium*. Biochem Biophys Res Commun 176:269–275.

PART III. APPLICATIONS IN CLEANUP AND BIOREMEDIATION

10

MICROBIOLOGICAL TREATMENT OF CHEMICAL PROCESS WASTEWATER

LAURENCE E. HALLAS

Wastewater Solutions Division,
LMII Corporation,
St. Louis, Missouri 63131

MICHAEL A. HEITKAMP

Environmental Sciences Center,
Monsanto Company,
St. Louis, Missouri 63166

1. INTRODUCTION

1.1. Scope of Issue

Management of chemical wastes has evolved into two distinct industrial operations: the *remediation* of past practices and the *continual treatment* of process wastes from current manufacturing. Both areas have received considerable attention in recent years as environmental issues have become more prevalent. This chapter focuses on the latter operation. It reviews some biological options that one chemical company has considered in treating organic compounds, particularly in current process wastewater.

Projected costs for domestic and worldwide pollution control have increased

Microbial Transformation and Degradation of Toxic Organic Chemicals, pages 349–387
© 1995 Wiley-Liss, Inc.

TABLE 10.1. Chemical Waste Survey[a]

Date	Wastewater (Million Tons per Year)	Percent Treated		
		On Site	Injection	Off Site
1985	163	90	6	4
1986	183	92	5	3
1987	165	90	6	4
1988	186	91	6	3
1989	188	90	6	4

[a]Six hundred twenty-seven facilities from 93 chemical companies (38% of the top 50 chemical companies in sales). Fifteen percent of the facilities did not generate any waste. Data from Chemical Manufacturers Association (1991).

considerably in the last two decades. In constant dollars, the U.S. Environmental Protection Agency (EPA) has found that costs in the United States have grown from $27B in 1972 to $90B in 1990, and they are expected to be $155B in 2000 (Reilly, 1990). Costs in Eastern Europe alone are projected to rise from $2B in 1991 to $11B by 1995.

In addition, amounts of wastes have remained relatively high in recent years. Table 10.1 presents survey data from 627 facilities representing 93 chemical companies that are members of the Chemical Manufacturers Association (1991). Aqueous wastes represented 97% of the total. Of that amount, 89% was either chemically or biologically treated. These data show the need for quality biological treatment technologies that can process high volumes of wastes.

1.2. Regulatory Issues

The large volumes of wastes have certainly not gone unnoticed. Regulatory guidelines and laws have been enacted for solids and emissions to air and water. A brief summary of these laws and how they impact wastewater treatment is in order.

The Clean Water Acts (CWA) had its origins in the Federal Water Pollution Control Act (FWPCA) of 1972 (Arbuckle et al., 1989). The CWA was enacted in 1977 and is implemented by the U.S. EPA. Changes were made in 1987, and reauthorization is expected in 1995 (Rotman, 1991). The CWA allows for setting minimum national effluent industrial standards. Its primary effect is to require permits (with enforceable limitations) for discharges into navigable waters. Limitations can be either technology or water quality based. These latter limitations are more stringent and are designated to protect highly used bodies of water. The permits are issued within 6 months under the National Pollutant Discharge Elimination System (NPDES), and authority for monitoring is usually relegated to the state. Permits focus on three general categories of pollutants:

1. Conventional (total organic carbon, total suspended solids, biological oxygen demand, and so forth)

2. Nonconventional (specific to the chemical process)

3. Toxic (listed toxics and priority pollutants)

Other legal guidelines and regulations can be incorporated into NPDES permits. These include

1. Organic Chemical, Plastic & Synthetic Fibers (OCPSF)

2. Pesticides Guidelines

3. Resource Conservation and Recovery Act (RCRA)

RCRA deserves some attention. It provides an incentive to reduce wastes and to pretreat hazardous wastewater (Case, 1989). It was enacted in 1976 and was amended in 1984 by the Hazardous and Solid Waste Amendments (HSWA).

RCRA can be incorporated into the permitting process since it applies to solid wastes and waste streams that exhibit certain toxic waste characteristics (ignitability, corrosivity, reactivity, and flammability) or are listed on specific hazardous waste lists. RCRA applies to active facilities and is designed to provide "cradle to grave" control of the defined waste (recall that inactive or abandoned sites are regulated under the Superfund Act of 1980).

RCRA classification has added a new level of severe restrictions to industrial operations. Treatment, storage, and disposal (TSD) facilities are permitted only after certain standards are met; e.g., surface impoundments that contain and treat defined wastes must be double lined, have leachate collection, leak detection, and groundwater monitoring systems. Sludge from such systems would be incinerated (controlled flame combustion). HSWA 1984 states that trial burns must be conducted to show destruction and removal efficiencies (Rotman and Young, 1991) of >99.9% of the specific constituents. Continuous monitoring and automated shutoff controls must be installed and emission standards will apply.

In summary, the chemical industry continues to generate millions of tons of process wastewater. More stringent federal guidelines are being promulgated, and billions of dollars are predicted to be spent for legal compliance. Therefore, it is becoming increasingly important to industry that they develop strategies first to minimize waste production and then to implement treatment technologies to eliminate the remaining wastes in wastewater in a cost-effective manner. This chapter outlines one corporation's attempt in the latter area. Background information is provided on common biotreatment strategies. Several examples are then given that highlight both the microbiology of biotreating industrial organic chemicals and the recent field advances in biotreatment.

2. BACKGROUND

2.1. Engineering Design

Biological treatment of organic wastes has been used for centuries. Composting, for example, is a natural way to enhance nutrient cycling. In this century, the science of wastewater biotreatment has become more developed since waste treatment plant operations became commonplace (Thayer, 1991). However, most plants were built in the 1960s and design was based on meeting simple water quality criteria like biological oxygen demand (BOD) and total organic carbon (TOC). Increased demand for performance has required extensive modifications of treatment practice and a proliferation of new biological technologies.

Biological reactor systems fall into two general categories—suspended growth and immobilized cell systems (Stensel and Strand, 1989). In either case, the principle is the same; i.e., to bring microbial biomass in contact with 1) organic compounds as a source of energy, 2) an electron acceptor (like oxygen or nitrate), and 3) appropriate nutrients for microbial growth. As the organics are degraded, a means to separate the treated liquor from the increased biomass must be provided (Stensel and Strand, 1989).

Suspended growth systems have been further classified into two main types: continuous flow (parallel and series) and semicontinuous batch. Immobilized (or fixed film) systems, on the other hand, can take many forms, including sparged packed beds, fluidized beds, rotating biological contactors, or even the centuries old trickling filter system. Distinctions in design can be based on several factors as listed in Table 10.2. Hybrids of both classifications exist as typified by suspended growth membrane reactors (Wilderer and Markl, 1989) and fixed growth-suspended reactor combinations (Kelly, 1991).

Immobilized bacteria systems configured as fluidized bed reactors (FBRs) offer technical advantages for the biotreatment of concentrated waste streams, volatile chemicals, and toxic chemicals. In FBRs, the biocarrier bed is fluidized by liquid recycle within the reactor. Since chemical wastes are injected into the recycle water, toxic chemicals are immediately diluted and microorganisms in the FBR are more resistant to direct chemical toxicity than many conventional treatment systems (Edwards and Heitkamp, 1992; Edwards et al., 1994). In addition, since the FBRs are usually oxygenated by supplying enriched oxygen into the recycle loop, a high level of microbial activity may be supported in the FBR with minimal airstripping of volatile chemicals (Stensel and Strand, 1989). An early review of FBRs said they were "the most significant development in the wastewater treatment field in the last 50 years" (Cooper and Atkinson, 1981). By 1990, over 25 field operations were

TABLE 10.2. Factors That Can Be Optimized in Suspended Growth and Immobilized Cell Systems

Factor	Example
Large waste volumes	Activated sludge
Enhanced kinetic rates	Series operation
Specific bacterial enrichment	Batch operation
Higher biomass/substrate ratio	Fixed film
High oxygen transfer	Fluidization
Settleability, adsorption	Activated carbon
Equalization, gas production	Anaerobic
Compact, space limitation	Semicontinuous, fixed film

TABLE 10.3. Methods of Primary Treatment
During Wastewater Equalization[a]

Method	Purpose
Screening	Particle sizing
Sedimentation, clarification	Solids removal
Flocculation	Solids concentration
Gravity separation	Oil–water partitioning
Neutralization	pH adjustment

[a]Modified from Eckenfelder et al. (1985).

documented with sizes ranging from 16,000 to 193,000 gallons (Sutton and Mishra, 1991).

Primary treatment ensues once the waste passes battery limits. The purpose of this step is to provide mixing time to wastewater in order to equalize composition. Several technologies can also be employed to make the waste more amenable to treatment (Table 10.3). An extensive review of the principles is given by Eckenfelder et al. (1985).

Secondary treatment system development is where key technological advances have occurred in recent years. This chapter focuses on aerobic treatment of aqueous wastes. But, as noted in Figure 10.1, both bio-oxidation of contaminated air (biofiltration) and anaerobic treatment technology are a large part of secondary treatment capabilities (the use of lagoons and ponds is not considered here due to their regulatory phaseout).

In aerobic systems, continuous flow activated sludge has been the mainstay technology. Waste is first mixed and aerated with microorganisms in a tank for a defined time. A second tank provides for separation (clarification) of the biomass from water with a greatly reduced organic (sometimes inorganic) content. The activated sludge process is mechanically well understood (Eckenfelder et al. 1985). Key operating variables are used to design and operate systems. Waste characterization is followed by reactor studies to define system organic loadings (F/M, or food/mass ratio). Rates of air delivery and oxygen transfer also play a role in full-scale design. Finally, the rate of clarification followed by a definition of the growth rate (and subsequent biomass wastage) are critical to the successful operation of the system. Addition of powdered activated carbon to the aeration tank has provided a successful variation in treating dilute but variable wastewater (Dietrich et al., 1988).

Semicontinuous operations have also been successful in industry. A sequencing batch reactor (SBR) is a time-oriented technique with waste flow, air/nutrient additions, and contact time determined by a strictly defined operational strategy (Irvine and Busch, 1979; Arora et al., 1985). A single tank (or tank sequence) is used in a regime of FILL, REACT, SETTLE, DRAW, IDLE. Cycle manipulation (e.g., air on/off through FILL and REACT or periodic nutrient additions) can create a unique microbial ecology for treating specific wastewater (Irvine et al., 1984; Irvine, 1989).

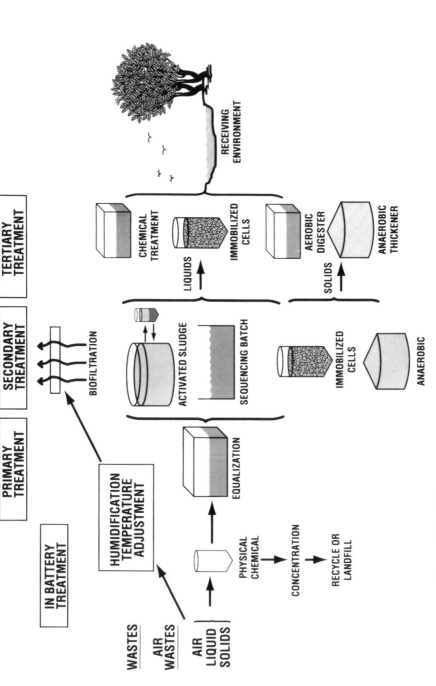

Figure 10.1. Schematic of general waste treatment technologies for industrial plant wastes.

Secondary treatment also includes microbial immobilization technologies (reviewed by Hao et al., 1991). This technology involves the attachment of chemical-degrading bacteria to inert biocarriers in a reactor system designed to support a high level of chemical-degrading activity by the immobilized microorganisms. Some of the simplest and oldest configurations of immobilized bacteria are trickling filters, rotating biological contactors, and submerged fixed film reactors. More recently, packed bed reactors (PBRs) and FBRs have been shown to achieve high performance for biotreatment of chemical industry wastes (Heitkamp et al., 1993; Edwards and Heitkamp, 1992; Edwards et al., 1994). In the PBRs, inert supports such as diatomaceous earth, hollow fibers, or porous silica serve as biocarriers for microbial attachment and growth (Stensel and Strand, 1989). These systems are oxygenated by compressed air and can be simple to operate in a compact area, and microbial selection is much easier to maintain (Eaton et al., 1989). However, diffusion kinetics can be slow, and high levels of microbial growth from concentrated wastewater can cause operational problems due to channeling. Lupton and Zupancie (1991) have shown success in utilizing PBRs to treat creosote-contaminated wastewater. Similarly, Portier and Fujisaki (1986) reported the biotreatment of chlorinated phenols by immobilized bacteria in a PBR.

Another category of wastewater processing is tertiary treatment of clarified water and waste sludge. Aqueous discharge from secondary treatment may not meet water quality standards. This may be due to high suspended solids or nutrient levels, as well as specific regulated components. Several options are available, including additional chemical treatment (e.g., filtration, coagulation, air stripping, carbon or chemical treatment, and ion exchange). Immobilization by trickling filters as well as polishing components with specific microbes on activated carbon or alginate are common (Hallas et al., 1992; Lin et al., 1989). Finally, sludge disposal has become a critical problem due to the high costs of incineration and limited landfill space for waste defined as hazardous. Thus, the oxidation of biomass through endogenous respiration or the anaerobic thickening of sludges has become a cost-effective alternative (Eckenfelder et al., 1985) for reducing these solid wastes.

2.2. Microbial Design

The key to the biological treatment of organic wastes is the microbial ecology of biological flocs in activated sludge. Floc formation is caused by bacterial excretions and allows for microbial community development (Butterfield, 1935; Unz and Davis, 1975; Eckenfelder et al., 1985). Several organic polymers are involved, but polysaccharides predominate in a slime layer. Both suspended growth (for clarification purposes) and immobilized cell systems show floc formation (Updike et al., 1969).

Microorganisms in flocs are widely varied. Aerobic and facultative anaerobic organisms predominate (Allen, 1944; Pike, 1975). Protozoa and bacteria make up most of the biomass, though small nematodes, insect larvae, yeasts, fungi, and even algae are found in the sludge. Figure 10.2 depicts the relative abundance of microbes as a function of carbon biomass ratio. As hydraulic residence time (HRT)

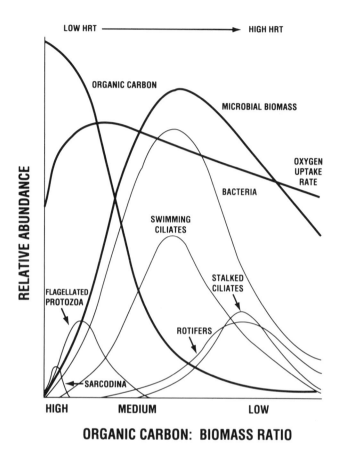

Figure 10.2. Microbial abundance in activated sludge. Note that biomass and bacterial abundance have been scaled to fit on the graph. (Adapted from Niedringhaus, 1982.)

increases, the ratio of organics to biomass (F/M) decreases; this can dramatically change the community structure. On a start up, F/M is high, encouraging biomass growth. *Sarcodina* populations (e.g., amoeba) flourish as they depend on high food levels due to food-gathering inefficiencies. As the biosystem matures, F/M ratios remain high and flagellate protozoa dominate. Bacterial numbers increase, and the highly motile flagellates use their grazing efficiency. Sludge is very light colored and slow to settle; flocculation has not begun, and the flagellates move throughout the water column. However, oxygen uptake rates are quite high.

Most biosystem operations maintain a steady-state operation between 0.1 and 0.2 F/M (BOD). In Figure 10.2, this is depicted in the center of the graph, where bacteria and free-swimming ciliates predominate. Floc formation (and thus sludge settling) is quite high as the community structure becomes more defined. Both microbial types use less energy in obtaining food, and grazing of bacteria by ciliates develops into a steady state (Eckenfelder et al., 1985). At the lower F/M, attached protozoa (stalked ciliates and rotifers) are apparent. At this stage, organic carbon limitations allow for a low BOD wastewater and a keen competition for organics. At some point, settling may be quite fast, but turbidity would be high as detritus levels increase. Indeed, an F/M (BOD) of <0.05 results in high endogenous respiration and destruction of the polysaccharide floc. Dispersed growth results, and water quality effluent deteriorates dramatically.

When general BOD removal is important, traditional waste treatment operations work well. When specific organic removals are required for NPDES permits, industrial biotreatment operations need to consider the microbes involved in waste treatment. Yet few studies have been done on the microbiology of activated sludge, particularly from industrial wastewater. Early work (Butterfield, 1935; McKinney and Horwood, 1952; McKinney and Weichlein, 1953) established that bacteria were responsible for organic biodegradation and water purification (flocculation) in the sludge. It has only been in the last 20 years that studies on microbial distribution and abundance have been undertaken. Dias and Bhat (1964) and a series of reports by Prakasam and Dondero (1967a–c, 1970a,b) incorporated domestic sewage extract into their enumeration media. They found that the numbers and types of microbes varied with the sewage composition and that artificial nutrient media tended to misrepresent the microbial composition. Lighthart and Oglesby (1969) described some of the first rapid microbial characterization tests for studying large numbers of activated sludge isolates. Benedict and Carlson (1971) and a literature review by Pipes (1980) have noted several bacterial species associated with activated sludge, including *Acinetobacter, Alcaligenes, Brevibacterium, Flavobacterium, Pseudomonas, Zooglea,* and several budding bacteria.

Few studies have been done with industrial activated sludge. Liao and Dawson (1975) monitored sludge-treating Kraft bleachery influent, and Takii (1977) studied sludge of industrial carbohydrates. They noted population shifts as the feed composition changed. In a numerical taxonomic study, Seiler and Blaim (1982) considered two Bayer AG biosystems and found *Alcaligenes, Pseudomonas,* and *Zooglea* as the predominant genera. While these studies used artificial media (not sewage extracts), the results suggested that system performance and microbial community structure were related.

Manipulating bacterial species has become an important part of industrial operations. One approach is to provide a constant source of bacteria to degrade specific compounds or compound classes. Cardinal and Stenstrum (1991) recently demonstrated the use of enricher reactors (operated as SBRs) that produced biomass that was seeded into activated sludge reactors. This bioaugmentation enhanced removals of naphthalene. Such operations allow for efficient BOD removals with selected degradation patterns to meet water quality permits.

A second approach for enhancing specific organic-degrading activities is to develop an understanding of the master variables controlling the microbial populations. Manipulating those controls (e.g., temperature, nutrient addition, organic loadings) along with a careful choice of reactor design has also been shown to be successful in establishing high specific microbial activities. Following are several case studies that consider both the microbiology of sludge as well as the use of different types of reactors to enrich for specific bacterial populations.

3. MICROBIOLOGY OF WASTEWATER TREATMENT

Acclimation is the adaptation of the microbial consortium to a change in reactor conditions. The reactor environment may be modified by the addition of new substrates or a change in physical parameters such as pH, temperature, operating strategies, and dissolved oxygen concentration. Acclimation, in the simplest case, then, is the time required for the development of significant numbers of bacteria that can metabolize the substrate(s) available (Busch, 1971). When the microbial population experiences a change in influent composition, the relative numbers of certain microorganisms shift in response to the new conditions.

The bacteria that predominate in activated sludge are the ones that have adapted most efficiently to the reactor conditions and feed components. They are able to compete most efficiently for the substrate and energy source (Irvine, 1989; Metcalf and Eddy, Inc., 1979). The amount of time required for acclimation is dependent on many factors. To utilize the substrate(s) present, the bacteria must possess the genetic information for specific catabolic enzymes. The location of the genes can also influence acclimation time; e.g., genes found on the bacterial chromosome can be constitutive (constantly present and independent of environmental conditions) or inducible (requiring an external factor to initiate enzyme synthesis). If a plasmid contains the gene, acclimation time may be shortened because of plasmid transfers to other microorganisms. Finally, the relative number of microorganisms capable of degrading the substrate and their growth rate determines acclimation time (Busch, 1971).

Acclimating activated sludge to degrade specific wastewater organics compounds can be accomplished through manipulating the environmental conditions. However, characterizing the microbes involved is a critical first step. This section describes the microbial ecology and key enrichment methods for three biosystems dedicated to oxidizing a herbicide (glyphosate), a chemical intermediate (nitrophenol), and an inorganic compound (ammonia).

3.1. Characterization of Glyphosate Degradation

N-phosphonomethylglycine (glyphosate) is a widely used, broad-spectrum herbicide manufactured by the Agricultural Group of the Monsanto Company. Several formulations have glyphosate as an active ingredient. The largest selling is Roundup™, which is the isopropylamine salt of glyphosate. These products have one of the largest shares of the herbicide market. Thus, it is important to understand the environmental fate of glyphosate.

All Monsanto facilities producing glyphosate use some form of activated sludge for treating process wastewater. Glyphosate is regulated at the federal (Pesticide Guidelines) and state (NPDES permitting) levels. To minimize its presence in discharge from manufacturing facilities, an extensive research program was developed. Microbial resistance and degradation patterns were described and activated sludge enumerated to characterize microbial interactions with glyphosate more fully. Sludge from three biosystems was used: I1 and I2 were industrial biosystems that treated several chemical process wastes; D1 was a biosystem that treated wastes from domestic sources.

3.1.1. Glyphosate Resistance/Degradation

Samples from all three biosystems were enumerated in a single experiment on a defined salt (Leadbetter and Foster, 1958) medium containing seven glyphosate concentrations. In all cases, as glyphosate levels increased, the microbial counts decreased. Figure 10.3 presents the percent counts enumerated at each glyphosate level when compared with plates containing no glyphosate. For example, at 300 mg/liter glyphosate, the percent resistant population at D1, I2, and I1 was 29%, 23%, and 58%, respectively. Based on the standard deviations, D1 and I2 were not different, but I1 had a significantly larger resistant population.

These studies were designed to determine if there was a correlation between glyphosate resistance and the degradation of glyphosate in the biosystems. No pattern was clear after several samplings. However, in these experiments plates were incubated for 48 hours before counting colonies. Figure 10.4 presents counts from D1, I2, and I1 after 96 hours incubation. As glyphosate concentrations increased, D1 counts decreased as previously noted. Yet I1 and I2 counts increased 60%–80% as glyphosate levels increased. A range of standard deviations (2%–16%) was noted in D1 and I1, though I2 was more variable (5%–30%). A replication of this experiment with D1 and I1 confirmed these trends.

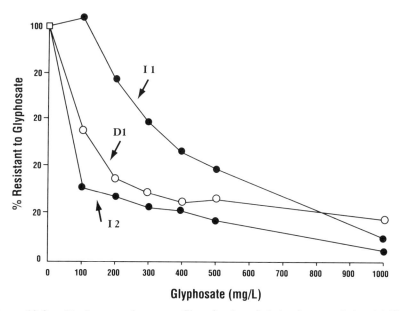

Figure 10.3. Glyphosate resistance profiles of activated sludge from two industrial (I1, I2) and one domestic (D1) biosystems. Plates were incubated for 48 hours and counted. Standard deviations ranged from 3% to 18%.

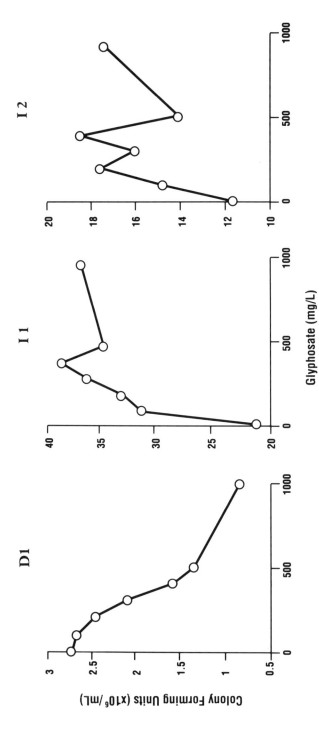

Figure 10.4. Microbial counts of I1, I2, and D1 activated sludge after 96 hours incubation. Standard deviations ranged from 3% to 18%.

It was clear that a population of slow-growing microorganisms was present in I1 and I2 that was not present in D1. It was possible that a portion of the population might be inhibited by glyphosate, recovering only after 48 hours incubation (and thus counted after 96 hours incubation). The dramatic increase in counts on glyphosate plates over plates containing no glyphosate strongly suggested that a population of microorganisms was either enhanced by the absence of fast-growing, glyphosate-sensitive microorganisms or glyphosate itself had a direct effect.

To determine if glyphosate could be used as a nutrient source, the industrial sludge samples were streaked on a defined medium. The colony-forming units (cfu) were determined after incubating for 5 days at 25°C. In this experiment, glyphosate was the sole source of phosphorus in the medium, and care was taken to remove all other contaminating phosphorus. Thus, it was assumed that the only colonies growing on these plates would be those utilizing glyphosate as a phosphorous source. On three separate sampling dates, I1 had the highest bacterial counts, averaging 23.5×10^6 cfu/ml. I2 bacterial counts averaged 5.48×10^6 cfu/ml. Control plates that did not contain glyphosate showed only a few pinpoint colonies. These results indicate that glyphosate increased total microbial numbers. The growth of some small colonies on the control plates containing no phosphorous source was presumably due to phosphorous carry-over from the aqueous inoculum or cellular endogenous reserves.

A single colony was picked at random. This isolate was streaked on fresh medium containing glyphosate, and a second colony was picked. This isolate was subsequently identified as a *Flavobacterium* sp. (Balthazor and Hallas, 1986). It was inoculated into a glyphosate broth as well as a control broth lacking glyphosate. The microbial rate of utilization of glyphosate was monitored (Fig. 10.5). Lag phase lasted for 25 hours. Slow growth then began as glyphosate was completely metabolized. Nearly stoichiometric amounts of aminomethylphosphonic acid (AMPA) were produced. Log growth lasted for 50 hours. About 25% of the AMPA was mineralized as indicated by the increase in PO_{4-3}. Other experiments (data not shown) indicated that AMPA could also be used as a sole phosphorous source. Furthermore, the presence of PO_4 did not affect glyphosate metabolism to AMPA but did inhibit AMPA degradation.

Rueppel et al. (1977) first identified a glyphosate-degrading activity (GDA) in flasks containing soil and water. The compound was metabolized to AMPA, which was subsequently mineralized. The *Flavobacterium* sp. described here also produced AMPA, and subsequent studies have noted isolates producing AMPA when using glyphosate as a sole carbon source (McAuliffe et al., 1990). Glyphosate metabolism is also known to occur by direct cleavage of the C–P bond and the intermediate production of sarcosine followed by glycine. This was first noted by Moore et al. (1983), Shinabarger and Braymer (1986), and Kishore and Jacob (1987) in a *Pseudomonas* sp. and subsequently in *Agrobacterium radiobacter* (Wackett et al., 1987), an *Arthrobacter* sp. (Pipke et al., 1987; Pipke and Amrhein, 1988), an *Alcaligenes* sp. (Lerbs et al., 1990), and several soil isolates (Quinn et al., 1988, 1989).

These two general methods for glyphosate degradation are characterized by the

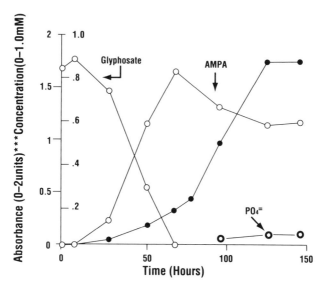

Figure 10.5.   *Flavobacterium* growth in the presence (●) of glyphosate. In the absence of glyphosate, growth was <0.4 absorbance. Glyphosate was transformed to AMPA. AMPA was partially mineralized to PO_4. (Modified from Balthazor and Hallas, 1986, with permission of the publisher.)

different intermediates produced (AMPA and sarcosine/glycine) as well as by the initial insensitivity of the AMPA pathway to phosphate. In a series of studies, Jacob et al. (1985, 1987, 1988) analyzed cells grown on ^{13}C, ^{15}N-labeled glyphosate using cross-polarization magic-angle spinning (CPMAS) and ^{13}C nuclear magnetic resonance (NMR). Figure 10.6 is a compilation of what they described. In the AMPA pathway, at least two enzymatic steps are involved: a cleavage of glyphosate to AMPA and a phosphonatase activity that mineralizes AMPA. The mechanism of carbon cleavage is unclear, with either one or two carbon units being produced. In the glycine pathway, a phosphonatase activity is followed by a sarcosine oxidase-dehydrogenase. There is published precedent for the more unique reactions (Quinn, 1989). LaNauze et al. (1970) have suggested the presence of a phosphonatase that

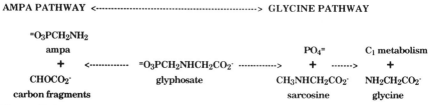

Figure 10.6.   The glycine and aminomethylphosphonic acid (AMPA) pathways for glyphosate biodegradation.

would cleave a C–P bond in both the AMPA and glycine pathway. Cook et al. (1978) have proposed similar enzyme step(s) operating for AMPA biodegradation.

At least one isolated microbe (a *Pseudomonas* sp. strain LBr) possesses both pathways (Jacob et al., 1988). It is capable of quickly metabolizing almost 20 mM of glyphosate, and about 5% of the compound is degraded via the glycine pathway. This microbe was isolated from industrial activated sludge treating high levels of glyphosate. Key to its performance, though, was its ability to degrade glyphosate even in the presence of inorganic phosphate. Schowanek and Verstraete (1990a,b) have suggested that phosphonate degradation in the presence of phosphate can occur in the environment. As noted below and in the next section, this uncoupling from phosphate regulation is a critical characteristic in the ability of the sludge to detoxify the herbicide.

3.1.2. Enumeration and Numerical Taxonomy of GDA

The next step was to characterize the relationship of GDA microbes with other members of the activated sludge consortia. Hallas et al. (1988) enumerated and then characterized microorganisms from the I1, I2, and D1 activated sludges. Nearly 400 microbial strains were identified and used to correlate GDA to 155 biochemical and morphological characteristics. Initially, several enumerations were done by microscopic examination and eight different media. These media were designed to select for 1) total microbial counts and counts of 2) gram-positive microorganisms, 3) gram-negative microorganisms, 4) pseudomonads, 5) *Bacillus*, 6) yeasts, 7) methanotrophs, and 8) formalotrophs.

Three subsamples from each biosystem were plated in triplicate for 5 months. While the direct counts enumerated all cells (viable and nonviable), the plating media enumerated any microbe that could grow 1) on agar plates containing glucose and yeast extract and 2) in the presence of the amendments unique to the selection media.

In addition, a dilution of process wastewater was added to industrial plates as a selective pressure (only the methylotroph and formalotroph media were exempted). This amendment consisted of several components including formaldehyde, formic acid, phenolics, and several phosphonates. It was assumed that this general amendment would inhibit the growth of transients while letting microbes grow that thrive in these biosystems (Hallas et al., 1988).

In general, the industrial biosystems had higher total counts than the domestic biosystem, but fewer pseudomonads and no yeasts. All biosystems had large populations of gram-positive and gram-negative bacteria as well as *Bacillus*, methylotrophs, and formalotrophs. Statistical comparisons were made between microbial counts and measures of biosystem performance. Enumerations of I1 and I2 tended to correlate with routine measurements currently done at these plants. Thus, high counts were associated with high BOD, COD, mixed liquor volatile suspended solids (MLVSS), and oxygen uptake rates, while lower counts were associated with high pH. Correlations of counts with removal efficiencies were not as clear cut, and no enumeration correlated with GDA.

Only four known references in the literature have considered broad groups of activated sludge microbes. Dias and Bhat (1964) found that gram-negative microbes predominated in their studies and Unz and Davis (1975) found low fungi (yeast) counts. Both studies were done with domestic activated sludge. Takii (1977) studied industrial carbohydrate waste biosystems and found gram-negative microbes predominated in low carbohydrate wastes and gram-positive microbes predominated in high carbohydrate wastes. Rogovskaya et al. (1959), Goud et al. (1985), and Seiler and Blaim (1982) also looked at industrial activated sludge (in this case the wastes from the chemical industry) and found that pseudomonads were the dominant genera.

The ultimate goal was to relate microbes with GDA to the other activated sludge microbes. However, glyphosate treatment efficiencies did not correlate with any of the plate counts. This could have provided a quick plating assay for detecting biosystem GDA efficacy. Further characterizations were then attempted using numerical taxonomic techniques.

Numerical taxonomy in microbiology has been applied in several ways. For example, it has been used to determine characterization keys for taxonomic groups (Molin and Turnstroem, 1982; Simidu and Tsukamoto, 1985) in order to assign diagnostic tests for specific intragroup differentiations. Cluster analysis has been used to probe the effect of environment parameters on microbial diversity (Bell et al., 1982; Hauxhurst et al., 1980; Quesada et al., 1983; West et al., 1984). Limited studies also have been done on the microbial diversity of activated sludge (Seiler et al., 1980; Seiler and Blaim, 1982; Takaii, 1977), though none of these have probed species diversity versus the degradation of specific substrates.

The numerical taxonomy of I1, I2, and D1 was begun with the isolation of almost 400 microbial strains. All strains were lyophilized for long-term storage, and a microbial strain library was constructed. A laboratory protocol was then developed for testing 155 microbial traits with commercially available microbial identification kits. These traits included cellular and colonial morphologies as well as biochemical analyses. All test traits were coded in binary form, and strain similarities were calculated by the unweighted pair-group arithmetic average clustering (UPGMA) using Simple Matching and Jaccard coefficients (Sneath, 1980; Walczak and Krichevsky, 1980).

Initial analyses noted that the activated sludge of all three systems contained bacterial populations that were not identifiable using the available diagnostic kits (Hallas et al., 1988). Mapping showed that clustering of strains reflected their biosystem origins. In total 5 D1, 12 I2, and 2 I1 clusters were described at the 83% similarity level. General characterizations could also be made. Industrial sludge was able to utilize a wider range of carbohydrates (facultatively and aerobically) than the domestic sludge (Table 10.4). This showed that the industrial biosystems were different from the domestic sludge and contained a wider range of genetic capabilities.

These general analyses were compared with an individual strain's ability to utilize glyphosate as a sole phosphorous source. Note that the ability to degrade glyphosate was not a selection criteria during the initial isolation of the strains. It

TABLE 10.4. Key Metabolic Characteristics Found in Two Industrial (I1, I2) and Domestic (D1) Activated Sludges

Occurrence (%)[a]	Biosystem	Characteristic[b]
100	I1	Glycerol, gly aminopeptidase, arg aminopeptidase, leu aminopeptidase
>90	I1	d-Arabitol, d-turanose, trehalose, saccharose, maltose, mannitol, inositol, d-glucose, adonitol, n-acetylglucosamine, o-glucosidase, glucosaminidase
	I1, I2	l-Fructose, arginine
	I1, D1	Alkaline phosphatase
	I2	Glycerol, gly aminopeptidase, arg aminopeptidase, leu aminopeptidase
>70	I2	Alkaline phosphatase
	D1	Arg aminpepeptidase
>70	D1	d-Glucose, arginine

[a]Percent occurrence of characteristic among isolated strains.
[b]Characterization is fermentative or enzymatic.

was found that all the GDA isolates metabolized glyphosate to AMPA; none mineralized the compound under the flask assay conditions. In addition, the GDA trait was entirely absent in the D1 strains. Thus, GDA is a trait that requires enrichment through selective pressure; it does not seem to occur spontaneously. Isolates with GDA did not cluster into a separate group; they were distributed along the whole dendogram as members of seven defined clusters of I1 and I2 subsets.

As with the enumeration data, an attempt was made to determine if any positive test reactions in the characterization protocol could be correlated with GDA as a quick bioassay. The data matrix was transposed and a cluster analysis done using the Jaccard coefficient. Figure 10.7 is a dendogram from this analysis. The Jaccard coefficient accounts for matching only positive test reactions within the data matrix. Thus, two tests reacting positive for a large number of the isolates link at a high similarity coefficient (low on the Y axis). Most tests linked at a similarity coefficient of <0.55. This is less than the >0.85 found when mapping the strains and indicated the uniqueness of the tests. Indeed, the dendrogram showed that GDA did not correlate to any other tests (similarity coefficient, <0.15). Clearly, the ability to degrade glyphosate was a trait not common in activated sludge microorganisms.

In summary, each activated sludge contained unique bacterial populations, with the industrial microbes capable of utilizing a wide range of carbohydrates. Numerical taxonomic characterization confirmed that there were several clusters within the sludges. GDA was found in only a small portion of the industrial clusters, and no GDA microbes were found in the domestic sludge. In addition, GDA did not correlate with any other characteristic tested, even though the GDA strains had a large phenotypic diversity. The study suggested that GDA is a trait induced by the selective pressures of the waste stream. Therefore, it is possible that domestic

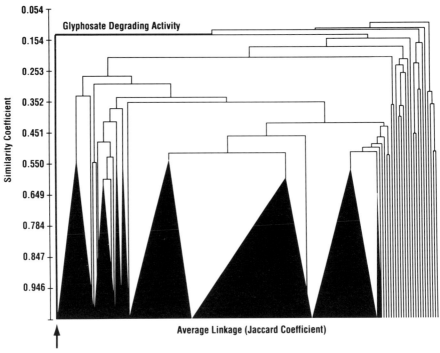

Figure 10.7. Linkage of Jaccard coefficient on 155 biochemical tests for 393 microbial strains. The characteristic for glyphosate degradation to AMPA (GDA) is the first test on the x axis (marked by an arrow).

sludge could contain bacteria capable of degrading glyphosate, but the enzymes needed are probably inducible.

3.2. Degradation of Chemical Mixtures by Immobilized Bacteria

Industrial waste streams often contain mixtures of organic chemicals, and the ability of biotreatment technologies to handle these mixtures successfully must be evaluated. In some cases, a particular mixture of organic chemicals may be completely compatible, and microbial degradation presents no problems. However, there is always concern that chemical mixtures may contain some toxic components that may inhibit microbial degradation. In addition, the formation of toxic chemical intermediates resulting from either co-metabolism or incomplete chemical degradation may also diminish reactor performance and, in extreme cases, could poison the system. In some cases, microorganisms having the ability to degrade a specific chemical may simply not utilize it in the presence of alternate, more desirable chemical substrates. These possibilities and concerns point to the need for specific testing of microbial performance in order to evaluate compatibility of chemicals found as mixtures in industrial waste streams. An additional benefit of this approach is the ability to obtain mass-balance accountability for chemicals degraded by microorganisms. These material-balance data are often useful for proving that chemical mineralization (as opposed to incomplete degradation, chemical adsorption, volatilization, and so forth) is occurring and in showing that significant levels of partially degraded, and potentially toxic, chemical intermediates are not accumulating in the system.

This approach was recently used in evaluating *Pseudomonas* strain PNP1 for use in the degradation of a waste stream containing aniline (ANL) and *p*-nitrophenol (PNP) as two of the principal components (Heitkamp et al., 1990b). *Pseudomonas* strain PNP1 was previously shown in flask studies to have exception PNP-degrading activity resulting in over 70% mineralization of a 70 mg/liter dose within 24 hours (Heitkamp et al., 1990a). Furthermore, this bacterium reduced PNP concentrations to below 50 μg/liter, produced less than 1% of volatile metabolites, incorporated 10.2% into cell biomass, and only left 5.3% of PNP residues as highly polar, water-soluble metabolites (Fig. 10.8A). Although this bacterium had exceptional and well-characterized PNP-degrading activity, its ability to degrade ANL as well as PNP in the presence of ANL had to be evaluated in order to determine its potential application to this waste stream.

Radiometric evaluations of ANL degradation and PNP degradation in the presence of ANL are shown in Figure 10.9A, 10.9B, respectively. The *Pseudomonas* sp. was found to mineralize 63% of a 70 mg/liter dose of ANL within 24 hours. Cellular residues accounted for an additional 25% of the total and aqueous residues were reduced to 8%. Chemical analyses of these aqueous residues by reversed-phase high-performance liquid chromatography showed the absence of ANL and the presence of highly polar water-soluble metabolites. No volatile metabolites of ANL were detected. It is noteworthy that the presence of ANL did not adversely affect the rate or extent of PNP degradation by this bacterium. The *Pseudomonas* sp. mineralized over 74% of PNP in the presence of ANL.

A comparison of the material balances for the degradation of PNP and ANL by *Pseudomonas* sp. PNP1 is presented in Figure 10.8. These studies show that *Pseudomonas* sp. PNP1 readily mineralizes ANL and PNP in the presence of ANL and indicate that this bacterium is a good candidate for use in further laboratory-scale feasibility studies to evaluate immobilized bacteria for biotreatment of this industrial waste stream.

A microbial consortium containing *Pseudomonas* sp. PNP1 was adapted in the laboratory to 2.8% salinity and evaluated for biotreatment of an industrial waste stream containing 50–300 mg/liter of ANL, 200–450 mg/liter of PNP, and 3,000–5,000 mg/liter of potassium formate (Heitkamp et al., 1990b). The microbial inoculum was immobilized on a Manville R635 diatomaceous earth biocarrier and tested in immobilized bacteria technology (IBT) columns as previously described (Heitkamp et al., 1990a). The experimental design consisted of monitoring the performance of the immobilized bacteria for removing ANL, PNP, and potassium formate from synthetic wastes pumped through the reactor during a 6 week period. The performance of the immobilized bacteria on this waste stream is shown in Figure 10.10. Some breakthrough of undegraded ANL or PNP was detected early in

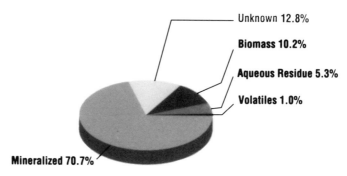

A p-NITROPHENOL DEGRADATION

PSEUDOMONAS SP. PNP1

Unknown 12.8%

Biomass 10.2%

Aqueous Residue 5.3%

Volatiles 1.0%

Mineralized 70.7%

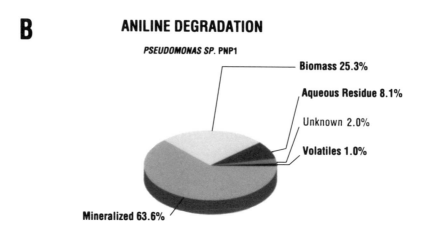

B ANILINE DEGRADATION

PSEUDOMONAS SP. PNP1

Biomass 25.3%

Aqueous Residue 8.1%

Unknown 2.0%

Volatiles 1.0%

Mineralized 63.6%

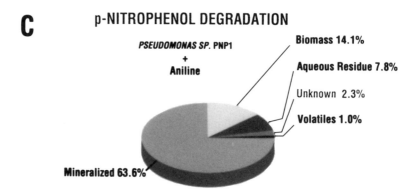

C p-NITROPHENOL DEGRADATION

PSEUDOMONAS SP. PNP1
+
Aniline

Biomass 14.1%

Aqueous Residue 7.8%

Unknown 2.3%

Volatiles 1.0%

Mineralized 63.6%

Figure 10.8. Mass-balance of ^{14}C residues of *p*-nitrophenol (**A**), aniline (**B**), and *p*-nitrophenol in the presence of aniline (**C**) for *Pseudomonas* sp. PNP1. Data for panel A taken from Heitkamp et al. (1990a).

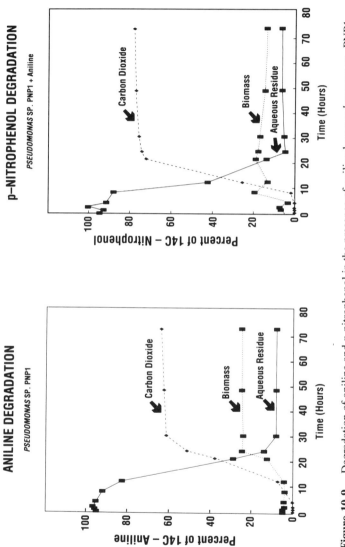

Figure 10.9. Degradation of aniline and *p*-nitrophenol in the presence of aniline by *pseudomonas* sp. PNP1.

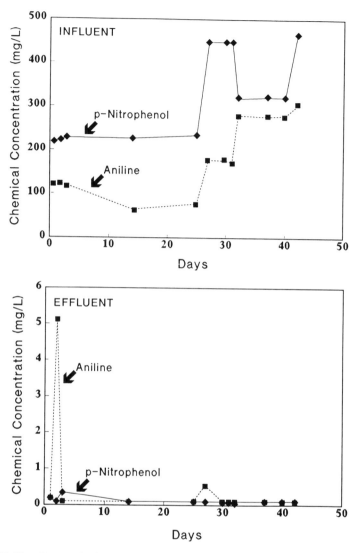

Figure 10.10. Removal of *p*-nitrophenol and aniline from a waste stream by bacteria immobilized on diatomaceous earth beads in a packed bed reactor. The waste stream also contained 3,000–5,000 mg/liter of potassium formate (90% degraded, data not shown) and 2.8% salinity.

the study and on day 28 when the concentrations of ANL and PNP in the feed were sharply increased. Overall, the IBT column consistently removed over 99% of chemicals in this waste stream for a period of 30 days at HRTs ranging from 5 to 8 hours. These results confirm the results of pure culture laboratory testing, with *Pseudomonas* sp. PNP 1 showing degradation of both ANL and PNP, and indicate that IBT using highly selected microbial strains is an effective alternative for the biotreatment of chemical mixtures in this waste stream.

3.3. Elemental Transformations (Ammonia)

Biological treatment of industrial wastewater primarily involves the detoxification or degradation of product active ingredients or key intermediates. However, mineralization of organics can lead to high inorganic levels in effluent discharges (Arbuckle et al., 1989). A good example is the production of ammonia in high strength nitrogenous wastes that could eutrophy receiving waters.

Figure 10.11 diagrams the basic steps in the nitrogen cycle. Ammonification of organic nitrogen to NH4$^+$-N can be accomplished by several microbial genera. Nitrification of NH4$^+$-N is a two-step process: *Nitrosomonas* sp. oxidizes NH_4 to NO_2^- and *Nitrobacter* sp. oxidizes NO_2^- to NO_3^-. Both are strict aerobes and autotrophic (i.e., their carbon source is CO_2), respiring the NH_4/NO_3^- for energy and reducing power. Denitrification, a common microbial trait, follows nitrification. It involves use of NO_3^- as an electron acceptor, resulting in the reduction of NO_3^- to gases when oxygen is unavailable.

The nitrogen cycle occurs in industrial biosystems. Verstraete et al. (1977) and Prakasam et al. (1978) have found that biosystem acclimation is possible. However, microbial enrichments can be hindered in three key ways. First, the process is sensitive to pH fluctuations, which can upset biosystems in the initial acclimation phase. Second, nitrifiers are slow growers. This requires longer residence times and higher sludge ages that most waste treatment operators may not be accustomed to. Finally, the toxicity of NH_2-N to *Nitrobacter* sp. (approximately 1 mg/liter) and NO_2-N itself (around 3 mg/liter) inhibits the system. High NH_3-N levels are usually considered the culprit in poisoning nitrification. Actually, the problem is the $NH_3:NH_4^+$ equilibrium. At high pHs, NH_3-N is favored. Yet, by lowering the pH to 6.0, the overall process slows down. Thus, the key to starting nitrification is keeping the steady-state NH_4-N (actually NH_3-N) levels low until the oxidation begins and the nitrifying population increases.

Microbiological ammonification, nitrification, and denitrification can be a reliable and cost-effective technology for removing NH_4-N from industrial waste streams. However, successful start up and operation require understanding the

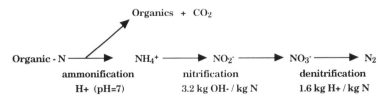

Figure 10.11. The nitrogen cycle as it can occur in activated sludge. In a single-stage system, pH demands are somewhat buffered.

microbiology of the system. The following example shows a case where organic and inorganic bioxidations were successfully combined.

A preliminary step in the glyphosate process is the synthesis of iminodiacetic acid (IDA). This interim process can produce wastewater with total nitrogen levels in excess of 500 mg/liter and carbon nitrogen levels of 4:1. The biotreatment of this waste would be more cost effective than physically stripping the high ammonia levels. Therefore, an investigation was conducted to determine if nitrifying micro-organisms were present in an activated sludge that would treat this waste. Ulti-mately, the study goal was to establish a lab scale "IDA" bioreactor that could ammonify and nitrify the wastewater in a single stage.

Initially, seeds from several sources were inoculated into a modification of the L-salts medium previously described in this chapter (Leadbetter and Foster, 1958). In this case, ammonium chloride (250 mg/liter) was the sole nitrogen (and energy) source. Also added were calcium carbonate chips (as a carbon source), a chelated iron solution (0.1 mL/liter; NaEDTA (0.14 g), iron sulfate septahydrate (0.5 g), sulfuric acid (0.05 ml) per 1 liter stock), and Cresol Red 25 ml/liter of a 0.005% stock). The Cresol Red was a pH indicator changing from red to yellow as the pH dropped. Potassium carbonate was added to raise the pH.

Several sources of microbes were used to isolate nitrifying microbes: 1) An aged aquarium detritus; 2) activated sludge from I2 activated sludge biosystems and 3) SBRs; and 4) activated sludge from a lab-scale bioreactor treating IDA wastewater, but not exhibiting nitrification. Flasks were incubated for several weeks, and the amount of potassium carbonate for neutralizing the media was monitored. As noted in Table 10.5, the IDA bioreactor and I2 SBR needed the largest amounts of base:

TABLE 10.5. The Isolation of a Nitrifying Activity in Several Sludge Seed Sources

		K_2CO_3 added (ml)		
Seed	Day	Actual	Cumulative	Comments
Aquarium	7	0.3		Small detritus
	11	0.3	0.6	
I2 Biosystem	7	0.3		Detritus, small gm coccoid cells
	10	1.0	1.3	
	11	0.4	1.7	
I2 SBR	4	0.6		Detritus, small gm coccoid cells
	5	1.0	1.6	
	6	0.7	2.3	
	11	0.7	3.0	
IDA bioreactor[a]	4	0.3		Large, coccoid and diplococ-
	5	1.3	1.3	coid gm cells
	6	0.7	2.3	
	7	0.6	2.9	
	8	0.3	3.2	

[a]The IDA bioreactor was enriched two more times. After 14 days, on the third run, total nitrogen was reduced 53% (63–29 mg/liter), NH_3-N was reduced 780% (62–18 mg/liter), NO_2-N increased 64× (0.05–3 mg/liter), and NO_3-N increased 37× (1–44 mg/liter).

almost $2\times$ higher than the I2 biosystem flask and $5\times$ higher than the aquarium flask.

Microscopic examinations of the active flasks revealed gram-positive coccoids and diploccoid cells that were similar to the cellular morphology of nitrifying bacteria. The IDA bioreactor culture was passed two more times in fresh media, resulting in a 10^9 dilution of the original population. Nitrogen analyses of spent media revealed significant reductions in total nitrogen and NH_3-N and increases in NO_2-N and NO_3-N (Table 10.5). These data confirmed that nitrifying populations were present in activated sludge and could be cultivated.

The next step was to establish ammonification and nitrification in a single-stage lab bioreactor. A plexiglass bioreactor (14 liters) was divided into a 10 liter aeration volume and a 4 liter clarifier volume (Eckenfelder et al., 1985). Initially, the HRT of the reactor was 10 days. Ammonification of the organic nitrogen produced levels of NH_3-N in excess of 300 mg/liter. After several months, an enrichment for nitrification was attempted by 1) increasing the HRT to 30 days (nitrifiers are slow growers), 2) adding HCO_3 (nitrifiers are autotrophs), and 3) maintaining a pH of 7–7.5 (toxic NH_3-N species less favored). Within 3 weeks, the bioreactor pH became erratic and dropped as low as 6.2. Potassium carbonate was substituted for HCO_3, and the pH was successfully maintained at 7.0. Figure 10.12 presents the nitrogen balance during these 2 months. NH_3-N levels began decreasing almost immediately, and NO_3-N levels began increasing. NO_2-N also accumulated for a short time. Close to 21 mM NH_3-N was available in the bioreactor. The data suggested that at least 5 mM (25%) accumulates as NO_3-N. The remaining NH_3-N was either used for growth or was denitrified to nitrogen gases.

Ultimately, the HRTs were lowered from 30 days to 5 days over 4 months. BOD, TOC, and total nitrogen treatment efficiencies were in excess of 95%, 97%, and 99%, respectively. The removal rates dropped slightly when HRTs were pushed to 5 days. Nevertheless, the treatment efficiencies produced acceptable effluent. Eventually, a 7 day HRT was used in design specification for this reactor. However, critical to the field scale unit would be the biological reduction of the remaining nitrate; denitrification in clarifiers can result in bulking and system upset. Still, this

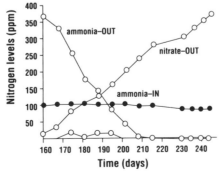

Figure 10.12. Ammonification and nitrification in a single-stage lab bioreactor. Some of the nitrate was not accounted for and is presumed denitrified to nitrogen gases.

study demonstrates that organic and inorganic biooxidations can be coupled when the system microbial ecology is better understood.

4. TECHNOLOGY ADVANCES IN WASTEWATER TREATMENT

The previous section focused on the microbiology of industrial wastewater biotreatment. With ever more stringent water quality guidelines, a better understanding of the microbial ecology is critical. In the last decade, advances in full-scale treatment systems have been made in two key areas. First, successful field trials have been documented where manipulating the microbiology of full-scale activated sludge biosystems produced measurable benefits in performance. This has helped waste treatment operators gain confidence in the multimillion liter biological processes they are trying to control. Second, novel hardware advances have been made that increase performance and may decrease costs. Examples of developments in these areas are described below.

4.1. Activated Sludge: Continuous Flow Reactors

Activated sludge has played a key role in the biotreatment of glyphosate process wastes. Stringent NPDES permits at process facilities require high treatment efficiencies of organic carbon (COD, BOD) as well as glyphosate. Establishing and maintaining GDA has been critical in keeping this active ingredient from discharge. Work described earlier noted the different microbial populations found in activated sludge treating glyphosate process wastewater (I1,I2) versus domestic activated sludge (D1). Industrial isolates were capable of utilizing a wider range of organics (including glyphosate), which suggested a more diverse microbial taxonomy. In addition, once industrial biosystems had established a GDA, they were able to maintain it even as organic carbon, glyphosate, and phosphate loading, temperature, and pH changed (Table 10.6). These observations suggested that the key to GDA was in initiating the activity itself.

As noted throughout this chapter, critical to developing special organic-

TABLE 10.6. **Range of Key Biotreatment Parameters Where Glyphosate Degradation Activity Was Maintained**

Parameter	Range[a]
Organic carbon	0.1–1.0%
Inorganic carbon	0–6.3 mM
Glyphosate	1.2–6.5 mM
Temperature	13°–40°C
pH	5–10

[a]Studies from lab batch and chemostat studies using pure and mixed cultures from industrial activated sludge.

degrading activities is understanding the master variables controlling the special microbial populations. However, maintaining bacteria in activated sludge—especially in 4–7 million liter tanks—is not easy and requires a higher level of communication with waste treatment operations. When the education is there, classic and simple microbial approaches can be successfully demonstrated in full-scale systems.

The first approach taken to establish GDA was to recycle sludge from the I2 biosystems aerobic digester. The digester's role in waste treatment is to reduce solids; it is always operated in a carbon-starved condition. GDA has always been detected in the digester whenever sampled. Thus, it was hypothesized that if low microbial populations with GDA were the reason no activity was detected in a downstream aeration basin, then seedings from the digester would initiate GDA. Table 10.7 presents the seeding schedule. Almost twice the volume of the biosystems was added as seed over a 6 week period. Other phosphonate treatment efficiencies increased to 99%. Effluent glyphosate concentrations also decreased but to levels only half of influent loadings. It was clear that augmenting digester microbes into the biosystem was not enough to establish GDA.

A second approach was taken using a specific process waste stream that contained high levels of glyphosate. This method of enrichment has been used in lab studies selecting for specific microbes. However, its use in large, field-scale systems has not been as common. Four feedings of the concentrated glyphosate wastewater were made over a 15-week period. The glyphosate increases were 4× to 9× the normal glyphosate loads (Table 10.7).

Effluent glyphosate levels dropped to undetectable limits, though sustained per-

TABLE 10.7. Digester Sludge and Glyphosate Feed Amendments to I2 Aeration Basins and the Response by Glyphosate-Degrading Microorganisms

Time (Weeks)	Seeding (Liters × 10⁵)[a]	Feed Added (Glyphosate, mg/liter)[b]	Flask Assay (Incubation, days)[c]
0	1.1	—	40
1	2.2	—	—
3	1.7	—	—
4	1.1	—	14
5	4.2	—	—
6	—	40	—
7	—	60	—
8	—	90	—
20	—	80	—
21	—	—	6

[a]The combined volume of the aeration basins was 5.7 million liters.
[b]The glyphosate process wastewater contained 3%–4% glyphosate. Higher glyphosate levels were measured within 48 hous of a feeding. Normal glyphosate levels were 8–12 mg/liter.
[c]Assay as described by Balthazor and Hallas (1986). A 100% GDA was the time taken to degrade glyphosate to 0 and accumulate stoichiometric amounts of AMPA.

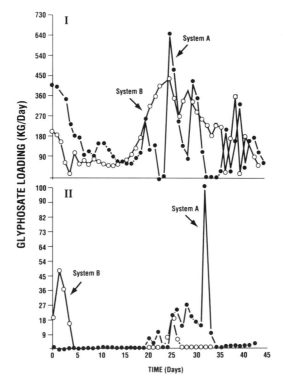

Figure 10.13.   Glyphosate levels in influent (I) and effluent (II) of I2 biosystems A (●) and B (○).

formance was not demonstrated for 4 months. To determine if this activity was due to higher populations of microbes possessing GDA, a flask assay (Balthazor and Hallas, 1986) was used to sample the biosystem (Table 10.7). The initial results indicated that, while GDA was not expressed in the biosystem, a very low level of activity was present (i.e., 100% GDA after 40 days). In 5 weeks (when the bioaugmentation ended) the flask assay noted a 14 day reaction time. After the test period ended, a 100% GDA was being demonstrated in 6 days. Thus, populations of specific microbes with GDA had grown to levels able to sustain GDA in the biosystems.

A third approach to initiating GDA in field units was attempted in the I1 biosystems. In this case, it was hoped that acclimation time could be significantly shortened by combining concentrated feed amendments with an increased HRT (i.e., lower F/M). Figure 10.13 depicts the glyphosate levels in the influent (Fig. 10.13, I) and effluent (Fig. 10.13, II) of two biosystems with volumes of 3.2 million liters. Initially, systems A and B were receiving 430 and 210 kg/day of glyphosate, respectively. Flow to system B was lowered by 50%–75%, and sludge glyphosate levels immediately dropped. After flow was restored to normal on day 6,

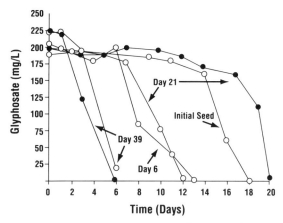

Figure 10.14. GDA flask assays for biosystems A (●) and B (○). Noted days are when samples from the sludges were taken.

GDA maintained effluent glyphosate at <1 kg/day from the 20–40 kg/day prior to the slowdown. On day 15, influent glyphosate was increased to both systems using waste from a concentrated glyphosate feed. Biosystem B maintained low effluent glyphosate levels even as influent glyphosate concentrations reached 450 kg/day. On day 24, system A performance began to deteriorate. The aeration tank was batched for 24 hours on days 24 and 33. Microbial response was immediate, and GDA was near 100% within days.

Flask assays were also conducted on samples from I1 biosystems during the field study (Fig. 10.14). As noted previously, a healthy sludge with GDA will biodegrade glyphosate in the bioassay within 6 days. System B's sludge was sampled before the study began, and it took 18 days to degrade glyphosate in the assay. After the flow slowdown, the sludge had matured to a 10–13 day "competency." By day 39, samples of the sludge were normal (i.e., glyphosate degradation was degraded after 6 days in the assay). In contrast, the biosystem A sludge inoculated into the assay on day 21 did not degrade glyphosate. This was just prior to the biosystems loss of GDA. After batching this system, the sludge response to GDA returned to normal.

These two field studies demonstrate several simple strategies for establishing GDA in continuous flow activated sludge aeration tanks. The challenge is twofold. First, manipulation of thousands of kilograms of sludge in millions of liters of aeration volume is difficult. Response time is slow, and any mistakes linger. Second, continuous flow activated sludge is an excellent technology for treating high BOD loadings, but populations are quite stable if near steady state kinetics are achieved. Changing the microbial ecology of activated sludge is better accomplished when operators have more control over operating strategies. This is described in the next section.

4.2. Activated Sludge: Sequencing Batch Reactors

There are many ways to classify waste treatment systems. As previously noted, continuous flow activated sludge systems are spatially oriented with wastewater flowing from aeration tanks to clarifiers. Semicontinuous operation like SBRs are a time-oriented process that takes place in a single tank. In many cases, SBR systems offer distinct advantages over traditional activated sludge.

For example, operational strategies can be used to more easily provide strong selective pressures. FILL/REACT phases can be manipulated (e.g., air delivery, nutrient addition) to subject microbes to a wide range of environmental conditions. The result is a high biological activity specifically tailored to operational need.

In the case of glyphosate, concentrated process streams provided an opportunity for selective biotreatment. Since glyphosate was the sole constituent that had to be removed. The I2 industrial facility had used incineration to remove glyphosate at considerable expense. SBR technology was an inexpensive alternative if microbes with GDA could be enriched in the sludge.

In preliminary batch studies (D.V.S. Murthy, University of Notre Dame, personal communication) SBR seed was used in three flasks treating 1) a dilution of the concentrated wastewater (BR1), 2) pure glyphosate alone (BR2), and 3) glyphosate plus 50 mg/L NH_3-N (BR3). Figure 10.15 notes the degradation of glyphosate over time. Glyphosate removal was fastest when NH_3-N was present, the slowest reaction time was noted in sludge treating the waste stream. That was partly due to a 24 hour lag period.

It was also interesting that NH_3-N levels in BR1 (initially 23 mg/liter) increased to 130 mg/liter during the lag period prior to decreasing for the remainder of the study. These data suggest that there is an inhibition of GDA during the lag phase when competing reactions (especially ammonification) are occurring. However, GDA is enhanced by NH_3-N. It is unclear if this is due to the direct involvement of nitrifiers or to the removal of toxic intermediates.

Other operational control studies have been done. Murthy et al. (1988) found that different filling strategies for a concentrated glyphosate waste stream led to different

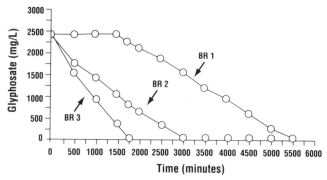

Figure 10.15. Glyphosate degradation in three batch flask reactors: BR1 (glyphosate wastewater), BR2 (glyphosate alone), BR3 (glyphosate plus NH_3-N).

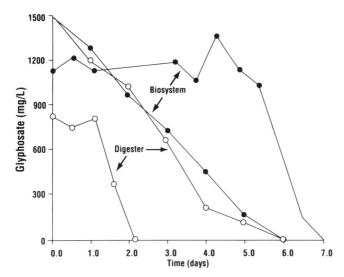

Figure 10.16. Glyphosate degradation in a 10 liter reactor using seed from the biosystem (●) and digester (○).

microbial selection. When a MIXED FILL strategy was used, glyphosate accumulated in the SBR. An AERATED FILL accumulated less glyphosate, but all of the herbicide was gone in both reactors at the end of REACT. The difference in reactors was the presence of denitrifiers in the MIXED FILL SBR. This promoted better sludge settling and gave a more stable reactor.

Field SBR studies were done starting with 10 liter units. A 150,000 liter SBR unit was then installed with initial work also done in a 30,000 liter compartment. In the 10 liter reactor, seed was added along with distilled water and wastewater, with final glyphosate levels reaching 200–400 mg/liter. The influent was added over 4–6 hours with aeration. The react phase lasted until the glyphosate disappeared. Figure 10.16 depicts glyphosate degradation in the 10 liter reactor with seed from either biosystem sludge or aerobic digester sludge. After the first run, both seeds degraded glyphosate with the digester seed 66% faster than the biosystem seed. A lag phase of 1 and 5 days was noted, respectively. No wasting of sludge occurred in these SBR studies, so the microbes were used through several cycles. By the fourth run, both seed sources degraded glyphosate in the 6 days.

For the larger scale studies (Fig. 10.17), a 24 hour cycle time was used: FILL (3–4 hours, air on), REACT (15 hours, air on), SETTLE (3 hours, air off), DRAW (2 hours, air off), IDLE (0–24 hours, air on). In the 30,000 liter reactor compartment, an acclimation of the biomass occurred with glyphosate degrading after 3 days in run 1 and 1–2 days in run 3. This is similar to the reaction time in the 10 liter reactors; initial experiments in the full-scale unit also followed the same trend, with glyphosate degradation occurring after 3 days in run 1. By run 15, only a 15 hour react phase was needed to degrade glyphosate.

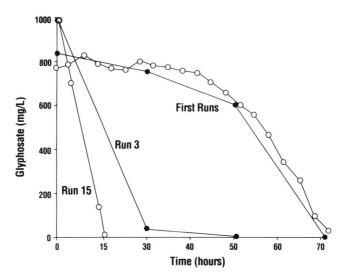

Figure 10.17. Glyphosate degradation in 30,000 (●) and 150,000 (○) liter SBR units.

These data support the premise that an SBR can be used to control specific microbes with GDA. In fact, glyphosate degradation rates up to 50 g/kg sludge/hour were realized in the study. Such activities are over 60 times better than traditional continuous flow systems. In any event, substantial savings in operational costs were realized at the I2 facility. Future industrial plants manufacturing glyphosate will also avoid installing incinerators—a multimillion dollar savings.

4.3. Immobilized Bacteria: Tertiary Biotreatment of Wastewater

One particularly attractive application of IBT is for the "polishing" or "final removal" of low concentrations of specific chemicals from high volume liquid waste streams. Examples of these types of waste stream include total plant effluents in which most organic chemicals have already been removed by a conventional biotreatment system or contaminated groundwater where a large aquifer may contain relatively low concentrations of recalcitrant chemicals. In either case, the ability of IBT to hold specific, chemical-degrading bacteria in a reactor at hydraulic flow rates that would wash them out of conventional bioreactors can result in very efficient removal of chemicals from high volume waste streams.

One example for this use of IBT is the removal of low concentrations of glyphosate from high volume liquid wastes. Table 10.8 presents a summary of results from a study in which a PBR inoculated with a microbial consortium having GDA was fed liquid wastes containing glyphosate (Heitkamp et al., 1992). Over 99% of initial glyphosate concentrations of around 400 mg/liter were removed at an HRT as short as 58 minutes. The flow of wastes into the PBR was then increased stepwise as the concentration of glyphosate was lowered to a target concentration of 50 mg/liter.

TABLE 10.8. Degradation of Glyphosate by Immobilized Bacteria

Day	Flow (ml/min)	Hydraulic Residence Time (min)	Glyphosate (mg/liter)[a]		Glyphosate[a] Removal (%)
			Influent[b]	Effluent[b]	
1–2	3	193	448	<2	>99
3–9	5	116	444	<2	>99
10–13	10	58	407	<2	>99
14–19	20	29	140	<2	>98
20–25	25	23	68	2.1	>96
26–35	30	19	49	8.8	82
36–56[c]	30	19	49	14.0	71
57–62[c]	30	19	51	13.0	75

[a]Glyphosate concentrations presented as the average measured during the presented time interval.
[b]Inorganic nitrogen added as 50 mg/liter of NH_4NO_3.
[c]IBT column was fluidized to remove excess biomass.

The peak performance of this PBR was achieved at an HRT of 23 minutes when over 96% of an average concentration of 68 mg/liter of glyphosate was removed. The PBR performance ranged from 71% to 82% removal at an HRT of 19 minutes, and repeated fluidization did not enhance performance by removing excessive biomass from the PBR. This laboratory study demonstrated that IBT was highly effective for removing low levels of glyphosate from high volume wastes.

A summary of the results from pilot-scale testing of an immobilized bacteria reactor to remove glyphosate from a total plant effluent is shown in Figure 10.18

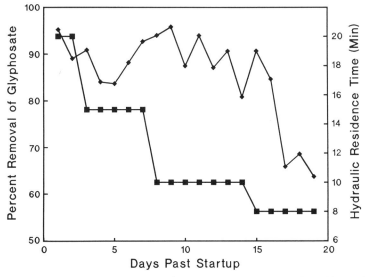

Figure 10.18. Glyphosate degradation (◆) by immobilized bacteria in a pilot-scale packed reactor at hydraulic residence times (■) ranging from 8 to 20 minutes.

(Hallas et al., 1992). This project utilized highly enriched glyphosate-degrading bacteria immobilized on diatomaceous earth beads in a 380-L packed bed reactor located at the glyphosate production facility. Hydraulic loading of the reactor was increased stepwise, and the performance was monitored at each set-point. These field results show that glyphosate removal rates ranging from 80% to 90% were achieved at HRTs as short as 10 minutes. Lowering the HRT to 8 minutes resulted in a consistent breakthrough of >30% undegraded glyphosate. Additional testing with this reactor showed that after a 21 day period of dormancy (no glyphosate in feed) the reactor was able to recover from a surge loading of glyphosate within 5 days (Hallas et al., 1992). This study demonstrated that immobilized bacteria in a packed bed reactor were highly effective for the tertiary removal of low levels of active herbicide from wastewater prior to discharge.

5. SUMMARY

This chapter presented an industrial perspective on the microbiological treatment of organic chemicals in wastes. The development of operational strategies that optimize existing conventional waste treatment systems as well as the successful development of new, high-performance waste treatment technologies both require a multidisciplinary research strategy. This strategy begins with basic research using microbiological and chemical methods to discover, characterize, evaluate, and optimize the performance of chemical-degrading microorganisms. This phase is followed by a progression from laboratory-scale testing to pilot-scale testing in order to demonstrate feasibility, further optimize operating conditions, and provide the engineering evaluation of performance needed to scale-up and operate a full-scale system successfully.

Ultimately, these systems will be required to withstand the harsh operating conditions and high performance expectations commonly required for industrial applications. Therefore, great care must be taken during development to evaluate biotreatment technologies under conditions that effectively simulate their final application. This approach provides a more realistic evaluation of system performance and can serve to identify key limiting factors leading to subsequent optimization in order to achieve even higher levels of performance and reliability.

REFERENCES

Allen LA (1944): The bacteriology of activated sludge. J Hyg 43:424–431.

Arbuckle NJG, Gordon T, Vanderver A, Randol RV (1989): Water pollution control. In Arbuckle NJ, et al. (eds): Environmental Law Handbook. Bethesda, MD: Government Institutes, Inc., 19th ed, pp 177–258.

Arora ML, Barth EF, Umphres MB (1985): Technology evaluation of sequencing batch reactors. J Water Pollut Cont Fed 57:867–875.

Balthazor TM, Hallas LE (1986): Glyphosate-degrading micro-organisms from industrial activated sludge. Appl Environ Microbiol 51:432–434.

Bell CR, Holder-Franklin MA, Franklin M (1982): Correlations between predominant heterotrophic bacteria and physicochemical water quality parameters in two Canadian Rivers. Appl Environ Microbiol 43:269–283.

Benedict RG, Carlson DA (1971): Aerobic heterotrophic bacteria in activated sludge. Water Res 5:1023–1030.

Busch AW (1971): Aerobic Biological Treatment of Wastewaters, Principles and Practice. Dallas: Oligodynamics Press, Inc.

Butterfield CT (1935): Studies of sewage purification. II. A zoogloea forming bacterium isolated from activated sludge. Public Health Rep 50:671–684.

Cardinal LJ, Stenstrum MK (1991): Enhanced biodegradation of polyaromatic hydrocarbons in the activated sludge process. Res J Water Pollut Cont Fed 63:950–957.

Case DR (1989): Resource conservation and recovery act (RCRA). In Arbuckle NJ, et al. (eds): Environmental Law Handbook, 10th ed. Bethesda, MD: Government Institutes, Inc., pp 563–605.

Chemical Manufacturers Association (1991): CMA Hazardous Waste Survey '89. Washington, DC: Chemical Manufacturers Association, Inc.

Cook AM, Daughton CG, Alexander M (1978): Phosphonate utilization by bacteria. J Bacteriol 133:85–90.

Cooper EF, Atkinson D (1981): Biological Fluidized Bed Treatment of Water and Wastewater. London: Ellis Horwood, Inc.

Dias FF, Bhat JV (1964): Microbial ecology of activated sludge. I. Dominant bacteria. Appl Microbiol 12:412–417.

Dietrich MJ, Copa WM, Chowdhury AK, Randall TL (1988): Removal of pollutants from dilute wastewater by the PACT trademark treatment process. Ind Prog 7:143–149.

Eaton D, Andrews S, Attaway H, Gallagher B (1989): Point source remediation of industrial wastewater using immobilized microbes. In Levandowski NJ (ed): Biotechnology Applications in Hazardous Waste Treatment. New York: Engin Foundation, Inc., pp 320–338.

Eckenfelder WW, Goronszy M, Quirk TP (1985): The activated sludge process: State of the art. In CP Straub (ed): CRC Critical Review in Environmental Control. Boca Raton, FL: CRC Press, Inc.

Edwards DE, Adams WJ, Heitkamp MA (1994): Laboratory-scale evaluation of aerobic fluidized bed reactors for the biotreatment of a synthetic, high-strength chemical industry waste stream. Water Environ Res. 66:70–83.

Edwards DE, Heitkamp MA (1992): Application of immobilized cell technology for biotreatment of industrial waste streams. In Industrial Environmental Chemistry: Waste Minimization in Industrial Processes and Remediation of Hazardous Wastes. New York: Plenum Press, pp 247–259.

Goud D, Parekh LJ, Ramakrishnan CV (1985): Bacterial profile of petrochemical industry effluents. Environ Pollut Ser A 39:27–37.

Hallas LE, Adams WJ, Heitkamp MA (1992): Glyphosate degradation by immobilized bacteria: Field studies with industrial wastewater effluent. Appl Environ Microbiol 58:1215–1219.

Hallas LE, Hahn EM, Korndorfer C (1988): Characterization of microbial traits associated with glyphosate biodegradation in industrial activated sludge. J Ind Microbiol 3:377–385.

Hao OJ, Chen JM, Davis AP, Al-Ghusain IA, Phull KK, Kim MH (1991): Biological fixed-film systems. Res J Water Poll Cont Fed 63:388–394.

Hauxhurst JD, Krichevsky MI, Atlas RM (1980): Numerical taxonomy of bacteria from the Gulf of Alaska. J Gen Microbiol 120:131–148.

Heitkamp MA, Adams WJ, Camel V (1993): Evaluation of five biocarriers as supports for immobilized bacteria: Comparative performance during high chemical loading, acid shocking, drying and heat shocking. J Environ Toxicol Chem 12:1013–1023.

Heitkamp MA, Adams WJ, Hallas LE (1992): Glyphosate degradation by immobilized bacteria: Laboratory studies showing feasibility for glyphosate removal from waste water. Can J Microbiol 38:921–928.

Heitkamp MA, Camel V, Reuter TJ, Adams WJ (1990a): Biodegradation of p-nitrophenol by immobilized bacteria. Appl Environ Microbiol 56:2967–2973.

Heitkamp MA, Camel V, Reuter TJ, Adams WJ (1990b): Use of immobilized bacteria for the degradation of aniline, *p*-nitrophenol and potassium formate in aqueous waste streams. In Morello JA, Domer JE (eds): Abstracts of the 90th Annual Meeting of the American Society for Microbiology. Washington, DC: American Society for Microbiology. Abstract Q-176, p 317.

Irvine RL (1989): Overview of biological processes as applied to hazardous waste problems. In Levandowski NJ (ed): Biotechnology Applications in Hazardous Waste Treatment. New York: Engineering Foundation, Inc., pp 7–26.

Irvine RL, Busch AW (1979): Sequencing batch biological reactors—An overview. J Water Poll Cont Fed 51:235–243.

Irvine RL, Sojka SA, Colaruotolo JE (1984): Enhanced biological treatment of leachates from industrial landfills. Hazard Wastes 1:123–135.

Jacob GS, Garbow J, Hallas LE, Kishore GM, Schaefer J (1988): Metabolism of glyphosate in *Pseudomonas* sp. strain LBr. Appl Environ Microbiol 54:2953–2958.

Jacob GS, Garbow JR, Schaefer J, Kishore GM (1987): Solid-state NMR studies of regulation of glyphosate and glycine metabolism in *Pseudomonas* sp. strain PG2982. J Biol Chem 262:1552–1557.

Jacob GS, Schaefer J, Stejskal EO, McKay RA (1985): Solid-state NMR determination of glyphosate metabolism in a *Pseudomonas* sp. J Biol Chem 260:5899–5905.

Kelly H (1991): Fixed growth reactor-suspended growth reactor combination. Water Eng Manag 138:27.

Kishore GM, Jacob GS (1987): Degradation of glyphosate by *Pseudomonas* sp. PG2928 via a sarcosine intermediate. J Biol Chem 262:12164–12168.

LaNauze JM, Rosenberg H, Shaw DC (1970): The enzymic cleavage of the carbon–phosphorus bond: Purification and properties of phosphonatase. Biochim Biophys Acta 212:332–350.

Leadbetter ER, Foster JW (1958): Studies on some methane-utilizing bacteria. Arch Mikrobiol 30:91–118.

Lerbs W, Stock M, Parthier B (1990): Physiological aspects of glyphosate degradation in *Alcaligenes* sp. strain GL. Arch Microbiol 153:146–150.

Liao CF-H, Dawson RN (1975): Microbiology of two-stage Kraft waste treatment. J Water Pollut Cont Fed 47:2384–2396.

Lighthart B, Oglesby RT (1969): Bacteriology of an activated sludge wastewater treatment plant. A guide to methodology. J Water Polut Cont Fed 41:267–281.

Lin JE, Hickey RF, Shen GJ, Wang HY (1989): Co-immobilized systems for the biodegradation of toxic chemicals. In Levandowski NJ (ed): Biotechnology Applications in Hazardous Waste Treatment. New York: Engineering Foundation, Inc., pp 351–362.

Lupton FS, Zupancie DM (1991): Removal of Phenols From Waste Water by a Fixed Bed Reactor. Patent Number 4,983,299. Washington, DC: United States Patent Office.

McAuliffe KS, Hallas LE, Kulpa CW (1990): Glyphosate degradation by *Agrobacterium radiobacter* isolated from activated sludge. J Ind Microbiol 6:219–222.

McKinney RE, Horwood MP (1952): Fundamental approach to the activated sludge process. I. Floc-producing bacteria. Sewage Ind Wastes 24:117–123.

McKinney RE, Weichlein RG (1953): Isolation of floc producing bacteria from activated sludge. Appl Microbiol 1:259–261.

Metcalf and Eddy, Inc. (1979): Wastewater Engineering Treatment, Disposal, Reuse, 2nd ed., New York: McGraw-Hill Book Co.

Molin G, Turnstroem, G (1982): Numerical taxonomy of psychotrophic pseudomonades. J Gen Microbiol 128:1249–1264.

Moore JK, Braymer HD, Larson AD (1983): Isolation of a *Pseudomonas* sp. which utilizes the phosphonate herbicide glyphosate. Appl Environ Microbiol 46:316–320.

Murthy DVS, Irvine RL, Hallas LE (1988): Principles of organism selection for the degradation of glyphosate in a sequencing batch reactor. In 43rd Annual Purdue Industrial Waste Conference. Ann Arbor, Michigan: Lewis Publishers, pp 267–274.

Niedringhaus EL (1982): Keeping track of the bugs. Water Pollut Cont Fed Highlights 10:7–11.

Pike LF (1975): Aerobic bacteria. In Curds CR, Hawkes HA (eds): Ecological Aspects of Used Water Treatment. London: Academic Press, Inc., vol 1, pp 1–63.

Pipes WO (1980): Microbiology of wastewater treatment. J Water Poll Cont Fed 52:1847–1853.

Pipke R, Amrhein N (1988): Degradation of the phosphonate herbicide glyphosate by *Arthrobacter atrocyaneus* ATCC 13752. Appl Environ Microbiol 54:1293–1296.

Pipke R, Schulz A, Amrhein N (1987): Uptake of glyphosate by an *Arthrobacter* sp. Appl Environ Microbiol 53:974–978.

Portier RJ, Fujisaki K (1986): Continous biodegradation and detoxification of chlorinated phenols using immobilized bacteria. Toxicity Assessment Int Q 1:501–513.

Prakasam TBS, Dondero NC (1967a): Aerobic heterotrophic bacterial populations of sewage and activated sludge. I. Enumeration. Appl Microbiol 15:461–467.

Prakasam TBS, Dondero NC (1967b): Aerobic heterotrophic bacterial populations of sewage and activated sludge. II. Method of characterization of activated sludge bacteria. Appl. Microbiol. 15:1122–1127.

Prakasam TBS, Dondero NC (1967c): Aerobic heterotrophic bacterial populations of sewage and activated sludge. III. Adaptation in a synthetic waste. Appl Microbiol 15:1128–1137.

Prakasam TBS, Dondero NC (1970a): Aerobic heterotrophic bacterial populations of sewage and activated sludge. IV. Adaptation of activated sludge to utilization of aromatic compounds. Appl Microbiol 19:663–670.

Prakasam TBS, Dondero NC (1970b): Aerobic heterotrophic bacterial populations of sewage and activated sludge. V. Analysis of population structure and activity. Appl Microbiol 19:671–680.

Prakasam TBS, Lue-Hing C, Loehr RC (1978): Nitrogen control in wastewater treatment systems by microbial nitrification and denitrification. In Schlessinger D (ed): Microbiology 1978. Washington, DC: American Society for Microbiology, pp 372–379.

Quesada E, Ventosa A, Rodriguez-Valera F, Megias L, Ramos-Cormenzana A (1983):

Numerical taxonomy of moderately halophilic gram-negative bacteria from hypersaline soils. J Gen Microbiol 129:2649–2657.

Quinn JP (1989): Carbon-phosphorus lyase activity—a novel mechanism of bacterial resistance to the phosphonic acid antibiotics. Lett Appl Microbiol 8:113–116.

Quinn JP, Peden JMM, Dick RE (1988): Glyphosate tolerance and utilization by the microflora of soils treated with the herbicide. Appl Microbiol Biotechnol 29:511–516.

Quinn JP, Peden JMM, Dick RE (1989): Carbon-phosphorus bond cleavage by gram-positive and gram-negative soil bacteria. Appl Microbiol Biotechnol 31:283–287.

Reilly W (1990): Aiming Before We Start: The Quiet Revolution in Environmental Policy. Speech delivered to the National Press Club on September 26,1990.

Rogovskaya ZI, Lazareva MF (1959): Intensification of biochemical purification of industrial sludge. I. Microbiological characteristics of activated sludges for the purification of various industrial effluents. Mikrobiologiya 28:565–573.

Rotman D (1991): Regulation drives upgrading. Chemical Week 5/15/91, 33–36.

Rotman D, Young I (1991): Hazardous waste: Shrinking options—Tough choices. Chemical Week 8/21/91 40–44.

Rueppel ML, Brightwell BC, Schaefer J, Marvel JT (1977): Metabolism and degradation of glyphosate in soil and water. J Agric Food Chem 25:517–528.

Schowanek D, Verstraete W (1990a): Phosphonate utilization by bacterial cultures in the presence of alternative phosphorus sources. Biodegradation 1:43–53.

Schowanek D, Verstraete W (1990b): Phosphonate utilization by bacterial cultures and enrichments from environmental samples. Appl Environ Microbiol 56:895–903.

Seiler H, Blaim H (1982): Population shifts in activated sludge from sewage treatment plants of the chemical industry: A numerical cluster analysis. Eur J Appl Microbiol Biotechnol 14:97–104.

Seiler H, Braatz R, Ohmayer R (1980): Numerical cluster analysis of the coryneform bacteria from activated sludge. Zbl Bakt Hyg I Abt Orig C 1:357–375.

Shinabarger DL, Braymer HD (1986): Glyphosate catabolism by *Pseudomonas* sp. strain PG2982. J Bacteriol 168:702–707.

Simidu U, Tsukamoto K (1985): Habitat segregation and biochemical activities of marine members of the family Vibrionaceae. Appl Environ Microbiol 50:781–790.

Sneath P (1980): Basic program for the most diagnostic properties of groups from an identification matrix of percent positive characters. Comp Geosci 6:21–26.

Stensel HD, Strand SE (1989): Reactor configurations in hazardous waste treatment. In Levandowski NJ (ed): Biotechnology Applications in Hazardous Waste Treatment. New York: Engineering Foundation, Inc., pp 47–62.

Sutton PM, Mishra PN (1991): Biological fluidized beds for water and waste water treatment. Water Environ Technol 24:52–60.

Takii S (1977): Bacterial characteristics of activated sludge treating carbohydrate wastes. Water Res 11:85–89.

Thayer AM (1991): Bioremediation: Innovative technology for cleaning up hazardous waste. Chem Eng News 8:23–44.

Unz RF, Davis JA (1975): Microbiology of combined chemical–biological treatment. J Water Pollut Contr Fed 47:183–194.

Updike SJ, Harris DR, Shrago E (1969): Microorganisms, alive and imprisoned in a polymer cage. Nature 224:1122–1124.

Verstraete W, Vanstaen R, Voets JP (1977): Adaptation to nitrification of activated sludge systems treating highly nitrogenous waters. J Water Pollut Cont Fed 22:1604–1608.

Wackett LP, Shames SL, Venditti CP, Walsh CT (1987): Bacterial carbon–phosphorus lyase: Products, rates, and regulation of phosphonic and phosphinic acid metabolism. J Bacteriol 169:710–717.

Walczak CA, Krichevsky MI (1980): Computer methods for describing groups from binary phenetic data: Preliminary summary and editing of data. Int J Syst Biotechnol 30:515–521.

West PA, Okpokwasili GC, Brayton PR, Grimes DJ, Colwell RR (1984): Numerical taxonomy of phenanthrene degrading bacteria isolated from the Chesapeake Bay. Appl Environ Microbiol 48:988–993.

Wilderer PA, Markl H (1989): Innovative reactor design to treat hazardous waste. In Levandowski NJ (ed): Biotechnology Applications in Hazardous Waste Treatment. New York: Engineering Foundation, Inc., pp 241–259.

BIOREMEDIATION OF
CHLOROPHENOL WASTES

MAX M. HÄGGBLOM

Center for Agricultural Molecular Biology,
Cook College,
Rutgers University,
New Brunswick, New Jersey 08903

RISTO J. VALO

Bioteam Corp.,
Maininkitie 21 A 6,
Espoo 02320, Finland

1. ENVIRONMENTAL CONTAMINATION BY CHLOROPHENOLS

1.1. Sources of Chlorophenols

Chlorinated phenols, primarily pentachlorophenol (PCP) and tetrachlorophenols (TeCP), have been widely used as broad-spectrum biocides in industry and agriculture since the 1920s. Because of their solubility in organic solvents and water solubility as sodium salts, chlorophenols have versatile uses. One of the major uses is to protect freshly sawn timber against sapstain fungi and also (frequently in combination with creosote) for long-term protection of wood poles, railway ties, and timber. Pentachlorophenol has also been used as a biocide in paints and oils and as a herbicide on rice fields (Cirelli, 1978; Kobayashi, 1978).

The annual production of PCP in 1984 was estimated at 35,000 to 40,000 tons (Korte, 1987). This estimate does not, however, include production in the former

---

Microbial Transformation and Degradation of Toxic Organic Chemicals, pages 389–434
© 1995 Wiley-Liss, Inc.

Eastern Block countries, and the annual worldwide production in 1970–1980 may have been closer to 90,000 tons (Detrick, 1977; Dougherty, 1978). The annual worldwide production of all chlorophenols was estimated to be as high as 200,000 tons, with approximately 80% used by the wood-preserving industry (Ahlborg and Thunberg, 1980). The technical chlorophenol formulations used for wood preservation varied from country to country. In the United States and Canada mainly PCP was used, while the chlorophenol formulation used in Finland and Sweden consisted of approximately 80% 2,3,4,6-TeCP, 5%–10% PCP, and 5%–10% 2,4,6-trichlorophenol (TCP).

The technical chlorophenol formulations contain several dimeric impurities, such as polychlorinated phenoxyphenols, dibenzo-*p*-dioxins, and dibenzofurans (Ahlborg and Thunberg, 1980; Humppi et al., 1984; Kitunen, 1990; Nilsson et al., 1978). Due to these highly toxic impurities as well as to the toxicity and persistence of chlorophenols in the environment, the use of chlorophenols has recently been restricted or banned in several countries, such as Sweden, Finland, Germany, and Japan, and the production and use of chlorophenols has hence decreased substantially after the mid-1980s. Chlorophenols are, however, still in use for wood preservation in several other countries.

There are a number of other sources of chlorophenols in addition to their direct use as biocides. Chlorinated phenols are intermediates in the synthesis of other biocides, such as the herbicides 2,4-dichlorophenoxyacetic acid (2,4-D) and 2,4,5-trichlorophenoxyacetic acid (2,4,5-T) (Nilsson et al., 1978). 2,4-Dichlorophenol and 2,4,5-trichlorophenol, respectively, are also microbial breakdown products of these herbicides. 2,4-D is one of the most commonly used herbicides in the world.

In addition to biocide use, other sources of chlorophenols can be identified. A wide range of chlorinated organic compounds, including chlorinated phenolics, are produced during chlorine bleaching of pulp (Suntio et al., 1988). Conventional chlorine bleaching of kraft pulp has been estimated to give rise to approximately 100–300 g of chlorinated phenolic compounds per ton of pulp, but these compounds represent only a few percent of the total discharge of chlorinated organics (Jokela and Salkinoja-Salonen, 1992; Salkinoja-Salonen et al., 1981). New bleaching processes have significantly reduced the levels of chlorinated phenolics (Jokela et al., 1993). Chlorophenols are analogously formed during chlorination of potable water containing humic material (Detrick, 1977). Chlorophenols are also generated during combustion of organic material in the presence of chloride, e.g., incineration of municipal solid waste or during burning of fresh wood, resulting in airborne distribution (Ahling and Lindskog, 1982; Paasivirta et al., 1985, 1990). Chlorophenols are thus ubiquitous pollutants and have, for example, been found in lake sediments remote from industrial sites (Salkinoja-Salonen et al., 1984). PCP was found in sediment layers older than 50 years, dating before industrial production, and presumably originated from forest fires. In addition, it is of interest to note that some chlorophenols are also produced biologically, e.g., 2,6-dichlorophenol, which is a sex pheromone of ticks (Berger, 1972; McDowell and Waladde, 1986). Chlorophenols are therefore not solely of anthropogenic origin.

1.2. Contamination of Soil and Groundwater

A common method used in Finland and other European countries for treatment of freshly sawn timber was by dip treatment, in which bundled lumber was submerged for a short time in a 1%–2% solution of sodium chlorophenolate in 50–100 m^3 basins (Kitunen, 1990; Valo, 1990). Before transporting the lumber to the storage area the chlorophenol solution was allowed to drip for a short time. The area around the treatment facility was frequently used as a short-term dripping area resulting in contamination of the soil. Another application technology, frequently used in Canada and the United States, is the use of either high- or low-pressure spraying hoods for continuous treatment of individual pieces of lumber (Krahn and Shrimpton, 1988). Pentachlorophenol was used as a 3%–6% solution in petroleum oil. Low-pressure spraying produces a large droplet size, and the lumber is wet after the spray box. In high-pressure treatment, where the droplet size is smaller, the lumber is dryer and not dripping. During the process PCP can escape with the liquid condensates and vapors as well as through spills. Treated lumber is usually stored outdoors with no protection against rain, promoting leaching of chlorophenols during rain fall (Krahn et al., 1987). Storm water runoff at sawmills and lumber export terminals was shown to cause significant contamination of an adjacent river in British Columbia (Krahn et al., 1987). Utility poles and railway ties treated with chlorophenols were shown to be a constant source of contamination via runoff and leaching (Wan, 1992).

The continued use of chlorophenols for wood preservation over several decades has frequently caused serious local contamination of the environment during normal operation as well as after accidental spills. In a study of Finnish sawmill sites the soil around wood-preserving facilities was shown to contain up to several grams of chlorophenols per kilogram of soil (Kitunen et al., 1985, 1987; Valo et al., 1984, 1985). The dimeric impurities polychlorinated phenoxy phenols and dibenzofurans were found to accumulate in the top soil, while vertical migration of chlorophenols occurred. The vertical distribution of chlorophenols was uneven, with certain layers of the soil accumulating more chlorophenols than others. Chlorophenols were detected in household wells 100–200 m from the preserving facility (Kitunen, 1990). Similar contamination of soil and groundwater has been detected elsewhere. At a treatment site in California PCP was found to migrate over 500 m in the aquifer (Goerlitz, 1992). In acidic groundwater the solubility and thus movement of PCP is more limited (Goerlitz et al., 1985). In addition to pH, also the organic matter content of soil influences migration of chlorophenols in soil and groundwater (Kitunen, 1990).

A large-scale contamination of an aquifer by chlorophenols was revealed in southern Finland in 1987 (Lampi et al., 1990, 1992b). This contamination was traced to originate from a nearby sawmill where chlorophenols had been used as an antisapstain agent for several decades. The volume of contaminated water was estimated to be 100,000 m^3, with the chlorophenol concentrations as high as 200 mg/liter. In 1987 the water supply of the community contained 70–140 μg/liter of chlorophenols and was subsequently closed, but prior to this the groundwater had been used by the community for over 20 years (Lampi et al., 1992a). Cancer incidence in the municipality indicated an excess of soft tissue sarcomas and non-Hodgkin's lymphomas (Lampi et al., 1992a). Five years after the contamination was discovered the chlorophenol concentration in the aquifer remains high, at 30–50 mg/liter.

In an accident in British Columbia 18,000 liters of a 7,500 mg/liter chlorophenol solution escaped from a dip treatment basin (Patterson and Liebscher, 1987). About 40% of the chlorophenol was recovered, but levels of 1–2 mg/liter remained in the groundwater. Six years later the chlorophenol contamination was still at a level of 0.3 mg/liter. Similar accidental spills of chlorophenols into a waterway led to serious contamination of surface waters and fish kills (Pierce and Victor, 1978; Renberg et al., 1983).

It is evident from these examples that wood-preserving sites have remained heavily contaminated with chlorophenols as well as the dimeric impurities for several years after discontinuation of chlorophenol use. Old and abandoned treatment sites will therefore pose a health risk for years to come.

2. MICROBIAL DEGRADATION AND TRANSFORMATION OF CHLOROPHENOLS

2.1. Microbial Isolates

It is apparent from the studies on contaminated wood-preserving sites that polychlorinated phenols are persistent in soil and groundwater. Self-cleansing seems to be slow, and the contamination may thus persist for decades. While degradation of chlorophenols may be slow in the environment, it is, however, possible to isolate microbes capable of utilizing PCP and other polychlorinated phenols as a source of carbon and energy. Indeed, the degradation of PCP and other chlorinated phenols has been extensively studied, and the degradation pathways are fairly well characterized. Several aerobic bacteria and fungi capable of degrading chlorophenols have been isolated, a number of which are listed in Table 11.1. In addition to the aerobic isolates, obligately anaerobic bacteria that reductively dechlorinate chlorophenols have been isolated. The list of strains in Table 11.1 is by no means comprehensive, but is intended more to illustrate the diversity of chlorophenol degrading microbes.

Table 11.1 also indicates the general mechanism for chlorophenol degradation by the different strains. The aerobic chlorophenol-degrading bacteria can be divided into two distinct groups based on their substrate specificity and the mechanism of chlorophenol degradation: (1) strains capable of degrading pentachlorophenol as well as tetra- and trichlorophenols and (2) strains that degrade mono- and dichlorophenols. Polychlorinated phenols are in general degraded by initial dechlorination, via hydroxylation and reductive dechlorinations, with ring cleavage occurring only after all or most of the chlorine substituents have been removed. Chlorinated p-hydroquinones are central intermediates in the degradation of penta-, tetra-, and

TABLE 11.1. Examples of Chlorophenol-Degrading Microbial Isolates

Strain	Substrates[a]	Mechanism[b]	Reference
Bacteria			
Pseudomonas sp. strain B13	CPs	A	Knackmuss and Hellwig (1978)
Pseudomonas sp. strain JS6	2,5-DCP	A	Spain and Gibson (1988)
Rhodococcus sp. strains An117 and An213	3-CP, 4-CP	A	Ihn et al. (1989)
Rhodococcus erythropolis 1cp	4-CP, 2,4-DCP	A	Gorlatov et al. (1989)
Azotobacter chroococcum strain MSB1	4-CP	A	Balajee and Mahadevan (1990)
Arthrobacter sp.	4-CP	—	Kramer and Kory (1992)
Alcaligenes sp. strain A7-2	4-CP	A	Schwien and Schmidt (1982)
Alcaligenes eutrophus JMP134	2,4-DCP (2,4-D)	A	Don et al. (1985), Pemberton et al. (1979), Pflug and Burton (1988)
Pseudomonas sp. strain NCIB 9340	2,4-DCP (2,4-D)	A	Evans et al. (1971)
Pseudomonas cepacia BRI6001	2,4-DCP (2,4-D)	A	Greer et al. (1990)
Pseudomonas cepacia	2,4-DCP (2,4-D)	A	Radjendirane et al. (1991)
Acinetobacter sp.	2,4-DCP (2,4-D)	A	Beadle and Smith (1982)
Xanthobacter sp. strain CP	2,4-DCP (2,4-D)	A	Ditzelmüller et al. (1989)
Flavobacterium sp. strain 50001	2,4-DCP (2,4-D)	A	Chaudry and Huang (1988)
Arthrobacter sp.	2,4-DCP (2,4-D)	A	Bollag et al. (1986)
Flavobacterium sp. strain MH	2,4-DCP (2,4-DP)	A	Horvath et al. (1990)
Pseudomonas cepacia strain AC1100	2,4,5-TCP (2,4,5-T), PCP	H	Karns et al. (1983), Sangodkar et al. (1988)
Nocardioides simplex strain 3E	2,4,5-TCP (2,4,5-T)	—	Golovleva et al. (1990)
Pseudomonas pickettii	2,4,6-TCP	H	Kiyohara et al. (1992)
Azotobacter sp. strain GP1	2,4,6-TCP	H	Li et al. (1991)
Streptomyces rochei 303	2,4,6-TCP, TeCPs, PCP	H	Golovleva et al. (1992)

(continued)

TABLE 11.1. (Continued)

Strain	Substrates[a]	Mechanism[b]	Reference
Mycobacterium (Rhodococcus) chlorophenolicum[c]			
Strain PCP-I	PCP, TeCPs, TCPs	H, R	Apajalahti et al. (1986), Apajalahti and Salkinoja-Salonen (1987a,b)
Strain CP-2	PCP, TeCPs, TCPs	H, R	Häggblom et al. (1988b, 1989b)
Strain CG-1	PCP, TeCPs, TCPs	H	Häggblom et al. (1988b)
Mycobacterium fortuitum strain CG-2	PCP, TeCPs, TCPs	H, R	Häggblom et al. (1988b), Nohynek et al. (1993), Uotila et al. (1992b)
Arthrobacter sp. strain ATCC 33790	PCP	H	Edgehill and Finn (1982), Schenk et al. (1989)
Arthrobacter sp. strain NC	PCP	—	Stanlake and Finn (1982)
Strain KC-3	PCP, 2,3,4,6-TeCP, 2,4,6-TCP	H, R	Chu and Kirsch (1972, 1973), Reiner et al. (1978)
Flavobacterium sp. strain ATCC 39723	PCP, TeCPs, TCPs	H, R	Saber and Crawford (1985), Steiert and Crawford (1986)
Pseudomonas sp.	PCP	—	Watanabe (1973)
Pseudomonas sp.	PCP	—	Suzuki (1977)
Pseudomonas spp.	PCP	—	Trevors (1982)
Pseudomonas sp. strain RA2	PCP	—	Radehaus and Schmidt (1992)
Pseudomonas sp. strain SR3	PCP	—	Middaugh et al. (1993)
Pseudomonas aeruginosa	PCP	—	Premalatha and Rajakumar (1994)

Organism	Substrate	Mechanism[b]	Reference
Pseudomonas saccharophila strains KF1, NKF1	2,4,6-TCP, 2,3,4,6-TeCP	—	Puhakka et al. (1994)
Desulfomonile tiedjei strain DCB-1	PCP, DCP-TeCPs	R	Mohn and Kennedy (1992b)
Strain DCB-2	2,4,6-TCP, DCPs	R	Madsen and Licht (1992)
Desulfitobacterium dehalogenans strain JW/IU-DC1	2,4-DCP	R	Utkin et al. (1994)
Strain 2 CP-1	2-CP	R	Cole et al. (1994)
Fungi			
Candida tropicalis	3-CP, 4-CP	A	Ivoilov et al. (1983)
Candida maltosa	3-CP, 4-CP	A	Polnisch et al. (1992)
Penicillium frequentans Bi 7/2	4-CP, CPs	A	Hofrichter et al. (1993, 1994)
Phanerochaete chrysosporium	PCP, 2,4,5-TCP, 2,4-DCP	O	Joshi and Gold (1993), Mileski et al. (1988), Valli and Gold (1991)
Phanerochaete sordida	PCP	—	Lamar et al. (1990b)
Mycena avenacea	PCP	O	Kremer et al. (1992)
Pleurotus cornucopiae	2,4,5-TCP, 2,4,6-TCP	—	Seeholzer-Nguyen and Hock (1991)

[a] 2,4-DP, 2-(2,4-dichlorophenoxy)propionic acid; PCP, pentachlorophenol; TeCP, tetrachlorophenol; TCP, trichlorophenol; DCP, dichlorophenol; CP, monochlorophenol; 2,4-D, 2,4-dichlorophenoxyacetic acid; 2,4,5-T, 2,4,5-trichlorophenoxyacetic acid; 2,4-DP, 2-(2,4-dichlorophenoxy)propionic acid.

[b] A, hydroxylation to chlorocatechol with dehalogenation after ring cleavage; H, hydroxylation to chlorohydroquinones, hydroxylation, and dechlorination; R, reductive dechlorination; O, oxidation to chlorobenzoquinone, followed by quinone reduction and methylation; —, not reported.

[c] First described as *R. chlorophenolicus* (Apajalahti et al., 1986), but transferred to the genus *Mycobacterium* as *M. chlorophenolicum* (Häggblom et al., 1994).

trichlorophenols. Mono- and dichlorinated phenols, on the other hand, are generally degraded via chlorinated catechols, with dechlorination after ring cleavage. The anaerobic bacterial isolates dechlorinate chlorophenols reductively but do not degrade the aromatic ring. In fungi, the lignin and manganese peroxidase systems are involved in degradation of polychlorinated phenols, while monochlorophenols are degraded via chlorocatechols. These different mechanisms for microbial degradation of chlorophenols are briefly discussed below. The emphasis is on PCP and other polychlorinated phenols, since these are the major contaminants from wood preservation.

2.2. Aerobic Chlorophenol Degradation

Several bacterial strains belonging to a variety of genera have been shown to degrade mono- and dichlorophenols (Table 11.1). The common feature of these bacteria is that they first oxidize the chlorophenols to chlorocatechols by a phenol hydroxylase (monooxygenase) with elimination of the chlorine substituent(s) after intra-diol cleavage of the aromatic ring (Häggblom, 1992; Reineke and Knackmuss, 1988). The degradation pathway for chlorinated phenoxyalkanoic acids converges with that of the corresponding chlorophenol after initial cleavage of the ether side chain (Fukumori and Hausinger, 1993). The degradation pathway for 2,4-dichlorophenoxyacetic acid via 2,4-dichlorophenol and 3,5-dichlorocatechol, followed by *ortho* ring cleavage has been extensively investigated (for reviews, see Häggblom, 1992; Rochkind-Dubinsky et al., 1987; Sangodkar et al., 1988).

Degradation of chlorocatechols seems to be the critical step in degradation of mono- and dichlorophenols, while the initial hydroxylation of chlorophenol is less specific. For example, several phenol-degrading strains of the genus *Rhodococcus* also hydroxylated mono-, di-, and trichlorophenols, but the resulting chloro-catechols were not degraded (Häggblom et al., 1989c). Similar results have been shown with several other bacteria, as well (Engelhardt et al., 1979; Spokes and Walker, 1974).

Metabolism of the chlorinated catechols via a "modified *ortho* cleavage pathway" is carried out by broad-specificity enzymes analogous to those oxidizing the nonchlorinated counterpart (Reineke and Knackmuss, 1988). Two types of isofunctional enzymes for catechol and chlorocatechol degradation have been found in *Pseudomonas* sp. B13 (Dorn and Knackmuss, 1978a,b; Schmidt and Knackmuss, 1980), as well as in other bacteria (Ghosal and You, 1989; Gorlatov et al., 1989; Pieper et al., 1988). Type II pyrocatechase and cycloisomerase have a broad substrate specificity and act on both chlorinated and nonchlorinated substrates. Chloride is eliminated spontaneously after ring cleavage.

Polychlorinated phenols are more resistant to microbial degradation. Bacteria capable of degrading PCP were first isolated in the early 1970s (Chu and Kirsch, 1972; Suzuki, 1977; Suzuki and Nose, 1971; Watanabe, 1973). It is, however, only recently that the degradation pathways (Fig. 11.1) of pentachlorophenol and other polychlorinated phenols have been elucidated. In contrast to mono- and dichlorophenols, polychlorinated phenols (penta-, tetra-, and trichlorophenols) undergo

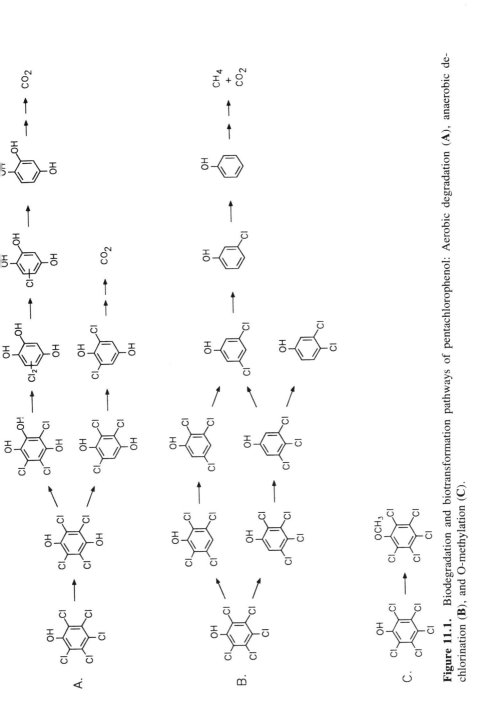

Figure 11.1. Biodegradation and biotransformation pathways of pentachlorophenol: Aerobic degradation (**A**), anaerobic de-chlorination (**B**), and O-methylation (**C**).

initial hydroxylation and dechlorination, with chlorinated *p*-hydroquinones as central intermediates (Häggblom, 1992). Accumulation of tetrachlorohydroquinone during PCP degradation was first demonstrated with the coryneform bacterium strain KC-3 (Reiner et al., 1978). Suzuki (1977) had identified both tetrachlorohydroquinone and tetrachlorocatechol in extracts of a PCP-degrading *Pseudomonas* strain. Chlorinated hydroquinones have since been identified as the initial metabolites during degradation of pentachlorophenol and other polychlorophenols by strains of *Mycobacterium chlorophenolicum* (Apajalahti and Salkinoja-Salonen, 1987b; Häggblom et al., 1988a,b), formerly classified as *Rhodococcus chlorophenolicus* but now assigned to the genus *Mycobacterium* (Häggblom et al., 1994; Briglia et al., 1994a); *Mycobacterium fortuitum* strain CG-2 (Häggblom et al., 1988b; Uotila et al., 1992b); *Arthrobacter* sp. strain ATCC 33790 (Schenk et al., 1989, 1990); *Flavobacterium* sp. strain ATCC 39723 (Steiert and Crawford, 1986); *Azotobacter* sp. strain GP1 (Li et al., 1991); *Pseudomonas pickettii* (Kiyohara et al., 1992); *Pseudomonas cepacia* strain AC1100 (Sangodkar et al., 1988); and *Streptomyces rochei* 303 (Golovleva et al., 1992).

Interestingly, the *Azotobacter* sp. strain GP1 appears to have both a chlorohydroquinone and a chlorocatechol pathway (Li et al., 1991). 2,4,6-Trichlorophenol was degraded via 2,6-dichlorohydroquinone, while monochlorophenols appear to be hydroxylated by a phenol-induced pathway to chlorocatechols. Recently, a derivative strain of *P. cepacia* AC1100 was constructed capable of degrading mixtures of 2,4-D and 2,4,5-T, with 2,4,5-T metabolized via a chlorohydroquinone pathway and 2,4-D metabolized via a chlorocatechol pathway (Haugland et al., 1990).

The mechanism of *para* hydroxylation has been the subject of some controversy and conflicting results. Partially this may be due to artifacts caused by different experimental conditions causing exchange of the hydroxyl label (Uotila, 1993; Uotila et al., 1992a). Depending on the extraction and derivatization methods used (especially acidic extraction) the chemical lability of the chlorohydroquinones may result in exchange of the hydroxylic oxygen in ^{18}O-labeling experiments.

Initial experiments with *M. chlorophenolicum* indicated that the hydroxyl group was derived from water, although the reaction required molecular oxygen (Apajalahti and Salkinoja-Salonen, 1987b). Uotila et al. (1992a), however, demonstrated that either H_2O or O_2 could serve as the oxygen source in *para* hydroxylation, depending on the environment of the enzyme. A membrane-bound P_{450} monooxygenase appears to be involved in PCP *para* hydroxylation by both *M. chlorophenolicum* strain PCP-I and *M. fortuitum* strain CG-2 (Uotila et al., 1992a,b, 1995). In the presence of sulfite or iodosobenzene the requirement for O_2 was circumvented, and *para* hydroxylation proceeded under anaerobic conditions (Uotila et al., 1992a,b, 1995). Attempts to purify the membrane-bound *para*-hydroxylating enzyme from *M. chlorophenolicum* have failed due to loss of activity at each purification step (Uotila, 1993).

para Hydroxylation of PCP by the *Flavobacterium* sp. is catalyzed by a different type of monooxygenation. Although initial experiments with cell extracts indicated that the hydroxyl group was derived from water (Steiert and Crawford, 1986), the hydroxylase has recently been purified and sequenced and shown to be a flavoprotein monooxygenase with a requirement for NADPH and O_2 (Orser et al., 1993b; Xun and Orser, 1991; Xun et al., 1992a,b). Hydroxylation of PCP by *Arthrobacter* sp. strain ATCC 33790 also required NADPH and O_2 (Schenk et al., 1989), although ^{18}O-labeling experiments were inconclusive on the origin of the hydroxy group (Schenk et al., 1990). The PCP-hydroxylating enzymes of the *Flavobacterium* sp. and *Arthrobacter* sp. are clearly distinct from those of the PCP-hydroxylating enzymes of *M. chlorophenolicum* and *M. fortuitum*. DNA hybridization experiments showed that the PCP-4 monooxygenase of the *Flavobacterium* strain was present in the PCP-degrading strains *Arthrobacter* sp. ATCC 33790 and *Pseudomonas* sp. SR3, but it was not found in *M. chlorophenolicum* strain PCP-I (Orser et al., 1993b).

Tetrachlorohydroquinone, the first intermediate during PCP degradation, appears to be degraded further by two different pathways, as illustrated in Figure 11.1A. In one pathway tetrachlorohydroquinone is reductively dechlorinated to 2,6-dichlorohydroquinone. This was first demonstrated with mutants of the coryneform bacterium strain KC-3, which accumulated tetra- and 2,6-dichlorohydroquinone during PCP metabolism (Reiner et al., 1978). Similarly, the *Flavobacterium* sp. degraded PCP via tetra- and 2,6-dichlorohydroquinone (Steiert and Crawford, 1986). Cell extracts of *Flavobacterium* sp. dechlorinated tetrachlorohydroquinone via trichlorohydroquinone to 2,6-dichlorohydroquinone, with glutathione serving as the reducing agent (Xun et al., 1992c). The tetrachlorohydroquinone reductive dehalogenase has recently been purified (Xun et al., 1992d) and the gene encoding the enzyme sequenced (Orser et al., 1993a). It is not known whether the remaining two chlorine substituents are removed by *Flavobacterium* sp. before or after ring cleavage.

The *M. chlorophenolicum* strains and *M. fortuitum* strain CG-2 degrade tetrachlorohydroquinone via a different pathway, involving a hydroxylation to trichloro-1,2,4-trihydroxybenzene followed by three reductive dechlorinations to 1,2,4-trihydroxybenzene (Fig. 11.1) (Apajalahti and Salkinoja-Salonen, 1987a; Häggblom et al., 1989b; Uotila et al., 1992b). The hydroxyl group is derived from water. All chlorine substituents are thus removed before cleavage of the aromatic ring. Dehalogenation–hydroxylation is catalyzed by cytoplasmic enzyme(s) and seems to have a nonspecific requirement for a reductant (Apajalahti and Salkinoja-Salonen, 1987a; Uotila et al., 1995, 1992b).

In contrast to bacterial PCP degradation, fungi degrade polychlorophenols by nonspecific oxygenations. *Phanerochaete chrysosporium* and a number of other white-rot fungi (Table 11.1) are able to degrade different chlorinated phenols by mechanisms involving the lignin-degrading system of the fungus. Nutrient-nitrogen limited cultures of *Phanerochaete chrysosporium* mineralized approximately 23% of ^{14}C-labeled PCP in 30 days (Mileski et al., 1988), with up to 40% mineralization observed in one study (Lin et al., 1990), but in many cases less than 10% of the carbon is mineralized by *Phanerochaete* strains (Lamar et al., 1990b). Extracellular lignin peroxidases were shown to oxidize PCP to tetrachloro-*p*-benzoquinone (Hammel and Tardone, 1988). Valli and Gold (1991) and Joshi and Gold (1993) presented a mechanism by which 2,4-dichlorophenol and 2,4,5-trichlorophenol are degraded via cycles of peroxidase-catalyzed oxidation and dechlorination to chloro-benzoquinones and subsequent quinone reduction and hydroquinone methylations.

2.3. Anaerobic Chlorophenol Degradation

While aerobic degradation of chlorophenols is well characterized, much less is known about how anaerobic microbes degrade chlorophenols. Reductive de-chlorination is a key reaction in anaerobic degradation of chlorophenols. Ide et al. (1972) were the first to suggest that chlorophenols were reductively dechlorinated in anaerobic soils. Kuwatsuka and Igarashi (1975) and Murthy et al. (1979) reported similar dechlorinations of PCP, suggesting preferential removal of *ortho* and *para* chlorines. Anaerobic dechlorination of chlorophenols has since been studied more closely with methanogenic enrichment cultures using sewage sludges, sediments, and soils as inoculum, and the dechlorination pathways have been well established (for reviews, see Häggblom et al., 1989b; Kuhn and Suflita, 1989; Mohn and Tiedje, 1992). Methanogenic degradation of chlorophenols is initiated by reductive dechlorination, in most cases with preferential removal of the *ortho* chlorine(s) (Boyd and Shelton, 1984; Hale et al., 1990, 1991; Hrudey et al., 1987a; Kohring et al., 1989a,b; Suflita and Miller, 1985; Zhang and Wiegel, 1990), but dechlorination pathways differ with different microbial consortia. A commonly observed transfor-mation sequence for PCP is via initial *ortho* dechlorinations producing 3,4,5-trichlorophenol followed by *para* dechlorination to 3,5-dichlorophenol (Abrahams-son and Klick, 1991; Mikesell and Boyd, 1985, 1986; Nicholson et al., 1992), but initial *meta* or *para* dechlorination of PCP has also been observed (Bryant et al., 1991; Hendriksen and Ahring, 1993; Hendriksen et al., 1992; Larsen et al., 1991). The main dechlorination pathways that have been observed for PCP are illustrated in Figure 11.1B, although a number of minor pathways can be discerned (Nicholson et al., 1992). Frequently, dechlorination is not complete, but di-, tri-, and tetra-chlorophenols accumulate (Abrahamsson and Klick, 1991; Armenante et al., 1992; Laren et al., 1991; Madsen and Aamand, 1991; Mohn and Kennedy, 1992a; Nicholson et al., 1992). However, complete mineralization of monochlorophenols and pentachlorophenol is a well-documented phenomenon (Boyd and Shelton, 1984; Boyd et al., 1983; Dietrich and Winter, 1990; Genthner et al., 1989b; Hägg-blom et al., 1993a; Hrudey et al., 1987b; Madsen and Aamand, 1992; Mikesell and Boyd, 1985, 1986).

A major obstacle in studying anaerobic dechlorination has been the difficulty in isolating dechlorinating microbes. To date, three obligately anaerobic bacteria that dechlorinate chlorophenols have been described (Madsen and Licht, 1992; Mohn and Kennedy, 1992b; Utkin et al., 1994). *D. tiedjei* strain DCB-1, originally isolated for its ability to dechlorinate 3-chlorobenzoate (DeWeerd et al., 1990; Shelton and Tiedje, 1984), was also shown to dechlorinate PCP and other chloro-phenols (Mohn and Kennedy, 1992b). 3-Chlorobenzoate was required as an inducer of dehalogenating activity, and only the *meta* chlorine of chlorophenols was re-moved. Another dehalogenating bacterium, strain DCB-2, was recently isolated for its ability to dechlorinate chlorophenols at the *ortho* position (Madsen and Licht, 1992). 2,4,6-Trichlorophenol, which was used as the substrate during enrichment, was also the preferred substrate by this *Clostridium*-like bacterium. The *para* chlorines were not removed, and *meta* dechlorination was observed only with 3,5-dichlorophenol. Recently Utkin et al. (1994) described the isolation of *Desulfitobacterium dehalogenans*, which reductively *ortho*-dechlorinates 2,4-dichlorophenol. Dietrich and Winter (Dietrich and Winter, 1990) reported on a 2-chlorophenol dechlorinating consortium consisting of three morphologically distinct microbes. A spirochaete-like organism was possibly the dechlorinating organism, since the mixed culture lost its dechlorinating capability when this organism was lost. Attempts to isolate the dechlorinating organism in pure culture failed. Although these anaerobic dehalogenating bacteria are not applicable to bioremediation of (poly)chlorophenol wastes,they have greatly contributed to our understanding of anaerobic reductive dechlorination. It is interesting to note that reductive dechlorination (of the chlorohydroquinone intermediates) is also central in degradation of PCP by strictly aerobic bacteria.

It has been suggested that the chlorinated aromatics are serving as electron acceptors for the anaerobic community (Mohn and Tiedje, 1992). Addition of auxiliary carbon sources frequently enhances dechlorination rates, either by acting as an electron donor for dechlorination or by serving as a growth substrate for the microbial population (Häggblom et al., 1993a; Hendriksen and Ahring, 1993; Hendriksen et al., 1992; Madsen and Aamand, 1992). Dechlorination of 3-chlorobenzoate by *D. tiedjei* DCB-1 was shown to provide energy that could be conserved by the bacterium (Dolfing, 1990; Mohn and Tiedje, 1990, 1991). Thermodynamic calculations indicate that reductive dechlorination of chlorophenols could also yield energy (Dolfing and Harrison, 1992), which would support the growth of anaerobic dehalogenating microorganisms in the environment.

Several studies have indicated that sulfate and other alternative electron acceptors inhibit reductive dechlorination by methanogenic communities (Häggblom et al., 1993a; Kohring et al., 1989b; Madsen and Aamand, 1991). This inhibition may involve interspecific competition for electron donors (Mohn and Tiedje, 1992). Alternatively, as in the case with *D. tiedjei* DCB-1, reduction of the sulfuroxyanions thiosulfate and sulfite competed for reducing equivalents and was favored over dechlorination (DeWeerd and Suflita, 1990; DeWeerd et al., 1991). Nonetheless, dechlorination of chlorophenols in the presence of sulfate (Genthner et al., 1989a; King, 1988; Kohring et al., 1989b), as well as degradation of chlorophenols coupled to sulfate reduction, has been demonstrated (Häggblom and Young, 1990; Häggblom et al., 1993b, and unpublished data). The different monochlorophenol isomers are indeed degraded under a variety of reducing conditions, including methanogenesis, sulfidogenesis, and iron reduction (Häggblom et al., 1993b; Kazumi et al., 1993). Under denitrifying conditions only 2-chlorophenol was shown to be degraded (Häggblom et al., 1993b). The different substrate specificities observed under each reducing condition suggested that distinct microbial populations were enriched. These studies indicated that chlorophenols can be degraded by diverse anaerobic microbial communities and that the availability of alternative electron acceptors will greatly influence the dechlorination and degradation of chlorophenols in anaerobic environments.

2.4. Biotransformation of Chlorophenols

It is important to keep in mind that loss of parent compound is insufficient evidence of biodegradation and that not all biotransformation reactions lead to ultimate degradation and mineralization of the compound. One environmentally important biotransformation reaction of chlorophenols is O-methylation (Fig. 11.1C). The chloroanisoles that are produced are more lipophilic than the corresponding chlorophenols and hence have a higher potential for bioaccumulation (Knuutinen et al., 1990; Neilson et al., 1984; Palm et al., 1991), and they also tend to be more resistant to biodegradation. Chloroanisoles are frequently found as biotransformation products in chlorophenol-contaminated soils (Behechti et al., 1988; Kuwatsuka and Igarashi, 1975; Lamar and Dietrich, 1990; Lamar et al., 1990a; Murthy et al., 1979), and they have also been detected in earthworms collected from these soils (Haimi et al., 1992; Knuutinen et al., 1990; Palm et al., 1991). The polychlorinated phenoxyphenols that are contaminants of the technical chlorophenols can also be O-methylated in soils (Valo and Salkinoja-Salonen, 1986b). Chlorinated anisoles formed by microbial O-methylation of chlorophenols have been identified as the cause of off-flavors in foods and beverages (Buser et al., 1982; Cserjesi and Johnson, 1972; Curtis et al., 1974; Engel et al., 1966; Tindale et al., 1989; Whitfield et al., 1985). For example, the contamination of dried fruit by 2,4,6-trichloroanisole and 2,3,4,6-tetrachloroanisole was traced to fiberboard boxes used in packaging (Tindale et al., 1989; Whitfield et al., 1985). These contained significant levels of the corresponding chlorophenols, which were shown to be O-methylated by fungi isolated from the packaging materials.

O-methylation of chlorophenols is catalyzed by a wide range of bacteria and fungi (for reviews, see Häggblom, 1990, 1992; Neilson et al., 1991) and may possibly serve as a detoxification mechanism, since the chloroanisoles are frequently less toxic to bacteria and fungi (Häggblom et al., 1988b; Ruckdeschel and Renner, 1986; Ruckdeschel et al., 1987). The O-methylating enzymes are fairly nonspecific and transform a range of halogenated phenolic compounds (Häggblom et al., 1988b, 1989a; Neilson et al., 1983, 1984). Interestingly, the capability to O-methylate chlorophenols was found in bacterial strains with no known previous exposure to chlorophenols as well as in strains enriched on chlorophenols (Häggblom et al., 1989a). This suggests that this trait is not necessarily the result of adaptation to chlorophenols.

Oxidative coupling of chlorophenols is another important transformation process in which phenol oxidases (peroxidases) catalyze the oxidation of phenols to phenoxy radicals, which then react by self-coupling or cross-coupling with other compounds to form dimers and polymers. Laccases from the fungi *Rhizoctonia practicola* and *Trametes versicolor* catalyze the polymerization of mono- and dichlorophenols to a mixture of oligomers (Dec and Bollag, 1990; Minard et al., 1981; Sjoblad and Bollag, 1977). Partial dechlorination may also occur during oxidation by laccases (Roy-Arcand and Archibald, 1991). Chlorophenoxy radicals produced by peroxidases may react to form highly toxic dimers such as polychlorinated dibenzo-*p*-dioxins, dibenzofurans, and diphenyl ethers (Minard et al., 1981;

Öberg et al., 1990; Svenson et al., 1989), indicating that biogenic formation of these compounds is possible, although it is not known to what extent this occurs in soils. Potential bioremediation technologies utilizing peroxidases should therefore be given critical consideration.

Cross-coupling of chlorophenols with other phenolics (Bollag and Liu, 1985; Bollag et al., 1988; Shuttleworth and Bollag, 1986) may explain the formation of humus- and soil-bound chlorinated residues in soils (Schmitzer et al., 1989). Chlorophenols bound to humic material were found to be only slowly released through microbial action (Dec et al., 1990), and the use of laccases has been proposed as a method for detoxifying and binding chlorophenol contaminants (Bollag, 1992; Shannon and Bartha, 1988). The polymeric chlorinated materials formed in soils seem to be very stable (Dec et al., 1990; Salkinoja-Salonen et al., 1995); however, recent studies have suggested that they can be taken up by earthworms and may thus enter the food chain (Salkinoja-Salonen et al., 1995). It is also not known whether the toxic dimers are formed by polymerization of chlorophenols in soils.

3. BIOREMEDIATION

There is considerable interest in developing microbiological methods for treatment of chlorophenol wastes. Alternative methods of remediation, such as incineration or absorption, are expensive due to the large volumes of contaminated soil or groundwater that need to be treated. In addition, many physical processes (such as sorption onto activated carbon or soil washing) only transfer the contaminant to another compartment, with no degradation of the compound. Bioremediation is attractive since complete mineralization of the contaminant to innocuous endproducts can be achieved. As seen from Table 11.1, numerous microbes capable of degrading polychlorinated phenols have been isolated, and knowledge of their physiology and other characteristics gives a strong base for bioremediation applications.

Biological treatment technologies for chlorophenol-contaminated soils include solid- and slurry-phase processes. Contaminated waters have been treated in activated sludge and other aerobic or anaerobic bioreactors, as well as special bioreactors with immobilized chlorophenol-degrading microbes. A number of laboratory studies have investigated these different microbial methods for degradation of chlorophenol wastes, but results from field work have rarely been reported. Our aim is to review some of the different treatment technologies that have been studied in the laboratory and show some examples of their implementation in the field.

3.1. Soil Treatment

Chlorophenol-degrading bacteria can frequently be isolated from contaminated soils, indicating that the potential for biodegradation exists. Watanabe (1977) showed that pentachlorophenol-decomposing bacteria were enriched 1,000-fold during annual applications of PCP to agricultural soil over a period of 3 years. Laboratory experiments also indicated an increase in the PCP-decomposing activity of the treated soils (Watanabe, 1978). Chlorophenol-contaminated soils in Finland contained 10^4–10^5 PCP-mineralizing microbes per gram of dry soil compared with 10^2 per gram of agricultural soil (Valo and Salkinoja-Salonen, 1986a). However, as is evident from the persistence of chlorophenols at contaminated sites, these microbes do not degrade the chlorophenols *in situ*, or the degradation rates are insufficient to handle the contamination.

A number of environmental factors may limit the biodegradability of chlorophenols in soils even when competent microbes are present (Salkinoja-Salonen et al., 1990). These include an unsuitable temperature (usually too low, since most of the isolated strains have temperature optima around $25°$–$35°C$), the absence (or in some cases presence) of oxygen, too high or too low a pH and redox potential, or the lack of essential nutrients. In addition, a high concentration of chlorophenols (which were used specifically because they are highly effective biocides and toxic to most organisms) or some other toxic chemical (e.g., heavy metals or creosote, which were frequently used as wood preservatives in conjunction with chlorophenols) may inhibit biological activity. The chlorophenols may in other cases be unavailable to the microbes, possibly tightly bound to humus or clays. This is exemplified in experiments in which freshly added PCP (e.g., ^{14}C-labeled PCP) was readily degraded, but the "old" chlorophenols were recalcitrant (Salkinoja-Salonen et al., 1990). The environmental conditions will be different at each site, and the main task of bioremediation is to optimize these conditions so that either indigenous or inoculated microbes can degrade the contaminants.

Biotreatment technologies for soil can be divided into two main categories; solid-phase or slurry-phase treatment. Slurry-phase bioremediation involves bioreactors that are operated under water-logged conditions, with mechanical mixing to maintain optimum conditions, while solid-phase treatment (e.g., landfarming or composting) is operated under unsaturated conditions (Morgan and Watkinson, 1989; Stroo, 1992). These two types of treatments differ in the bioavailability of substrates and access of oxygen, and thus pose different demands on the microbial populations (Briglia et al., 1994b).

The use of inoculation in bioremediation has several constraints, a main restricting factor being the survival and competitiveness of the inoculants, in addition to the metabolic capabilities of the strains (Pritchard, 1992; Salkinoja-Salonen et al., 1990), and there are only a few well-documented reports showing successful application in large scale. A number of laboratory studies, however, have examined the activity of PCP-degrading pure cultures when inoculated into soil (Table 11.2). Most of these have been performed on a very small scale of 50–100 g of soil and in the laboratory, although a few experiments have included outdoor field tests.

One of the first studies involved inoculation of 10^6 PCP-utilizing *Arthrobacter* sp. cells per gram of soil that had been.treated with 20 mg/kg PCP. The inoculation resulted in 90% removal of PCP in 4 days, which corresponded to a reduction of the half-life to less than 1 day compared with 2 weeks in the uninoculated control (Edgehill and Finn, 1983b). Soil inoculation was also effective in an outdoor study with similarly treated 37 liter test plots (Edgehill and Finn, 1983b).

Crawford and Mohn (1985) investigated the effectiveness of the PCP-degrading

TABLE 11.2. Use of Inocula To Enhance Degradation of Chlorophenols in Soils

Inoculum	Scale, Time	Inoculum Size	Conditions	CP Concentration Beginning	CP Concentration End	Reference
Arthrobacter sp.	50 g, 6 days	10^5/g	Aerobic, 30°C	20 mg/kg	2 mg/kg	Edgehill and Finn (1983b)
Arthrobacter sp.	37 liters, 12 days	10^8/liter	Aerobic, 8°–16°C, outdoor test plot	182 mg/liter	28 mg/liter	Edgehill and Finn (1983b)
Arthrobacter sp.	50 g, 2 days	10^6/g	Aerobic, room temperature, clay-soil	15 mg/kg	0.6 mg/kg	Edgehill (1994)
Flavobacterium sp.	100 g, 10 days	10^7/g	Aerobic, 30°C, 60% mineralization of [^{14}C]PCP	100 mg/kg	13 mg/kg	Crawford and Mohn (1985)
Flavobacterium sp.	100 g, 100 days	10^7/g	Aerobic, 30°C, multiple inoculations at 20–40 day intervals	298 mg/kg	59 mg/kg	Crawford and Mohn (1985)
Flavobacterium sp.	40 ml, 60 hours	10^{11}/liter	Aerobic, 21°C, 25% wt/vol soil slurry with glucose supplementation	75 mg/liter	2 mg/liter	Topp and Hanson (1990)
Flavobacterium sp.	20 g, 210 days	10^6/g	Aerobic, 25°C, inoculation at days 40 and 168; 68% mineralization of [^{14}C]PCP	175 mg/kg	21 mg/kg	Seech et al. (1991)
CP enrichment	1.5 m³, 120 days	10^5/g	Aerobic, 15°–30°C, outdoor compost, nutrient supplements, no enhancement of CP degradation over uninoculated control	280 mg/kg	20 mg/kg	Valo and Salkinoja-Salonen (1986a)
Mycobacterium chlorophenolicum	1 liter, 68 days	10^9/g	Aerobic, 20°C, 30%–40% of [^{14}C]PCP mineralizaton in 40 days	120 mg/kg	20 mg/kg	Valo and Salkinoja-Salonen (1986a)

(continued)

TABLE 11.2. (Continued)

| Inoculum | Sale, Time | Inoculum Size | Conditions | CP Concentration | | Reference |
				Beginning	End	
Mycobacterium chlorophenolicum	1 liter, 68 days	10^9/g	Aerobic, 20°C, 30%–40% of [^{14}C]PCP mineralizaton in 40 days	120 mg/kg	20 mg/kg	Valo and Salkinoja-Salonen (1986a)
Mycobacterium chlorophenolicum	50 g, 130 days	10^8/g	Aerobic, 22°C, immobilized on polyurethane, 40%–60% mineralization of [^{14}C]PCP in 129 days	23 mg/kg	<1 mg/kg	Middeldorp et al. (1990)
Mixed culture	100 ml, 23 days	10^6/ml	Aerobic, 25°C, 5%–40% wt/vol soil slurry inoculated with mixed culture after 13 days	50–250 mg/liter	<2 mg/liter	Mahaffey and Sanford (1990)
Mixed culture	94 m³, 30 days	10^7/ml	Aerobic reactor, 20% wt/vol soil slurry	100 mg/kg	<0.5 mg/kg	Mahaffey and Sanford (1990)
Phanerochaete chrysosporium	24 g, 60 days		Aerobic, 39°C, sterilized soil, <2% mineralization of [^{14}C]PCP in 60 days	50 mg/kg	1 mg/kg	Lamar et al. (1990a)
Phanerochaete spp.	370 kg, 46 days		Aerobic, outdoor 10°–22°C, addition of wood chips and peat, 80% decrease of PCP, extensive O-methylation	300–400 mg/kg	<100 mg/kg	Lamar and Dietrich (1990)
Phanerochaete spp.	9 m², 2 tons, 56 days		Aerobic, outdoor plots, 55%–89% depletion of PCP. Inoculum added as grain-sawdust mixture	570–1010 mg/kg	74–490 mg/kg	Lamar et al. (1993)
Sludge enrichment	50 g, 28 days	5 g/kg	Anaerobic, >60% of PCP recovered as reductive dechlorination products (TeCP, TCP, DCP)	10–30 mg/kg	<0.5 mg/kg	Mikesell and Boyd (1988)

Flavobacterium sp. in remediation of contaminated soil. The addition of 10^7 *Flavobacterium* sp. cells/g of three different soils (loam, clay, and sand) resulted in 60% mineralization of the added PCP (100 mg/kg, wet weight) in 10 days. The rate of mineralization was slightly faster at PCP concentrations of 10 and 50 mg/kg, while 500 mg/kg PCP was not mineralized. Inoculation of the *Flavobacterium* sp. to industrially contaminated soil was less effective. Four inoculations of 10^7–10^8 cells/g of soil at 20–40 day intervals were required to decrease the PCP concentration from 300 to 60 mg/kg. The residual PCP level of 10–50 mg/kg was difficult to eliminate (Crawford and Mohn, 1985). In another study, inoculation with the *Flavobacterium* sp. stimulated degradation of PCP in soil, resulting in a decrease of PCP from 175 to 21 mg/kg in 210 days (Seech et al., 1991). A major constraint to successful soil inoculation with *Flavobacterium* seems to be the rapid decline of the inoculum. The viability of *Flavobacterium* sp. in soil was shown to be low, and inoculated populations declined by over 90% within 3 days (Topp and Hanson, 1990). Although the *Flavobacterium* is an effective PCP degrader in liquid culture and, as typical of gram-negative organisms, tolerates high PCP concentrations (Gonzales and Hu, 1991), it appears to be less effective for soil treatment. An alternative method to overcome the poor survival of the *Flavobacterium* sp. in soil is physical washing of the contaminated soil followed by microbial treatment of the PCP-containing process water (Pflug and Burton, 1988).

The use of the nocardioform bacterium *Mycobacterium* (*Rhodococcus*) *chlorophenolicum* for soil bioremediation has been extensively studied. *M. chlorophenolicum* was shown to survive for lengthy periods when inoculated into soil (Briglia et al., 1990). Immobilization of the bacteria onto a porous solid support (bark chips, polyurethane foam) was found to increase the viability of the inoculum in soil. In one experiment, *M. chlorophenolicum* immobilized onto polyurethane foam and introduced into a peaty soil with 750 mg/kg PCP to give approximately 10^9 cells/g of soil (dry weight) remained active for over 200 days (Briglia et al., 1990, 1994b). Electron microscopic investigation revealed the presence of viable *M. chlorophenolicum*–like cells in the polyurethane particles almost 1 year after inoculation. In another experiment a single inoculum of 10^5–10^8 *M. chlorophenolicum* cells/g of peaty soil and as little as 500 cells/g of sandy soil initiated mineralization of [^{14}C]PCP (Middeldorp et al., 1990). These soils had no detectable PCP-mineralizing microbes prior to inoculation, and PCP was persistent in uninoculated soil. A viable population of *M. chlorophenolicum* can thus be established in contaminated soils, and the inoculum is effective in degrading chlorophenols at the high levels found at polluted sites.

If sufficient numbers of chlorophenol-degrading microbes are already present in soil, further inoculation will not enhance degradation. For example, in a field experiment the addition of 10^5 PCP degraders per gram of soil to a 1.5 m^3 compost did not enhance remediation over uninoculated controls, since the total number of indigenous PCP degraders was already 4×10^5/g soil (Valo and Salkinoja-Salonen, 1986a). *M. chlorophenolicum* has been used to inoculate composts of volumes up to 100 m^3 for treatment of contaminated soils (Table 11.3). In these cases, however, controls with no inoculation were not included, and similar composts receiving no

TABLE 11.3. Results of Composting of Chlorophenol-Contaminated Soils

Compost[a]	Volume (m³)	Soil Type	Treatment Time (Months)	Chlorophenol Conc.[b] (mg/kg Dry Weight)
A	50	Sand; 19% organic matter	0	212
			16	16
B[c]	300	Peat, sawdust; 75% organic matter	0	9,000
			45	730
C[d]	10	Peat, sawdust; 75% organic matter	0	5,500
			25	70
D	200	Sand, silt; low organic matter	0	360
			24	50
E[e]	200	Sand, silt, clay; low organic matter	0	790
			11	110
			24	16
			34	10
F[f]	70	Sawdust; 80% organic matter	0	8,520
			13	4,950
			23	84
			32	18

[a]Five different soils were treated on site at sawmills in Finland.
[b]Sum of 2,4,6-TCP, 2,3,4,6-TeCP, and PCP determined from a composite sample of 20–50 subsamples of soil as described by Valo and Salkinoja-Salonen (1986a).
[c]Soil pH initially 4.7; neutralized by repeated additions of lime to maintain pH above 6.5.
[d]Soil from same location as windrow B. Inoculated with chlorophenol-degrading enrichment culture attached on sawdust.
[e]Inoculated with *Mycobacterium chlorophenolicum* to 10^6 cells/g wet weight.
[f]Inoculated with *Mycobacterium chlorophenolicum* to 10^6 cells/g wet weight at 0 and 13 months.

inoculation and relying on activation of the indigenous microbial population proved to be effective, as well. Inoculation with chlorophenol-degrading microbes is important, however, if the number of indigenous chlorophenol degraders is low. Laboratory studies indicated that for most soils more than 10^7 active bacteria were needed per gram of soil to ensure effective mineralization (Briglia et al., 1994b).

The use of composting for treatment of chlorophenol-contaminated soils has been studied in the laboratory and in field tests (Valo, 1990; Valo and Salkinoja-Salonen, 1984, 1986a) and this technology is currently being implemented for bioremediation of several contaminated sawmill sites in Finland. Initial pilot studies (see Table 11.2) indicated that conditions required for the activity of chlorophenol-degrading microbes could be created in field composts, which led to 80% removal of chlorophenols (initial concentration 200–300 mg/kg) within 4 months (Valo and Salkinoja-Salonen, 1986a). Laboratory studies with soil from the pilot composts demonstrated that removal of chlorophenols was due to mineralization with negligible loss due to volatilization. O-methylation, which is catalyzed by a number of soil bacteria and fungi (Häggblom, 1990, 1992), accounted for less than 2% of chlorophenol loss and was thus insignificant compared with degradation. It is important to note that the main dimeric impurities of technical chlorophenols, polychlorinated phenoxyphenols, were recalcitrant and were not degraded during composting (Valo and Salkinoja-Salonen, 1986a).

The windrow composting method has since been used to remediate a number of former sawmill sites (Table 11.3). The chlorophenol-contaminated soil was composted on site, within the sawmill area, close to the original location. The chlorophenol contamination ranged from 200 to 9,000 mg/kg soil dry weight. The composting area was lined with plastic, which was covered with a 5–10 inch layer of gravel to protect the plastic lining from machinery. The maximum dimensions of the windrows were width, 6 m; height, 3 m; and length, up to 50 m. The treated soil volume ranged from 10 to 300 m^3. Softwood bark was added to the contaminated soil as a bulking agent to improve aeration in the compost windrows. Drain pipes (5–10 cm in diameter) were installed in some of the composts, but this proved later to be unnecessary, as there was no indication of anoxic conditions in the middle of the compost heaps. The soil was supplemented with inorganic nutrients, containing 250 mg/kg of nitrogen and 40 mg/kg of phosphorus. Some of the composts were inoculated either with a chlorophenol-degrading mixed culture or with a culture of *M. chlorophenolicum*. Finally, the windrows were covered with plastic lining to prevent leaching of chlorophenols by rain. Gas exchange through the surface of the windrow was maintained by leaving an air layer below the plastic lining. The windrows were mixed completely once or twice a year, during which nutrients were added and the pH and moisture adjusted if needed, and samples were taken for chlorophenol analysis (Valo, 1990). The excavation of chlorophenol-contaminated soil and construction of compost heaps containing a total soil volume of 2,000 m^3 at a sawmill site in Finland is illustrated in Figure 11.2.

In Finland, where the maximum temperature of surface soil reaches about 16°C in summer and falls below 0°C for 4–6 months in the winter, maintaining a favorable temperature for microbial activity is imperative for effective treatment. When the contaminated soil was built into a compost the temperature was maintained at 5°–15°C above ambient (Valo, 1990; Valo and Salkinoja-Salonen, 1986a). However, no measures were taken to maintain the temperature during winter, thus reducing the effective treatment time to about 6 months of the year.

The results of six different composts of contaminated soils are compiled in Table 11.3. The effectiveness of treatment varied in the different soils, and in some soils the chlorophenols seemed to be biologically unavailable (Salkinoja-Salonen et al., 1990). Soils containing up to 9,000 mg/kg of chlorophenols could be remediated, and treatment levels below 20 mg/kg were obtained after a composting period of 20–30 months. The indigenous population of chlorophenol-degrading microorganisms could thus be effectively stimulated to remediate the sites.

Soils contaminated with both chlorophenols and creosote pose a more complex problem for bioremediation. Both solid-phase and slurry-phase systems for treatment of PCP- and creosote-contaminated soil have been evaluated in bench scale (Mueller et al., 1991a,b), but in neither case was degradation of PCP demonstrated, although some of the lower molecular weight creosote components were biodegraded.

Figure 11.2. Construction of a 2,000 m² chlorophenol compost: Excavation of chlorophenol-contaminated soil at sawmill (**A,B**), mixing contaminated soil with bark chips as a bulking agent (**C**), and completed windrows (**D**).

Full-scale bioremediation of a wood treatment site contaminated with PCP and polynuclear aromatic hydrocarbons has been reported (Compeau et al., 1991). Seventeen thousand cubic yards of soil with an average PCP concentration of 410 mg/kg was treated in an 8 acre land treatment unit. During a treatment period of 13 weeks the level of PCP was reduced to 150 mg/kg. At another site, soil washing

Figure 11.2. (*Continued*)

combined with slurry-phase biodegradation was demonstrated (Compeau et al., 1991; Mahaffey and Sanford, 1990). The wash solution was treated in 94 m³ slurry bioreactors inoculated with 10⁷ cells/ml of a PCP-degrading mixed culture. Laboratory studies verified that the consortium mineralized PCP to CO_2 with release of chloride. The slurry treatment resulted in removal of 370 mg/kg PCP on slurry solids to a level below 0.5 mg/kg in 14–21 days. Treatment of the soil, without the washing step, was less effective. At a 20% wt/vol solids loading rate and an inoculum of 10^7 cells/ml, degradation of PCP to a level below 0.5 mg/kg took 30 days (Mahaffey and Sanford, 1990).

In situ subsurface bioremediation of PCP-contaminated soil has also been evaluated. An aboveground water treatment system combined with nutrient addition and reinjection of the water to the subsurface was studied in bench scale and in pilot scale on site (Fu and O'Toole, 1990). The aboveground treatment system consisted of UV/ozone chemical oxidation and activated carbon adsorption. Depletion of PCP in soil after groundwater pumping was observed in both the laboratory study and on site, but controls demonstrating that PCP loss was due to biodegradation were lacking. Although the chloride concentration in groundwater increased during continuous recycling, suggesting biological transformation of PCP, no data showing the mass balance of transformation was presented, and the study is inconclusive.

Since the white-rot fungi *Phanerochaete* spp. are capable of transforming and degrading chlorinated aromatic compounds, their use for remediation of PCP-contaminated soil has been investigated. Inoculation of *Phanerochaete chrysosporium* into sterilized soils with 50 mg/kg PCP resulted in an average PCP decrease of 98% in 56 days compared with 43% in uninoculated controls (Lamar et al., 1990a). However, ^{14}C-labeling studies indicated that only 2% of PCP was mineralized to CO_2. Both *Phanerochaete chrysosporium* and *Phanerochaete sordida* were shown to O-methylate PCP nearly stoichiometrically to pentachloroanisole (Lamar et al., 1990b). Although liquid cultures of the *Phanerochaete* spp. were shown to mineralize pentachloroanisole to some extent, it persisted in soil (Lamar et al., 1990b). PCP was also extensively O-methylated in wood chips inoculated with *Phanerochaete chrysosporium* and *Phanerochaete sordida,* but inoculation with another white-rot fungus, *Trametes hirsuta,* resulted in 27% mineralization of PCP (Lamar and Dietrich, 1992). The fungus *Lentinus edodes* transformed PCP in soil with pentachloroanisole as a significant product (Okeke et al., 1993). The use of *Phanerochaete chrysosporium* and *Phanerochaete sordida* inoculations for soil treatment was tested in a field study in plots of $1 \times 1 \times 0.35$ m with approximately 370 kg (dry weight) of soil, which indicated that PCP was *depleted* (Lamar and Dietrich, 1990). However, similar to the laboratory study, extensive O-methylation occurred with over 10% of PCP converted to pentachloroanisole. The extent of volatilization of pentachloroanisole at the field site was not investigated. Similar observations were made in a larger scale field study (Lamar et al., 1993). Separate 9 m² plots were inoculated with *Phanerochaete chrysosporium*, *Phanerochaete sordida* and *Trametes hirsuta* with different treatments. Depletion of PCP varying from 15% to 89% was observed, but the extent of PCP degradation was not determined. Up to 10% of PCP accumulated as pentachloroanisole (Lamar et al., 1993). In addition, the potential for formation of polychlorinated dioxins and dibenzofurans by fungal peroxidases remains (Öberg et al., 1990; Svenson et al., 1989) and should be addressed. Fungal treatment of chlorophenol-contaminated sites should therefore be considered with caution. It is clearly important to make the distinction between *depletion* and *degradation* of chlorophenols during biological treatment.

4.2. Water Treatment

A variety of bioreactor types have been investigated for treatment of chlorophenol-contaminated waters. These have been operated under either aerobic or anaerobic conditions (and in sequence) and include trickling filters, rotating disk reactors, fluidized bed systems, fixed film and sludge blanket reactors, activated sludge plants, and aerated lagoons. The use of immobilized specialized strains has also been investigated.

Several studies have indicated that chlorophenols are degraded in laboratory and full-scale activated sludge treatment of wastewater, with removal efficiencies ranging from under 20% to over 90% (Chudoba et al., 1989; Edgehill and Finn, 1983a; Ettala et al., 1992; Hickman and Novak, 1984; Kirk and Lester, 1988). Other types of aerobic reactors utilizing mixed cultures of acclimated biomass have also been used (Dierberg et al., 1987; Etzel and Kirsch, 1974; Klecka and Maier, 1985; Moos et al., 1983; Nevalainen et al., 1993; Seidler et al., 1986; Tokuz, 1991). In many of these studies, however, controls were lacking, and a distinction between biodegradation and removal through sorption cannot always be made.

Puhakka and coworkers (see Järvinen and Puhakka, 1994; Mäkinen et al., 1993; Puhakka and Järvinen, 1992) studied the use of aerobic fluidized bed bioreactors for treatment of chlorophenol-contaminated simulated groundwater in laboratory scale. A mixed culture of polychlorophenol-degrading microbes originally enriched from activated sludge of a municipal wastewater treatment plant was immobilized on a celite carrier (Puhakka and Järvinen, 1992). The reactors were fed a mixture containing 2,4,6-TCP, 2,3,4,6-TeCP, and PCP at loading rates of 217 mg/liter/day and operated at a temperature of 20°–28°C. Over 90% of the chlorophenols were removed at a hydraulic retention time of 5 hours during a 12 day experiment. Inorganic chloride release corresponded well with the amount fed as chlorophenol-bound chlorine, and also total organic carbon removal agreed well with chlorophenol removal (Puhakka and Järvinen, 1992). In another experiment, chlorophenol removal efficiency was 99% or more at a hydraulic retention time of 5 hours and a loading rate of 78–445 mg/liter/day over a 3 month monitoring period (Mäkinen et al., 1993). Of specific interest is that the fluidized reactors could be operated at a temperature below 10°C with the same chlorophenol removal efficiency, indicating that high rate degradation of chlorophenols at groundwater temperatures is feasible (Järvinen and Puhakka, 1994; Järvinen et al., 1994). Fluidized bed bioremediation of chlorophenol-contaminated groundwater at low temperatures is currently being evaluated on a larger scale at the municipality of Kärkölä in Finland (J. Puhakka, personal communication).

Several of the chlorophenol-degrading bacteria listed in Table 11.1 show promise for use in bioremediation of chlorophenol-contaminated waters. Edgehill and Finn (1983a) showed that addition of a PCP-metabolizing *Arthrobacter* sp. to activated sludge resulted in immediate removal of PCP, whereas without inoculation a 1 week acclimation period was required before onset of activity. The use of immobilized bacteria or fungi for water treatment is of special interest. Immobilization of microbes has been shown to increase their tolerance against toxic compounds (Balfanz and Rehm, 1991; O'Reilly and Crawford, 1989; Portier and Fujisaki, 1986; Suzuki, 1977; Valo et al., 1990) and, by co-immobilizing the microbes with a sorptive material, chlorophenols can be concentrated from dilute waters. Two extensively studied systems involve the PCP-degrading *M. chlorophenolicum* and *Flavobacterium* strains.

Pignatello et al. (1983) observed that PCP degradation in experimental streams occurred largely by microbes attached to rock surfaces, which led them to design a bioreactor using an immobilized biofilm covering rock surfaces (Brown et al., 1986; Frick et al., 1982). The *Flavobacterium* sp. strain ATCC 39723 isolated from an enrichment culture of sediment from the experimental streams (Saber and Crawford, 1985) was used in further work. Polyurethane-immobilized cells of the *Flavobacterium* sp. were capable of degrading PCP at initial concentrations of up to 300 mg/liter (Crawford et al., 1990; O'Reilly and Crawford, 1989). Supplementary carbon sources greatly stimulated PCP degradation by the *Flavobacterium* sp. (Gonzales and Hu, 1991; Topp et al., 1988). In a continuous-culture reactor a PCP degradation rate was 3.5–4 mg/day per gram of foam (O'Reilly and Crawford, 1989). Glutamate (0.5 g/liter) was, however, required as a supplementary carbon source to maintain activity.

A pilot scale treatment system inoculated with the *Flavobacterium* sp. was evaluated for its effectiveness in treatment of PCP-contaminated groundwater (Stinson et al., 1991). The treatment system consisted of a three-stage fixed-film reactor (approximately 2 m^3) and was operated at flow rates of 1–5 gal/min (220–1,130 liters/hr). The bioreactor achieved approximately 96% removal of PCP at a flow rate of 5 gal/min (19 liters/hr) and an influent PCP concentration of 27 mg/liter, producing an effluent of 1 m/liter PCP before activated carbon polishing.

The severe case of groundwater contamination at Kärkölä, Finland, prompted us to study the use of *M. chlorophenolicum* for treatment of chlorophenol-contaminated groundwater. Previous work by Apajalahti and Salkinoja-Salonen (1984) and by Valo et al. (1985) indicated that a biofilter consisting of a mixed bacterial population immobilized onto soft wood bark chips was effective in degrading PCP. The bark chips sorbed PCP reversibly and thus detoxified the medium and allowed PCP degradation at concentrations of 200 μM and higher, which would otherwise have been inhibitory to the bacteria. *M. chlorophenolicum* strain PCP-I, which was isolated from an enrichment culture of this biofilter (Apajalahti et al., 1986), was sensitive to the toxicity of PCP and did not grow in media with a PCP concentration higher than 50 μM (Apajalahti and Salkinoja-Salonen, 1986). Immobilization of the bacterium onto a solid support, such as polyurethane, increased its tolerance against chlorophenols (Valo et al., 1990), similar to what was seen with the *Flavobacterium* sp. (O'Reilly and Crawford, 1989) and with other bacteria (Balfanz and Rehm, 1991; Portier and Fujisaki, 1986; Siahpush et al., 1992). In a trickling filter the immobilized *M. chlorophenolicum* degraded chlorophenols at feed concentrations of up to 500 μM (130 mg/liter) (Valo et al., 1990). Figure 11.3 shows a micrograph of the polyurethane material seeded with *M. chlorophenolicum* after 2 months of operation.

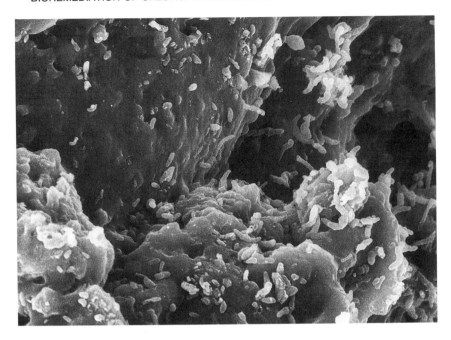

Figure 11.3. Scanning electron micrograph of polyurethane bed material seeded with *Mycobacterium chlorophenolicum* after 2 months of operation (Valo et al., 1990). (Micrograph kindly provided by Eeva-Liisa Nurmiaho-Lassila.) Bar = 10 μm.

In Finland and in other northern countries a principal factor limiting *in situ* biodegradation is the low temperature of groundwater, typically below 6°C year round. Heating the groundwater to 25°–30°C, the optimal temperature of most chlorophenol-degrading bacteria, would be an added expense in biological treatment. We found that we could avoid this by operating the *M. chlorophenolicum* biofilter in two stages (Valo et al., 1990). Since the activated carbon–polyurethane matrix that we used for immobilizing the bacterium effectively adsorbs chlorophenols, we could charge the filter by feeding chlorophenol-contaminated (penta-, 2,3,4,6-tetra-, and 2,4,6-trichlorophenol) water at 4°C. After charging the filter with chlorophenols, the temperature was elevated to 25°C, and one bed volume of water or buffer was recycled in the biofilter. This enabled the immobilized bacteria to dehalogenate and mineralize the chlorophenols that had been sorbed onto the polyurethane. Degradation of the chlorophenols proceeded in the absence of a supplementary carbon source. After the chlorophenols were degraded, as evidenced by release of inorganic chloride, the charge cycle was repeated. The biofilter remained active for over 4 months of repeated charging at 4°C and subsequent biodegradation stages at 25°C. No supplementary carbon sources were necessary to maintain activity, which is important if the groundwater is to be returned to the aquifer.

Treatment of chlorophenol-contaminated groundwater by immobilized *M. chlo-*

rophenolicum was tested further in two pilot-scale bioreactors of 50 liters and 2 m³, respectively (Valo and Hakulinen, 1990). The contaminated groundwater was from the earlier described site at Kärkölä, in southern Finland. During a 3 month trial period the 2 m³ reactor, containing 1 m³ of polyurethane (with immobilized *M. chlorophenolicum*) was operated at 20°C at a feeding rate of 700–3,000 liters/day to treat 60 m³ of contaminated water with a chlorophenol concentration of 25–40 mg/liter. The degradation rate was 25–55 g/m³/day, giving an effluent chlorophenol concentration of 0.002–0.075 mg/liter. Analysis of inorganic chloride indicated that chlorophenol removal corresponded with complete dechlorination.

Immobilization of a number of other microbial pure cultures for water treatment has been evaluated as well. *Pseudomonas* and *Arthrobacter* strains were immobilized on chitin, glass, or cellulose material and utilized in degradation of a mixture of phenol and chlorophenols (Portier and Fujisaki, 1986). The *Alcaligenes* strain A7-2 immobilized by either entrapment or adsorption was able to degrade 4-CP and mixtures of monochlorophenols and phenol (Balfanz and Rehm, 1991; Menke and Rehm, 1992; Westmeier and Rehm, 1985). *Arthrobacter* sp. strain ATCC 33790 immobilized into alginate was effective in degrading PCP (Lin and Wang, 1991). Recently, Golovleva et al. (1993) investigated the degradation of chlorophenols by *Streptomyces rochei* strain 303 immobilized on a polycaproamide fiber carrier. The bioreactor was operated continuously at a flow rate of 0.08/hr and a 2,4,6-trichlorophenol concentration of 1 g/liter without loss of activity for 2.5 months. At a lower inflow concentration of 150 mg/liter the bioreactor was operated for 11 months without any additional carbon source (Golovleva et al., 1993).

In another study, the PCP-degrading *Pseudomonas* strain SR3 was inoculated in combination with other polyaromatic hydrocarbon–degrading bacterial strains to a stirred tank reactor that was evaluated for treatment of PCP- and creosote-contaminated groundwater (Mueller et al., 1993). Results from a 1.2 liter bench-scale reactor indicated that 70% of PCP was degraded, while in a 454 liter pilot-scale reactor less than 25% of PCP was degraded during a 14 day operation period. The poor performance in the pilot reactor was attributed to low cell viability and poor induction of the inoculum.

There has been considerable interest in the potential of using *Phanerochaete chrysosporium* and other white-rot fungi for treatment of pulp bleaching effluents containing chlorinated organic compounds. A rotating biological contactor with immobilized *Phanerochaete chrysosporium* that was initially developed for decolorization of kraft bleach plant effluents was later shown also to dechlorinate high- and low-molecular weight components of the effluents, including chlorophenols (Huynh et al., 1985; Pellinen et al., 1988). A number of other white-rot fungi have been shown to be effective in removing the chlorinated organics of bleach plant effluents, as well (Bergbauer et al., 1991; Wang et al., 1992), and fungal biofilm reactors have also been tested for degradation of PCP and other chlorophenols (Lewandowsky et al., 1990; Lin et al., 1991; Venkatadri et al., 1992). A co-immobilized system using activated carbon and the white-rot fungus *Phanerochaete chrysosporium* was tested for its ability to degrade PCP (Lin et al., 1991). In such a system 95% of PCP was removed within 25 hours; however, only 10% of PCP was mineralized over a period of 300 hours, although the mineralization rate was higher than with nonimmobilized cells. In a biofilm membrane reactor of *Phanerochaete chrysosporium* physical binding to fungal cells accounted for up to 75% of PCP removal (Venkatadri et al., 1992). The extent of PCP degradation was not determined. As in soil treatment with *Phanerochaete chrysosporium*, a drawback with fungal technologies is that the mechanism of chlorophenol degradation is poorly understood and frequently does not result in complete mineralization.

It is interesting to note that while anaerobic dechlorination is lately receiving increasing attention, the first studies on biological treatment of chlorophenol-contaminated waters specifically involved anaerobic reactors. Hakulinen and Salkinoja-Salonen (1981, 1982) demonstrated that an anaerobic fluidized bed reactor was effective in dechlorinating chlorinated phenols in pulp bleaching effluents. These studies also involved another, now popular, concept, sequential anaerobic and aerobic treatment, in which anaerobic dechlorination yields metabolites that are more amenable to aerobic degradation. Recently, sequential anaerobic–aerobic degradation of 2,4,6-TCP was demonstrated in a process consisting of an anaerobic reactor that dechlorinated 2,4,6-TCP to 4-CP, which was subsequently mineralized in an aerobic reactor (Armenante et al., 1992).

Reductive dechlorination of pentachlorophenol has been demonstrated in anaerobic fixed film and sludge blanket reactors (Hendriksen and Ahring, 1992, 1993; Hendriksen et al., 1992; Krumme and Boyd, 1988; Mohn and Kennedy, 1992a; Woods et al., 1989; Wu et al., 1993), and methanogenic granules with high dechlorinating activity have been developed (Wu et al., 1993). It is important to note that reductive dechlorination is not always complete, but may result in accumulation of persistent mono- and dichlorinated phenols and is therefore a limitation for anaerobic treatment of chlorophenol-contaminated waters.

4. CONCLUSIONS

Successful bioremediation must result in the mineralization of the hazardous compounds to CO_2 and water or in their detoxification to innocuous endproducts. It is pivotal to make the distinction between *degradation* and *depletion* when evaluating, and particularly before implementing, a treatment technology, since a number of biotransformation reactions can produce more recalcitrant and hazardous compounds. The effectiveness of a proposed bioremediation strategy should first be demonstrated by laboratory and pilot studies.

The examples given in this chapter demonstrate that a number of different strategies for biological remediation of chlorophenol wastes can successfully be implemented, and removal of the chlorophenols can result in complete mineralization to innocuous endproducts with release of chloride. Bioremediation of contaminated soil has been demonstrated in full scale at several sites. For example, composting was shown to be a highly effective bioremediation method for chlorophenol-contaminated soils. Treatment can be performed on site with low costs of maintaining composts during the treatment period. Treatment of chlorophenol-contaminated water has been demonstrated with several different bioreactor types involving both aerobic or anaerobic bacteria, as well as immobilized pure cultures. Several of these show promise for treatment of contaminated groundwater. The choice of the appropriate (bio)technology for soil or groundwater remediation will depend on the site characteristics and on the nature of the contaminants, and in some cases chemical/physical treatment may be needed in combination with bioremediation.

Although chlorophenol degradation at contaminated sites is slow, indigenous microbial populations can effectively be stimulated to bioremediate chlorophenol-contaminated soils. However, the use of specific inocula for treatment of contaminated soils may have a number of advantages. If the levels of indigenous chlorophenol-degrading microbes are low or nonexistent, inoculation will be needed to promote biodegradation. In addition, by inoculating a known microbe with a characterized degradation pathway, unwanted biotransformation reactions (such as O-methylation catalyzed by a variety of soil fungi and bacteria) can be minimized. Survival of the inoculant is central to its effective use, and large differences in the survival characteristics of different PCP-degrading bacteria have been shown. Immobilization or microencapsulation of the inoculant can improve its survival and activity in soil. Successful development of bioremediation technologies will require a much better understanding of the survival and activities of introduced microbes.

ACKNOWLEDGMENTS

We are grateful to Mirja Salkinoja-Salonen, Jussi Uotila, Jaakko Puhakka, and their coworkers for many helpful discussions and for providing us with manuscripts prior to their publication.

We acknowledge financial support from the U.S. Environmental Protection Agency and the Office of Naval Research.

REFERENCES

Abrahamsson K, Klick S (1991): Degradation of halogenated phenols in anoxic natural marine sediments. Mar Pol Bull 22:227–233.

Ahlborg UG, Thunberg TM (1980): Chlorinated phenols: Occurrence, toxicity, metabolism, and environmental impact. CRC Crit Rev Toxicol 7:1–35.

Ahling B, Lindskog A (1982): Emission of chlorinated organic substances from combustion. In Hutzinger O, Frei RW, Merian E, Pocchiari F (eds): Chlorinated Dioxins and Related Compounds. Oxford: Pergamon Press, pp 215–225.

Apajalahti JHA, Kärpänoja P, Salkinoja-Salonen MS (1986): *Rhodococcus chlorophenolicus* sp. nov., a chlorophenol-mineralizing actinomycete. Int J Syst Bacteriol 36:246–251.

Apajalahti JHA, Salkinoja-Salonen MS (1984): Absorption of pentachlorophenol (PCP) by bark chips and its role in microbial PCP degradation. Microb Ecol 10:359–367.

Apajalahti JHA, Salkinoja-Salonen MS (1986): Degradation of chlorinated phenols by *Rhodococcus chlorophenolicus*. Appl Microbiol Biotechnol 25:62–67.

Apajalahti JHA, Salkinoja-Salonen MS (1987a): Complete dechlorination of tetrachlorohydroquinone by cell extracts of pentachlorophenol-induced *Rhodococcus chlorophenolicus*. J Bacteriol 169:5125–5130.

Apajalahti JHA, Salkinoja-Salonen MS (1987b): Dechlorination and *para*-hydroxylation of polychlorinated phenols by *Rhodococcus chlorophenolicus*. J Bacteriol 169:675–681.

Armenante PM, Kafkewitz D, Lewandowski G, Kung C-M (1992): Integrated anaerobic–aerobic process for the biodegradation of chlorinated aromatic compounds. Environ Progr 11:113–122.

Balajee S, Mahadevan A (1990): Utilization of chloroaromatic substances by *Azotobacter chroococcum*. Syst Appl Microbiol 13:194–198.

Balfanz J, Rehm H-J (1991): Biodegradation of 4-chlorophenol by adsorptive immobilized *Alcaligenes* sp. A 7-2 in soil. Appl Microbiol Biotechnol 35:662–668.

Beadle CA, Smith RW (1982): The purification and properties of 2,4-dichlorophenol hydroxylase from a strain of *Acinetobacter* species. Eur J Biochem 123:323–332.

Behechti A, Ballhorn L, Freitag D (1988): Verhalten von 2,4-Dichlorophenol, 2,4,6-Trichlorophenol und Pentachlorophenol bie der Kompostierung in einer standardisierten Laboranlage, Chemosphere 17:2433–2440.

Bergbauer M, Eggert C, Kraepelin G (1991): Degradation of chlorinated lignin compounds in a bleach plant effluent by the white-rot fungus *Trametes versicolor*. Appl Microbiol Biotechnol 35:105–109.

Berger RS (1972): 2,6-Dichlorophenol, sex pheromone of the lone star tick. Science 177:704–705.

Bollag J-M (1992): Decontaminating soil with enzymes. Environ Sci Technol 26:1876–1881.

Bollag J-M, Helling CS, Alexander M (1968): 2,4-D metabolism. Enzymatic hydroxylation of chlorinated phenols. J Agric Food Chem 16:826–828.

Bollag J-M, Liu S-Y (1985): Copolymerization of halogenated phenols and syringic acid. Pestic Biochem Physiol 23:261–272.

Bollag J-M, Shuttleworth KL, Anderson DH (1988): Laccase-mediated detoxification of phenolic compounds. Appl Environ Microbiol 54:3086–3091.

Boyd SA, Shelton DR (1984): Anaerobic biodegradation of chlorophenols in fresh and acclimated sludge. Appl Environ Microbiol 47:272–277.

Boyd SA, Shelton DR, Berry D, Tiedje JM (1983): Anaerobic biodegradation of phenolic compounds in digested sludge. Appl Environ Microbiol 46:50–54.

Briglia M, Eggen RIL, van Elsas DJ, de Vos WM (1994a): Phylogenetic evidence for transfer of pentachlorophenol-mineralizing *Rhodococcus chlorophenolicus* PCP-I to the genus *Mycobacterium*. Int J Syst Bacteriol 44:494–498.

Briglia M, Middeldorp PJM, Salkinoja-Salonen MS (1994b): Mineralization performance of *Rhodococcus chlorophenolicus* strain PCP-1 in contaminated soil simulating on site conditions. Soil Biol Biochem 26:377–385.

Briglia M, Nurmiaho-Lassila EL, Vallini G, Salkinoja-Salonen M (1990): The survival of the pentachlorophenol-degrading *Rhodococcus chlorophenolicus* PCP-1 and *Flavobacterium* sp. in natural soil. Biodegradation 1:273–281.

Brown EJ, Pignatello JJ, Martinson MM, Crawford RL (1986): Pentachlorophenol degradation: A pure bacterial culture and an epilithic microbial consortium. Appl Environ Microbiol 52:92–97.

Bryant FO, Hale DD, Rogers JE (1991): Regiospecific dechlorination of pentachlorophenol by dichlorophenol-adapted microorganisms in freshwater, anaerobic sediment slurries. Appl Environ Microbiol 57:2293–2301.

Buser H-R, Zanier C, Tanner H (1982): Identification of 2,4,6-trichloroanisole as a potent compound causing cork taint in wine. J Agric Food Chem 30:359–362.

Chaudry GR, Huang GH (1988): Isolation and characterization of a new plasmid from a *Flavobacterium* sp. which carries the genes for degradation of 2,4-dichlorophenoxyacetate. J Bacteriol 170:3897–3902.

Chu JP, Kirsch EJ (1972): Metabolism of pentachlorophenol by an axenic bacterial culture. Appl Microbiol 23:1033–1035.

Chu J, Kirsch EJ (1973): Utilization of halophenols by a pentachlorophenol metabolizing bacterium. Dev Ind Microbiol 14:264–273.

Chudoba J, Albokova J, Lentge B, Kümmel R (1989): Biodegradation of 2,4-dichlorophenol by activated sludge microorganisms. Water Res 23:1439–1442.

Cirelli DP (1978): Patterns of pentachlorophenol usage in the United States of America—An overview. In Rao KR (ed): Pentachlorophenol. Chemistry, Pharmacology, and Environmental Toxicology. New York: Plenum Press, pp 13–18.

Cole JR, Cascarelli AL, Mohn WW, Tiedje JM (1994): Isolation and characterization of a novel bacterium growing via reductive dehalogenation of 2-chlorophenol. Appl Environ Microbiol 60:3536–3542.

Compeau GC, Mahaffey WD, Patras L (1991): Full-scale bioremediation of contaminated soil and water. In Sayler GS, Fox R, Blackburn JW (eds): Environmental Biotechnology for Waste Treatment. New York: Plenum Press, pp 91–109.

Crawford RL, Mohn WW (1985): Microbial removal of pentachlorophenol from soil using a *Flavobacterium*. Enzyme Microb Technol 7:617–620.

Crawford RL, O'Reilly KT, Tao H-L (1990): Microorganism stabilization for *in situ* degradation of toxic chemicals. In Kamely D, Chakrabarty A, Omenn GS (eds): Biotechnology and Biodegradation. Houston: Gulf Publishing Co., pp 203–211.

Cserjesi AJ, Johnson EL (1972): Methylation of pentachlorophenol by *Trichoderma virgatum*. Can J Microbiol 18:45–49.

Curtis RF, Dennis C, Gee JM, Gee MG, Griffiths NM, Land DG, Peel JL, Robinson D (1974): Chloroanisoles as a cause of musty taint in chickens and their microbiological formation from chlorophenols in broiler house litters. J Sci Food Agric 25:811–828.

Dec J, Bollag J-M (1990): Detoxification of substituted phenols by oxidoreductive enzymes through polymerization reactions. Arch Environ Contam Toxicol 19:543–550.

Dec J, Shuttleworth KL, Bollag J-M (1990): Microbial release of 2,4-dichlorophenol bound to humic acid or incorporated during humification. J Environ Q 19:546–551.

Detrick RS (1977): Pentachlorophenol, possible sources of human exposure. Forest Products J 27:13–16.

DeWeerd KA, Concannon, F, Suflita JM (1991): Relationship between hydrogen consumption, dehalogenation, and the reduction of sulfur oxyanions by *Desulfomonile tiedjei*. Appl Environ Microbiol 57:1929–1934.

DeWeerd KA, Mandelco L, Tanner RS, Woese CR, Suflita JM (1990): *Desulfomonile tiedjei* gen. nov. and sp. nov., a novel anaerobic dehalogenating, sulfate reducing bacterium. Arch Microbiol 154:23–30.

DeWeerd KA, Suflita JM (1990): Anaerobic aryl reductive dehalogenation of halobenzoates by cell extracts of "*Desulfomonile tiedjei*." Appl Environ Microbiol 56:2999–3005.

Dierberg FE, Goulet NA Jr, DeBusk TA (1987): Removal of two chlorinated compounds from secondary domestic effluent by a thin film technique. J Environ Q 16:321–324.

Dietrich G, Winter J (1990): Anaerobic degradation of chlorophenol by an enrichment culture. Appl Microbiol Biotechnol 34:253–258.

Ditzelmüller G, Loidl M, Streichsbier F (1989): Isolation and characterization of a 2,4-dichlorophenoxyacetic acid-degrading soil bacterium. Appl Microbiol Biotechnol 31:93–96.

Dolfing J (1990): Reductive dechlorination of 3-chlorobenzoate is coupled to ATP production and growth in an anaerobic bacterium, strain DCB-1. Arch Microbiol 153:264–266.

Dolfing J, Harrison BK (1992): Gibbs free energy of formation of halogenated aromatic compounds and their potential role as electron acceptors in anaerobic environments. Environ Sci Technol 26:2213–2218.

Don RH, Weightman AJ, Knackmuss H-J, Timmis KN (1985): Transposon mutagenesis and cloning analysis of the pathway for degradation of 2,4-dichlorophenoxyacetic acid and 3-chlorobenzoate in *Alcaligenes eutrophus* JMP134(pJP4). J Bacteriol 161:85–90.

Dorn E, Knackmuss H-J (1978a): Chemical structure and biodegradability of halogenated aromatic compounds. Substituent effects on 1,2-dioxygenation of catechol. Biochem J 174:85–94.

Dorn E, Knackmuss H-J (1978b): Chemical structure and biodegradability of halogenated aromatic compounds. Two catechol 1,2-dioxygenases from a 3-chlorobenzoate-grown pseudomonad. Biochem J 174:73–84.

Dougherty RC (1978): Human exposure to pentachlorophenol. In Rao KR (ed): Pentachlorophenol: Chemistry, Pharmacology, and Environmental Toxicology. New York: Plenum Press, pp 351–361.

Edgehill RU (1994): Pentachlorophenol removal from slightly acidic mineral salts, commercial sand, and clay soil by recovered *Arthrobacter* strain ATCC 33790. Appl Microbiol Biotechnol 41:142–148.

Edgehill RU, Finn RK (1982): Isolation, characterization and growth kinetics of bacteria metabolizing pentachlorophenol. Eur J Appl Microbiol Biotechnol 16:179–184.

Edgehill RU, Finn RK (1983a): Activated sludge treatment of synthetic wastewater containing pentachlorophenol. Biotechnol Bioeng 25:2165–2176.

Edgehill RU, Finn RK (1983b): Microbial treatment of soil to remove pentachlorophenol. Appl Environ Microbiol 45:1122–1125.

Engel C, deGroot AP, Weurman C (1966): Tetrachloroanisol: A source of musty taste in eggs and broilers. Science 154:270–271.

Engelhardt G, Rast HG, Wallnöfer PR (1979): Cometabolism of phenol and substituted phenols by *Nocardia* spec. DSM 43521. FEMS Microbiol Lett 5:377–383.

Ettala M, Koskela J, Kiesilä A (1992): Removal of chlorophenols in a municipal sewage treatment plant using activated sludge. Water Res 26:797–804.

Etzel JE, Kirsch EJ (1974): Biological treatment of contrived and industrial wastewater containing pentachlorophenol. Dev Ind Microbiol 16:287–295.

Evans WC, Smith BSW, Fernley HN, Davies JI (1971): Bacterial metabolism of 2,4-dichlorophenoxyacetate. Biochem J 122:543–551.

Frick TD, Crawford RL, Martinson M, Chresand T, Bateson G (1988): Microbiological cleanup of groundwater contaminated by pentachlorophenol. In Omenn GS (ed): Environmental Biotechnology. Reducing Risks From Environmental Chemicals Through Biotechnology. New York: Plenum Press, pp 173–191.

Fu JK, O'Toole R (1990): Biodegradation of PCP contaminated soils using in situ subsurface bioreclamation. In Akin C, Smith J (eds): Gas, Oil, Coal, and Environmental Biotechnology II. Chicago: Institute of Gas Technology, pp 145–169.

Fukumori F, Hausinger RP (1993): *Alcaligenes eutrophus* JMP134 "2,4-dichlorophenoxyacetate monooxygenase" is an α-ketoglutarate-dependent dioxygeanse. J Bacteriol 175:2083–2086.

Genthner BRS, Price WA II, Pritchard PH (1989a): Anaerobic degradation of chloroaromatic compounds in aquatic sediments under a variety of enrichment conditions. Appl Environ Microbiol 55:1466–1471.

Genthner BRS, Price WA II, Pritchard PH (1989b): Characterization of anaerobic dechlorinating consortia derived from aquatic sediments. Appl Environ Microbiol 55:1472–1476.

Ghosal D, You I-S (1989): Operon structure and nucleotide homology of the chlorocatechol oxidation genes of plasmids pJP4 and pAC27, Gene 83:225–232.

Goerlitz DF (1992): A review of studies of contaminated groundwater conducted by the U.S. Geological Survey Organics Project, Menlo Park, California, 1961–1990. In Lesage S, Jackson RE (eds): Groundwater contamination and analysis at hazardous waste sites. New York: Marcel Dekker, pp 295–355.

Goerlitz DF, Troutman DE, Godsy EM, Franks BJ (1985): Migration of wood-preserving chemicals in contaminated groundwater in a sand aquifer at Pensacola, Florida. Environ Sci Technol 19:955–961.

Golovleva LA, Pertsova RN, Evtushenko LI, Baskunov BP (1990): Degradation of 2,4,5-trichlorophenoxyacetic acid by a *Nocardioides simplex* culture. Biodegradation 1:263–271.

Golovleva LA, Zaborina OE, Arinbasarova AY (1993): Degradation of 2,4,6-TCP and a mixture of isomeric chlorophenols by immobilized *Streptomyces rochei* 303. Appl Microbiol Biotechnol 38:815–819.

Golovleva LA, Zaborina O, Pertsova R, Baskunov B, Schurukhin Y, Kuzmin S (1992): Degradation of polychlorinated phenols by *Streptomyces rochei* 303. Biodegradation 2:201–208.

Gonzales JF, Hu W-S (1991): Effect of glutamate on the degradation of pentachlorophenol by *Flavobacterium* sp. Appl Microbiol Biotechnol 35:100–104.

Gorlatov SN, Mal'tseva OV, Shevchenko VI, Golovleva LA (1989): Degradation of chlorophenols by a culture of *Rhodococcus erythropolis*. Mikrobiologiya 58:802–806 [English translation 58:647–651].

Greer CW, Hawari J, Samson R (1990): Influence of environmental factors of 2,4-dichlorophenoxyacetic acid degradation by *Pseudomonas cepacia* isolated from peat. Arch Microbiol 154:317–322.

Häggblom MM (1990): Mechanisms of bacterial degradation and transformation of chlorinated monoaromatic compounds. J Basic Microbiol 30:115–141.

Häggblom MM (1992): Microbial breakdown of halogenated aromatic pesticides and related compounds. FEMS Microbiol Rev 103:29–72.

Häggblom MM, Apajalahti JHA, Salkinoja-Salonen MS (1988a): Hydroxylation and dechlorination of chlorinated guaiacols and syringols by *Rhodococcus chlorophenolicus*. Appl Environ Microbiol 54:683–687.

Häggblom MM, Janke D, Middeldorp PJM, Salkinoja-Salonen MS (1989a): O-methylation of chlorinated phenols in the genus *Rhodococcus*. Arch Microbiol 152:6–9.

Häggblom MM, Janke D, Salkinoja-Salonen MS (1989b): Hydroxylation and dechlorination of tetrachlorohydroquinone by *Rhodococcus* sp. strain CP-2 cell extracts. Appl Environ Microbiol 55:516–519.

Häggblom MM, Janke D, Salkinoja-Salonen MS (1989c): Transformation of chlorinated phenolic compounds in the genus *Rhodococcus*. Microb Ecol 18:147–159.

Häggblom MM, Nohynek LJ, Palleroni NJ, Kronqvist K, Nurmiaho-Lassila E-L, Salkinoja-Salonen MS, Klatte S, Kroppenstedt RM (1994): Transfer of polychlorophenol-degrading *Rhodococcus chlorophenolicus* (Apajalahti et al., 1986) to the genus *Mycobacterium* as *Mycobacterium chlorophenolicum* comb. nov. Int J Syst Bacteriol 44:485–493.

Häggblom MM, Nohynek LJ, Salkinoja-Salonen MS (1988b): Degradation and O-methylation of polychlorinated phenolic compounds by *Rhodococcus* and *Mycobacterium* strains. Appl Environ Microbiol 54:3043–3052.

Häggblom MM, Rivera MD, Young LY (1993a): Effects of auxiliary carbon sources and electron acceptors on methanogenic degradation of chlorinated phenols. Environ Toxicol Chem 12:1395–1403.

Häggblom MM, Rivera MD, Young LY (1993b): Influence of alternative electron acceptors on the anaerobic biodegradability of chlorinated phenols and benzoic acids. Appl Environ Microbiol 59:1162–1167.

Häggblom MM, Young LY (1990): Chlorophenol degradation coupled to sulfate reduction. Appl Environ Microbiol 56:3255–3260.

Haimi J, Salminen J, Huhta V, Knuutinen J, Palm H (1992): Bioaccumulation of organochlorine compounds in earthworms. Soil Biol Biochem 24:1699–1703.

Hakulinen R, Salkinoja-Salonen M (1981): An anaerobic fluidised-bed reactor for the treatment of industrial wastewater containing chlorophenols. In Cooper PF, Atkinson B (eds): Biological Fluidised Bed Treatment of Water and Wastewater. Chichester: Ellis Horwood Ltd., pp 374–382.

Hakulinen R, Salkinoja-Salonen M (1982): Treatment of pulp and paper industry wastewaters in an anaerobic fluidised bed reactor. Process Biochem 17:18–22.

Hale DD, Rogers JE, Wiegel J (1990): Reductive dechlorination of dichlorophenols by nonadapted and adapted microbial communities in pond sediments. Microb Ecol 20:185–196.

Hale DD, Rogers JE, Wiegel J (1991): Environmental factors correlated to dichlorophenol dechlorination in anoxic freshwater sediments. Environ Toxicol Chem 10:1255–1265.

Hammel KE, Tardone PJ (1988): The oxidative 4-dechlorination of polychlorinated phenols is catalyzed by extracellular fungal lignin peroxidases. Biochemistry 27:6563–6568.

Haugland RA, Schlemm DJ, Lyons RP III, Sferra PR, Chakrabarty AM (1990): Degradation of the chlorinated phenoxyacetate herbicides 2,4-dichlorophenoxy acetic acid and 2,4,5-trichlorophenoxyacetic acid by pure and mixed bacterial cultures. Appl Environ Microbiol 56:1357–1362.

Hendriksen HV, Ahring BK (1992): Metabolism and kinetics of pentachlorophenol transformation in anaerobic granular sludge. Appl Microbiol Biotechnol 37:662–666.

Hendriksen HV, Ahring BK (1993): Anaerobic dechlorination of pentachlorophenol in fixed-film and upflow anaerobic sludge blanket reactors using different inocula. Biodegradation 3:399–408.

Hendriksen HV, Larsen S, Ahring BK (1992): Influence of a supplemental carbon source on anaerobic dechlorination of pentachlorophenol in granular sludge. Appl Environ Microbiol 58:365–370.

Hickman GT, Novak JT (1984): Acclimation of activated sludge to pentachlorophenol. J Water Pollut Contr Fed 56:364–369.

Hofrichter M, Günther T, Fritsche W (1993): Metabolism of phenol, chloro- and nitrophenols by the *Penicillium* strain *Bi 7/2* isolated from a contaminated soil. Biodegradation 3:415–421.

Hofrichter M, Bublitz F, Fritsche W (1994): Unspecific degradation of halogenated phenols by the soil fungus *Penicillium frequentans* Bi 7/2. J Basic Microbiol 34:163–172.

Horvath M, Ditzelmüller G, Loidl M, Streichsbier F (1990): Isolation and characterization of a 2-(2,4-dichlorophenoxy) propionic acid-degrading soil bacterium. Appl Microbiol Biotechnol 33:213–216.

Hrudey SE, Knettig E, Daignault SA, Fedorak PM (1987a): Anaerobic biodegradation of monochlorophenols. Environ Technol Lett 8:65–76.

Hrudey SE, Knettig E, Fedorak PM, Daignault SA (1987b): Anaerobic semi-continuous culture biodegradation of dichlorophenols containing an ortho chlorine. Water Pollut Res J Can 22:427–436.

Humppi T, Laitinen R, Kantolahti E, Knuutinen J, Paasivirta J, Tarhanen J, Lahtiperä M, Virkki L (1984): Gas chromatographic-mass spectrometric analysis of chlorinated phenoxyphenols in the technical chlorophenol formulation Ky-5. J Chromatogr 291:135–144.

Huynh V-B, Chang H-M, Joyce TW, Kirk TK (1985): Dechlorination of chloro-organics by a white-rot fungus. Tappi J 68:98–102.

Ide A, Niki Y, Sakamoto F, Watanabe I, Watanabe H (1972): Decomposition of pentachlorophenol in paddy soil. Agric Biol Chem 36:1937–1944.

Ihn W, Janke D, Tresselt D (1989): Critical steps in degradation of chloroaromatics by rhodococci. III. Isolation and identification of accumulating intermediates and dead-end products. J Basic Microbiol 29:291–297.

Ivoilov VS, Karasevich YN (1983): Monochlorophenols as substrates of the enzymes of preparatory phenol metabolism in the yeast *Candida tropicalis*. Mikrobiologiya 52:956–961.

Järvinen KT, Melin ES, Puhakka JA (1994): High-rate bioremediation of chlorophenol contaminated ground water at low temperatures. Environ Sci Technol 28:2387–2392.

Järvinen KT, Puhakka JA (1994): Bioremediation of chlorophenol contaminated ground water. Environ Technol 15:823–832.

Jokela JK, Laine M, Salkinoja-Salonen M (1993): Effect of biological treatment on halogenated organics in bleached kraft pulp mill effluents studied by molecular weight distribution analysis. Environ Sci Technol 27:547–557.

Jokela JK, Salkinoja-Salonen M (1992): Molecular weight distributions of organic halogens in bleached kraft pulp mill effluents. Environ Sci Technol 26:1190–1197.

Joshi DK, Gold MH (1993): Degradation of 2,4,5-trichlorophenol by the lignin-degrading basidiomycete *Phanerochaete chrysosporium*. Appl Environ Microbiol 59:1779–1785.

Karns JS, Kilbane JJ, Duttagupta S, Chakrabarty AM (1983): Metabolism of halophenols by 2,4,5-trichlorophenoxyacetic acid-degrading *Pseudomonas cepacia*. Appl Environ Microbiol 46:1176–1181.

Kazumi J, Häggblom MM, Young LY (1993): Biodegradation of monochlorinated and nonchlorinated aromatic compounds under iron (III) reducing conditions. In American Society for Microbiology 93 General Meeting, Atlanta, GA, 1993. Washington, DC: American Society for Microbiology, abstract Q-370, p 414.

King GM (1988): Dehalogenation in marine sediments containing natural sources of halophenols. Appl Environ Microbiol 54:3079–3085.

Kirk PWW, Lester JN (1988): The behaviour of chlorinated organics during activated sludge treatment and anaerobic digestion. Water Sci Technol 20(11/12):353–359.

Kitunen VH (1990): The Use and Formation of CPs, PCPPs and PCDDs/PCDFs in Mechanical and Chemical Wood Processing Industries. Ph.D. Thesis, University of Helsinki.

Kitunen V, Valo R, Salkinoja-Salonen M (1985): Analysis of chlorinated phenols, phenoxyphenols and dibenzofurans around wood preserving facilities. Int J Environ Anal Chem 20:13–28.

Kitunen VH, Valo RJ, Salkinoja-Salonen MS (1987): Contamination of soil around woodpreserving facilities by polychlorinated aromatic compounds. Environ Sci Technol 21:96–101.

Kiyohara H, Hatta T, Ogawa Y, Kakuda T, Yokoyama H, Takizawa N (1992): Isolation of *Pseudomonas pickettii* strains that degrade 2,4,6-trichlorophenol and their dechlorination of chlorophenols. Appl Environ Microbiol 58:1276–1283.

Klecka GM, Maier WJ (1985): Kinetics of microbial growth on pentachlorophenol. Appl Environ Microbiol 49:46–53.

Knackmuss H-J, Hellwig M (1978): Utilization and cooxidation of chlorinated phenols by *Pseudomonas* sp. B 13. Arch Microbiol 117:1–7.

Knuutinen J, Palm H, Hakala H, Haimi J, Huhta V, Salminen J (1990): Polychlorinated phenols and their metabolites in soil and earthworms of sawmill environment. Chemosphere 20:609–623.

Kobayashi K (1978): Metabolism of pentachlorophenol in fishes. In Rao KR (ed): Pentachlorophenol: Chemistry, Pharmacology, and Environmental Toxicology. New York: Plenum Press, pp 89–105.

Kohring G-W, Rogers JE, Wiegel J (1989a): Anaerobic biodegradation of 2,4-dichlorophenol in freshwater lake sediments at different temperatures. Appl Environ Microbiol 55:348–353.

Kohring G-W, Zhang X, Wiegel J (1989b): Anaerobic dechlorination of 2,4-dichlorophenol in freshwater sediments in the presence of sulfate. Appl Environ Microbiol 55:2735–2737.

Korte F (1987): Lehrbuch der Ökologischen Chemie, 2nd ed. Stuttgart: George Thieme Verlag.

Krahn PK, Shrimpton JA (1988): Stormwater related chlorophenol releases from seven wood protection facilities in British Columbia. Water Pollut Res J Can 23:45–54.

Krahn PK, Shrimpton JA, Glue RD (1987): Assessment of Storm Water Related Chlorophenol Releases From Wood Protection Facilities in British Columbia. Regional Program Report 87-14. Environment Canada Conservation and Protection, Environmental Protection Pacific and Yukon Region.

Kramer CM, Kory MM (1992): Bacteria that degrade *p*-chlorophenol isolated from a continuous culture system. Can J Microbiol 38:34–37.

Kremer S, Sterner O, Anke H (1992): Degradation of pentachlorophenol by *Mycena avenacea* TA 8480—Identification of initial dechlorinated metabolites. Z Naturforsch 47c:561–566.

Krumme ML, Boyd SA (1988): Reductive dechlorination of chlorinated phenols in anaerobic upflow bioreactors. Water Res 22:171–177.

Kuhn EP, Suflita JM (1989): Dehalogenation of pesticides by anaerobic microorganisms in soils and groundwater—A review. In Reactions and Movement of Organic Chemicals in Soils. SSSA Special Publication No. 22. Madison: Soil Science Society of America, pp 111–180.

Kuwatsuka S, Igarashi M (1975): Degradation of PCP in soils. II. The relationship between the degradation of PCP, the properties of soils, and the identification of the degradation products of PCP. Soil Sci Plant Nutr 21:405–414.

Lamar RT, Dietrich DM (1990): In situ depletion of pentachlorophenol from contaminated soil by *Phanerochaete* spp. Appl Environ Microbiol 56:3093–3100.

Lamar RT, Dietrich DM (1992). Use of lignin-degrading fungi in the disposal of pentachlorophenol-treated wood. J Ind Microbiol 9:181–191.

Lamar RT, Evans JW, Glaser JA (1993): Solid-phase treatment of a pentachlorophenol-contaminated soil using lignin-degrading fungi. Environ Sci Technol 27:2566–2571.

Lamar RT, Glaser JA, Kirk TK (1990a): Fate of pentachlorophenol (PCP) in sterile soils inoculated with the white-rot basidiomycete *Phanerochaete chrysosporium:* Mineralization, volatilization and depletion of PCP. Soil Biol Biochem 22:433–440.

Lamar RT, Larsen MJ, Kirk TK (1990b): Sensitivity to and degradation of pentachlorophenol by *Phanerochaete* spp. Appl Environ Microbiol 56:3519–3526.

Lampi P, Hakulinen T, Luostarinen T, Pukkala E, Teppo L (1992a): Cancer incidence following chlorophenol exposure in a community in southern Finland. Arch Environ Health 47:167–175.

Lampi P, Tolonen K, Vartiainen T, Tuomisto J (1992b): Chlorophenols in lake bottom sediments: A retrospective study of drinking water contamination. Chemosphere 24:1805–1824.

Lampi P, Vartiainen T, Tuomisto J, Hesso A (1990): Population exposure to chlorophenols, dibenzo-*p*-dioxins and dibenzofurans after a prolonged ground water pollution by chlorophenols. Chemosphere 20:625–634.

Larsen S, Hendriksen HV, Ahring BK (1991): Potential for thermophilic (50°C) anaerobic dechlorination of pentachlorophenol in different ecosystems. Appl Environ Microbiol 57:2085–2090.

Lewandowsky GA, Armenante PM, Pak D (1990): Reactor design for hazardous waste treatment using a white rot fungus. Water Res 24:75–82.

Li D-Y, Eberspächer J, Wagner B, Kuntzer J, Lingens F (1991): Degradation of 2,4,6-trichlorophenol by *Azotobacter* sp. strain GP1. Appl Environ Microbiol 57:1920–1928.

Lin J-E, Wang HY (1991): Degradation of pentachlorophenol by non-immobilized, immobilized and co-immobilized *Arthrobacter* cells. J Ferm Bioeng 72:311–314.

Lin J-E, Wang HY, Hickey RF (1990): Degradation kinetics of pentachlorophenol by *Phanerochaete chrysosporium*. Biotechnol Bioeng 35:1125–1134.

Lin J-E, Wang HY, Hickey RF (1991): Use of coimmobilized biological systems to degrade toxic organic compounds. Biotechnol Bioeng 38:273–279.

Madsen T, Aamand J (1991): Effects of sulfuroxy anions on degradation of pentachlorophenol by a methanogenic enrichment culture. Appl Environ Microbiol 57:2453–2458.

Madsen T, Aamand J (1992): Anaerobic transformation and toxicity of trichlorophenols in a stable enrichment culture. Appl Environ Microbiol 58:557–561.

Madsen T, Licht D (1992): Isolation and characterization of an anaerobic chlorophenol-transforming bacterium. Appl Environ Microbiol 58:2874–2878.

Mahaffey WR, Sanford RA (1990): Bioremediation of pentachlorophenol contaminated soil: bench scale to full scale implementation. In Akin C, Smith J (eds): Gas, Oil, Coal, and Environmental Biotechnology II. Chicago: Institute of Gas Technology, pp 117–143.

Mäkinen PM, Theno TJ, Ferguson JF, Ongerth JE, Puhakka JA (1993): Chlorophenol toxicity removal and monitoring in aerobic treatment: Recovery from process upsets. Environ Sci Technol 27:1434–1439.

McDowell PG, Waladde SM (1986): 2,6-Dichlorophenol in the tick *Rhicephalus appendiculatus* Neumann. A reappraisal. J Chem Ecol 12:69–81.

Menke B, Rehm H-J (1992): Degradation of mixtures of monochlorophenols and phenol as substrates for free and immobilized cells of *Alcaligenes* sp. A7-2. Appl Microbiol Biotechnol 37:655–661.

Middaugh DP, Resnick SM, Lantz SE, Heard CS, Mueller JG (1993): Toxicological assessment of biodegraded pentachlorophenol: Microtox and fish embryos. Arch Environ Contam Toxicol 24:165–172.

Middeldorp PJM, Briglia M, Salkinoja-Salonen MS (1990): Biodegradation of pentachlorophenol in natural soil by inoculated *Rhodococcus chlorophenolicus*. Microb Ecol 20:123–139.

Mikesell MD, Boyd SA (1985): Reductive dechlorination of the pesticides 2,4-D, 2,4,5-T and pentachlorophenol in anaerobic sludges. J Environ Q 14:337–340.

Mikesell MD, Boyd SA (1986): Complete reductive dechlorination and mineralization of pentachlorophenol by anaerobic microorganisms. Appl Environ Microbiol 52:861–865.

Mikesell MD, Boyd SA (1988): Enhancement of pentachlorophenol degradation in soil through induced anaerobiosis and bioaugmentation with anaerobic sewage sludge. Environ Sci Technol 22:1411–1414.

Mileski GJ, Bumpus JA, Jurek MA, Aust SD (1988): Biodegradation of pentachlorophenol by the white rot fungus *Phanerochaete chrysosporium*. Appl Environ Microbiol 54:2885–2889.

Minard RD, Liu S-Y, Bollag J-M (1981): Oligomers and quinones from 2,4-dichlorophenol. J Agric Food Chem 29:250–252.

Mohn WW, Kennedy KJ (1992a): Limited degradation of chlorophenols by anaerobic sludge granules. Appl Environ Microbiol 58:2131–2136.

Mohn WW, Kennedy KJ (1992b): Reductive dehalogenation of chlorophenols by *Desulfomonile tiedjei* DCB-1. Appl Environ Microbiol 58:1367–1370.

Mohn WW, Tiedje JM (1990): Strain DCB-1 conserves energy for growth from reductive dechlorination coupled to formate oxidation. Arch Microbiol 153:267–271.

Mohn WW, Tiedje JM (1991): Evidence for chemiosmotic coupling of reductive dechlorination and ATP synthesis in *Desulfomonile tiedjei*. Arch Microbiol 157:1–6.

Mohn WW, Tiedje JM (1992): Microbial reductive dehalogenation. Microbiol Rev 56:482–507.

Moos LP, Kirsch EJ, Wukasch RF, Grady CPL Jr (1983): Pentachlorophenol biodegradation—I. Aerobic Water Res 17:1575–1584.

Morgan P, Watkinson RJ (1989): Microbiological methods for the cleanup of soil and ground water contaminated with halogenated aromatic compounds. FEMS Microbiol Rev 63:277–300.

Mueller JG, Lantz SE, Blattmann BO, Chapman PJ (1991a): Bench-scale evaluation of alternative biological treatment processes for the remediation of pentachlorophenol- and creosote-contaminated materials: Solid-phase bioremediation. Environ Sci Technol 25:1045–1055.

Mueller JG, Lantz SE, Blattmann BO, Chapman PJ (1991b): Bench-scale evaluation of alternative biological treatment processes for the remediation of pentachlorophenol- and creosote-contaminated materials: Slurry-phase bioremediation. Environ Sci Technol 25:1055–1061.

Mueller JG, Lantz SE, Ross D, Colvin RJ, Middaugh DP, Pritchard PH (1993): Strategy using bioreactors and specially selected microorganisms for bioremediation of groundwater contaminated with creosote and pentachlorophenol. Environ Sci Technol 27:691–698.

Murthy NBK, Kaufman DD, Fries GF (1979): Degradation of pentachlorophenol (PCP) in aerobic and anaerobic soil. J Environ Sci Health B14:1–14.

Neilson AH, Allard A-S, Hynning P-A, Remberger M (1991): Distribution, fate and persistence of organochlorine compounds formed during production of bleached pulp. Toxicol Environ Chem 30:3–41.

Neilson AH, Allard A-S, Hynning P-A, Remberger M, Landner L (1983): Bacterial methylation of chlorinated phenols and guaiacols: Formation of veratroles from guaiacols and high-molecular-weight chlorinated lignin. Appl Environ Microbiol 45:774–783.

Neilson AH, Allard A-S, Reiland S, Remberger M, Tärnholm A, Viktor T, Landner L (1984): Tri- and tetra-chloroveratrole, metabolites produced by bacterial O-methylation of tri- and tetra-chloroguaiacol: An assessment of their bioconcentration potential and their effects on fish reproduction. Can J Fish Aquat Sci 41:1502–1512.

Neilson AH, Lindgren C, Hynning P-Å, Remberger M (1988): Methylation of halogenated phenols and thiophenols by cell extracts of gram-positive and gram-negative bacteria. Appl Environ Microbiol 54:524–530.

Nevalainen I, Kostyal E, Nurmiaho-Lassila E-L, Puhakka JA, Salkinoja-Salonen MS (1993): Dechlorination of 2,4,6-trichlorophenol by a nitrifying biofilm. Water Res 27:757–767.

Nicholson DK, Woods SL, Istok JD, Peek DC (1992): Reductive dechlorination of chlorophenols by a pentachlorophenol-acclimated methanogenic consortium. Appl Environ Microbiol 58:2280–2286.

Nilsson C-A, Norström Å, Andersson K, Rappe C (1978): Impurities in commercial products related to pentachlorophenol. In Rao KR (ed): Pentachlorophenol: Chemistry, Pharmacology, and Environmental Toxicology. New York: Plenum Press, pp 313–324.

Nohynek LJ, Häggblom MM, Palleroni NJ, Kronqvist K, Nurmiaho-Lassila E-L, Salkinoja-Salonen M (1993): Characterization of a *Mycobacterium fortuitum* strain capable of degrading polychlorinated phenolic compounds. Syst Appl Microbiol 16:126–134.

Öberg LG, Glas B, Swanson SE, Rappe C, Paul KG (1990): Peroxidase-catalyzed oxidation of chlorophenols to polychlorinated dibenzo-p-dioxins and dibenzofurans. Arch Environ Contam Toxicol 19:930–938.

Okeke BC, Smith JE, Paterson A, Watson-Craik IA (1993): Aerobic metabolism of pentachlorophenol by spent sawdust culture of "Shiitake" mushroom (*Lentinus edodes*) in soil. Biotechnol Lett 15:1077–1080.

O'Reilly KT, Crawford RL (1989): Degradation of pentachlorophenol by polyurethane-immobilized *Flavobacterium* cells. Appl Environ Microbiol 55:2113–2118.

Orser CS, Dutton J, Lange C, Jablonski P, Xun L, Hargis M (1993a): Characterization of a *Flavobacterium* glutathione S-transferase gene involved in reductive dechlorination. J Bacteriol 175:2640–2644.

Orser CS, Lange CC, Xun L, Zahrt TC, Schneider BJ (1993b): Cloning, sequence analysis, and expression of the *Flavobacterium* pentachlorophenol-4-monooxygenase gene in *Escherichia coli*. J Bacteriol 175:411–416.

Paasivirta J, Hakala H, Knuutinen J, Otollinen T, Särkkä J, Welling L, Paukku R, Lammi R (1990): Organic chlorine compounds in lake sediments. III. Chlorohydrocarbons, free and chemically bound chlorophenols. Chemosphere 21:1355–1370.

Paasivirta J, Heinola K, Humppi T, Karjalainen A, Knuutinen J, Mäntykoski K, Paukku, R, Piilola T, Surma-Aho K, Tarhanen J, Welling, L, Vihonen H, Särkkä J (1985): Polychlorinated phenols, guaiacols and catechols in environment. Chemosphere 14:469–491.

Palm H, Knuutinen J, Haimi J, Salminen J, Huhta V (1991): Methylation products of chlorophenols, catechols and hydroquinones in soil and earthworms of sawmill environments. Chemosphere 23:263–267.

Patterson RJ, Liebscher HM (1987): Laboratory simulation of pentachlorophenol/phenate behaviour in an alluvial aquifer. Water Pollut Res J Can 22:147–155.

Pellinen J, Joyce TW, Chang H-M (1988): Dechlorination of high-molecular-weight chlorolignin by the white-rot fungus *P. chrysosporium*. Tappi J 71:191–194.

Pemberton JM, Corney B, Don RH (1979): Evolution and spread of pesticide degrading ability among soil micro-organisms. In Timmis KN, Pühler A (eds): Plasmids of Medical, Environmental and Commercial Importance. Amsterdam: Elsevier/North-Holland Biomedical Press, pp 287–299.

Pflug AD, Burton B (1988): Remediation of multimedia contamination from the wood-preserving industry. In Omenn GS (ed): Environmental Biotechnology. Reducing Risks From Environmental Chemicals Through Biotechnology. New York: Plenum Press, pp 193–201.

Pieper DH, Reineke W, Engesser K-H, Knackmuss H-J (1988): Metabolism of 2,4-dichlorophenoxyacetic acid, 4-chloro-2-methylphenoxyacetic acid and 2-methylphenoxyacetic acid by *Alcaligenes eutrophus* JMP 134. Arch Microbiol 150:95–102.

Pierce RH Jr, Victor DM (1978): The fate of pentachlorophenol in an aquatic ecosystem. In Rao KR (ed): Pentachlorophenol: Chemistry, Pharmacology, and Environmental Toxicology. New York: Plenum Press, pp 41–52.

Pignatello JJ, Martinson MM, Steiert JG, Carlson RE, Crawford RL (1983): Biodegradation and photolysis of pentachlorophenol in artificial freshwater streams. Appl Environ Microbiol 46:1024–1031.

Polnisch E, Kneifel H, Franzke H, Hofmann KH (1992): Degradation and dehalogenation of monochlorophenols by the phenol-assimilating yeast *Candida maltosa*. Biodegradation 2:193–199.

Portier RJ, Fujisaki K (1986): Continuous biodegradation and detoxification of chlorinated phenols using immobilized bacteria. Toxicity Assessment 1:501–513.

Premalatha A, Rajakumar GS (1994): Pentachlorophenol degradation by *Pseudomonas aeruginosa*. World J Microbiol Biotechnol 10:334–337.

Pritchard PH (1992): Use of inoculation in bioremediation. Curr Opin Biotechnol 3:232–243.

Puhakka JA, Herwig RP, Koro PM, Wolfe GV, Ferguson JF (1994): Isolation and partial characterization of bacteria from chlorophenol degrading biofilm enrichment. Appl Microbiol Biotechnol (in press).

Puhakka JA, Järvinen K (1992): Aerobic fluidized-bed treatment of polychlorinated phenolic wood preservative constituents. Water Res 26:765–770.

Radehaus PM, Schmidt SK (1992): Characterization of a novel *Pseudomonas* sp. that mineralizes high concentrations of pentachlorophenol. Appl Environ Microbiol 58:2879–2885.

Radjendirane V, Bhat MA, Vaidyanathan CS (1991): Affinity purification and characterization of 2,4-dichlorophenol hydroxylase from *Pseudomonas cepacia*. Arch Biochem Biophys 288:169–176.

Reineke W, Knackmuss H-J (1988): Microbial degradation of haloaromatics. Annu Rev Microbiol 42:263–287.

Reiner EA, Chu J, Kirsch EJ (1978): Microbial metabolism of pentachlorophenol. In Rao KR (ed): Pentachlorophenol: Chemistry, Pharmacology and Environmental Toxicology. New York: Plenum Press, pp 67–81.

Renberg L, Marell E, Sundström G, Adolfsson-Erici M (1983): Levels of chlorophenols in natural waters and fish after an accidental discharge of a wood-impregnating solution. Ambio 12:121–123.

Rochkind-Dubinsky ML, Sayler GS, Blackburn JW (1987): Microbiological Decomposition of Chlorinated Aromatic Compounds. New York: Marcel Dekker.

Roy-Arcand L, Archibald FS (1991): Direct dechlorination of chlorophenolic compounds by laccases from *Trametes* (*Coriolus*) *versicolor*. Enzyme Microb Technol 13:194–203.

Ruckdeschel G, Renner G (1986): Effects of pentachlorophenol and some of its known and possible metabolites on fungi. Appl Environ Microbiol 51:1370–1372.

Ruckdeschel G, Renner G, Schwarz K (1987): Effects of pentachlorophenol and some of its known and possible metabolites on different species of bacteria. Appl Environ Microbiol 53:2689–2692.

Saber DL, Crawford RL (1985): Isolation and characterization of *Flavobacterium* strains that degrade pentachlorophenol. Appl Environ Microbiol 50:1512–1518.

Salkinoja-Salonen M, Middeldorp P, Briglia M, Valo R, Häggblom M, McBain A (1990): Cleanup of old industrial sites. In Kamely D, Chakrabarty A, Omenn GS (eds): Biotechnology and Biodegradation, vol. 4: Advances in Applied Biotechnology Series. The Woodlands, TX: Portfolio Publishing Company, pp 347–367.

Salkinoja-Salonen M, Saxelin M-L, Pere J, Jaakkola T, Saarikoski J, Hakulinen R, Koistinen O (1981): Analysis of toxicity and biodegradability of organochlorine compounds released into the environment in bleaching effluents of kraft pulping. In Keith LH (ed): Advances in the Identification and Analysis of Organic Pollutants in Water. Ann Arbor, MI: Ann Arbor Science, vol 2, pp 1131–1164.

Salkinoja-Salonen M, Uotila J, Jokela J, Laine M, Saski E (1995): Organic halogens in the environment: Studies of environmental biodegradability and human exposure. Submitted.

Salkinoja-Salonen MS, Valo R, Apajalahti J, Hakulinen R, Silakoski L, Jaakkola T (1984): Biodegradation of chlorophenolic compounds in wastes from wood-processing industry. In Klug MJ, Reddy CA (eds): Current Perspectives in Microbial Ecology. Proceedings of the Third International Symposium on Microbial Ecology, 7–12 August, 1983. Washington, DC: American Society for Microbiology, pp 668–676.

Sangodkar UMX, Aldrich TL, Haugland RA, Johnson J, Rothmel RK, Chapman PJ, Chakrabarty AM (1989): Molecular basis of biodegradation of chloroaromatic compounds. Acta Biotechnol 9:301–316.

Sangodkar UMX, Chapman PJ, Chakrabarty AM (1988): Cloning, physical mapping and expression of chromosomal genes specifying degradation of the herbicide 2,4,5-T by *Pseudomonas cepacia* AC1100. Gene 71:267–277.

Schenk T, Müller R, Lingens F (1990): Mechanism of enzymatic dehalogenation of pentachlorophenol by *Arthrobacter* sp. strain ATCC 33790. J Bacteriol 172:7272–7274.

Schenk T, Müller R, Mörsberger F, Otto MK, Lingens F (1989): Enzymatic dehalogenation of pentachlorophenol by extracts from *Arthrobacter* sp. strain ATCC 33790. J Bacteriol 171:5487–5491.

Schmidt E, Knackmuss H-J (1980): Chemical structure and biodegradability of halogenated aromatic compounds. Conversion of chlorinated muconic acids into maleoylacetic acid. Biochem J 192:339–347.

Schmitzer J, Bin C, Scheunert I, Korte F (1989): Residues and metabolism of 2,4,6-trichlorophenol-^{14}C in soil. Chemosphere 18:2383–2388.

Schwien U, Schmidt E (1982): Improved degradation of monochlorophenols by a constructed strain. Appl Environ Microbiol 44:33–39.

Seech AG, Trevors JT, Bulman TL (1991): Biodegradation of pentachlorophenol in soil: The response to physical, chemical, and biological treatments. Can J Microbiol 37:440–444.

Seeholzer-Nguyen B, Hock B (1991): Biodegradation of trichlorophenols by fungi. Angew Botanik 65:219–227.

Seidler JJ, Landau M, Dierberg FE, Pierce RH (1986): Persistence of pentachlorophenol in a wastewater-estuarine aquaculture system. Bull Environ Contam Toxicol 36:101–108.

Shannon MJR, Bartha R (1988): Immobilization of leachable toxic coil pollutants by using oxidative enzymes. Appl Environ Microbiol 54:1719–1723.

Shelton DR, Tiedje JM (1984): Isolation and partial characterization of bacteria in an anaerobic consortium that mineralizes 3-chlorobenzoic acid. Appl Environ Microbiol 48:840–848.

Shuttleworth KL, Bollag J-M (1986): Soluble and immobilized laccase as catalysts for the transformation of substituted phenols. Enzyme Microb Technol 8:171–177.

Siahpush AR, Lin J-E, Wang HY (1992): Effects of adsorbents on degradation of toxic organic compounds by coimmobilized systems. Biotechnol Bioeng 39:619–628.

Sjoblad RD, Bollag J-M (1977): Oxidative coupling of aromatic pesticide intermediates by a fungal phenol oxidase. Appl Environ Microbiol 33:906–910.

Spain JC, Gibson DT (1988): Oxidation of substituted phenols by *Pseudomonas putida* F1 and *Pseudomonas* sp. strain JS6. Appl Environ Microbiol 54:1399–1404.

Spokes JR, Walker N (1974): Chlorophenol and chlorobenzoic acid co-metabolism by different genera of soil bacteria. Arch Microbiol 96:125–134.

Stanlake GJ, Finn RK (1982): Isolation and characterization of a pentachlorophenol-degrading bacterium. Appl Environ Microbiol 44:1421–1427.

Steiert JG, Crawford RL (1986): Catabolism of pentachlorophenol by a *Flavobacterium* sp. Biochem Biophys Res Commun 141:825–830.

Stinson MK, Skovronek HS, Chresand TJ (1991): EPA SITE demonstration of BioTrol aqueous treatment system. J Air Waste Manag Assoc 41:228–233.

Stroo HF (1992): Biotechnology and hazardous waste treatment. J Environ Q 21:167–175.

Suflita JM, Miller GD (1985): Microbial metabolism of chlorophenolic compounds in ground water aquifers. Environ Toxicol Chem 4:751–758.

Suntio LR, Shiu WY, Mackay D (1988): A review of the nature and properties of chemicals present in pulp mill effluents. Chemosphere 17:1249–1290.

Suzuki T (1977): Metabolism of pentachlorophenol by a soil microbe. J Environ Sci Health B12:113–127.

Suzuki T, Nose K (1971): Decomposition of pentachlorophenol in farm soil, Part 2: PCP metabolism by a microorganism isolated from soil [in Japanese]. Noyaku Seisan Gijutsu 26:21–24.

Svenson A, Kjeller L-O, Rappe C (1989): Enzyme-mediated formation of 2,3,7,8-tetrasubstituted chlorinated dibenzodioxins and dibenzofurans. Environ Sci Technol 23:900–902.

Tindale CR, Whitfield FB, Levingston SD, Nguyen THL (1989): Fungi isolated from packaging materials: Their role in the production of 2,4,6-trichloroanisole. J Sci Food Agric 49:437–447.

Tokuz RY (1991): Biotreatment of hazardous organic wastes using rotating biological contactors. Environ Prog 10:198–204.

Topp E, Crawford RL, Hanson RS (1988): Influence of readily metabolizable carbon on pentachlorophenol metabolism by a pentachlorophenol-degrading *Flavobacterium* sp. Appl Environ Microbiol 54:2452–2459.

Topp E, Hanson RS (1990): Factors influencing the survival and activity of a pentachlorophenol-degrading *Flavobacterium* sp. in soil slurries. Can J Soil Sci 70:83–91.

Trevors JT (1982): Effect of temperature on the degradation of pentachlorophenol by *Pseudomonas* species. Chemosphere 11:471–475.

Uotila J (1993): Dehalogenases for Polyhalogenated Aromatic Compounds in *Rhodococcus chlorophenolicus* PCP-1 and *Mycobacterium fortuitum* CG-2. Ph.D. Thesis. Helsinki, Finland: University of Helsinki, Department of Applied Chemistry and Microbiology.

Uotila JS, Kitunen VH, Apajalahti JHA, Salkinoja-Salonen MS (1992a): Environment-dependent mechanism of dehalogenation by *Rhodococcus chlorophenolicus* PCP-1. Appl Microbiol Biotechnol 38:408–412.

Uotila JS, Kitunen VH, Coote T, Saastamoinen T, Salkinoja-Salonen MS, Apajalahti JHA (1995): Metabolism of halohydroquinones in *Rhodococcus chlorophenolicus* PCP-I. Biodegradation (in press).

Uotila JS, Kitunen VH, Saastamoinen T, Coote T, Häggblom MM, Salkinoja-Salonen MS (1992b): Characterization of aromatic dehalogenase activities of *Mycobacterium fortuitum* strain CG-2. J Bacteriol 174:5669–5675.

Uotila JS, Salkinoja-Salonen MS, Apajalahti JHA (1991): Dechlorination of pentachlorophenol by membrane bound enzymes of *Rhodococcus chlorophenolicus* PCP-I. Biodegradation 2:25–31.

Utkin I, Woese C, Wiegel J (1994): Isolation and characterization of *Desulfitobacterium dehalogenans* gen. nov., sp. nov., an anaerobic bacterium which reductively dechlorinates chlorophenolic compounds. Int J Syst Bacteriol 44:612–619.

Valli, K, Gold MH (1991): Degradation of 2,4-dichlorophenol by the lignin-degrading fungus *Phanerochaete chrysosporium*. J Bacteriol 173:345–352.

Valo R (1990): Occurrence and Metabolism of Chlorophenolic Wood Preservative in the Environment. Ph.D. Thesis. Helsinki, Finland: University of Helsinki, Department of General Microbiology.

Valo R, Apajalahti J, Salkinoja-Salonen M (1985): Studies on the physiology of microbial degradation of pentachlorophenol. Appl Microbiol Biotechnol 21:313–319.

Valo RJ, Häggblom MM, Salkinoja-Salonen MS (1990): Bioremediation of chlorophenol containing simulated ground water by immobilized bacteria. Water Res 24:253–258.

Valo R, Hakulinen R (1990): Bakteerit puhdistamaan kloorifenoleilla likaantunutta pohjavettä. Kemia-Kemi 17:811–813.

Valo R, Kitunen V, Salkinoja-Salonen MS, Räisänen S (1984): Chlorinated phenols as contaminants of soil and water in the vicinity of two Finnish sawmills. Chemosphere 13:835–844.

Valo R, Kitunen V, Salkinoja-Salonen MS, Räisänen S (1985): Chlorinated phenols and their derivatives in soil and groundwater around wood-preserving facilities in Finland. Water Sci Technol 17:1381–1384.

Valo R, Salkinoja-Salonen M (1984): Kemiallisesti likaantuneen maan puhdistaminen mikrobien avulla. Kemia-Kemi 11:948–949.

Valo R, Salkinoja-Salonen M (1986a): Bioreclamation of chlorophenol-contaminated soil by composting. Appl Microbiol Biotechnol 25:68–75.

Valo R, Salkinoja-Salonen M (1986b): Microbial transformation of polychlorinated phenoxy phenols. J Gen Appl Microbial 32:505–517.

Venkatadri R, Tsai S-P, Vukanic N, Hein LB (1992): Use of a biofilm membrane reactor for the production of lignin peroxidase and treatment of pentachlorophenol by *Phanerochaete chrysosporium*. Hazard Waste Hazard Mat 9:231–243.

Wan MT (1992): Utility and railway right-of-way contaminants in British Columbia: Chlorophenols. J Environ Q 21:225-231.

Wang S-H, Ferguson JF, McCarthy JL (1992): The decolorization and dechlorination of kraft bleach plant effluent solutes by use of three fungi: *Ganoderma lacidum, Coriolus versicolor,* and *Hericium erinaceum.* Holzforschung 46:219–233.

Watanabe I (1973): Isolation of pentachlorophenol decomposing bacteria from soil. Soil Sci Plant Nutr 19:109–116.

Watanabe I (1977): Pentachlorophenol-decomposing and PCP-tolerant bacteria in field soil treated with PCP. Soil Biol Biochem 9:99–103.

Watanabe I (1978): Pentachlorophenol (PCP) decomposing activity of field soils treated annually with PCP. Soil Biol Biochem 10:71–75.

Westmeier F, Rehm HJ (1985): Biodegradation of 4-chlorophenol by entrapped *Alcaligenes* sp. A 7-2. Appl Microbiol Biotechnol 22:301–305.

Whitfield FB, Nguyen THL, Shaw KJ, Last JH, Tindale CR, Stanley G (1985): Contamination of dried fruit by 2,4,6-trichloroanisole and 2,3,4,6-tetrachloroanisole adsorbed from packaging materials. Chem Ind 661–663.

Woods SL, Ferguson JF, Benjamin MM (1989): Characterization of chlorophenol and chloromethoxybenzene biodegradation during anaerobic treatment. Environ Sci Technol 23:62–68.

Wu W-M, Bhatnagar L, Zeikus JG (1993): Performance of anaerobic granules for degradation of pentachlorophenol. Appl Environ Microbiol 59:389–397.

Xun L, Orser CS (1991): Purification and properties of pentachlorophenol hydroxylase, a flavoprotein from *Flavobacterium* sp. strain ATCC 39723. J Bacteriol 173:4447–4453.

Xun L, Topp E, Orser CS (1992a): Confirmation of oxidative dehalogenation of pentachlorophenol by a *Flavobacterium* pentachlorophenol hydroxylase. J Bacteriol 174:5745–5747.

Xun L, Topp E, Orser CS (1992b): Diverse substrate range of a *Flavobacterium* pentachlorophenol hydroxylase and reaction stoichiometries. J Bacteriol 174:2898–2902.

Xun L, Topp E, Orser CS (1992c): Glutathione is the reducing agent for the reductive dehalogenation of tetrachloro-*p*-hydroquinone by extracts from a *Flavobacterium* sp. Biochem Biophys Res Commun 182:361–366.

Xun L, Topp E, Orser CS (1992d): Purification and characterization of a tetrachloro-*p*-hydroquinone reductive dehalogenase from a *Flavobacterium* sp. J Bacteriol 174:8003–8007.

Zhang X, Wiegel J (1990): Sequential anaerobic degradation of 2,4-dichlorophenol in freshwater sediments. Appl Environ Microbiol 56:1119–1127.

12

BIOLOGICAL TREATMENT OF CHLORINATED ORGANICS

PETER ADRIAENS

Department of Civil and Environmental Engineering, University of Michigan, Ann Arbor, Michigan 48109

TIMOTHY M. VOGEL

Department of Safety and the Environment, Rhone-Poulenc Industrialisation, 69153 Decines-Charpieu, France

1. INTRODUCTION

Biological treatment of wastes has been utilized for centuries, whether or not intentionally. The ubiquitous presence of microorganisms has allowed for the evolution of populations acclimated to many, if not most, types of hazardous wastes. Our understanding of the interactions between microbes and waste compounds results from studies of either axenic microorganisms and pollutants or mixed microbial communities and mixed wastes. In either case, an increasing body of knowledge exists that addresses the capabilities of microorganisms to degrade organic and inorganic pollutants.

The outcome of each degradation process, as shown in laboratory-scale experiments, depends on microbial (e.g., biomass concentration, population diversity, enzyme specificities), substrate (e.g., molecular structure, solubility, concentration), and a range of environmental (e.g., pH, temperature, moisture content, E_h,

Microbial Transformation and Degradation of Toxic Organic Chemicals, pages 435–486
© 1995 Wiley-Liss, Inc.

availability of electron acceptors and carbon and energy sources) parameters. These parameters may effect an acclimation period of the microorganisms to the substrate or environment. To apply these results in pilot or full-scale engineered systems, additional knowledge of mainly physical–chemical factors, such as sorption and desorption processes, volatilization and advection, and multiphasic movement and partitioning, is required. The relative importance of these factors obviously depends on the system of interest and on the spatial and temporal variability of the waste to be biodegraded.

Thus, a critical decision in biotreatment process selection pertains to the different types of wastes it is intended to be able to treat. This is exemplified by the application of an activated sludge treatment system developed for domestic sewage to handle industrial chlorinated solvents. The activated sludge system with its intense aeration would surely strip most volatile compounds into the air without biologically degrading them. It will become apparent from this discussion that the application of biotechnology for the biological destruction of recalcitrant, synthetic, and, at times, toxic components associated with environmentally hazardous wastes is not immediately transferrable from our knowledge of conventional waste treatment process engineering, which generally is intended for "easily metabolizable" or labile substrates. Logically, then, the approach often pursued is to start with laboratory-scale experiments to define potential treatment schemes, prior to expanding the chosen design to the pilot-scale and investing in full-scale biotreatment systems. This chapter focuses on a discussion of the biological treatment of chlorinated organic compounds, whose "disappearance" in many current treatment systems may not be biologically mediated.

Chlorinated organic compounds are among the most ubiquitous anthropogenic chemicals found in the environment (Vogel et al., 1987; Kuhn and Suflita, 1989a; Neilson, 1990). They have been (and many still are) produced either intentionally, as is the case with chlorinated solvents and herbicides, chlorophenols, and polychlorinated biphenyls, or are formed as byproducts during industrial processes, an exemplified by polychlorinated dibenzo-p-dioxins or dibenzofurans. Though most chlorinated organic compounds are xenobiotics, evidence exists for the natural production of chlorinated compounds in marine environments. Aliphatic (e.g., tetrachloroethylene [PCE] and trichloroethylene [TCE]) and aromatic (e.g., p-dichlorobenzene, chlorophenols) chlorinated organics are among the most prevalent hazardous chemical contaminants found in groundwater (Westrick et al., 1984) and at CERCLA sites (U.S. EPA, 1990c). Thus, the ubiquity and known or assumed carcinogenicity of organohalides in the environment has prompted research into the development of safe and practical remediation techniques.

Traditional clean-up techniques employ physical processes such as pumping or venting techniques followed by concentration on activated carbon or air stripping in the case of volatile compounds or incineration in the case of higher molecular weight contaminants. Whereas pump-and-treat methods may be successful for removal of nonsorbing lesser chlorinated compounds, their time-efficiency (and cost-effectiveness) decreases rapidly with increasing chlorine content and molecular size due to increasing mass transfer limitations. As the compounds are generally not destroyed unless a post-treatment step is included, *in situ* or on site biological degradation of these contaminants has the potential to offset the above restrictions. Nationally coordinated initiatives such as the U.S. EPA Biosystems Technology Development Program, as well as independent industrial and university-sponsored research worldwide, have addressed the potential for site-directed bioremediation systems. This research has resulted in the development of laboratory-, pilot-, and field-scale applications for treatment of soils, sediments, and liquid wastes contaminated with chlorinated organic compounds (reviewed by Morgan and Watkinson, 1989).

The decision for development of any bioremediation system for a given type of contamination relies primarily on information obtained from laboratory studies, with the exception perhaps of petroleum hydrocarbons. A limited number of field-scale studies, such as for chlorinated solvents, chlorophenols, and polychlorinated biphenyls, are emerging and provide site-specific information on 1) the types and capabilities of microbial populations present, 2) the acclimation period required before degradation commences, 3) the extent of degradation that can be expected, and 4) the environmental prameters that affect the degradation potential.

2. BIODEGRADATION OF CHLORINATED ORGANIC COMPOUNDS

Considerable research over the last 20 years has demonstrated that most chlorinated organic compounds can at least be transformed, and in some cases mineralized, by microorganisms in spite of the recent historical contamination of these compounds in the environment. This would indicate that microbial evolutionary genetics has resulted in the development of novel metabolic capabilities and, furthermore, that inter- and intrageneric dissemination of genetic material (mainly plasmid encoded) must have occurred (Chakrabarty, 1982; Sayler et al., 1990; van der Meer et al., 1992). In this chapter, the genetics of chlorinated hydrocarbon metabolism will only be evaluated to the extent that it pertains to the discussion on microorganism selection for biotreatment processes.

The immense potential for the application of biotreatment, however, cannot be fully appreciated without complete understanding of the microbiological and biochemical principles that underlie metabolic abilities and activities. Elucidation of these factors will then aid in the assessment of natural detoxification processes and in the design and improvement of biological treatment technologies. The following section describes some of the specific degradative abilities reported to date under different environmental conditions, with an emphasis on their potential applicability in biotreatment.

2.1. Physiology of Biodegradative Microorganisms

One of the primary variables affecting the activity of heterotrophic bacteria is the ability and availability of reduced organic materials to serve as electron donors (and energy sources). Whether a contaminant serves as an effective energy source for a

TABLE 12.1. Abbreviations, Chemical Names, and Chlorine to Carbon Ratios of Common Alkyl and Aryl Halide Contaminants

Abbreviation	Chemical Name(s)	Chlorine Carbon
Alkyl		
PCE	Perchloroethylene, tetrachloroethylene	2
TCE	Trichloroethylene, 1,1,1-trichloroethylene	1.5
c-DCE	*cis*-Dichloroethylene, 1,1-dichloroethylene	1
t-DCE	*trans*-Dichloroethylene, 1,2-dichloroethylene	1
VC	Vinyl chloride, monochloroethylene	0.5
PCA	Perchloroethane, hexachloroethane	3
CT	Carbon tetrachloride, tetrachloromethane	4
CF	Chloroform, trichloromethane	3
DCM	Dichloromethane	2
DDT	Dichloro-diphenyl-trichloroethane	0.4
Aryl		
PCB	Polychlorinated biphenyl	0.08–0.8
PCDD/F	Polychlorinated dibenzo-*p*-dioxin/furan	0.08–0.7
HCB	Hexachlorobenzene	1
DCB	Dichlorobenzene	0.3
CB	Chlorobenzene	0.17
PCP	Pentachlorophenol	0.8
DCP	Dichlorophenol	0.3
CP	Chlorophenol	0.17
CBA	Chlorobenzoic acid	0.17–0.8

given microorganism is a function of the average oxidation state of the carbon in the material. In general, higher oxidation states correspond to lower energy yields and thus provide less energetic incentive for an organism to degrade it. The oxidation level of chlorinated organic compound is affected by the number of (halogen) substituents. Thus, the relative oxidation state of the organic carbon present in a range of alkyl and aryl halides can be expressed as the ratio of the number of chlorines to the number of carbons in the molecule (Table 12.1). Thus, tetra-chloromethane (CT) and hexachlorobenzene (HCB) represent the most oxidized alkyl (No. Cl/No. C = 4) and aryl (No. Cl/No. C = 1) halides, respectively. The implication of relative oxidation state of organic carbon with respect to the deha-logenation mechanisms is discussed later.

The biodegradation of chlorinated organic compounds can be divided into three types of metabolic processes: 1) the compound is able to serve as a sole source of carbon and energy to a single organism or a sequence of microorganisms, 2) the compound has to be degraded via co-metabolic processes by virtue of the lack of "recognition" by the enzymes involved and thus serves not as a growth substrate, or 3) the substrate concentration is too low to support a microbial population, and secondary substrate utilization will ensue. Obviously, the first type would be the most favorable for bioremediation purposes. However, when the substrate concentration present is insufficient to sustain a microbial population, the compound will be degraded via secondary metabolism, and the population grows on a more abundant substrate. Alternatively, the substrate concentration may be insufficient to induce the appropriate enzymes, and the compound will have to be degraded via co-metabolism (Horvath, 1972; Stirling and Dalton, 1981).

2.2. Co-Metabolic Processes

Co-metabolism has been recognized to be a ubiquitous metabolic (detoxification) process, pertinent to the aerobic transformation of most halogenated organic compounds in the environment, even though many chlorinated compounds have been shown to serve as sources of carbon and energy under laboratory conditions (Table 12.2). Hence, many reactions catalyzed by oxidative enzymes are of a co-metabolic nature, i.e., the chlorinated substrates are only fortuitously oxidized by enzymes induced on alternative (often the nonchlorinated analog, i.e., analog enrichment) substrates (Table 12.3). Since in these cases the chlorinated compounds must compete with the inducing compounds for the active sites of the enzyme, transformation often results in endproducts that are not easily further metabolized. Aerobic PCB transformation is thus enhanced by the addition of biphenyl, which induces the

TABLE 12.2. Biological Processes and Environmental Conditions Under Which Organohalides May Be Transformed by Bacteria

	Examples	
Process	Alkyl Halides	Aryl Halides
Primary substrates		
Aerobic	Chlorinated aliphatic acids, vinyl-chloride, 1,2-dichloroethane	Chlorobenzene; mono-, di-, and trichlorobenzoates (BA); mono-, di, and pentachlorophenols; monochlorobiphenyls (BP); chlorinated phenoxyherbicides
Anaerobic	Dichloromethane	Chlorobenzenes; mono- and di-chlorophenols
Co-metabolism (nongrowth substrates)		
Oxidations	Trichloroethylene, dichloro-ethylene, vinylchloride; chloro-form; chloroacetates	Di- and trichloroBA; di- through pentachlorinated BPs; mono-, di-, and tetrachlorinated dioxins and furans; chloronaphthalene
Reductions	DDT; carbon tetrachloride, dichlo-roethane 1,1,1-trichloroethane; tetra-, tri-, and dichloro-ethylenes; chloroform	Di-, tri-, tetra-, and hexachloroben-zenes; chlorobenzoates; chloro-anilines; di- through hepta-chlorinated BPs; heptachlorinated dioxin/furan; lindane

Modified from Adriaens and Hickey (1993).

TABLE 12.3. Physical, Chemical, and Biological Information Required for Prediction of Organic Contaminant Mobility and Biotransformation

Parameter	Environmental/Physical Compartment		
Hydraulic	Contaminant source Location Amount (concentration) Rate of release	Reactor environment Temperature Location Volume Depth Pumping rates	Hydrogeologic environment Extent of aquifer and aquitard Characteristics of aquifer Hydraulic gradient Groundwater flow rate
Sorption	Distribution coefficient Characteristic Partitioning	Solids characteristics Organic carbon content Clay content Particle size distri- bution	Contaminant characteristics Octanol/water partition coeff. Solubility
Chemical	Solvent characteristics Ionic strength pH NO_3^-, SO_4^{2-}, O_2 Potential inhibitors	Solids characteristics Potential catalysis Heavy metals Clays	Contaminant characteristics Potential products Concentration
Biological	Water characteristics Ionic strength pH Macro and trace nutrients Substrate concentra- tion Organism concentra- tion, distribution, type	Solids characteristics Particle size distribution Active bacteria	Contaminant characteristics Enzyme specificity Potential microbial meta- bolites Potential product toxicity Monod rate constants Molecular structure

biphenyl dioxygenases (Ahmed and Focht, 1973; Furukawa, 1982; Brunner et al., 1985; Kohler et al., 1988). An example of co-metabolism where the natural growth substrate is structurally unrelated to the chlorinated aromatic compound (i.e., no analog enrichment) is methane monooxygenase-catalyzed hydroxylations of aromatic compounds (Higgins et al., 1980; Green and Dalton, 1989; Adriaens and Grbić-Galić, 1994).

2.3. Microbial Interactions During Organochloride Degradation

Whether chlorinated organic compounds will undergo biological transformation or mineralization depends not only on substrate interactions such as co-metabolism and secondary metabolism but also on the different types of microbial mechanisms that evolved to deal with the carbon–chlorine bond(s) present in the xenobiotic (Reineke, 1984). In the soil and aqueous environment, an extensive array of microbial interactions has been recognized that are responsible for the production of either nonobligatory or highly intergrated obligatory relationships. These relationships, in turn, enable microbial populations to maintain a flexibility toward environmental changes, such as the introduction of xenobiotic compounds. Whereas interactions have been described to occur at the ecological and physiological levels (Grbić-Galić, 1990), increasing evidence from artificial laboratory communities and natural microbial populations indicates that genetic interactions are at least partly responsible for this dynamism of microbial associations (Reanney et al., 1983; Focht, 1988; van der Meer et al., 1992a).

Since mixed microbial populations have a greater possibility of degrading chlorinated compounds due to their larger catabolic gene pool, the complete biodegradation of xenobiotics in nature by a single organism is the exception rather than the rule. This pertains especially to the aerobic degradation of chlorinated hydrocarbons, which requires a number of oxidases and hydrolases, enzymes each with their own substrate specificities. This is exemplified for the degradation of PCBs, which requires inducible biphenyl dioxygenases for the initial co-metabolic transformation steps, resulting in the formation of chlorobenzoate intermediates. Since biphenyl dioxygenases do not recognize the intermediates, a different set of (inducible) oxygenases is required to degrade these substrates further, thus accomplishing mineralization of the PCB (Adriaens et al., 1989). Similarly, mineralization of highly chlorinated compounds has been recognized to be accomplished only by sequential anaerobic dehalogenation, followed by aerobic oxidative degradation. Alternatively, a community of microorganisms may be required to remove toxic intermediates generated during degradation of the xenobiotic. Such is the case during anaerobic degradation of halogenated compounds, where build-up of hydrogen by one of the methanogenic community members is growth inhibitory to acetogens and thus needs to be scavenged. As the microorganisms need to work together for complete degradation to occur, these types of community interactions have been called *concerted metabolism*.

Ultimately, the adaptation of a community to a novel environmental niche may result in genetic exchange, resulting in one microorganism harboring the combined catabolic gene pool for complete degradation of the pollutant (Focht, 1988; van der Meer et al., 1992a). The likelihood of genetic exchange to occur may be very significant when 1) pathway encoding genes are located on appropriate transfer systems such as plasmids and 2) a significant environmental selection pressure is exercised. Sayler et al. (1990) reviewed the environmental significance of plasmids encoding chlorinated hydrocarbon degradation in the disemination of catabolic pathways. It has indeed been shown that many aerobic microorganisms isolated from different locations carry similar operons for the same chlorinated organic compounds, including PCBs, chlorobenzoates, chlorocatechols, and 2,4-D (van der Meer et al., 1992a).

3. BIOLOGICAL DEHALOGENATION REACTIONS

From a chemical viewpoint, there are two major types of microbially mediated reactions involved in the removal of halogen substituents from organic molecules.

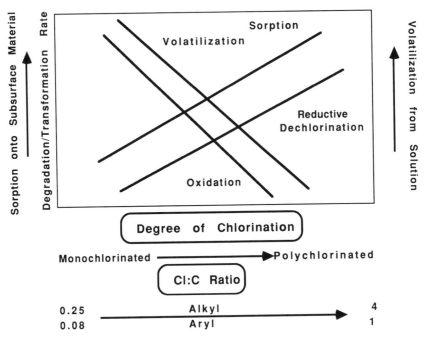

Figure 12.1. Relative trends of oxidative and reductive dehalogenation rates, sorption onto subsurface soils, and volatilization rates over the spectrum of alkyl and aryl chlorination. (Modified from Vogel et al., 1987).

Oxidation and reduction reactions involve the transfer of electrons from or to the halogenated compound, respectively. Due to the electronegative character of halogen substituent groups, highly halogenated compounds are more oxidized than the lesser chlorinated isomers and thus are less susceptible to oxidative reactions; rather, they will undergo reductive reactions for thermodynamic reasons. The dependency of oxidative and reductive reactions on the relative oxidation state of alkyl and aryl halides is plotted in Figure 12.1.

Oxidative reactions usually do not directly cause loss of a halogen substituent, except when a hydrolytic dehalogenation mechanism (i.e., a dehalogenase) is involved. In this case, the halogen is replaced via nucleophilic substitution with hydroxy groups from the (aqueous) medium. Chlorinated aromatic compounds, however, often undergo several oxidation step(s) prior to (spontaneous) dechlorination due to steric mechanisms (Dorn and Knackmuss, 1978a,b; Neilson, 1990; Häggblom, 1992). Under anaerobic conditions, reductive dehalogenation is the dominant mechanism for halogen removal (Mohn and Tiedje, 1992). These reductive reactions involve the transfer of electrons to the halogenated compound, resulting in the replacement of a halogen with a hydrogen atom. Within a family of chlorinated compounds, the ease of reductive dechlorination is generally proportional to the degree of chlorination. For example, tetrachloromethane (carbon tetrachloride) has the highest oxidation-reduction potential of the chloromethanes. It thus represents the most favorable electron acceptor within this family due to its most negative value in change of Gibbs free energy.

3.1. Aerobic Dehalogenation

3.1.1. Physiology

Under aerobic conditions, most reactions are dependent on the presence of molecular oxygen, whether as a terminal acceptor for the electrons released during oxidation of organic carbon or as a reactant in processes mediated by oxygenases. Microbial oxidations of halogenated compounds are generally accepted to be catalyzed by enzymes that have evolved for catabolism of natural substrates but that exhibit low substrate specificities (van der Meer et al., 1992). A range of oxygenases has been recognized that, depending on their functional differences and activated oxygen species, have been shown to oxidize a whole range of chlorinated organic compounds (Table 12.3). While numerous examples of aerobic biodegradation have been elucidated, almost all involve the oxidation of the molecule without the removal of a chlorine atom during the first step.

Aerobic bacteria employ oxygenases that insert one (monooxygenase) or two (dioxygenase) oxygen atoms into the molecule via an electrophilic attack on an unsubstituted carbon-atom, resulting in the formation of (a) carbon–hydroxyl bond(s). In other cases, the oxygen can electrophilically attack the electrons in the π bond(s) of an alkene or in aromatic rings. While both aromatic and aliphatic compounds may serve as substrates for monooxygenases, dioxygenase substrates are generally limited to aromatic compounds. A special type of oxygenase, found in methanotrophic bacteria, is the methane monooxygenase (MMO), which has an unusually low substrate specificity (Fox and Libscomb, 1990). Both a particular (membrane-bound) MMO and a soluble form of the enzyme exist, and their relative expression is dependent on the copper concentration in the environment. Although both forms of the enzyme have been demonstrated to oxidize chlorinated ethylenes, only the latter form was able to attack aryl halides (Burrows et al., 1984; Oldenhuis et al., 1989). The physiology and substrate specificities of co-metabolic, alkyl halide–oxidizing enzymes is discussed elsewhere in this volume. Finally, the peroxide-dependent (per)oxidases, which to date have only been reported in lignin-degrading white-rot fungi, have received increasing attention for their role in the oxidation, dehalogenation, and ring cleavage of chlorinated aromatic compounds.

Other reactions pertinent to the degradation of chlorinated compounds that can be catalyzed by aerobes but that have been shown to be independent of molecular oxygen are dehalogenations mediated by highly inducible hydrolytic aryl dehalogenases and alkyl halidohydrolases. When a microorganism posseses dehalogenases (or halidohydrolases), the halogenated compound (aryl or alkyl halide), provided it is present at concentrations sufficient to sustain a microbial population, will usually serve as a source of carbon and energy.

3.1.2. Mechanisms of Aerobic Dechlorination

In the case of chlorinated aromatic compounds (CAH), two main pathways have been described: The ring is dechlorinated either prior to or after ring fission.

In the first case, a (coenzyme A [CoA]) dependent hydrolytic dehalogenase replaces a chlorine with a hydroxyl from water, whereafter aromatic monoxygenases oxidize the ring to form a dihydroxy intermediate that can then be further cleaved by dioxygenases, as is the case with, for example, chlorobenzoates and chlorophenols (Karns et al., 1983; Marks et al., 1984; Adriaens, et al., 1989). Aerobic dechlorination by either reduced glutathione-dependent or CoA-dependent dehalogenases has been demonstrated to occur with alkyl (Slater et al., 1979; Kohler-Staub and Leisinger, 1985; Kohler-Staub and Kohler, 1989) and aryl (Scholten et al., 1991; Copley and Crooks, 1992) halides, respectively. In either case, the enzyme first activates, then dechlorinates the compound by nucleophilic substitution of the chlorine(s) with hydroxy group(s), thus rendering it more accessible to further aerobic metabolism. Some of the chlorines on highly chlorinated CAH such as pentachlorophenol may be reductively removed, even under aerobic conditions (Crawford and Mohn, 1985).

Alternatively, after initial oxidation by dioxygenases to form chlorinated dihydrodiols, the aromatic ring can be cleaved either between or adjacent to either hydroxyl substituent (*ortho-* or *meta*-ring fission) by ring-cleaving dioxygenases and is subsequently dehalogenated (e.g., chlorobenzenes, chlorocatechols) Dorn and Knackmuss, 1978a,b; Reineke, 1984; Hickey and Focht, 1990). Due to the vast number of intermediates formed from aerobic transformation and degradation of CAH, potential problems arise when aerobic processes are considered for bioremediation of CAH. Inhibitory substrate interactions between product, cometabolic substrate, and growth substrate have been shown to present further impediment to complete mineralization of the CAH (Dorn and Knackmuss, 1978; Adriaens and Focht, 1991; Hernandez et al., 1991).

Peroxidases activate the aromatic nuclei via the formation of highly active radicals by single electron extractions and have a very low substrate specificity. These enzymes have been shown to mediate hydroxylations, hydrations, and dehalogenations of aryl halides, as has been demonstrated with chlorobenzenes, pentachlorophenol, PCBs, and DDT, for example. The MMO, on the other hand, oxidizes chlorinated ethylenes and halogenated aromatic compounds via epoxidation of the double bond. This enzyme has only been shown to dechlorinate alkyl halides (e.g., TCE, DCE) and chlorinated aliphatic acids (e.g., trichloroacetate).

3.2. Anaerobic Dehalogenation

3.2.1. Physiology

Reductive dechlorination has been recognized as the only environmental biotransformation process for many highly chlorinated alkyl (Vogel and McCarty, 1985, Vogel et al., 1987; Fathrepure and Vogel, 1991) and aryl (Bosma et al., 1988; van der Meer et al., 1992; Nies and Vogel, 1990; Quensen et al., 1990; Adriaens and Grbić-Galić, 1993; Mohn and Tiedje, 1992) halides of concern (Table 12.2). Whereas several axenic cultures have been isolated that can grow anaerobically on alkyl halides as a sole source of carbon and energy, only in two cases have isolates been obtained that are able to utilize aryl halides (i.e., 3-chlorobenzo-ate, 2-chlorophenol). In all other cases, mixed microbial populations were enriched on the chlorinated substrates, resulting in growth and in methane and chloride production. Alternatively, halogenated aliphatic and aromatic compounds often only serve as alternative electron sinks.

Though the actual mechanism of reductive dehalogenation is still unclear, factors that influence the process include the type of electron donor (Nies and Vogel, 1990), the presence of specific electron acceptors (Petrovskis et al., 1994), temperature, and substrate availability to the microorganisms. The most effective substrate (electron donor) is apparently dependent on the microbial population involved rather than on the chemical structure of the compound to be reduced. Electron acceptor conditions affect the population composition of microbial communities and have been correlated with the occurrence of reductive dehalogenation activity. Thus, it was found that methanogenic conditions (no exogenous electron acceptor but carbon dioxide) are conducive to dechlorinating activity of aryl halides such as PCBs, both in the laboratory and in the environment (Brown et al., 1987; Quensen et al., 1988; 1990; Nies and Vogel, 1990, 1991). Whereas alkyl halides have been shown to be dechlorinated under methanogenic, sulfidogenic, and denitrifying conditions, aryl halides have only been demonstrated to undergo dechlorination under methanogenic and sulfidogenic conditions. The presence of sulfate has been demonstrated to slow down dechlorinating activity on aryl halides, including chlorobenzo-ates, chlorophenols, and chlorophenoxyacetates (Gibson and Suflita, 1986; Hägg-blom et al., 1993). Thus, reductive dehalogenation depends, at least in part, on the electron donor requirements and the efficiency with which electrons can be directed to dehalogenation.

Until recently, reductive dehalogenation was generally considered a nonenergy-yielding detoxification mechanism. Hence, reductive dehalogenation can be re-garded as a form of "anaerobic co-metabolism," whether or not the compound is further degraded. Since reductive dechlorination is generally accepted to be a co-metabolic process, except in a few isolated cases (Dolfing, 1990; Cole et al., 1994), no benefit is derived by the microorganism, making enrichment and adapta-tion difficult. Though the reason is still unclear, inocula from previously PCB-contaminated sediments require shorter acclimation times to dechlorinated PCBs than those derived from pristine sediments. Stimulation of dehalogenating activity has been observed after amendment with 1) additional electron donors (e.g., metha-nol, acetate), 2) reducing agents or a reduced environment, and, recently, 3) after addition of freshly spiked halogenated compounds (e.g., PCBs, bromobiphenyls).

3.2.2. Mechanism of Anaerobic Dechlorination

With few exceptions, anaerobic transformation of halogenated compounds is initi-ated by the removal of the bulky halogens via reductive dehalogenation. This reaction is more exergonic, and usually more rapid, with the more highly chlori-nated members of a homologous series than with the lesser chlorinated compounds (Dolfing and Harrison, 1992; Nies, 1993). The rate and occurrence of dehalogenation are strongly dependent on the environmental conditions and chlorinated compounds present.

Reductive dehalogenation of aryl halides is thought to involve two one-electron reduction steps, resulting in the removal of a halogen substituent and the formation of an intermediate aryl halide radical, which abstracts a proton from water to complete the reaction (Nies and Vogel, 1991). In all reported examples of biologically mediated reductive dehalogenation, the halogen is released as a halide. Dehalogenation appears to be the rate-limiting step during anaerobic aryl halide biotransformation, as the aromatic ring can only be further degraded once all the halogen substituents are removed. Reductive dehalogenation has been observed with chlorinated single ring structures (e.g., benzenes, benzoates, phenols), fused rings (e.g., naphthalenes) and carbon–carbon linked (e.g., PCBs and ether-linked (e.g., furans, dioxins) rings (Toussaint et al., 1992; Adriaens and Grbić-Galić, 1992; Wittich, 1992). Depending on the system, chlorines have been removed, resulting in the accumulation of up to monochlorinated aromatic compounds when the compound becomes too reduced for further reductive dechlorination. Steric factors have also been shown to affect the dechlorination of some compounds; generally, the *ortho*-substituted isomers are removed the slowest.

Alkyl halide dechlorination has been demonstrated for methanes (tetrachloro-, dichloro-), ethanes (hexachloro-, tetrachloro-, dichloro-), ethylenes (perchloro-, trichloro-) and DTT and has recently been extensively reviewed (Vogel et al., 1987; Mohn and Tiedje, 1992). Contrary to what would thermodynamically be predicted, some aryl and alkyl halides were shown to dechlorinate completely. All three monochlorophenol isomers were dechlorinated to phenol under methanogenic conditions (Boyd and Shelton, 1984; Hrudey et al., 1987); similarly, tetrachloroethylene (PCE) has been shown to dechlorinate to ethylene and reduce to ethane (de Bruin et al., 1992).

4. ENVIRONMENTAL PARAMETERS/CHARACTERISTICS

That the degradation of chlorinated organic compounds is dependent not only on microbial physiology and molecular structural limitations is apparent from the fact that, even though most chlorinated compounds can at least be transformed by some type of microbial consortium or axenic culture, many of these same compounds remain unaltered in the environment. Thus, the importance of environmental physical and chemical limitations and how they affect microbial degradation of organohalides has to be recognized. The physical, chemical, and biological limitations thought to play a significant role during degradation processes are qualitatively summarized in Table 12.4. Since recent reviews have described the relation between environmental characteristics and microbial biodegradative activity (Focht, 1988; Morgan and Watkinson, 1989; Grbić-Galić, 1990; Adriaens and Hickey, 1993), only those parameters that may be subjected to manipulation and optimization are described here (see Sections 5 and 6, below).

The importance of physical separation of contaminant and microorganism, as

TABLE 12.4. Examples of Enzymes Involved in Aerobic Cometabolic Degradation of Chlorinated Compounds

Enzyme	Substrate Range	Dehalogenation	Requirements
Alkyl oxygenases			
Methane monooxygenase (soluble, particulate)	Alkyl and aryl	Alkyl, yes; aryl, no	Cu^{2+}-dependence; methane, formate
Alkyl monooxygenase	Alkyl	Yes	Reducing power (NAD[P]H), inducers (isoprene, propane)
Ammonia monooxygenase	Alkyl	Yes	Ammonia
Aryl oxygenases			
Aryl monooxygenase	Aryl, alkyl (TCE)	Alkyl, yes; aryl, no	Inducers (phenol, toluene, 2,4-D)
Ring dioxygenase	(Chlorinated) ring(s)	Yes/no	Inducible
Extra- and intradiol dioxygenase	(Chloro-)catechols	Yes	Inducible
Peroxidase	Aryl	Yes	H_2O_2-dependent, nitrogen-starvation
Dehalogenases[a]			
Alkyl	Alkyl	Yes	Coenzyme A, inducible
4-Chlorobenzoyl	Chlorobenzoates	Yes	Coenzyme A, inducible

[a]Dehalogenases are active under aerobic and anaerobic conditions; the hydroxyl is derived from water, instead of from atmospheric oxygen.

effected by sorption processes onto inorganic and organic matrices, may be one of the least well understood yet most significant phenomena in biological treatment processes. Where sorption onto soil material, biomass, or organic colloids may be characteristic of natural environments (*in situ*), the surfaces of inert organics, such as activated carbon or dead and inactive cells, may be dominant sorption sites in bioreactors. Contaminant characteristics such as hydrophobicity and volatility provide insight into the relative tendency of the chemical to sequester in environmental compartments. Moreover, organohalides containing hydroxy or carboxy substituents (e.g., chlorophenolics, 2,4-D) may be incorporated into humic material due to chemical coupling reactions with humic components. These sorption mechanisms are believed either to render the organic inaccessible to the microorganisms, or at least to limit the rate of biodegradation due to diffusion and desorption phenomena.

These issues are critical in the interpretations of the effects of chemical structure on the apparent rate of biodegradation of chlorinated compounds. Highly chlorinated compounds tend to sorb more strongly to organic matter than lesser chlorinated contaminants. On the other hand, lesser chlorinated compounds tend to be more volatile (Fig. 12.1). Thus, in biological treatment systems, chlorinated compounds may disappear faster than they are biodegraded due to physical removal or may be degraded slower than predicted due to bioavailability limitations. This simplification may be misleading in the case where highly chlorinated organohalides tend to sorb stronger but dechlorinate more rapidly than the lesser chlorinated isomers under active methanogenic conditions.

The influence of the chemical characteristics of the localized environment, and of the chemical itself, has a direct effect on the microorganisms. The chemical composition of the solution that contains chlorinated compounds can be responsible for the inhibition of microbial growth, changes in the microbial community, and possibly death of microbial cells. Thus, the probable efficacy of biological treatment schemes, based on extrapolation of near-optimum chemical conditions in the laboratory, may be misleading. For example, the dissolved oxygen concentration and presence of other inorganic or organic potential electron acceptors will be critical to determining which microbial community is dominant. Clearly then, any contaminated environment should be analyzed not only for the pollutant(s) present but also for all the chemical parameters that may be directly related to microbial activity. These include, but are not restricted to, pH, ionic strength of the water, presence of heavy metals, and of potential electron acceptors (Table 12.4). Screening tests that evaluate microbial activity and environmental factors limiting (or even inhibiting) microbial metabolism thus prove indispensable in determining the potential for *in situ* bioremediation processes or whether *ex situ* approaches should be considered.

5. BIOREMEDIATION: PROCESSES AND BENCH-SCALE RESEARCH

5.1. Introduction

The implication of a microbial degradation process that is dependent on the primary substrate concentration, but not on the halogenated organic concentration (i.e., cometabolism), is that there is no limit to the final concentration of the halogenated organic substance. When the chlorinated compound serves as a primary substrate, fewer organisms would survive when its supply became low, and a threshold concentration may exist below which dechlorination is expected to cease (McCarty, 1988). As a secondary substrate, the halogenated organic substance can continue to be reduced by a large, healthy population of bacteria grown on the primary substrate until the halogenated solvent has been completely degraded.

The actual environmental contamination is often a complex mixture of chemicals. Therefore, technologies aimed at bioremediation of halogenated compounds should achieve complete destruction of all hazardous chemicals. Based on kinetic and metabolic characteristics of aerobic and anaerobic bacteria, two-stage biological processes consisting of an initial anaerobic dechlorination of highly chlorinated chemicals followed by aerobic degradation of the partially dechlorinated metabolites may effectively treat wastes containing complex mixtures of chlorinated organics. The conditions present, as well as the structure of the chlorinated compounds, dictate to a large extent the transformations that will predominate and the expected products that will accumulate.

The most important factor influencing biological degradation is whether the necessary organisms are present. This should be determined before a full-scale remediation scheme is begun by sampling the contaminated site and conducting laboratory-scale treatability studies (van der Meer et al., 1992b). If the required microorganisms are present but inactive or at too low concentrations, *in situ* or *ex situ* stimulation will be necessary. If the required microorganisms are not present, bioremediation may be enhanced by inoculating the soil or bioreactor with laboratory isolates. The success of inoculation is governed primarily by the competitive exclusion principle, i.e., the inoculant will survive when it has a function (niche) in the environment not previously occupied by indigenous bacteria (Focht, 1988). Thus, both selection pressure by the contaminant (i.e., concentration) and general growth kinetics of the inoculant in soil need to be considered. Microbial substrates and electron acceptors are the next factors to be considered. In a given location, the biological transformations will be oxidations, mediated by aerobic microorganisms, as long as oxygen is present. Once oxygen has been depleted, alternative electron acceptors such as nitrate and sulfate will be used. Finally, anaerobic (even methanogenic) microorganisms will dominate and reductions will take place. The different laboratory-scale approaches taken to enhance organohalide transformation and mineralization are summarized in Table 12.5 and will be discussed on a compound-specific basis. The array of bioreactor configurations described in Table 12.5 are represented in Figure 12.2.

5.2. Chlorinated Solvents

In remediation of groundwater contaminated with chlorinated solvents, the choice of treatment method should be based on which type of reaction will be most rapid and whether the expected products are less hazardous than the original halogenated contaminant. For example, reduction of PCE leads to the production of vinylchloride, which is a known carcinogen. However, the above discussion suggests an efficient means of converting halogenated compounds to nonhazardous compounds by using sequential anaerobic and aerobic treatments. Thus, PCE would be reduced to trichloroethylene (TCE) and dichloroethylene (DCE) under anaerobic conditions, followed by an aerobic environment in which the TCE and DCE would be oxidized to carbon dioxide. Overall, degradation rates could be maximized and the accumulation of vinylchloride prevented (Fathepure and Vogel, 1991). Whereas few studies have focused on *in situ* processes, bench-scale bioreactors have been developed and applied to simulate groundwater and soil bioreclamation from chlorinated solvents.

A study conducted by Dooley-Dana et al. (1989) demonstrated the feasibility of *in situ* degradation of low concentrations of PCE and TCE. A laboratory aquifer simulator was filled with soil, through which amended groundwater was circulated. Naturally occurring bacteria were simulated by controlled addition of nutrients. Methanogenic conditions were subsequently established within 2 weeks. During

TABLE 12.5. Comparison of Bench-Scale Process Designs for Transformation and Mineralization of Alkyl and Aryl Halides

Process Design/ Application	Compound[a]	Removal Efficiency (%)	Amendment/Inoculum	Reference
Activated sludge, aerobic				
Acclimated	PCP	98	Cell residence time, 5–10 days; Hydr. res. time (HRT), 25 hours	Edgehill and Finn (1983a)
	Kaneclor 500	45–82	No intermediates detected; HRT, 5–32 days	Kaneko et al. (1976)
Nonacclimated	PCP	7–11	Sorption significant	Blackburn et al. (1984)
	Aroclor 1221	81 ± 6	No lesser chlorinated PCBs or chloro-benzoate intermediates detected	Tucker et al. (1975)
	Aroclor 1242	26 ± 16		
	Aroclor 1254	15 ± 38		
Bioreactor, aerobic				
Suspended	TCE	95	Acclimation period required	Phelps et al. (1990)
	TCE	99	Preinduction with phenol	Nelson et al. (1988)
	PCP	>99	Two-stage inoculation (*Pseudomonas* sp. SR3)	Mueller et al. (1993)
Biofilm	TCE	92	TCE provided in gas phase; no inducers required; *Pseudomonas cepacia*	Shields et al. (1993), Shields and Reagin (1992)
	TCE	>99	Intermediates not detected	Pflug and Burton (1988)
	DCM	>99	Based on chloride appearance; inocula-tion with DCM degrader	Stucki (1990)
	CB	91–96	Inoculation with distillery waste	Bouwer and McCarty (1985)
	DCB	71–99	Acclimation period, 20–500 days	
	TCB	75–95	Removal dependent on acetate concen-tration	
	PCP	99	Based on disappearance; inoculated with *Flavobacterium*	Frick et al. (1988)
	TCP, PCP	99.9	Inoculation with rhodococci; mineraliza-tion (Cl^-, $[^{14}C]CO_2$)	Valo et al. (1990)

	DCP	99.9	Rotating Biological Contactor; *Phanero-chaete crysosporium*	Tabak et al. (1991)
	3,4-PCB	>99	Based on metabolites detected	Adriaens and Focht (1990)
	4,4'-PCB	63	Mineralization, 7%–11%	
	3,3',4,4'-PCB	32	*Acinetobacter* spp.; inducer; biphenyl-vapors	
Bioreactor, anaerobic				
Biofilm	CF	70	Acclimated inoculum	Bouwer and McCarty (1983)
	CT	73		
	TCA	84		
	PCE	92		
	DCM	100	HRT: 0.5–2 days; inoculum growth on DCM	Freedman and Gossett (1991)
Stirred tank reactor	PCP	99.9	HRT, 10–40 days; acclimated to PCP; sorption and volatilization insignificant	Guthrie et al. (1984)
Upflow sludge	PCP	94–97	Based on metabolites detected	Woods et al. (1989)
blanket	2,3,4,5-CP	65–95	Efficiency dependent on concentration	
	2,4,6-CP	95–97	*Ortho-* and *meta*-dechlorination	
	2,4-CP	80–87	No dechlorination of monochlorophenols	
	2,3-CP	74–81		
	2,6-CP	90–92	Anaerobic municipal sludge	
	CP	66–95	Acclimated (2 years) sludge	Krumme and Boyd (1988)
	TCP	23–71	Mineralization dependent on substrate loading; HRT, 2–5 days	
	PCP	35		
Sequential processes				
Washing/anaerobic	PCP	99	Based on 80% extracted conversion to mono-CPs	Khodadoust et al. (1993)
Anaerobic/aerobic	PCE	96	Lesser chlorinated intermediates detected	Fathepure and Vogel (1991)
	CF	83		

(*continued*)

TABLE 12.5. (*Continued*)

Process Design/Application	Compound[a]	Removal Efficiency (%)	Amendment/Inoculum	Reference
	HCB	94	Activity dependent on carbon source provided	Anid et al. (1993)
	Aroclor 1242	80	Change in homolog pattern; Hudson River sediments	Hakulinen and Salkinoja-Salonen (1982)
	CP-mixture	93	Nonacclimated inoculum	
	PCP	100	Mineralization (CO_2, Cl^-)	
Solid-Phase Processes				
Soil inoculations	PCP	80–87	*Flavobacterium* inoc. (repeat); mineralization dependent on soil type, moisture, temperature	Crawford and Mohn (1985)
	PCP	45–65	*Rhodococcus chlorophenolicus*; dependent on soil texture, initial PCP conc., and inoculum size; *o*-methylation (volatilization), <2%	Middeldorp et al. (1990)
	PCP	100	Induced anaerobiosis; sewage sludge amendment	Mikesell and Boyd (1988)
	PCP	67	Based on $^{14}CO_2$ recovery; *Flavobacterium*; soil moisture, 60%	Seech et al. (1991)
	PCP		*Arthrobacter*	Edgehill and Finn (1983b)
	2,4,5-T		*P. cepacia*	Chatterjee et al. (1982)
	CB, DCB		*Pseudomonas* sp.	van der Meer et al. (1987)
	3-CBA	100	*P. alcaligenes* C-O; dependent on inoculum size	Focht and Shelton (1987)

Process	Compound	%	Description	Reference
	Aroclor 1242	48–49	*Acinetobacter* strain P6; biphenyl (BP) added	Brunner et al. (1985); Focht and Brunner (1985)
	Aroclor 1242	72	Inoculation with chlorobenzoate degraders, BP	Hickey et al. (1993)
	PCP	88–91	Formation of anisoles; addition of peat; *Phanerochaete* spp.	Lamar and Dietrich (1990)
Composting	PCP	55–72 (soils) 1–10 (sediments)	"Landfarming chambers" Amended/unamended; moisture, 8%–12%	Mueller et al. (1991a)
	PCP	98	*R. chlorophenolicus*	Valo and Salkinoja-Salonen (1986)
Bioslurry	Aroclor 1242	3	Amended with straw, sludge	Brunner et al. (1985)
	PCP	11 (soil) 23 (sediment)	Bioreactor design (1.5L) DO, 90%; solids, 2.1%–2.7%	Mueller et al. (1991b)
	PCP	6–97	*Flavobacterium* inoculation; degradation dependent on C-source, copper/chromate/arsenate conc.	Topp and Hanson (1990)
Aquifer simulator	CF	90	Stimulated with CH_4	Janssen et al. (1991)
	TCE	44–69	Stimulation with CH_4 and O_2; dependent on groundwater velocity; no product monitoring	Hicks et al. (1991)

[a]For list of abbreviations, ses Table 12.1.

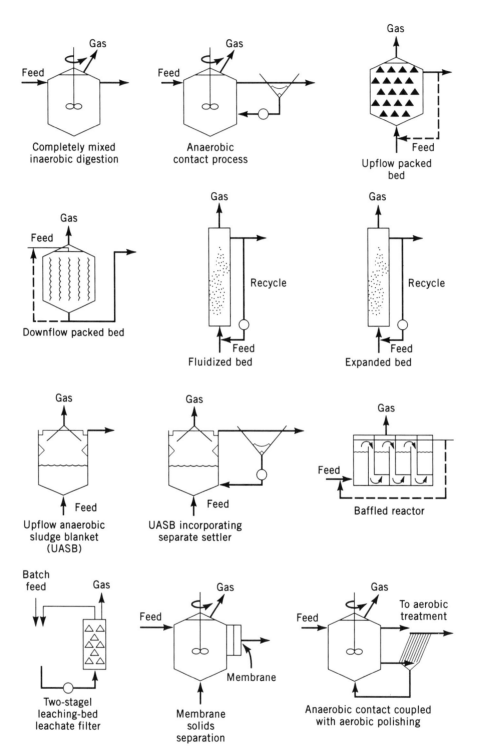

Figure 12.2. Typical reactor configurations used in bench-scale treatment of chlorinated organic compounds. (Reproduced from Speece, 1983, with permission of the publisher.)

this period, PCE and TCE were dechlorinated to both *cis-* and *trans*-DCE. Upon introduction of oxygen, oxidative degradation of DCE was initiated by the methanotrophs present. Similarly, Dolan and McCarty (1994) were able to stimulate methanotrophic activity on TCE, both DCE isomers, and vinylchloride in laboratory columns containing aquifer samples from St. Joseph (MI). These studies suggest that aquifer conditions could be manipulated *in situ* to achieve complete destruction of problematic chemicals in the same manner as in soil column reactors.

Laboratory-scale bioreactor systems have been developed for the aerobic (Bouwer and McCarty, 1985; Phelps et al., 1990; Nelson et al., 1988; Rittmann and McCarty, 1980; Stucki, 1990; Shields et al., 1993), anaerobic (Bouwer and McCarty, 1983, 1985; Cobb and Bouwer, 1991), and sequential (Fathepure and Vogel, 1991) degradation of chlorinated solvents (Fig. 12.2). Their relative efficiencies in solvent removal, generally measured by product appearance, are given in Table 12.5. These studies have demonstrated that the lesser chlorinated compounds (e.g., dichloromethane, vinylchloride) are readily degraded under aerobic conditions, whereas PCE, TCA, and CT persist. The latter are reduced quite rapidly under methanogenic conditions. Simple organics (e.g., acetate, methane) were required as the primary substrates during organohalide degradation, except for dichloromethane, which has been shown to support microbial growth. Aerobic degradation of the chlorinated alkenes (TCE, DCE, vinylchloride) was dependent on the presence of methanotropic bacteria in the bioreactor and on the addition of inducing substrates for methane monooxygenase activity (methane, phenol). Competitive substrate inhibition of aerobic enzymes excluded, no inhibitory effects of substrates or products were observed during the degradation experiments.

In a novel approach to aerobic TCE degradation, Shields et al. (1993) used a nonrecombinant derivative of *Pseudomonas cepacia* G4, an organism previously shown to express an inducible monooxygenase (toluene *ortho*-monooxygenase, Tom), which is able to co-oxidize TCE (Shields and Reagin, 1992). The derivative, which constitutively expresses Tom (no need for inducing substrates), was applied in a fixed film vapor phase bioreactor to degrade air-entrained TCE. The biofilm, maintained on lactate, was able to degrade 90% of TCE concentrations as low as 130 μg/liter.

5.3. Chlorinated Aromatics

5.3.1. Chlorobenzenes

Especially the 1,3- and 1,4-isomers have been demonstrated to be ubiquitous groundwater and CERCLA site contaminants. Hexachlorobenzene is sequentially reduced to tri- and dichlorobenzenes in batch microcosms using anaerobic sewage sludge (Tiedje et al., 1987; Fathepure et al., 1988). Mineralization of chlorobenzenes under anaerobic conditions has not been demonstrated. Aerobic chlorobenzene degradation is limited by the number of chlorines on the benzene ring; up to tetrachlorinated benzenes have been shown to degrade via chlorocatechol intermediates (Häggblom, 1992).

Reductive dechlorination of all di- and trichlorobenzene isomers to monochlorobenzene was demonstrated in column reactors containing anaerobic sediments from the Rhine River (Kuhn et al., 1985; Bosma et al., 1988; van der Meer et al., 1990). Enrichment culture derived from these sediments dechlorinated chlorobenzenes in the presence of simple aliphatic acids, sugars, or alcohols as electron donors for the methanogenic populations. Mono- through trichlorobenzenes were shown to be degraded by aerobic acetate-grown biofilms, after inoculum acclimation periods ranging from 20 to 500 days (Bouwer and McCarty, 1985). Using a two-stage sequential anaerobic–aerobic set-up, Fathepure and Vogel (1991) demonstrated dehalogenation of hexachlorobenzene to di- and trichlorobenzenes, followed by aerobic mineralization of these intermediates to carbon dioxide and nonvolatile products. This column study resulted in a total sequential metabolism of 94%, but was shown to be dependent on the primary carbon source added: glucose < methanol < acetate.

5.3.2. Chlorophenols and Phenoxy Herbicides

Degradation of this type of contaminants has recently been reviewed (Häggblom, 1992) and constitutes a chapter in this volume. Anaerobic dehalogenation and mineralization of chlorophenols has been observed under methanogenic and sulfidogenic conditions, though the rates of dechlorination were slower under sulfidogenic conditions (Häggblom et al., 1993). Whereas pentachlorophenol was dehalogenated to mainly di- and trichlorophenols, all monochlorophenols have been demonstrated to dechlorinate to phenol in anaerobic sludge (Ide et al., 1972; Murthy et al., 1979; Boyd and Shelton, 1984; Hrudey et al., 1987). Under aerobic conditions, two types of bacterial isolates have been found: those with a preference for lesser chlorinated phenols and strains that preferentially attack highly chlorinated phenols (Häggblom, 1992). Additionally, different species of the lignin-degrading fungi *Phanerochaete* have been demonstrated either to transform or to mineralize PCP by means of nonspecific lignin peroxidases, which are expressed under nutrient-limiting conditions. The main products resulting from transformation are pentachloroanisoles (i.e., volatilization) and nonextractable soil-bound residues (i.e., humification) (Ryan and Bumpus, 1989).

Laboratory feasibility studies for degradation of chlorophenols have mainly focused on pentachlorophenol degradation in soils and aqueous systems. The degradation of liquid chlorophenol-containing wastes was investigated using either aerobic or anaerobic bioreactors of different configurations (Table 12.5). Since indigenous microorganisms often did not exhibit activity to chlorophenols, most bench- and pilot-scale designs involve a separate acclimation period of the inocula. Acclimation periods ranging from a few days up to years have been reported, depending on the inoculum and primary substrate used, the experimental design (aerobic or anaerobic), and the chlorophenol isomer degraded. In the case of aerobic reactors, PCP-degrading isolates, such as *Flavobacterium* (Frick et al., 1988) or *Pseudomonas* sp. SR3 (Mueller et al., 1993) were added to biofilm or suspended culture reactors to enhance chlorophenol degradation. Stability and activity of the added PCP degraders were dependent on the competition of the bacteria in the waste stream for the primary growth substrate (e.g., acetate), the concentration of PCP in the waste (selection pressure), and the hydraulic retention time.

Three types of treatment designs for PCP-contaminated soils were evaluated: soil inoculation, composting techniques, and bioslurries. Inoculations with strains of *Flavobacterium* (Crawford and Mohn, 1985; Seech et al., 1991), *Rhodococcus chlorophenolicus* (Middeldorp et al., 1990), and *Arthrobacter* (Edgehill and Finn, 1983b) were shown to enhance the removal of PCP. However, the competitiveness of the inocula differed widely: Whereas *Rhodococcus* inocula sustained activity for months after inoculation, repeated inoculations of *Flavobacterium* were required for substantial removal of PCP.

Recently, inoculation of lignin-degrading fungi has received a great deal of attention with respect to their potential for *in situ* treatment of aryl halides in general and PCP in particular (Lamar and Dietrich, 1990; Lamar et al., 1990; Glaser et al., 1993). Laboratory- and pilot-scale studies involve inoculum preparation, using wood or aspen chips, followed by soil inoculation (top 30 cm) and soil mixing (rototilling) to equalize the PCP concentration. The soil is amended with a carbon source, such as peat moss (applied at approx. 2% dry weight), and the moisture is adjusted to field capacity (and maintained at minimum 20%). In laboratory-scale studies, the loss of PCP via mineralization and volatilization was found to be negligible, a small fraction of the PCP (8%–13%) was methylated to penta-chloroanisole, and the major fraction was converted to nonextractable soil-bound products.

Composting and bioslurry techniques proved to be effective only when PCP degraders (or anaerobic inocula) were added and when the concentration of copper/arsenate/chromium compounds often associated with wood preserving wastes was sufficiently low (Salkinoja-Salonen et al., 1984; Topp and Hanson, 1990). The lower removal efficiency observed by Mueller et al. (1991a) may thus be explained by the inability of indigenous soil and sediment microorganisms to metabolize PCP. Using a combined physical–biological approach, Khodadoust et al. (1993) describe a soil-washing technique of PCP-contaminated soils with an ethanol–water mixture, followed by anaerobic treatment. The soil extract was fed into an expanded bed granular activated carbon (GAC) anaerobic bioreactor. Equimolar conversion of PCP into monochlorophenols was the predominant degradation process by mixed cultures rather than mineralization to carbon dioxide and methane. Speitel et al. (1989) investigated the simultaneous aerobic biodegradation and desorption of 2,4-dichlorophenol and PCP in granular activated carbon columns (GAC) by acclimated cultures. It was found that sorption affected the rate of degradation of 2,4-DCP and that biodegradation of PCP was affected by slow microbial kinetics.

Since phenoxy herbicides (e.g., 2,4-D, 2,4,5-T) are degraded aerobically via chlorophenol and chlorocatechol intermediates, laboratory treatability studies would follow the same approach as described for chlorophenols. Reports on the degradation of 2,4-D or 2,4,5-T in bioreactors are scarce. Repeated inoculations of a 2,4,5-T–contaminated soil with a *Pseudomonas cepacia* strain enhanced its degradation, but the inoculum was outcompeted as soon as the contaminant disappeared

(Chatterjee et al., 1982; Kilbane et al., 1983). This would be expected, as the function of this organism in the soil environment disappears with the substrate it is intended to degrade.

5.3.3. Polychlorinated Biphenyls

Dechlorination of Aroclor 1242 (average of four chlorines), Aroclor 1254 (average of five chlorines), and Aroclor 1260 (average of six chlorines) was first demonstrated in the laboratory by Quensen et al. (1988, 1990) using Hudson River (NY) sediment. All mixtures dehalogenated to biphenyl isomers with an average chlorine substitution of two to three chlorines, the latter albeit slowly. Investigations on the influence of the type of growth substrate (i.e., electron donor) on the rate of dechlorination revealed that methanol and glucose were better substrates than acetone and acetate (Nies and Vogel, 1990). Samples not amended with organic carbon, yet supplied with the same nutrient concentrations as the aforementioned, did not affect dechlorination within the time monitored. Ye et al. (1992) reported on the influence of heat and ethanol treatment on anaerobic dechlorination of PCBs. Their results indicated that treated cultures retained their dechlorinating activity, albeit less extensively than untreated controls. Sulfate-reducing conditions were shown not to mediate PCB dechlorination and was demonstrated for Aroclor 1242 in two different lake sediments (Häggblom et al., 1993).

Aerobically, PCB co-metabolism is generally limited to congeners with four to five chlorines and results in the accumulation of a range of oxidized (chlorinated) intermediates such as chlorobenzoates (Adriaens et al., 1991). Aerobic chlorine removal does not occur until either 1) after ring fission of the biphenyl molecule or the chlorocatechol intermediates or 2) after dehalogenase activity on the chlorobenzoate intermediates. Aerobic PCB degradation is limited by induction of the proper microorganisms and degradative enzymes by analog substrates (Brunner et al., 1985; Hickey et al., 1993), by the biomass density of the required microorganisms (Brunner et al., 1985), and by the possible inhibitory substrate interactions that may ensue from the generation of chlorinated analogs (Adriaens and Focht, 1991).

These fundamental processes have been assessed for their practical applicability in a number of bench-scale studies (Adriaens and Focht, 1990; Anid et al., 1991, 1993). Using a defined axenic coculture with the combined catabolic ability to mineralize *meta*- and *para*-substituted PCBs aerobically, Adriaens and Focht (1990) described the transformation and mineralization of 4,4'- and 3,4-dichlorobiphenyl and 3,4,3',4'-tetrachlorobiphenyl in a fixed-film column reactor degradation study. Their results indicated that, whereas competition of both strains for the same growth substrate (i.e., benzoate) during co-metabolism of the PCBs (and chlorobenzoates) may result in washout of the slower growing bacterium, diffusion of the aerobic PCB metabolites to the chlorobenzoate degraders is likely the rate-limiting process for complete PCB mineralization.

Alternatively, the metabolism of ^{14}C-labeled PCBs (Aroclor 1242 mixture) in soil was shown to be greatly enhanced (48%–49%) by the addition of biphenyl as a growth substrate and by inoculation of _Acinetobacter_ sp. strain P6, a PCB co-metabolizing strain (Brunner et al., 1985). Organic amendments of the soil with straw and sludge, anaerobic incubations, or a combination of both treatments in no case resulted in more than 3% mineralization of the amount of ^{14}C-labeled Aroclor 1242 added. Hickey et al. (1993) amended the same soil with chlorobenzoate utilizers, a PCB co-metabolizer, and a combination of both. It was found that the ^{14}C-Aroclor 1242 was mineralized most extensively when the soil was amended only with chlorobenzoate utilizers, suggesting that the indigenous microflora was able to co-metabolize the PCBs and that chlorobenzoate metabolism is the rate-limiting step during aerobic Aroclor 1242 mineralization. This finding was corroborated by Focht and Shelton (1987), who compared the disappearance of [^{14}C]-3-CB in culture with that in soil. 3-Chlorobenzoate was refractile to attack in soil by the indigenous microflora, but it was completely mineralized upon inoculation of a chlorobenzoate utilizer.

Recently, Lajoie et al. (1992) developed the concept of field application vectors (FAVs), where a temporary environmental niche is created for a microorganism in which relevant (foreign) degradative genes are cloned. This concept has been applied to the co-metabolism of PCBs using a _Pseudomonas paucimobilis_ strain in which the biphenyl (_bph_) operon (encodes the biphenyl dioxygenases) was cloned and its expression linked to the degradation of a nonionic detergent (Lajoie et al., 1993). Although a significant drop in biphenyl dioxygenase expression was observed in nonsterile environments after five generations, the concept of FAVs does present a potential for _in situ_ degradation of PCBs. The coupling of gene expression to a temporary, nontoxic, inexpensive, and easy to create niche may help alleviate regulatory concerns with the release of genetic engineered microorganisms in the environment.

Anid et al. (1991, 1993) describe a bench-scale river model to examine design parameters that would be important in _in situ_ bioremediation of PCBs, particularly the transition from anaerobic to aerobic conditions. Hudson River sediments (0.8 ton) were spiked with 300 ppm of Aroclor 1242, and methanol-amended anaerobic medium was recirculated over the sediments. At the end of the anaerobic phase, the congener profile of sediment PCBs shifted to an increase in mono- and dichlorobiphenyls and a decrease in more chlorinated homologs. Subsequently, the sediments were amended with a recirculating hydrogen peroxide solution as an oxygenation agent to stimulate for aerobic microorganisms. The congener profile showed a decrease in the lesser chlorinated congeners, as could be expected from batch studies, but no chlorobenzoate intermediates were identified. This sequential process thus demonstrates that either _in situ_ or _ex situ_ remediation processes of PCBs would be feasible.

5.4. Summary

The application of aerobic and anaerobic bench-scale treatment designs toward the degradation of a range of aryl and alkyl halides in either liquid or solid medium has been described. From a microbial physiological point of view, the onset of degradation in most of these systems was shown to be dependent on at least one of the following requirements:

1. An acclimation period prior to biodegradative activity
2. An (inducing) growth substrate during co-metabolism
3. The (repeated) inoculation of specialized laboratory strains

The review also indicated that removal efficiencies are often based on substrate disappearance and do not differentiate between transformation and mineralization of the chlorinated organic and physical–chemical processes. For example, many different enzymes need to be induced and react in sequence during aerobic cometabolic degradation processes for complete degradation to occur. Thus, an extensive literature review and/or batch (i.e., microcosm or flask) treatability studies are prerequisites to investigate the appearance and accumulation of any recalcitrant metabolites. These studies should also be undertaken to assess removal due to sorption and volatilization. Only when all of the substrate removal can be accounted for can the efficiency of any bench-scale study be accurately calculated and reasonable extrapolations be made to pilot- and field-scale remediation systems.

Clearly, the application of the bench-scale studies presented here to groundwater remediation or any engineered biological treatment process requires more information than provided thus far (see Table 12.5). The results of microbial degradation of hazardous waste studies should be useful for understanding the occurrence and transformations of these compounds in artificial and natural environments. The next step is the rational design of biotreatment processes for the complete or environmentally relevant destruction of hazardous wastes. This design requires complete evaluation of the potential hazardous compounds, their potential products, the health risks, and the microbiology for both current and future wastes.

6. BIOREMEDIATION STRATEGIES: FIELD-SCALE APPLICATIONS

6.1. Introduction

Field-scale bioremediation studies are primarily, but not exclusively, extensions of bench- and pilot-scale investigations on the biodegradability of specific contaminants and on the parameters that can be modified to enhance biodegradation. Thus, the general criteria that will determine the application of either *in situ* or on site bioremediation processes of soil and groundwater are based on 1) the general biodegradability and physical–chemical properties of the contaminant, 2) the prevailing environmental conditions, and 3) the quantity (or volume) of the soil (water) that needs to be treated. Smaller contaminated areas (plumes), surface spills, easily leachable or volatile compounds, and contaminants that require specialized bacteria for biodegradation are generally good candidates for *ex situ* treatment in bioreactors. Examples of both *in situ* and on site remediation technologies will be provided and discussed in light of their intrinsic characteristics, removal efficiency, and

TABLE 12.6. Comparison of Bioremediation Technology Performance on Organohalide Contaminated Soils

Treatment Technology	Compound	Representative Range of Removal (%)	Comments
Physical–Chemical			
Aeration	Chlorinated solvents	99.9	Limited by soil types and to volatile compounds; limited application
Soil flushing/ washing	Organohalides	75–89	Proven effective; higher cost than bioremediation alone; not effective on clay or silt
In situ bioremediation			
Bioventing	TCE	20–30	Limited by soil permeability
	VC	95	Effectiveness on TCE< DCE<VC
	t-DCE	80–85	Up to 20% due to microbial activity
	c-DCE	40	
Inoculation/ stimulation	Aroclor 1242	37–52	Oxygen, biphenyl, nutrients
	Aroclor 1260	>20	Bromobiphenyl addition
	PCP	15–89	Dependent on fungus and inoculum-loading
	Chloroalkenes	90	Phenol induction
Ex situ bioremediation			
Bioreactors	TCE	90	TCE-co-metabolizer added; methane stimulation
	PCP	99	*Flavobacterium* added; dependent on extraction efficiency from soils
Composting	Freons	94	Applicable to wide range of organohalides
	TCE, semi-volatiles	89–99	
	PCP	70–80	Indigenous organisms
Bioslurry	TCE	86–100	Nutrient addition
	PCP	90–99	Treatment of residuals
Landfarming	PCP	80	Dependent on concentration; nutrients no effect

general applicability for the degradation of chlorinated organics (Table 12.6). For a more extensive description of the technologies and the specifications of their site-specific application, the reader is referred to separate volumes on the topic (Hinchee and Olfenbuttel, 1992a,b; Nyer, 1992; Hinchee et al., 1994a,b: Lee et al., 1988; Norris et al., 1994; Morgan and Watkinson, 1989).

The soil is a complex system, consisting of four phases: gas (15%–35%), water (15%–35%), inorganic solids (38%–45%), and organic solids (5%–12%). By volume, gas and water comprise about 50% of the pore spaces in soil (McFarland et al., 1991). The organic contaminant will thus, depending on its solubility and volatilization, distribute in certain proportions in the fluid phases and partition between the various soil solid phases. In the case of chlorinated solvents (aromatic and aliphatic), the contaminant, which is denser than water (DNAPL), tends to get trapped in the soil pores. For heavier, less soluble chlorinated organics such as PCBs, phenols, and chlorinated herbicides, partitioning in the sediment, aquitard, or soil organic matter tends to be the major process of distribution.

Under field conditions, the rate and extent of remediation is generally limited by mass transfer limitations, with respect to both the contaminant and the potential electron donors, electron acceptors, and nutrients (Chapelle, 1993). Successful remediation systems will thus depend on a combination of physical–chemical methods to ensure 1) susceptibility of the contaminant to transformation reactions and 2) chemical optimization of the soil–sediment environment for the proper degradation reactions and biodegradation methods. Depending on the specific properties of the organic contaminant and the soil characteristics, different remediation schemes will then be favored.

6.2. *In Situ* Stimulation

The success of *in situ* biodegradation processes is often dependent on the degree to which desired degradation processes or required physiological groups of microorganisms can be stimulated. This, in turn, will be dictated by the type of contaminant and by the prevalent redox conditions in the contaminated environment. Depending on whether oxidative–reductive processes need to be stimulated or the required catabolic activity is lacking, either oxygen and/or additional electron donors or microorganisms will need to be introduced.

Since oxygen is often the limiting factor in aerobic bioremediation processes, different approaches have been considered to deliver oxygen to the soil: injection of aerated water, air sparging, venting, and injections with hydrogen peroxide or nitrate. Where air sparging and the different injections are mainly geared toward the saturated zone (Fig. 12.3B–D), venting finds its application in the unsaturated zone (Fig. 12.3A). Recently, bioventing received attention for its potential to enhance biodegradative activity in soil: Air is injected directly into the vadose zone and circulated via air extraction wells. As such, two goals are accomplished: The oxygen is delivered in close proximity to the (aerobic) microorganisms, and the air may enhance solubilization and redistribution of the nonaqueous phase liquids (NAPLs) in the soil environment. Most information pertaining to bioventing is based on observations made at field sites contaminated with either petroleum hydrocarbons (Baehr et al., 1989) or aviation (JP4) fuels (Hinchee et al., 1991; Miller et al., 1992; Dupont et al., 1992; Kampbell and Wilson, 1991). The effectiveness of this technique to enhance organohalide (DNAPLs, denser than water) remediation has not been fully tested. Air sparging techniques are currently being developed for field applications at hydrocarbon-contaminated sites (Hinchee, 1994). Controlled additions of hydrogen peroxide, on the other hand, have been shown to enhance effectively the degradation of chlorinated organics. An approach where nitrate is supplemented to a given site as an alternative electron acceptor has shown potential for organohalide degradation.

More complex types of biostimulation are required when the chlorinated contaminant can only be degraded either via (sequential) co-metabolic processes (see Section 2, above, and Table 12.1) or when the conditions present are not conducive to complete mineralization. For example, methanogenic conditions in aquifers might result in the production of lesser halogenated compounds, but possibly not complete mineralization of these compounds. Indeed, the results presented here indicate that vinylchloride might be a product of TCE transformation under methanogenic conditions in aquifers. Fortunately, vinylchloride can be degraded more rapidly under aerobic conditions. Thus, aquifer remediation might involve two stages of operation either in sequence in the same part of the aquifer or in series radial from a contamination site (Fig. 12.4). The first stage would be a methanogenic zone, where dehalogenation of the polyhalogenated compounds would occur. The second would be an aerobic zone, where the mono- and dihalogenated compounds would be oxidized to nonhalogenated compounds, such as carbon dioxide.

When the required enzymes or microbial populations are not present, either innocuous inducing substrates will need to be added or laboratory-isolated microorganisms with specific degradative abilities will need to be inoculated. The addition of specific microorganisms is currently limited to naturally isolated (i.e., not genetically engineered) or indigenous microorganisms. This approach has been described earlier, and its difficulties have been reviewed recently (Focht, 1988).

6.2.1. Methanotrophic Oxidation

Stimulation of aerobic (methanotrophic) microorganisms in subsurface spills of TCE and DCE has been met with success, resulting in the biodegradation of both solvents. Natural attenuation of TCE *cis-* and *trans-*DCE, and vinylchloride by autochthonous methanotrophic bacteria in the Moffett aquifer has been demonstrated to occur after external stimulation of these microorganisms with oxygen and methane (Roberts et al., 1990; Semprini et al., 1990, 1992) or phenol (Hopkins et al., 1993). The essential features of the process include the introduction of oxygen to provide an environment suitable for the aerobic methanotrophs and a carbon source to support growth and the generation of reducing power and to induce enzyme activity (Roberts et al., 1990; Henry and Grbić-Galić, 1990, 1991). Generally, the rate of degradation was dependent on the level of chlorination of the solvent, in the order vinylchloride > *cis-*DCE > *trans-*DCE > TCE. It was demonstrated that the biotransformation of TCE, *trans-*DCE, and vinylchloride was competitively inhibited by methane. Upon addition of alternative electron donors such as formate or methanol, inhibition ceased. However, formate does not induce the MMO and may stimulate other types on microorganisms that do not participate in solvent oxidation. Since transformation could only be temporarily achieved, the issue of primary substrate addition and stimulation of methanotrophic bacteria needs to be addressed further. Current field studies on chloroalkene degradation with phenol-induced microorganisms have resulted in 90% removal of TCE concentrations of up to 1 mg/liter (Hopkins et al., 1993). Overall, the technology appears to be limited by soil permeability and heterogeneity and the requirement of stimulation of methanotrophs that are present at very low concentrations.

More highly chlorinated solvents, such as tetrachloroethylene (PCE) or hexachloroethane (PCA), have not been shown to degrade aerobically and will thus first need to be reductively dechlorinated under anaerobic conditions prior to aerobic (methanotrophic) degradation. Conceptually, a two-zone plume interception strategy is envisioned in Figure 12.4(A), whereby nutrients and simple organic electron donors are provided to stimulate the methanogenic populations. Nutrients, oxygen, and methane will need to be provided to stimulate the methanotrophic zone downstream from the methanogenic zone.

6.2.2. Aerobic PCB Oxidation

Polychlorinated biphenyls currently present in the anaerobic (mainly methanogenic) Hudson River sediments have been demonstrated to have undergone extensive dechlorination. Of all the mono- and dichlorobiphenyls present in the sediment (62%–73% of total), 55%–60% were o- or o,p-chlorinated. Since these congeners are susceptible to aerobic degradation, Harkness et al. (1993) conducted a field study on *in situ* stimulation of aerobic PCB biodegradation. Thus, biphenyl, oxygen (as hydrogen peroxide), and inorganic nutrients were added to isolated plots of Hudson River bottom. Upon stimulation, the indigenous aerobic microorganisms were able to degrade 37%–55% of these lightly chlorinated PCBs, resulting in the production of o-, p-, or o,p-substituted mono- and dichlorobenzoates. Repeated inoculation with a purified PCB–co-metabolizing microorganism failed to enhance biodegradative activity, presumably due to competition with acclimated indigenous aerobic bacteria.

6.2.3. Anaerobic Reductive Dechlorination of PCB

Contrary to that in Hudson River sediments, Aroclor 1260 present in Woods Pond sediments has not undergone extensive reductive dechlorination, suggesting that methanogenic conditions alone were not conducive for dechlorination. It was found that the addition of freshly spiked PCB and polybrominated biphenyls stimulated extensive *meta*-dechlorination of the sediment-bound Aroclor. In a field-scale study undertaken to investigate their *in situ* stimulatory effect, DeWeerd and Bedard (1993) added 2- and 2,2′-dibromobiphenyl to sediment plots that were physically isolated in "caissons." Whereas the PCBs did not dechlorinate in unamended sediments, debromination of the added congeners to biphenyl stimulated dechlorination of the Aroclor mixture, as was found in the laboratory (DeWeerd and Bedard, 1992). Current studies focus on alternative stimulating substrates, such as halobenzoates (DeWeerd and Bedard, 1993).

6.2.4. Fungal Inoculation Technology

Recently a 2 month field study was initiated at the Brookhaven Wood Preserving facility to evaluate the potential of bioremediation of pentachlorophenol-containing wastes using lignin-degrading fungi (EPA, 1991; Glaser et al., 1993). Bench-scale and *in situ* pilot scale studies (Lamar and Dietrich, 1990) have shown that suboptimal growth conditions, an additional carbon source, and preinoculation of the fungus on a wood chip support resulted in an 88%–91% decrease of PCP in 6.5 weeks. In the field study, soil from a waste sludge pile was sieved (2.5 cm) and homogenized, wood chips were added, and the plots (3×3 m, 4 tons of soil) were inoculated with three fungal species. The plots were then irrigated and rototilled on a weekly basis. This treatment resulted in 15%–89% transformation of PCP, depending on the fungus and inoculum loading used; the greatest removal of PCP was achieved in plots inoculated with *P. sordida*. Limitations of this method include the high cost of the inoculation technique, the low degree of mineralization of PCP, and the potential toxicity and instability of the soil-bound residues.

6.2.5. *Anaerobic Degradation of PCE*

A full-scale integrated demonstration of reductive dechlorination of PCE under anaerobic conditions is underway at a site in Germany. At this site, air, soil, and groundwater are contaminated with PCE. The current treatment affects both the surface soil and the groundwater through different applications of basic processes. The air treatment involves passing the gas through activated carbon for sorbing the PCE. Soil reclamation is performed by placing the soil in long, covered composting piles, sealed against oxygen diffusion. The system depends entirely on indigenous anaerobic bacteria that are stimulated by nutrient addition and induced anaerobiosis. This process is currently being characterized to determine the relative rate of PCE degradation and the major metabolites produced.

PCE-contaminated water at the site is treated initially in an anaerobic fixed bed bioreactor, which exhibits similar reductive dechlorination as observed in the soil environment. This system also requires nutrients to maintain the active anaerobic processes. The biogas produced in conjunction with active anaerobic reductive dechlorination may contain PCE due to its fugacity and is treated separately by activated carbon filters. After anaerobic treatment, the water is intended to be treated by an aerobic biological process.

Currently, an *in situ* approach for this sequential process is attempted by addition of nutrients to maintain both active anaerobic and aerobic degradative activities within the indigenous microbiota. The approach follows basic principles that have been described earlier and that in practice do not rely on the use or selection of a specific microorganism. A combination of this process with aerobic microbial degradation of the metabolites formed may lead to a practical technology for complete removal of PCE.

6.3. Natural Bioremediation

Recently, evidence has been presented for the potential of natural (i.e., biodegradation without external stimulation) bioremediation of organics in both aerobic and anaerobic aquifers and in anaerobic river and lake sediments. The best documented cases in groundwater concern plumes containing aromatic hydrocarbons (i.e.,

BTEX), which have been shown to attenuate in aerobic aquifers (Hadley and Armstrong, 1991; Klec'ka et al., 1990), or in anaerobic groundwater under nitrate-respiring (Wilson et al., 1990, 1993), sulfate-reducing (Acton and Barker, 1992; Thierin et al., 1992), or methanogenic (Godsy et al., 1992; Wilson et al., 1993) conditions. In an attempt to quantify the relative fraction of electron scavenging by the different electron acceptors present, Wilson et al. (1993) found that nitrate respiration was the dominant metabolic process in a BTEX-contaminated aquifer. Evidence for natural biodegradation of chlorinated compounds is limited to anaerobic groundwaters.

6.3.1. Anaerobic Natural Bioremediation

Much remains to be learned about the mechanisms of dechlorination, about which microbes are responsible, and about the physiological conditions that encourage this type of detoxification. However, knowledge gained over the last decade provides a basis for initiating testing of bioremediation schemes. An example of the potential complexity of determining transformation pathways is represented by a TCA-contaminated groundwater supply. Here, a knowledge of TCA transformation pathways under anaerobic and aerobic conditions and a qualitative understanding of the relative transformation rates of TCA and its intermediates is most useful (Fig. 12.1). Detectable concentrations of 1,1-DCE (5 μg/liter) should be observed where a significant concentration of TCA (about 120 μg/liter or more) has been present in water for a period approaching a year or longer, assuming a temperature of about 20°C and no significant advection, dispersion, or sorption. Obviously, advection, dispersion, and sorption will occur and would tend to reduce the apparent 1,1-DCE concentration.

If the microorganisms capable of mediating the reductive dehalogenation of TCA are active, 1,1-DCA would be produced following contamination of groundwater with TCA. Chloroethane (CA) might not be detected if its abiotic transformation rate is faster than the reductive dehalogenation of DCA to CA. Where 1,1-DCE is formed, the presence of vinylchloride might also be observed if proper anaerobic microbial conditions prevail. However, if the initial TCA was about 120 μg/liter and the subsequent concentration of 1,1-DCE formed abiotically within 1 year is about 5 μg/liter, then possible vinylchloride production might be difficult to detect. Since all of these intermediates would be subject to partitioning onto aquifer solids, their concentrations would be further reduced, and perhaps also their activities. This may complicate somewhat the interpretations of measurements made for these contaminants in aquifer interstitial fluid. However, a knowledge of what intermediates to expect and why should be very useful. Different retardation factors governing movement of this group of compounds as a result of their different partition coefficients, analytical detection limits, presence of suitable bacteria, environmental conditions, and time will affect the actual observed co-occurrences. However, the group of contaminants associated with the abiotic and biotic transformation of TCA are commonly found together in groundwater. For example, in a co-occurrence study by the U.S. EPA, 1,1-DCE and 1,1-DCA were noted in many (21% and 36%, respectively) of the same well waters where TCA was detected.

During site characterization of a TCE spill in St. Joseph, Michigan, both *cis*- and *trans*-DCE as well as vinylchloride were found downgradient of and in a larger perimeter around the two main spill areas, indicating that reductive dechlorination had occurred (Keck, 1988). Similarly, *trans*-DCE, vinylchloride, and mono-chlorobenzene have been detected in TCE- and DCE-contaminated groundwater at the Wurtsmith Air Force Base (MI). *trans*-DCE was generally detected at lower depths and in higher concentrations than TCE (Stark et al., 1983). Evidence for *in situ* transformation of PCE to ethene and ethane in the presence of acetate and methanol was reported in anaerobic groundwater below a chemical transfer facility (Toronto, Canada). Though the transformation primarily proceeded to *cis*-DCE, vinylchloride, and ethene, low concentrations of ethane were found in areas contaminated with methanol (Major et al., 1991). Laboratory investigations using core samples taken from the site indicated the predominant presence of methanogens and sulfate-reducing bacteria, and confirmed their ability to transform PCE when acetate and/or methanol were added.

Environmental reductive dechlorination of PCBs has been observed to occur *in situ* in anaerobic freshwater, estuarine, and marine sediments (Brown et al., 1987; Tiedje et al., 1993; Alder et al., 1993). Gas chromoatograms of original Aroclor mixtures were compared with those of the sediment samples and indicated that the highly chlorinated congeners (three to six chlorines) had decreased significantly, with a concurrent increase in lesser chlorinated biphenyls. The rate and extent of dechlorination and the dechlorination pattern have been shown to be site-dependent and are affected by the microbial population involved and by the type of PCB mixture (e.g., Aroclor) present. Thus, Aroclor 1242 (42% chlorine, w/w) present in Hudson River sediment was shown to be dechlorinated to mono-, di-, and tri-chlorobiphenyls. On the contrary, Silver Lake (MA) sediments containing Aroclor 1260 (60% chlorine, w/w) accumulated mainly tri- and tetrachlorinated biphenyls. New Bedford Harbor, an estuarine sediment historically contaminated with Aroclors 1242 and 1254 (54% chlorine, w/w), showed di-, tri-, and tetrachlorobiphenyls accumulating.

6.3.2. Aerobic Natural Bioremediation

Under aerobic conditions, many chlorinated organic contaminants can only be degraded via a co-metabolic process during growth of the microorganism on a primary substrate, resulting in a partial degradation or transformation of the pollutant. Due to their relative abundance in aquifers, methanotrophic bacteria can potentially have a significant influence on catalyzing transformations of xenobiotic organic compounds (Higgins et al., 1980). Aerobic co-metabolic processes generally rely on unspecific enzyme systems that fortuitously oxidize nongrowth substrates. In the case of methanotrophs, the MMO is induced during growth on methane, but is able to oxidize chlorinated solvents such as TCE, DCE, and vinylchloride. Since the organic pollutant is incompletely oxidized, this reaction yields no benefit to the microorganism (Horvath, 1972). Moreover, since the primary substrate and the contaminant compete for the same enzyme (i.e., the same reaction), the rate and extent of co-metabolic transformation is governed by competitive inhibition mechanisms (McCarty, 1988; Adriaens and Focht, 1991). The phenomenon of competitive inhibition by methane in co-metabolism of chlorinated solvents (TCE, DCE), has been shown for all methanotrophic systems examined to date. Hence, depending on the affinity of the enzyme for methane and the chlorinated contaminant (Ks) and on their respective concentrations, methane will interfere more or less efficiently with aryl halide transformation (and vice versa) if both compounds are simultaneously present.

6.4. *Ex Situ* Bioremediation Technologies for Solid Wastes

When the hydrogeological characteristics, the general contaminant parameters, or the zone of contamination is not favorable to implement *in situ* treatment technologies, the soils may have to be treated after excavation and separate containment. Three main categories of *ex situ* solid waste treatment systems are bioslurries, composting, and landfarming technologies (Fig. 12.3E–F). Whereas the latter two are based on spreading the contaminated soil on specially constructed sites, followed by cultivation methods and nutrient addition, bioslurry treatment relies on maintaining an aqueous slurry of the soil. Their applicability is primarily based on the size of the contaminated area, the type of contamination, and the cost to remediate the soil. Many different types of solid waste reactor technologies are being developed or are in the process or field application stage (EPA, 1992b). Even though most field demonstrations have focused on petroleum hydrocarbon decontamination, the basic concepts of these systems should eventually be applicable to the degradation of chlorinated wastes (EPA, 1992b) once the microbial physiological and physical chemical impediments pertaining to these wastes are overcome. Examples where solid phase technologies have been successfully applied are given in Table 12.6.

6.3.3. Bioslurries

Bioreactor (bioslurry) treatment processes are conducted with the waste materials in an aqueous slurry that contains approximately 20%–50% solids. The optimal percent solids is determined based on the types and concentrations of contaminants and on the difficulty of keeping the particles in suspension. Thus, whereas lesser soluble chlorinated organics require a higher ratio of water to solids, soils contaminated with chlorinated solvents are mixed at a 50% ratio (Just and Stockwell, 1991). The bioslurry is amended with hydrogen peroxide and nutrients, seeded with microbial cultures when necessary, and treated as a continuously mixed batch system.

This method is geared toward the treatment of high concentrations of organic contaminants in soils and sludges, and the success of its application is primarily based on vendor claims (EPA, 1992b). The removal efficiency of TCE- and pentachlorophenol (PCP)–contaminated soils via bioslurries has been demonstrated to be in the range of 86%–100% for TCE (ENSITE, 1990) and >99% for PCP (EPA, 1990a). Bioslurry treatment of PCP-contaminated soils was demonstrated in a combined soil washing and screening process (Compeau et al., 1991). The slurry resulting from washing contained 20% solids (less than 60 mesh fraction), was

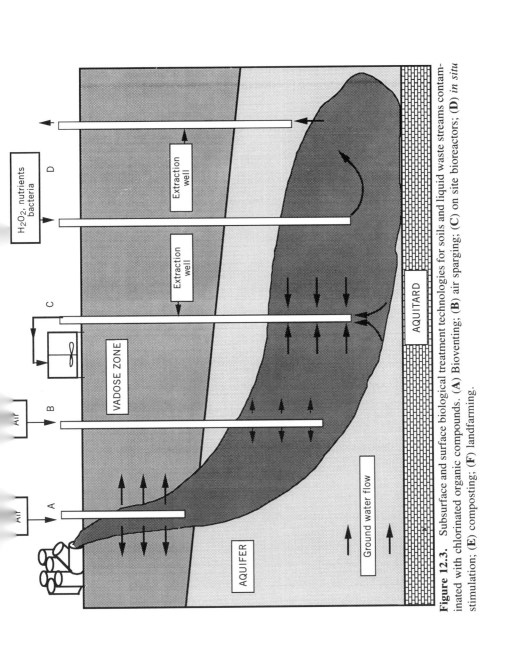

Figure 12.3. Subsurface and surface biological treatment technologies for soils and liquid waste streams contaminated with chlorinated organic compounds. (**A**) Bioventing; (**B**) air sparging; (**C**) on site bioreactors; (**D**) *in situ* stimulation; (**E**) composting; (**F**) landfarming.

(E)

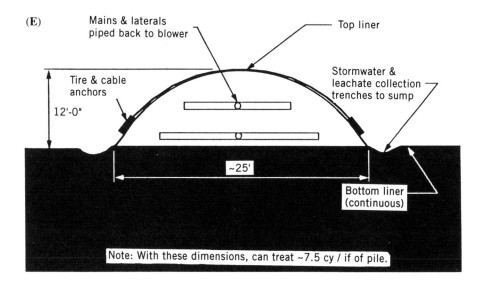

Mains & laterals piped back to blower

Top liner

Tire & cable anchors

Stormwater & leachate collection trenches to sump

12'-0"

~25'

Bottom liner (continuous)

Note: With these dimensions, can treat ~7.5 cy / lf of pile.

(F)

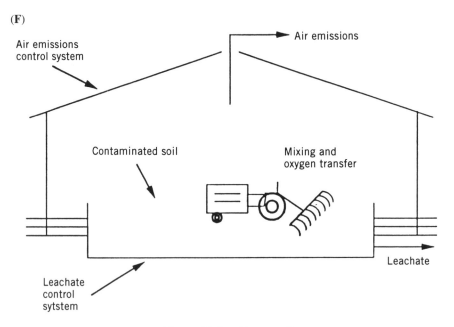

Air emissions control system

Air emissions

Contaminated soil

Mixing and oxygen transfer

Leachate

Leachate control sytstem

Figure 12.3. (*Continued*)

inoculated with a PCP-degrading consortium at 10^7 cells/ml slurry, and amended with nutrients at a C:N:P ratio of 25:8:1. Thus, a removal efficiency of >99% was obtained in 2 weeks.

Bioslurry treatment processes produce soil residues that need to be de-watered, and potential air emissions and wastewaters must be monitored. The solids may still require additional treatment for additional organic or metal contaminants prior to disposal. Moreover, due to extensive mixing, separation, and de-watering systems, and the addition of nutrients, bioslurry remediation systems tend to be more costly than other bioremediation systems (Nyer, 1992).

6.3.4. Composting

Composting processes are typically conducted in long piles (windrows) or in reactors. The contaminated soil is mixed with highly biodegradable bulking material (e.g., wood chips, straw) at a ratio of 9 to 1 (bulking to waste). Aeration is accomplished by turning the windrows/piles or by forced aeration via a grid of perforated pipes. Alternatively, a composting reactor mixes the waste and bulking material by mechanical means. Moisture, nutrients, temperature, and microbial populations are enhanced as found necessary based on pilot-scale studies.

Composting processes have been successfully applied to treat PCP-contaminated wood processing sludges. In a field-scale (50–300 m³) composting study of chlorophenol-contaminated soils, Valo and Salkinoja-Salonen (1986) stimulated indigenous microorganisms to enhance the degradation. To distinguish disappearance by aeration from that by biodegradation, the formation of hydroxylated and methylated metabolites was monitored. Addition of *Rhodococcus chlorophenolicus,* a versatile chlorophenol degrader, did not enhance the degradation of PCP, as it did in bench-scale studies (Middeldorp et al., 1990).

6.3.5. Landfarming

Landfarming has mainly been used for bioreclamation of soils containing different types of mineral oil, and soils contaminated with polycyclic aromatic hydrocarbons and PCP. The basic layout consists of a layered system, where the contaminated soil is spread out over a clean sand drainage layer placed on an impermeable liner. The purpose of the sand layer is to allow for drainage of soil leachates, which are collected separately during the treatment. Over the course of the treatment, the soil will be tilled on a regular basis to promote homogenization and oxygen availability. Generally, the success of this technology will depend on the presence of microorganisms, oxygen, nutrients, and the bioavailability of the compound (Harmsen, 1991).

Bioremediation of soil contaminated with wood-preserving wastes containing creosote and PCP was successfully demonstrated. Landfarming of soils contaminated with 400–1,000 ppm of PCP was demonstrated on an 8 acre plot (Compeau et al., 1991). After pH adjustments, the time required for 50% reduction in PCP was shown to be dependent on the contaminant concentration and varied from 8 to 12 weeks. Hutzler et al. (1989) demonstrated the reduction of PCP by 95% in a 5 month field study. A reduction of 80% was accomplished within 10–20 weeks by Harmsen (1991) using contaminated soil from a small wood-conserving factory on an experimental landfarm. In this study the addition of nutrients or moisture had no effect on the rates of disappearance. The concentration of the PCP in the leachates remained below 1 μg/liter during the entire period and thus required no further treatment. The appearance of degradation products was not monitored during either study.

6.5. *Ex Situ* Technologies for Liquid and Gaseous Wastes

The application of bioreactors for the treatment of liquid domestic wastes is an established technology, and its fundamentals constitute the basis for application on wastes contaminated with chlorinated organics. Two basic design concepts, suspended and fixed-film bioreactors, have addressed the problems of bringing the contaminant in close contact with the bacteria and of either optimizing nutrient addition in general, or oxygen transfer rates in the case of aerobic processes.

Finally, the success of *ex situ* technologies for decontamination of soils and groundwater will be dependent on the efficiency of the processes that deliver the liquid or gaseous waste to the bioreactor, such as pump-and-treat methods (saturated zone), soil vapor extraction (SVE) technologies (unsaturated zone), or soil liquid extraction (surficial soil). Vacuum extraction is the most commonly applied remedial technology for contaminant removal in the unsaturated zone (EPA, 1990a,b, 1992b). Its application for NAPLs with densities greater than water, such as chlorinated compounds, often requires groundwater pump down prior to extraction, since the contaminant may have moved to capillary and saturated zones (Johnson et al., 1990). Volatility of the contaminant (vapor pressure greater than 0.5 mmHg) and soil type (porosity) represent the most important characteristics for determining the effectiveness of SVE. Thus, the average removal of TCE from sandy to clayey soil types was 62% (Pederson and Curtis, 1991). Alternatively, soil washing techniques have been applied for the removal of chlorinated hydrocarbons from soil. Efficiencies of up to 98% have been reported, dependent on the organic carbon content of the soil (EPA, 1990b).

Once the contaminant is physically removed, the waste stream can either be subjected to further physical treatment such as air stripping or carbon sorption combined with biological treatment or treated directly in bioreactors. The reactor configuration depends on the parameters obtained during treatability studies of the chlorinated organic contaminant present and on site-specific considerations.

6.5.1. *Aerobic Fluidized–Fixed Bed Reactors*

Aerobic reactor technology has been used primarily for remediation of groundwaters contaminated with chlorinated solvents such as TCE or aryl halides such as PCP. Immobilized biofilm reactors inoculated with the PCP-degrading *Flavobacterium* strain were shown to treat PCP and creosote-contaminated groundwater effectively to acceptable levels of discharge on a continuous basis (Frick et al., 1988; Pflug and Burton, 1988). The Biotrol Aqueous Treatment System (BATS) first conditions (nutrient addition, pH adjustments) the contaminated well water prior to feeding it in a sequential aerobic reactor system. Thus, a single pass through the reactor results in 99% removal, without the build-up of chlorinated intermediates. Alternatively, PCP-contaminated soils have been treated after excavation and liquid extraction techniques. The process water is then treated using the BAT system. Effectiveness of this technology was shown to be dependent on the extraction efficiency of PCP from the soil (up to 90%, depending on soil organic matter content).

Few reports exist on the treatment of chlorinated solvents in bioreactor systems. Recently, removal of TCE in air from a field-scale groundwater air stripping system was demonstrated in an aerobic biofilm reactor seeded with a TCE-co-metabolizing bacterium. Removal efficiencies averaged 90% (Envirogen, 1992). Full-scale feasibility of chlorinated solvent oxidation using methane-stimulated methanotrophic bacteria such as *Methylosinus trichosporium* has been demonstrated in pilot-scale studies (EPA, 1992a).

6.5.2. Sequential Anaerobic–Aerobic Bioreactor Processes

The full-scale anaerobic biodegradation approach of PCE described earlier constitutes the first step in a full-scale sequential system. The second step involves the aerobic degradation of the partially dechlorinated metabolites. This aerobic process could in principle mirror the technologies discussed for aerobic biotreatment of chlorinated organic compounds. At a site in Germany, Umweltschutz Nord (Warrelman and Schulz-Berendt, 1992) is treating partially dechlorinated PCE-contaminated soil and groundwater with the mixed aerobic microbial communities found at the site. The microorganisms are provided with nutrients and a source of oxygen (e.g., hydrogen peroxide) during soil treatment (composting), water treatment in reactors, or subsurface groundwater treatment. The sequential process was either spatially or temporally separated. For the soil treatment, the process is temporally sequential, by allowing air to diffuse in the composting piles after the anaerobic phase is completed. For water and subsurface groundwater treatment, the process will be spatially sequential as the anaerobic biofilm reactor is followed by an aerobic system.

6.6. Summary

It was demonstrated that the choice of any given type of bioremediation strategy for chlorinated organic compounds, whether *in situ* or *ex situ,* is highly dependent on 1) microbiological parameters, 2) compound-specific characteristics, 3) site-specific characteristics, 4) the nature and location of the spill, and 5) questions of an economic, toxicological, and demographic nature. *Ex situ* technologies of solid and liquid waste streams may, due to their highly controlled nature and potential for process optimization, result in a greater efficiency of removal of the chlorinated organic (Table 12.6). However, this advantage over *in situ* technologies is offset by the higher energy requirements for 1) extracting the solids by means of vapor or liquid extraction technologies, 2) mixing the solids and slurries with nutrients, or 3)

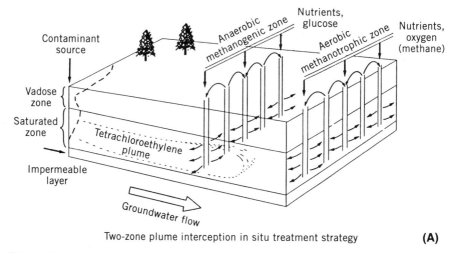

Figure 12.4. Innovative treatment technologies for sites contaminated with highly chlorinated compounds. (**A**) Sequential anaerobic/aerobic *in situ* stimulation; (**B**) Biofilter Trench Design; (**C**) *In Situ* Modular Biocassette. (A, reproduced from EPA, 1992b, with permission of the publisher; C, reproduced from Mayer, 1992, with permission of the publisher.)

displacing the liquid or solid waste by pump-and-treat or soil excavation techniques. The cost of *in situ* technologies, on the other hand, will be primarily determined by the cost and efficiency of supplying oxygen to the contaminated site.

Even though the cost of bioremediation technologies may be favorable relative to physical–chemical methods, often the biological process alone may not be able to reduce the concentrations of the chlorinated organic to the required clean-up level, or at least not within a required time frame (Nyer, 1992). Novel sequential remediation techniques are currently being evaluated to offset these limitations (Fig. 12.4) (The Bioremediation Report, 1992). However, since bioremediation results in the destruction rather than physical displacement of the chlorinated pollutant, requirements for site clean-up may have to be reevaluated. Recently, the U.S. EPA ruled that contaminant removal was an insufficient criterion during bioremediation of PCB-contaminated sites. Instead, the appearance of chlorobenzoate metabolites needed to be demonstrated (EPA, 1992a). Thus, the priorities of required clean-up levels and cost for remediation will need to be evaluated on a site-specific basis in light of the immediate potential for harm to the surrounding population and/or the resources it draws on.

7. CONCLUSIONS

This chapter indicates that bioremediation of solid and liquid wastes contaminated with chlorinated organic compounds has tremendous potential. Its importance as an innovative technology is underscored by the U.S. EPA Bioremediation Field Initia-

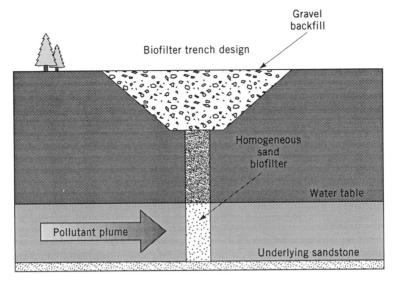

(B)

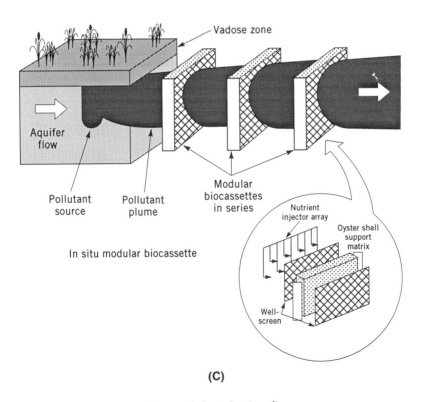

(C)

Figure 12.4. (*Continued*)

tive, which addresses the research needs to help foster implementation of this technology at contaminated sites. As of December 1991, 124 sites were considered for either *in situ* or on site bioremediation, and 10 more were under consideration, most of which are in the predesign stage (Kovalick, 1992). Of these, almost 50% of the sites were contaminated with chlorinated compounds, mainly in the form of solvents or creosote (contains pentachlorophenol). Though encouraging, the application of bioremediation technologies for chlorinated organic compounds is still in its infancy.

Clearly, the potential to employ microorganisms for the degradation of chlorinated organics exists and is currently being developed into practical engineering processes. The relative progress in this area of research has been phenomenal, as in the mid-1970s many scientists believed that highly chlorinated organic compounds would be resistant to biodegradation, even though at that time considerable work had been done on chlorinated compounds such as pesticides (DDT). Most of this work concerned aerobic microorganisms and focused on the mechanisms microorganisms used to circumvent the difficulties of dealing with carbon–chloride bonds. With the observations of anaerobic reductive dechlorination of highly chlorinated compounds, a process more favorable at increasing chlorine numbers, the basic foundation was laid for the complete biotreatment of any chlorinated organic compound. The difficulties currently being faced deal with process development and such issues as bioavailability, concentration-dependent toxicity, relative rates of degradation, and sequential processes. Once these hurdles are overcome with future research, the application of microbial processes should prove increasingly successful with respect to the biotreatment of chlorinated organic compounds.

REFERENCES

Acton DW, Barker JF (1992): *In situ* biodegradation potential of aromatic hydrocarbons in anaerobic ground water. J Contam Hydrol 9: 325–352.

Adriaens P, Focht DD (1990): Continuous coculture degradation of selected polychlorinated biphenyl congeners by *Acinetobacter* spp. in an aerobic reactor system. Environ Sci Technol 24:1042–1049.

Adriaens P, Focht DD (1991): Evidence for inhibitory substrate interactions during cometabolism of 3,4-dichlorobenzoate by *Acinetobacter* sp. strain 4CB1. FEMS Microbiol Ecol 85:293–300.

Adriaens P, Grbić-Galić D (1992): Reductive dehalogenation of highly chlorinated dioxins and dibenzofurans: Retention of the 2,3,7,8-sequence. Dioxin '92 Conference, Tampere, Finland (Abstr.).

Adriaens P, Grbić-Galić D (1994): Cometabolic transformation of mono- and dichloro- and chlorohydroxybiphenols by methanotrophic groundwater isolates. Environ Sci Technol 28:1325–1330.

Adriaens P, Fu Q, Grbić-Galić D (1994): Bioavailability and transformation of highly chlorinated dibenzo-*p*-dioxins and dibenzofurans in anaerobic soils and sediments. Environ Sci Technol (In review).

Adriaens P, Hickey WJ (1993): Physiology of biodegradative microorganisms. In Stoner DL (ed): Biotechnology for Hazardous Waste Treatment. Boca Raton, FL: Lewis Publishers, pp 45–71.

Adriaens P, Huang C-M, Focht D (1991): Biodegradation of PCBs by aerobic microorganisms. In Baker RA (ed): Organic Substances and Sediments in Water. Chelsea, MI: Lewis Publishers, Inc., vol I, pp 311–327.

Adriaens P, Kohler H-PE, Kohler-Staub D, Focht DD (1989): Bacterial dehalogenation of chlorobenzoates and co-culture biodegradation of 4,4′-dichlorobiphenyl. Appl Environ Microbiol 55:887–892.

Alder AC, Haggblom MM, Oppenheimer SR, Young LY (1993): Reductive dechlorination of polychlorinated biphenyls in anaerobic sediments. Environ Sci Technol 27:530–539.

Ahmed M, Focht DD (1973):Degradation of polychlorinated biphenyls by two species of *Achromobacter*. Can J Microbiol 19:47–52.

Anid PJ, Nies L, Vogel TM (1991): Sequential anaerobic–aerobic biodegradation of PCBs in the river model. In Hinchee RF, Olfenbuttel RF (eds): On-Site Bioreclamation. Boston: Butterworth-Heinemann, pp 428–436.

Anid PJ, Ravest-Webster BP, Vogel TM (1993): Effect of hydrogen peroxide on the biodegradation of PCBs in anaerobically dechlorinated river sediments. Biodegradation 4:241–249.

Baehr AL, Hoag GE, Marley MC (1989): Removing volatile contaminants from the unsaturated zone by inducing advective air-phase transport. J Contam Hydrol 4:1–26.

Blackburn JW, Troxler WL, Sayler GS (1984): Prediction of the fates of organic chemicals in a biological treatment process—An overview. Environ Prog 3:163–176.

Bosma TNP, van der Meer JR, Schraa G, Tros ME, Zehnder AJB (1988): Reductive dechlorination of all trichloro- and dichlorobenzene isomers. FEMS Microbiol Ecol 53:223–229.

Bouwer EJ, McCarty PL (1983): Transformation of 1- and 2-carbon halogenated aliphatic organic compounds under methanogenic conditions. Appl Environ Microbiol 45:1286–1294.

Bouwer EJ, McCarty PL (1985): Utilization rates of trace halogenated compounds in acetate-supported biofilms. Biotechnol Bioeng 27:1564–1571.

Boyd SA, Shelton DR (1984): Anaerobic biodegradation of chlorophenols in fresh and acclimated sludge. Appl Environ Microbiol 47:272–277.

Brown JF Jr, Bedard DL, Brennan MJ, Carnahan JC, Feng H, Wagner RE (1987): Polychlorinated biphenyl dechlorination in aquatic sediments. Science 236:709–712.

Brunner W, Sutherland FH, Focht DD (1985): Enhanced biodegradation of polychlorinated biphenyls in soil by analog enrichment. J Environ Q 14:324–328.

Burrows KJ, Cornish A, Scott D, Higgins IG (1984): Substrate specificities of the soluble and particulate methane monooxygenases of *Methylosinus trichosporium* OB3b. J Gen Microbiol 130:3327–3333.

Chakrabarty AM (1982): Genetic mechanisms in the dessimination of chlorinated compounds. In Chakrabarty AM (ed): Biodegradation and Detoxification of Environmental Pollutants. Boca Raton, FL: CRC Press, pp 127–141.

Chapelle F (1993): Ground Water Microbiology and Geochemistry. New York: Wiley Interscience.

Chatterjee DK, Kilbane JJ, Chakrabarty AM (1982): Biodegradation of 2,4,5-trichlorophenoxyacetic acid in soil by a pure culture of *Pseudomonas cepacia*. Appl Environ Microbiol 44:514–516.

Cobb GD, Bower EJ (1991): Effects of electron acceptors on halogenated organic compound biotransformations in a biofilm column. Environ Sci Technol 25:1068–1074.

Cole JR, Cascarelli AL, Mohn WW, Tiedje JM (1994): Isolation and characterization of a novel bacterium growing via reductive dehalogenation of 2-chlorophenol. Appl Environ Microbiol 60:3536–3542.

Compeau GC, MaHaffey WD, Patras L (1991): Full-scale bioremediation of contaminated soil and water. In Sayler GS, Fox R, Blackburn JW (eds): Environmental Biotechnology for Waste Treatment. New York: Plenum Press, pp 91–111.

Copley SD, Crooks GP (1992): Enzymic dehalogenation of 4-chlorobenzoyl coenzyme A in *Acinetobacter* sp. strain 4-CB1. Appl Environ Microbiol 58:1385–1387.

Crawford RL, Mohn WW (1985): Microbial removal of pentachlorophenol from soil using a *Flavobacterium*. Enzyme Microbiol Technol 7:617–620.

de Bruin WP, Kotterman MJJ, Posthumus MA, Schraa G, Zehnder AJB (1992): Complete biological reductive transformation of tetrachloroethene to ethane. Appl Environ Microbiol 58:1996–2000.

DeWeerd KA, Bedard DL (1993): Stimulation of anaerobic microbial dechlorination of PCBs with halogenated benzoic acids in Woods Pond sediment slurries. In Annual Meeting of the American Society for Microbiology. Washington, DC: American Society for Microbiology, abstract Q147.

Dolan ME, McCarty PL (1994): Factors affecting transformation of chlorinated aliphatic hydrocarbons by methanotrophs. In Hinchee RE, Leeson A, Semprini L, Ong SK (eds): Bioremediation of Chlorinated Polycyclic Aromatic Hydrocarbon Compounds. Ann Arbor, MI: Lewis Publishers, pp 303–309.

Dolfing J (1990): Reductive dechlorination of 3-chlorobenzoate is coupled to ATP production and growth in an anaerobic bacterium, strain DCB-1. Arch Microbiol 153:264–266.

Dolfing J, Harrison BK (1992): Gibbs free energy of formation of halogenated aromatic compounds and their potential role as electron acceptors in anaerobic environments. Environ Sci Technol 26:2213–2218.

Dooley-Dana M, Fogel S, Findley M (1989): Abstracts of the International Association of Water Pollution Research and Control. p A14.

Dorn EM, Knackmuss H-J (1978a): Chemical structure and biodegradability of halogenated aromatic compounds: Two catechol 1,2-dioxygenases from a 3-chlorobenzoate grown pseudomonad. J Biochem 174:73–84.

Dorn EM, Knackmuss H-J (1978b): Chemical structure and biodegradability of halogenated aromatic compounds: Substituent effects on 1,2-dioxygenation of catechol. J Biochem 174:85–94.

Dupont RR, Doucette WJ, Hinchee RE (1992): Assessment of *in situ* bioremediation potential and the application of bioventing at a fuel-contaminated site. In Hinchee RE and Olfenbuttel RF (eds): *In Situ* Bioremediation. Boston: Butterworth-Heinemann, pp 262–282.

Edgehill RU, Finn RK (1983a): Activated sludge treatment of synthetic wastewater containing pentachlorophenol. Biotechnol Bioeng 25:2165–2176.

Edgehill RU, Finn RK (1983b): Microbial treatment of soil to remove pentachlorophenol. Appl Environ Microbiol 45:1122–1125.

ENSITE (1990): Case history on Channel Gateway Development site. Report on Pilot testing and Full Scale Remediation.

ENVIROGEN News, October 28, 1992, Lawrenceville, NJ.

EPA (1990a): EPA/540/2-90/016, Engineering Bulletin—Slurry Biodegradation. Washington, DC: EPA.

EPA (1990b): EPA/540/2-90/017, Engineering Bulletin—Soil Washing Treatment. Washington, DC: EPA.

EPA (1990c): EPA/540/2-90/011 Subsurface Contamination Reference Guide. Washington, DC: EPA.

EPA (1991): EPA/540/2-91/027, Bioremediation in the Field, No. 4. Washington, DC: EPA.

EPA (1992): EPA/540/N-92/004, Bioremediation in the Field, No. 7. Washington, DC: EPA.

EPA (1992b): EPA/540/R-92/077, The Superfund Innovative Technology Evaluation Program: Technology Profiles, 5th ed. Washington, DC: EPA.

Fathepure BZ, Tiedje JM, Boyd SA (1988): Reductive dechlorination of hexachlorobenzene to tri- and dichlorobenzenes in anaerobic sewage sludge. Appl Environ Microbiol 54:327–330

Fathepure BZ, Vogel TM (1991): Complete degradation of polychlorinated hydrocarbons by a two-stage biofilm reactor. Appl Environ Microbiol 57:3418–3422.

Focht DD (1988): Performance of biodegradative microorganisms in soil: Xenobiotic chemicals as unexploited metabolic niches. In Omenn GS (ed): Environmental Biotechnology: Reducing Risks From Environmental Chemicals Through Biotechnology. New York: Plenum Press. pp 15–29.

Focht DD, Shelton D (1987): Growth kinetics of *Pseudomonas alcaligenes* C–O relative to inoculation and 3-chlorobenzoate metabolism in soil. Appl Environ Microbiol 53:1846–1849.

Focht DD, Brunner W (1985): Kinetics of biphenyl and polychlorinated biphenyl metabolism in soil. Appl Environ Microbiol 50:1058–1063.

Fox BG, Lipscomb JD (1990): Methane monooxygenase: A novel biological catalyst for hydrocarbon oxidation In Hamilton G, Reddy C, Madyastha KM (eds): Biological Oxidation Systems. Orlando, FL: Academic Press, vol 1, pp 367–388.

Freedman DL, Gossett JM (1991): Biodegradation of dichloromethane in a fixed-film reactor under methanogenic conditions. In Hinchee RE, Olfenbuttel RF (eds): On-Site Bioreclamation Processes for Xenobiotic and Hydrocarbon Treatment. Boston: Butterworth-Heinemann, pp 113–134.

Frick TD, Crawford RL, Martinson M, Chresand T, Bateson G (1988): Microbiological cleanup of groundwater contaminated by pentachlorophenol. In Omenn GS (ed): Environmental Biotechnology: Reducing Risks From Environmental Chemicals Through Biotechnology. New York: Plenum Press, pp 173–191.

Furukawa K (1982): Microbial degradation of polychlorinated biphenyls (PCB's). In Chakrabarty AM (ed): Biodegradation and Detoxification of Environmental Pollutants. CRC Press, pp 33–57.

Gibson SA, Suflita JM (1986): Extrapolation of biodegradation results to groundwater aquifers: Reductive dehalogenation of aromatic compounds. Appl Environ Microbiol 56:1825–1832.

Glaser JA, Lamar RL, Davis MW, Dietrich DM (1993): Application of wood-degrading fungi to treat contaminated soils. In Symposium on Bioremediation of Hazardous Wastes:

Research, Development, and Field Evaluations. EPA/600/R-93/054. Washington, DC: EPA.

Godsy EM, Goerlitz DF, Grbić-Galić D (1992): Methanogenic degradation kinetics of phenolic compounds in aquifer-derived microcosms. Biodegradation 2:211–221.

Grbić-Galić D (1990): Anaerobic microbial transformation of nonoxygenated aromatic and alicyclic compounds in soil, subsurface, and freshwater sediments. In Bollag JM, Stotzky G (eds): Soil Biochemistry. New York: Marcel Dekker, vol 6, pp 117–189.

Green J, Dalton H (1989): Substrate specificity of soluble methane monooxygenase, mechanistic implications. J Biol Chem 264:17698–17703.

Guthrie MA, Kirsch EJ, Wukasch RF, Grady CPL Jr (1984): Pentachlorophenol biodegradation II: Anaerobic. Water Res 18:451–461.

Hadley PW, Armstrong R (1991): Where's the benzene—Examining California ground water quality surveys. Ground Water 29:35–40.

Häggblom MM (1992): Microbial breakdown of halogenated aromatic pesticides and related compounds. FEMS Microbiol Rev 103:29–72.

Häggblom MM, Rivera MD, Young LY (993): Influence of alternative electron acceptors on the anaerobic biodegradability of chlorinated phenols and benzoic acids. Appl Environ Microbiol 59:1162–1167.

Hakulinen R, Salkinoja-Salonen M (1982): Treatment of pulp and paper industry wastewaters in an anaerobic fluidised bed reactor. Process Biochem 17:18–22.

Harkness MR, McDermott JB, Abramowicz DW, Salvo JJ, Flanagan WP, Stephens ML, Mondello FJ, May RJ, Lobos JH, Carroll KM, Brennan MJ, Bracco AA, Fish KM, Warner GL, Wilson PR, Dietrich DK, Lin DT, Morgan CB, Gately WL (1993): *In situ* stimulation of aerobic PCB biodegradation in Hudson River sediments. Science 259:503–507.

Harmsen J (1991): Possibilities and limitations of landfarming for cleaning contaminated soils. In Hinchee RE, Olfenbuttel RF (eds): On-Site Bioreclamation. Boston: Butterworth-Heinemann, pp 255–272.

Henry SM, Grbić-Galić D (1990): Effect of mineral media on trichloroethylene oxidation by aquifer methanotrophs. Microbiol Ecol 20:151–169.

Henry SM, Grbić-Galić D (1991): Influence of endogenous and exogenous electron donors and trichloroethylene oxidation toxicity on trichloroethylene oxidation by methanotrophic cultures from a groundwater aquifer. Appl Environ Microbiol 57:236–244.

Hernandez BS, Higson FK, Kondrat R, Focht DD (1991): Metabolism of and inhibition by chlorobenzoates in *Pseudomonas putida* P111. Appl Environ Microbiol 57:3361–3366.

Hickey WJ, Focht DD (1990): Degradation of mono-, di-, and trihalogenated benzoic acids by *Pseudomonas aeruginosa* JB2. Appl Environ Microbiol 56:3842–3850.

Hickey WJ, Searles DB, Focht DD (1993): Enhanced mineralization of polychlorinated biphenyls in soil inoculated with chlorobenzoate-degrading bacteria. Appl Environ Microbiol 59:1194–1200.

Hicks DD, Reddell DL, Coble CG (1991): Methanotrophic removal of TCE in an aquifer simulator. In Hinchee RE, Olfenbuttel RF, (eds): On-Site Bioreclamation. Boston: Butterworth-Heinemann, pp 437–443.

Higgins IJ, Best DJ, Hammond RC (1980): New findings in methane-utilizing bacteria highlight their importance in the biosphere and their commercial potential. Nature 286:561–564.

Hinchee RE (ed) (1994): Air Sparging for Site Remediation. Ann Arbor, MI: Lewis Publishers.

Hinchee RE, Leeson A, Semprini L, Ong SK (eds) (1994a): Bioremediation of chlorinated and polycyclic aromatic hydrocarbon compounds. Ann Arbor, MI: Lewis Publishers.

Hinchee RE, Olfenbuttel RF (eds) (1992a): *In Situ* Bioremediation. Boston: Butterworth-Heinemann.

Hinchee RE, Olfenbuttel RF (eds) (1992b): On Site Bioremediation. Boston: Butterworth-Heinemann.

Hinchee RE, Anderson DB, Metting FB Jr, Sayles GD (eds) (1994b): Applied Biotechnology for Site Remediation. Ann Arbor, MI: Lewis Publishers.

Hinchee RE, Downey DC, Dupont RE, Aggarwal PK, Miller RN (1991): Enhancing biodegradation of petroleum hydrocarbons through soil venting. J Hazard Materials 27:315–325.

Hopkins G, Semprini L, McCarty PL (1993): Field evaluation of phenol for cometabolism of chlorinated solvents. In Abstracts of Symposium on Bioremediation of Hazardous Wastes: Research, Development, and Field Evaluations. EPA/600/R-93/054.

Horvath RS (1972): Microbial cometabolism and the degradation of organic compounds in nature. Bacteriol Rev 36:146–155.

Hrudey SE, Knettig E, Daignault SA, Fedorak PM (1987): Anaerobic biodegradation of monochlorophenols. Environ Technol Lett 8:65–76.

Hutzler NJ, Baillod CR, Schaepe PA (1989): Biological reclamation of soils contaminated with pentachlorophenol. In Proceedings of the Sixth National Conference on Hazardous Wastes and Hazardous Materials, p 361.

Ide A, Niki F, Sakamoto F, Watanabe I, Watanabe H (1972): Decomposition of pentachlorophenol in paddy soil. Agric Biol Chem 36:1937–1944.

Janssen DB, van den Wijngaard AJ, van der Waarde JJ, Oldenhuis R (1991): Biochemistry and kinetics of aerobic degradation of chlorinated aliphatic hydrocarbons. In Hinchee RE, Olfenbuttel RF (eds): On-Site Bioreclamation. Boston: Butterworth-Heinemann, pp 92–113.

Johnson PC, Kemblowski MW, Colthart JD (1990): Quantitative analysis for the cleanup of hydrocarbon-contaminated soils by *in-situ* soil venting. Ground Water 28:403–412.

Just SR, Stockwell KJ (1991): A comparison of the effectiveness of emerging *in-situ* technologies and traditional *ex-situ* treatment of solvent-contaminated soils. In Tedder JW (ed): Emerging Technologies for Hazardous Waste Management.

Kampbell DH, Wilson JT (1991): Bioventing to treat fuel spills from underground storage tanks. J Hazard Materials 28:75–80.

Kaneko M, Morimoto K, Nambu S (1976): The response of activated sludge to a polychlorinated biphenyl (KC-500). Water Res 10:157–163.

Karns JS, Kilbane JJ, Duttagupta S, Chakrabarty AM (1983): Metabolism of halophenols by 2,4,5-trichlorophenoxyacetic acid-degrading *Pseudomonas cepacia*. Appl Environ Microbiol 46:1176–1181.

Khodadoust AP, Wagner JA, Suidan MT, Safferman SI (1993): Treatment of PCP-contaminated soils by washing with ethanol/water followed by anaerobic treatment. In Symposium on Bioremediation of Hazardous Wastes: Research, Development, and Field Evaluations, EPA/600/R-93/054. Washington, DC: EPA.

Kilbane JJ, Chatterjee DK, Chakrabarty AM (1983): Detoxification of 2,4,5-

trichlorophenoxyacetic acid from contaminated soil by *Pseudomonas cepacia.* Appl Environ Microbiol 45:1697–1700.

Klec'ka GM, Davis JW, Gray DR, Madsen SS (1990): Natural bioremediation of organic contaminants in ground water: Cliffs-Dow Superfund Site. Ground Water 28:534–543.

Kohler-Staub D, Kohler H-PE (1989): Microbial degradation of b-chlorinated four-carbon aliphatic acids. J Bacteriol 171:1428–1434.

Kohler-Staub D, Leisinger T (1985): Dichloromethane dehalogenase of *Hyphomicrobium* sp. strain DM2. J Bacteriol 162:676–681.

Kovalick WW (1992): Use of innovative technologies (especially bioremediation) in the U.S. Preprints of International Symposium on Soil Contamination Using Biological Processes (Karlsruhe, Germany). Frankfurt am Main, Germany: Schön and Wetzel GmbH, pp 259–266.

Krumme ML, Boyd SA (1988): Reductive dechlorination of chlorinated phenols in anaerobic upflow bioreactors. Water Res 22:171–177.

Kuhn EP, Colberg PJ, Schnoor JL, Wanner O, Zehnder AJB, Schwartzenbach RP, (1985): Microbial transformation of substituted benzenes during infiltration of river water to ground water: Laboratory column studies. Environ Sci Technol 19:961–967.

Kuhn EP, Suflita JM (1989): Dehalogenation of pesticides by anaerobic microorganisms in soils and groundwater—A review. In Sawhney BL, Brown K (eds): Reactions and Movement of Organic Chemicals in Soils. Madison, WI: Soil Science Society of America and American Society for Agronomy, pp 111–180.

Lajoie CA, Chen S-Y, Oh K-C, Strom PF (1992): Development and use of field application vectors to express nonadaptive foreign genes in competitive environments. Appl Environ Microbiol 58:655–663.

Lajoie CA, Zylstra GJ, DeFlaun MF, Strom PF (1993): Field application vectors for bioremediation of soils contaminated with polychlorinated biphenyls. Appl Environ Microbiol 59:1749–1751.

Lamar RT, Dietrich DM (1990): *In situ* depletion of pentachlorophenol from contaminated soil by *Phanerochaete* spp. Appl Environ Microbiol 56:3093–3100.

Lamar RT, Glaser JA, Kirk TK (1990): Fate of pentachlorophenol (PCP) in sterile soils inoculated with *Phanerochaete crysosporium:* Mineralization, volatilization, and depletion of PCP. Soil Biol Biochem 22:433–440.

Lee MD, Thomas JM, Borden RC, Bedient PB, Ward CH, Wilson JT (1988): Biorestauration of aquifers contaminated with organic compounds. CRC Crit Rev Environ Control 18:29–89.

Major DW, Hodgkins EW, Butler BJ (1991): Field and laboratory evidence of in situ biotransformation of tetrachloroethene to ethene at a chemical transfer facility in North Toronto. In Hinchee RE, Olfenbuttel RF (eds): On-Site Bioreclamation. Boston: Butterworth-Heinemann, pp 147–171.

Marks TS, Wait R, Smith ARW, Quirk AV (1984): The origin of the oxygen incorporated during the dehalogenation/hydroxylation of 4-chlorobenzoate by an *Arthrobacter* sp. Biochem Biophys Res Comm 124:669–674.

Mayer WS (ed) (1982): The Bioremediation Report 1. Santa Rosa, CA: Cognis, Inc.

McCarty PL (1988): Bioengineering issues related to *in situ* remediation of contaminated soils and groundwater. In Omenn GS (ed): Environmental Biotechnology: Reducing Risks From Environmental Chemicals Through Biotechnology. New York: Plenum Press, pp 143–162.

McFarland MJ, Sims RC, Blackburn JW (1991): Use of treatability studies in developing remediation strategies for contaminated soils. In Sayler GS, Fox R, Blackburn JW (eds): Environmental Biotechnology for Waste Treatment. New York: Plenum Press, pp 163–175.

Middeldorp PJM, Briglia M, Salkinoja-Salonen MS (1990): Biodegradation of pentachlorophenol in natural soil by inoculated *Rhodococcus chlorophenolicus*. Microbiol Ecol 20:123–139.

Mikesell MD, Boyd SA (1988): Enhancement of pentachlorophenol degradation in soil through enhanced anaerobiosis and bioaugmentation with anaerobic sewage sludge. Environ Sci Technol 22:1411–1414.

Miller RN, Vogel CC, Hinchee RE (1992): A field-scale investigation of petroleum hydrocarbon biodegradation in the vadose zone enhanced by soil venting at Tyndall AFB, Florida. In Hinchee RE, Olfenbuttel RF (eds): *In Situ* Bioremediation. Boston: Butterworth-Heinemann, pp 282–302.

Mohn WW, Tiedje JT (1992): Microbial reductive dechlorination. Microbiol Rev 56:482–507.

Morgan P, Watkinson RJ (1989): Microbiological methods for the cleanup of soil and groundwater contaminated with halogenated organic compounds. FEMS Microbiol Rev 63:277–300.

Mueller JG, Lantz SE, Blattmann BO, Chapman PJ (1991a): Bench-scale evaluation of alternative biological treatment processes for the remediation of pentachlorophenol- and creosote-contaminated materials: Solid-phase bioremediation. Environ Sci Technol 25:1045–1055.

Mueller JG, Lantz SE, Blattmann BO, Chapman PJ (1991b): Bench-scale evaluation of alternative biological treatment processes for the remediation of pentachlorophenol- and creosote-contaminated materials: Slurry-phase bioremediation. Environ Sci Technol 25:1055–1061.

Mueller JG, Lantz SE, Ross D, Colvin RJ, Middaugh DP, Pritchard PH (1993): Strategy using bioreactors and specially selected microorganisms for bioremediation of groundwater contaminated with creosote and pentachlorophenol. Environ Sci Technol 27:691–698.

Neilson AH (1990): The biodegradation of halogenated organic compounds. J Appl Bacteriol 69:445–470.

Nelson MJK, Pritchard PH, Bourquin AW (1988): Preliminary development of a bench-scale treatment system for aerobic degradation of trichloroethylene. In Omenn GS (ed): Environmental Biotechnology: Reducing Risks From Environmental Chemicals Through Biotechnology. New York: Plenum Press, pp 203–209.

Nies L, Vogel TM (1990): Effects of organic substance on dechlorination of Aroclor 1242 in anaerobic sediments. Appl Environ Microbiol 56:2612–2617.

Nies L, Vogel TM (1991): Identification of the proton source for the microbial reductive dechlorination of 2,3,4,5,6-pentachlorobiphenyl. Appl Environ Microbiol 57:2771–2774.

Nies LF (1993): Microbial and Chemical Reductive Dechlorination of Polychlorinated Biphenyls and Chlorinated Benzenes. Ph.D. Dissertation, The University of Michigan.

Norris RD, Hinchee RE, Brown R, McCarty PL, Semprini L, Wilson JT, Kampbell DH, Reinhard M, Bouwer EJ, Borden RC, Vogel TM, Thomas JM, Ward CH (1994): Handbook of Bioremediation. Ann Arbor MI: Lewis Publishers.

Nyer EK (1992): Groundwater Treatment Technology, 2nd ed. New York: Van Nostrand Reinhold.

Oldenhuis R, Ruud LJM, Vink JM, Janssen DB, Witholt B (1989): Degradation of chlorinated aliphatic hydrocarbons by *Methylosinus trichosporium* OB3b expressing soluble methane monooxygenase. Appl Environ Microbiol 55:2819–2826.

Pedersen TA, Curtis JT (1991): Soil vapor extraction technology. Pollution Technology Review, No 204. Park Ridge, NJ: Noyes Data Corporation.

Petrovskis EA, Vogel TM, Adriaens P (1994): Effects of electron acceptors and donors on transformation of tetrachloromethane by *Shewanella putrefaciens* MR1. FEMS Microbiol Lett 121:357–364.

Pflug AD, Burton MB (1988): Remediation of multimedia contamination from the wood-preserving industry. In Omenn GS (ed): Environmental Biotechnology: Reducing Risks From Environmental Chemicals Through Biotechnology. New York: Plenum Press, pp 193–201.

Phelps TJ, Niedzielski JJ, Schram RM, Herbes SE, White DC (1990): Biodegradation of trichloroethylene in continuous-recycle expanded-bed bioreactors. Appl Environ Microbiol 56:1702–1709.

Quensen JF III, Boyd SA, Tiedje JM (1990): Dechlorination of four commercial polychlorinated biphenyl mixtures (Aroclors) by anaerobic microorganisms from sediments. Appl Environ Microbiol 56:2360–2369.

Quensen JF III, Tiedje JM, Boyd SA (1988): Reductive dechlorination of polychlorinated biphenyls by anaerobic microorganisms from sediments. Science 242:752–754.

Reanney DC, Gowland PC, Slater HJ (1983): Genetic interactions among microbial communities. In Slater JH, Wittenbury R, Wimpenney JWT (eds): Microbes in Their Natural Environments. Cambridge: Cambridge University Press, pp 379–421.

Reineke W (1984): Microbial degradation of halogenated aromatic compounds. In Gibson DT (ed): Microbial Degradation of Organic Compounds. New York: Marcel Dekker, pp 319–360.

Rittmann BE, McCarty PL (1980): Utilization of dichloromethane by suspended and fixed-film bacteria. Appl Environ Microbiol 39:1225–1226.

Rittmann BE, Valocchi AJ, Odencrantz JE, Bae W (1988): *In-situ* Bioremediation of Contaminated Ground Water. Illinois Hazardous Waste Research Information Center, HWRIC RR 031.121.

Roberts PV, Hopkins GD, Mackay DM, Semprini L (1990): A field evaluation of *in situ*-biodegradation of chlorinated ethenes: Part 2, results of biostimulation and biotransformation experiments. Ground Water 28:591–604.

Ryan TP, Bumpus JA (1989): Biodegradation of 2,4,5-trichlorophenoxyacetic acid in liquid culture and in soil by the white rot fungus *Phanerochaete crysosporium*. Appl Microbiol Biotechnol 31:302–307.

Salkinoja-Salonen M, Valo R, Apajalahti J, Hakulinen R, Silakoski R, Jaakkola T (1984): Biodegradation of chlorophenolic compounds in wastes from the wood processing industry. In Klug MJ, Reddy CA (eds): Current Perspective in Microbial Ecology. Washington DC: American Society for Microbiology, pp 668–676.

Sayler GS, Hopper SW, Layton AC, King JMH (1990): Catabolic plasmids of environmental and ecological significance. Microbiol Ecol 19:1–21.

Scholten JD, Chang KH, Babbitt PC, Charest H, Sylvestre M, Dunaway-Mariano D (1991):

Novel enzymic hydrolytic dehalogenation of a chlorinated aromatic. Science 253:182–185.

Seech AG, Trevors JT, Bulman TL (1991): Biodegradation of pentachlorophenol in soil: The response to physical, chemical, and biological treatments. Can J Microbiol 37:440–444.

Semprini L, Grbić-Galić D, McCarty PL, Roberts PV (1992): Methodologies for Evaluating *In-Situ* Bioremediation of Chlorinated Solvents, EPA/600/R-92/042. Washington, DC: EPA.

Semprini L, Roberts PV, Hopkins GD, McCarty PL (1990): A field Evaluation of *In-Situ* Biodegradation of Chlorinated Ethenes: Part 2, results of biostimulation and biotransformation experiments. Ground Water 28:715–727.

Shields MS, Reagin MR (1992): Selection of a *Pseudomonas cepacia* strain constitutive for the degradation of trichloroethylene. Appl Environ Microbiol 58:3977–3983.

Shields MS, Reagin MR, Gerger R, Schaubhut R, Campbell R, Somerville C, Pritchard PH (1993): Field demonstration of a constitutive TCE-degrading bacterium for the bioremediation of TCE. In Symposium on Bioremediation of Hazardous Wastes: Research, Development, and Field Evaluations. EPA/600/R-93/054. Washington, DC: EPA.

Slater JH, Lovatt D, Weightman AJ, Senior E, Bull AT (1979): The growth of *Pseudomonas putida* on chlorinated aliphatic acids and its dehalogenase activity. J Gen Microbiol 114:125–136.

Speece RE (1983): Environ Sci Technol 17:416A–427A.

Speitel GE Jr, Lu CJ, Turakhia M, Zhu X-J (1989): Biodegradation of trace concentrations of substituted phenols in granular activated carbon columns. Environ Sci Technol 23:68–74.

Stark JR, Cummings TR, Twenter FR (1983): Groundwater Contamination at Wurtsmith Air Force Base, Michigan. U.S. Geological Survey: Water Resources Investigations, Report 83-4002.

Stirling DI, Dalton H (1981): Fortuitous oxidations by methane-utilizing bacteria. Nature 291:169–170.

Stucki G (1990): Biological decomposition of dichloromethane from a chemical process effluent. Biodegradation 1:221–228.

Tabak HH, Glaser JA, Strohofer S, Kupferle MJ, Scarpino P, Tabor MW (1991): Characterization and optimization of organic wastes and toxic organic compounds by a lignolytic white rot fungus in bench-scale bioreactors. In Hinchee RE, Olfenbuttel RF (eds): On-Site Bioreclamation. Boston: Butterworth-Heinemann, pp 341–365.

Thierrin J, Davis GB, Barber C, Patterson BM, Pribac F, Power TR, Lambert M (1992): Natural degradation rates of BTEX compounds and naphthalene in a sulfate reducing ground water environment. Abstracts of *In-Situ* Bioremediation Symposium '92, Niagara-on-the-Lake, Canada.

Tiedje JM, Quensen JF III, Chee-Sanford J, Schimmel JP, Boyd SA (1993): Microbial reductive dechlorination of PCBs. Biodegradation 4:231–241.

Tiedje JM, Boyd SA, Fathepure BZ (1987): Anaerobic degradation of chlorinated aromatic hydrocarbons. Dev Ind Microbiol 27:117–127.

Topp E, Hanson RS (1990): Factors influencing the survival and activity of a pentachloro-phenol-degrading *Flavobacterium* sp. in soil slurries. Can J Soil Sci 70:83–91.

Toussaint M, Commandeur LCM, Parsons JR (1992): Reductive dechlorination of 1,2,3,4-tetrachloro-dibenzo-p-dioxin by a bacterial consortium isolated from Lake Ketelmeer sediment: Preliminary results. Preprints of International Symposium on Soil Decontamination Using Biological Processes. Frankfurt am Main, Germany: Schön and Wetzel GmbH, pp 578–585.

Tucker ES, Saeger VW, Hicks O (1975): Activated sludge primary biodegradation of polychlorinated biphenyls. Bull Environ Contam Toxicol 14:705–713.

Valo RJ, Häggblom MM, Salkinoja-Salonen MS (1990): Bioremediation of chlorophenol-containing simulated ground water by immobilized bacteria. Water Res 24:253–258.

Valo RJ, Salkinoja-Salonen M (1986): Bioreclamation of chlorophenol-contaminated soil by composting. Appl Microbiol Biotechnol 25:68–75.

van der Meer JR, de Vos WM, Harayama S, Zehnder AJB (1992a): Molecular mechanisms of genetic adaptation to xenobiotic compounds. Microbiol Rev 56:677–694.

van der Meer JR, Bosma TNP, de Bruin WP, Harms H, Holliger C, Rijnaarts HHM, Tros ME, Schraa G, Zehnder AJB (1992b): Versatility of soil column experiments to study biodegradation of halogenated compounds under environmental conditions. Biodegradation 3:265–285.

van der Meer JR, Roelofsen W, Schraa G, Zehnder AJB (1987): Degradation of low concentrations of dichlorobenzenes and 1,2,4-trichlorobenzene by *Pseudomonas* sp. P51 in nonsterile soil columns. FEMS Microbiol Ecol 45:333–341.

Vogel TM, Criddle CS, McCarty PL (1987): Transformations of halogenated aliphatic compounds. Environ Sci Technol 21:722–736.

Vogel TM, McCarty PL (1985): Biotransformation of tetrachloroethylene to trichloroethylene, dichloroethylene, vinyl chloride, and carbon dioxide under methanogenic conditions. Appl Environ Microbiol 49:1080–1083.

Warrelman J, Schulz-Berendt V (1992): Development of biological technologies for degradation of CKW contaminants at the model location Eppelheim. Preprints of International Symposium on Soil Contamination Using Biological Processes (Karlsruhe, Germany). Frankfurt am Main, Germany: Schön and Wetzel GmbH, pp 806–810.

Westrick JJ, Mello JW, Thomas RG (1984): The ground water supply survey. J Am Water Works Assoc 76:52–59.

Wilson BH, Wilson JT, Kampbell DH, Bledsoe BE, Armstrong JM (1990): Biotransformation of monoaromatic and chlorinated hydrocarbons at an aviation gasoline spill site. J Geomicrobiol 8:225–240.

Wittich RM (1992): Aerobic and anaerobic degradation and transformation/dehalogenation of low-chlorinated dibenzofurans and dibenzo-*p*-dioxins. Preprints of International Symposium on Soil Decontamination Using Biological Processes. Frankfurt am Main, Germany: Schön and Wetzel GmbH, pp 284–291.

Woods SL, Ferguson JF, Benjamin MM (1989): Characterization of chlorophenol and chloromethoxybenzene degradation during anaerobic treatment. Environ Sci Technol 23:62–68.

Ye D, Quensen JF III, Tiedje JM, Boyd SA (1992): Anaerobic dechlorination of polychlorobiphenyls (Aroclor 1242) by pasteurized and ethanol-treated microorganisms from sediments. Appl Environ Microbiol 58:1110–1114.

13

FIELD TREATMENT OF COAL TAR-CONTAMINATED SOIL BASED ON RESULTS OF LABORATORY TREATABILITY STUDIES

MARGARET FINDLAY
SAMUEL FOGEL
Bioremediation Consulting, Inc., Newton, Massachusetts 02159

LEE CONWAY
Northeast Utilities Service Company, Hartford, Connecticut 06141

ARTHUR TADDEO
Unitec Environmental Consultants, Inc., Honolulu, Hawaii 96819

1. INTRODUCTION

Coal tar is a byproduct of the coal gasification process used widely between 1820 and 1950 to produce a natural gas substitute known as *town gas*. It has been estimated that during that time 11 billion gallons were produced (Eng, 1985), a large portion of which was discarded as a waste product at more than 1,500 gas manufacturing facilities located throughout the United States. Coal tar contains monoaromatics and polynuclear aromatic hydrocarbons (PAHs), which have been

Microbial Transformation and Degradation of Toxic Organic Chemicals, pages 487–513
© 1995 Wiley-Liss, Inc.

placed on the EPA priority pollutant list (U.S. GPO, 1978) because of their toxicity, mutagenicity, and tendency to bioconcentrate (Anderson, 1979; Atlas, 1981; Wolfe, 1977).

This chapter concerns a site at which coal tar had been released to a wetland and had been subsequently covered with aquatic sediment. Although the hydrocarbon constituents of coal tar are biodegradable, natural biodegradation did not occur at this site since most of the tar was located in a concentrated submerged hydrophobic strata that was apparently impervious to the necessary oxygen. Since these conditions would also make *in situ* treatment difficult, it was decided that excavation and biological treatment by land application would be the most cost-effective approach to remediation. The work described here includes the laboratory treatability study that defined the biodegradation process and determined the optimal conditions for treatment of the site material, followed by the field pilot project, which demonstrated the full-scale process and documented the extent of biodegradation.

2. ANALYTICAL METHODS FOR TREATABILITY AND FIELD SAMPLES

Percent humus was determined as weight loss at 400°C after extraction with methylene chloride. Grain size analysis was carried out according to methods described by the American Society for Testing and Materials (1981). Suspended solids were analyzed according to EPA Method 160.2

Soils were prepared for metal analysis according to EPA Method 3050, and metals were analyzed as follows: Hg, Method 245.1; Pb, Method 239.2; Zn, Cr, Cu, Ni, Method 200.7; As, Method 206.2; Cd, Method 213.2. Leachate and run-off samples were analyzed for metals according to EPA Method 600-4/79-020 (U.S. EPA, 1979). During the treatability study, soil was prepared for N and P analysis by EPA Method 3050, and analysis for nitrate N, ammonia N, and P were done according to EPA Methods 353.3, 350.2 and 365.3, respectively (U.S. EPA, 1979, 1982). During the field study, nitrogen analyses were carried out on a water extract of the soil. Ammonia nitrogen was analyzed according to EPA Method 350.2 or by field test kit model SL-AN (La Motte Chemical Co, Chestertown, MD).

Volatile organic analysis was performed by purge and trap gas chromatography according to EPA Method 602. Soil samples were analyzed for PAH by first extracting by shaking with 1:1 dichloromethane:acetone according to modified EPA Method 3550. Leachate and run-off samples were extracted for PAH according to EPA Method 3510. Analysis for PAH was carried out by gas chromatography with flame ionization detection (GC/FID) according to EPA Method 8100 (U.S. EPA, 1982). The chromatographic system used a capillary DB5 column 0.25 mm $\times$ 30 m, a temperature program of 8°C per minute between 40° and 290°C, an HP model 5880 GC, Model 7672 autosampler, and Level 4 integrator. Identification of PAH was by comparison of retention times to those of standards, and quantitation was by the internal standard method using o-terphenyl.

Field air samples were analyzed on an AID portable gas chromatograph (Analytical Instrument Development Inc, Avondale, PA) model 210 GC with 10.4 eV PID, calibrated with BTEX and naphthalene standards. Air samples (1 cc) were taken from above the test plots using a Gastight syringe (Hamilton Co, Reno, NV) and injected into the GC.

3. LABORATORY TREATABILITY INVESTIGATION

3.1. Treatability Materials and Methods

3.1.1. Site Samples: Contaminated Material and Treatment Area Soil

Samples were collected at the discharge point of the former tar outfall pipe. Three borings were advanced to a depth of 9 feet using a hollow-stem auger and split-spoon sampler, obtaining continuous samples. The cores contained 1 foot of organic sediment, 2 feet of black heavily contaminated tarry material, and 6 feet of coarse sand with residual tar contamination. These layers were mixed to give a composite sample representing the material to be excavated and treated. In addition to the sample of contaminated material, a soil sample from the intended treatment area was also obtained. Both contaminated and treatment soils were sieved through 3 mm to remove stones and debris.

3.1.2. Culturing and Enumeration of PAH-Degrading Bacteria

PAH-degrading bacterial culture were obtained from tar-contaminated lakebed soil by enrichment with PAH. The culture medium was a mineral salts media published previously (Fogel et al., 1986) except that $NaNO_3$ was replaced by ammonium sulfate at 777 mg/liter. Cultures were grown with shaking in 50 ml of this media in closed 160 ml serum bottles, receiving either crystals of naphthalene, phenanthrene, or pyrene or 1 g of coal tar. The headspace of the bottles was supplemented with 10 cc oxygen every 5 days, and subculturing was carried out every 14 days. Bacteria in PAH-degrading cultures were enumerated by plating on nobel agar containing mineral salts, with naphthalene crystals in the cover of an inverted petrie dish.

3.1.3. Mineralization of Radiolabeled Naphthalene

Three 20 g dry weight treatment mixtures that contained 1% tar were assembled by mixing 1 part tar-contaminated soil and 1 part treatment soil, 0.067 μCi [^{14}C]naphthalene, and nonradiolabeled naphthalene to a final concentration of 150 ppm. Microorganisms in one of the mixtures were killed by thoroughly mixing in 2% weight mercuric chloride. Each treatment mixture was placed in a sealed biometer flask (see Section 3.2.9) having a side-arm flask containing CO_2-free KOH solution to absorb the radiolabeled carbon dioxide produced by biodegradation of the radiolabeled naphthalene. The KOH solution was removed at intervals, and an

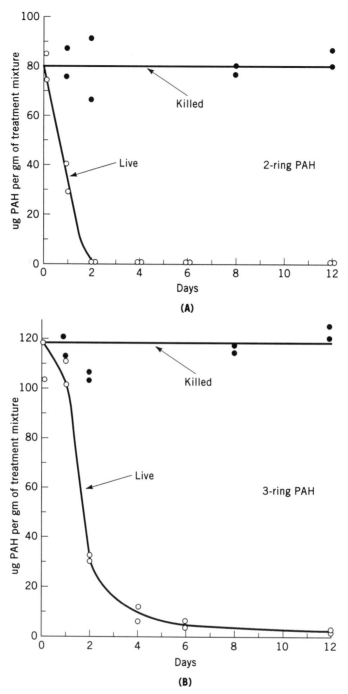

Figure 13.1. Two identical laboratory treatment mixtures containing 2 parts tar-contaminated material and 1 part soil, one mixture killed with $HgCl_2$. Treatment conditions are described in the text. Samples were analyzed at intervals for three 2-ring and four 3-ring PAHs by GC/FID. Closed circles, killed; open circles, live. (**A**) 2-ring PAH. (**B**) 3-ring PAH.

aliquot was mixed with Aquasol scintillant and counted in a Packard Model 2425 Tricarb Liquid Scintillation Spectrometer, using the channels ratio method to correct for quenching.

3.1.4. Trapping and Quantitation of Volatilized Hydrocarbons

A treatment mixture, 44 g dry weight, 1:1 mix of tar soil and treatment soil was placed in a 25 mm × 30 cm glass column (Ace Glass, Vineland, NJ), and moist air was passed over the top of the soil as illustrated in Section 3.2.10, then through a series of two 1.2 g Carbotrap columns (Supelco, Inc., Bellefonte, PA) to absorb volatilized hydrocarbons. An identical treatment mixture was extracted immediately after preparation to give starting PAH concentrations. After 4 days of biodegradation, the treatment mixture in the column was extracted and analyzed to determine the extent of biodegradation. Each sorbent was transferred to a 4 ml Teflon-capped vial and was extracted for 10 minutes with 3.0 ml of carbon disulfide that contained 50 ng/ml of OTP standard. One milliliter of each extract was concentrated to 100 μl by gentle evaporation with nitrogen, and 1 μl was analyzed by GC/FID to determine the amount volatilized during biodegradation.

3.2. Treatability Results

3.2.1. Properties of Tar-Contaminated and Treatment Area Soils

Properties of the tar soil and treatment soil are given in Table 13.1, and concentrations of selected PAH in each are given in Table 13.2.

3.2.2. Coal Tar Biodegradation Treatment Simulation

Laboratory Treatment Mixture. A treatment mixture was constructed by mixing 2 parts tar-contaminated material with one part soil from the proposed treatment area. Fertilizer was added to give 122 ppm nitrogen and 28 ppm PO_4 on a dry weight basis, in addition to the amounts naturally present in the soils. pH was adjusted to 7.0 by the addition of agricultural ground limestone. PAH-degrading

TABLE 13.1. Characteristics of Treatability Site Samples

	Treatment Area Soil	Contaminated Soil
Sand/silt/clay	80/13/8	80/16/5
Humus, % dry wt	30%	2%
Moisture holding	1.9 g water/g dry	0.2 g water/g dry
pH	7.1	5.1
Lime to get pH to 7	—	0.5%
N (NH_3 + NO_3) dry wt	39 ppm	46 ppm
Total P, dry wt	650 ppm	390 ppm
K, dry wt	870 ppm	350 ppm

TABLE 13.2. PAH and VOC in Treatability Site Samples (ppm Dry Weight)

	Tar-Contaminated Soil		Treatment Soil	
	Single	Sum	Single	Sum
1 Ring				
Toluene EB and xylenes	0.41			
2 Rings				
Naphthalene	67		<0.5	
2-Methylnaphthalene	73	200	<0.5	
1-Methylnaphthalene	71		<0.5	
3 Rings				
Acenaphthene	84		<0.7	
Fluorene	38	230	<1	5
Phenanthrene	84		5	
Anthracene	24		<1	
4 Rings				
Fluoranthene	20		9	
Pyrene	26	64	9	25
Benzo[a]anthracene	9		3	
Chrysene	9		4	
5 Rings				
Benzo[b]fluoranthene	2		2	
Benzo[k]fluoranthene	4	12	3	8
Benzo[α]pyrene	6		3	
6 Rings				
Indenopyrene	2	4	<3	
Benzoperylene	2		<3	

bacteria isolated from the site and cultured in the laboratory were added to give 1 million bacteria per gram dry weight of treatment mixture. The moisture content was adjusted to 55% of the moisture-holding capacity of the mixture. For the 2:1 mix, this amounted to 0.42 g water per g dry mix. For an experimental control, an identical mixture was prepared, and the bacteria were killed by thoroughly mixing in 2% mercuric chloride.

The Live and Killed treatment mixtures were placed in opaque open containers to a 6 inch depth and incubated at room temperature. The desired moisture content was maintained by adding water to maintain constant weight. Aeration by tilling was simulated by emptying the test mix into a bowl and thoroughly mixing three times per week. These conditions are summarized in Table 13.3.

Laboratory Biodegradation Results.   Duplicate samples were removed from both the Live and Killed treatment mixtures at intervals and analyzed for PAH according to the GC/FID method described above. The values for the selected 2-ring and 3-ring PAH (listed in Table 13.2) were summed and are presented

TABLE 13.3.   2 : 1 Tar Soil : Treatment Soil Laboratory Mixture

2 : 1 Weight ratio	Tar-contaminated soil : treatment soil
Added fertilizer	122 ppm N, 28 ppm PO_4 (dry wt)
pH	7.0
Added bacteria	1 Million PAH degraders/g dry soil
Moisture content	30%
Soil depth	6 Inches, open container
Aeration	Mixing three times per week

separately in Figure 13.1A and 13.1B. These data show very good agreement between duplicate samples and that the concentrations of 2- and 3-ring compounds did not change significantly in the killed control mixture during the 12 days of incubation. As illustrated in Figure 13.1A, the 2-ring PAH were more than 50% degraded within 24 hours and completely degraded in 2 days. The 3-ring PAH, as shown in Figure 13.1B, did not degrade significantly during the first 24 hours, but were more than 50% degraded by day 2 and 90% degraded by day 4.

The gas chromatographic fingerprints obtained by analysis of day 12 samples, from both the Live and Killed treatment mixtures, are presented in Figure 13.2. Comparison of the Live and Killed chromatograms shows that the 2-ring PAHs (naphthalene and the methyl naphthalenes), as well as the 3-ring PAHs (acenaphthene, fluorene, phenanthrene, and anthracene), were biodegraded to nondetectable levels. Other unidentified peaks in this area of the chromatogram, which are substituted PAHs and alkanes, were also degraded. Of the 4-ring PAHs, only fluoranthene was extensively degraded by day 12, while pyrene, benzo[α]anthracene, and chrysene were only slightly degraded by this time.

3.2.3. Effect of Added Site Bacteria

Treatment Conditions.   An experiment was carried out to determine the effect of enhancing the bacterial population in the treatment mixtures by adding cultured site bacteria. Three treatment mixtures similar to those in the previous experiment were prepared (but having 1:1 tar soil:treatment soil). One was killed by adding 2% mercuric chloride, one contained the naturally occurring concentration of PAH-degrading bacteria, and the third received added PAH-grown site bacteria, 1 million per dry gram of treatment mixture.

Results.   At 3 days, the 2-ring PAHs in the mixture containing added bacteria were 97% degraded, and those in the mixture containing the native levels of bacteria were only 86% degraded, a significant but small difference. The 3-ring PAHs also degraded faster in the treatment mixture containing added bacteria, but the difference decreased from 25% at 3 days to 11% at 5 days, to 5% at 7 days of treatment, indicating that native bacteria responded quickly to the provision of optimal conditions. The data are summarized in Table 13.4.

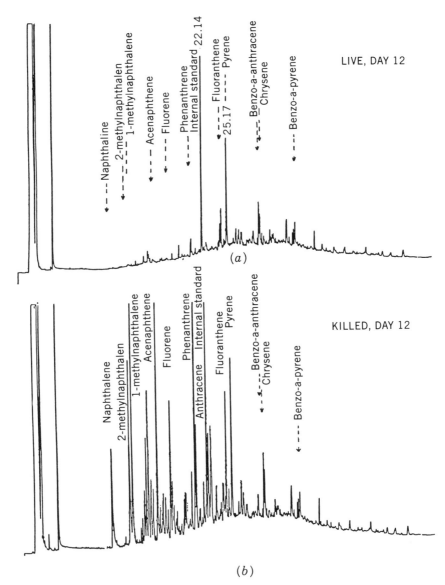

Figure 13.2. Gas chromatographic analysis of day 12 samples from the experiment depicted in Figure 13.1. (*a*) Live test mixture (major peak is the GC internal standard). (*b*) Killed test mixture.

3.2.4. Effect of Moisture Content

Treatment mixtures (1:2 tar soil:treatment soil) were assembled as described above except that different amounts of water were added to achieve 25%, 50%, and 75% of moisture-holding capacity, and the 2- and 3-ring PAHs were analyzed at inter-

TABLE 13.4. Effect of Added Site Bacteria

		Percent Degraded		
No. of Rings	Days	Native Levels	Added Bacteria	Difference
2	3	86	97	11
3	3	45	70	25
3	5	56	67	11
3	7	73	78	5

vals. The rates of biodegradation at 25% and 50% of moisture-holding capacity were similar, but that at 75% moisture-holding capacity was significantly lower. The results are summarized in Table 13.5 and indicate that a moisture content of 30%–60% of capacity is satisfactory.

3.2.5. Effect of Fertilizer Concentration

The treatment mixtures contained a significant amount of native readily available nitrogen, about 40 ppm. Native total phosphorus was about 500 ppm, but *readily available* phosphorus was not determined (see Table 13.1). An experiment was carried out to determine the relative effect of added "low" (30 ppm N, 7 ppm PO_4) and "moderate" (115 ppm N, 27 ppm PO_4) concentrations of fertilizer. The treatment mixtures (1:2 tar soil:treatment soil) were prepared similarly to those described above. The results are summarized in Table 13.6. Each entry represents the average of two treatment mixtures. The degradation of 2-ring PAHs, which was complete in 2 days, was not affected by the additional fertilizer. However, the degradation of 4-ring PAHs, requiring 8 days, was stimulated by the higher fertilizer concentration, indicating that fertilizer would be needed for the field remediation.

3.2.6. Effect of pH

Treatment mixtures had a natural pH slightly below 7. A treatment mixture (1:2 tar soil:treatment soil) was assembled as described above and was brought to a pH of 7

TABLE 13.5. Effect of Moisture Content: Percent PAH Degraded in 8 Days[a]

	Percent Moisture-Holding Capacity		
No. of Rings	25%	50%	75%
2	99	99	61
3	92	94	33

[a]Average of duplicates.

TABLE 13.6. Effect of Nitrogen and Phosphate Addition

		Percent Degraded	
No. of Rings	Days	N = 30 ppm PO$_4$ = 7 ppm	N = 115 ppm PO$_4$ = 27 ppm
2	2	100	100
3	5	86	95
4	8	25	35

with ground limestone. A second mixture having a pH of 6 was prepared by the addition of aluminum sulfate, and the treatment mixtures were analyzed for PAH at intervals. The data, summarized in Table 13.7, indicate that biodegradation of PAH was slightly slower at pH 6.

3.2.7. Effect of Loading Rate

The biodegradation in treatment mixtures containing different loading rates of tar-contaminated soil in treatment soil were studied, representing 1:2, 1:1, and 2:1 mixtures of tar to treatment soil. Samples were analyzed for PAH at intervals. The data, summarized in Table 13.8, indicated that the percent biodegradation was slightly slower at the higher loading level but that the overall amount of PAH degraded was higher at that level. The mixture having the highest loading rate, however, was slightly more difficult to work with during the initial mixing stage and had a reduced moisture-holding capacity.

3.2.8. Effect of Aeration Frequency

The treatment mixtures in the preceding experiment were aerated by mixing every 2 days. Additional duplicate treatment mixtures were set up at the same time but were not aerated after initial assembly. Analysis of these mixtures after 6 days indicated that the rate of PAH degradation was similar to those that had been aerated more frequently. The data are summarized in Table 13.9.

TABLE 13.7. Effect of pH

		Percent Degraded	
No. of Rings	Days	pH 6	pH 7
2	1	71	86
3	4	88	94

TABLE 13.8. Effect of Loading Rate

No. of Rings	Days	Percent Degraded[a]		
		1 : 2[b]	1 : 1	2 : 1
2	1	76	72	61
3	4	94	93	90

[a]Average of duplicates.
[b]Ratio Tar Soil : Treatment Soil.

3.2.9. Mineralization of Naphthalene

To demonstrate that the loss of naphthalene was due to complete mineralization to carbon dioxide and water rather than to a partially degraded intermediate, [14C]naphthalene was added to treatment mixtures (1:1 tar soil:treatment soil), and these were placed in sealed biometer flasks (Fig. 13.3A) having KOH in the side-arm flask to absorb the $^{14}CO_2$ produced from the biodegradation of the radiolabeled naphthalene. The KOH was sampled at various times and the amount of $^{14}CO_2$ determined by scintillation counting. Fig. 13.3B shows the accumulated $^{14}CO_2$ from the two live treatment mixtures and the killed control. The live mixtures released 50% of the added ^{14}C. Since aerobic microorganisms incorporate roughly one atom of carbon into biomass for each atom of carbon released as CO_2, the data are consistent with the idea that 100% of the naphthalene was mineralized.

3.2.10. Measurement of Volatilization During Biodegradation

Volatile losses from treatment mixtures containing killed bacteria indicate the upper limit of volatilization that could occur during biodegradation. Under conditions of rapid biodegradation, however, volatilization losses would be much less than in a killed control. This experiment was designed to measure directly the volatilization from a live treatment mixture using the optimal conditions of moisture, loading rate, and pH determined in earlier experiments. A treatment mixture was assembled (1:1 tar soil:treatment soil), and a portion was analyzed to determine zero time PAH concentrations. The remainder was placed to a depth of 6 inches in a sealed cylin-

TABLE 13.9. Effect of Aeration Frequency

Tar Soil : Treatment Soil	Percent Degraded		
	1 : 2	1 : 1	2 : 1
3-Ring PAHs			
Aerated day 2, analyzed day 4	94	93	90
No aeration, analyzed day 6	>98	97	96

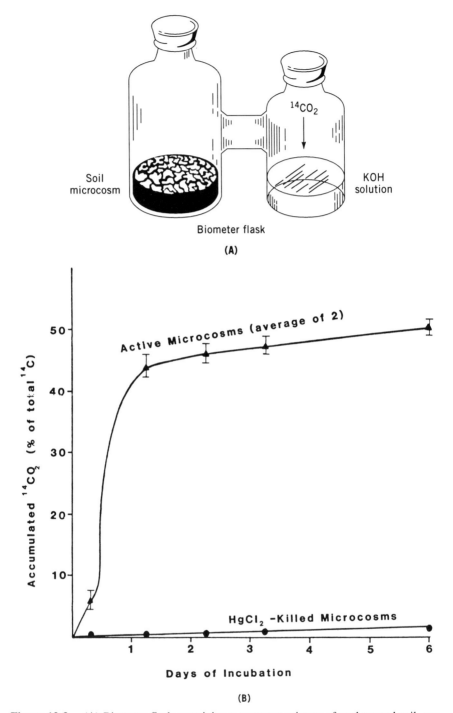

Figure 13.3. (A) Biometer flask containing a treatment mixture of coal tar and soil, prepared as described in the text, containing radiolabeled naphthalene, with KOH solution in the side-arm flask to trap $^{14}CO_2$. (B) Cumulative evolution of $^{14}CO_2$ from the experiment depicted in A. Triangles, average of duplicate live treatment mixtures; circles, killed treatment mixture.

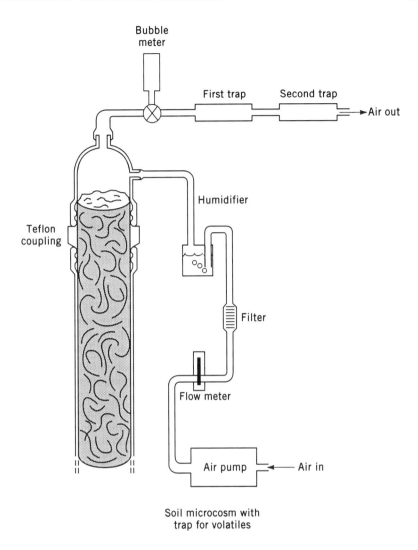

Figure 13.4.   Experimental set-up for the measurement of volatile emissions from a coal tar treatment mixture undergoing biodegradation.

der, and moist air was passed over the soil surface and then through a series of two sorbent traps to capture volatilized compounds (Fig. 13.4). At the end of the 4 day experimental period, a sample was taken from the treatment mixture for PAH analysis, and the traps were desorbed and analyzed for PAH. The results are presented in Table 13.10.

The analysis of the treatment mixture indicated that 100% of the 2-ring PAHs and 91% of the 3-ring PAHs had disappeared due to both biodegradation and

TABLE 13.10. Direct Measurement (μg PAH/g Treatment Mix) of PAH Volatilized From Treatment Mixture During Biodegradation

No. of Rings	Initial PAH in Soil	Final PAH in Soil	PAH in Trap No. 1	PAH in Trap No. 2	Total Trapped	Percent Volatilized
2	30	0.5	.51	.02	.53	2%
3	14	1.2	.11	—[a]	.13[b]	1%[b]
4	14	16	.05	<.01	.05	0.3%

[a]No data.
[b]Estimate.

volatilization and that the concentration of the 4-ring PAHs had not changed significantly during the 4 days. Of the trapped 2-ring PAHs, 96% were in the first trap, indicating that these compounds were trapped efficiently. Dividing the total trapped 2-ring PAHs, 0.53 μg/g treatment mix, by the initial amount in the treatment mix, gives 2% as the amount volatilized. For the 3-ring PAHs, based on the amount in the first trap, about 1% volatilized. Data for 4-ring PAHs indicate that they were also efficiently trapped and that only 0.3% were volatilized. These data are consistent with the data from killed treatment mixtures such as shown in Figure 13.1, which indicated that volatilization of 2- and 3-ring PAHs from a 0 to 6 inch soil depth is not a significant fraction compared with biodegradation of these compounds under the conditions employed in this study. For the 2-ring PAHs, Table 13.10 indicates that biodegradation is 50 times faster than volatilization.

3.3. Treatability Discussion and Conclusions

The treatability study demonstrated that soil contaminated with coal tar could be rapidly treated by mixing at approximately a 1:1 ratio with soil from the intended treatment area to create a friable mixture having good porosity and drainage properties, bringing the pH to 7, maintaining the moisture content within 25% to 50% of the moisture-holding capacity of the mixture, adding supplementary nitrogen and phosphate fertilizer, and aerating by mixing.

In a typical biodegradation experiment, 2-ring PAHs were reduced from a total concentration of 80 ppm to nondetectable amounts in 2 days, and 3-ring PAHs were reduced from about 120 ppm to 5 ppm in 6 days, while 4-ring PAHs did not degrade extensively in the 12 day treatment time. Numerous other compounds, assumed to be alkanes and substituted 2- and 3-ring PAHs, were also degraded.

The soil from the intended treatment area contained an unusually high organic component, 30%, and when mixed with the tar-contaminated soil caused the formation of a friable mixture having good moisture-holding ability. This "treatment" soil contained 40% as much 4-ring PAHs and 65% as much 5-ring PAHs as the tar-contaminated soil, indicating prior contamination with tar, and probably contained an acclimated population of tar-degrading bacteria.

Experiments comparing biodegradation in treatment mixtures having native bacteria with those having added cultured site PAH degraders showed that stimulation due to added bacteria decreased from 25% at 3 days to 5% at 7 days. This result indicated that, for the field project, native bacteria would be sufficiently stimulated by the provision of optimal conditions to carry out the remediation without added cultured bacteria.

Experiments measuring biodegradation in treatment mixtures having different moisture contents and different pH indicated that optimal conditions included a wide range of moisture, from 25% to over 50% of moisture-holding capacity, and a pH range from 6 to 7. Experiments with different concentrations of added fertilizer indicated that, although the soils had significant concentrations of nitrogen and phosphorus, the higher fertilizer concentration resulted in a slightly increased rate of biodegradation during an 8 day test. It was concluded that during field remediation, which would be carried out for a longer time period, fertilizer should be added at regular intervals.

4. FIELD PILOT DEMONSTRATION

4.1. Field Methods

4.1.1. Test Plan

The test plan involved the study of seven test plots constructed near the site of contamination. One plot was maintained as a control and had no tar added. On day 0, six plots received 1.5 inch applications of tar-contaminated soil, mixed into the top 6 inches of soil. On day 27, four plots received a second application of tar-contaminated soil, and on day 55 two plots received a third application.

4.1.2. Plot design

Each plot, 6 × 9 feet and 2 feet deep, was excavated and graded for leachate collection at one end; then the bottom and sides were lined with 6 mil polyethylene. Backfill consisted of a lower drainage layer of sandy soil and an upper 1 foot layer of the organic topsoil (treatment soil) described above for the treatability study, sloped for collection of run off. Before application to the test plots, the treatment soil was thoroughly mixed in a pile by extensive reworking with a backhoe. PAH and metal analysis of pilot test treatment soil is given in Tables 13.11 and 13.12. Test plot detail is given in Figure 13.5. Test plots were separated by 4 feet.

4.1.3. Excavation and Application of Tar-Contaminated Soil

The tar-contaminated soil was excavated to a depth of 8 feet from the wetland by backhoe, drained briefly, then placed in a 12 × 12 × 1 foot mixing box, where it was homogenized in batches by rototilling. Potassium phosphate and ammonium sulfate were mixed in to add 27 ppm phosphate and 120 ppm nitrogen. Batches were mixed in a pile with a backhoe to provide a homogeneous source of contaminated

TABLE 13.11. PAH and VOC (ppm (Dry Weight) GC/FID Analysis) in Pilot Study Treatment Soil and Tar-Contaminated Soil

	Tar-Contaminated Soil		Treatment Soil	
	Single	Sum	Single	Sum
1 Ring				
Toluene EB and xylenes	0.41			
2 Rings				
Naphthalene	292		<0.5	
2-Methylnaphthalene	268	757	<0.5	
1-Methylnaphthalene	197		<0.5	
3 Rings				
Acenaphthene	185		1	
Fluorene	121	658	1	16
Phenanthrene	276		11	
Anthracene	76		3	
4 Rings				
Fluoranthene	71		20	
Pyrene	99	243	21	72
Benzo[α]anthracene	32		16	
Chrysene	41		15	
5 Rings				
Benzo[b]fluoranthene	6		10	
Benzo[k]fluoranthene	11	27	10	30
Benzo[a]pyrene	10		10	
6 Rings				
Indenopyrene	3	6	5	1
Benzoperylene	3		6	

soil for the test plots. This material was black and had a stiff consistency and a strong organic odor. PAH analysis of pilot stock contaminated soil showed significantly higher concentrations than those in the sample used for the laboratory study (Table 13.2). The values for the pilot soils are given in Table 13.11. Metals are given in Table 13.12. The stock contaminated soil was then applied to test plots and rototilled thoroughly into the top 6 inches of soil.

TABLE 13.12. Metals (ppm, Dry Weight) in Treatment and Tar-Contaminated Soils

	As	Cd	Cr	Cu	Ni	Pb	Hg	Zn
Treatment	2	1	7	54	2	150	0.3	70
Tar-Contam.	4	10	250	1300	90	190	1	610

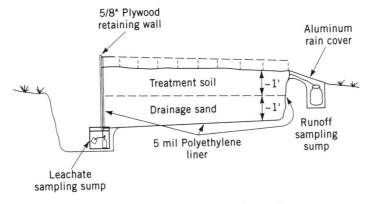

Figure 13.5. Test plot detail. Vertical section, 9 foot dimension.

4.1.4. Collection of Soil Samples

Eight to ten core samples, 6 inches deep, were obtained from each test plot during every sampling event. These were composited to produce one sample per plot. The soil was then passed through a 4 mm sieve to remove occasional stones and organic debris to ensure reproducible analyses. Samples were preserved by mixing in 2% mercuric chloride, transported on ice, and stored under refrigeration or freezing.

4.1.5. Leachate and Run-Off Collection

Leachate was directed into an evacuated 500 ml Tedlar bag, with overflow to a glass jar. Samples to be analyzed for volatiles were taken from the Tedlar bag, and samples for metals were taken from the jar. Run-off was directed by surface grade to one end of the plot and was funneled by aluminum flashing to an amber collection bottle.

4.1.6. Soil Aeration by Tilling, and Moisture Maintenance

Rototilling to a depth of 6 inches was carried out on a weekly basis. Each plot was tilled three times in two directions. After tilling, the surface was graded and sloped toward the run-off collection system. The moisture content of the plots was maintained between 35% and 65% of the moisture holding capacity by adding water as necessary.

4.2. Field Pilot Test Results and Discussion

4.2.1. Test Plot Soil pH

Table 13.13 presents data on test plot soil pH and limestone additions. Treatment soil had a pH of 7.1, but the contaminated soil was pH 5.1. Thus, plots having a greater amount of contaminated soil required slightly more limestone to maintain

TABLE 13.13. Test Plot Tar Applications and pH Maintenance (10 weeks)

Plots	Tar Applications			Total Limestone (4 Applications; lb/100 lb soil)	Average pH (9 Weekly Analyses)
	Day 0	Day 27	Day 55		
A	0″	0″	0″	0.3	7.2
B + C	1.5″	0″	0″	2.0	6.6
D + E	1.5″	1.5″	0″	2.3	6.7
F + G	1.5″	1.5″	1.5″	3.3	6.8

the desired pH of about 6.7. The control plot, having no contaminated soil, required only 0.3% limestone and maintained an average pH of 7.2. The test plots required from 2% limestone (for 1.5 inches of tar) to 3.3% limestone (for 4.5 inches of tar).

4.2.2. Test Plot Available Nitrogen and Phosphate

The treatment soil and contaminated soil initially had about 40 ppm of available nitrogen and an undetermined amount of organically bound nitrogen. These soils also contained several hundred ppm of total phosphorus, but an undetermined amount of water-soluble phosphorus (Table 13.1). Thus the initial fertilizer applications were relatively conservative. By day 40, the available nitrogen in all plots was relatively low, about 14 ppm, and the four plots that had received a second application of tar received additional nitrogen and phosphate. By day 62, after two plots had received a third application of contaminated soil, the concentration of available nitrogen was again low, and all plots received application of nitrogen and phosphate, greater amounts for plots having received a greater total amount of tar. The data for available nitrogen and concentrations of added nitrogen and phosphate are presented in Table 13.14. The concentration of available phosphate in the soil was not determined.

4.3. PAH Concentrations in the Test Plots

4.3.1. Presentation of Data

The biodegradation of 2-ring, 3-ring and 4-ring PAH is shown on separate axes in Figure 13.6, for the two plots that received three applications of tar-contaminated

TABLE 13.14. Test Plots: Available N (ppm) and Fertilizer Addition (ppm)

Total Tar (Inches)	Day 0, Add. N, PO$_4$	Day 40, Avail. N	Day 41, Add. N, PO$_4$	Day 62, Avail. N	Day 69, Add. N, PO$_4$
0	0, 0	5	0, 0	7	0, 0
1.5	30, 7	—	0, 0	14	240, 27
3.0	30, 7	14	240, 27	14	240, 27
4.5	30, 7	14	240, 27	10	480, 54

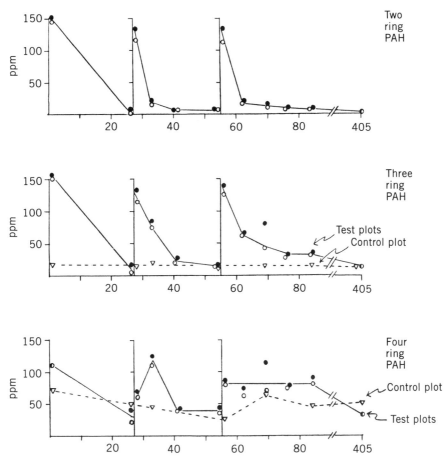

Figure 13.6.   Test plot biodegradation of 2-, 3-, and 4-ring PAHs. Solid lines, PAH concentrations in 10-fold composite samples collected from duplicate 9 × 6 foot test plots that received three applications of 1.5 inches of tar-contained soil on August 21, September 17, and October 15. Dashed lines, PAH concentrations in 10-fold composite samples collected from the control plot that did not receive application of tar but that had significant concentrations of 3- and 4-ring PAHs. The PAHs analyzed are the eleven 2-, 3-, and 4-ring compounds listed in Table 13.11.

soil. Figure 13.6 also contains data for the control plot, which received no application of tar but which contained high levels of 3-ring and 4-ring PAHs. Figure 13.7 presents an overview of the biodegradation, showing a plot of the sum of the 11 indicator PAHs, while Figure 13.8 gives the GC/FID fingerprints showing all compounds present in the test plots.

The PAH concentrations for the three-application plots are also presented in Table 13.15 and those for the control plots are presented in Table 13.16. The percent biodegradation achieved at the end of the active treatment period, on No-

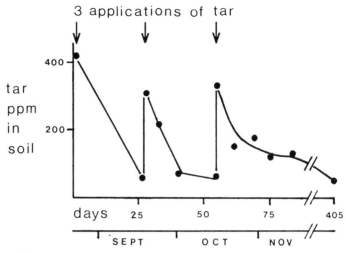

Figure 13.7.   Summary of the test plot data presented in Figure 13.6. Each data point represents the sum of the 11 compounds in Figure 6, and the average of the two test plots.

vember 13 (day 84), and by the end of the second season (no tilling or fertilizer) on day 405, are presented in Table 13.17.

4.3.2.   Control Plot

The 3- and 4-ring PAHs in the control plot, which had been decreasing gradually during the first 55 days of the field test, showed unexpected increases during late October. The concentration of 4-ring PAHs, for example, increased from 25 to 63 ppm, and 2-ring PAHs appeared in the samples. Similar increases were documented for test plots (data not shown) that had received only one or two applications of tar and were being monitored during this time. It is speculated that tar-contaminated particulates from the near-by tar disposal site were deposited on the test plots and control plot. Such deposits, which might have occurred during August or September, could have been rapidly degraded, but those during late October resulted in accumulation.

4.3.3.   Plots With Three Applications of Tar

The data for 2-, 3-, and 4-ring PAHs in the test plots are discussed separately below. In general, the biodegradation in the first 2 weeks was less extensive for the third application than for the second, most likely due to the cold late October weather. It is likely that the first application degraded as rapidly as the second, but 2 week samples were not analyzed.

4.3.4.   2-Ring PAHs

After the first application on August 21, the 2-ring PAHs degraded from 150 to 2 ppm in 26 days. For the second application on day 27 (September 17), degradation

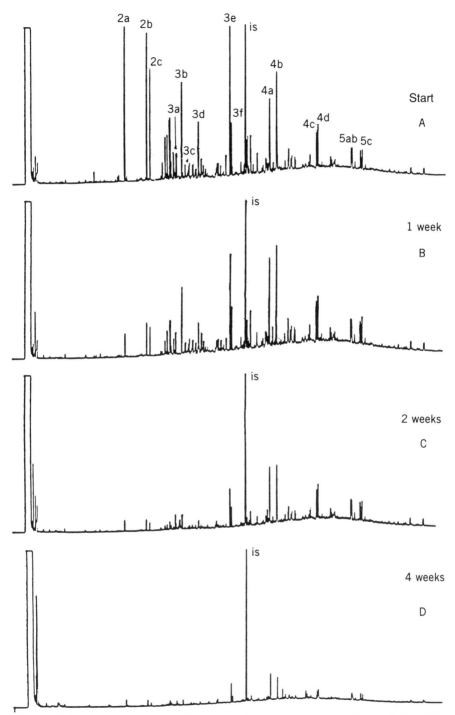

Figure 13.8. Gas chromatograph fingerprints from gas chromatographic analysis of 10-fold composite samples from a test plot having an application of 1.5 inches of tar, sampled at day 0 and 1, 2 and 4 weeks and analyzed under identical conditions. The major constituent at 1, 2, and 4 weeks is the GC internal standard. The compounds quantitated were 2a, naphthalene; 2b, 2-methylnaphthalene; 2c, 1-methylnaphthalene; 3b, acenahthene; 3d, fluorene; 3e, phenanthrene; 3f, anthracene; 4a, fluoranthene; 4b, pyrene; 4c, benzanthracene; and 4d, chrysene.

TABLE 13.15.   PAH Biodegradation in Plots With 3 Applications of Tar-Contaminated Soil (ppm Dry Weight, GC/FID Analysis, Average of 2 Plots)[a]

No. of Rings	First Application		Second Application		Third Application			
	Day 0	Day 26	Day 27	Day 41	Day 55	Day 69	Day 84	Day 405
2	150	2	124	8	124	16	11	3
3	155	12	122	22	132	61	33	14
4	110	30	66	42	81	98	85	33

[a]Data are for sums of selected PAH listed in Table 13.11.

was from 133 to 8 ppm in 14 days, and for the third application (October 15), 2-ring PAHs degraded from 125 to less than 25 ppm in 14 days. By the end of active treatment on day 84 (November 13), 29 days after the third application, 2-ring PAHs had decreased further, and after an additional year of passive biodegradation the 2-ring PAHs had decreased to 3 ppm, a concentration still above that in the control plot.

4.3.5.   3-Ring PAH

After the first application, the 3-ring PAHs degraded from 155 to 12 ppm in 26 days, after the second application from 122 to 22 ppm in 14 days, and after the third application from 132 to 33 ppm in 29 days. After a year of passive biodegradation, the concentration was 14 ppm, the same as that in the control plot.

4.3.6.   4-Ring PAH

The treatment soil had a high concentration of 4-ring PAHs, about 72 ppm. The first application of tar to the test plots increased this concentration to only 110 ppm. During the first application period, the 4-ring PAHs in the test plot decreased from 110 to 30 ppm, while in the control plot a smaller decrease from 72 to 50 ppm was observed. The greater biodegradation in the test plot can be attributed to the presence of fertilizer as well as to the stimulation of the bacterial population by the added 2-ring PAHs.

After the second application, the 4-ring PAHs decreased only 36% in 2 weeks, and those in the control plot decreased 26%. After the third application, on October

TABLE 13.16.   PAH Concentrations in Control Plot, ppm Dry Weight (Tilled, No Fertilizer, No Added Tar)[a]

No. of Rings	Days							
	0	26	27	41	55	69	84	405
2	<0.5	<0.5	—	<0.5	<0.5	2	6	0.2
3	19	15	—	18	8	14	17	13
4	72	50	—	37	25	63	47	51

[a]Data are for sums of selected PAH listed in Table 13.11; ppm dry wt, GC/FID analysis.

TABLE 13.17. Percent PAH Degraded at 84 and 405 Days

	No. of Rings	Total PAH Treated (ppm)	Percent Degraded	
			84 Days	405 Days
Control plot	2	—	—	—
	3	19	10	32
	4	72	35	29
One application	2	150	99	99
	3	155	92	95
	4	110	62	83
Two applications	2	258	99	99
	3	257	95	96
	4	141	74	82
Three applications	2	388	97	99
	3	380	91	96
	4	186	54	82

15, the 4-ring PAHs did not decrease during the 29 day active treatment period. After an additional year of passive biodegradation, however, the 4-ring PAHs decreased from 85 to 33 ppm, a value significantly below that in the control plot (51 ppm).

4.3.7. Plots Receiving Only One or Two Applications of Tar

Plots that received only one application of tar (on day 0), or only two applications (days 0 and 27), were treated and sampled for the full 84 days. Although the detailed data are not presented here, these plots showed decreases in PAH concentrations during the first two application periods similar to those shown here for the plots that received three applications.

4.3.8. Percent Biodegradation

The percent biodegradation was calculated by dividing the concentration remaining in each PAH category by the total amount treated, which included the amount applied in the three applications plus any PAHs already present in the treatment soil. These values were calculated for plots having one, two, and three applications of tar and are presented in Table 13.17. The values calculated for the end of the treatment period on November 13 (day 84) are considered to represent incomplete treatment of the third application, as the cold weather probably slowed the rate of biodegradation. The values obtained by sampling after the following warm season, on day 405, are similar for all test plots. In general, 2-ring PAHs were degraded 99%; 3-ring, 96%; and 4-ring, 82%.

4.4. GC/FID Fingerprints

Figure 13.8 presents GC/FID fingerprint data from soil extracts prepared in an identical manner from samples taken on days 0, and 1, 2, and 4 weeks after application (second application period). In addition to the 11 major constituents quantitated in the tables, about 50 additional peaks are visible, all of which decreased markedly during the 4 week treatment period. It can also be seen that the 5-ring PAHs decrease significantly after the second week. Other compounds, probably substituted PAHs such as dimethylnaphthalenes, as well as alkanes, also degraded significantly during the 4 weeks.

4.5. Leachate PAHs

Leachate accumulated in the leachate collection systems on five occasions during the 84 day treatment period. As some of the collection systems did not operate at times, only a total of 14 samples were collected from five test plots and 1 from the control plot. Since the concentrations of PAHs in the samples did not correlate with the number of applications of tar, or the number of days since application, the values were averaged, giving concentrations of 2-, 3-, and 4-ring PAHs in the leachate of 0.2, 0.7, and 0.9 ppb, respectively. These data are presented in Table 13.18.

To estimate the significance of these concentrations, assume 10 inches of rain during the treatment period, 50% of which passes through the soil, giving 620 liters of leachate containing a total of 0.5 mg of 4-ring PAHs. The average amount of 4-ring PAHs treated was 150 ppm × 680 kg soil = 100 g. Thus the leachate represents 0.0005% of the amount treated.

4.6. Leachate Metals

Metal concentrations in the treatment soil and in the tar-contaminated soil were given in Table 13.12. The contaminated soil contained about 10 times as much cadmium and zinc and over 25 times as much chromium, copper, and nickel as did the treatment soil. Leachate samples for metal analysis for both test and control plots were available only on days 54 and 84. These data are presented in Table 13.19. The test plot data represent ranges for three plots having 3 inches of tar by day 54, and for two with 3 inches and two with 4.5 inches by day 84. Safe drinking water levels (CFR 40, part 141.11, and 143.3, July 19, 1979) are included for comparison.

Cadmium and zinc concentrations in the soil mixture were calculated based on the amounts present in the tar and treatment soil to be three- to five-fold higher in

TABLE 13.18. PAH Concentrations (ppb, GC/FID Analysis) in Leachate[a]

		Test Plots	
No. of Rings	Control Plot	Range	Average
2	<0.1	<0.1 to 2.1	0.2
3	<0.1	<0.1 to 2.6	0.7
4	<0.2	<0.2 to 6.6	0.9

[a]Average of 14 samples, days 5–84, five plots.

TABLE 13.19. Metal Concentrations (ppb) in Leachate

	As	Cd	Cr	Cu	Ni	Pb	Zn	Hg
Control plot	<8	2	<25	20	150	<5	160	—
	<8	<1	<25	30	130	<6	20	<0.4
Test plots	<8	2	<25	40	280	<7	190	<0.2
		5		60	580		2,000	40
Drinking water	50	10	50	1,000	—	50	5,000	2

the test plots relative to the control plots, and these increases are reflected in the increase of those elements in the leachate. Nickel and copper would be increased about 10- and 20-fold, respectively, in the test plots, but only slight increases of these metals in the leachate were measured relative to the control plot. None of the metals in the leachate exceeded the drinking water standards except mercury, which exceeded the standard by 20-fold in two samples.

4.7. Run-Off Metals and PAH

Run-off collected from one of the test plots on day 4 was high in suspended solids, about 26 g per liter. Analysis for total and dissolved metals demonstrated that 95%–99% of the metals were associated with the particulate matter and therefore could be readily removed in a settling basin. The concentrations of the dissolved metals were all below the drinking water levels except Cd, which exceeded the standard by only 30%. PAHs in the unfiltered test plot run-off on day 4 were about 30 ppb each for 2-, 3-, and 4-ring PAHs. Although dissolved and particulate PAHs were not determined, it is likely that the PAHs are also associated with the solids and would be easily removed by settling.

4.8. Air Emissions

A T.I.P. detector measured 7–30 ppm total organic compounds in the air 15 feet from the pit during excavations and 10–60 ppm closer to the pit. This value is less than the NIOSH 8 hour exposure limit for coal tar of 100 ppm (400 mg/m³) (US Dep of Health, 1985). The NIOSH safety limit of 10 ppm for naphthalene would also be met. A field gas chromatograph was used to determine the sum of aromatic compounds in the air near the test plots during the initial application and tilling. These values averaged 10 mg/m³ and did not exceed 20 mg/m³ of air. NIOSH safety limits for toluene, xylene, and ethylbenzene of 100 ppm each (435 mg/m³) were therefore met.

5. CONCLUSIONS

5.1. Extent of Treatment

The data for tar applied to test plots in August, September, and October show that the rate and extent of biodegradation decreased with decreasing temperature.

For the August application, 2-, 3-, and 4-ring PAHs degraded rapidly. For the September application, 2- and 3-ring PAHs degraded rapidly but 4-ring PAHs degraded slowly. For the October application, only the 2-ring PAHs degraded rapidly, 3-ring slowly, and 4-ring not at all. Samples collected at the end of the following summer, however, demonstrated that the same percentage of treatment was achieved for plots that had received a third application in October as those having only one application in August. Thus, the limiting factor during the first season for the third application was the cold weather and not the loading rate of contaminated material on the test plots.

The ultimate percent treatment for plots having one, two, or three applications of tar-contaminated soil was 99%, 96%, and 82% for 2-, 3-, and 4-ring PAHs, respectively. It is likely that more extensive treatment of the 4-ring PAHs would have occurred if tilling, fertilizer, and pH maintenance had been carried out during the second season.

5.2. Recommended Application Frequency and Land Requirement

Based on the results obtained for the August and September applications, it was concluded that efficient treatment could be carried out between April and September 31 and that extensive degradation can be obtained within 2 weeks of application. The most efficient use of a treatment site would be to apply tar at 2 week intervals, resulting in about 10 applications per season. At 1.5 inches per application, this would be 1.25 cubic feet of contaminated soil per square foot of treatment site per season, allowing 5,000 cubic yards of contaminated soil to be treated on 1 acre of land in three seasons. An additional small area would be needed for a run-off settling pond.

The repeated application of tar would result in an increased amount of leachate. Data from the test plots indicated that after passing through only 2 feet of soil the leachate met drinking water standards for most metals, and less than 0.001% of applied PAHs could be detected. Treatment of leachate from repeated applications of tar could be achieved by providing a greater depth to groundwater. Analysis of run-off showed that most of the metals were associated with particulates. Even at higher rates of tar application, metals in run-off, and probably also PAHs, would be easily recovered in a settling pond. Limited data on emission of volatile mono-aromatics and naphthalene indicate that NIOSH safety limits could be attained within 30 feet of the excavation and incorporation processes.

5.3. Alternative Treatment Options

The process described here involved the biodegradation of coal tar constituents at a concentration of about 650 ppm total hydrocarbons by gas chromatography (TH-GC). Much higher concentrations of coal tar can be tolerated and biodegraded by bacteria. Taddeo et al. (1989) described compost treatment of coal tar at a starting concentration of 6,400 ppm TH-GC, and Findlay and Dooley (1990) reported biodegradation of a light coal tar in an aqueous emulsion at a staring concentration of 200,000 ppm TH-GC. Thus, in land application of coal tar, the limiting factor in determining application rate is not bacterial tolerance, but the physical properties of the tar/soil mixture.

At high concentrations of tar, the mixture is difficult to aerate by tilling and has hydrophobic properties. Compost treatment involves combining sludges such as coal tar with a bulking agent to provide improved handling properties, air permeability, and moisture holding capacity in order to allow treatment of high concentrations of contaminants. Composting has been used successfully to treat relatively high concentrations of petroleum production sludge (Fyock et al., 1991). While composting has the advantage of requiring less land area, the mixing equipment required is relatively expensive. Where land is available for treatment, soil application has the advantage of simplicity and the use of low-cost tilling equipment.

REFERENCES

Anderson JW (1979): An assessment of knowledge concerning the fate and effects of petroleum hydrocarbons in the marine environment. In Vernberg WB, Thurberg FP, Calabrese A, Vernberg FJ (eds): Marine Pollution. New York: Academic Press.

Atlas RM (1981): Microbiological degradation of petroleum hydrocarbons: An environmental perspective. Microbiol Rev 45:180–209.

Eng R (1985): Survey of town gas and by-product production and location in the U.S. (1880–1950). NTIS PB8J-173813.

Federal Register 40 CFR 143.3 (1979): Secondary maximum contaminant levels. July 19, 1979, pp 42, 198.

Findlay M, Dooley M (1990): Bioreactor treatment of liquid coal tar. Institute of Gas Technology 3rd International Symposium on Gas, Oil and Environmental Biotechnology. December 1990, New Orleans.

Fogel MM, Taddeo A, Fogel S (1986): Biodegradation of chlorinated ethenes by methane-utilizing bacteria. Appl Environ Microbiol 51:720–724.

Fyock O, Nordrum S, Fogel S, Findlay M (1991): Pilot scale composting of petroleum production well sludge. 3rd Annual Symposium on Environmental Protection in the Energy Industry. December 12, 1991, University of Tulsa.

NIOSH Pocket Guide to Chemical Hazards (1990): U.S. Dept. of Health and Human Services, U.S. GPO, Pub 90-117, Washington, DC.

Taddeo A, Findlay M, Danna M, Fogel S (1989): Field demonstration of a forced aeration composting treatment for coal tar. HMCRI Superfund Conference, December 1989, Washington, DC, pp 1–22.

U.S. EPA (1979): Methods for Chemical Analysis of Water and Wastes. EPA/4-79-020, rev ed 1983. OH: Cincinnati, EPA/EMSL.

U.S. EPA (1982): Test Methods for Evaluating Solid Waste—Physical/Chemical Methods, SW-846, 2nd ed. Washington, DC: EPA, Office of Solid Waste.

U.S. GPO Federal Register (1978): 34:4103–4109.

Wolfe DE (ed) (1977): Fate and Effects of Petroleum Hydrocarbons in Marine Organisms and Ecosystems. New York: Pergamon Press.

14

IN SITU PROCESSES FOR BIOREMEDIATION OF BTEX AND PETROLEUM FUEL PRODUCTS

GENE F. BOWLEN

Envirogen, Inc., Lawrenceville, New Jersey 08648

DAVID S. KOSSON

Department of Chemical and Biochemical Engineering, Rutgers, The State University of New Jersey, Piscataway, New Jersey 08855

1. INTRODUCTION: SCOPE OF THE PROBLEM

Contamination of soils and groundwater by various petroleum hydrocarbons is widespread throughout the United States and much of the world. The use and storage of various fuels, including gasoline, kerosene, diesel fuel, various heating oils (Nos. 2, 3, 4, and 6), and jet fuels (JP4, and JP5), has resulted in surface spills and leakage from numerous underground storage tanks. Leakage from many of the tanks has been identified as one of the primary sources of groundwater contamination (Cheremisinoff et al., 1983). The number of underground gasoline tanks has been estimated at 1.4 million, with 85% of the tanks constructed of steel with no corrosion protection (Feliciano, 1984). Feliciano (1984) indicated that 75,000–100,000 of the tanks are leaking their contents into the surrounding soils and groundwater. Recent

Microbial Transformation and Degradation of Toxic Organic Chemicals, pages 515–542
© *1995 Wiley-Liss, Inc.*

TABLE 14.1.   BTEX Production Capacity[a]

Hydrocarbon	Production (Billions of Gallons)			
	1989	1988	1980	1979
Benzene	1.58	1.59	1.56	1.67
Toluene	0.81	0.87	na	1.01
p-Xylene	0.76	0.78	0.53	0.65
Xylene	0.81	0.76	na	na

[a]Data are from Anonymous (1981) and Reisch (1990). na, not available.

legislation has mandated the replacement of many of the older, poorly designed tanks, and reductions in the number of leaking tanks may have occurred in recent years.

Benzene, Toluene, Ethylbenzene, and Xylene (BTEX) and naphthalene were observed in various types of soil samples in 1966 (Simonart and Batistic, 1966). The source of the contamination was not clear, but, considering the amounts of BTEX produced annually, it is likely that significant quantities find their way into the environment (Table 14.1) Annual production of benzene exceeds 1.6 billion gallons each year with toluene and xylene production in excess of 750 million gallons of each compound (Anonymous, 1981; Reisch, 1990). BTEX is a major component in unleaded gasoline and, therefore, is a ubiquitous presence throughout much of the world. BTEX and methyl-*tert*-butyl ether (MTBE) are the most water-soluble fractions in gasoline and therefore the most mobile in an aquifer system.

Additional sources of petroleum contamination include accidental spills and the long-term practice of landfilling industrial and petroleum wastes. Leachate from landfills often contains considerable amounts of dissolved BTEX. Two leachate plumes, which were examined in detail, contained many aromatic compounds, including all the BTEX components, in concentrations ranging from 47 to 820 μg/liter (Reinhard and Goodman, 1984) and from 3,800 to 41,000 μg/liter (Dienemann et al., 1987).

Once BTEX has entered an aquifer, biological and abiotic transformation, volatilization, and sorption are the three principal removal mechanisms. Recent studies indicated that BTEX migration was only slightly retarded by partitioning in low organic content aquifers and not irreversibly sorbed (Patrick et al., 1985). Biological transformation and volatilization are the major removal mechanisms for BTEX contamination of soils and groundwater.

Remediation of BTEX contaminated soils historically focused on either excavation and disposal or treatment of contaminated soils, or pumping of groundwater to remove dissolved contaminants from the aquifer. Both approaches have been found to be extremely costly. In addition, excavation can be disruptive to other site activities. Recent emphasis has focused on the development of *in situ* remediation processes to reduce remediation costs, site disruption, and, in some cases, remediation time frames. *In situ* remediation has several basic approaches. Contaminated groundwater has been remediated through the introduction of additional oxygen or alternate electron acceptors, or nutrients, including nitrogen, phosphorus, and in some cases potassium. Indigenous microbial populations most frequently have been appropriate for remediation of petroleum hydrocarbons once locally limiting environmental conditions (oxygen, nutrients, pH, and so forth) have been altered. Additions of oxygen and nutrients are usually accomplished through groundwater recirculation coupled with above-ground aeration or the use of injection and/or infiltration galleries. Remediation of vadose zone soils has been accomplished either through nutrient injection or infiltration, or through the use of techniques designed to increase air flow through the soil system (bioventing). Remediation of surficial soils has been accomplished through soil tilling with concurrent addition of nutrients.

The objective of this chapter is to review the current state of the art for *in situ* bioremediation of soils and groundwater contaminated with BTEX and petroleum fuels. Process design considerations and limitations are discussed, with a focus on implementation issues that have been encountered.

2. PROPERTIES OF BTEX AND OTHER FUELS

Petroleum hydrocarbon products consist of blends of distillate fractions from the processing of crude oil. The fractions include aliphatic, aromatic, and asphaltic compounds obtained during the refining process. Aliphatic compounds can be divided further into straight-chain and branched alkanes and cycloalkanes. The various fuels can be separated by the boiling range used to obtain the fuel fraction. The boiling ranges used to obtain the more common fuels range from 23° to 320°C (Table 14.2). The various jet fuels are blends of naphtha, gasoline, and kerosene.

TABLE 14.2. Properties of Petroleum Distillate Fractions[a]

Fuel	Distillate Range	Chemical Characteristics	
Naphtha	$\geq 10\% = <175°C$		
	$\geq 95\% = <240°C$		
Kerosene	180°–320°C	C_{10}–C_{16} fraction	
Diesel fuel		C_{10}–C_{18} fraction	
(No. 2)			
Gasoline	23°–204°C	C_5–C_9 fraction	High volatility, high aromatic content
Blended Fuel			
JP4		Highest volatility and hydrogen content and lowest viscosity of aviation fuels	Mixture components: naphtha, gasoline, and kerosene
JP5		Higher molecular weight fraction	Mixture components: naphtha, gasoline, and kerosene

[a]Data are from Riser-Roberts (1992).

The BTEX components in the fuel are the lower molecular weight aromatic fractions commonly obtained from petroleum fractionation and hydrocarbon reforming. The relatively high aqueous solubilities of benzene (1,750 mg/liter) and toluene (525 mg/liter) are a major consideration from an environmental standpoint. Entry into the groundwater will be followed by rapid movement of a significant portion of the release. The adsorption of benzene and toluene has been determined to be significantly less than xylene isomers, which would contribute further to the greater mobility of these compounds. The xylene isomers also migrate, but at a slower rate than either benzene or toluene.

The densities of all the BTEX compounds and the majority of other fuel components are less than water, so that any BTEX present at levels above solubility would be dispersed in the soil pore structure or form a floating layer within the groundwater system. The high vapor pressure of benzene and, to a lesser extent, that of remaining BTEX compounds could produce a significant gas phase transfer within the vadose zone and possibly into the atmosphere.

3. REGULATORY CONCERNS FOR BTEX AND FUEL OILS

The permitted level for benzene in drinking water, established by the New Jersey Department of Environmental Protection, is 2 μg/liter and the following maximum contaminant levels have been proposed: benzene 1 μg/liter, toluene 1,000 μg/liter, and total xylenes 40 μg/liter (New Jersey Register, 1992). BTEX concentration standards for soil have also been proposed and are divided into surface and subsurface soil standards. The proposed standards for surface soils are: benzene 3 mg/kg soil; toluene, 1,000 mg/kg soil; and total xylenes, 360 mg/kg soil (New Jersey Register, 1992). The proposed standards for subsurface soils are benzene, 1 mg/kg soil; toluene, 500 mg/kg soil; and total xylenes, 10 mg/kg soil (New Jersey Register, 1992). The threat to human health from low levels of BTEX contamination in drinking water is substantial, with a one-in-a-million cancer risk for lifetime exposure to benzene concentrations of 0.67 μg/liter in drinking water (Federal Register, 1984).

4. REQUIREMENTS FOR *IN SITU* BIOLOGICAL SOLUTIONS TO PETROLEUM CONTAMINATION

Bioremediation of subsurface contamination requires the presence of microorganisms within the contaminated site capable of utilizing the carbon, present as contaminant, for cell growth and energy. For biological activity to occur the microorganisms, typically bacteria, require nutrients, such as ammonia, nitrate, and phosphate, and an electron acceptor. For petroleum-contaminated sites, oxygen has usually been the electron acceptor. However, anoxic or anaerobic systems are possible with nitrate, sulfate, iron, manganese, carbon dioxide, and other organic compounds available to function as the electron acceptor.

Implementation of an *in situ* bioremediation system requires several initial steps to characterize the subsurface contamination and physical characteristics. The steps, including implementation of the system, should include (Thomas and Ward, 1989; Wilson et al., 1990; Lee et al., 1984, 1988):

1. Site characterization
2. Free product recovery (if necessary)
3. Laboratory feasibility tests for microbial and nutrient evaluation
4. Laboratory treatability tests for evaluation of bioremediation potential, utilizing site samples
5. Pilot-scale operation (if necessary)
6. System design
7. Operation
8. Monitoring and closure

The extent of the contaminant plume and the concentration ranges of the various contaminants need to be defined to determine the scope of the problem. Definition of the contaminant plume also should include evaluation of the nature of contaminant–soil interactions. Contaminants may be present dissolved in the soil solution (aqueous) phase, sorbed to the soil, or dispersed in the organic phase. Evaluation of soil chemistry should include determination of the soil pH and buffer capacity, moisture content, cation exchange capacity, particle size distribution, organic matter, total Kjeldahl nitrogen (TKN), and available phosphate. Evaluation of groundwater chemistry should include measurement of pH, temperature, dissolved oxygen, total organic carbon (TOC), total inorganic carbon, nitrate, chloride, sulfate, total dissolved solids, and dissolved iron. These parameters will assist in evaluating what the current limiting factors for natural biodegradation are, and in determining necessary nutrient, pH, or other needed system modifications. TOC can be an important parameter because it reflects the total concentration of dissolved substrate that may be present, in addition to anticipated hydrocarbon contamination. Microbial activity and biodegradation rates can be a strong function of soil moisture content in unsaturated soils. Optimum moisture has been reported to be between 30% and 90% relative saturation (Dibble and Bartha, 1979a) and approximately 50% relative saturation (Chiang and Petkowsky, 1993). Elevated concentrations of dissolved iron may lead to subsequent flow difficulties with an aerobic groundwater recirculation system resulting from iron oxidation and precipitation *in situ* (McAllister et al., 1993; Chiang and Petkowsky, 1993).

Soil permeability must be determined and hydraulic conductivity should be 10^{-4} cm/sec or greater to allow transport of nutrients and electron acceptor through the contaminated zone (Thomas and Ward, 1989). Because of subsurface heterogeneity, the site characterization should include multiple sampling points. The physical characterization of the site also should include groundwater information, such as flow rate and direction, and depth to groundwater, with seasonal fluctuations. Temperature, including seasonal variations, of the soil and groundwater can limit the potential for bioremediation and should be determined. Low temperatures decrease microbial activity and can cause viscosity increases in the oil, reduced volatility, and increased solubility of toxic short-chain alkanes resulting in increased lag periods before biodegradation (Atlas and Bartha, 1972; Leahy and Colwell, 1990).

The design and operation of free product recovery systems are essential to minimize the source of contamination. Recovery systems can include soil vapor extraction, which can remove much of the volatile portion of the contamination. Recovery of product by physical means can account for as much as 90% of the contaminant (Brown et al., 1985) but often recoveries of only 30%–60% are eco- nomically feasible (Yaniga and Mulry, 1985). Free product recovery minimizes the bioremediation system requirements, such as oxygen and nutrient transfer, allowing smaller systems and/or shorter treatment periods.

5. MICROBIAL POPULATIONS AND ENUMERATION

Feasibility tests for petroleum contamination remediation evaluate the microbial and nutrient characteristics of the subsurface soils and groundwater. Most frequently, the indigenous microbial populations in a contaminated area are capable of biode- grading petroleum-based contaminants, but population density or activity is often limited by one or more local environmental factors. Enumeration of total microbial populations typically is determined by serial dilution of site samples,then plating on one or more non-selective growth media. Contaminant-specific microbial enumera- tion is determined utilizing a selective culture medium with the contaminant as the sole source of carbon. Plate count techniques do not provide complete quantitation of microbial populations, with results estimated to vary from <1% to as much as 50% efficiency (Lee et al., 1988; Riser-Roberts, 1992). Soil microbial numbers typically range from 10^3 to 10^8 colony-forming units (cfu)/g or ml of sample (Hardaway et al., 1991; Frankenberger et al., 1989).

Petroleum-degrading bacteria usually increase in number following prolonged exposure to the contaminant (Atlas, 1984, 1986, 1991). A site contaminated with JP5 fuel contained two to three orders of magnitude greater bacteria than the surrounding uncontaminated soil (Ehrlich et al., 1985). Initial contamination causes a reduction in microbial numbers, with chemical toxicity being assumed to be the cause of the decrease in bacterial colonies (Odu, 1972). The process of natural acclimation, or adaptation, occurs over time as only the bacteria that can utilize the contaminant as carbon source remain viable. Three explanations for the adaptation process are (Leahy and Colwell, 1990; Spain et al., 1980; Spain and vanVeld, 1983):

1. Induction and/or depression of specific enzymes
2. Genetic variability with altered metabolic capabilities
3. Selected enrichment for bacteria with desired metabolic capabilities

Bacteria and fungi capable of degrading petroleum hydrocarbons are reported to be ubiquitous in nature. Numerous researchers have identified petroleum-degrading bacteria in soil and in freshwater and marine environments (Atlas, 1991; Bossert and Bartha, 1984; Cooney, 1984; Gibson and Subramanian, 1984). The most prevalent hydrocarbon-degrading bacteria in soil and aquatic environments are *Achromobacter, Acinetobacter, Alcaligenes, Arthrobacter, Bacillus, Flavobacterium, Nocardia,* and *Pseudomonas* spp. and the coryneforms (Leahy and Colwell, 1990). *Pseudomonas* spp. are reported to be the most widespread and to have the broadest capability to degrade pollutants, with *Corynebacterium* spp. having the widest application for heterocyclic compounds (Kobayashi and Rittmann, 1982).

Several direct and indirect measurements may be used as indicators of *in situ* microbial activity, including bacterial cell counts from soil and groundwater samples (Lee et al., 1988). Elevated carbon dioxide and reduced oxygen concentrations in soil gas samples are indicative of aerobic degradation, and relative concentrations of methane can be used as a relative indicator of methanogenic activity. Oxygen depletion coupled with elevated concentrations of dissolved iron in groundwater can be indicative of active iron-reducing bacterial populations.

6. CHEMICAL/NUTRIENT REQUIREMENTS

Feasibility tests for the remediation of petroleum contamination evaluate other environmental characteristics such as pH, nutrient concentrations, available oxygen or other electron acceptors, soil moisture content, and salinity. Microbial populations typically require near-neutral pH conditions for optimal activity. Variations in pH from 6.5 to 8.0 or 8.5 are usually acceptable. A soil pH of 7.8 has been reported as the optimum for petroleum degradation (Dibble and Bartha, 1979b), but each system should be evaluated individually. No effect on toluene degradation has been observed for pH values between 5.0 and 7.5 (Armstrong et al., 1991). Bowlen (1992) investigated the effect of solution pH on the biodegradation of BTX under denitrifying conditions. Biodegradation was most rapid at pH values between 7.0 and 8.0, but significant contaminant removal occurred at pH 6.5. Lower pH values resulted in inhibition of degradation activity, but activity generally was restored when the pH was adjusted to greater than 6.5. Soils with pH values outside of the optimal range may be modified with various nutrient amendments. Biodegradation rates for gasoline in an acidic soil, pH at 4.5, increased by nearly twofold when the pH of the soil was adjusted to 7.4 (Verstraete et al., 1976). Basic soils can be treated with nitrogen fertilizers, such as ammonium sulfate or ammonium nitrate, to reduce the soil pH value. Acidic soils can be treated with a liming technique utilizing various forms of $CaCO_3$ or $MgCO_3$. Use of divalent cations also minimized disaggregation of the soils, which can result when monovalent cations such as sodium, i.e., Na_2CO_3, are utilized. Many petroleum process wastes are highly acidic and may require significant amendments prior to treatment.

Many soil systems are severely nutrient limited due to the high concentration of carbon present. Microbial communities require nitrogen, typically as ammonia or nitrate, and phosphorus, as phosphate, for continued cellular growth. A general rule of thumb for nutrients is to maintain a C:N:P ratio of 100:13:3 to 100:10:1, respectively. In most cases the required nutrient loading would be done on a systematic basis to prevent plugging of the injection system by biomass growth or accumulation of salts (Frankenberger, 1988). A typical implementation strategy is to alternate oxygen and nutrient injections so that bacterial growth in the immediate vicinity of the injection well is minimized.

Several recent reports, primarily from bioventing applications, have indicated that much lower nutrient ratios may be sufficient for hydrocarbon degradation in the subsurface environment (Hinchee et al., 1989; Miller, 1990). Effective stimulation of bioremediation has been reported with C:N ratios of 200:1 and C:P ratios of 1,000:1 (Wang and Bartha, 1990). Significant reductions of hydrocarbons have been observed in the presence of mineral inorganic nutrients. Analysis of the TKN and total phosphorus in the soils has indicated that sufficient nutrients may be bound within the organic fraction of the soil to support the observed petroleum degradation (Miller, 1990). At subsurface locations, with limited bacterial growth, the possibility of nutrient recycling from decaying biomass also exists. Efficient recycling of nutrients could allow slow, continual utilization of the carbon contaminant without large influxes of nitrogen or phosphorus, and with little or no increase in viable bacterial populations.

Nitrogen sources include various ammonium or nitrate salts, including ammonium nitrate, ammonium sulfate, urea, and diammonium phosphate (Frankenberger, 1988; Frankenberger and Karlson, 1991; Aggarwal et al., 1991a,b). Additions of nitrate compounds are limited by regulatory concerns with nitrate concentrations, as nitrogen, limited to 10 mg/liter in groundwater. Urea contains the highest percentage of ammmonia (46%) and when applied to soils hydrolyzes to ammonia and carbon dioxide. Urea has been reported as an acceptable nitrogen source for remediation of petroleum hydrocarbons (Wang and Bartha, 1990) and also has been indicated to be preferred over nitrate salts (I. Bossert, personal communication, 1993). However, petroleum hydrocarbons, especially refined oils, have been reported to inhibit the hydrolysis of urea (Frankenberger, 1988). Because of the potential for hydrolytic inhibition inorganic forms of nitrogen most frequently have been used, with ammonia generally recommended over nitrate. Nitrogen has also been added in the gaseous phase as anhydrous ammonia (Dineen et al., 1989). Various nutrient application rates and formulations, especially for landfarming treatments, were recently reviewed (Frankenberger and Karlson, 1991).

Phosphate sources are typically orthophosphate salts, i.e., sodium, ammonium, or potassium. Various polyphosphate formulations are utilized, particularly when stabilization of hydrogen peroxide is required (Aggarwal, 1991b). Actual orthophosphate requirements tend to exceed theoretical values due to high levels of adsorption to soil and precipitation by multivalent cations such as iron or calcium. Various water-soluble organic phosphorous compounds such as potassium diphenyl-, calcium diethyl- or glycerol-phosphate, and potassium diphenyl pyrophosphate are available (Frankenberger and Karlson, 1991). Four polyphosphate compounds, pyrophosphate, tripolyphosphate, and trimetaphosphate and orthophosphate, were evaluated as nutrient sources in soils, concurrent with hydrogen peroxide treatment (Aggarwal et al., 1991a,b). Sorption of the polyphosphate formulations was minimal but pyrophosphate and tripolyphosphate rapidly decomposed to orthophosphate, which then sorbed to the soils. Trimetaphosphate sorbed less to the soil, had a higher solubility, and hydrolyzed slower to orthophosphate. The authors do not recommend any phosphate additions to soils with a high calcium content due to nearly complete precipitation of the added phosphate.

Screening of nutrient requirements is recommended to assess the potential for nutrient binding to site-specific soils. Hardaway et al. (1991) used a screening test where nutrient solutions were added to a soil slurry at a liquid to solid ratio of 1:2, allowed to react for 4 hours, and the residual nutrient concentration in the aqueous phase was measured. Addition of potassium also has been reported to enhance mineralization of toluene in contaminated groundwater (Armstrong et al., 1991). However, additions of monovalent cations should be minimized and balanced with the addition of divalent cations to avoid disaggregation of the soil structure. Maintenance of a sodium absorption ratio (SAR) of less than 2.0 is recommended for injection of nutrients.

7. DEVELOPMENT OF *IN SITU* PROCESSES

7.1. Remedial Options—Addition of Electron Acceptor

Many elements, such as nutrients, pH control, or hydraulic control of contaminants during aquifer remediation, need to be addressed to provide an aerobic *in situ* process. However, the paramount concern is how to get sufficient oxygen, serving as electron acceptor, to the contaminated area. Several processes have been utilized, including reinjection of oxygenated water to aquifers utilizing air, pure oxygen, or hydrogen peroxide to provide the oxygen source. Recently, air or oxygen sparging directly into the saturated zone has been utilized to enhance both stripping of volatile components into the vadose zone and biostimulation through increased dissolved oxygen concentrations. The technique is commonly called *biosparging*. Treatment of the vadose zone has utilized either a bioventing approach or percolation/flooding of the unsaturated zone with oxygen-rich groundwater.

Anoxic processes have been less frequently applied, but may include the injection of nitrate and various forms of sulfate into saturated subsurface systems. Nitrate enhances denitrification processes whereas sulfate serves as an electron sink for sulfidogenic processes to treat petroleum fuel contamination.

7.2. Aerobic Approaches

7.2.1. *Landfarming*

Landfarming is an above ground process where contaminated soils are provided with required nutrients and spread out in thin layers that can be tilled by various mechanical means. Tilling provides oxygen and ensures the thorough dispersal of added moisture and nutrients. Landfarms were originally operated with contaminated soils and wastes added directly onto the ground, with no provisions for leachate control or vapor emissions. Present day operations require containment of soils with leachate collection systems and above ground enclosures for vapor control. Petroleum wastes were historically treated by landfarming techniques (Frankenberger and Karlson, 1991). At one time it was estimated that nearly half of petroleum oily wastes were treated at landfarm operations (Huddleston, 1979).

7.2.2. Oxygen Sparging/Hydrogen Peroxide

Enhancement of *in situ* bioremediation through injection of oxygenated groundwater and usually nutrients was first demonstrated at a site contaminated with gasoline from a leaking pipe (Raymond et al., 1975). Following free product recovery, the bioremediation was initiated by batch-wise injection of nutrients and air sparging within the injection wells. Nutrients consisted of inorganic ammonium and phosphate salts. Within 1 year of termination of the process, no gasoline was detected in the site groundwater (Lee et al., 1988).

Numerous applications have approximated the procedure implemented by Jamison et al. (1975). Oxygen addition systems have been varied in an attempt to increase the transfer of oxygen into the subsurface. Air injection has been replaced with pure oxygen to increase theoretical dissolved oxygen concentrations from 8–10 mg/liter with air, to 40–50 mg/liter with pure oxygen. The expense of pure oxygen for injection has been minimized by on site generation utilizing zeolite catalysts (Prosen et al., 1991). Oxygen has been added directly to groundwater, by sparging or in above ground systems prior to reinjection of groundwater. Several treatment scenarios have been investigated at a JP4 spill site in Traverse City (MI). The investigations, performed by several researchers from EPA Kerr Laboratories and others, have included air injection, hydrogen peroxide injection, and nitrate addition to stimulate *in situ* biorestoration (Wilson et al., 1987, 1989, 1990; Ostendorf and Kampbell, 1990, 1991; Hutchins and Wilson, 1991; Huling et al., 1991; Hutchins, 1991; Hutchins et al., 1991; Lawes, 1991; Barenschee et al., 1991; Lee and Raymond, 1991; Wilson and Ward, 1988). Chiang and Petkowsky (1993) reported a 91% reduction in BTEX mass in a contaminated aquifer in less than one year, through recirculation of groundwater with above ground re-oxygenation. Groundwater was reinjected at several depths to achieve uniform distribution in the contamination plume. Contaminant transport modeling was used to demonstrate a reduction of the remediation time required by several years.

Dissolved iron is typically found in many ground waters when dissolved oxygen levels are low. Oxygenation, by air, oxygen, or peroxide addition, while increasing dissolved oxygen content also can cause precipitation of the dissolved iron as it is changed from Fe^{2+} to Fe^{3+} forms. Iron precipitate formed in groundwater by oxygenation, typically with hydrogen peroxide, is removed via filtration prior to reinjection of the groundwater. Other chemical treatments include pH adjustment with caustic or lime and addition of flocculating agents, such as $FeCl_3$, which may be required to form a precipitate suitable for high flow rate filtration.

Hydrogen peroxide addition has been used at many sites to provide significant increases in dissolved oxygen. Peroxide additions up to 500 mg/liter have been utilized to increase dissolved oxygen concentrations. At a site contaminated with up to 1,500 mg diesel fuel/kg soil, injections of nutrients and hydrogen peroxide over a 6 month period reduced hydrocarbon concentrations to background levels (Frankenberger et al., 1989). At a site where biofilms were plugging injection wells, hydrogen peroxide was added at 100 mg/liter to enhance hydrocarbon removal and to control biofilm growth (Lee et al., 1988). Additional case histories of peroxide additions are discussed in the review by Lee et al. (1988).

The success rate of peroxide addition has been limited by several problems. Hydrogen peroxide, at levels as low as 100–200 mg/liter, can be toxic to microorganisms. Acclimation of subsurface microorganisms has been accomplished by initiating treatment at 20–50 mg/liter and slowly increasing the concentration to operational levels of 100–500 mg/liter. Peroxide can decompose rapidly to gaseous oxygen in the presence of iron or enzymatic catalysts such as catalase (Aggrawal et al., 1991a). The rapid degassing can lead to formation of large air bubbles that can block pores in aquifer material (Lee et al., 1988). As discussed previously, iron oxidation in the presence of peroxide can lead to precipitation and plugging of injection wells or aquifer formations.

Laboratory studies examined several additives reported to stabilize peroxide solutions with soils from Eglin AFB (Aggrawal et al., 1991a). Batch and continuous flow experiments evaluated trimetaphosphate (TMP), citric acid, perborate, boric acid, catechol, and fluoride and found that only TMP and citric acid provided any significant improvement in peroxide stability. During the follow-up field evaluation, peroxide solutions (500 mg/liter) were injected through injection galleries. The peroxide solutions were stabilized with the phosphate nutrient mixture, Restore 375. Following several months of injection, no detectable elevation in hydrogen peroxide was observed even at a monitoring well only 2 feet from the point of injection (Hinchee et al., 1991a).

7.2.3. Bioventing

Volatilization of BTEX from subsurface soil systems can be accomplished by soil vapor extraction (SVE). Biological transformation and SVE recently have been combined as bioventing, where the vapor flow rates from the SVE system are reduced to provide sufficient oxygen to the subsurface, while minimizing volatilization and mobilization of the organic contaminants toward the surface. The concentration of oxygen in air, 20.9 vol%, provides a highly efficient oxygen transfer mechanism. Oxygen concentrations in groundwater are limited to the solubility of oxygen in water. Dissolved oxygen concentrations are as follows: air, 8–10 mg/liter; pure oxygen, 40–50 mg/liter; hydrogen peroxide, 500–1000 mg/liter if completely utilized under ideal conditions (see Section 7.2.2.). Even with the most optimistic peroxide conditions, air provides approximately 200 times more oxygen per unit of transfer medium to the subsurface.

Bioventing is ideally suited for the treatment of the vadose zone contaminated with middle distillate, low volatility fuels, such as diesel or jet fuels or SVE residuals from gasoline fuel spills. The majority of lighter weight products in gasoline can effectively and economically be removed from vadose zone soils with conventional SVE systems. Adaptation of SVE systems from removal systems to oxygen transfer systems was initially proposed in the early 1980s from laboratory studies at the Texas Research Institute (Hoeppel, 1991). Laboratory studies indicated that bioremediation was responsible for more than one-third of the gasoline removal observed during soil venting. Several researchers suggested utilization of SVE to provide subsurface oxygen (Hoeppel, 1991) and Ely and Heffner (1988) obtained a patent for utilization of SVE systems to enhanced bioremediation.

Bioventing system hardware requirements are similar to SVE systems, with injection and extraction wells or trenches connected via piping networks to a blower or vacuum pump. The size of the bioventing system blower can be less than SVE systems due to the lower flow rate requirements. In addition, off-gas collection/treatment systems requirements are substantially reduced or eliminated due to decreases in vapor transport to the surface. The placement of extraction wells is typically outside of the zone of contamination to enhance the biological treatment of contaminants within the subsurface and to minimize the transport of volatilized compounds to the surface (Hoeppel, 1991; Dupont, 1993).

Bioventing demonstrations have been carried out at several Air Force sites around the country over the past few years and are now well-documented in the literature (Dupont, 1993; Hinchee et al., 1989; Miller et al., 1991; Sayles et al., 1992; Brown et al., 1991; Hinchee and Arthur, 1991; Dupont et al., 1991; Hinchee et al., 1991b,c; Urlings et al., 1991). Environmental conditions for implementation have ranged from Florida to Alaska and soils depths have varied from near surface to approximately 100 feet below ground surface. An *in situ* respirometry method has demonstrated oxygen uptake and carbon dioxide production from contaminated subsurface soils (Hinchee et al., 1991c; Hinchee and Ong, 1992). Operating requirements have varied between sites with no additional nutrients required at Tyndall AFB (Miller, 1990) to Hill AFB where moisture and nutrient additions greatly enhanced the biodegradation rates (Dupont, 1993). At the Hill AFB site, laboratory studies indicated that nutrient addition was required, but field data indicated that the greatest improvement in removal rates was observed after moisture addition, and nutrient additions may not have been required. The improvement observed with moisture addition may be the result of relatively dry subsurface soils following a year of high volume SVE at the site. Organically bound nitrogen was the primary source of nutrients at Tyndall AFB (Miller, 1990). Total nitrogen concentrations, inorganic and organically bound, should be determined in the subsurface soils as a better indicator of nutrient requirements.

Additional operating parameters have been developed during the various site demonstrations. Intermittent operation of the blower systems, i.e., 45 min/day, has provided sufficient oxygen to promote biodegradation without the need for surface containment or treatment of off-gas (Dupont, 1993). Bioventing in cold climates in shallow soils (2–7 ft) has been demonstrated at Eielson AFB in Fairbanks, Alaska. The active warming system applied warm water (35°C) through soaker hoses placed 2 feet below ground surface (Sayles et al., 1992). Winter soil temperatures were typically at or above 10°C in the actively warmed soils and −20°C in the unheated control soils. The heated soils maintained significant biological activity throughout the winter season.

To implement a bioventing system, additional information needs to be collected during the site characterization. A soil gas survey should be conducted with gas samples taken at a variety of depths throughout the suspected area of contamination. The gas samples should be analyzed for contaminant, oxygen, and carbon dioxide concentrations. The samples provide baseline data for assessing the removal efficiency of the system and for establishing any oxygen deficiency within the site soils. If oxygen is already present, then additions of more oxygen will not be beneficial (Hinchee and Ong, 1992). The permeability of the vadose zone soils to air flow and the radius of influence of venting wells must be determined to provide a full-scale bioventing design. Permeability is a function of grain size, soil homogeneity, porosity, and moisture content. A site should be permeable enough to allow a minimum of one soil gas exchange per week (Hinchee and Ong, 1992). The radius of influence is the maximum distance from an injection well where significant pressure or vacuum differences can be determined. Optimal placement of wells and the sizing of blower equipment typically have been determined empirically, although subsurface mathematical modeling recently has been applied to some sites (Marley et al., 1992).

An *in situ* respirometry method has demonstrated oxygen uptake and carbon dioxide production from contaminated subsurface soils (Hinchee and Ong, 1992). The respiration test monitors oxygen consumption and carbon dioxide production in contaminated and background locations after oxygenating the subsurface. An inert gas tracer is added to measure air diffusion characteristics within the soils. The oxygenation process is typically operated for 24 hours and then the soil vapor phase monitored for approximately 72 hours. Oxygen utilization is apparent within the 3 day monitoring period in the contaminated soils but not in the background soils. Carbon dioxide production has not proven to be a reliable monitor of biological activity at the sites investigated. At Tyndall AFB, the oxygen utilization rates determined by the respiration test were consistent with values obtained during long-term operation of a nearby pilot test (Miller, 1990; Hinchee and Ong, 1992).

Several configurations combining injection and extraction wells have been demonstrated at various AFB locations. The simplest and lowest cost installation utilizes only injection well(s) placed within the contaminated zone where air is injected into the subsurface. While a low cost alternative, the system provides no control over migration of vapors. Other alternatives, enabling better control of fugitive vapors, provide for injection within the contaminated zone with an extraction well located within clean soil. In theory hydrocarbons from the contaminated off-gas have been degraded before reaching the extraction well. Another alternative provides extraction from within the contaminated area with reinjection of the off-gases into clean soils outside the zone of contamination. The reinjected air is then drawn back through the contaminated zone in a recirculation loop (Hinchee and Ong, 1992).

7.2.4. Biosparging

An extension of soil vapor extraction and bioventing, referred to as *biosparging* or *air sparging,* has emerged in recent years to extend the volatilization and biological treatment of subsurface VOCs by sparging air or oxygen into contaminated soils within the saturated zone. The added oxygen may stimulate biological activity in the saturated zone or may help transfer the contaminant to the vadose zone where biological activity and/or vapor extraction can occur. The contaminants within the saturated zone partition into the advective air phase, effectively simulating an *in situ* air stripping system. Biosparging has been described as a "crude air stripper in the subsurface" (Brown and Fraxedas, 1991).

Biosparging addresses one of the major limitations to the bioventing technique by expanding application to the saturated zone. Enhanced oxygen transfer in saturated media has previously been accomplished by groundwater drawdown followed by implementation of the bioventing system. Groundwater drawdown requires the removal of large volumes of groundwater, which can be difficult and costly and often leads to treatment of the extracted water. De-watering cannot be applied when contamination exists at any significant depth within the saturated zone.

Biosparging systems are often combined with bioventing or SVE systems because the contamination of fuel products that have reached the saturated zone is spread between the vadose zone, the capillary fringe area, and the upper few feet of the saturated zone (Marley and Hoag, 1990). The remedial system must address each of these areas. Control of volatile components sparged from the saturated zone can be accomplished in the vadose zone system by biological means or by above ground collection. Bioventing or SVE systems designs can accomplish control of saturated zone emissions while remediating vadose zone contamination.

System designs to date have been empirically based. Major design parameters include contaminant type, extent of contamination, gas injection pressures and flow rates, site geology, bubble geometry, injection interval (areal and vertical), and radius of influence (Marley et al., 1992; Martin et al., 1992). Contaminants with high volatility and low aqueous solubility, such as the lighter petroleum products, can be removed from the saturated subsurface by dissolution from the soil or liquid phase and displacement into the vapor phase for transport to the surface. Residual, less volatile compounds, that are major components of the fuel oils and their products, would be treated through enhanced biological activity due to increased dissolved oxygen content. These and other design aspects were also discussed by Griffin et al. (1991).

The gas injection pressure required to force air into a saturated formation is a combination of the static water head and the "air entry pressure" (Marley et al., 1992). Air entry pressure is the pressure needed to overcome capillary forces and displace water from the saturated soil. Approximately 1 psi of air pressure is required for every 2.3 feet of hydraulic head (Brown, 1993). Moderate increases in injection pressure, beyond the required air entry pressure, can increase the horizontal component of air movement, but large increases can cause vertical channeling along well implacements or excessive spreading of contaminants (Marley et al., 1992; Brown, 1993).

Site geology is often considered the predominant design parameter. Many of the limitations to biosparging can be linked to heterogeneities within the site geology. Homogeneous, course-grained soils are considered optimal for biosparging applications (Marley et al., 1992). Air entry pressures are lowest in coarse media, and air distribution extends evenly in horizontal and vertical directions. Horizontal flow is important to establish a significant and consistent radius of influence. Vertical flow provides transport out of the contaminated area under controlled conditions. Soil heterogeneities, such as a clay lens, can disrupt the flow patterns and cause spreading of nondegraded contaminants, possibly beyond the control range of the surface extraction systems. Heterogeneities can be areas with only slightly reduced permeabilities, as the sparged gases will follow a path of least resistance through a formation. Once a flow channel has been opened, it requires less air pressure to remain open than to open a new channel. If channeling occurs, the area of influence for the treatment system may be reduced substantially. Contaminated areas located directly above areas of reduced permeability may be highly contaminated due to the downward movement of contaminants, but largely untreated due to air flow only around the periphery of the contaminated area. A horizontal stratum of increased permeability can lead to spreading of contaminants along the preferential flow path of higher permeability. Volatilized contaminants can migrate significant distances along high permeability strata and move beyond the radius of influence of the vadose zone control system. A radius of influence of greater than 60 feet has been reported in a stratified soil (Marley et al., 1992; Martin et al., 1992). To reduce and/or control the risk of channelized flow, a thorough geological evaluation of the treatment area is required. Boring logs from any borehole placed at the site should be complete with no unevaluated intervals. Pilot tests should evaluate the potential for channelized flow (Brown, 1993).

Site materials must be permeable to the flow of air. Horizontal permeability is typically much greater than vertical permeability, which helps to increase the radius of influence. Permeability limits outlined by Brown (1993) vary depending on the ratio of horizontal to vertical permeability. If the horizontal–vertical permeability ratio is low (<2:1), then sparging can be implemented in soil systems with hydraulic conductivities of 10^{-5} cm/sec or greater. When the ratio is high (>3:1), permeabilities must be greater than 10^{-4} cm/sec for sparging to be successful.

Vertical well screen intervals are typically short, 1–3 feet, because air enters the formation at the shallowest possible depth; entry pressures increase with increasing depth. Placing well screen intervals deeper in the formation increases the radius of influence, but requires greater air pressures to overcome the increased hydraulic head. In homogeneous material, air injection patterns approximate a parabola so that only small increases in depth of injection result in significantly increased radii of influence.

Sparge depth is limited to a range of 4–30 feet below the water table (Brown, 1993). At depths less than 4 feet, the radius of influence is too small to be economically feasible; the 4 foot limit realistically can be increased by several feet. At 30 feet, control of the air flow patterns becomes too difficult to manage safely; horizontal migration of contaminants beyond the range of the vadose zone extraction system poses too great a risk. In addition, at 30 feet the operating costs are higher due to increased hydraulic head; air injection pressures are at least 10 psi greater than at the lower limit of 4 feet.

The placement of sparging wells is dependent on each of the factors discussed above, and on the location and extent of contamination at the site. The depth of contamination below the water table and the required radius of influence determine the vertical placement of the sparge wells. The number of sparge wells and their spacing is determined by the areal extent of contamination and the radius of influence obtained at the various well locations.

The primary concern about air sparging systems is loss of control of the injected air. Loss of control can be a result of lateral spreading of contaminant due to flow around a low permeability area, flow along a high permeability area, or a poorly designed vadose zone vapor extraction system. At one pilot demonstration, VOCs were observed in monitoring wells 60 feet from the injection wells (Marley et al., 1992; Martin et al., 1992). The cause of significant lateral migration was believed to be a combination of lithological characteristics, including the micaceous structure of sands, lateral channeling, and fine-grained silt layers. The complexity of flow patterns in the subsurface indicates that pilot tests should be required at sites to ensure control of vapor migration during full-scale remediation.

Alternative sparging methods have been discussed recently in the literature. Newer applications include "well/trench sparging"(WTS), where air is injected into an open well or trench (Pankow et al., 1993), or *"in situ* sparing" (ISS), where air is injected directly into saturated porous media through a well screen, discussed previously (Marley and Hoag, 1990; Marley et al., 1992; Brown and Fraxedas, 1991; Brown, 1993). ISS is suggested when the contamination is present in a defined area of the vadose zone and migrates downward into the groundwater within a confined area. WTS has been proposed for treatment of groundwater in areas where the zone of contamination is dispersed and not well defined. Long-term plume control of contaminated aquifers would be provided by installing either a trench across the width of the contaminated flow or a cut-off wall with a series of sparge gates within the wall. The sparge gates would comprise 5%–20% of the cut-off wall to prevent excessive spreading of the plume behind the cutoff wall (Pankow et al., 1993).

The trench approach applies to areas where the source of contamination cannot be located or effectively remediated. To minimize the movement of the contaminated groundwater off site, the control trench would be installed on the down gradient side of the site. The contaminated groundwater, typically moving off site at rates of 0.015–0.5 m/day, would pass slowly through the sparge trench, where volatile contaminants would be stripped from the groundwater to the surface for either release to the atmosphere or treatment. Significant oxygenation of groundwater would stimulate the biological removal of any remaining organic contaminants. Physical removal is the primary goal of this technology, and the degree of biological enhancement is unclear.

Case histories detailing the implementation of air sparging systems have concentrated on volatilization of contaminants from groundwater into the vadose zone with subsequent capture by an SVE system. As with bioventing, proof of enhancement of biological activity will be difficult without substantial monitoring of dissolved oxygen levels in the saturated zone. Contaminants have included chlorinated VOCs such as trichloroethylene (TCE) and BTEX from various gasoline spills (Ardito and Billings, 1990; Brown et al., 1991; Marley et al., 1992; Martin et al., 1992). At a gasoline site in New Mexico, air sparging replaced a pump-and-treat system and resulted in removal of greater than 90% of the dissolved benzene within 9 months. Installation costs for the pump-and-treat and the air sparging systems were comparable, and the air sparging system saved $9,000/month in operating costs (Ardito and Billings, 1990).

7.2.5. Biofiltration

Biofiltration for the treatment of vapor phase contamination, especially hydrocarbons, has been a recent development within the United States. The simplest form of biofiltration is a soil bed with a horizontal network of perforated pipe placed 2 to 3 feet below ground surface (Bohn, 1992). The vapor phase contaminants are passed from the piping into the soil and utilized as a carbon source by microorganisms present in the soil. The system can be compared with the process described in a bioventing system. A similar system, which could be described as passive bioventing, has been demonstrated at the Traverse City site, where analysis of the soil gas in the vadose zone above the contaminated soil has demonstrated significant reductions in hydrocarbon concentrations as the soil gas migrates upward through the vadose zone (Ostendorf and Kampbell, 1990, 1991). The vadose zone soils, rich in nutrients, were approximately 4 m thick above an extended zone of capillary fringe hydrocarbon contamination.

Reactor based configurations for biofiltration utilize a variety of packing materials, including soil, peat wood bark, and other bulking agents. The bulking agents may be required to minimize pressure drops during vapor flow through porous packed media. Efficient operation requires abundant microorganisms, nutrients, moisture, and pH buffering capacity. An alternative configuration for vapor phase treatment is the biotrickling filter, which operates by recirculating liquid media over packing either co- or countercurrent to the vapor flow (Dharmavaram, 1991). Operational parameters, including temperature, bed moisture, and pH, need to be carefully controlled if consistent removal efficiencies are to be maintained in the system.

Treatment of vapor phase hydrocarbons in the field has been limited, but several bench-scale investigations have indicated that the volatile fractions of various hydrocarbons can be degraded in biofilters. Biofilters and soil beds were utilized to degrade aromatics such as styrene and toluene (Ottengraf et al., 1986). Removal of the volatile portion of gasoline and middle distillate fuels, such as aviation and jet fuels, has been demonstrated in biofilters using soil, carbon, diatomaceous earth, and various combinations as support media (Hodge et al., 1991).

7.3. Anaerobic Approaches

Aerobic biodegradation of aromatic hydrocarbons has been widely studied, but much less attention has been given to the anaerobic biodegradation of aromatic hydrocarbons. The first aromatic compounds shown to be anaerobically biodegraded all had at least one oxygen or nitrogen substituent associated with the benzene ring Grbić-Galić and Young, 1985). The biodegradation of benzoates, cresols, phenols, and various aromatic acids has been reviewed by several authors (Evans, 1977; Sleat and Robinson, 1984; Berry et al., 1987; Berwanger and Barker, 1988; Ball et al., 1991). The benzene ring, without any oxygen substituents, is highly resistant to biodegradation. Batch studies by Horowitz et al. (1988), with lake sediment and municipal digester sludge cultures, showed no degradation of toluene. Schink (1985) also observed no removal of benzene, toluene, or xylene (no indication of which isomers) during batch studies under methanogenic conditions. The first evidence of possible biodegradation of the xylene isomers was observed in two different, contaminated anaerobic aquifers. In an aquifer, including trace amounts of BTEX (820 μg/liter for the *m*- and *p*-xylenes) (Patrick et al., 1985), most of the aromatic compounds showed a reduction in concentration proportional to the distance traveled from the landfill site. However, the xylene isomers showed a much greater reduction over the same distance. An additional aquifer study described the anaerobic degradation of benzene, toluene, and smaller amounts of the xylenes (Battermann and Werner, 1984). The reductions of BTEX were seen after injection of nitrate into the contaminated aquifer.

Laboratory biodegradation studies have been carried out under methanogenic, denitrifying, sulfidogenic, and iron-reducing conditions, using both batch and continuous flow column methodologies. Only limited field demonstrations have utilized anaerobic techniques, and case studies are presented at the end of the following laboratory sections.

7.3.1. Biodegradation by Denitrification

Denitrification has been shown to be an effective process for biodegradation in anoxic systems with nitrate ion available as the terminal electron acceptor. The nitrate is reduced, through a series of steps, to nitrite, nitric oxide, nitrous oxide, and finally dinitrogen, which allows a concomitant oxidation of the carbon substrate. Several studies indicated complete degradation of toluene and the xylenes by denitrifying bacteria (Zeyer et al., 1986; Kuhn et al., 1985, 1988). Laboratory columns were packed with aquifer material taken from a river/groundwater infiltration site. The *p*- and *m*-xylenes were removed within the first 47.7 cm of the soil column, and the *o*-xylene was completely removed within 87.7 cm in the column. Nitrate requirements for the denitrification pathway were demonstrated by nitrate dependence and substrate removal techniques.

Subsequent column studies were the basis for calculating the degradation rate constant of *m*-xylene as 0.45/hour (Zeyer et al., 1986): 80% of the ^{14}C-ring-labeled *m*-xylene was oxidized to [14]CO_2 with 6% as acid-extractable metabolites, 3% nonextractable metabolites, and 11% unrecovered and assumed to be incorporated into the biomass. Additional experiments showed rapid mineralization of *m*-xylene and toluene, with nitrous oxide replacing nitrate as the electron acceptor (Kuhn et al., 1988). The acclimated microorganisms also were able to degrade oxidized intermediates but were unable to degrade unsubstituted aromatics or methylated cycloalkanes. Benzoate was suggested as an intermediate formed during toluene metabolism by isotope dilution experiments (Dolfing et al., 1990).

The biodegradation of the BTEX compounds was investigated in anaerobic batch denitrification studies (Major et al., 1988). The aquifer material used was from a shallow sand aquifer that had been exposed to BTEX during a previous natural gradient field injection experiment (Patrick et al., 1985). The BTEX disappearance was observed in the presence of various electron acceptors, including anoxic nitrate, nitrate and oxygen, and oxygen, but very little disappearance was seen under strict anaerobic conditions with no nitrate present. Denitrification was confirmed by using acetylene blockage of nitrate reductase and measuring the nitrous oxide production. Reduction of the nitrate levels resulted in a stoichiometric decrease in BTEX removal.

Recent studies have demonstrated the degradation of toluene and m-xylene and the transformation of o-xylene by several mixed denitrifying cultures (Evans et al., 1991a). Parallel studies have isolated a pure culture, referred to as T1, that mineralized toluene at concentrations of half saturation (3 mM) with a concomitant reduction in nitrate (Evans et al., 1991b). A subculture of the original mixed culture, with only m-xylene (100 μM) as substrate, completely degraded the m-xylene within 1 week. The acclimated m-xylene culture degraded as much as 500 μM of substrate. Complete carbon, electron, and nitrogen mass balances were established for the toluene and the m-xylene cultures. The mass balances accounted for 80%–90% of the expected constituents, with slightly greater than one-half of the original carbon being converted to carbon dioxide.

The degradation of toluene and xylene isomers was reported in batch and column studies under anoxic denitrifying conditions (Bowlen, 1992). No removal of benzene was observed during any of the experiments. Toluene was rapidly degraded in field and laboratory experiments in an aquifer material (Barbaro et al., 1992). Field injection of nitrate was carried out over an extended period (100–300 days), and toluene was transformed within the first 5 m from the injection well. The remaining aromatic compounds were not degraded in the field experiments. In laboratory microcosms, the xylene isomers and ethylbenzene concentrations were reduced in the presence of nitrate. Benzene was recalcitrant under all experimental conditions.

A field-scale treatment of BTEX contamination by injection of nitrate into a contaminated aquifer has been described for a site in Germany (Battermann and Werner, 1984; Geldner, 1987). Laboratory and field results indicated that all BTEX compounds could be removed following addition of nitrate. Geldner (1987) states that benzene, toluene, and the xylene isomers were degraded sequentially in microcosms and in the field application. Injections of 500 mg/liter of nitrate were initiated at the start of the field project, and no detectable BTEX compounds were observed after 3 months of operation.

Field demonstrations in the United States have been limited, due to concerns over the subsurface injection of any substantial concentration of nitrate and mixed results on the biodegradability of benzene (the BTEX component of greatest concern) at the laboratory scale. The Traverse City site in Michigan has utilized both aerobic and denitrifying processes to reduce residual JP4 concentrations in subsurface soils. Injection concentrations of nitrate were limited to the regulatory limit of 10 mg/liter as nitrogen (Hutchins, 1991b,c). Substantially greater amounts of nitrate were consumed, approximately 10×, than could be accounted for by the aromatic removal stoichiometry. The excess nitrate usage may have resulted from degradation of alternative carbon substrates.

7.3.2. Biodegradation by Methanogenisis

Toluene and benzene were anaerobically biodegraded by mixed methanogenic cultures (Grbić-Galić and Vogel, 1987) derived from ferulic acid-degrading sewage sludge enrichments. The serum bottle studies used unlabeled and ^{14}C-labeled substrate, with concentrations ranging from 120 to 2,400 mg/liter. Toluene, with the methyl group labeled, produced a high percentage of $^{14}CO_2$, indicating that most of the methyl group carbon was transformed to CO_2 and not to CH_4. Much lower levels of $^{14}CO_2$ were generated from the ring-labeled benzene and toluene. Gas chromatography–mass spectrometry analysis of the intermediates indicated a degradation pathway of either ring hydroxylation or methyl oxidation, resulting in the formation of phenol, cresol, and aromatic acids. Further investigation using ^{18}O-labeled water, indicated that water was the source of oxygen for the initial hydroxylation step in the degradation pathway of benzene and toluene (Grbić-Galić, 1986; Grbić-Galić and Vogel, 1987; Vogel and Grbić-Galić, 1986).

The biotransformation of several alkylbenzenes has been observed in methanogenic aquifer material (Wilson et al., 1986). Benzene (613 µg/liter), toluene (547 µg/liter), ethylbenzene (269 µg/liter) and o-xylene (257 µg/liter) disappeared from the active cultures but not from the autoclaved inactive cultures. Toluene removal occurred rapidly (15% remaining after 20 weeks), while biodegradation of the remaining compounds occurred after prolonged lag times (>75% remaining after 20 weeks). By 120 weeks each compound was reduced to 3% or less of the original concentration. ^{14}C-labeled toluene tracer studies indicated $^{14}CO_2$ as the principle degradation product.

7.3.3. Biodegradation Involving Sulfate Reducing Bacteria

Recent evidence has indicated that toluene and the xylene isomers can be biodegraded under sulfidogenic conditions in subsurface soils contaminated with fuel products. Aquifer solids from Seal Beach, California, and subsurface soils from Maryland and Oklahoma have demonstrated degradation of alkylbenzenes (Edwards et al., 1991, 1992; Haag et al., 1991; Beller et al., 1992; Sewell et al., 1991; Reinhard et al., 1991). Acclimation patterns for the alkylbenzenes in the Seal Beach samples were sequential, with toluene removed within 40 days, p-xylene in 72 days, and o-xylene in 104 days. Further additions of the compounds were degraded without evidence of a lag period. Once acclimated to toluene or p- and o-xylene the microcosms degraded subsequent additions to m-xylene without a lag period.

Subsurface soils contaminated with aviation fuel demonstrated complete removal of toluene, but not benzene or ethylbenzene (Beller et al., 1992). o- and p-xylene were included only in the initial stages of the experiment, and no further information about xylene degradation was given. Addition of amorphous iron hydroxide reduced the lag period before initiation of activity and enhanced the degradation rate. One hypothesis advanced to explain the role of the iron indicated a reaction between the hydrogen sulfide and the iron to produce an iron sulfide precipitate.

7.3.4. Biodegradation Involving Transition Metals

Ferric ion has been investigated as an alternative microbial electron acceptor for the oxidation of toluene and possibly other aromatic compounds (Lovley et al., 1989). A crude oil–contaminated glacial outwash aquifer in Minnesota was investigated.

The disappearance of the aromatic compounds was linked to a reduction of Fe(III) to Fe(II) in the sediments by an organism designated as GS-15. Alternative electron acceptors (nitrate and sulfate) were at <20 μM, and the methane concentration was $<$μM, thereby limiting the possibility of these pathways.

Serum bottle studies utilizing the sediments as inoculum and 600 μM of toluene as sole carbon source showed $>98\%$ removal of the toluene after 45 days incubation. Heat-sterilized controls showed no loss of toluene and no reduction of Fe(III).

8. CONCLUSIONS

Contamination of groundwater and soils by petroleum hydrocarbons is widespread. *In situ* bioremediation has been demonstrated as an effective remediation technology. Aerobic biodegradation processes utilizing indigenous bacterial populations have been employed most frequently, although use of alternative electron acceptors may have practical application. Natural biodegradation processes are most often limited by the absence of terminal electron acceptor (oxygen), nutrients (nitrogen and phosphorus), and to a lesser extent pH. The most successful approach for treatment of surficial soils has been landfarming. Treatment of vadose zone soils has been most readily achieved through soil venting (*"bioventing"*) to increase air circulation in the subsurface. Treatment of contaminated aquifers has been most readily accomplished through recirculation of oxygenated groundwater. The success of hydrogen peroxide injection has been severely limited by rapid peroxide decomposition and precipitation of metals in soils. While *in situ* bioremediation strategies have been successfully demonstrated, the engineering design basis for implementation is severely limited and needs extensive further development.

REFERENCES

Aggarwal PK, Hinchee RE (1991): Monitoring *in situ* biodegradation of hydrocarbons by using stable carbon isotopes. Environ Sci Technol 25:1176–1180.

Aggarwal PK, Means JL, Downey DC, Hinchee RE (1991a): Use of hydrogen peroxide as an oxygen source for *in situ* biodegradation, Part II. Laboratory studies. J Hazard Mat 27:301–314.

Aggarwal PK, Means JL, Hinchee RE (1991b): Formulation of nutrient solutions for *in situ* biodegradation. In Hinchee RE, Olfenbuttel RF (eds): *In Situ* Bioreclamation: Applications and Investigations for Hydrocarbon and Contaminated Site Remediation. Stoneham, MA: Butterworth-Heinemann, vol 1, pp 51–66.

Anonymous (1981): Tough decade ahead for aromatics. Chemical Eng News 59:18–20.

Ardito CP, Billings JF (1990): Alternative remediation strategies: The subsurface volatilization and ventilation system. In Proceedings of The Conference on Petroleum Hydrocarbons and Organic Chemicals in Ground Water: Prevention, Detection and Restoration. NWWA. pp 281–296.

Armstrong AQ, Hodson RE, Hwang HM, Lewis DL (1991): Environmental factors affecting toluene degradation in ground water at a hazardous waste site. Environ Toxicol Chem 10:147–158.

Atlas RM (ed) (1984): Petroleum Microbiology. New York: Macmillan.

Atlas RM (1986): Microbial degradation of petroleum hydrocarbons: An environmental perspective. Microbial Rev 45:180–209.

Atlas RM (1991): Bioremediation of fossil fuel contaminated soils. In Hinchee RE, Olfenbuttel RF (eds): *In Situ* Bioreclamation: Applications and Investigations for Hydrocarbon and Contaminated Site Remediation. Stoneham, MA: Butterworth-Heinemann, vol 1, pp 14–33.

Atlas RM, Bartha R (1972): Biodegradation of petroleum in seawater at low temperatures. Can J Microbiol 18:1851–1855.

Ball HA, Reinhard M, McCarty PL (1991): Biotransformation of monoaromatic hydrocarbons under anoxic conditions. In Hinchee RE, Olfenbuttel RF (eds): *In Situ* Bioreclamation: Applications and Investigations for Hydrocarbon and Contaminated Site Remediation. Stoneham, MA: Butterworth-Heinemann, vol 1, pp 458–462.

Barbaro JR, Barker JF, Lemon LA, Mayfield CI (1992): Biotransformation of BTEX under anaerobic, denitrifying conditions: Field and laboratory observations. J Contam Hydrol 11:245–272.

Barenschee ER, Bochem P, Helmling O, Weppen P (1991): Effectiveness and kinetics of hydrogen peroxide and nitrate-enhanced biodegradation of hydrocarbons. In Hinchee RE, Olfenbuttel RF (eds): *In Situ* Bioreclamation: Applications and Investigations for Hydrocarbon and Contaminated Site Remediation. Stoneham, MA: Butterworth-Heinemann, vol 1, pp 103–124.

Battermann G, Werner P (1984): Elimination of underground hydrocarbon contamination by microbial degradation. Abwasser/Wasser 125:366–373.

Beller HR, Grbić-Galić D, Reinhard M (1992): Microbial degradation of toluene under sulphate-reducing conditions and the influence of iron on the process. App Environ Microbiol 58:786–793.

Berwanger DJ, Barker JF (1988): Aerobic biodegradation of aromatic and chlorinated hydrocarbons commonly detected in landfill leachates. Water Pollut Res J Can 23:460–475.

Berry DF, Francis AJ, Bollag JM (1987): Microbial metabolism of homocyclic and heterocyclic aromatic compounds under anaerobic conditions. Microbiol Rev 51:43–59.

Bohn H (1992): Consider biofiltration for decontaminating gases. Chemical Eng Prog 88: 34–40.

Bossert I, Bartha R (1984): The fate of petroleum in soil ecosystems. In Atlas RM (ed): Petroleum Microbiology. New York: Macmillan, pp 434–476.

Bowlen GF (1992): Anaerobic Biodegradation of Benzene, Toluene and the Xylenes. Ph.D. Dissertation. Rutgers, The State University of New Jersey.

Brown R (1993): Treatment of petroleum hydrocarbons in groundwater by air sparging. In *In situ* bioremediation of ground water and geological material: A review of materials, EPA/600/R-93/124. Washington, DC: EPA.

Brown RA, Dey JC, McFarland WE (1991): Integrated site remediation combining groundwater treatment, soil vapor extraction, and bioremediation. In Hinchee RE, Olfenbuttel RF (eds): *In Situ* Bioreclamation: Applications and Investigations for Hydrocarbon and Contaminated Site Remediation. Stoneham, MA: Butterworth-Heinemann, vol 1, pp 444–449.

Brown RA, Longfield JY, Norris RD, Wolfe GE (1985): Enhanced bioreclamation: Designing a complete solution to groundwater problems. Presented at Water Pollution Control Fed. Ind. Waste Symposium, Kansas City, MO.

Brown RA, Fraxedas R (1991): Air sparging—extending volatilization to contaminated aquifers. In Symposium on Soil Venting. R.S. Kerr Environmental Research Laboratory, Houston, TX.

Cheremisinoff PN, Casana JG, Ouellette RP (1983): Underground storage tank control. Pollut Eng 18.

Chiang CY, Petkowsky PD (1993): An enhanced bioremediation system at a central production facility—system design and data analysis. In Society of Environ. Tox. and Chemistry, 14th Annual mtg., Houston, TX.

Cooney JJ (1984): The fate of petroleum pollutants in fresh waterecosystems. In Atlas RM (ed): Petroleum Microbiology. New York: Macmillan, pp 399–434.

Dharmavaram S (1991): Biofiltration—A lean emissions abatement technology. In 84th Annual Air & Waste Management Meeting & Exhibition, Vancouver, BC.

Dibble JT, Bartha R (1979a): Effect of environmental parameters on the biodegradation of oil sludge. Appl Environ Microbiol 37:729–739.

Dibble JT, Bartha R (1979b): Rehabilitation of oil-inundated agricultural land: A case history. Soil Sci 128:56–60.

Dienemann EA, Sikkema S, Kosson DS, Ahlert RC (1987): Evaluation of sequential anaerobic/aerobic packed bed bioreactors for treatment of a Superfund leachate. AIChE National Meeting, New York, NY, November.

Dineen D, Slater JP, Hicks P, Holland J, Clendening LD (1989): *In situ* biological remediation of petroleum hydrocarbons in unsaturated soils. In Kostecki PT, Calabrese EJ (eds): Petroleum Contaminated Soils. Chelsea, MI: Lewis, vol 3, pp 177–187.

Dolfing J, Zeyer J, Binder-Eicher P, Schwarzendach RP (1990): Isolation and characterization of a bacterium that mineralizes toluene in the absence of molecular oxygen. Arch Microbiol 154:336–341.

Dupont RR (1993): Fundamentals of bioventing applied to fuel contaminated sites. Environ Prog 12:45–53.

Dupont RR, Doucette W, Hinchee RE (1991): Assessment of *in situ* bioremediation potential and the application of bioventing at a fuel contaminated site. In Hinchee RE, Olfenbuttel RF (eds): *In Situ* Bioreclamation: Applications and Investigations for Hydrocarbon and Contaminated Site Remediation. Stoneham, MA: Butterworth-Heinemann, vol 1, pp 262–282.

Edwards EA, Wills LE, Grbić-Galić D, Reinhard M (1991): Anaerobic degradation of toluene and xylene: Evidence for sulphate as the terminal electron acceptor. In Hinchee RE, Olfenbuttel RF (eds): *In Situ* Bioreclamation: Applications and Investigations for Hydrocarbon and Contaminated Site Remediation. Stoneham, MA: Butterworth-Heinemann, vol 1, pp 463–470.

Edwards EA, Wills LE, Reinhard M, Grbić-Galić (1992): Anaerobic degradation of toluene and xylene: Evidence for sulphate as the terminal electron acceptor. Appl Environ Microbiol 58:794–800.

Ehrlich GG, Schoeder RA, Martin P (1985): Microbial Populations in a Jet-Fuel Contaminated Shallow Aquifer at Tustin, California. U.S. Geological Society, Open-file Report 85-335. Prepared in cooperation with the U.S. Marine Corps.

Ely DL, Heffner DA (1988): Process for *in situ* biodegradation of hydrocarbon contaminated soil. U.S. Patent Office, Patent No. 4,765,902, August 23.

Evans PJ, Mang DT, Young LY (1991a): Degradation of toluene and *m*-xylene and transformation of *o*-xylene by denitrifying enrichment cultures. Appl Environ Microbiol 57: 450–454.

Evans PJ, Mang DT, Kim KS, Young, LY (1991b): Anaerobic degradation of toluene by a denitrifying bacterium. Appl Environ Microbiol 57:1139–1145.

Evans WC (1977): Biochemistry of the bacterial catabolism of aromatic compounds in anaerobic environments. Nature 270:17–22.

Federal Register 49(114):24334, 1984.

Feliciano D (1984): Leaking underground storage tanks: A potential environmental problem. Congressional Research Service Report. U.S. Library of Congress, Washington, DC.

Frankenberger WT Jr (1988): Use of urea as a nitrogen fertilizer in bioreclamation of petroleum hydrocarbons in soil. Bull Environ Contam Toxicol 40:66–68.

Frankenberger WT Jr, Emerson KD, Turner DW (1989): *In situ* bioremediation of an underground diesel fuel spill: A case history. Environ Manage 13:325–332.

Frankenberger WT Jr, Karlson U (1991): Bioremediation of seleniferous soils. In Hinchee RE, Olfenbuttel RF (eds): On-site bioreclamation: Applications and investigations for hydrocarbon and contaminated site remediations. Stoneham, MA: Butterworth-Heinemann, vol 2, pp 239–254.

Geldner P (1987): Stimulated *In Situ* Biodegradation of Aromatic Hydrocarbons. EPA/600/ 9-87/018F. Washington, DC: EPA.

Gibson DT, Subramanian V (1984): Microbial degradation of aromatic hydrocarbons. In Gibson DT (ed): Microbial Degradation of Organic Compounds. New York: Marcel Dekker.

Grbić-Galić D (1986): *O*-demethylation, dehydroxylation, ring-reduction and cleavage of aromatic substrates by *Enterobacteriaceae* under anaerobic conditions. J App Bacteriol 61:491–497.

Grbić-Galić D, Vogel TM (1987): Transformation of toluene and benzene by mixed methanogenic cultures. Appl Environ Microbiol 53:254–260.

Grbić-Galić D, Young LY (1985): Methane fermentation of ferulate and benzoate: Anaerobic degradation pathways. Appl Environ Microbiol 50:292–297.

Griffin CJ, Armstrong JM, Douglass RH (1991): Engineering design aspects of an *in situ* soil vapor remediation system (sparging). In Hinchee RE, Olfenbuttel RF (eds): *In Situ* Bioreclamation: Applications and Investigations for Hydrocarbon and Contaminated Site Remediation. Stoneham, MA: Butterworth-Heinemann, vol 1, pp 517–522.

Haag F, Reinhard M, McCarty PL (1991): Degradation of toluene and *p*-xylene in anaerobic microcosms: Evidence for sulphate as a terminal electron acceptor. Environ Toxicol Chem 10:1379–1389.

Hardaway KL, Katterjohn MS, Lang CA, Leavitt ME (1991): Feasibility and other considerations for use of bioremediation in subsurface areas. In Sayler GS (ed): Environmental Biotechnology for Waste Treatment. New York: Plenum Press.

Hinchee RE, Arthur M (1991): Bench scale studies of the soil aeration process for bioremediation of petroleum hydrocarbons. Appl Biochem Biotechnol 28/29:901–906.

Hinchee RE, Ong SK (1992): A rapid *in situ* respiration test for measuring aerobic biodegradation rates of hydrocarbons in soil. J Air Waste Manage Assoc 42:1305–1312.

Hinchee RE, Downey DC, Aggarwal PK (1991a): Use of hydrogen peroxide as an oxygen source for *in situ* biodegradation. Part I. Field studies. J Hazard Mat 27:287–299.

Hinchee RE, Downey DC, Dupont RR, Arthur M, Miller RN, Aggarwal P, Beard T (1989): Enhanced biodegradation through soil venting. Final Report No. SSPT 88–427. Prepared for HQ AFESC/RDV by Battelle Columbus, Columbus, OH.

Hinchee RE, Downey DC, Dupont RR, Aggarwal PK, Miller RN (1991b): Enhancing biodegradation of petroleum hydrocarbons through soil venting. J Hazard Mat 27:315–325.

Hinchee RE, Ong SK, Hoeppel RE (1991c): A field treatability test for bioventing. Presented at the Air & Waste Management 84th Annual Meeting, Vancouver, BC.

Hodge DS, Medina VF, Islander RL, Devinny JS (1991): Treatment of hydrocarbon fuel vapors in biofilters. Environ Technol 12:655–662.

Hoeppel RE, Hinchee RE, Arthur MF (1979): Bioventing soils contaminated with petroleum hydrocarbons. J Ind Microbiol 8:141–146.

Horowitz A, Shelton DR, Cornell CP, Tiedje JM (1982): Anaerobic degradation of aromatic compounds in sediments and digested sludge. Dev Ind Microbiol 23:433–444.

Huddelston RL (1979): Solid-waste disposal: Landfarming. Chemical Eng 86:119–124.

Huling SG, Bledsoe BE, White MV (1991): The feasibility of utilizing hydrogen peroxide as a source of oxygen in bioremediation. In Hinchee RE, Olfenbuttel RF (eds): *In situ* Bioreclamation: Applications and Investigations for Hydrocarbon and Contaminated Site Remediation. Stoneham, MA: Butterworth-Heinemann, vol 1, pp 83–102.

Hutchins SR (1991): Optimizing BTEX biodegradation under denitrifying conditions. Environ Toxicol Chem 10:1437–1448.

Hutchins SR, Downs CW, Wilson JT, Smith GB, Kovacs DA (1991): Effect of nitrate addition on biorestoration of fuel-contaminated aquifer: Field demonstration. Ground Water 29:571–580.

Hutchins SR, Wilson JT (1991): Laboratory and field studies on BTEX biodegradation in a fuel-contaminated aquifer under denitrifying conditions. In Hinchee RE, Olfenbuttel RF (eds): *In situ* Bioreclamation: Applications and Investigations for Hydrocarbon and Contaminated Site Remediation. Stoneham, MA: Butterworth-Heinemann, vol 1, pp 157–172.

Jamison VW, Raymond RH, Hudson JO (1975): Biodegradation of high-octane gasoline in ground water. Dev Ind Microbiol 16:305–312.

Kobayashi M, Rittmann BF (1982): Microbial removal of hazardous organic compounds. Environ Sci Technol 16:170A–183A.

Kuhn EP, Colberg PJ, Schnoor JL, Wanner O, Zehnder AJB, Schwarzenbach RP (1985): Microbial transformations of substituted benzenes during infiltration of river water to groundwater: Laboratory column studies. Environ Sci Technol 19:961–968.

Kuhn EP, Zeyer J, Eicher, Schwarzenbach RP (1988): Anaerobic degradation of alkylated benzens in denitrifying laboratory aquifer columns. Appl Environ Microbiol 54:490–496.

Lawes BC (1991): Soil-induced decomposition of hydrogen peroxide. In Hinchee RE, Olfenbuttel RF (eds): *In Situ* Bioreclamation: Applications and Investigations for Hydrocarbon and Contaminated Site Remediation. Stoneham, MA: Butterworth-Heinemann, vol 1, pp 143–156.

Leahy JG, Colwell RR (1990): Microbial degradation of hydrocarbons in the environment. Microbiol Rev 54:305–315.

Lee MD, Raymond RL Sr (1991): Case history of the application of hydrogen peroxide as an oxygen source for *in situ* bioreclamation. In Hinchee RE, Olfenbuttel RF (eds): *In Situ* Bioreclamation: Applications and Investigations for Hydrocarbon and Contaminated Site Remediation. Stoneham, MA: Butterworth-Heinemann, vol 1, pp 429–436.

Lee MD, Thomas JM, Borden RC, Bedient PB, Ward CH, Wilson JT (1988): Biorestoration of aquifers contaminated with organic compounds. CRC Crit Rev Environ Control 18:29–89.

Lee MD, Wilson JT, Ward CH (1984): Microbial degradation of selected aromatics in a hazardous waste site. Dev Ind Microbiol 25:557–565.

Lovley DR, Baedecker MJ, Lonergan DJ, Cazzarelli IM, Phillips EJP, Siegel DI (1989): Oxidation of aromatic contaminants coupled to microbial iron reduction. Nature 339:297–300.

Major DW, Mayfield CI, Barker JF (1988): Biotransformation of benzene by denitrification in aquifer sand. Ground Water 26:8–14.

Marley MC, Hazebrouck DJ, Walsh MT (1992): The application of *in situ* air sparging as an innovative soils and ground water remediation technology. Ground Water Monit Rev pp 137–145.

Marley MC, Hoag GE (1990): The application of vapor extraction and air sparging technologies to achieve cleanup at a gasoline spill site. Presented at the A&WMA International Specialty Conference "How Clean is Clean?" Boston, MA: November 7–9.

Martin LM, Sarnelli RJ, Walsh MT (1992): Pilot-scale evaluation of groundwater air sparging: Site-specific advantages and limitations. National R&D Conference on Control of Hazardous Materials, San Francisco, CA, February 4–6.

McAllister PM, Chiang CY, Ettinger RA, Salanitro JP (1993): Field demonstration of an enhanced bioremediation system at a service station. Society of Environmental Toxicology and Chemistry, 14th Annual Meeting, Houston, TX.

McGinnis GD, Matthews J, Pope D, Kerr RS (1991): Bioremediation studies at a northern California Superfund site. Presented at the Air & Waste Management 84th Annual Meeting, Vancouver, BC.

Mihelic JR, Luthy RG (1988a): Degradation of polycyclic aromatic hydrocarbons under various redox conditions in soil–water systems. Appl Environ Microbiol 54:1182–1187.

Mihelic JR, Luthy RG (1988b): Microbial degradation of acenaphthalene and naphthalene under denitrification conditions in soil–water systems. Appl Environ Microbiol 54:1188–1198.

Miller RN (1990): A Field Scale Investigation of Enhanced Petroleum Hydrocarbon Biodegradation in the Vadose Zone Combining Soil Venting as an Oxygen Source With Moisture and Nutrient Addition. Ph.D. Dissertation. Utah State University.

Miller RN, Hinchee RE, Vogel CC (1991): A field investigation of petroleum hydrocarbon biodegradation in the vadose zone enhanced by soil venting at Tyndall AFB, Florida. In Hinchee RE, Olfenbuttel RF (eds): *In Situ* Bioreclamation: Applications and Investigations for Hydrocarbon and Contaminated Site Remediation. Stoneham, MA: Butterworth-Heinemann, vol 1, pp 283–302.

Mueller JG, Lantz SE, Thomas RL, Kline EL, Chapman PJ, Middaugh DP, Pritchard PH, Calvin RJ, Rozich AP, Ross D (1991): Bioremediation of the American Creosote Works Superfund site, PansAacola, Florida. Presented at the Air & Waste Management 84th Annual Meeting, Vancouver, BC.

New Jersey Register (1992): 24:389–390 (February 3, 1992).

Odu CTI (1972): Microbiology of soils contaminated with petroleum hydrocarbons. I. Extent of contamination and some soil and microbial properties after contamination. J Inst Petrol 58:201–208.

Ostendorf DW, Kampbell DH (1990): Bioremediated soil venting of light hydrocarbons. Hazard Waste Hazard Mat 7:319–334.

Ostendorf DW, Kampbell DH (1991): Biodegradation of hydrocarbon vapors in the unsaturated zone. Water Resources Res 27:453–462.

Ottengraf SPP, et al (1986): Biological elimination of volatile zenobiotic compounds in biofilters. Bioprocess Eng 1:61.

Pankow JF, Johnson RL, Cherry JA (1993): Air sparging in gate wells in cutoff walls and trenches for control of plumes of volatile organic compounds (VOCs). Ground Water 31:654–663.

Patrick GC, Ptacek CJ, Gillham RW, Barker JF, Cherry JA, Major D, Mayfield CI, Dickhout RD (1985): The behavior of Soluble Petroleum Product Derived Hydrocarbons in Groundwater, Phase I. PACE Report No. 85-3. University of Waterloo: Petroleum Association for Conservation of the Canadian Environment, Institute for Groundwater Research.

Prosen BJ, Korreck WM, Armstrong JM (1991): Design and preliminary performance results of a full-scale bioremediation system utilizing an on-site oxygen generator system. In Hinchee RE, Olfenbuttel RF (eds): *In Situ* Bioreclamation: Applications and Investigations for Hydrocarbon and Contaminated Site Remediation. Stoneham, MA: Butterworth-Heinemann, vol 1, pp 523–528.

Raymond RL, Jamison VW, Hudson JO (1975): Beneficial Stimulation of Bacterial Activity in Ground Waters Containing Petroleum Products. API Publication No. 4427. Washington, DC: American Petroleum Institute.

Reinhard M, Wills LE, Ball HA, Harmon T, Phipps DW, Ridgeway HF, Eisman MP (1991): A field experiment for the anaerobic biotransformation of hydrocarbon compounds. In Hinchee RE, Olfenbuttel RF (eds): *In Situ* Bioreclamation: Applications and Investigations for Hydrocarbon and Contaminated Site Remediation. Stoneham, MA: Butterworth-Heinemann, vol 1, pp 487–495.

Reinhard M, Goodman NL (1984): Occurrence and distribution of organic chemicals in two leachate plumes. Environ Sci Technol 18:953–960.

Reisch MS (1990): Top 50 chemicals production slowed markedly last year. Chemical Eng News 68:11–15.

Riser-Roberts E (1992): Bioremediation of Petroleum Contaminated Sites. Boca Raton, FL: C.K. Smoley and CRC Press, Inc.

Sayles GD, Hinchee RE, Brenner RC, Vogel CM, Miller RN (1992): *In situ* bioventing: Two U.S. EPA and Air Force sponsored field studies. U.S. EPA/Air & Waste Management Association International Symposium, pp 207–216.

Schink B (1985): Degradation of unsaturated hydrocarbons by methanogenic enrichment cultures. FEMS Microbiol Ecol 31:69–77.

Sewell GW, Russell HH, Gibson SA (1991): Anaerobic biotransformation of toluene under sulphate-reducing conditions. Poster presented at the R.S. Kerr Ground Water Research Seminar, Oklahoma City, OK.

Simonart P, Bastistic L (1966): Aromatic hydrocarbons in soil. Nature 212:1461–1462.

Sleat R, Robinson JP (1984): The bacteriology of anaerobic degradation of aromatic compounds—A review. J Appl Bacteriol 57:381–394.

Spain JC, Pritchard PH, Bourquin AW (1980): Effects of adaptation on biodegradation rates in sediment/water cores from estuarine and freshwater environments. Appl Environ Microbiol 40:726–734.

Spain JC, vanVeld PA (1983): Adaptation of natural microbial communities to degradation of xenobiotic compounds: Effects of concentration, exposure time, inoculum, and chemical structure. Appl Environ Microbiol 45:428–435.

Stuart BJ, Bowlen GF, Kosson DS (1991): Competitive sorption of benzene, toluene and the xylenes onto soil. Environ Prog 10:104–109.

Thomas JM, Ward CH (1989): *In situ* biorestoration of organic contaminants in the subsurface. Environ Sci Technol 23:760–766.

Urlings LCGM, Spuy F, Coffa S, van Vree HBRJ (1991): Soil vapor extraction of hydrocarbons: *In situ* and on-site biological treatment. In Hinchee RE, Olfenbuttel RF (eds): *In Situ* Bioreclamation: Applications and Investigations for Hydrocarbon and Contaminated Site Remediation. Stoneham, MA: Butterworth-Heinemann, vol 1, pp 321–336.

Verstraete W, Vanloocke R, BeBorger R, Verlinde A (1976): Modelling of the breakdown and the mobilization of hydrocarbons in unsaturated soil layers. In Sharpley JM, Kaplan AM (eds): Proceedings of the 3rd International Biodegradation Symposium. Applied Science, London, England: pp 99–112.

Vogel TM, Grbić-Galić D (1986): Incorporation of oxygen from water into toluene and benzene during anaerobic fermentative transformation. Appl Environ Microbiol 52:200–202.

Wang X, Bartha R (1990): Effects of bioremediation on residues, activity and toxicity in soil contaminated by fuel spills. Soil Biol Biochem 22:501–505.

Wang X, Yu X, Bartha R (1990): Effects of bioremediation on polycyclic aromatic hydrocarbon residues in soil. Environ Sci Technol 24:1086–1089.

Wilson BH, Smith GB, Rees JF (1986): Biotransformations of selected alkylbenzenes and halogenated aliphatic hydrocarbons in methanogenic aquifer material: A microcosm study. Environ Sci Technol 20:997–1002.

Wilson JT, Lee MD, Ward CH (1987): *In situ* biorestoration as a ground water remediation technique. J Hazard Materials 14:71–82.

Wilson JT, Leach LE, Michalowski J, Vandegrift JS, Callaway R (1989): *In Situ* Bioremediation of Spills From Underground Storage Tanks: New Approaches for Site Characterization, Project Design, and Evaluation of Performance. EPA/600/2-89/042. Washington, DC: EPA, pp 1–56.

Wilson JT, Hutchins SR, Downs WC, Douglas RH, Newman WA, Hendrix DJ (1990): *In Situ* Biorestoration of a Fuel Spill. EPA/600/9-90/041. Washington, DC: EPA.

Wilson JT, Ward CH (1988): Opportunities for biorememdiation of aquifers contaminated with petroleum hydrocarbons. J Ind Microbiol 27:109–116.

Yaniga PM, Mulry J (1985): Accelerated aquifer restoration: *In situ* applied techniques for enhanced free product recovery/absorbed hydrocarbon reduction via bioreclamation. In Proceedings of the NWWA/API Conference on Petroleum Hydrocarbons and Organic Chemicals in Ground Water—Prevention, Dectection and Restoration, Worthington, OH. p 421.

Zeyer J, Kuhn EP, Schwarzenbach RP (1986): Rapid mineralization of toluene and 1,3-dimethylbenzene in the absence of molecular oxygen. Appl Environ Microbiol 52:944–947.

PART IV. FUTURE TRENDS

≡ 15

DEGRADATIVE GENES IN THE ENVIRONMENT

TAMAR BARKAY

Microbial Ecology and Biotechnology, U.S. Environmental Protection Agency, Gulf Breeze, Florida 32561

SYLVIE NAZARET

Laboratoire d'Ecologie Microbienne du Sol, U.A. CNRS, Université Lyon I, Villeurbanne, F-69622, France

WADE JEFFREY

Center for Environmental Diagnostics and Bioremediation, University of West Florida, Pensacola, Florida 32514

1. INTRODUCTION

Recent advances in the application of molecular biological tools in microbial ecology (Sayler and Layton, 1990; Steffan and Atlas, 1991; Pichard and Paul, 1993) have opened opportunities to address, at a greater depth and precision than ever before, questions pertaining to microbial processes in nature. This development has the potential to create a bridge between knowledge gained by studies at organismal biochemical and genetic levels that have been proceeding in the laboratory ever since the beginning of modern-day microbiology and microbially driven environmental processes that are critical to ecosystem homeostasis. Biodegradation and its application in bioremediation of toxic compounds stand to benefit from molecular

Microbial Transformation and Degradation of Toxic Organic Chemicals, pages 545–577
© 1995 Wiley-Liss, Inc.

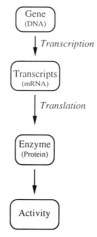

Figure 15.1. The flow of genetic information in gene expression. Boxes indicate products; italics describe processes; arrows depict direction of information flow.

studies of environmental processes. This is because the essence of recent efforts aimed to using microbes in environmental management is the application of processes studied in the test tube to complex and variable environments. The success of this ambitious undertaking depends on the expression of specialized genetic systems, or on the free flow of genetic information from genes to gene products (Fig. 15.1) in microbial populations subjected to numerous biotic and abiotic interactions in the environment.

Traditionally, biodegradation and bioremediation have been studied using a "black box" approach, whereby the disappearance of substrates and appearance of intermediates indicate degradation, and the evolution of $^{14}CO_2$ from radiolabeled substrates indicates mineralization. By correlating such observations with changes in microbial measurements (e.g., increased counts of active organisms, consumption of electron acceptors, increase in biomass of microbial predators) the biological origin of the degradation process is sometimes revealed (Madsen, 1991). In practice, activities are enhanced *in situ* by treatments that stimulate the entire microbial community, such as addition of fertilizers (Pritchard and Costa, 1991), or those that enhance specific populations known to degrade the target compound (Semprini et al., 1990). Concepts, approaches, and methods that promote understanding of factors controlling expression of biodegradative pathways *in situ* have not been developed. Yet it is exactly this understanding that is needed to attain the full potential of bioremediation. Here we propose that tools to stimulate remedial treatments can be obtained by developing an understanding of how the environment modulates gene expression of biodegradative genes.

We present experimental approaches, methods, and early results pertaining to various processes that constitute the flow of information in gene expression in prokaryotes as they occur in the environment. We divide the chapter into subsections relating to processes (transcription, translation, activity of gene products) and products (DNA template, mRNA transcript, protein) in gene expression (Fig. 15.1).

We have taken the view that the microbial community in a given environment is an entity comprised of many individual cells (Sonea, 1989, 1991). Because the activity of interest, as it occurs in an environmental sample, is the product of a mixed community that cannot be partitioned into specific active populations without the bias that is introduced by culturing microbes, approaches and methods that allow for detection of processes and products in gene expression *in situ* are stressed.

The chapter presents a brief description of the processes of gene expression, with recent literature references (as of summer 1993) provided for the interested reader. We do not intend to review the literature comprehensively; rather, potential benefits to bioremediation strategies as might arise from understanding the molecular events that lead to biodegradation *in situ* are highlighted.

2. THE GENE LEVEL

Catabolic genes that enable degradation of contaminants (such as substituted aliphatic and aromatic compounds, PAHs and PCBs) have evolved in many bacterial species inhabiting contaminated environments (water, sediment, and soil). The study of microbes isolated from these environments form the basis of our knowledge of the metabolism, genetics, and ecology of biodegradation. However, the population dynamics and activities of these organisms within complex microbial communities remain poorly understood primarily due to lack of methods for the detection of specific populations and their activities *in situ*. It is in this area that molecular biological tools are of assistance. Among these, gene probes and hybridization techniques provide high sensitivity and specificity by using the nucleic acid sequence of the gene of interest as a molecular marker. A biodegrading microorganism may be identified by the presence of a catabolic gene, whether or not this gene is expressed.

2.1. DNA–DNA Hybridization

DNA–DNA hybridization for the detection of target genes involves annealing a labeled gene probe to a complementary nucleotide sequence (Sambrook et al., 1989). Single-stranded DNA or RNA probes may vary from an entire genome, through a plasmid, operonic, or gene size, to an oligonucleotide sequence of 15–20 nucleotides. Probes may be labeled radioactively with ^{32}P, ^{3}H, or ^{35}S or by non-isotopic reporters (hapten, biotin, chemiluminescent dyes, and others) and hybridized with the target DNA that is usually bound to a solid support (nitrocellulose or nylon membrane). Nucleic acids of varying purity, including those obtained from environmental samples, may be used in hybridization. Factors affecting hybridization include temperature, contact time, salt concentration and ionic strength of hybridization and washing buffers, the degree of homology between the hybridizing sequences, and the length and concentration of the target and probe sequences (Berent et al., 1985). Varying these factors during hybridization and posthybridization washes controls the degree of homology that is required for probe–target binding (stringency; Beltz et al., 1983). Conditions of high stringency (requiring high homology) would detect highly conserved sequences, whereas a lower stringency permits detection of more divergent sequences.

Little attention has been paid to stringency conditions in applications of DNA–DNA hybridization in environmental studies, which is surprising in light of the popularity of this technique and the appreciated divergence of genes specifying similar functions (van der Meer et al., 1992). Mercury resistance (*mer*) genes have been one of the most frequently used gene probes. Barkay et al. (1990) investigated why a high proportion of mercury resistant bacteria were consistently found to be *mer* negative (Barkay and Olson, 1986; Barkay, 1987; Barkay et al., 1989b). They hybridized a probe comprised of *merA* from Tn*501* under varied stringency conditions. The results showed that when tested at low, but not at high, stringency, all isolates hybridized to the probe and led to the conclusion that *mer* genes with a high degree of sequence divergence encoded a common molecular mechanism for Hg(II) resistance in aerobic heterotrophic bacteria (Barkay et al., 1990). The same conclusion was also reached by Rochelle et al. (1991), who observed a wide distribution of *mer* among unselected and Hg(II)-resistant water and sediment isolates, using four *mer* probes from different gram-negative and -positive bacteria. The authors emphasized the necessity of using gene probes derived from a variety of origins, various hybridization conditions, and multiple detection techniques to ensure a reliable estimation of genotype distribution.

2.1.1. *Colony Hybridization*

Colony hybridization is the simplest application of nucleic acid hybridization for detection of specific target genes in bacteria isolated from the environment. Colonies grown on nonselective or selective media are transferred to a membrane (alternatively, colonies can be grown directly on a filter) and lysed by alkaline or enzymatic treatments prior to probing for the target gene. A modified method combining DNA–DNA hybridization with the most probable number (MPN) approach has been developed to quantitate specific populations that are present at low densities (Frederickson et al., 1988). It involves a serial dilution of the original sample and microbial growth for a few days, before filtration onto membranes and subsequent hybridization.

Unlike selective cultivation techniques that require several enrichment steps, colony hybridization is rapid. In addition, it avoids false identification of degraders due to growth on medium contaminants. Since colony hybridization can be performed with bacteria grown on nonselective media that support abundant growth, detection of specific genotypes is achieved regardless of their expression (Bogdanova et al., 1992). Colony hybridization may, therefore, improve estimates of abundance of a given genotype as compared with classic methods. Whether the detected genes specify the corresponding activities needs to be verified, since the genes could be mutated or deleted (Walia et al., 1990; Tebbe et al., 1992). Like all techniques that require culturing, colony hybridization leaves a vast untapped fraction of the community unaccounted for and consequently may bias surveys of specific genes. This is of particular concern when frequencies of catabolic genes are related to biodegradative activities (see Section 2.3.).

2.1.2. Hybridization to Nucleic Acids Isolated From the Environment

Methods for hybridization of nucleic acids extracted and purified from environmental samples have been applied to ensure a representative sampling of the genetic diversity among the studied communities. Various strategies have been developed to extract nucleic acids from environmental matrices. Extraction from water requires a biomass concentration step (filtration) and subsequent lysis of cells (Fuhrman et al., 1988; Somerville et al., 1989; Giovannoni et al., 1990; Barkay et al., 1991). With sediment and soil samples, two strategies have been employed: 1) separation of bacterial cells from soils or sediment particles by successive centrifugation steps (Fægri et al., 1977; Bakken, 1985) or by cation exchange resin (Jacobsen and Rasmussen, 1992) prior to lysis of the cell suspension and nucleic acid purification (Torsvik, 1980; Holben et al., 1988; Jansson et al., 1989; Pillai et al., 1991); and 2) direct lysis in soil–sediment, extraction, and subsequent purification of nucleic acid fractions (Ogram et al., 1987; Olson, 1991). Evaluation of the two approaches (Steffan et al., 1988) showed that direct extraction resulted in higher recovery with reduced specificity, since contaminant DNA, such as eukaryotic or extracellular DNA, was present in the extract. Estimates of nucleic acids recovered from soil, sediment, and water samples range from 50% to 90% (Steffan and Atlas, 1988). Recovery is affected by clay particles that adsorb nucleic acids (Lorenz and Wackernagel, 1987; Romanowski et al., 1991), whereas humic acids in organic matter co-purify with nucleic acids during extraction (Ogram et al., 1987). Different purification procedures have been described that yield high purity DNA appropriate for further enzymatic treatments or hybridization with full efficiency and specificity (Boom et al., 1990; Tsai and Olson, 1992b; Picard et al., 1992; Jacobsen and Rasmussen, 1992).

2.1.3. Amplification of Hybridization Signals

Environmental samples often do not contain sufficient numbers of the target microorganisms, and consequently their nucleic acids, to allow detection by DNA–DNA hybridization. Hybridization signals can be amplified to increase sensitivity by using multiple probes for each target or by hybridizing a second probe to the primary probe (Fahrlander and Klausner, 1988). However, the most common approach to increasing sensitivity of hybridization is the polymerase chain reaction (PCR) (Saiki et al., 1985), whereby a large amount of the target DNA sequence is generated. PCR involves melting to convert double-stranded to single-stranded DNA, annealing oligonucleotide primers to two regions flanking the target sequence, and extension of the primed sequence by DNA polymerase. The specificity of the amplified region is determined by the sequences of the oligonucleotide primers. The amount of target DNA is exponentially increased by repeating the cycle of melting, annealing, and DNA synthesis (Saiki et al., 1985).

PCR has a very wide use in all branches of biological sciences, including numerous applications in environmental microbiology (Steffan and Atlas, 1991; Bej and Mahbubani, 1992). The first demonstration of PCR analysis in microbial ecology employed repeat sequences (Steffan and Atlas, 1988) or highly conserved genes such as 16S rDNA and *nifH* (encoding a subunit of the nitrogenase enzyme; Simonet et al., 1990).

Amplification by PCR requires knowledge of the target sequence to synthesize oligonucleotide primers. Because of a lack of strict conservation of DNA sequences among divergent genes (Barkay et al., 1990; Rochelle et al., 1991), primers may be too specific for detection of genes specifying a certain function in indigenous populations. PCR has been used to increase sensitivity of hybridization with nucleic acids from environmental samples (Steffan and Atlas, 1988; Bej et al., 1991; Pillai et al., 1991; Tsai and Olson, 1992a; Picard et al., 1992; Bruce et al., 1992; Neilson et al., 1992; Tsai et al., 1993; Herrick et al., 1993).

2.2. Abundance of Degradative Genes in the Environment

The application of molecular techniques in environmental microbiology gives us tools to investigate the impacts of contaminants on microbial community structure and function. This is achieved by following changes in abundance of specific taxonomic groups or catabolic genotypes in response to the presence of contaminants in the environment.

2.2.1. Genotypic Surveys of Indigenous Microbial Communities

Detection and enumeration of degrading or detoxifying genotypes in contaminated versus noncontaminated sites constitute the initial steps in relating a certain genotype to biodegradation or biodetoxification. Genotypic screening is carried out by hybridization to either bacterial colonies or DNA extracts. Results are usually compared with those of more traditional techniques such as enrichment and growth in selective and differential media. Degradative genotypes surveyed by this approach are listed in Table 15.1.

The distribution of the mercury resistance operon (*mer*) in environmental samples, as determined by DNA–DNA hybridization, has been extensively studied. The *merA* gene encodes a mercuric reductase that transforms Hg(II) to the volatile Hg⁰, and the *merB* specifies a lyase that degrades organomercury to Hg(II) and a reduced organic moiety (Silver and Walderhaug, 1992; Summers, 1992). The *mer* operon has been detected in a large variety of gram-negative and gram-positive bacterial isolates (Barkay and Olson, 1986; Olson et al., 1989; Rochelle et al., 1991) and in nucleic acids extracted from environmental samples (Barkay et al., 1989a, 1991; Bruce et al., 1992).

High abundance of heavy metal resistant genotypes (resistance to Cd^{2+}, Co^{2+}, Zn^{2+}, Hg^{2+}) in contaminated versus control soils was also reported by Diels and Mergeay (1990) and of polychlorinated biphenyl (PCB) degradation genotypes by Walia et al. (1990). More than 80% of bacteria isolated from a PCB-contaminated site, and less than 1% of bacterial colonies isolated from a control site, hybridized with the PCB probe. Using nucleic acid extracts, Bruce et al. (1992) recently reported amplification of *mer* in contaminated soil samples, and Herrick et al. (1993) amplified *nahAc* (specifying naphthalene dioxygenase) sequences from con-

TABLE 15.1. Genotypes Detected in Environmental Samples Using Gene Probing

Detected Genotype	Probe	References
Degradation of		
Naphthalene	Entire plasmid	Sayler et al. (1985), Blackburn et al. (1987)
	nahAB	Tebbe et al. (1992)
	nahAc	Herrick et al. (1993)
Toluene	Entire plasmid	Sayler et al. (1985), Jain et al. (1987)
	xylE	Morgan et al. (1989), Nüßlein et al. (1992)
3-Chlorobenzoate	*dpb*	Fulthorpe and Wyndham (1989, 1992)
	4-Methyl-enelactone isomerase gene	Nüßlein et al. (1992)
4-Methylbenzoate	4-Methyl-enelactone isomerase gene	Nüßlein et al. (1992)
2,4-Dichlorophenoxy acetic acid	*tfdB*	Neilson et al. (1992)
	tfdABCDEF	Holben et al. (1992)
Polychlorinated biphenyls	*cbpABCD*	Walia et al. (1990)
	cbpC	Walia et al. (1990)
Resistance to		
Mercury	*mer* operon	Barkay (1987, 1989b, 1991), Bruce et al. (1992)
	merA	Barkay et al. (1990), Diels and Mergeay (1990), Rochelle et al. (1991), Tebbe et al. (1992)
	merB	Rochelle et al. (1991), Tebbe et al. (1992)
	merR	Rochelle et al. (1991)
Cadmium, cobalt, zinc	*czc*	Diels and Mergeay (1990)

taminated aquifer samples and analyzed their diversity by restriction enzyme patterns.

The number of probe-positive strains often exceeds corresponding phenotypic enumerations. Thus, Diels and Mergeay (1990) reported that the number of *czc* (specifying resistance to Co^{2+}, Zn^{2+}, and Cd^{2+}) positive bacteria grown on a nonselective medium exceeded the number of zinc-resistant bacteria by 70%. A similar percentage was shown for the recovery of *merA*-positive and Hg(II)-resistant bacteria (Tebbe et al., 1992).

Genotypic surveys can identify environmental interactions that affect functioning of a specific molecular mechanism. For example, Barkay et al. (1991) showed a lower abundance of *mer* genes in DNA fractions extracted from mercury-contaminated sediment communities than from water column samples, although the sediment contained several orders of magnitude higher mercury concentrations. Mercury in sediments may be present in biologically unavailable forms that would not enrich for resistant populations, suggesting that some *mer*-mediated reactions may be ineffective in sediments. Holben et al. (1992) compared the abundance of 2,4-dichlorophenoxyacetic acid (2,4-D) degrading genotypes and phenotypes in soils that had been treated with 2,4-D for 42 years with those of control unimpacted soils. The 2,4-D history of the soils had no apparent effect on 2,4-D degraders, and similar levels of *tfd* genes (specifying 2,4-D degradation) were found in both soil types. The authors suggest that this absence of *in situ* acclimation was due to infrequent applications of low concentrations of 2,4-D (1 ppm) that did not promote persistence of degrading populations at levels above those observed in control soils. Obviously, hypotheses formulated by results of genotypic screens need to be explicitly tested by an appropriate experimental design.

2.2.2. Genotypic Enrichment in Microcosms

An increase in the abundance of a genotype in controlled exposures in the laboratory and comparisons with unexposed samples that are treated similarly confirm results of genotypic surveys (see Section 2.2.1.) and clearly indicates that the detected genotype (and, likely, the mechanism it specifies) plays a role in the response of the indigenous community to the target compound. This approach was used to demonstrate the role of *mer* in response to Hg(II) in communities of fresh and saline waters (Barkay, 1987). Using a *mer* probe and a high stringency of hybridization, the authors found homologous DNA sequences in 50% of resistant strains from two Hg(II) acclimated freshwater communities but in only 12% of strains representing two acclimated saline communities.

DNA sequences homologous to the NAH7 plasmid, specifying the degradation of naphthalene, in the native community were increased significantly upon exposure to elevated concentrations of naphthalene as a sole carbon source (Sayler et al., 1985). More recently, Holben et al. (1992) monitored the response of soil microbial communities to treatments with 2,4-D. Both 2,4-D–degrading phenotypes (monitored by MPN) and genotypes (DNA–DNA hybridization with soil extracts) were increased following exposure to 2,4-D.

2.3. Relationship Between Gene Abundance and Gene Activity in the Environment

The importance of a specific genotype to contaminant degradation can be evaluated by relating its abundance to its corresponding activity. DNA–DNA hybridization has been used by several investigators to relate abundance of degradative genes to disappearance of pollutants. These studies used either field samples or microcosm incubations. Such studies often demonstrate good correlations among the concentration of the test compound, the degradative genotype, and the degradation rate. This approach has been used with genotypes specifying catabolism of substituted aromatic compounds in sediment microcosms exposed to synthetic oil (Sayler et al., 1985), naphthalene degradation in activated sludge (Blackburn et al., 1987), degradation of 3-chlorobenzoate (3Cba) in lake microcosms (Fulthorpe and Wyndham, 1989), and reduction of Hg(II) in a contaminated freshwater pond (Barkay et al., 1991).

Several important factors that affect activities *in situ* can be studied by relating genotypes to activities. The dynamics of specific processes and how they are controlled by interactions in the environment can be followed. For example, working with an introduced 3Cba and 4-methyl benzoate (4MB) degrading *Pseudomonas* sp. (see Section 2.4.), Nüßlein et al. (1992) observed a lag period of several days in activated sludge microcosms, although the same strain degraded these substrates without delay *in vitro*. The authors suggest that other growth substrates available in the microcosm were preferentially used and that biodegradation of 3Cba and 4MB was initiated only after these were depleted.

The diversity of biodegrading populations in mixed communities can be evaluated by obtaining restriction enzyme patterns of sequences that encompass target genes. Considerations regarding stringency of hybridization (see Section 2.1.) are critical for this approach. A diversity in naphthalene dioxygenase sequences, revealed by hybridization patterns to an *nahAc* probe (from plasmid NAH7), was reported in communities from a contaminated aquifer (Herrick et al., 1993). Holben et al. (1992) hybridized several *tfd* genes with DNA extracted from soil samples that had been acclimated to 2,4-D, to suggest that 2,4-D degraders belonged to a single population. Furthermore, hybridization signals were obtained with *tfdA* and *tfdB*, but not with *tfdC, D, E,* or *F,* suggesting that, in the soil, degradation of 3,5-dichlorocatechol proceeded by a pathway dissimilar to the one that is encoded by the classic 2,4-D degradation plasmid, pJP4. Thus, coupling DNA–DNA hybridization with activity measurements can indicate if a known pathway is involved in degradation *in situ*. Obviously, strains with new or modified degradation pathways can then be isolated from the studied environmental samples.

The meaning of detecting a DNA sequence homologous to a known biodegradative gene is rather limited. Not only is detection affected by the lack of sequence conservation (see above), it does not differentiate expressed from nonexpressed genes, contributing to erroneous relationships between genotype abundance and activity. Thus, complementary approaches that address expression of degradative genes by looking at their transcription and translation (see Sections 3–6) are needed.

2.4. Using DNA–DNA Hybridization to Track Genetically Engineered Microorganisms in the Environment

2.4.1. Establishment of Degrading Populations in the Environment

The introduction of either genetically engineered (GEMs) or native microorganisms to contaminated environments has been used to accelerate bioremediation *in situ*. The success of such introductions depends on the survival and subsequent activity of the newly introduced species. These, in turn, depend on biotic and abiotic factors that are variable in time and space. DNA–DNA hybridization with genetic markers that distinguish released microorganisms from indigenous flora has proven a useful tool in tracking newly introduced organisms and their recombinant DNA in a variety of environments.

The maintenance of four introduced bacterial strains in groundwater aquifer microcosms that had been experimentally contaminated with toluene, chlorobenzene, or styrene was followed, using colony hybridization (Jain et al., 1987). The introduced genotypes were stable at approximately $10^5/g$ (wet weight) of microcosm material for as long as 8 weeks. Interestingly, probe-positive colonies were detected in samples from control microcosms to which contaminants were not added. These results suggest that newly introduced microorganisms can survive and be maintained in the absence of selective substrates, under conditions with no obvious competitive advantage.

Nüßlein et al. (1992) tested the survival and activity of two *Pseudomonas* sp. strains in activated sludge microcosms. One of the strains was capable of degrading 3Cba and 4MB, and the other degraded toluene and 4-ethylbenzoate (4EB). When these strains were introduced at initial densities of 10^6-10^7 cfu/ml, population densities declined and stabilized at 10^4-10^5 cfu/ml in both contaminated and control microcosms. On the other hand, the presence of 3Cba affected survival of *Alcaligenes* sp. BR60 in 3Cba-supplemented lake water microcosms (Fulthorpe and Wyndham, 1989). The strain survived for 53 days in supplemented microcosms, but declined to an undetected level in the control microcosms after 41 days. Furthermore, numbers of 3Cba-degrading genotypes, determined by the MPN–DNA hybridization approach, were positively correlated with the concentration of 3CBA and its uptake and degradation rates (Fulthorpe and Wyndham, 1989). Other studies showed that the introduction of degrading microorganisms did not significantly increase the degradation rates over uninoculated controls that contained indigenous degradative genotypes (Jain et al., 1987; Nüßlein et al., 1992).

Amplification with PCR, because of its improved sensitivity, has been most useful in facilitating detection of GEMs in the environment. Steffan and Atlas (1988) used the repeat sequence (15–20 copies per genome) of *Pseudomonas cepacia* AC1100 as a target to detect one cell against a background of 10^{11} nontarget microorganisms in sediment samples. Similar sensitivity was achieved for detection of Tn5 (*nptII* probe specifying kanamycin resistance) in soil samples using a "double PCR" method (Pillai et al., 1991). Bej et al. (1991) detected one cell in 100 ml water by amplifying DNA from filter-bound cells that had been lysed by repeated freezing and thawing. Amplification of *tfdB* in *Alcaligenes eutrophus* was also reported (Neilson et al. 1992).

2.4.2. Gene Probing in Biotechnology Risk Assessment

Safety concerns require that questions pertaining to the survival of newly introduced strains, their mobility in the environment, their activity, and the potential for gene transfer to indigenous bacteria be addressed. The later is of concern mainly with regard to GEMs. Few studies have evaluated transport and dispersal of GEMs in the environment, and, to the best of our knowledge, none were directly concerned with degradative genotypes. However, experiments conducted on released microorganisms (*Rhizobium* spp. and *Pseudomonas* spp.) for agricultural applications (Madsen and Alexander, 1982) showed that mobility in soil is dependent on water flow, the soil matrix, and the presence of plants (Parker et al., 1986; Trevors et al., 1990).

Gene tracking is useful in following dissemination of recombinant DNA from GEMs to indigenous microbial flora. Establishment of recombinant genes in indigenous microbes may enhance the survival and expression of these genes and, thus, their potential to cause deleterious effects in the environment. Transfer in the environment between introduced strains is well documented (Levy and Miller, 1989). Transfer by interspecies or intergeneric conjugation has been followed most extensively (Krasovsky and Stotsky, 1987; Trevors and Starodub, 1987; Nüßlein et al., 1992), but mobilization (McLure et al., 1989; Henschke and Schmidt, 1990; Top et al., 1990) and transduction events have also been described (Saye et al., 1987).

Potential of environmental isolates to receive recombinant DNA has been shown (Schilf and Klingmüller, 1983; Genthner et al., 1988), but transfer to indigenous microbes is harder to demonstrate. DNA probing is a useful tool to distinguish transconjugants from background indigenous organisms. Both Smit et al. (1991) and Barkay et al. (1993) used hybridizations with eukaryotic marker sequences to confirm the presence of indigenous transconjugants. In a freshwater ecosystem, Fulthorpe and Wyndham (1992) showed the presence of the 3Cba catabolic transposon Tn*5271* in strains with different characteristics (colony morphology, pigmentation, growth, and motility) from that of *Alcaligenes* sp. BR60, suggesting a role of plasmid transfer and/or transposition during community adaptation to 4-chloroanilline.

3. TRANSCRIPTION

3.1. The Transcription Process

Transcription is the mechanism whereby information is transferred from the DNA sequence to RNA. In prokaryotes this process is well understood; transcription proceeds through three distinct stages, including initiation, elongation, and termination. Products of this process are functional RNA species (rRNA and tRNA) and mRNA, which serve as an intermediate in the flow of genetic information, conserving the code based on which proteins are subsequently synthesized by translation. Whereas rRNA and tRNA are quite stable, mRNA is very labile, with half-lives as short as minutes (Lewin, 1985).

Initiation of transcription involves the binding of RNA polymerase to a promotor region in the DNA sequence, the unwinding of the supercoiled DNA template upstream of the binding site, followed by the formation of a ternary complex (establishment of the first phosphodiester bond; Collado-Vides et al., 1991). Two consensus sequences upstream of the promoter are essential for identification of a promotor by, and binding of, RNA polymerase, respectively (McCLure, 1985). Elongation of the initiated nascent RNA molecule proceeds by the movement of the polymerase along the DNA template (at a rate of about 40 nucleotides/sec in actively growing *Escherichia coli* [Gotta et al., 1991]) and the release of a 5' to 3' RNA chain that is complementary to the coding strand of the DNA template and similar to the noncoding strand. Transcription terminates at sites that are predetermined by the sequence of the growing RNA chain, and termination is affected by termination and antitermination factors (Platt, 1986).

Gene expression is largely regulated at the transcription level by controlling initiation and termination events. Elaborate regulatory circuits have evolved in bacteria in response to the conflicting demands of having to respond to the intermittent presence of various growth substrates (requiring carriage of genetic information for many alternative catabolic pathways) in environments with scarce energy resources (necessitating energy conservation). Thus, strategies that turn on specific genes when the substrates for the enzyme that they specify are present and turn them off upon substrate depletion have been selected during evolution. In addition, gene expression is controlled by the rate of transcript degradation. The short half-lives of prokaryotic transcripts provide an important regulatory function. When a protein is no longer required, transcription of the mRNA specifying that protein is turned off, and shortly thereafter pre-existing transcripts are degraded and translation of the encoded protein is terminated. Thus, to maintain protein synthesis, mRNA must be continually produced (Freifelder, 1983). Two examples for gene expression controlled by mRNA stability are expression of the arsenical resistance operon (Owolabi and Rosen, 1990) and the differential half-lives of transcripts encoded by the plasmid maintenance controlling element, *parB*, in the plasmid R1 (Gerdes et al., 1988).

3.2. Induction of Biodegradative Functions *In Situ*

With rare exceptions (e.g., Neidle et al., 1989), most biodegradative pathways whose regulation has been studied to date are positively regulated. Transcription is induced by activators following their interaction with substrates or intermediates of the transcribed pathway (van der Meer et al., 1992). Because biodegradation processes are specified by inducible pathways, their utilization in remediation strategies by either indigenous or introduced organisms depends on concentrations of inducers (Janke, 1987) and their bioavailability *in situ* (Ogram et al., 1985; Morris and Pritchard, 1994). Bioremediation is an option only if concentrations that induce gene expression are above those that satisfy levels required by regulatory agencies. Induction conditions have been defined for many biodegradative pathways (Reineke, 1988; van der Meer et al., 1992; Chaudhry and Chapalamadugu, 1991; Dagley, 1986). For some, details of regulatory circuits have also been elucidated (Ramos et al., 1987; Schell, 1990; Neidle et al., 1989). At present we do not know the implications of these studies in the induction of degradative activities *in situ*.

An evaluation of induction *in situ* could be performed by considering concentrations of substrates known to induce biodegradative pathways from studies with pure cultures. Such studies are commonly performed with millimolar concentrations of substrates intended to support growth. However, in the presence of other growth substrates, contaminants may be degraded at much lower concentrations. Available examples indicate that threshold-inducing concentrations are at the micromolar range (Sokol and Howell, 1981; Reber, 1982; Lechner and Straube, 1984; Janke, 1987). Application of results from pure culture studies to degradation in the environment requires care because, *in situ,* different populations within the community may degrade pollutants by different pathways at different concentrations (Rubin et al., 1982; Alexander, 1985; Hwang et al., 1989). It is with molecular approaches that we can begin to understand how specific pathways are induced *in situ.*

3.2.1. Biosensors To Detect Induction of Specific Pathways

Newly developed bioluminescence-based biosensors are a powerful tool to demonstrate induction of, and thus potential biodegradation by, specific pathways *in situ* (King et al., 1990; Heitzer et al., 1992; Selifonova et al., 1993). These biosensors can also distinguish the bioavailable from the inert contaminant because, by definition, they would only respond to bioavailable inducers. Biosensors are comprised of a receptor, the regulatory element (operator/promoter [O/P] sequences and other essential regulatory functions) that controls expression of the biodegradative operon, cloned upstream from the bioluminescence (*lux*) reporter functions. Thus, transcription of *lux* ensues when the biodegradative operon is induced. Luminescence genes, isolated from a variety of bacteria, contain genes specifying the subunits of a luciferase (*luxAB*) and a fatty acid reductase enzyme complex (*luxCDE*) that is involved in the production of long chain fatty acids (Meighen, 1988; Hastings et al., 1985). Biosensors for detecting gene expression in the environment all contain the intact *luxCDABE* operon (Shaw and Kado, 1986). Biosensors to detect naphthalene and salicylate (King et al., 1990; Heitzer et al., 1992; Johnston and Sayler, 1992), toluene and xylene (Kuo et al., 1992), ionic mercury (Selifonova et al., 1993), and organomercurial compounds (Taylor et al., 1992) have been prepared. Their response, at certain inducer concentrations, is quantitative (Heitzer et al., 1992; Selifonova et al., 1993). However, because luminescence is highly sensitive to reaction conditions (Hastings et al., 1985), the response can only be considered semiquantitative (Selifonova et al., 1993).

We recently used *mer–lux* biosensors to demonstrate that the *mer* operon is induced in a mercury-contaminated freshwater pond and may play a role in the reduction of mercury *in situ.* Furthermore, a reduction in mercury bioavailability during retention (≈ 8 hours) of the contaminated water in the pond was noted by comparing analytically measured mercury with the biosensor's response. Water discharged to the pond contained 20.3 ± 0.15 nM total mercury and induction level of *mer–lux* fusions corresponded to ≈ 20 nM Hg(II). Total mercury in a sample collected at the pond's outlet was 7.35 ± 0.15 nM, and the level reported by *mer–lux* was 1–2 nM Hg(II). Thus, not only did total mercury concentration decline during retention in the pond, there was an even greater decrease in the bioavailable form of mercury (Selifonova et al., 1993). We propose that biosensors in combination with analytical measurements of contaminant concentrations could be used to study factors that control induction of biodegrading pathways in the environment. Detection of bioluminescence is possible in complex environmental samples such as biofilms (Mittelman et al., 1992) and soils (Rattray et al., 1990). Other reporter systems (β-galactosidase, catechol-2, 3-dioxygenase, and others) are widely used in studies of transcription regulation. Unlike these methods that require cell collection and enzyme assays, bioluminescence allows for *in situ* real-time measurements. Because recovery of cells for these assays from environmental samples is not always possible, *lux* is essentially the only genetic system currently available to study bioavailability in the environment.

3.2.2. Constitutively Expressed Biodegradative Pathways

Biodegradation processes that do not depend on induction can be obtained by modifying regulatory circuits using genetics and recombinant DNA technologies. For example, Shields and Reagin (1992) recently described a mutated *P. cepacia* G4 that constitutively degrade trichloroethylene. Strategies to obtain constitutively expressed degradative pathways include 1) mutated activator genes that no longer require interaction with the inducer to turn on transcription: Zhou et al. (1990) isolated mutants of *xylS* (specifying a positive regulator that induces the lower operon of the *meta* cleavage pathway of the TOL plasmid pWWO) that did not require the presence of benzoates or their analogs for induction. 2) Structural genes cloned downstream from a strong constitutive promotor: A transposon vector (Herrero et al., 1990) mediated insertion of *merTPAB* in the chromosome of *P. putida* KT2440 resulted in overexpression of the mercuric reductase enzyme and elevated resistance to organic and inorganic mercury (Horn et al., 1994). 3) Inactivation or elimination of a repressor: The *mer* operon is both positively and negatively regulated by the *merR* gene product. Inactivation of *merR* results in a low level constitutive expression of the operon that can be magnified to levels similar to those of the induced wild-type operon by increasing the copy number of *mer* (Hamlett et al., 1992). Similarly, inactivation of *catM* results in constitutive expression of the *cat* structural genes (β-ketoadipate pathway) in *Acinetobacter calcoaceticus* (Neidle et al., 1989). How these modifications affect performance of specialized strains in biotreatments is presently unknown.

4. THE TRANSCRIPT LEVEL

Interest in detecting *in situ* microbial gene expression has increased recently. For several years researchers have been identifying and isolating specific gene sequences from a variety of environments (see Section 2). The ecological importance of these genes depends on the extent of their *in situ* expression. Detection of transcripts is an indication that specific gene systems, such as those specifying biodegradation, in indigenous bacteria are expressed in the environment. Furthermore, *in situ* transcription could be correlated with environmental factors to study ecological phenomena and their effects on biodegradative gene expression and, by inference, biodegradation.

The first to introduce transcript detection to microbial ecological studies were

Tsai and Olson (1990), who examined the effects of varied Hg(II) and CH$_3$Hg(I) concentrations and temperature on accumulation of transcripts specifying mercury transformations in 20 mercury-resistant strains isolated from the environment. Although transcription of *mer* had been studied earlier (Lund et al., 1986; Heltzel et al., 1990), no attention had been paid to its ecological ramifications.

4.1. *In Situ* Measurements of Transcripts

Methods to study gene expression at the transcription level *in situ* have been developed in the past few years. These represent two distinctly different approaches. The first follows transcription from specific promoters under *in situ* conditions using biosensors (see Section 3.2.1). Reporter genes such as *lux* (King et al., 1990) and *xylE* (specifying catechol 2,3-dioxygenase; Morgan et al., 1989) have been used for this purpose. This approach is utilized in laboratory and microcosm studies whereby effects of various environmental parameters on the transcription process *in situ* can be investigated.

The second approach examines gene expression by detecting specific transcripts by probing mRNA fractions that are extracted from environmental samples. This approach relies on intact indigenous genes and their promoters' activities. Several methods to detect transcripts *in situ* have recently been developed. All make use of the acid-guanidinium isothiocyanate extraction (GITC) of Chomczynski and Sacchi (1987). Tsai et al. (1991) described a soil RNA extraction method that, using inoculated soil samples, recovered less than 40% of the RNA relative to extraction of a pure culture with the same cell density. Pichard and Paul (1991) examined diurnal patterns of ribulose-1,5-bisphosphate carboxylase (RuBisCO) expression in marine waters and were the first to recover specific transcripts from the indigenous microbial community. They improved previous methods by including nuclease digestion controls and probing with both sense and antisense probes to distinguish between RNA and DNA hybridization signals. Messenger RNA was extracted by lysing cells in a mini bead-beater in the presence of the GITC-denaturing solution. Most recently, Pichard and Paul (1993) described an approach that relates mRNA transcripts to gene dose as a sensitive indicator of gene expression. They followed transcription of *xylE* controlled by bacteriophage lambda P$_R$ and the temperature-sensitive repressor cI$_{857}$ that was maintained in a *Vibrio* sp. By normalizing *xylE* mRNA to *xylE* DNA, they determined whether measured changes in *xylE* transcript dosage were due to variations in transcriptional activity, *xylE* gene copy, or host cell density. Catechol-2,3-dioxygenase activity lagged behind transcription by approximately 1 hour.

The ability to detect indigenous transcripts may be limited by the biomass carrying the target gene. Pichard and Paul (1991, 1993) monitored transcription in cultured cells or detected RuBisCO transcripts in natural populations. RuBisCO is the most abundant protein in the living world (Ellis, 1979) and probably the gene whose expression in the environment should be the easiest to detect. Other genes that are present at lower levels would require biomass concentration from a large volume of environmental samples. With aquatic samples, this may be prohibited by the size of the filters (25 mm diameter) that are used in the mini bead-beater, the time required to pass a large volume through the filters, or clogging of the filters before sufficient biomass has been collected. Thus, development of alternative cell collection techniques to allow the processing of larger volumes of water was necessary. Jeffrey et al. (1994) have recently developed a procedure to extract mRNA from large flat filters (142 mm diameter) or cartridge-type filters (e.g., Sterivex, Millipore Corp.). Filters carrying cells are frozen by dry ice or liquid nitrogen treatments to stabilize mRNA. Frozen filters may be stored for extended periods until extraction. RNA is obtained by GITC/phenol/chloroform extraction following cell lysis by boiling in the presence of a detergent and an RNase inhibiter (Jeffrey et al., 1994). The rest of the procedure is as described by Pichard and Paul (1991, 1993). This procedure was used to monitor transcript accumulation during acclimation of a freshwater community to 1 mg/liter Hg(II) (Fig. 15.2). Accumulation of *merA* transcripts (specifying mercuric reductase) preceded measurable growth of indigenous bacteria (Fig. 15.2A). When nutrients (10 mg/liter yeast extract) were added together with Hg(II), acclimation was accelerated with *merA* transcription followed by an increase in cell counts (Fig. 15.2B). This method was used to relate *merA* transcript abundance with Hg(II) reduction/volatilization by the indigenous microbial community of a mercury-contaminated freshwater pond in Oak Ridge, TN (Nazaret et al., 1994).

However, even with recent improvements of methods for the recovery of transcripts from environmental samples (Pichard and Paul, 1993; Jeffrey et al., 1994), this approach remains limited by the abundance of the target mRNA *in situ*. Methods that increase the sensitivity of mRNA detection are, therefore, needed.

4.2. Future Developments in Detection of Transcripts *In Situ*

The potential applications of gene expression studies in environmental microbiology are currently limited by the lack of understanding of transcription *in situ*. Knowledge in this area is dependent on availability of methods that would enable studying basic concepts such as rates of transcript synthesis and degradation *in situ* and how they are affected by availability of growth substrates and other ecological factors.

At least two critical methodologies, one to improve sensitivity of transcript detection and another to enable precise quantitation of transcripts in environmental samples, are currently available. Sensitivity of transcript detection can be improved by PCR amplification (see Section 2.2.2) of DNA produced by reverse transcriptase using target mRNA as a template (Goblet et al., 1989; Erlich et al., 1991; Myers and Gelfend, 1991). Viral, ribosomal, and eukaryotic mRNAs (Kawasaki, 1990; Noonan and Robinson, 1988), and more recently bacterial mRNAs (Mahbubani et al., 1991), have been detected by PCR. Selenska and Klingmüller (1992) recently reported amplification and detection of *nif* (specifying nitrogen fixation) and Tn5 DNA and transcripts from soils.

Quantitation of transcripts present at low abundance is achieved by including known concentrations of standards in PCR amplification reactions. Delidow et al. (1989) and Becker-André and Hahlbrock (1989) have reported methods to measure specific mRNA from rat tissue and potato cells, respectively. Delidow et al. (1989)

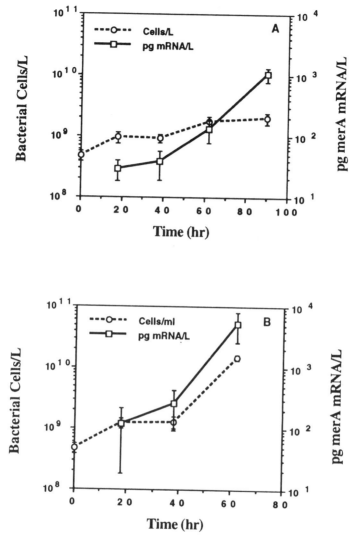

Figure 15.2. The relationship between cell growth and *merA* transcript production. (A) Freshwater community from Fort Pickens Pond, Pensacola Beach, FL, was diluted 1:10 in 0.2 μm filtered pond water and exposed to 1 mg/liter Hg(II). (B) Identical to A except that 10 mg/liter yeast extract was added. Samples were removed at different times after exposure for direct microscopic counts and for determinations of *mer* mRNA levels (Jeffrey et al., 1994).

relied on amplification of a dilution series of the target. Becker-André and Hahl-brock (1989) included a dilution series of *in vitro*–generated transcripts that differed from the target by a single nucleotide. Because a novel restriction site was generated by this substitution, the ratio of DNA amplified from the target mRNA to DNA amplified from the known amount of standard RNA was determined after restriction endonuclease digestion and separation by gel electrophoresis. As has been the case with DNA probing and PCR amplification, where applications in environmental studies lagged behind their development by molecular biologists, it is only a matter of time before quantitation of mRNA is accomplished with samples isolated from the environment.

5. TRANSLATION

Translation is the process whereby genetic information that is carried in the mRNA transcript is transferred to the functional unit, the protein. This is a well-understood process described in numerous text books (for example, Lewin, 1985).

6. THE PROTEIN LEVEL

Proteins are the final product in gene expression. The synthesis of intact catalytically active proteins in the environment is needed for commencement of biodegradative processes and the elimination of contaminants. Detection of organisms producing specific proteins, and of these proteins and their activities, could reveal processes and interactions that affect transformations of target compounds *in situ*. Although approaches for the detection of proteins in environmental samples, mostly by immunoassays (Bohlool and Schmidt, 1980; Ward, 1990) and enzymatic activities (Smith et al. 1992), have long been employed in microbial ecology, these approaches have not been widely used in studies on biodegradation and bioremediation. However, they have received recent attention (Strategies and Mechanisms for field Research in Environmental Bioremediation, ASM Colloquium, San Antonio, TX, January 7–9 1993) and are likely to be more extensively used in the near future.

6.1. Immunoassays for the Detection of Proteins in Environmental Matrices

Immunoassays, especially when coupled with fluorescent labels, are widely applied in microbial ecology. Their attractiveness lies in the high specificity of the antigen–antibody recognition (Campbell et al., 1964; Goldman, 1968; Kawamura, 1977) that is provided by the ability of the mammalian immune system to recognize foreign molecules. The development of the hybridoma technology for the production of monoclonal antibodies (Köhler and Milstein, 1975; Harlow and Lane, 1988) improved the ability to produce large quantities of specific antibodies.

Immunoassays in microbial ecology are used to observe specific microbes in complex environments from which their selective isolation is difficult (Bohlool and Schmidt, 1980), to study spatial and taxonomic diversity (Muyzer et al., 1987), to quantitate cell associated enzymes (Ward, 1990), and to track viable but nonculturable bacteria in the environment (Colwell et al., 1985; Brayton et al., 1987).

Specific antibodies have been prepared for methylotrophs (Reed and Dugan, 1978), phycoerythrin-containing organisms (Campbell and Iturriaga, 1988), ammonia- and nitrite-oxidizing bacteria (Ward and Carlucci, 1985; Ward et al., 1989), nitrogen-fixing *Klebsiella pneumoniae* (Schmidt and Hayasaka, 1985), heterotrophic bacteria (Dahle and Laake, 1982), picoplankton cyanobacteria and eukaryotic phytoplankton (Campbell et al., 1983; Shapiro and Campbell, 1988), and nitrogen-fixing organisms (Currin et al., 1990). Diurnal patterns in synthesis and degradation of nitrogenase in *Trichodesmium thiebautii* were studied by Capone et al. (1990) with antibodies generated against polypeptides expressed from fragments of a *nif* gene from *Trichodesmium* spp. (Zehr et al., 1990).

Immunoassays have been mostly used to study the role of specific organisms and enzymes in geochemical and nutrient cycles. This approach could assist in optimizing biodegradative activities *in situ* by following the persistence of degrading organisms and the expression of their degrading enzymes. Indeed, fluorescent antibodies specific to surface determinants and to flagellar components detected biodegrading strains of *P. putida* 2440 (Ramos-González et al., 1992) and *P. putida* PaW340 (Morgan et al., 1991), respectively, in lakewater microcosms. Enzyme-linked immunosorbant assays (ELISA) were used to detect C230 in microcosms containing a *P. putida* host carrying a recombinant *xylE* plasmid (Morgan et al., 1989) and to study the role of *Thiobacillus ferrooxidans* in desulfurization of coal (Muyzer et al., 1987). Thus, immunoassays are likely to become a useful tool in research addressing bioremediation *in situ*.

6.2. Enzyme Assays *In Situ*

Enzyme assay is another experimental tool that is commonly used by microbial ecologists, yet it has had very limited applications in biodegradation studies. These assays serve as a tool in understanding fundamental ecological processes. Topics, such as soil productivity (Burns, 1982), corrosion of steel (Bryant et al., 1991), and element and nutrient cycling (Smith et al., 1992), have been addressed by measuring enzyme activities. In addition, inhibition of *in situ* enzymatic activities are used as an indication of toxicity (Bitton and Koopman, 1986).

The major problem with *in situ* enzyme assays is the complexity of the test matrix that makes interpretation a rather difficult task. Thus, most authors (Johnen and Drew, 1977; Greaves et al. 1980; Alexander, 1977; Burns, 1982) caution that such assays should only be used in conjunction with other tests that more directly assess substrates, products, and microbial activities. It is as a part of a comprehensive monitoring scheme that enzyme assays may contribute to improved understanding of biodegradation *in situ*. Morgan et al. (1989) related C230 activities in lake microcosms to abundance of the *xylE* gene and to levels of the C230 protein. These authors, however, did not indicate whether they followed degradation of catechol (or its substituted derivatives) to the corresponding semialdehydes.

Employing both approaches to study gene expression (see Section 6.1) may allow examination of factors that control enzyme activities *in situ* by comparing abundance of a protein with its enzymatic activity. Data manipulations to determine ratios between two enzymes or relationships between enzymes and other chemical parameters (e.g., pigments, inducer, substrate, and so forth), may enhance the power of measuring enzymes by their physical presence or activities *in situ* (Ward, 1990).

7. CHEMICAL ACCLIMATION OF MICROBIAL COMMUNITIES

Acclimation, or adaptation, of microbial communities to biodegradable compounds are processes that occur between exposure and initiation of degradation. Acclimation takes between hours and weeks, and it results in more efficient transformations of the target compound. It is the outcome of various processes and interactions within the microbial community in its natural habitat. Some of these processes constitute steps in gene expression resulting in the production of enzymes that catalyze chemical transformations. Here we outline these processes, provide pertinent examples, and suggest how our perspective on gene expression in the environment may assist in exploiting the acclimation phenomena for the degradation of contaminants by indigenous microorganisms in the environment.

Several mechanisms have been proposed for the acclimation process (Spain et al., 1980; Wiggins et al., 1987; Barkay and Pritchard, 1988; van der Meer et al., 1992):

1. Enrichment of pre-existing degrading populations: This is a major mechanism of acclimation (Wiggins et al., 1987; Chen and Alexander, 1989). One implication of enrichment is that genes specifying the biodegradative reactions are multiplied in the community genome (Barkay et al., 1989a; Holben et al., 1992). Viewing the microbial community as a whole entity (Sonea 1989, 1991), enrichment could be considered a gene amplification process. Another implication of enrichment is a change in community structure when degrading populations outgrow populations that are unable to utilize the contaminant as a substrate. For example, a reduction in community diversity indices was noted during acclimation to Hg(II) (Barkay, 1987). This change may have dire consequences in terms of community function and ecosystem stability. Molecular tools may be used to measure changes in community structure *in situ*, without culturing of specialized populations. Characterization of community structure by 16S rRNA-targeted phylogentic probes (Tsien et al., 1990; Devereux et al., 1992) and phospholipid ester-linked fatty acids (Federle et al., 1983; Rajendran et al., 1992) are well established. New promising approaches include DNA–DNA hybridization (Lee and Fuhrman 1990; 1991), and DNA-reannealing patterns indicative of GC content (B. Holben and A. Ogram, personal communications).

2. Induction of degradative genes: This mechanism is likely to contribute to acclimation of the community as a whole, similar to its role in adaptation on the organismal level (see Section 3). Experimental approaches to distinguish induction from increases in activities that are due to enrichment are now

feasible with the recent development of methods for the detection of transcription and translation products in environmental samples (see Sections 4 and 6). This research has already been initiated with the work of Pichard and Paul (1993) and Jeffrey et al. (1994), who observed increases in *xylE* and *merA* (Fig. 15.2) transcripts, respectively, that preceded increases in biomass in microcosm studies. Further refinement of approaches and methods to relate gene expression *in situ* with degradation/transformation rates of the test compound, and microbial growth and metabolism are likely to result in defining the role of enzyme induction in acclimation.

3. Genetic change by mutations and gene exchange: To date, this mechanism of acclimation has not been distinguished from enrichment (van der Meer et al., 1992). Preliminary findings suggest an increased rate of mutation *in situ* (Miller et al., 1989), but the possible role of this phenomenon in acclimation has not been addressed. Because the topic of this chapter is the vertical transfer of genetic information in gene expression, we have not discussed processes of horizontal gene transfer among microorganisms in the environment. This has been an area of active research in recent years (Levy and Miller, 1989) due to concerns with the safety of recombinant DNA in the environment (Halvorson et al., 1985). Whereas gene transfer among indigenous microbes has been observed (Smit et al., 1991; Fulthorpe and Wyndham, 1992), the data suggest that it is a rather rare event, occurring in specialized environments, that could go unnoticed in the absence of selection for the newly created phenotype (Barkay et al., 1995). Rare as it may be, horizontal gene exchange may be critical to acclimation by increasing the diversity of the acclimated community (see above) and thus ensuring stability of the ecosystem.

Acclimation has been investigated for more than 20 years (Atlas and Bartha, 1972; Alexander, 1985). Early work resulted in the formulation of hypotheses regarding the molecular and biochemical events of acclimation. Spain et al. (1980) suggested "three ways by which adaptation can occur upon exposure of the population to a new substrate: (i) induction or depression of specific enzymes, . . . (ii) selection of new metabolic capabilities produced by genetic changes, and (iii) increase in the number of organisms able to catalyze a particular transformation." It seems that these hypotheses could now be tested using molecular-based experimental approaches to the study of microbial communities in their natural habitat.

8. SUMMARY

Biodegradation and its application in bioremediation hold promise for efficient and cost effective means for cleaning contaminated environments. Optimization of conditions for the expression of biodegradative pathways in contaminated sites is essential for the realization of this promise. This goal could be facilitated by understanding the processes and interactions that influence gene expression in the environment. By

TABLE 15.2.   Potential Contributions of Studies Employing Molecular Approaches to Biotreatments of Contaminated Sites

Experimental Approach[a]	Assistance to Bioremediation by
Genotypic surveys (Section 2.2)	Defining environmental interactions that affect a specific genotype. Isolating strains with alternative pathways for degradation
Relating gene abundance to *in situ* activities (Section 2.3)	Determining diversity of degrading populations and of mechanisms involved in degradation
Tracking introduced biodegrading genotypes (Section 2.4)	Verifying success of environmental introductions (establishment of degrading populations)
Luminescent biosensors (Section 3.2.1)	Verifying that conditions *in situ* allow induction of a specific biodegradative pathway. Studying factors that control bioavailability of contaminants
Detection of mRNA transcripts (Section 4.1)	Providing an indication that a biodegradative pathway is expressed *in situ*
Immunoassay to detect specific organisms and their protein products (Section 6.1)	Evaluating spatial distribution of degrading populations *in situ*. Detecting the physical presence of degrading enzyme
Enzyme activities in environmental samples (Section 6.2)	Verifying enzyme activities *in situ*

[a]Entries in parentheses refer to the section in this chapter where each specific approach is discussed.

applying approaches and methodologies that examine ecological events at their molecular level *in situ,* the transformation of biodegradation from a trial and error science ("black box": contaminants in, degradation products out) to a refined fully controlled technology would be accelerated.

The flow of genetic information is based on a chain of events, leading from the gene, where genetic information is stored and restored to an active gene product. Disruption at any juncture in this chain of events would result in a failure of the remedial treatment. Thus, molecular studies on the various processes and products of gene expression in the environment would yield information that can assist in understanding, and consequently manipulating, remediation *in situ* (Table 15.2). At the gene level, detection of biodegradation genes in indigenous and introduced microbes reveals the diversity of degrading populations and pathways, identifies strains with novel catabolic pathways, and follows the establishment of introduced and indigenous biodegrading species in treatment sites. Molecular tools to follow transcription assist in identifying conditions conducive for induction of catabolic pathways (using biosensors) and confirm that biodegradative genes are indeed transcribed *in situ*. Detection of enzymes and their activities in the treated site matrix ensures that the gene products are made and are active.

The obvious choice of model systems to study gene expression in the environment is catabolic pathways whose genetics and biochemistry are well understood. Such systems are available in the *mer* operon, specifying detoxification of mercurial compounds (Summers, 1992; Silver and Walderhaug, 1992) and the catabolic pathway for the degradation of toluene and xylenes specified by the *xyl* operon of the TOL plasmid (Ramos et al., 1987; van der Meer et al., 1992). Indeed, many of the studies described in this chapter used molecular tools that had been developed for *mer* and *xyl*. Such tools to study *in situ* gene expression of the catechol 2,3-dioxygenase gene (*xylE*) of TOL include gene probes (Morgan et al., 1989), bioluminescence sensors (Kuo et al., 1992), transcript detection (Pichard and Paul, 1993), and ELISA-based assays for C230 (Morgan et al., 1989). Detection of *mer* at the gene (Barkay, 1987), transcription (Selifonova et al., 1993), and transcript (Tsai and Olson, 1990; Jeffrey et al., 1993) levels have been described. These systems could therefore be used as prototype pathways to study gene expression of degradation and detoxification operons in contaminated environments.

ACKNOWLEDGMENTS

Gratitude is extended to our colleagues at the Microbial Ecology and Biotechnology Branch of the EPA Environmental Research Laboratory in Gulf Breeze: Richard Eaton, Pam Morris, Jim Mueller, Rick Cripe, and Matt Hoch, for providing literature references and advice; Juan L. Ramos, for sending copies of his numerous papers; and Pam Morris, Joe Lepo, and Bob Burlage, for their critical review of the manuscript. This chapter was partially supported by cooperative agreement CR 818676-01-0 between the Environmental Protection Agency and the University of West Florida (W.H.J.).

REFERENCES

Alexander M (1977): Introduction to Soil Microbiology. New York: John Wiley & Sons.

Alexander M (1985): Biodegradation of organic chemicals. Environ Sci Technol 18:106–111.

Atlas RM, Bartha R (1972): Biodegradation of petroleum in seawater at low temperatures. Can J Microbiol 18:1851–1855.

Bakken LR (1985): Separation and purification of bacteria from soil. Appl Environ Microbiol 49:1482–1487.

Barkay T (1987): Adaptation of aquatic microbial communities to Hg^{2+} stress. Appl Environ Microbiol 53:2725–2732.

Barkay T, Gillman M, Leibert C (1990): Genes encoding mercuric reductases from selected gram-negative aquatic bacteria have a low degree of homology with *merA* of transposon Tn*501*. Appl Environ Microbiol 56:1695–1701.

Barkay T, Kroer N, Rasmussen LD, Sørensen SJ (1995): Conjugal transfer at natural population densities in a microcosm simulating an estuarine environment. FEMS Microbiol Ecol (in press).

Barkay T, Liebert C, Gillman M (1989a): Environmental significance of the potential for *mer*(Tn*21*)-mediated reduction of Hg^{2+} to Hg^0 in natural waters. Appl Environ Microbiol 55:1196–1202.

Barkay T, Liebert C, Gillman M (1989b): Hybridization of DNA probes with whole-community genome for detection of genes that encode microbial responses to pollutants: *mer* genes and Hg²⁺ resistance. Appl Environ Microbiol 55:1574–1577.

Barkay T, Liebert C, Gillman M (1993): Conjugal gene transfer to aquatic bacteria detected by the generation of a new phenotype. Appl Environ Microbiol 59:807–814.

Barkay T, Olson BH (1986): Phenotypic and genotypic adaptation of aerobic heterotrophic sediment bacterial communities to mercury stress. Appl Environ Microbiol 52:403–406.

Barkay T, Pritchard H (1988): Adaptation of aquatic microbial communities to pollutant stress. Microbiol Sci 5:165–169.

Barkay T, Turner RR, VandenBrook A, Liebert C (1991): The relationships of Hg(II) volatilization from a freshwater pond to the abundance of *mer* genes in the gene pool of the indigenous microbial community. Microb Ecol 21:151–161.

Becker-André M, Hahlbrock K (1989): Absolute mRNA quantification using the polymerase chain reaction (PCR). A novel approach by a PCR aided transcript titration assay (PATTY). Nucleic Acids Res 22:9437–9446.

Bej AK, Mahbubani MH (1992): Applications of the polymerase chain reaction in environmental microbiology. PCR Methods Applic 1:151–159.

Bej AK, Mahbubani MH, Dicesare JL, Atlas RM (1991): Polymerase chain reaction–gene probe detection of microorganisms by using filter-concentrated samples. Appl Environ Microbiol 57:3529–3534.

Beltz GA, Jacobs KA, Eickbush TH, Cherbas PT, Kafotos FC (1983): Isolation of multigene families and determination of homologies by filter hybridization methods. Methods Enzymol 100:266–285.

Berent SL, Mahmoudi M, Torczynski RM, Bragg PW, Bollon AP (1985): Comparison of oligonucleotide and long DNA fragments as probes in DNA and RNA dot, Southern, Northern colony and plaque hybridizations. BioTechniques 3:208–220.

Bitton G, Koopman B (1986): Biochemical tests for toxicity screening. In Bitton G, Dutka BJ (eds): Toxicity Testing Using Microorganisms. Boca Raton: CRC Press, Inc., vol I, pp 27–55.

Blackburn JW, Jain RK, Sayler GS (1987): Molecular microbiol ecology of a naphthalene-degrading genotype in activated sludge. Environ Sci Technol 21:884–890.

Bogdanova ES, Mindlin SZ, Pakrová E, Kocur M, Rouch DA (1992): Mercuric reductase in environmental gram-positive bacteria sensitive to mercury. FEMS Microbiol Lett 97:95–100.

Bohlool BB, Schmidt EL (1980); The immunofluorescence approach in microbial ecology. Adv Microb Ecol 4:203–241.

Boom R, Sol CJA, Salimans MMM, Jansen CL, Wertheim-van Dillen PME, van der Noordaa J (1990): Rapid and simple method for purification of nucleic acids. J Clin Microbiol 28:495–503.

Brayton PR, Tamplin ML, Huq A, Colwell RR (1987): Enumeration of *Vibrio cholerae* 01 in Bangladesh waters by fluorescent-antibody direct viable count. Appl Environ Microbiol 53:2862–2865.

Bruce KD, Hiorns WD, Hobman JL, Osborn AM, Strike P, Ritchie DA (1992): Amplification of DNA from native populations of soil bacteria by using the polymerase chain reaction. Appl Environ Microbiol 58:3413–3416.

Bryant RD, Jansen W, Boivin J, Laishley EJ, Costerton JW (1991): Effect of hydrogenase and mixed sulfate-reducing bacterial populations on the corrosion of steel. Appl Environ Microbiol 57:2804–2809.

Burns RG (1982): Enzyme activity in soil: Location and a possible role in microbial ecology. Soil Biol Biochem 14:423–427.

Campbell DH, Garvey JS, Cremer NE, Sussdorf DH (1964): Methods of Immunology. New York: W.A. Benjamin.

Campbell L, Carpenter EJ, Iacono VJ (1983): Identification and enumeration of marine chroococcoid cyanobacteria by immunofluorescence. Appl Environ Microbiol 46: 553–559.

Campbell L, Iturriaga R (1988): Identifiction of *Synechococcus* spp. in the Sargasso Sea by immunofluorescence and fluorescence excitation spectroscopy performed on individual cells. Limnol Oceanogr 33:1196–1201.

Capone DG, O'Neil JM, Zehr J, Carpenter EJ (1990): Basis for diel variation in nitrogenase activity in the marine planktonic Cyanobacterium *Trichodesmium thiebautii*. Appl Environ Microbiol 56:3532–3536.

Chaudhry GR, Chapalamadugu S (1991): Biodegradation of halogenated organic compounds. Microbiol Rev 55:59–79.

Chen S, Alexander M (1989): Reasons for the acclimation for 2,4-D biodegradation in lake water. J Environ Qual 18:153–156.

Chomczynski P, Sacchi N (1987): Single-step method of RNA isolation by acid guanidinium isothiocyanate–phenol–chloroform extraction. Anal Biochem 162:156–159.

Collado-Vides J, Magasanik B, Gralla JD (1991): Control site location and transcriptional regulation in *Escherichia coli*. Microbiol Rev 55:371–394.

Colwell RR, Brayton PR, Grimes DJ, Roszak DB, Huq SA, Palmer LM (1985): Viable but non-culturable *Vibrio cholerea* and related pathogens in the environment: Implication for release of genetically engineered microorganisms. BioTechnology 3:817–820.

Currin CA, Paerl HW, Suba GK, Alberte R (1990): Immunofluorescence detection and characterization of N_2 fixing microorganisms from aquatic environments. Limnol Oceanogr 35:59–71.

Dagley S (1986): Biochemistry of aromatic hydrocarbon degradation in pseudomonads. In Sokatch JR, Ornston LN (eds): The Bacteria: A Treatise on Structure and Function. Orlando: Academic Press, pp 527–555.

Dahle AB, Laake M (1982): Diversity dynamics of marine bacteria studied by immunofluorescent staining on membrane filters. Appl Environ Microbiol 43:169–176.

Delidow BC, Peluso JJ, White BC (1989): Quanitative measurement of mRNAs by polymerase chain reaction. Gene Anal Tech 6:120–124.

Devereux R, Kane MD, Winfrey J, Stahl DA (1992): Genus- and group-specific hybridization probes for determinative and environmental studies of sulfate-reducing bacteria. Syst Appl Microbiol 15:601–609.

Diels L, Mergeay M (1990): DNA probe-mediated detection of resistant bacteria from soils highly polluted by heavy metals. Appl Environ Microbiol 56:1485–1491.

Ellis RJ (1979): The most abundant protein in the world. Trends Biochem Sci 4:241–244.

Erlich HA, Gelfand D, Sninsky JJ (1991): Recent advances in the polymerase chain reaction. Science 252:1643–1651.

Fægri A, Torsvik VL, Goksöyr J (1977): Bacterial and fungal activities in soil: Separation of bacteria and fungi by a rapid fractionated centrifugation technique. Soil Biol Biochem 9:105–112.

Fahrlander PD, Klausner A (1988): Amplifying DNA probe signals; A "Christmas tree" approach. Biol Technol 6:1165–1168.

Federle TW, Hullar MA, Livingston RJ, Meeter DA, White DC (1983): Spatial distribution of biochemical parameters indicating biomass and community composition of microbial assemblies in estuarine mud flat sediments. Appl Environ Microbiol 45:58–63.

Fredrickson JK, Bezdicek DF, Brockman FJ, Li SW (1988): Enumeration of Tn5 mutant bacteria in soil by using a most-probable-number-DNA hybridization procedure and antibiotic resistance. Appl Environ Microbiol 54:446–453.

Freifelder D (1983): Molecular Biology. Boston: Jones and Bartlett.

Fuhrman JA, Comeau DE, Hagström Å, Chan AM (1988): Extraction from natural planktonic microorganisms of DNA suitable for molecular biological studies. Appl Environ Microbiol 54:1426–1429.

Fulthorpe RR, Wyndham RC (1989): Survival and activity of a 3-chlorobenzoate-catabolic genotype in a natural system. Appl Environ Microbiol 55:1584–1590.

Fulthorpe RR, Wyndham RC (1992): Involvement of a chlorobenzoate-catabolic transposon, Tn5271, in community adaptation to chlorobiphenyl, chloroaniline, and 2,4-dichlorophenoxyacetic acid in a freshwater ecosystem. Appl Environ Microbiol 58:314–325.

Genthner FJ, Chatterjee P, Barkay T, Bourquin AW (1988): Capacity of aquatic bacteria to act as recipients of plasmid DNA. Appl Environ Microbiol 54:115–117.

Gerdes K, Helin K, Christensen OW, Løbner-Olesen A (1988): Translational control and differential RNA decay are key elements regulating postsegregational expression of the killer protein encoded by the parB locus of plasmid R1. J Mol Biol 203:119–129.

Gilbert MP, Summers AO (1988): The distribution and divergence of DNA sequences related to the Tn21 and Tn501 mer operons. Plasmid 20:127–136.

Giovannoni SJ, Britschgi TB, Moyer CL, Field KG (1990): Genetic diversity in Sargasso Sea bacterioplankton. Nature 345:60–63.

Goblet C, Prost E, Whalen RG (1989): One-step amplification of transcripts in total RNA using the polymerase chain reaction. Nucleic Acids Res 17:2144.

Goldman M (1968): Fluorescent Antibody Methods. New York: Academic Press.

Gotta SL, Miller OL Jr, French SL (1991): rRNA transcription rate in Escherichia coli. J Bacteriol 173:6647–6649.

Greaves MD, Poole NJ, Domsch KH, Jagnow G, Verstraete W (1980): Recommended Tests for Assessing the Side-Effects of Pesticides on the Soil Microflora. Agricultural Research Council Weed Research Organization, Oxford, Technical Report No. 59.

Halvorson HO, Pramer D, Rogul M (1985): Engineered Organisms in the Environment: Scientific Issues. Washington, DC: American Society for Microbiology.

Hamlett NV, Landale EF, Davis BH, Summers AO (1992): Roles of the Tn21 merT, merP, and merC gene products in mercury resistance and mercury binding. J Bacteriol 174:6377–6385.

Harlow E, Lane D (1988): Antibodies: A Laboratory Manual. Cold Spring Harbor, NY: Cold Spring Harbor Laboratory.

Hastings JW, Potrikus CJ, Gupta SC, Kurfürst M, Makemson JC (1985): Biochemistry and physiology of bioluminescent bacteria. Adv Microbiol Physiol 26:235–291.

Heitzer A, Webb OF, Thonnard JE, Sayler GS (1992): Specific and quantitative assessment of naphthalene and salicylate bioavailability by using a bioluminescent catabolic reporter bacterium. Appl Environ Microbiol 58:1839–1846.

Heltzel A, Lee IW, Totis PA, Summers AO (1990): Activator-dependent preinduction binding of σ-70 RNA polymerase at the metal-regulated mer promoter. Biochemistry 29:9572–9584.

Henschke RB, Schmidt FRJ (1990): Plasmid mobilization from genetically engineered bacteria to members of the indigenous soil microflora *in situ*. Curr Microbiol 20:105–110.

Herrero M, de Lorenzo V, Timmis KN (1990): Transposon vectors containing non-antibiotic resistance selection markers for cloning and stable chromosomal insertion of foreign genes in gram-negative bacteria. J Bacteriol 172:6557–6567.

Herrick JB, Madsen EL, Batt CA, Ghiorse WC (1993): Polymerase chain reaction amplification of naphthalene-catabolic and 16S rRNA gene sequences from indigenous sediment bacteria. Appl Environ Microbiol 59:687–694.

Holben WE, Jansson JK, Chelm BK, Tiedje JM (1988): DNA probe method for the detection of specific microorganisms in the soil bacterial community. Appl Environ Microbiol 54:703–711.

Holben WE, Schroeter BM, Calabrese VGM, Olsen RH, Kukor JK, Biederbeck VO, Smith AE, Tiedje JM (1992): Gene probe analysis of soil microbial populations selected by amendment with 2,4-dichlorophenoxyacetic acid. Appl Environ Microbiol 58:3941–3948.

Horn JM, Brunke M, Deckwer W-D, Timmis KN (1994): *Pseudomonas putida* strains which constitutively overexpress mercury resistance for biodetoxification of organomercurial pollutants. Appl Environ Microbiol 60:357–362.

Hwang H-M, Hodson RE, Lewis DL (1989): Microbial degradation kinetics of toxic organic chemicals over a wide range of concentrations in natural aquatic systems. Environ Toxicol Chem 8:65–74.

Jacobsen CS, Rasmussen OF (1992): Development and application of a new method to extract bacterial DNA from soil based on separation of bacteria from soil with cation-exchange resin. Appl Environ Microbiol 58:2458–2462.

Jain RK, Sayler GS, Wilson JT, Houston L, Pacia D (1987): Maintenance and stability of introduced genotypes in groundwater aquifer material. Appl Environ Microbiol 53:996–1002.

Janke D (1987): Use of salicylate to estimate the "threshold" inducer level for *de novo* synthesis of the phenol-degrading enzymes in *Pseudomonas putida* strain H. J Basic Microbiol 27:83–89.

Jansson JK, Holben WE, Tiedje JM (1989): Detection in soil of a deletion in an engineered DNA sequence by using gene probes. Appl Environ Microbiol 55:3022–3025.

Jeffrey WH, Nazaret S, Von Haven R (1994): Improved method for recovery of mRNA from aquatic samples and its application to detection of *mer* expression. Appl Environ Microbiol 60:1814–1821.

Johnen BG, Drew EA (1977): Ecological effects of pesticides on soil microorganisms. Soil Sci 123:319–324.

Johnston WH, Sayler GS (1992): Maintenance and stability of *nah–lux* bioluminescent reporter strains and plasmids. Abstr Annu Meet Am Soc Microbiol p 388.

Kawasaki ES (1990): Amplification of RNA. In Innis MA, Gelfand DH, Sninsky JJ, White TJ (eds): PCR Protocols: A Guide to Methods and Applications. San Diego: Academic Press, pp 399–406.

Kawaura A Jr (1977): Fluorescent Antibody Techniques and Their Applications. Baltimore: University Park Press.

King JMH, DiGrazia PM, Applegate B, Burlage R, Sanseverino J, Dunbar P, Larimer F, Sayler GS (1990): Rapid, sensitive bioluminescent reporter technology for naphthalene exposure and biodegradation. Science 249:778–781.

Köhler G, Milstein C (1975): Continuous cultures of fused cells secreting antibody of predefined specificity. Nature 256:495–497.

Krasovsky VN, Stotzky G (1987): Conjugation and genetic recombination in *Escherichia coli* in sterile and nonsterile soil. Soil Biol Biochem 19:631–638.

Kuo C-T, Webb O, Burlage R, Palumbo A (1992): Effect of iron on light output from a *xyl-lux* bioluminescent reporter strain, RB1401. Abstr Annu Meet Am Soc Microbiol p 388.

Lechner U, Straube G (1984): Influence of substrate concentration on the induction of amidases in herbicide degradation. Z allgemeine Mikrobiol 24:581–584.

Lee S, Fuhrman JA (1990): DNA hybridization to compare species compositions of natural bacterioplankton assemblages. Appl Environ Microbiol 56:739–746.

Lee S, Fuhrman JA (1991): Species composition shift of confined bacterioplankton studied at the level of community DNA. Mar Ecol Progr Ser 79:195–201.

Levy SB, Miller RV (1989): Gene Transfer in the Environment. New York: McGraw-Hill.

Lewin B (1985): Genes II, 2nd ed. New York: John Wiley and Sons.

Lorenz MG, Wackernagel W (1987): Adsorption of DNA to sand and variable degradation rates of adsorbed DNA. Appl Environ Microbiol 53:2948–2952.

Lund PA, Ford SF, Brown NL (1986): Transcriptional regulation of the mercury-resistance genes of transposon Tn*501*. J Gen Microbiol 132:465–480.

Madsen EL (1991): Determining in situ biodegradation: Facts and challenges. Environ Sci Technol 25:1663–1673.

Madsen EL, Alexander M (1982): Transport of *Rhizobium* and *Pseudomonas* through soil. Soil Sci Soc Am J 46:557–560.

Mahbubani MH, Bej AK, Miller RD, Atlas RM, DiCesare JL, Haff LA (1991): Detection of bacterial mRNA using polymerase chain reaction. BioTechniques 10:48–49.

McClure NC, Weightman AJ, Fry JC (1989): Survival of *Pseudomonas putida* UWC1 containing cloned catabolic genes in a model activated sludge unit. Appl Environ Microbiol 55:2627–2634.

McLure WR (1985): Mechanism and control of transcription initiation in prokaryotes. Annu Rev Biochem 54:171–204.

Meighen EA (1988): Enzymes and genes from the *lux* operons of bioluminescent bacteria. Annu Rev Microbiol 42:151–176.

Miller RV, Kokjohn TA, Sayler GS (1989): Environmental and molecular characterization of systems which affect genome alteration in *Pseudomonas aeruginosa*. In Silver S, Chakrabarty AM, Iglewski B, Kaplan S (eds): *Pseudomonas:* Biotransformations, Pathogenesis, and Evolving Biotechnology. Washington, DC: American Society for Microbiology, pp 252–268.

Mittelman MW, King JMH, Sayler GS, White DC (1992): On-line detection of bacterial adhesion in a shear gradient with bioluminescence by a *Pseudomonas fluorescens (lux)* strain. J Microbiol Methods 15:53–60.

Morgan JAW, Winstanley C, Pickup RW, Jones JG, Saunders JR (1989): Direct phenotypic and genotypic detection of a recombinant pseudomonad population released into lake water. Appl Environ Microbiol 55:2537–2544.

Morgan JAW, Winstanley C, Pickup RW, Saunders JR (1991): Rapid immunocapture of *Pseudomonas putida* cells from lake water by using bacterial flagella. Appl Environ Microbiol 57:503–509.

Morris PJ, Pritchard PH (1994): Concepts in improving polychlorinated biphenyl bio-availability to bioremediation strategies. In Hinchee RE, Leeson A, Semprini L, Ong SK (eds): Bioremediation of Chlorinated and Polycyclic Aromatic Hydrocarbon Compounds. Boca Raton: Lewis Publishers, pp 359–367.

Muyzer G, de Bruyn AC, Schmedding DJM, Bos P, Westbroek P, Kuenen GJ (1987): A combined immunofluorescence–DNA–fluorescence staining technique for enumeration of *Thiobacillus ferrooxidans* in a population of acidophilic bacteria. Appl Environ Microbiol 53:660–664.

Myers TW, Gelfand DH (1991): Reverse transcription and DNA amplification by a *Thermus thermophilus* DNA polymerase. Biochemistry 31:7661–7666.

Nazaret S, Jeffrey WH, Saouter E, Von Haven R, Barkay T (1994): *mer* A gene expression in aquatic environments measured by mRNA production and Hg(II) volatilization. Appl Environ Microbiol 60:4059–4065.

Neidle EL, Hartnett C, Ornston LN (1989): Characterization of *Acinetobacter calcoaceticus catM*, a repressor gene homologous in sequence to transcriptional activtor genes. J Bacteriol 171:5410–5421.

Neilson JW, Josephson KL, Pillai SD, Pepper IL (1992): Polymerase chain reaction and gene probe detection of the 2,4-dichlorophenoxyacetic acid degradation plasmid, pJP4. Appl Environ Microbiol 58:1271–1275.

Noonan KE, Robinson IB (1988): mRNA phenotyping by enzymatic amplification of randomly primed cDNA. Nucleic Acids Res 16:10366.

Nüßlein K, Maris D, Timmis K, Dwyer DF (1992): Expression and transfer of engineered catabolic pathways harbored by *Pseudomonas* spp. introduced into activated sludge microcosms. Appl Environ Microbiol 58:3380–3386.

Ogram AV, Jessup RE, Ou LT, Rao PSC (1985): Effect of sorption on biological degradation rates of (2,4-dichlorophenoxy) acetic acid in soils. Appl Environ Microbiol 49:582–587.

Ogram A, Sayler GS, Barkay T (1987): The extraction and purification of microbial DNA from sediments. J Microbiol Methods 7:57–66.

Olson BH (1991): Tracking and using genes in the environment. Environ Sci Technol 25:604–611.

Olson BH, Lester JN, Cayless SM, Ford S (1989): Distribution of mercury resistance determinants in bacterial communities of river sediments. Water Res 23:1209–1217.

Owolabi JB, Rosen BP (1990): Differential mRNA stability controls relative gene expression within the plasmid-encoded arsenical resistance operon. J Bacteriol 172:2367–2371.

Parke JL, Moen R, Rovira AD, Bowen GD (1986): Soil water flow affects rhizosphere distribution of a seed-borne biological control agent, *Pseudomonas fluorescens*. Soil Biol Biochem 18:583–588.

Picard C, Ponsonnet C, Paget E, Nesme X, Simonet P (1992): Detection and enumeration of bacteria in soil by direct DNA extraction and polymerase chain reaction. Appl Environ Microbiol 58:2717–2722.

Pichard SL, Paul JH (1991): Detection of gene expression in genetically engineered microorganisms and natural phytoplankton populations in the marine environment by mRNA analysis. Appl Environ Microbiol 57:1721–1727.

Pichard SL, Paul JH (1993): Gene expression per gene dose: A specific measure of gene expression in aquatic microorganisms. Appl Environ Microbiol 59:451–457.

Pillai SD, Josephson KL, Bailey RL, Gerba CP, Pepper IL (1991): Rapid method for processing soil samples for polymerase chain reaction amplification of specific gene sequences. Appl Environ Microbiol 57:2283–2286.

Platt T (1986): Transcription termination and the regulation of gene expression. Annu Rev Biochem 55:339–372.

Pritchard PH, Costa CF (1991): EPA's Alaska oil spill bioremediation project. Environ Sci Technol 25:372–379.

Rajendran N, Suwa Y, Urushigawa Y (1992): Microbial community structure in sediments of a polluted Bay as determined by phospholipid ester-linked fatty acids. Mar Pollut Bull 24:305–309.

Ramos JL, Mermod N, Timmis KN (1987): Regulatory circuits controlling transcription of TOL plasmid operon encoding *meta*-cleavage pathway for degradation of alkylbenzoates by *Pseudomonas*. Mol Microbiol 1:293–300.

Ramos-González M-I, Ruiz-Cabello F, Brettar I, Garrido F, Ramos JL (1992): Tracking genetically engineered bacteria: Monoclonal antibodies against surface determinants of the soil bacterium *Pseudomonas putida* 2440. J Bacteriol 174:2978–2985.

Rattray EAS, Prosser JI, Killham K, Glover LA (1990): Luminescence-based nonextractive technique for *in situ* detection of *Escherichia coli* in soil. Appl Environ Microbiol 56:3368–3374.

Reber HH (1982): Inducibility of benzoate oxidizing cell activities in *Acinetobacter calcoaceticus* strain Bs 5 by chlorobenzoates as influenced by the position of chlorine atoms and the inducer concentration. Eur J Appl Microbiol Biotechnol 15:138–140.

Reed WM, Dugan PR (1978): Distribution of *Methylomonas methanica* and *Methylosinus trichosporium* in Cleveland Harbour as determined by an indirect fluroescent antibody-membrane filter technique. Appl Environ Microbiol 35:422–430.

Reineke W (1988): Microbial degradation of haloaromatics. Annu Rev Microbiol 42:263–287.

Rochelle PA, Wetherbee MK, Olson BH (1991): Distribution of DNA sequences encoding narrow- and broad-spectrum mercury resistance. Appl Environ Microbiol 57:1581–1589.

Romanowski G, Lorenz MG, Wackernagel W (1991): Absorption of plasmid DNA to mineral surfaces and protection against DNase I. Appl Environ Microbiol 57:1057–1061.

Rubin HE, Subba-Rao RV, Alexander M (1982): Rates of mineralization of trace concentrations of aromatic compounds in lake water and sewage samples. Appl Environ Microbiol 43:1133–1138.

Saiki RK, Scharf S, Faloona F, Mullis KB, Horn GT, Erlich HA, Arnheim N (1985): Enzymatic amplification of β-globin genomic sequences and restriction site analysis for diagnosis of sickle cell anemia. Science 230:1350–1354.

Sambrook J, Fritsch EF, Maniatis T (1989): Molecular Cloning: A Laboratory Manual, 2nd ed. Cold Spring Harbor, NY: Cold Spring Harbor Laboratory.

Saye DJ, Ogunseitan O, Sayler GS, Miller RV (1987): Potential for transduction of plasmids in a natural freshwater environment: Effect of plasmid donor concentration and a natural microbial community on transduction in *Pseudomonas aeruginosa*. Appl Environ Microbiol 53:987–995.

Sayler GS, Layton AC (1990): Environmental application of nucleic acid hybridization. Annu Rev Microbiol 44:625–648.

Sayler GS, Shields MS, Tedford ET, Breen A, Hooper SW, Sirotkin KM, Davis JW (1985): Application of DNA–DNA colony hybridization to the detection of catabolic genotypes in environmental samples. Appl Environ Microbiol 49:1295–1303.

Schell MA (1990): Regulation of the naphthalene degradation genes of plasmid NAH7: Example of a generalized positive control system in *Pseudomonas* and related bacteria. In Silver S, Chakrabarty AM, Iglewski B, Kaplan S (eds): *Pseudomonas:* Biotransformations, Pathogenesis, and Evolving Biotechnolgoy. Washington, DC: American Society for Microbiology, pp 165–176.

Schilf W, Klingmüller W (1983): Experiments with *Escherichia coli* on the dispersal of plasmids in environmental samples. Recomb DNA Tech Bull 6:101–102.

Schmidt MA, Hayasaka SS (1985): Localization of nitrogen-fixing *Klebsiella* sp. isolated from the root rhizomes of the seagrass *Halodule wrightii* Aschers. Bot Mar 28:437–442.

Selenska S, Klingmüller W (1992): Direct recovery and molecular analysis of DNA and RNA from soil. Microb Releases 1:41–46.

Selifonova O, Burlage R, Barkay T (1993): Bioluminescent sensors for detection of bioavailable Hg(II) in the environment. Appl Environ Microbiol 59:3083–3090.

Semprini L, Roberts PV, Hopkins GD, McCarty PL (1990): A field evaluation of in-situ biodegradation of chlorinated ethenes: Part 2, results of biostimulation and biotransformation experiments. Ground Water 28:715–727.

Shapiro LP, Campbell L (1988): Immunochemical characterization of the ultraphytoplankton. EOS 69:1099.

Shaw JJ, Kado CI (1986): Development of a *Vibrio* bioluminescence gene-set to monitor phytopathogenic bacteria during the ongoing disease process in a nondisruptive manner. BioTechnology 4:560–564.

Shields MS, Reagin MJ (1992): Selection of a *Pseudomonas cepacia* strain constitutive for the degradation of trichloroethylene. Appl Environ Microbiol 58:3977–3983.

Silver S, Walderhaug M (1992): Gene regulation of plasmid-and chromosome-determined inorganic ion transport in bacteria. Microbiol Rev 56:195–228.

Simonet P, Normand P, Moiroud A, Bardin R (1990): Identification of *Frankia* strains in nodule by polymerase chain reaction products with strain specific oligonucleotide probes. Arch Microbiol 153:235–240.

Smit E, van Elsas JD, van Veen JA, de Vos WM (1991): Detection of plasmid transfer from *Pseudomonas fluorescens* to indigenous bacteria in soil by using bacteriophage øR2f for donor counterselection. Appl Environ Microbiol 57:3482–3488.

Smith DC, Simon M. Alldredge AL, Azam F (1992): Intense hydrolytic enzyme activitiy on marine aggregates and implications for rapid particle dissolution. Nature 359:139–142.

Sokol W, Howell JA (1981): Kinetics of phenol oxidation by washed cells. Biotechnol Bioeng 23:2039–2049.

Somerville CC, Knight IT, Straub WL, Colwell RR (1989): Simple, rapid method for direct isolation of nucleic acids from aquatic environments. Appl Environ Microbiol 55:548–554.

Sonea S (1989): A new look at bacteria. ASM News 55:584–585.

Sonea S (1991): Bacterial evolution without speciation. In Margulis L, Fester R (eds): Symbiosis as a Source of Evolutionary Innovation. Cambridge, MA: MIT Press, pp 95–105.

Spain JC, Pritchard PH, Bourqin AW (1980): Effects of adaptation on biodegradation rates in sediment/water cores from estuarine and freshwater environments. Appl Environ Microbiol 40:726–734.

Steffan RJ, Atlas RM (1988): DNA amplification to enhance detection of genetically engineered bacteria in environmental samples. Appl Environ Microbiol 54:2185–2191.

Steffan RJ, Atlas RM (1991): Polymerase chain reaction: Applications in environmental microbiology. Annu Rev Microbiol 45:137–161.

Steffan RJ, Goksøyr J, Bej AK, Atlas RM (1988): Recovery of DNA from soils and sediments. Appl Environ Microbiol 54:2908–2915.

Summers AO (1992): Untwist and shout: A heavy metal–responsive transcriptional regulator. J Bacteriol 174:3097–3101.

Taylor J, Frackman S, Langley KM, Rosson RA (1992): Luminescent biosensors for detection of inorganic and organic mercury. Abstr Annu Meet Am Soc Microbiol p 389.

Tebbe CC, Ogunseitan OA, Rochelle PA, Tsai Y-L, Olson BH (1992): Varied responses in gene expression of culturable heterotrophic bacteria isolated from the environment. Appl Microbiol Biotechnol 37:818–824.

Top E, Mergeay M, Springael D, Verstraete W (1990): Gene escape model: Transfer of heavy metal resistance genes from *Escherichia coli* to *Alcaligenes eutrophus* on agar plates and in soil samples. Appl Environ Microbiol 56:2471–2479.

Torsvik VL (1980): Isolation of bacterial DNA from soil. Soil Biol Biochem 12:15–21.

Trevors JT, Starodub ME (1987): R-plasmid transfer in nonsterile agricultural soil. Syst Appl Microbiol 11:223–227.

Trevors JT, van Elsas JD, van Overbeek LS, Starodub M-E (1990): Transport of a genetically engineered *Pseudomonas fluorescens* strain through a soil microcosm. Appl Environ Microbiol 56:401–408.

Tsai Y-L, Olson BH (1990): Effects of Hg^{2+}, CH_3-Hg^+, and temperature on the expression of mercury resistance genes in environmental bacteria. Appl Environ Microbiol 56:3266–3272.

Tsai Y-L, Olson BH (1992a): Detection of low numbers of bacterial cells in soils and sediments by polymerase chain reaciton. Appl Environ Microbiol 58:754–757.

Tsai, Y-L, Olson BH (1992b): Rapid method for separation of bacterial DNA from humic substances in sediments for polymerase chain reaction. Appl Environ Microbiol 58:2292–2295.

Tsai Y-L, Palmer CJ, Sangermano LR (1993): Detection of *Escherichia coli* in sewage and sludge by polymerase chain reaction. Appl Environ Microbiol 59:353–357.

Tsai Y-L, Park MJ, Olson BH (1991): Rapid method for direct extraction of mRNA from seeded soils. Appl Environ Microbiol 57:765–768.

Tsien HC, Bratina BJ, Tsuji K, Hanson RS (1990): Use of oligonucleotide signature probes for identification of physiological groups of methylotrophic bacteria. Appl Environ Microbiol 56:2858–2865.

van der Meer JR, de Vos WM, Harayama S, Zehnder AJB (1992): Molecular mechanisms of genetic adaptation to xenobiotic compounds. Microbiol Rev 56:677–694.

Walia S, Khan A, Rosenthal N (1990): Construction and applications of DNA probes for detection of polychlorinated biphenyl-degrading genotypes in toxic organic-contaminated soil environments. Appl Environ Microbiol 56:254–259.

Ward BB (1990): Immunology in biological oceanography and marine ecology. Oceanography, April 30–35.

Ward BB, Carlucci AF (1985): Marine ammonia- and nitrite-oxidizing bacteria: Serological diversity determined by immunofluorescence in culture and in the environment. Appl Environ Microbiol 50:194–201.

Ward BB, Glover HE, Lipschultz F (1989): Chemoautotrophic activity and nitrification in the oxygen minimum zone of Peru. Deep-Sea Res 36:1031–1051.

Wiggins BA, Jones SH, Alexander M (1987): Explanations for the acclimation period preceding the mineralization of organic chemicals in aquatic environments. Appl Environ Microbiol 53:791–796.

Zehr JP, Limberger RJ, Ohki K, Fujita Y (1990): Antiserum to nitrogenase generated from an amplified DNA fragment from natural populations of *Trichodesmium* spp. Appl Environ Microbiol 56:3527–3531.

Zhou L, Timmis KN, Ramos JL (1990): Mutations leading to constitutive expression from the TOL plasmid *meta*-cleavage pathway operon are located at the C-terminal end of the positive regulator protein XylS. J Bacteriol 172:3707–3710.

16

RISK ASSESSMENT FOR TOXIC CHEMICALS IN THE ENVIRONMENT

DAVID W. GAYLOR

National Center for Toxicological Research, U.S. Food and Drug Administration, Jefferson, Arkansas 72079

1. INTRODUCTION

During the past decade the process of risk assessment has become more formalized. The risk assessment process is generally considered to contain four components: hazard identification, exposure assessment, dose–response measurements, and quantitative risk characterization (estimation). Hazard identification is the process of establishing the potential for toxic effects in humans resulting from exposure to a substance. Exposure assessment is the process of determining the extent of exposure (dose rate and duration of exposure by inhalation, ingestion, and dermal absorption). Dose–response relationships are obtained from epidemiological data or animal bioassay experiments in order to establish the relationship between risk (probability of disease or a toxic effect) and dose for various exposure conditions. The final step of quantitative risk characterization is the integration of the available data to estimate risk for various exposure conditions. This final step often necessitates the use of assumptions required to extrapolate the results obtained at one set of exposure conditions to other exposure conditions and from results in experimental laboratory animals to humans.

Microbial Transformation and Degradation of Toxic Organic Chemicals, pages 579–601
© *1995 Wiley-Liss, Inc.*

580

The focus of this chapter is on quantitative risk assessment. Hence, emphasis is on the development of dose–response relationships to provide numerical estimates of risk as a function of dose. Attention is given to the assumptions required to extrapolate from high experimental dose levels to low environmental exposure levels. The uncertainties in extrapolating from animal results to humans is also emphasized. Risk estimation often is a very uncertain process, but the regulation of the use of substances generally cannot be postponed until accurate and precise risk estimates are achievable. The purpose of this chapter is not to justify the process of risk assessment but rather to discuss some of the facets of the current status of risk assessment. Much of the discussion is on cancer risk estimation because this area has received more attention than other diseases. Some discussion of risk assessment for neurotoxic and developmental effects is included.

2. HAZARD IDENTIFICATION

It is impossible even to introduce the subject of hazard identification. Large portions of the fields of biology and medicine are concerned with the study of toxicity. Structure–activity relationships are being developed to predict toxicity. Members of certain chemical classes generally produce similar biological effects. Genotoxic agents are often considered to have a high potential for carcinogenicity (e.g., alkalating agents). Growth hormones are often considered to have a high potential to produce developmental defects in offspring. The discussion in this section is limited to a few statistical data analysis problems that can have a profound affect on determining if a real (reproducible) effect is demonstrated from a set of experimental data.

One of the major problems associated with a chronic animal bioassay to test for carcinogenicity is that many animals will die before they may have an opportunity for tumor development. This censoring of observations poses no problem if the incidence of death due to nontumor causes is not related to the dose of the substance under study. However, if differential censoring occurs across dose groups, then statistical procedures are employed to adjust for differential longevity, which affects the numbers of animals at risk. Peto et al. (1980) present a detailed discussion of the problems. The following sections present a summary of the statistical tests for carcinogenicity proposed by Peto et al. (1980).

2.1. Statistical Tests for Carcinogenicity

To determine if substances pose a potential carcinogenic risk to humans, experiments are conducted in laboratory animals. Significantly higher tumor rates in treated animals than in controls are taken as evidence for carcinogenicity. If biases of allocation, animal husbandry, necropsy, and pathology are avoided, there are three possible causes for differences in tumor rates between treated and control groups of animals: 1) differences in the lengths of time animals are at risk for the different groups, 2) chance biological variation, or 3) real carcinogenic effects of the substance under test. The purpose of statistical analysis is to adjust for differences in longevity and to ascertain the magnitude of chance variation. If the differences in age-adjusted tumor rates are too large to be attributed plausibly to chance variation, then differences are attributed to real carcinogenic effects.

The necessity for age adjustments is illustrated by the following example. Suppose the liver tumor rate in a particular strain of mice is 5% at 20 months and 10% by 24 months. Consider a 2 year experiment in which most of the treated animals only survive for 20 months. An observation of equal tumor rates of 10 out of 100 animals with liver tumors in each group might be misinterpreted as a noncarcinogenic effect of the treatment. Due to the shorter longevity in the treated group, only five animals are expected to have tumors by 20 months; hence the tumor rate in the treated group is actually double the rate in the control animals at 20 months.

The appropriate method for adjusting the longevity depends on how tumors are discovered. "Death rate" (actuarial) methods are appropriate for *fatal* tumors, and a slight modification is appropriate in a *mortality-independent* context for "onset rates" of directly observable skin or palpable tumors. For *incidental* nonfatal tumors discovered upon necropsy at the time of scheduled sacrifices or at the time of death of an animal due to a cause other than the tumor type under investigation, a different "prevalence rate" method is required for correcting for longevity. Prevalence rates from sacrificed animals overestimate the probability of dying with a tumor at a particular time, t, because they would have times to death longer than t. On the other hand, prevalence rates at time, t, may underestimate the proportion of animals that die from other causes with the tumor present because more of these animals might produce tumors later.

Since different statistical methods are required for fatal and incidental tumors, it is necessary that pathologists attempt to distinguish between these types of tumors. For some animals there will be errors of judgment. The more accurate the distinction between incidental and fatal tumors, the more reliable will be the adjustments for longevity. Obviously, all tumors discovered at the time of a scheduled sacrifice are incidental regardless of their stage of development. Tumors that result in the death of their hosts are classified as *fatal* tumors. Tumors that are observed upon necropsy for an animal that has died of some unrelated cause are classified as *incidental*.

Misclassifying incidental tumors as fatal tumors tends to make the groups with high intercurrent mortality appear more carcinogenic than they really are. Conversely, misclassifying fatal tumors as incidental tumors tends to make the groups with high intercurrent mortality appear less carcinogenic than they really are. Peto et al. (1980) recommend a more flexible classification: fatal, probably fatal, probably nonfatal, and nonfatal. The effects of misclassification can be established by including and excluding the probably fatal tumor with the fatal tumors and by including and excluding the probably nonfatal tumors with the nonfatal tumors. Peto et al. (1980) contend that in their experience the effect of tumor misclassification has been unimportant.

A particular tumor type may be primarily nonfatal, primarily fatal, or may contain a large portion of animals of both types. Most likely, the proportion of fatal tumors at a particular site will change with age. If there are several tumors in an organ (e.g., liver cell hepatomas), then they must be taken collectively as fatal or nonfatal. Although it may be difficult to determine the exact cause of death in an animal, it generally may be easier to determine whether a particular tumor contributed to death, which is all that is needed for an appropriate statistical analysis. What is needed is a classification of a tumor at the time of necropsy, not a prediction of whether a tumor would eventually cause death.

2.2. Statistical Tests for Carcinogenicity; Equal Longevity

In the special case when there are no material differences between dose groups with respect to intercurrent mortality (i.e., death due to causes other than the tumor type under investigation), no adjustment for longevity is needed. In this case, unadjusted tumor rates can be compared. For the simplest case of testing for the equality of tumor rates in a control group and a treated (dosed) group of animals, Fisher's Exact Test often is used. Consider the following experimental results for the number of animals diagnosed as positive or negative for a particular characteristic (tumor in this instance).

	No. of Animals	
	Negative	Positive
Controls	a	b
Treated	c	d

The probability, $p(d)$, of obtaining b positive animals out of $(a + b)$ control animals and d positive animals out of $(c + d)$ treated animals is

$$p(d) = \frac{(a + b)!(c + d)!(a + c)!(b + d)!}{a!b!c!d!(a + b + c + d)!}$$

where $a! = 1 \times 2 \times \cdots \times a$, for $a > 0$ and $0! = 1$. The probability of obtaining a result of d or more positive by chance out of $(b + d)$ positives in $(c + d)$ treated animals is

$$p = p(d) + p(d + 1) + \cdots + p(b+d)$$

This probability, p is the familiar statistical significance level of obtaining d or more positives when in fact there is no difference in tumor rates between the control and treated animals. If p is small, e.g., less than 0.05, it is stated that there is a statistically significant effect of treatment on increasing the tumor rate. If there is no treatment effect, the probability is p that this is a false–positive result. One can require a smaller p value before concluding that the treatment had an effect in order to reduce the chance of obtaining a false–positive result, but this diminishes the probability of detecting true treatment effects (statistical power of the test). This topic is beyond the scope of this chapter. Power of tests is discussed in many statistical texts (see, e.g., Snedecor and Cochran, 1980).

To illustrate the calculation of p using Fisher's Exact Test, suppose 5 out of 20 treated animals exhibited a tumor and 1 out of 25 control animals were positive. Then

$$p = \frac{20! \ 25! \ 39! \ 6!}{15! \ 5! \ 24! \ 1! \ 45!} + \frac{20! \ 25! \ 39! \ 6!}{14! \ 6! \ 25! \ 0! \ 45!} = 0.052$$

The probability of a difference in tumor rates this large due to chance alone when there is no treatment effect is $p = 0.052$. Of course, it is never known whether a treatment truly influences tumor production. If it is stated that these data demonstrate an effect of the treatment on tumor production, the probability that this is a false–positive statement is $p = 0.052$ if there is no treatment effect.

In the case where an experiment consists of $k > 1$ dose groups, it may be of interest to determine if the treatment results in increasing disease (tumor) rates with increasing doses. That is, it may be important to test whether the data reflect a dose–response relationship. If no correction for unequal nontumor mortality among dose groups is required, the Cochran-Armitage Test for trend among proportions, corrected for a small tumor count, frequently is used (see e.g., Thomas et al., 1977). If x_i out of n_i animals are positive at a dose level of d_i, the statistical test for a positive trend is given by calculating the standard normal deviate $Z = b/s$, where

$$b = \frac{\Sigma x_i d_i - (\Sigma x_i)(\Sigma n_i d_i)/\Sigma n_i - (d_k - d_{k-1})/2}{\Sigma n_i d_i^2 - (\Sigma n_i d_i)^2/\Sigma n_i}$$

$$s = \left[\frac{\Sigma x_i (\Sigma n_i - \Sigma x_i)/(\Sigma n_i)^2}{\Sigma n_i d_i^2 - (\Sigma n_i d_i)^2/\Sigma n_i} \right]^{1/2}$$

The probability, p, of a standard normal deviate greater than Z is the statistical significance level. That is, the probability of stating the treatment produced a positive dose–response when in fact the observed trend is due to chance alone is approximately p. Again, p represents the approximate probability of a false–positive result when there is no treatment effect.

To illustrate the Corrected Cochran-Armitage Test for trend, consider the following experimental results:

| Dose (d_i) | No. of Animals | | $x_i d_i$ | $n_i d_i$ | $n_i d_i^2$ |
	Positive (x_i)	Examined (n_i)			
0	0	50	0	0	0
1	0	49	0	49	49
2	1	48	2	96	192
4	3	50	12	200	800
Total(Σ)	4	197	14	345	1,041

$$b = \frac{14 - 4(345)/197 - (4 - 2)/2}{1{,}041 - (345)^2/197} = 0.0137$$

$$s = \left[\frac{4(197 - 4)/(197)^2}{1{,}041 - (345)^2/197}\right]^{1/2} = 0.0067$$

$$Z = 0.0137/0.0067 = 2.03$$

The approximate probability, p, of obtaining a trend with a slope of $b = 0.0137$ due to chance alone is equal to the probability of a normal deviate larger than $Z = 2.03$. From standard normal tables (see, e.g., Snedecor and Cochran, 1980), the probability of exceeding $Z = 2.03$ is $p = 0.021$. That is, the statistical significance level of the observed trend (probability of concluding a trend exists when in fact there is no treated effect) is $p = 0.021$.

When nontumor mortality differs across dose groups, it is necessary to make adjustments in the numbers of animals at risk for developing incidental or fatal tumors. The actuarial methods required for those adjustments are beyond the scope of this chapter, but are discussed extensively in the literature (e.g., Peto et al., 1980; Kodell et al. 1982, 1983, and 1986).

The values of b and s depend on the actual units in which dose is measured, but the test statistic $Z = b/s$ does not depend on the units of measure, and hence the p value is independent of the units of measure. For example, it does not matter whether dose is expressed as ppm or percent.

The test given is for a linear trend in the doses. Departures from linearity of the dose–response will contribute to the standard deviation, s, and weaken the sensitivity of the test to detect a trend. Thus, a transformation of the dose scale may result in a more nearly linear dose–response relationship. For a concave dose–response log dose may provide a more sensitive test; e.g., for doses of 10, 100, and 1,000, the values of D in the calculation would be set equal to 1, 2, and 3, respectively, with $D = 0$ for the controls. For a convex dose–response, $\sqrt{\text{dose}}$ or some other root of dose may provide a reasonably straight dose response line.

In the special case of comparing two groups, the trend case can also be used. The computations are simplified by using $D = 0$ for the controls and $D = 1$ for the treated group. This test is equivalent to the test of homogeneity of tumor rates across groups.

Recall that the analyses up to this point are appropriate only in the case of *similar nontumor mortality* in each group. Under this condition proportionate numbers of fatal and incidental tumors are expected in groups at any point in time. There is no need therefore to distinguish between fatal and incidental tumors when nontumor mortality is similar among groups. In this case, tests can be performed on the crude cumulative number of animals with tumors.

2.3. Multiple Comparisons (Tests)

One of the problems associated with the interpretation of experiments in which there are many possible biological endpoints is the problem of multiple comparisons (tests) of treated groups with controls. Suppose it is decided to assign some importance to results that are statistically significant at the 5% level (i.e., the probability that the result is a false–positive is less than 0.05, often designed by $p < 0.05$). The chance for a false–positive result increases as the number of comparisons increases. For example, if a bioassay produces 20 independent endpoints each with a false–positive probability of 0.05, then on the average such a bioassay would produce one false–positive result. Fortunately, a typical bioassay for carcinogenicity does not provide an opportunity for this many false–positive results; nonetheless the problem of multiple comparisons must be considered. Up to 50 tissues may be examined in a chronic bioassay. Analyses of the data are generally performed on tumor rates for each tissue site separately. Occasionally, similar tumors for several sites may be combined, e.g., those of leukemia. Sometimes analyses are performed for malignant tumors and for benign and malignant tumors combined. Thus, there are several opportunities for false–positive results in a chronic bioassay for carcinogenicity.

Most bioassays utilize two or more dose level groups and control (undosed) animals. Generally, the tumor rate for each dose group is compared with the control tumor rate. This offers additional opportunities for false–positive results. False–positive rates are generally controlled for multiple dose comparisons by use of the Bonferroni correction (Miller, 1966). If there are k dose groups compared with the control, the overall false–positive rate is limited to p by only considering an individual comparison statistically significant at the p/k level. For example, if two groups are compared with controls and the overall false–positive is to be limited to $p < 0.05$, then a comparison of a dosed group with the controls must have a false–positive rate of $p < 0.025$ to be considered statistically significant.

Bioassays for carcinogenicity frequently are conducted in both sexes of two rodent species. It is possible that the increase in tumors at any particular tumor site may not achieve statistical significance, but may be close in both sexes of one or both species or in both species for one or both sexes. To increase the sensitivity of this bioassay to detect carcinogenicity, the test statistics may be combined to test for carcinogenicity in one or both species for both sexes combined or for one or both sexes for both species combined. That is, for each tumor site in bioassays conducted in both sexes of mice and rats, statistical tests may be performed for male mice, female mice, male rats, female rats, all males, all females, all mice, all rats, and all animals. Obviously, these tests are not independent, but they provide a method of increasing the sensitivity of bioassay screens that are exploratory in nature.

As the number of statistical tests on a set of data increases, the possibility of detecting a potential carcinogenic effect may increase, but the probability of false–positive results may also increase. Accompanying any statistical test is the possibility that relatively rare chance events result in a significantly higher tumor rate in treated animals when there is no real carcinogenic effect, i.e., a false–positive result. This can only happen if the spontaneous background tumor rate in control animals is high enough to produce enough excess tumors to be statistically significant. Fears et al. (1977) show that at least five more spontaneous tumors would have to arise in a group of 50 treated animals than in 50 control animals in order to produce a false-positive result at the $p \leq 0.05$ significance level. That is, there must be at least 5 out of 50 additional spontaneous tumors in the treated group. This generally can only occur at those few tumor sites with the higher spontaneous background tumor rates, as those sites with near-zero background rates will have negligible probabilities of producing false–positive results. Fears et al. (1977) show false–positive rates of 0.05, 0.03, 0.10, and 0.13 for B6C3F$_1$ male and female mice and in Fischer 344 male and female rats, respectively, when performing statistical tests for 21 tissue sites in groups of 50 animals. The overall false–positive rate is

$$1 - (1 - 0.05)(1 - 0.03)(1 - 0.10)(1 - 0.13) = 0.28$$

Utilizing the Bonferroni correction, the overall false–positive rate for this two-species, both-sexes bioassay can be reduced to approximately the $p \le 0.05$ level by requiring the significance level to reach $p \le 0.01$ for any particular test involving common tumors and $p \le 0.05$ for rare tumors (Haseman, 1984).

If the bioassay is being conducted to test specifically for an increase of tumors at a particular site, in a given sex and species, then no correction in the significance level for this specific hypothesis is required. However, any other test of an exploratory nature conducted at other tumor sites will require more stringent significance levels.

2.4. Historical Tumor Rates

Historical data on tumor rates among control animals are sometimes used for comparisons with tumor rates in dosed animals. In most chronic bioassays, the appearance of two or three tumors in a treated group would never achieve statistical significance. For example, ignoring time to tumor considerations, when no tumors are observed in 50 control animals, tumors in at least 5 of 50 treated animals are required to be considered statistically significant at the $p < 0.05$ level by Fisher's Exact Test (Sokal and Rohlf, 1969, p 595). However, three rare tumors may be considered to represent a statistically significant occurrence. The rareness of the tumor may be established from an accumulation of results in control animals. If no tumors had occurred in 85 control animals, then 3 of 50 tumors in the treated group would be statistically significant at the $p < 0.05$ level.

Common tumors present a more difficult situation. It is tempting to attempt to utilize the information on control animals. Even when control animals are utilized only from other studies conducted at the same laboratory within a period of a few months, the use of historical controls is questionable. Tumor rates may vary from experiment to experiment because of differences in the genetic source of the animals, pathologists, interpretation of pathological definitions of tumors, feed nutrients and contaminants, environmental conditions, aggressiveness of the removal of moribund animals, and so forth. Thus, the concurrent controls provide the best scientific basis for comparisons. If the values for the historical controls differ significantly from the concurrent controls, then experimental conditions may have differed, and only the concurrent controls should be used. If the historical and concurrent controls have similar tumor rates, use of the historical controls is justified, but generally little difference is noted.

Tarone (1982) and Hoel and Yanagawa (1986) give methods for incorporating historical tumor rates for control animals into statistical tests for carcinogenicity. Haseman et al. (1984) discuss issues involved in the use of historical control data. They conclude that the most appropriate control group is the concurrent controls, but there are instances for rare tumors or marginally significant tumors where the use of historical controls may be useful.

2.5. Statistical Tests for Reproductive and Developmental Effects

Reproductive and developmental bioassays provide data with quite different statistical properties than tumorigenicity studies. In reproductive and developmental studies, parent animals are exposed to toxic substances and the biological effects are observed in the offspring. One important aspect of this type of experiment is that the experimental unit is the litter. Hence, the sample size for statistical tests is the number of litters and not the number of fetuses or offspring. Some measures of reproduction are pregnancies per mating and the number of implants in a litter. Some measures of development are litter size, ratio of the number of fetuses with a particular type of malformation to the number of implants or live fetuses in a litter, and fetal weight.

Measurements obtained on a continuum, e.g., fetal weight, can generally be treated by classic statistical procedure (Snedecor and Cochran, 1980). Measurements involving counts, e.g., litter size, can usually be analyzed by classic statistical methods by using the square root of the count for each litter. Measures that are proportions or percentages, e.g., ratio of the number of normal fetuses to the number of live fetuses per litter, can sometimes be analyzed by classic statistical methods by using the arc sine of the square root of the proportion. The use of classic statistical methods requires that the measurements are approximately normally distributed with approximately equal variances. When the transformed data do not meet these requirements, nonparametric statistical methods can be used (see, e.g., Siegel, 1956).

3. EXPOSURE ASSESSMENT

Exposure assessment is concerned with estimating the concentrations of toxic substances in the air, water, food, drugs, cosmetics, and commercial and industrial products to which humans are exposed. Measurements may be made directly in various media or calculated from mathematical models describing the dispersion of toxic substances in space over time. To estimate risks it is necessary to estimate the number of people exposed under various conditions. Let N_i represent the number of people exposed to a concentration, C_i, of a toxic substance. If p_i is the probability of a toxic effect in an individual at a concentration of C_i, then the expected number of cases at this concentration is $X_i = p_i N_i$. For the whole population, the total number of cases expected is

$$X = p_1 N_1 + p_2 N_2 + \cdots + p_k N_k = \Sigma p_i N_i$$

where k is the number of different concentrations. As discussed later, risk at low doses is likely to be proportional to dose. Also, in the absence of data, low-dose linearity often is assumed. That is, $p_i = b\,D_i$, where b is the proportionality constant (slope). Also at low exposure levels, dose may be proportional to the concentration of the toxic substance in the environmental carrier. That is, $D_i = b'C_i$, where b' is the proportionality constant. Under these conditions, $p_i = b\,D_i = b\,b'C_i = B\,C_i$, where B is the proportionality constant relating risk to concentration. The number of individuals expected to produce toxic effects under these conditions is

$$X = \Sigma\,p_i\,N_i = \Sigma\,B\,C_i\,N_i = B\,\Sigma\,C_i\,N_i$$

The probability of risk, i.e, the proportion of individuals with the toxic effect, is

$$p = X/N = B\,\Sigma\,C_i\,N_i/N$$

where $N = \Sigma\,N_i$ the total number of individuals in the population at risk. Since the average concentration for exposed individuals is $\bar{C} = \Sigma\,C_i\,N_i/N$, the average risk per individual is

$$p = B\,\Sigma\,C_i\,N_i/N = B\bar{C}$$

Thus, when risk is proportional to concentration, it is only necessary to estimate the average concentration in order to estimate the expected risk for the whole population. Raising or lowering the average concentration by a factor will raise or lower the risk by that same factor. Under these conditions it is not necessary actually to know the distribution of concentration in the environment.

If the regulatory strategy is to eliminate exposures above a certain level to control risk, then it is necessary to know the distribution of exposure in the population in order to estimate the degree and effect of regulation. If $p(C_i)$ represents the probability of a toxic effect at a concentration of C_i in a subpopulation of N_i individuals, the average risk to an individual in the population is

$$p = \Sigma\,N_i \cdot p(C_i)/N$$

The effect of regulation that changes the $p(C_i)$ or N_i can be estimated from this formula. For example, consider the following population of 1 million individuals:

Subpopulation	No. of Individuals (N_i)	Concentration (C_i)	Probability of Toxicity ($p[C_i]$)
1	300,000	0 ppm	0
2	300,000	1	1×10^{-6}
3	200,000	2	4×10^{-6}
4	100,000	3	9×10^{-6}
5	100,000	4	16×10^{-6}

The average risk in this population is

$$p = (300,000 \times 1 \times 10^{-6} + 200,000 \times 4 \times 10^{-6} + 100,000 \times 9 \times 10^{-6} + 100,000 \times 16 \times 10^{-6})/10^6 = 3.6 \times 10^{-6}$$

If the exposure in subpopulation 5 were eliminated entirely, the average population risk is reduced to 2.0×10^{-6}. Or, if the maximum concentration in all subpopulations is reduced to 2 ppm at which the risk is 4×10^{-6}, the average population risk is

$$p = (300,000 \times 1 \times 10^{-6} + 400,000 \times 4 \times 10^{-6})/10^6 = 1.9 \times 10^{-6}$$

Likewise, the effect of other regulatory strategies could be estimated.

The effect of exposures by multiple routes or mixtures of toxic substances are discussed later. The science, engineering, and technology of exposure assessment is a very complex subject that is beyond the scope of this chapter.

4. DOSE–RESPONSE AND RISK ESTIMATION

4.1. Introduction

The final step in risk assessment is to combine the information from the exposure assessment with dose–response relationships in order to provide estimates of disease risk for various conditions. Risk is defined as the probability, on a scale of 0 to 1, that an individual will develop a disease by a given time for a specified set of exposure conditions. If the risk estimate is for a group (population) of individuals, then the risk also may be expressed as the proportion or percentage of individuals developing a disease.

The most relevant estimate of risk for exposure to a toxic substance is obtained from humans exposed for the same conditions. Such information seldom exists. Disease rates in human populations are generally difficult to ascertain, and accurate exposure data seldom are available. Humans are exposed to many toxic substances making it difficult to identify the particular toxic substance and to separate its effect from other exposures. Humans are relatively mobile, making it difficult to obtain medical information. If human dose–response information is available, the data often arise from occupational groups or groups of individuals exposed to higher levels than those existing in the usual environment. Hence, risk estimates for environmental exposure levels would require extrapolation to lower exposures.

Since adequate human dose–response information seldom is available, most risk estimates are based on animal bioassay data. These experiments are generally conducted using a few dose levels and at most a few dozen animals per dose. Doses higher than those encountered by humans are generally used in order to elicit potential toxic effects in relatively small numbers of animals. If we are concerned about a level of risk on the order of 1 in 1,000, it would take several thousand animals to detect and measure this level of risk with precision. Resources are not generally available for such studies. Disease risks on the order of 10% usually can

be detected with high probability. Thus, most risk estimates are based on experiments in which the proportions of animals with disease exceed 10%. This leads to
two major problems in risk assessment: high to low dose extrapolation and animal to
human extrapolation. Some of the problems involved in animal to human extrapolation are discussed in later sections. The remainder of this section is devoted to
discussing dose–response models and low dose extrapolation.

4.2. Cancer Risk Estimation

Many dose–response models proposed for cancer risk estimation are discussed by
Krewski and Brown (1981). These models can be divided into two general groups:
distribution and mechanistic.

The distributional models assume that each individual in a population has a dose
level above which a tumor will appear. The collection of these doses produces a
population distribution function. The probability that a tumor occurs at a particular
dose is the proportion of animals that produce tumors below that dose. Various
statistical models have been used to describe these distributions, e.g., log-probit,
logistic, and extreme value. In recent years the trend has been to develop mechanistic models for low dose extrapolation; thus distributional models are not discussed
further.

Mechanistic models attempt to describe the carcinogenic process in terms of
initiation and cell proliferation (promotion). A multistage model was suggested by
Armitage and Doll (1961) that expresses the probability, p, of tumor by a specified
time with continuous exposure to a carcinogen at a dose, d, as

$$p = 1 - e^{-(a_1 + b_1 d)(a_2 + b_2 d) \cdots (a_k + b_k d)}$$

This model assumes that there are k stages to the carcinogenic process, and $a_i \geq 0$
represents the spontaneous background rate for the ith stage. It is assumed that there
is a linear relationship between the rate and dose with a non-negative slope, $b_i \geq 0$.
Howe and Crump (1982) provide a computer program for fitting a modified multistage model

$$P = 1 - e^{-(q_0 + q_1 d + q_2 d^2 + \cdots)}$$

where the $q_i \geq 0$. Currently, this model is commonly used by the U.S. Environmental Protection Agency for low dose extrapolation (Anderson et al., 1983). Although
this model may fit the data reasonably well in the experimental dose range it may be
inaccurate in the low dose range. Other models may also fit the data equally well in
the experimental dose range, but provide extremely different estimates of risk at low
doses (Food and Drug Administration Advisory Committee, 1971). Crump (1982)
discusses the futility of choosing the best model from bioassay data. Since point
estimates depend on the model selected and the best model cannot be determined,
point estimates are essentially meaningless. At low doses the modified multistage
model is dominated by the linear term so that

$$p \cong q_0 + q_1 d$$

for small doses. Hence, the term *linearized multistage* is often applied to this model. The excess tumor rate above the spontaneous background rate is

$$(p - q_0) \cong q_1 d$$

If q_1^* represents an upper confidence limit estimate of the linear term, $q_1^* d$ provides an upper limit estimate of excess risk at a dose of d. For all dose–response models that are convex at low doses (curving upward), it is felt that $q_i^* d$ provides a plausible upper bound for all point estimates from convex models. The lower bound estimate is zero. This allows for a threshold dose below which no tumors would appear in any animals. Arguments for low dose linearity are presented in a later section. The multistage model has been criticized for not directly including the effect of dose on the cell proliferation rate. The effect of dose on cell proliferation is included in the proportion of animals with tumors at any point in time, but this effect is not accounted for separately.

Moolgavkar and Knudson (1981) and Moolgavkar and Venzon (1979) propose a two-stage initiation-promotion model to describe the carcinogenic process. This model assumes that normal cells are transformed (mutated) to initiated cells, followed by a birth–death process (proliferation) of initiated cells, which can be transformed to a malignant state. Let

N = The number of normal cells susceptible to transformation
M_0 = The net transition rate to initiated cells
M_1 = The net transition rate from initiated to malignant cells
B = The net birth (proliferation) rate of initiated cells

If these parameters are nearly constant over time and $p(t)$ is small, in the absence of competing mortality the probability, p, of a tumor by time, t, is approximately

$$p(t) = 1 - e^{-NM_0 M_1 (e^{Bt} - Bt - 1)/B^2}$$

If it is assumed that the transition and proliferation rates are linearly related to a target tissue dose, x, then

$$p(x,t) = 1 - e^{-(a + bx)(c + dx)[\exp(g + hx)t - (g + hx)t - 1]/(g + hx)^2}$$

as given by Thorslund et al. (1987). This model contains six parameters. In general it is not possible to estimate all of these from standard bioassay data. Specialized experiments are required. Research and experience with this model presently are inadequate to determine the general utility of this model. Again, this model approaches linearity at low doses.

4.3. Intermittent Exposure to Carcinogens

In chronic bioassays animals are usually dosed at a constant dose through the duration of a study. Humans may be exposed to a carcinogen intermittently due to occasional use of a substance, during various occupational exposures or accidents, or at variable environmental exposure levels. It would be unusual to be exposed to a constant level of a carcinogen throughout life.

A simplifying assumption often is made that risk is proportional to total dose and age at exposure is ignored. For example, the estimated tumor risk for an exposure for a fraction (1/F) of a lifetime is the lifetime tumor estimate divided by F where the daily dosage rate is the same.

Kodell et al. (1987) discuss the effect on tumor rates for intermittent exposure assuming an underlying multistage process. The most extreme case is the exposure to the total lifetime dose at one instance as opposed to that same total dose apportioned out equally over the total lifetime. Early life exposure to a late-acting carcinogen or late life exposure to an initiator may have little effect on the lifetime tumor rate. In such instances, the risk estimate based on a constant lifetime exposure may grossly overestimate the cancer risk. On the other hand, early life exposure to an initiator or late life exposure to a late-acting carcinogen may have a larger effect than that same total dose spread over a lifetime. In such instances, Kodell et al. (1987) show that instantaneous exposures may produce up to k times the cancer rate estimated from the same total dose apportioned out constantly over a lifetime, where k is the number of stages in a multistage process. If time to tumor information is available, it may be possible to estimate k. For a multistage model the tumor rate is expected to increase with the k*th* power of time. Doll (1971) indicates that the number of stages generally varies from three to six. In the absence of other information, a rule of thumb would be to increase the cancer risk estimate for an instantaneous exposure by a factor of six in order to obtain an upper limit for an estimate based on the same total dose spread over a lifetime.

Chen et al. (1988) investigated the effect of instantaneous exposure assuming the Moolgavkar-Knudson-Venzon two-stage-proliferation model. It appears for this model that a cancer risk estimate may need to be multiplied by a factor of 10 in order to allow for increases in risk based on the same total dose spread evenly over a lifetime. Again, the effect on cancer risk may be small depending on the age it occurs.

Gaylor (1988) surveyed the literature for cases where short-term exposures could be compared to lifetime exposures at the same dosage rate. This search only produced a limited database of 10 studies. For most of these studies the short-term exposures were for at least one-fourth of the duration of the study. Examples for single exposures were not included. Within this very limited database, estimates of short-term incidence were generally less than 10 times the estimate based on average daily lifetime dose.

4.4. Nonparametric Extrapolation

Nonparametric means that a model fit to the bioassay data is not used for low dose extrapolation. A model may be used to describe the dose–response relationship in the experimental dose range and to estimate the uncertainty in the experimental results, but with nonparametric extrapolation risk estimates outside the experimental dose range are not obtained by extending the model.

Gaylor and Kodell (1980) propose fitting a model to dose–response data in order to estimate the upper confidence on the excess risk above background at the lowest experimental dose level and then extrapolating linearly from this point to the origin (zero excess risk at zero dose). If p_u represents the upper confidence limit on the excess risk at the lowest experimental dose, d_L, then excess risk at any lower dose, d, is estimated to be less than

$$p \le \frac{p_u}{d_L} \cdot d$$

That is, p_u/d_L establishes the smallest low dose slope compatible with the experimental data. Gaylor and Kodell (1980) showed that for 14 sets of bioassay data covering a range of dose–response shapes and spontaneous background rates that the nonparametric procedures generally gave upper limits on low dose risk estimates that are within a factor of two of those obtained by extrapolating with the modified multistage model. These 14 data sets also included nontumor endpoints.

The above procedure is expected to provide a plausible upper limit on low dose risk estimates in those situations where the true low dose response is linear or sublinear (convex). Farmer et al. (1982) pointed out that for low values of risk at the lowest experimental dose, the estimate of the upper confidence limit on the excess risk, P_u, at the lowest dose may be somewhat influenced by the choice of the model fit to the data in the experimental dose range. To reduce the dependency of the linear low dose extrapolation on the choice of the model used to estimate P_u, Farmer et al. (1982) suggested that the low dose extrapolation should start at the lowest dose or the dose giving an estimate of the excess risk of 1%, whichever is larger. The above procedure can be applied to both tumor and nontumor endpoints. No assumptions are made that are unique to carcinogenesis. Arguments for low dose linearity, other than conservatism, are presented later.

Krewski et al. (1984) suggested linear extrapolation between the origin and the upper confidence limit on the lowest dose that displayed a higher proportion of animals with tumors than the controls. Krewski et al. (1986) modified their procedure to consider extrapolation from all doses that do not demonstrate a statistically significant increase in the tumor rate above the control animals. The lowest slope from the upper confidence limit on the excess risk for these doses is used for low dose extrapolation. No model is fit to the bioassay data and the higher doses that are more likely to produce tumors due to physiological stresses are not used in low dose extrapolation. Krewski et al. (1984) showed that the nonparametric extrapolation is generally within a factor of two of upper limits of excess risk obtained from the modified multistage model. Apparently little is gained by making the additional assumption of a multistage process.

4.5. Reproductive and Developmental Risk

As discussed previously, data for reproductive and developmental effects pose special problems in that substances are administered to one or both parents and the results are measured in the offspring. Thus, the litter is the experimental unit.

Generally, the regulation of substances that produce reproductive or developmental defects is accomplished by applying safety factors to no observed adverse effect levels (NOAEL). Such procedures do not provide risk estimates. Risks are assumed to be zero. Problems associated with the safety factor approach are discussed in a later section.

For reproductive and developmental effects, Kimmel and Gaylor (1988) suggest first estimating the effective dose that produces an excess risk in 10% of the animals, denoted by ED_{10}. The reason for the choice of this risk level is that it is about the lowest risk that can be detected with a moderate number of animals, typically about 20 litters per dose group. The ED_{10} can often be estimated by fitting a model to the bioassay. Since little or no extrapolation is required to estimate the ED_{10}, the choice of a model is not critical. Any model that provides a reasonable fit of the bioassay data can be used. A lower confidence limit estimate, LED_{10}, accounts for the experimental uncertainty. A discussion of statistical procedures to estimate confidence limits for reproductive and developmental data is beyond the scope of this chapter. Haseman and Kupper (1979) discuss the analysis of data from teratological experiments.

The LED_{10} gives an estimate of the dose at which the excess risk is expected to be less than 0.1. If this dose is divided by a factor of F, then the risk at the LED_{10}/F is expected to be less than $0.1/F$ if the dose–response curve is linear or sublinear (convex). Obviously, if the LED_{10}/F is below a threshold dose below which no animals produce any defects, then the risk is zero. For example, if the $LED_{10} = 100$ mg/kg of body weight per day and $F = 1,000$, the risk at 0.1 mg/kg of body weight per day is expected to be less than $0.1/1,000 = 0.0001$ (1 in 10,000). Crump (1984) proposed a similar procedure.

4.6. Neurotoxic Risk

Gaylor and Slikker (1990) describe a four-step process for estimating risk due to neurotoxic agents:

1. Establish a mathematical dose–response relationship between a measure of a biological effect and dose (preferably dose in the target tissue)
2. Determine the distribution (variability) of individual measurements of biological effects about average values
3. Establish an adverse, or abnormal, level of a biological effect in unexposed (control) individuals
4. Combine the information from the first three steps to estimate the risk (proportion of individuals exceeding an adverse or abnormal level of a biological effect) as a function of dose

Neurotoxic risk can be estimated from the first three steps. The first step provides an estimate of the average value for a particular dose level. If the distribution of the variability of measurements about the average value can be established, then the probability (risk) that an individual will fall outside the normal range for unexposed individuals can be estimated (Slikker and Gaylor, 1990).

5. MECHANISMS OF TOXICITY

5.1. Threshold Doses, Safety Factors, and Low Dose Linearity

Toxicologists generally assume that for all biological effects other than mutagenic carcinogens, threshold doses exist below which no toxic effects occur in any individual. This is generally based on the observation that it is impossible to distinguish differences in biological effects between animals treated at low doses and untreated control animals. The problem with this argument is that it is impossible to determine whether the risk is actually zero or if the statistical resolving power to detect small risks is inadequate. Every bioassay experiment has a statistical limit of detection depending on the number of animals used. For example, observing no effects in a large experiment with 100 animals does not mean that this dose is below a threshold dose and without risk. If the true risk were 0.03, there is a 5% chance that no effects will be observed in a random sample of 100 animals. That is, the upper 95% confidence limit estimate of a risk based on the observation of no effects in 100 animals is 0.03. No effects in an extremely large sample of 1,000 animals only indicates with 95% confidence that the true risk is less than 0.003, and so on.

Threshold doses cannot be established from animal bioassay data. In some cases it may be possible to show with pharmacokinetic studies that the active toxic substance does not get to the target tissue site at low doses. Even Michaelis-Menten processes are nonthreshold processes. In some cases, subcellular experiments may be able to show that a toxic event does not occur at low doses. In general, thresholds cannot be demonstrated and are assumed. This assumption has led to a regulatory process of dividing a NOAEL by a safety factor (SF) to establish a safe dose. The NOAEL is generally taken as the largest dose for which the effect is not statistically significant at the $p \le 0.05$ level. This is not equivalent to saying that this is a no risk dose. A safety factor of 10 is often assumed to allow for an increased sensitivity of humans compared with animals, and another safety factor of 10 accounts for variation in sensitivities among individuals (Lehman and Fitzhugh, 1954). The dose equal to the NOAEL/100 is assumed to pose no or negligible risk. This procedure serves as the basis for the regulation of most toxic substances. The procedure may be adequate for minor toxic effects, but its properties need to be investigated when serious toxic effects, even though rare, may be produced.

Gaylor (1989) showed that the estimates of the risks at the NOAEL for 10 sets of data varied from 2% to 6%. That is, the premise that risk at the NOAEL is close to zero in the experimental animal under the conditions of the test generally is false. If risk were proportional to dose for these data sets, the risks at the NOAEL/100 would actually vary from 0.0002 to 0.0006. Furthermore, there are many instances where interspecies and intraspecies sensitivities are greater than 10. There is no guarantee that the deficiency in a safety factor of 10 for one factor will be compensated for by the other factor.

The safety factor approach does not reward good experimentation. A precise experiment with many animals will generally be able to detect lower levels of risk at lower doses than poorer experiments. Hence, NOAELs will tend to be lower for better experiments, and the level of use allowed by the NOAEL/SF will be more stringent. An additional safety factor may be imposed for inadequate experiments, but "inadequate" is not usually defined and the application of an additional safety factor is very uneven.

Perhaps the strongest arguments for lack of threshold doses and for low dose linearity are expressed by Crump et al. (1976), Peto (1978), and Hoel (1980). They state that if a substance augments a toxic process that is already occurring in untreated animals or individuals, then the addition of even a small amount of a toxic substance will be accompanied by a small increase in risk. This assumes that the administration of a substance results in an increase in the toxic agent at the target tissue site. Even if a threshold dose exists, the presence of toxic effects in untreated animals or individuals indicates that exogenous or endogenous factors have already caused the threshold dose to be surpassed. Hence, the addition of any dose that augments a toxic process that is already occurring will produce additional risk. Since a short segment of any continuous curve can be approximated by a straight line segment, low dose linearity is expected in such instances. The question is not so much whether low dose linearity exists, but rather over what dose range is a linear approximation appropriate? Since most toxic effects appear in untreated animals or individuals, it is difficult to argue that a threshold dose occurs below which no individual will exhibit a toxic effect unless it can be shown that the substance administered acts through an entirely different process than any existing processes that may be producing the toxic effects.

Note that the above arguments for low dose linearity apply not only to mutagenic carcinogens but to all toxic effects. It is generally accepted that dose–response curves for mutagenic carcinogens may be linear at low doses. More impetus is given to this assumption since the formation of DNA adducts are often predictive of tumors and the relationship between administered dose of a carcinogen and DNA adduct formation frequently is linear at low doses (Poirier and Beland, 1987; Beland, 1989).

5.2. Interspecies Dose Scaling

Dose has been used up to this point in a very general sense. The U.S. Food and Drug Administration usually scales dose across species on a body weight basis. That is, equal toxic effects are assumed in humans and animals when the dose administered is expressed as milligram per kilogram of body weight per day. The U.S. Environmental Protection Agency generally scales dose across species on a surface area basis for carcinogens. Surface area is calculated as body weight to the 2/3 power. That is, dose is expressed in terms of mg/(kg body weight)$^{2/3}$ per day. Scaling on a surface area basis results in higher estimated risks in humans than body weight scaling by about a factor of six higher when extrapolation is from rats and about a factor of 12 higher for mice. Risk estimates based on the concentration of the substances in the food or water result in risk estimates between body weight and surface area dose scaling. Some investigators argue that various physiological factors correlate better across species when body weight to the 3/4 power is used. Andersen (1987) suggests that body weight scaling should be used for active stable metabolites, surface area scaling if the parent compound is carcinogenic, and the reciprocal of surface area if the active carcinogen is a reactive metabolite. Considerable research needs to be undertaken before patterns on interspecies dose scaling may develop.

Presumably interspecies extrapolations would improve as one moves from the administered dose to the effective dose at the target tissue site. When relative absorption rates between humans and animals are measured, risk estimates can be adjusted accordingly. In a few cases physiologically based pharmacokinetic models may be developed. Such procedures require estimates of many parameters and generally require extensive experimentation. Even if the toxic agent can be measured in the target organ, questions remain as to whether the most appropriate dose is the maximum concentration reached, area under the concentration-time curve, or some other measure. Questions also remain as to whether the amount of toxic substance in the target organ scales best across species on a weight, concentration, or surface area basis. Considerable research is required in this area before patterns develop for interspecies scaling of dose that best predict toxic effects.

5.3. Mixtures

Humans probably are exposed to more than one carcinogen at a time. The question arises as to what the total carcinogen risk is for a mixture of carcinogens. Gibb and Chen (1986) investigated this problem assuming an underlying multistage process. They show that an additive effect is expected, at low doses, for carcinogens that affect the same stage of a multistage process. Carcinogens acting on different stages have a multiplicative effect. At low doses, multiplication of relative risks near one will result approximately in the addition of the excess risks. For example, if the relative risk is 1.001 for each of three different stages affected, the over-all relative risk is $(1.001)^3$, which is approximately equal to 1.003. These results hold for the multistage model. If there is synergism or antagonism between two or more components in a mixture, the predictability of the total risk is likely to be poor. Experiments to study the interaction among components of a mixture will have to be conducted on high enough doses to observe effects in moderate numbers of animals. Again, these interactions might not occur at lower doses. At the present time, there generally appears to be no good alternative to the assumption of additive risks for the carcinogenic components in a mixture.

When estimates of the risks for individual components of a complex mixture, e.g., diesel exhaust, are not available, a bioassay is required at different concentrations of the mixture. Then, estimates of risk for the total mixture can be obtained. The shortcoming of this approach is obtaining representative samples of a mixture for testing.

6. EXPERIMENTAL DESIGN

The bioassays conducted by the National Toxicology Program to detect carcinogens currently administer chemicals at the maximum tolerated dose (MTD), 1/2 MTD, 1/4 MTD, and controls in both sexes of the $B6C3F_1$ mouse and Fischer 344 rat.

Fifty animals are generally used for each dose-species-sex group. The primary purpose of this experiment is to indicate if a chemical has carcinogenic tendencies. Portier and Hoel (1983) and Gaylor et al. (1985) show that this experimental design is also generally good for low dose linear extrapolation regardless of the background tumor rates and the shape of the dose–response curve. Little is gained by adding lower dose levels or more animals at the lower doses.

There is considerable criticism of the MTD. It is argued that tumors may be produced at the MTD due to physiological stress. This issue has been discussed previously. Low dose extrapolation primarily depends on the results at the lower doses. If the results at the higher doses are questionable, the data from these doses could be discarded before performing any low dose risk estimation.

7. UNCERTAINTY IN CANCER RISK ESTIMATES

It is often assumed that estimates of cancer risk overestimate the actual risk because conservative procedures are generally used that tend to overestimate risk for various steps of a risk assessment. Procedures that may be conservative are linear low dose extrapolation, upper confidence limits on the experimental tumor incidence, use of the most sensitive set of data (tumor site-species-strain-sex), interspecies dose scaling on a surface area basis, assuming 100% absorption in humans, use of extreme exposure levels for humans, and use of benign tumors.

Allen et al. (1988) compared the carcinogenic potency for 23 chemicals for which suitable data existed for both animals and humans. For those chemicals that are carcinogenic for both animals and humans, cancer risk estimates are about the same on the average for animals and humans where dose scaling is based on body weight and benign tumors are included. This analysis does not indicate that tumor risks are grossly overestimated for human carcinogens. In fact, animal data frequently underestimate the human risk. It is encouraging that predictions are more or less correct on the average, but individual predictions may be high or low by a factor of 100.

Low dose linear extrapolation is often criticized for being too conservative, but, as discussed previously, there are several arguments for low dose linearity. Also, Bailar et al. (1988) show that a high percentage of animal bioassays are supralinear. That is, the one-hit model is not conservative enough in these cases. This analysis is based on data collected at the MTD and 1/2 MTD. It is unknown if linear or sublinear relationships are more prevalent at low doses. It does not appear that low dose linearity is necessarily conservative. The nonlinear dose–response curves often observed may be due in part to the pharmacokinetics at high doses.

There can be little objection to using upper confidence limits on tumor rates observed in an experiment. Confidence limits account for the uncertainty inherent in any experiment. Upper limits generally are only two to three times higher than the point estimate of the tumor rate in the experimental dose range. The use of upper limits only allows for the uncertainty in the experimental results and does not impose undue conservatism on cancer risk estimates.

The general practice in cancer risk estimation is to select the tumor site in the species and sex that gives the highest estimate of risk. The justification for this is presumably to allow for the more sensitive humans. The study by Allen et al. (1988) indicates that the use of the mean sensitivity in animals tends to estimate the carcinogenic risk in humans.

In the absence of information about the relative absorption in animals and humans, equal absorption is assumed. If information on relative absorption is available, then this can be factored into dose estimates.

The inclusion of benign tumors does not necessarily increase the estimate of risk. Inclusion of benign tumors generally will increase the background tumor rate, and the result may be an increase or decrease in the risk.

Factors that affect cancer risk estimates tend to be multiplicative. Using a worst case scenario for each step of a cancer risk will likely produce an overestimate of the risk for a human carcinogen. However, depending on the procedures used, cancer risk estimates are not necessarily overly conservative.

A number of issues involved in risk assessment have been discussed in this chapter. Obviously, it is not possible to cover all aspects of risk assessment and certainly not possible to discuss even a few aspects in depth. Risk assessment approaches have only begun to be formalized. Information is being collected that no doubt will influence risk assessments for individual chemicals and collectively will continue to alter procedures used in risk assessment. Unlike most areas of science, direct measurement and observation generally are not possible. There will always be some uncertainty in estimating low dose effects from high dose experiments and in predicting risk in humans from animal data. Only additional research can reduce these uncertainties.

REFERENCES

Allen BC, Crump KS, Shipp AM (1988): Correlation between carcinogenic potency of chemicals in animals and humans. Risk Analysis 8:531–544.

Andersen ME (1987): Tissue dosimetry in risk assessment, or what's the problem here anyway? In Pharmacokinetics in Risk Assessment: Drinking Water and Health. Washington, DC: National Academy Press, vol 8, pp 9–22.

Anderson EL, and the Carcinogen Assessment Group of the U.S. Environmental Protection Agency (1983): Quantitative approaches in use to assess risk. Risk Analysis 3:277–295.

Armitage P, Doll R (1961): Stochastic models for carcinogenesis. In LeCam LM, Neyman J (eds): Proceedings of the Fourth Berkeley Symposium on Mathematical Statistics and Probability. Berkeley, CA: University of California Press, vol 4, pp 19–38.

Bailar JC, Crouch EAC, Shaikh R, Spiegelman D (1988): One-hit models of carcinogenesis: Conservative or not? Risk Analysis 485–497.

Beland FA (1989): Metabolic activation of aromatic amine carcinogens *in vitro* and *in vivo*. J Univ Occup. and Environ. Health 11:387–397.

Chen JJ, Kodell RL, Gaylor DW (1988): Using the biological two-stage model to assess risk from short-term exposures. Risk Analysis 8:223–230.

Crump KS (1982): Designs for discriminating between binary dose response models with applications to animal carcinogenicity experiments. Comm Stat (A) 11:375–393.

Crump KS (1984): A new method for determining allowable daily intakes. Fund Appl Toxicol 4:854–871.

Crump KS, Hoel DG, Langley CH, Peto R (1976): Fundamental carcinogenic processes and their implications for low dose risk assessment. Cancer Res 36:2973–2979.

Doll R (1971): The age distributuion of cancer: Implications for models of carcinogenesis. J R Stat Soc (A) 134:133–166.

Farmer JH, Kodell RL, Gaylor DW (1982): Estimation and extrapolation of tumor probabilities from a mouse bioassay with survival/sacrifice components. Risk Analysis 2:27–34.

Fears TR, Tarone RE, Chu KC (1977): False–positive and false–negative rates for carcinogenicity screens. Cancer Res 37:1941–1945.

Food and Drug Administration Advisory Committee on Protocols for Safety Evaluation (1971): Panel on carcinogenesis report on cancer testing in the safety evaluation of food additives and pesticides. Toxicol Appl Pharmacol 20:419–438.

Gaylor DW (1988): Risk assessment: Short-term exposure at various ages. In Woodhead AD, Bender MA, Leonard RC (eds): Phenotypic Variation in Populations: Relevance to Risk Assessment. New York: Plenum Press, pp 173–176.

Gaylor DW (1989): Quantitative risk analysis for quantal reproductive and developmental effects. Environ Health Perspect 79:243–246.

Gaylor DW, Kodell RL (1980): Linear interpolation algorithm for low-dose risk assessment of toxic substances. J Environ Pathol Toxicol 4:305–312.

Gaylor DW, Kodell RL, Chen JJ (1985): Experimental design of bioassays for screening and low dose extrapolation. Risk Analysis 5:9–16.

Gaylor DW, Slikker W Jr (1990): Risk assessment for neurotoxic effects. NeuroToxicology 11:211–218.

Gibb HJ, Chen CW (1986): Multistage model interpretation of additive and multiplicative carcinogenic effects. Risk Analysis 6:167–170.

Haseman JK (1984): Statistical issues in the design, analysis and interpretation of animal carcinogenicity studies. Environ Health Perspect 58:385–392.

Haseman JK, Huff J, Boorman GA (1984): Use of historical control data in carcinogenicity studies in rodents. Toxicol Pathol 12:126–135.

Haseman JK, Kupper LL (1979): Analysis of dichotomous response data from certain toxicological experiments. Biometrics 35:281-293.

Hoel DG (1980): Incorporation of background in dose-response models. Fed Proc 39:73–75.

Hoel DG, Yanagawa T (1986): Incorporating historical controls in testing for a trend in proportions. J Am Stat Asso 81:1095–1099.

Howe RB, Crump KS (1982): GLOBAL82: A computer program to extrapolate quantal animal toxicity data to low doses. Prepared for the Office of Carcinogenic Standards, OSHA, U.S. Department of Labor, Contract 41USC252C3.

Kimmel CA, Gaylor DW (1988): Issues in qualitative and quantitative risk analysis for developmental toxicology. Risk Analysis 8:15–20.

Kodell RL, Gaylor DW, Chen JJ (1986): Standardized tumor rates for chronic bioassays. Biometrics 42:867–873.

Kodell RL, Gaylor DW, Chen JJ (1987): Using average lifetime dose rate for intermittent exposures to carcinogens. Risk Analysis 7:339–345.

Kodell RL, Haskin MG, Shaw GW, Gaylor DW (1983): CHRONIC: A SAS procedure for statistical analysis of carcinogenesis studies. J Stat Comput Simulation 16:287–310.

Kodell RL, Shaw GW, Johnson AM (1982): Nonparametric joint estimators for disease resistance and survival functions in survival/sacrifice experiments. Biometrics 38:43–58.

Krewski DA, Brown CC (1981): Carcinogenic risk assessment: A guide to the literature. Biometrics 37:353–366.

Krewski DA, Brown CC, Murdoch D (1984): Determining "safe" levels of exposure: Safety factors or mathematical models? Fund Appl Toxicol 4:383–394.

Krewski DW, Murdoch D, Dewanji A (1986): Statistical modelling and extrapolation of carcinogenesis data. In Moolgavkar SH, Prentice RL (eds): *Modern Statistical Methods in Chronic Disease Epidemiology.* New York: John Wiley and Sons, pp 259–282.

Lehman AJ, Fitzhugh OG (1954): 100-Fold margin of safety. Q Bull Assoc Food Drug Officials 18:33–35.

Miller RG (1966): Simultaneous Statistical Inference. New York: McGraw-Hill.

Moolgavkar SH, Knudson AG (1981): Mutaton and cancer: A model for human carcinogenesis. J Natl Cancer Inst 66:1037–1052.

Moolgavkar SH, Venzon DJ (1979): Two-event models for carcinogenesis: Incidence curves for childhood and adult tumors. Math Biosci 47:55–77.

Peto R (1978): Carcinogenic effects of chronic exposure to very low levels of toxic substances. Environ Health Perspect 22:155–161.

Peto R, Pike MC, Day NE, Gray RG, Lee PN, Parish S, Peto J, Richards S, Wahrendorf J (1980): Guidelines for Simple, Sensitive Significance Tests for Carcinogenic Effects in Long-Term Animal Experiments. IARC Monographs, Supplement 2. Lyon, France: International Agency for Research on Cancer.

Poirier MC, Beland FA (1987): Determination of carcinogen-induced macromolecular adducts in animals and humans. Prog Exp Tumor Res 31:1–10.

Portier C, Hoel DG (1983): Optimal design of the chronic bioassay. J Toxicol Environ Health 12:1–9.

Siegel S (1956): Nonparametric Statistics for the Behavioral Sciences. New York: McGraw-Hill.

Slikker W Jr, Gaylor DW (1990): Biologically-based dose–response model for neurotoxicity risk assessment. Korean J Toxicol 6:205–213.

Snedecor GW, Cochran WG (1980): Statistical Methods, 7th ed. Ames: Iowa State University Press.

Sokal RR, Rohlf FJ (1969): Biometry: The Principles and Practice of Statistics in Biological Research. San Francisco, CA: W.H. Freeman.

Tarone RE (1982): The use of historical control information in testing for a trend in proportions. Biometrics 38:215–220.

Thomas DG, Breslow N, Gart JJ (1987): Trend and homogeneity analyses of proportions and life table data. Comput Biomed Res 10:373–381.

Thorslund TW, Brown CC, Charnley G (1987): Biologically motivated cancer risk models. Risk Analysis 7:109–119.

≡ 17

HAZARDOUS CHEMICALS AND BIOTECHNOLOGY: PAST SUCCESSES AND FUTURE PROMISE

BURT D. ENSLEY
MARY F. DeFLAUN

Envirogen, Inc.,
Lawrenceville, New Jersey 08648

1. INTRODUCTION

One of the most exciting aspects in the application of biotechnology to the control of hazardous chemicals is the fact that this technology is just beginning to produce results. The use of microorganisms to control wastes such as municipal sewage is taken for granted, but until recently the application of biotechnology to control hazardous wastes has been generally assumed to be impractical. These wastes were too toxic or too concentrated or too recalcitrant for microorganisms to be effective. These assumptions are now changing. From the dramatic examples of bioremediation such as at the beaches in Prince William Sound, to more quiet revolutions taking place in plant operations all over the United States where wastes are being treated and destroyed as they are being generated, biotechnology is becoming the cost-effective solution to many hazardous waste problems.

Microbial Transformation and Degradation of Toxic Organic Chemicals, pages 603–629
© *1995 Wiley-Liss, Inc.*

This change in emphasis toward biotechnology solutions has not been without resistance from scientists and engineers who are more comfortable with traditional chemical and physical means of treatment. In addition, new applications of biotechnology have had to overcome erroneous dogmas about microorganisms and enzymes used to degrade hazardous chemicals. It has only been recently that some notions about microorganisms such as their inability to grow deep in the subsurface or their lack of activity against airborne contaminants and highly chlorinated compounds have been dispelled. It is our belief that one of the largest obstacles still facing the application of biotechnology in hazardous waste treatment is the ignorance and fear many people have about microorganisms: that these creatures can get out of control and wreak havoc, that they are ineffective in most applications involving hazardous chemicals, and that they are still so poorly understood that their use in real world systems is not yet practical.

2. PAST SUCCESSES OF BIOTECHNOLOGY IN HAZARDOUS CHEMICALS TREATMENT

Clearly one of the most successful bioremediation projects undertaken in terms of environmental restoration and enhanced public awareness of bioremediation was the stimulation of indigenous microorganisms to degrade oil on contaminated beaches in Prince William Sound. The dramatic removal of oil from beaches by the application of fertilizer convinced important government officials, including EPA chief William Riley, that bioremediation was a viable alternative for at least some hazardous waste contamination problems. The fertilizer Inipol EAP 22, produced by Elf Aquitane, and a slow-release granular fertilizer (Customblen) were applied to oiled beaches during the summers of 1989 and 1990 (Pritchard and Costa, 1991).

These fertilizer applications stimulated the indigenous microorganisms to degrade the oil contaminating the beaches at rates approximately two- to threefold above those on beaches that did not receive fertilizer. Even greater (three- to fivefold) stimulation of biological degradation activity was observed on other beaches, and the stimulation continued into the beach itself to a depth of at least 7–8 cm. In addition to the strikingly obvious removal of oil on the beaches, this approach was shown to be an environmentally safe clean-up technology with no adverse environmental side effects or release of undegraded oil or oil residues (Lindstrom et al., 1991). In many ways this application was a triumph for biotechnology; while it was far from the most sophisticated methods used in biotreatment, the high visibility given the disaster, the positive results obtained in the face of considerable skepticism, and the clearly evident clean up of the treated area make this a highly successful bioremediation project. The enhanced public and official awareness of the benefits of bioremediation accrued to the entire industry due to the efforts of a few scientists working on this project under extremely adverse conditions.

3. THE PROMISE OF BIOTECHNOLOGY AND OBSTACLES TO FUTURE APPLICATIONS

The revolution in biotechnology applications to hazardous chemical treatment does not just lie in biotechnology itself. Certainly new scientific methods developed by the biotechnology industry can be applied to microorganisms and enzymes that treat hazardous waste. But equally exciting is the discovery of new applications using existing technology and microorganisms. These new applications include states of matter only recently regarded as treatable by microorganisms and the growing use of biotechnology in geographical areas that were previously politically unavailable to Western technology.

In a 1987 article entitled "Biotechnology and Hazardous Waste Disposal: An Unfulfilled Promise," Robert Nicholas described the reasons that biotechnology as a science and an industry had not seen more of an impact in hazardous waste treatment. One obstacle to broader application is the complex mixture of many wastes. This heterogeneity demands that many microorganisms, each with an activity against one of the components in the waste, be used in a system to detoxify the mixture completely. It is stated in the above treatise that little work on using microorganisms to treat complex waste mixtures is available and also that the wastes have high toxicity.

Little commercial or academic research aimed toward resolving problems associated with the biodegradation of hazardous waste had been undertaken when Mr. Nicholas' article appeared in 1987. Much has changed in the intervening years. More academic interest and funding by government agencies has been directed toward solving problems of bioremediation. Companies have been funded by venture capitalists with the idea that the time for applying biotechnology and its tools to hazardous waste treatment as a practical enterprise has come. This is very promising for the future of bioremediation as a useful discipline, since practical applications usually accompany rapid growth in scientific research and discovery.

Another difficulty in biotechnology applications is that microorganisms will not grow on some defined substrates at concentrations that are frequently of regulatory concern (such as below 10 ppb). These low concentrations, while still potentially toxic to humans or capable of reducing environmental quality, will not support the growth of degradative microorganisms. The issue of organism survival under environmental conditions also mitigates against the use of some microorganisms that are capable of degrading many hazardous wastes.

An important issue limiting biotechnology applications to hazardous waste treatment is that of regulation at the federal level by the Environmental Protection Agency and at the regional and local levels by other government bodies including local legislatures and town councils. There is a good deal of concern regarding the use of biotechnology in the environment. This concern centers around the potential harm that can be caused by genetically engineered or nonindigenous microorganisms, including further deterioration in an already fragile or marginal environment that has been impacted by chemical pollution. This fear in many ways arises from the problem itself. Chemical pollution is a direct result of developing technology, and the concern that application of further technology, such as biotechnology, will make matters worse is hard to counter.

At the same time, much of this fear is in fact irrational. Most microorganisms are not dangerous, including genetically engineered ones. While extreme positions on this subject can be taken for personal or intellectual advantage, genetically engineered microorganisms are almost all entirely benign and can only have a positive environmental impact. Also nonindigenous microorganisms are introduced into new environments every time the wind blows or birds and insects or other creatures pass over, under, or through the environment. The insistence by regulators from the state of Alaska that no nonindigenous organisms be used to treat the beaches of Prince William Sound may have seemed prudent at the time. The wisdom of this decision can be brought into perspective by considering that 11,000 nonindigenous organisms—human beings, each carrying billions of bacteria—were mobilized from all over the United States and a few foreign countries and sent to clean up the beaches in Alaska. Many of these people certainly introduced their own nonindigenous microorganisms onto the beaches by any number of different mechanisms. It is entirely possible that microorganisms originally degrading oil on the contaminated beaches of Normandy, France, found their way to Prince William Sound by the action of tide and currents, wind, ships, or human beings and have now established themselves on those beaches. It is ironic that, if isolated, these cultures would now be considered native "Alaskan" microorganisms.

The promising aspect in the regulatory arena is that common sense and reality are slowly asserting themselves. Regulations forbidding or limiting the use of nonindigenous or even genetically engineered microorganisms are easing and thus reducing the financial risk of attempting to improve the degradative potential of microorganisms. As common sense prevails, less regulatory oversight will be demanded for the use of these microorganisms, and prudent, intelligent application of specific, rationally designed microorganisms will be permitted as long as their use does not reduce environmental quality or increase health risks. This more positive regulatory atmosphere is one of the most promising developments for the future of biotechnology applications in hazardous waste treatment. It is also promising for the future of the environment, since nonindigenous or genetically engineered microorganisms may often be the best or only hope for improving some impacted environments in our lifetime.

4. VOLATILE (AIRBORNE) CONTAMINANTS

To date there has been a limited amount of activity in the treatment of volatile or airborne contaminants by microorganisms. Vapor phase treatment has not been widely regarded as practical because dissolving the contaminant into an aqueous phase containing the microorganisms can limit the degradation rate. However, it has recently been demonstrated that volatile contaminants can be efficiently degraded in vapor by microorganisms. As the number of compounds known to be degradable by microorganisms continues to grow and the activity of these organisms against volatile contaminants becomes more widely demonstrated, vapor phase treatment will be a practical method for controlling emissions from factories, and it may even be able to reduce contaminants in the lower atmosphere.

The use of microorganisms in a variety of bioreactor systems to degrade volatile hazardous emissions in air is becoming more widespread. Volatile compounds such as the chlorinated solvents, paint strippers, lower molecular weight alkanes, and aromatic hydrocarbons can be degraded in vapor by microorganisms. This phenomenon has been taken for granted in many laboratories for years. Microorganisms are often grown on vapors of various compounds both on solid media and in liquid cultures.

Extensive degradation of volatile organics can be achieved using specialized microbial cultures. We have recently observed the destruction of vaporous chlorinated solvents such as trichloroethylene (TCE) in excess of 95% using a gas lift reactor (Ensley and Kurisko, 1992). Other hazardous volatile compounds such as dichloroethylene and the carcinogen vinyl chloride are also readily degraded by these types of microbial cultures. Most of the recent studies appearing in the literature on the degradation of chlorinated solvents include degradation of volatilized materials. Unless steps are taken to prevent volatilization, compounds such as TCE, when added to the sealed vial systems used in laboratory analysis, reach an equilibrium between the aqueous phase and the head space above it. To achieve complete destruction, the microorganism must degrade both the material dissolved in the water and the material in the head space air above it. In laboratories and pilot systems, vaporous hazardous materials are being degraded in bioreactor systems efficiently and rapidly.

Hydrocarbon fuel vapors can be treated in an apparatus called a *biofilter* containing diatomaceous earth as a solid support that is seeded with microorganisms and kept at a high level of humidity. A stream of air containing either JP5 jet fuel or diesel fuel with high-molecular-weight hydrocarbons is passed through the biofilters. These simple reactors easily biodegrade the hydrocarbons to give outlet concentrations well below regulatory discharge limits, establishing that simple reactor designs can be used to treat hazardous contaminants in the vapor phase (Hodge et al., 1991; Medina et al., 1991). The removal of propane, isobutane, and butane from a waste air stream has also been demonstrated using microorganisms growing on the surface of soil in a reactor. This soil bioreactor reduced hydrocarbon concentrations in the air by at least 90%, again demonstrating the efficacy of microorganisms in treating these airborne contaminants (Kampbell et al., 1987). Xenobiotic hazardous waste carried as vapors in the air have also been treated with a biofilter reactor. Volatile chlorinated organics such as dichloromethane and aromatic compounds such as benzene and styrene have been degraded in bioreactors containing specifically enriched cultures (Ottengraf and van den Oever, 1983; Ottengraf et al., 1986).

A specific example that shows considerable promise for the treatment of volatile contaminants is the biological destruction of certain chlorofluorocarbons before they reach the stratosphere. The reductive dehalogenation of chlorofluorocarbons such as CFC-113 has been reported (Semprini et al., 1990a,b). In addition, we have recently discovered that methane-oxidizing microorganisms will oxidize some of the hydrochlorofluorocarbons (HCFCs) and hydrofluorocarbons (HFCs) proposed as CFC replacements (DeFlaun et al., 1992). A preliminary report that ammonia-oxidizing microorganisms catalyze a similar reaction (Hyman et al., 1992) suggests the use of broad substrate specificity oxygenases for the destruction of HCFCs in groundwater and at recycling and treatment facilities. Contaminated water treatment may not be the only application. These organisms have a relatively high rate of activity against HCFCs and HFCs, and it would be interesting to speculate whether the introduction as aerosols of these degradative microorganisms into the lower atmosphere, at altitudes that are still warm enough to support biological activity, would prevent some of the HCFCs in the lower atmosphere from rising to the stratosphere and destroying the ozone. It has been consistently predicted that all of the CFCs and HCFCs currently in the atmosphere will eventually rise to the strato-sphere and cause ozone destruction (Ramanathan, 1975). There is approximately a 10 year lag period between the time that these compounds are manufactured and the time they reach the stratosphere (Zurer, 1992). We may be able to accelerate the removal of HCFCs in the lower atmosphere by the application of microorganisms. While it seems that the cost of such an approach may be prohibitive, in fact it may be trivial compared with the economic impact of ozone destruction in the strato-sphere. Introduction of aerosols of the HCFC-oxidizing bacteria would be unlikely to cause any adverse environmental impact. The HCFCs would be oxidized to simple molecules such as CO_2 and chloride and fluoride ions, which would be removed from the atmosphere by precipitation. These ammonia and methane-oxidizing organisms are safe, they have no known pathogenic activity, and experi-ments on the degradative activity of microorganism aerosols released into the lower atmosphere could be highly fruitful.

In the future, manufacturing and other industries will be a reduced source of airborne contamination. Emissions and exhausts will be passed through bioreactors where the contaminants will be eliminated from the air by specific degradative bacteria. Contamination already existing in the atmosphere may also be treatable by extending the use of microbial aerosols to include lower atmosphere applications.

5. BIOREMEDIATION OF CONTAMINATED AQUIFERS

A growing application of bioremediation is the *in situ* treatment of contaminated groundwater and aquifer solids. There are literally hundreds of sites across the United States where important groundwater resources have been contaminated by hazardous chemicals. The current technology for treating these sites is to pump the contaminated water to the surface and treat it there by physical or chemical means. Unfortunately, this ignores much of what is known about soil and solid chemistry and the interaction between aquifer solids and organic chemicals. Water is not a good solvent for these compounds, and it has recently come to light that it could take hundreds of years to remove organic contaminants by pumping water to the surface (Mott, 1992). This is unacceptable, and alternatives must be found or these water resources will have to be abandoned. Clearly, abandoning fresh water because it has been contaminated by man and forever preventing the use of this scarce and declining resource cannot be the best solution.

Economic and regulatory issues associated with excavation of soil and the growing realization that pump-and-treat methods are not remediating aquifers within a reasonable time period have spurred interest and research in *in situ* methods. There are two approaches to bioremediation *in situ:* stimulation of the indigenous microbial activity, and bioaugmentation or the introduction of degradative bacteria. There are enthusiastic proponents and strong opposition to both of these approaches. Biostimulation of the indigenous population by adjusting nutrients, pH, temperature, and oxygen is an attractive option from a regulatory standpoint; however, an insufficient or nonexistent degradative population would hinder this approach (Broholm et al., 1991). One of the most common arguments against introducing degradative bacteria is that nonindigenous strains cannot compete with resident microbial populations. In most applications, however, it will not be necessary for the degradative bacteria to survive for an extended period of time. With the introduction of sufficient numbers of highly active bacteria, remediation may be completed before the decline of the introduced species.

One problem associated with pumping bacteria to depth in soils or aquifers is clogging of the injection well associated with adhesion of the introduced bacteria to the soil matrix around the well. Stimulation and growth of indigenous microorganisms when nutrients and oxygen are introduced to the subsurface can also cause plugging of the material surrounding the injection well. In studies designed to stimulate different populations of TCE-degrading organisms *in situ,* it was found that pulse injections of nutrients and oxygen rather than continuous injection helped to alleviate plugging problems (Semprini et al., 1990a).

Recent work has shown considerable potential in using microorganisms to treat contaminated water and the surrounding aquifer solids in place. This approach could result in markedly lower cost and more rapid clean-up scenarios. Researchers at Moffet Field have demonstrated that indigenous microorganisms, when properly stimulated, will degrade groundwater contaminants. Proper stimulation can mean the addition of gases such as methane or aromatic compounds to induce microbial oxygenases that are active against pollutants such as TCE. Biostimulation of methanotrophic populations *in situ* by the addition of methane and oxygen resulted in 20% degradation of TCE, 50% of *cis*-DCE, 85% of *trans*-DCE, and 95% of vinyl-chloride. Phenol and oxygen additions have stimulated indigenous organism to degrade up to 90% of the TCE added (Semprini and McCarty, 1992).

Similar results have recently been obtained by workers at the Savannah River laboratory. This aquifer may contain one of the most massive groundwater contamination episodes in the United States. Again, stimulating indigenous microorganisms through the addition of methane and oxygen to soil columns appears to cause biological destruction of contaminants in groundwater (Enzien et al., 1992).

The second approach for this type of application is the direct addition of microorganisms to the subsurface to effect treatment of groundwater contaminants. The *in situ* biodegradation of contaminated groundwater by adding catalytically proficient bacteria has been demonstrated in a field trial (Nelson et al., 1990). The TCE-degradative microorganism *Pseudomonas cepacia* strain G-4 was injected into contaminated groundwater along with nutrients and oxygen. After the injection, TCE concentrations *in situ* were observed to decline from a high of 3,000 ppb to a mean of 78 ppb over a 20 day period. This preliminary demonstration is very promising in that microorganisms are the only catalyst that can be added underground to treat contaminants directly in place. Their small size and high activity combined with the large numbers that can be added make this process competitive compared with any alternative.

A method to encapsulate degradative microorganisms in microbeads small enough for *in situ* aquifer remediation has been developed using a low pressure ultrasonic nozzle. Microbeads of alginate, agarose, or polyurethane (2–50 μm in diameter) containing >50% wt/vol of 100% active cells were produced with a *Flavobacterium* capable of degrading PCP. In addition to enhancing penetration into an aquifer, this size bead permits adequate transfer of oxygen into the interior, ensuring survival and activity of the bacteria. Cells thus encapsulated degraded PCP at the same rate as free cells and displayed a shorter lag period prior to degradation, probably because the effects of PCP toxicity are reduced (Stormo and Crawford, 1992). Additional carbon sources incorporated into these beads may allow for sustained survival of organisms *in situ*. Encapsulation may also increase cell loading capacities and rates of microbial production in bioreactors.

The practicality of *in situ* applications of degradative bacteria has been further supported by the discovery that microorganisms can survive and be isolated as deep as several thousand feet underground. The Deep Probe studies identified microorganisms surviving in geological formations at depths that were regarded previously to be beyond those suitable for growth of microorganisms (Balkwill, 1989; Fliermans and Hazen, 1991). These findings suggest that even deep contaminated aquifers can be restored by the introduction or stimulation of biodegradative microorganisms. As with many other biotechnology applications, the introduction of nonindigenous microorganisms into polluted aquifers will itself become a political issue. Opinion makers and regulators must be reassured that adding the organisms will not further deteriorate the quality of the aquifer. Public fears, including the sometimes irrational ones, must be allayed by scientists able and willing to prove that what they are doing causes no further harm. Beyond this, the high cost of treating deep aquifers by any approach could be a deterrent, but simply abandoning these precious fresh water resources for our grandchildren to deal with is not an acceptable alternative.

5.1. *In Situ* Soil Remediation Applications

New application methods include the combination of physical and biological processes for the *in situ* treatment of certain hazardous wastes in soil. The combination of stimulated bioremediation and the physical technique of soil venting has been termed *bioventing*. This technique has been used to remediate hydrocarbon contamination in the vadose zone of soils and groundwater. Soil venting both physically removes volatiles and provides the limiting nutrient oxygen for biodegradation of light components in materials such as hydrocarbons, kerosene, jet fuel, and gasoline. This *in situ* remediation technology has been used to treat contaminated sites both in the United States and in Europe (Hoeppel et al., 1991). Bioventing provides a source of oxygen to the microorganisms as fresh air is drawn into a biovented area to enhance contaminant degradation rates and also causes the vaporization of lighter contaminant fractions such as hydrocarbons. Vapor phase biodegradation has been shown to occur and can take place *in situ*. Direct degradation of dissolved hydrocarbons as well as degradation in the vapor phase can be stimulated by bioventing. The development and increasing popularity of bioventing techniques is yet another example of the strength of combining different technologies in application science and bioremediation.

The optimal temperature for maximum microbial activity is usually warmer than the average soil temperature; therefore, degradation rates may be enhanced by warming the soil. Stimulation of microbial degradation by soil warming was accomplished at a site containing jet fuel (JP4)–contaminated unsaturated soil at Eielson Air Force Base in Alaska. Bioventing was used at this site because soil gas measurements indicated that low oxygen concentrations limited biological activity. In addition to oxygen limitations caused by a high degree of contamination (6,000–40,000 ppm total hydrocarbons), the cold temperatures at this site also reduced microbial activity. To warm the soil, soaker hoses placed 2 feet below the surface delivered 35°C water at a rate of 1 gal/min/50 ft square test plot. This was sufficient to maintain the temperature in this soil at ∼12°C from September until January, while the temperature of the control plot dropped from 10 ° to −2°C. *In situ* respirometry tests conducted in December indicated a biodegradation rate in the actively heated plot at least twice that of the control plot (Sayles et al., 1992).

A number of different methods of warming soil to stimulate microbial activity for bioremediation have been proposed. The effectiveness of warming the soil by pumping warm water through it has been demonstrated in the Eilson AFB study. However, this method may only be effective in conjunction with bioventing, which would prevent anaerobiosis from occurring in the soil. Another method being tested in Alaska involves enhancing solar warming of the soil with a black plastic mulch. Conceivably solar heating could also be effected by other methods. Bioventing and vapor extraction technologies could serve a dual purpose by drawing warm air through the soil, volatilizing contaminants while stimulating biological activity. Another potential method of soil warming is the use of radio waves to generate heat. Radio frequency technology uses underground radio antennas to supply energy and has been demonstrated to be effective in heating unsaturated soil (sand) in Utah to greater than 100°C to volatilize contaminants over large areas of the subsurface. It is possible that the energy generated can be controlled sufficiently to provide uniform heating of soil to ∼30°C to promote biodegradation (R. Kasevich, personal communication).

5.2. Modifying the Oxidative State of the Environment

Another key ingredient for stimulating microbial activity *in situ* is oxygen. The addition of oxygen, however, is an inefficient process due to its low solubility, losses associated with injection, and reaction with inorganic species. Hydrogen peroxide is also used, but is toxic to microorganisms and can form precipitates with inorganic species such as ferric iron. In addition, air, oxygen, or hydrogen peroxide has to be added continuously to sustain high microbial metabolic activity. Oxygen bubbling out of solution can also cause well plugging.

Significant enhancement of bioremediation may be possible with the addition of an oxygen-releasing compound. Chemical oxygen delivery or peroxygen technology is being promoted for use in bioremediation; CaO_2 and MgO_2 have been demonstrated to be effective in enhancing microbial activity in contaminated soil. Both of these oxygen-releasing compounds can be delivered in a solid pelleted form that makes application very simple for surface soil contamination. In this type of application, MgO_2 is more stable than CaO_2 and therefore releases oxygen more slowly and has a longer residence time. MgO_2 also has less of a pH effect on the soil. CaO_2 is very basic and can shift the soil environment to a very high pH if applied too liberally. A method has been developed to produce MgO_2 in stable, but very small particles. This technology produces MgO_2 particles with a maximum size of 44 μm and an average size of 2 μm (S. Koenigsburg, personal communication). In this size range, these particles may be capable of penetrating aquifer solids and delivering molecular oxygen deep into the saturated zone. This technology may prove to be more effective than other oxygen delivery methods currently in use.

Nitrate has been investigated as an alternative to oxygen for *in situ* bioremediation. Unlike oxygen, nitrate is very soluble in water and does not form precipitates in groundwater with high iron content. Using nitrate results in anaerobic biodegradation associated with nitrate reduction and denitrification. The *in situ* application of nitrate as the terminal electron acceptor was investigated in the anaerobic biodegradation of a shallow water table aquifer contaminated with JP4 jet fuel. Laboratory microcosm studies using contaminated and uncontaminated aquifer material indicated that under anaerobic conditions with the addition of nitrate a number of monoaromatic hydrocarbon components of jet fuel, with the exception of benzene, were degraded (Hutchins et al., 1991b). Under field conditions, where oxygen is not excluded, benzene, which degrades rapidly under aerobic conditions, was also eliminated (Hutchins et al., 1991a). Rate constants based on projected nitrate demand indicated that a 3,400 m^2 area of an aquifer contaminated with petroleum hydrocarbons could be remediated in 210 days (Hutchins, 1992). The amount of nitrate or nitrite in the aquifer will have to be carefully controlled so that these compounds do not become a contamination problem themselves.

The discovery that a number of the most recalcitrant hazardous wastes, including the ozone-depleting chlorofluorocarbons (Lovely and Woodward, 1992), can be degraded by anaerobic microorganisms may lead to the increased use of *in situ* modification of the oxidative state of the environment. The utility of anaerobic treatment of CFCs and other hazardous wastes is limited except in certain oxygen-starved environments such as some landfills. Anaerobic conditions created in excavated soil contaminated with the herbicide dinoseb (2-*sec*-butyl-4,6-dinitrophenol) promoted the complete degradation of this teratogenic compound (Kaake et al., 1992). Anaerobiosis in the dinoseb-contaminated soil was created very economically in open containers by the addition of a starchy potato-processing byproduct and flooding with phosphate buffer. This pretreatment stimulated the consumption of both oxygen and nitrate from the soils and resulted in the establishment of an anaerobic microbial consortium. The idea of creating an anaerobic environment in soil is also being applied to PCB contamination. Anaerobic dechlorination in this case is also being induced by simply flooding the soil (Baker, 1992). Changing the oxidative state on the environment *in situ*, however, can be very risky. The transition from an aerobic to an anaerobic environment may mobilize toxic components such as heavy metals previously sorbed to the soil matrix.

The role of anaerobic *in situ* biodegradation in the conversion of TCE to vinylchloride and finally to environmentally benign compounds such as ethene and ethane has been examined in a sand aquifer (Vogel and McCarty, 1985; McCarty and Wilson, 1992). Depletion of organic matter in sections of the aquifer where ethene was present indicated that the organic material served as the electron donor during this transformation. These results suggest that addition of organic matter *in situ* may drive the reductive dechlorination of TCE beyond the more toxic vinylchloride to much less toxic products. The reductive dehalogenation of TCE has also been stimulated *in situ*, using indigenous microorganisms, by the addition of toluene as an electron donor (Sewell and Gibson, 1991). Complete reductive dehalogenation in sand columns of tetrachloroethylene to ethane in the presence of lactate has also recently been documented (Mayer et al., 1989; DeBruin et al., 1992).

6. DEGRADATION OF HAZARDOUS CHEMICALS SORBED TO SOILS AND SEDIMENTS

Many hazardous chemicals such as polychlorinated biphenyls (PCBs) and polycylic aromatic hydrocarbons (PAHs) are very insoluble in water, and their biological degradation is extremely slow. This becomes even more pertinent if the contaminants have been "weathered" or "aged" and are strongly sorbed to the soil or sediment matrix. They are then not readily degradable by most microorganisms, and their bioavailability is poor. Unfortunately some of these compounds are so toxic at such low concentrations (and tend to amplify up the food chain) that poor bioavailability does not completely reduce their environmental hazards. Bioremediation has not been widely regarded as a viable treatment method for these types of wastes because of the slow and incomplete degradation displayed to date. However, recent advances in the application of aerobic microorganisms have shown that substantial degradation of PCBs and PAHs can be achieved using processes that enhance the availability of these compounds to microorganisms.

In addition to aerobic oxidation processes, there have been discoveries that microorganisms can promote reductive dehalogenation of PCBs to less chlorinated metabolites under natural conditions. A study done in the Hudson River indicated that natural biodegradative processes may preclude the destructive and expensive dredging of the river sediments (Brown et al., 1987). Heavily contaminated soils and sediments may now be treatable by biological processes on site or in slurry reactors or treatment lagoons. These methods will contain the contaminants on the site, destroy them without reducing environmental quality, and provide a means of cleaning the sites at a cost substantially below the cost of methods currently in use.

In a number of cases, studies of highly recalcitrant wastes have found the best degradation using a combination of anaerobic and aerobic treatments. Natural anaerobic treatment of PCBs in Hudson River sediments occurred over time with the indigenous microorganisms dechlorinating the higher chlorinated PCB congeners. An *in situ* field study performed in enclosed caissons provided oxygen, nutrients, and agitation to these sediments. Stimulation of the indigenous aerobic PCB-degrading microorganisms was sufficient to accelerate the aerobic biodegradation of the lower chlorinated congeners. Control caissons were agitated but there were no other additions resulting in a considerably lower rate of degradation (Abramowicz et al., 1992).

The more recalcitrant or complex a waste is, the greater the likelihood that more than one degradative microorganism will be necessary for complete mineralization. Often enrichments for particular degradative abilities will yield a mixed culture of organisms that upon further isolation to pure culture do not exhibit the same pattern of degradation. Waste from the wood-preserving industry containing creosote and pentachlorophenol (PCP) is a very recalcitrant contaminant of soil and groundwater. Creosote is a complex mixture of constituents, including high molecular weight PAHs. A two-stage bioreactor system has been designed for the treatment of this combination in waste water. The design of this system was based on the activity of a bacterial consortium that utilized a number of the creosote constituents, plus two different organisms capable of utilizing PCP and fluoranthene as sole sources of carbon and energy. The 13 member bacterial consortium was used in the first stage and the two pure cultures in the second stage of the bioreactor. Under field conditions with a contaminated groundwater feed containing 1,000 ppm creosote, 98% of all monitored creosote constituents and ~70% of the PCP was biodegraded (Mueller et al., 1993).

The use of more sophisticated biological reactor systems and soil washing devices are at least partially effective in treating soils contaminated with compounds such as pentachlorophenol and some PCBs (Compeau et al., 1991). This approach will be expanded in the types of contaminants, the degree of contamination, and the number of sites treatable. This lower cost remediation strategy bodes well for both the environment and American industry. The clean up of contaminated sites has become a major liability, and the use of inexpensive microbial processes means that many sites can be cleaned without bankrupting the responsible parties. This also applies to clean-up solutions for developing countries such as Mexico and Brazil and provides a lower cost approach for the clean up of terribly polluted areas in

Eastern Europe and the former Soviet Union. These countries do not have nor will have in the future the considerable funds necessary to clean up their serious pollution problems using conventional technology. The application of effective biotechnology processes at low cost means that these sites can be cleaned and the environment restored in areas where it would otherwise be economically impractical.

7. BIOLEACHING

The science of geomicrobiology or the ability of bacteria to transform metals in the environment is not new. Naturally occurring sulfide-oxidizing bacteria have been used for several decades in the mining industry to solubilize metals from low-grade ore. The activity of sulfate-reducing bacteria, on the other hand, can be used to convert metals and heavier organics to a lower oxidation state, often less toxic product. Methods for removing heavy metals and radioactive waste from water and soil involve either making the material more soluble for recovery from the liquid phase or making it less soluble for precipitation. In many cases this type of transformation can be effective *in situ,* but there are situations where these natural degradative processes will not be effective without innovative engineering solutions. The variety of bacteria that can be used to recover and recycle heavy metals before they become environmental contaminants may prevent further pollution from these sources. It appears that even contamination by radioactive metals can be bioremediated. Recently, an organism was isolated that has the ability to remove highly soluble uranium from water by precipitation, even at concentrations as high as 2,000 ppm (Lovely et al., 1992a,b).

An *in situ* pilot project designed to reduce hexavalent chromium to chromium(III) biologically in soil used two different types of bacteria to effect this transformation. The addition of molasses as a carbon source promoted the formation of organic acids by facultative bacteria. These organic acids plus the addition of sulfate stimulated the indigenous sulfate-reducing population to produce H_2S, which reduces the chromium from its soluble highly toxic form to insoluble and less toxic chromium(III) (DeFilippi and Lupton, 1992).

8. THE POTENTIAL OF GENETICALLY ENGINEERED MICROORGANISMS IN HAZARDOUS WASTE TREATMENT

Improving biocatalysts used in hazardous chemical degradation has recently emphasized the use of modern molecular biology techniques that directly intervene in the genetic makeup of microorganisms to create new and desirable properties. There has been considerable success and even more promise in the use of these methods to make unique and heretofore unavailable microbial systems for the degradation of even the most recalcitrant hazardous waste.

8.1. Using Genetically Engineered Microorganisms to Improve Degradation Rates

One important factor limiting the bioremediation of sites contaminated with certain hazardous wastes is the slow rate of degradation characteristic of these compounds. Materials such as PCBs and PAHs are degraded very slowly by microorganisms. This slow rate of removal limits the practicality of using bacteria in remediating contaminated sites. If the site will take decades to clean up using microorganisms while the contamination continues to spread, a faster approach is necessary for biology to be considered at all. This is an area where genetic engineering can make a marked improvement. It is no secret that molecular biology techniques can be used to increase the level of a particular protein or enzyme or series of enzymes in a bacterial cell. It has also been established that in most instances if the level of an enzyme in the cell has increased then the rate of the reaction it catalyzes increases as well.

For example, the rate of naphthalene metabolism has been markedly increased by the use of genetic engineering methods. The levels of the enzyme naphthalene dioxygenase in a recombinant *Escherichia coli* are significantly higher than those measured in the original host microorganism (Ensley et al., 1988). The actual level of naphthalene degradation by this genetically modified organism is shown in Figure 17.1. Under these optimized conditions—high concentrations of naphthalene in a stirred tank reactor with high biomass levels (5 g per liter dry weight of cells)—the genetically engineered *E. coli* oxidized naphthalene at an initial rate of 5 g per liter of culture per hour. This high rate of metabolism could probably never be repeated in the field, nor would it be necessary. Degradation of PAHs by naphthalene dioxygenase in heavily contaminated sites under environmental conditions that reduce the degradation rate to a hundred-fold lower than those observed with this genetically modified organism in the laboratory would still result in complete metabolism of the contaminants within a few days.

Genes involved in the degradation of PCBs have been cloned and expressed in microorganisms such as *E. coli* (Mondello, 1989). The degradation of PCBs by the genetically engineered strains displayed rates approximately equivalent to those observed with the wild-type strain LB400. It is likely that these preliminary degra-

---

Figure 17.1. Growth and naphthalene oxidation by a recombinant *E. coli*. An *E. coli* harboring a plasmid containing the strong P_L promoter and the naphthalene dioxgenase genes was grown in a 1 liter stirred tank fermentor with LB + glucose. After several hours of growth at a temperature that restricts expression of the P_L promoter, the temperature was reduced and naphthalene dioxygenase enzyme activity measured (Ensley et al., 1982). When the levels of naphthalene dioxygenase in the recombinant *E. coli* had reached a maximum, 5 g of powdered naphthalene was added to the fermenter. The degradation of the naphthalene was monitored by withdrawing samples at 15 minute intervals and measuring the production of *cis*-naphthalene dihydrodiol. The concentration of remaining naphthalene was measured by gas chromatographic analysis using a flame ionization detector. The data from *cis*-diol accumulation and naphthalene disappearance were consistent with initial degradation rates of 5 g of naphthalene oxidized per hour.

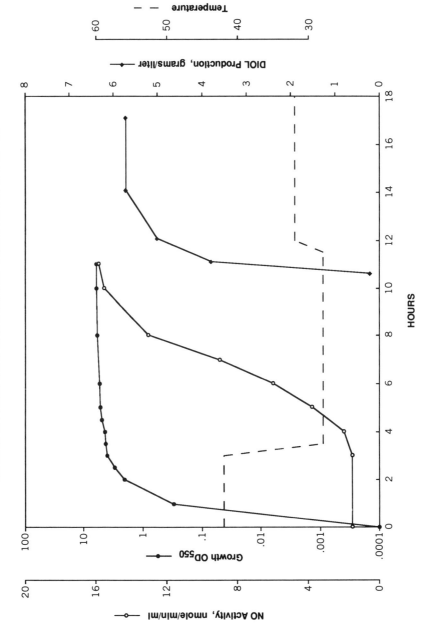

GROWTH AND NAPHTHALENE OXIDATION BY *E. coli*

dation rates will improve with further genetic manipulation of this metabolic pathway in *E. coli*. Strong promoter systems can be used to stimulate expression of the genes and other genetic modifications made so that higher levels of expression can be obtained in *E. coli* or other environmentally acceptable host microorganisms. Eventually, a superior, high-rate PCB-degrading strain may result from these efforts, and its application could alter the practicality and economics of using microorganisms to clean up PCB contaminated sites.

8.2. Broadened Substrate Specificity

One response to the complex mixture of wastes present at some contaminated sites is use of microorganisms that are active against more than one compound. Enzymes with broad substrate specificity and microorganisms with degradative competence against a range of substrates offer the possibility that one or a few microbial cultures can degrade all of the important wastes in a complex mixture. An example of a broad substrate specificity enzyme active against complex and recalcitrant hazardous wastes is an enzyme involved in pentachlorophenol metabolism (Xun and Orser, 1991). This broad substrate specificity enzyme system will attack a wide range of substituted phenols and removes hydrogen, nitro, amino, and cyano groups on the benzene ring (Xun et al., 1992). Such broad substrate range enzymes cloned into the appropriate host organism with a strong promoter system will provide practitioners of bioremediation with catalytic systems that can attack many compounds in complex mixtures present at contaminated sites.

A combination of techniques, including random mutagenesis and gene cloning with plasmids and transposons, has been used to construct microorganisms with broad substrate specificity against a variety of chloroaromatics (Fig. 17.2) (Timmis et al., 1988). This work also illustrates that the same approaches can be used to achieve several objectives at once: The new organism has a pathway engineered to avoid synthesis of toxic (to the organism) intermediates, the metabolic pathway displays activity against a much broader range of substrates than it did originally, and a series of recruited enzymes form the basis of a complete biochemical pathway for a family of chloroaromatics found in process waters. This example of engineering an organism to increase the range of substrates attacked also anticipated the biochemical consequences of broadening the substrate range to include methyl-substituted catechols as intermediates. These compounds are dead-end metabolites in most cells because they are not substrates for the *ortho*-cleavage catechol metabolic pathway. Use by the microorganisms of the alternate *meta*- cleavage pathway to degrade the methyl catechols was precluded by the simultaneous presence of chloro-substituted catechols, which are suicide substrates that will inactivate the *meta*-cleavage enzyme. Introduction of an isomerase gene into the engineered pathway permitted the simultaneous degradation of molecules containing either a methyl or chloro group at the *para* position. This approach emphasizes the usefulness of altering metabolic flow to avoid an intermediate that is toxic to the biochemical machinery.

The above constructs are a prototype for genetic enhancement methods used to

Substrate	Alteration	Source of genes or change

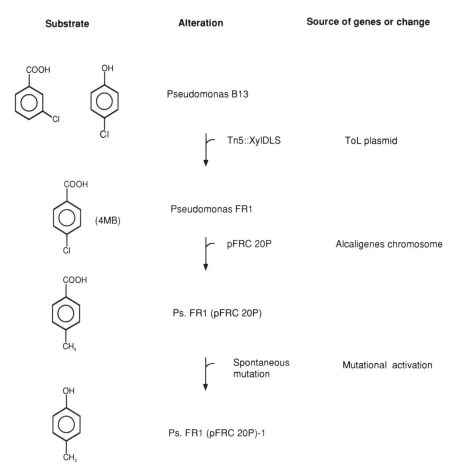

Figure 17.2. Increasing the degradative ability of microorganisms using genetic engineering techniques. The substrate range of *Pseudomonas* B13 was increased by the recruitment of genes encoding several enzymes from different organisms. The transposon Tn5 containing TOL genes D, L, and S was introduced into the chromosome by mating to produce *Pseudomonas* strain FR1 having additional activity against 4-chloro and 4-methyl benzoate. The 4-methyl benzoate was only partially degraded because it is not a substrate for the *ortho*-cleavage pathway. The introduction of an isomerase from the *Alcaligenes* chromosome on a plasmid permitted the complete metabolism of 4-methyl benzoate. Finally, isolation of a strain containing a spontaneous mutation in the initial oxidation pathway permitted the microorganism to degade *para*-cresol.

improve the properties of hazardous waste degrading microorganisms. Other examples of these techniques are appearing with greater frequency. Recently the cloning, nucleotide sequencing, and expression of the gene encoding the enzyme haloalkane dehalogenase was described (Janssen et al., 1989). This enzyme already has a broad substrate specificity, with activity against chloromethanes, chloroethane, dichloroethane, dichloropropane, and bromo and iodoalkanes (Keuning et al., 1985). It was found that this enzyme could be efficiently expressed from its own promoter sequences in a variety of gram-negative bacteria, including halocarboxilic acid degraders such as *Pseudomonas* sp. GJ1, *Xanthomonas autotrophicus XD,* and Pseudomonas sp. GJ31 (Janssen et al., 1984; Oldenhuis et al., 1989). Introduction of this plasmid-encoded haloalkane dehalogenase permitted *Pseudomonas* GJ1 to degrade 1,2-dichloroethane with complete dechlorination and allowed *X. auto-trophicus* XD and *Pseudomonas* GJ31 to utilize 1-chlorobutane as a sole carbon and energy source for growth. The use of this single new gene permitted the already existing biochemical pathways of these organisms to metabolize the reaction products and thus considerably broaden the substrate range of the cultures. Other such examples of enzyme recruitment will appear as this method of genetically altering catabolic enzyme systems becomes better understood and applied.

8.3. Using Genetically Engineered Microorganisms to Degrade Low-Level Contaminants

The use of molecular biology techniques provides a distinct advantage when microorganisms are needed that can degrade contaminants to extremely low concentrations because the contaminants are highly toxic, carcinogenic, or adversely impact the environment at low concentrations. Many natural microorganisms display a threshold effect and are not capable of using certain compounds as growth substrates below a critical concentration (Boethling and Alexander, 1979a,b). This shortcoming can be overcome through the use of genetic methods to introduce new promoter systems that direct the synthesis of degradative pathways by mechanisms that are independent of the substrate concentration.

The microbial degradation of TCE by a genetically engineered *E. coli* to extremely low levels has been reported (Winter et al., 1989). The recombinant microorganism contained a toluene monooxygenase enzyme system under the control of the temperature-inducible P_L promoter or the IPTG-inducible *lac* promoter. Once induced, these cultures would degrade TCE to extremely low levels. A 99.99% reduction in TCE was noted when provided to the culture at initial concentrations of 20 ppm. It would be difficult to use a natural culture to obtain this extensive degradation since the natural inducers and cosubstrates for TCE-oxidizing microorganisms are also competitive inhibitors of TCE degradation (Folsom et al., 1990). All of these compounds and TCE are substrates for the same enzyme, and their effectiveness as competitive inhibitors increases with decreasing TCE concentrations. Therefore, altering enzyme regulation by molecular biology techniques so that the natural substrate and inducer is no longer necessary is one approach to obtaining extensive degradation of recalcitrant hazardous wastes.

As other bacterial degradation systems for environmental pollutants or carcinogens such as PCBs or PAHs are identified, the complete mineralization of these molecules may require genetic manipulation to ensure that the proper enzymes are synthesized regardless of the concentration of the contaminant in the environment.

In this way, the microorganisms can be used as biological catalysts to facilitate the degradation of the target compounds without necessarily having to use the compounds as sole carbon or energy sources for growth or induction of enzyme synthesis.

8.4. The Use of Microorganisms Resistant to Environmental Stress

A barrier to the use of microorganisms in bioremediation is adverse environmental conditions present at the site. These conditions, such as the presence of toxic heavy metals or high concentrations of organic contaminants, can prevent indigenous microorganisms from growing and cause immediate mortality of organisms introduced to the site, significantly reducing the practicality of bioremediation.

New discoveries are expanding the number of sites that can be treated by microorganisms. An example of this is the recent isolation of microorganisms that are resistant to extremely high concentrations of organic solvents. Toluene is an organic solvent that is highly toxic and kills most microorganisms at concentrations as low as 0.1%. Toluene is a component of gasoline and many other fuels, and the use of microorganisms to clean up heavily contaminated areas by these materials are limited because of direct toxicity. Recently, a toluene-resistant microorganism has been isolated that will grow in the presence of high concentrations of organic solvents such as toluene, cyclohexene, styrene, and xylene (Inoue and Horikoshi, 1989). This microorganism has been identified as a strain of *Pseudomonas putida* (Inoue et al., 1991). *P. putida* strains have commonly been used as host organisms for genetic engineering experiments because broad host range gram-negative plasmids can be transferred into these cells by conjugation with *E. coli* or by electroporation. The use of such solvent-resistant microorganisms as hosts for degradative pathways could allow biodegradation at even highly contaminated sites where indigenous microorganisms, and indeed most microorganisms, could not hope to survive.

Some solvent-resistant organisms also display degradative activity against contaminants such as toluene or xylene and could be used to degrade these contaminants at high concentrations in the environment without any genetic intervention at all (Cruden et al., 1992). A particular strain of *P. putida* can grow with aromatic substrates such as *p*-xylene at concentrations as high as 20%, a concentration at which the aromatic substrate forms a separate liquid phase in the system. The discovery that this strain of *P. putida* utilizes the TOL metabolic pathway for aromatic degradation and displays significant hybridization with TOL-derived DNA further strengthens the contention that the use of solvent-resistant organisms as hosts for degradative pathways is a sound strategy for the biotreatment of highly contaminated sites.

There are other microorganisms in the microbiologists' arsenal that are relatively resistant to conditions or unusually high ionic strength, low or high pH, the presence of heavy metals such as mercury, lead, chromium and arsenic, and other conditions that would normally be regarded as prohibitive to life or the use of microorganisms as a remediation alternative. The combination of genetic engineering techniques to introduce desired degradative phenotypes into host strains that display tolerance or resistance to adverse environmental conditions is a powerful tool for bioremediation. This approach will see increasing application in the future as the practicality and desirability of this approach to treating hazardous waste becomes more widely accepted both by the practitioners of remediation and by regulators and the public at large.

8.5. *In Situ* Genetically Engineered Microorganisms

One argument against the application of nonindigenous organisms in the environment is their inability to compete with the established microbial population. Therefore, providing degradative organisms with a selective advantage over the indigenous population would help to establish these organisms temporarily in the environment. An approach that has been taken is to enrich for a host organism that can grow on a highly selective substrate and then provide that host with the appropriate degradative pathways. One such field application vector was developed with a host, *Pseudomonas paucimobilis* IGP4, that could grow in soil amended with 1% of the detergent Igepal CO-720. This host also contained a plasmid into which PCB-degradative genes had been cloned. The detergent substrate did not provide for significant growth of the indigenous population, but did allow proliferation of *P. paucimobilis* IGP4 when introduced into nonsterile soil (Lajoie et al., 1992). The Igepal may also serve to increase the bioavailability of the PCB by emulsification, but this effect has not yet been conclusively demonstrated. In studies with soils contaminated with Aroclor 1242 and amended with 1% Igepal these genetically modified strains were able to effect considerable degradation of the PCB congeners over a 48 day period (Lajoie et al., 1995).

One of the major technical obstacles to successful *in situ* site remediation is providing metabolically capable microorganisms throughout the contaminated region of the subsurface. Many types of motile bacteria that are metabolically desirable for use in degrading contaminants may have only limited mobility through subsurface soils. In addition, soil porosity also limits the passage of microorganisms through the subsurface. One study calculated that the expected normal attenuation in a sandy aquifer would preclude a nongrowing bacterial population from being transported more than 100 m through aquifer sediments at Cape Cod (Harvey et al., 1989). Thus subsurface soil properties and natural adhesive properties of microorganisms limit microbial access to the contaminants. The development of adhesion-deficient variants of degradative bacteria may provide for an effective strategy for *in situ* bioremediation. It has been shown that stable adhesion-deficient variants of soil isolates can be enriched in a sand column assay (DeFlaun et al., 1990). The adhesion-deficient phenotype of these isolates in the sand column assay is also predictive of their adhesion to soil and seeds (DeFlaun et al., 1994). Nonadhesive variants might be capable of traveling much farther through aquifer solids and preclude plugging problems associated with injecting bacteria into aquifers.

9. ASSESSMENT OF BIOREMEDIATION POTENTIAL OF CONTAMINATED SITES

Another exciting application of biotechnology techniques to hazardous waste problems is the adaptation of methods originally developed for disease diagnosis and drug monitoring. Immunoassays have been used to screen a wide range of conditions indicative of disease. A particular type of immunoanalysis called *enzyme-linked immunosorbent assay* (ELISA) has been adapted to detect certain organic contaminants. The challenge in developing immunoassays for organic contaminants is that most of these materials are too small to be recognized by mammalian immune systems. However, methods have been developed to conjugate specific contaminants such as PCBs and PAHs to larger molecules, including serum albumin (Carter, 1992).

Immunoassays used to detect the presence of hazardous wastes can be a significant advantage in remediation activities. Classic chemical methods of analysis for compounds such as aromatic hydrocarbons or PCBs are time consuming and expensive. These analyses can take from 4 to 6 weeks and cost from $200 to $2,000 per sample. While analyzing soil contaminated with the small hydrophobic molecules that are characteristic of many hazardous wastes does not appear to be an attractive application for immunoassays, the feasibility of this approach has been demonstrated for molecules such as the dioxins (Vanderlaan et al., 1988). Because these immuno test kits can be used directly in the field and give results within a few minutes, they make on-site sampling and analysis practical and should permit the practitioner to identify and define the extent and scope of contamination at a particular site within a few days. In addition, the degradation of hazardous materials by microorganisms can be rapidly analyzed on a real-time basis using such assays and the reduced cost and simplicity associated with immunoassays means that more samples can be taken, statistically valid sample numbers can be generated at relatively low cost, and the progress of the remediation activity can be monitored as it is going on.

As immunoassay kits for hazardous waste become more widely accepted and used, many previously untested sites will come under scrutiny and more thorough and complete analysis of contamination will become common place. It is not inconceivable that very simple dipstick-type tests for the most important contaminants such as the PCBs, petroleum hydrocarbons, dioxins, and pentachlorophenol may become as simple to use as home pregnancy test kits. Prospective buyers of real estate, real estate agents, or environmental consultants could use these tests to evaluate whether a piece of property is contaminated before a decision is made to buy. Likewise, owners of industrial sites can test for unsuspected contamination themselves and have any positive results confirmed by more conventional analysis such as GC or GC/MS. While these immunoassays may never completely replace more expensive and conventional analytical methods, they certainly will be useful as an adjunct or supplement to GC analysis.

An important consideration when analyzing a contaminated site for bioremediation potential is the availability of the contaminant to the microorganisms. As discussed previously, weathering and aging can result in hydrophobic contaminants becoming tightly bound to the soil and affect the ability of the bacteria to degrade them. A highly sensitive biological system for the detection of biologically available hazardous waste has been developed. Certain bacteria respond to the presence of contaminants in the environment by inducing the synthesis of degradative enzymes. The biological machinery involved in this degradative pathway induction can be exploited to provide a signal that the contaminant is present.

The gene-encoding bacterial luciferase (a light-emitting protein) has been cloned downstream from a regulated promoter system that induces the synthesis of the naphthalene-degradative pathway. When a culture of these organisms is exposed to naphthalene or salicylate, the bacteria synthesize luciferase and emit light: a sensitive and easily measured indicator of contamination (Burlage et al., 1990; King et al., 1990). In tests of this system utilizing media and soil extracts, the light response was linear and quantitative with naphthalene detected to concentrations as low as 45 ppb. Although the bioluminescence response was quenched in soil slurries, naphthalene could still be detected by the reporter strain (Heitzer et al., 1992). Classical methods for the detection of contaminants in soil can identify and quantify pollutants but cannot be used to predict their degradability. Assays for bioavailability of contaminants provide much more pertinent information relative to assessing the biotreatability of a site.

Theoretically, this system could be readily adapted to detect many different types of environmental contaminants as long as there was an inducible enzyme pathway for degradation and the regulatory machinery responded to low contaminant concentrations. The insertion of the light-producing luciferase gene into the pathway would cause the bacteria to "light up" in the presence of the contaminant. This simple method offers relatively high selectivity, although induction of a metabolic pathway is not always discriminating and a few nonhazardous molecules could interfere by causing a positive reaction. A putatively positive sample could then be further analyzed by more conventional methods once it has been identified. These assay methods give us new tools to monitor the presence and destruction of hazardous materials in the environment at a lower cost and more rapidly than the current sophisticated and sensitive, but time-consuming and expensive methods available.

We have attempted in this chapter to introduce several concepts that are important to the development of biotechnology applications in hazardous waste treatment. Because biotreating many contaminated areas is such a complicated endeavor, many techniques we have outlined, plus a few we have probably overlooked, will have to be integrated into a working system for a successful remediation. Microbiology, biochemistry, chemical engineering, molecular biology, analytical chemistry, mechanical engineering, hydrogeology, and other important disciplines all make a vital contribution to the treatment of hazardous materials, especially in the environment. *In situ* treatment processes, perhaps the most cost-effective approach, require a highly sophisticated understanding of the interactions between the contaminant and the water, the solids, the microbes, and oxygen and other nutrients at a minimum.

It is a good time to be a scientist interested in bioremediation, to feel a renewed interest and activity compared with a more sedate recent past. It appears that more efforts will meet with success now that there is a better understanding of what it takes to practice bioremediation and new tools have become available to the biological and engineering practitioners of this discipline. For industry, bioremediation has shaken off its reputation of being an unreliable approach enthusiastically recommended by individuals who often promised more than they could deliver. The ethical standards of industrial bioremediation activities have been raised as the quality of the science and the degree of public acceptance have increased. More widespread attention to this area has brought breakthroughs in identifying processes and new microorganisms that degrade a wider range of contaminants than ever before. The use of genetically engineered microorganisms in hazardous waste treatment will certainly see the light of day within the next few years. The adaptation of these powerful biological tools to the hazardous waste treatment arena is very promising. Equally exciting is the new collaboration between various disciplines to provide a path from initial discoveries in the laboratory to a cleaner and safer environment. None of us could ask for more in our subject of study or as a source of livelihood.

REFERENCES

Abramowicz DA, Harkness MR, McDermott JB, Salvo JJ (1992): *In Situ* Hudson River Research Study: A Field Study on Biodegradation of PCBs in Hudson River Sediments. Final Report, February 1992, General Electric Company Research and Development.

Baker NC (1992): GE researchers gaining ground in PCB cleanup. Environ Today 3:1–35.

Balkwill DL (1989): Numbers, diversity, and morphological characteristics of aerobic, chemoheterotrophic bacteria in deep subsurface sediments from a site in South Carolina. Geomicrobiol J 7:33–52.

Boethling RS, Alexander M (1979a): Effect of concentration of organic chemicals on their biodegradation by natural microbial communities. Appl Environ Microbiol 37:1211–1216.

Boethling RS, Alexander M (1979b): Microbial degradation of organic compounds at trace levels. Environ Sci Technol 13:989–991.

Broholm K, Christensen TH, Jennsen BJ (1991): *In situ* degradation of chlorinated solvents by stimulating indigenous organisms with methane and air. Environ Technol 12:279–289.

Brown JF, Wagner RE, Feng H, Bedard DL, Brennan MJ, Carnahan JC, May RJ (1987): Environmental dechlorination of PCBs. Environ Toxicol Chem 6:579–593.

Burlage RS, Sayler GS, Larimer F (1990): Monitoring of naphthalene catabolism by bioluminescence with *nah–lux* transcriptional fusions. J Bacteriol 172:4749–4757.

Carter KR (1992): On-site screening speeds sample analysis. Pollut Eng 36–38.

Compeau GC, Mahaffey WD, Patras L (1991): Full-scale bioremediation of contaminated soil and water. In Sayler GS (ed): Environmental Biotechnology for Waste Treatment. New York: Plenum Press, pp 91–122.

Cruden DL, Wolfram JH, Rogers RD, Gibson DT (1992): Physiological properties of a

Pseudomonas strain which grows with *p*-xyrene in a 2-phase (organic-aqueous) medium. Appl Environ Microbiol 58:2723–2729.

DeBruin WP, Kotterman MJJ, Posthunus MA, Schraa G, Zehnder AZB (1992): Complete biological reductive transformation of tetrochloroethene to ethene. Appl Environ Microbiol 58:1996–2000.

DeFilippi LJ, Lupton FS (1992): Bioremediation of chromium (VI) contaminated sold residues uing sulfate reducing baceria. In I and EC Special Symposium, American Chem. Society, Atlanta, GA, pp 117–120.

DeFlaun MF, Ensley BD, Steffan RJ (1992): Biological oxidation of hydrochlorofluorocarbons (HCFCs) and hydrofluorocarbons (HFCs) by a methanotrophic bacterium. Bio/ Technology (in press).

DeFlaun MF, Marshall B, Kulle EP, Levy SB (1994): TnS-insertion mutants of *Pseudomonas fluorescens* defective in adhesion to soil and seeds. Appl Environ Microbiol 60:2637–2642.

DeFlaun MF, Tanzer AS, McAteer AL, Marshall B, Levy SB (1990): Development of an adhesion-deficient mutant of *Pseudomonas fluorescens*. Appl Environ Microbiol 56:112–119.

Ensley BD, Gibson DT, Laborde AL (1982): Oxidation of naphthalene by a multicomponent enzyme system from *Pseudomonas* sp. strain NCIB 9816. J Bacteriol 149:948–954.

Ensley BD, Kurisko P (1992): Field trials of chlorinated solvents degradation by specialized microbial cultures. In Abstracts of the Annual Meeting of the Society for Industrial Microbiology, 9–14 August, San Diego, California, S37, p 64.

Ensley BD, Osslund TD, Joyce M, Simon MJ (1988): Expression and complementation of naphthalene dioxygenase activity in *Escherichia coli*. In Hagedorn, Hanson SR, Kunz DA (eds): Microbial Metabolism and the Carbon Cycle. New York: Harwood Academic Publishers, pp 437–450.

Enzien MV, Picardal FW, Hazen TC, Arnold RG (1992): Effects of trichloroethylene, and methane exposure on microbial community dynamics in a sediment column. In Abst. Ann. Mtg. Washington, DC: American Society for Microbiology, abstract N-25.

Fliermans CB, Hazen TC (1991): Proceedings of the First International Symposium on Microbiology of the Deep Subsurface. Aiken, SC: WSRC Information Services.

Folsom BR, Chapman PJ, Pritchard PH (1990): Phenol and trichloroethylene degradation by *Pseudomonas cepacia* G4: Kinetics and interactions between substrates. Appl Environ Microbiol 56:1279–1285.

Harvey RW, George LH, Smith RL, LeBlanc DR (1989): Transport of microspheres and indigenous bacteria through a sandy aquifer: Results of natural- and forced-gradient tracer experiments. Environ Sci Technol 23:51–56.

Heitzer A, Webb OF, Thonnard JE, Sayler GS (1992): Specific and quantitative assessment of naphthalene and salicylate bioavailability by using a bioluminescent catabolic reporter bacterium. Appl Environ Microbiol 58:1839–1846.

Hodge DW, Medina VF, Ilander RL, Devinny JS (1991): Treatment of hydrocarbon fuel vapors in biofilters. Environ Technol 12:655–662.

Hoeppel RE, Hinchee RE, Arthur MF (1991): Bioventing soils contaminated with petroleum hydrocarbons. J Ind Microbiol 8:141–146.

Hutchins SR (1992): Use of nitrate to bioremediate a pipeline spill at Park City, Kansas: Projecting from a treatability study to full-scale remediation. Abstr. EPA's Biosys. Technol. Dev. Program, Chicago, IL.

Hutchins SR, Downs WC, Wilson JT, Smith GB, Kovacs DA, Fine DD, Douglas RH, Hendrix DJ (1991a): Effect of nitrate addition on biorestoration of fuel-contaminated aquifer: Field demonstration. Ground Water 29:571–580.

Hutchins SR, Sewell GW, Kovacs DA, Smith GA (1991b): Biodegradation of aromatic hydrocarbons by aquifer microorganisms under denitrifying conditions. Environ Sci Technol 25:68–76.

Hyman MR, Ensign SA, Rasche ME, Arp DJ (1992): Degradation of hydrochlorofluorocarbons (HCFCs) by *Nitrosomonas europaea*. In Abstr. Ann. Mtg. Washington, DC: American Society of Microbiology, abstract k-49.

Inoue A, Amomoto MJ, Horikoshi K (1991): *Pseudomonas putida* which can grow in the presence of toluene. Appl Environ Microbiol 57:1560–1562.

Inoue A, Horikoshi K (1989): A *Pseudomonas* that thrives on high concentrations of toluene. Nature 338:264–266.

Janssen DB, Pries F, Van der Ploeg J, Kazemier B, Terpstra P, Witholt B (1989): Cloning of 1,2-dichloroethane degradation genes of *Xanthobacter autotrophicus* GJ10 and expression and sequencing of the dhl A gene. J Bacteriol 171:6791–6799.

Janssen DB, Scheper A, Witholt B (1984): Biodegradation of 2-chloromethane and 1,2-dichloroethane by pure bacterial cultures. Prog Ind Microbiol 20:169–178.

Kaake RH, Roberts DJ, Stevens TO, Crawford RL, Crawford DL (1992): Bioremediation of soils contaminated with herbicide 2-*sec*-butyl-4, 6-dinitrophenol (Dinoseb). Appl Environ Microbiol 58:1683–1689.

Kampbell DH, Wilson JT, Reed HW, Stocksdale TT (1987): Removal of volatile aliphatic hydrocarbons in a soil bioreactor. J Am Pollut Control Assoc 37:1236–1240.

Keuning S, Jenssen DB, Witholt B (1985): Purification and characterization of hydrolytic haloalkane dehalogenase from *Xanthobacter autotrophicus* GJ10. J Bacteriol 163:635–639.

King JMH, DiGrazia PM, Applegate B, Burlage R, Sanserverino J, Dunbar P, Larimer F, Sayler GS (1990): Rapid, sensitive bioluminescent reporter technology for naphthalene exposure and biodegradation. Science 249:778–781.

Lajoie CA, Chen SY, Oh KC, Strom PF (1992): Development and use of field application vectors to express non-adaptive foreign genes in competitive environments. Appl Environ Microbiol 58:655–663.

Lajoie CA, Zylstra GJ, DeFlaun MF, Strom PF (1995): Development of field application vectors for *in situ* bioremediation of soils contaminated with polychlorinated biphenyls. Manuscript in preparation.

Lindstrom JE, Prince RC, Clark JC, Grossman MJ, Yeager TR, Braddock JF, Brown EJ (1991): Microbial populations and hydrocarbon biodegradation potentials in fertilized shoreline sediments affected by the T/V Exxon Valdez oil spill. Appl Environ Microbiol 57:2514–2522.

Lovely DR, Phillips EJP, Widman PK (1992a): Bioremediation of uranium-contaminated waters and soils with microbial U (VI) reduction. In Abstr. Ann. Mtg. Washington, DC: American Society for Microbiology, abstract Q289.

Lovely DR, Phillips EJP, Widman PK (1992b): Enzymatic reduction of uranium by *Desulfovibrio desulfuricans*. In Abstr. Ann. Mtg. Washington, DC; American Society for Microbiology, abstract Q288.

Lovely DR, Woodward JC (1992): Consumption of freons CFC-11 and CFC-12 by anaerobic sediments and soils. Environ Sci Technol 26:925–929.

Mayer KP, Grbić-Galić D, Semprini L, McCarty PL (1988): Degradation of trichloroethylene by methanotrophic bacteria in a laboratory column of saturated aquifer material. Water Sci Technol 20:175–178.

McCarty P, Wilson JT (1992): Natural anaerobic treatment of a TCE plume: St. Joseph, Michigan, NPL site. In Abstr. EPA's Biosys. Technol. Develop. Program, Chicago, IL.

Medina HV, Islander RL, Devinny JS (1991): Treatment of hydrocarbon fuel vapors in biofilters. Environ Technol 12:655–662.

Mondello FJ (1989): Cloning and expression in *Escherichia coli* of *Pseudomonas* strain LB400 genes encoding polychlorinated biphenyl degradation. J Bacteriol 171:1725–1732.

Mott RM (1992): Aquifer restoration under CERCLA: New realities and old myths. Environ Rep 92:1301–1304.

Mueller JG, Colvin RJ, Lantz SE, Middaugh DP, Ross D, Pritchard PH (1993): Strategy using bioreactors and specially-selected microorganisms for bioremediation of ground water contaminated with creosote and pentachlorophenol. Environ Sci Technol In press.

Nelson MJ, Kinsella JV, Montoya T (1990): *In situ* biodegradation of TCE contaminated groundwater. Environ Prog 9:190–196.

Nicholas RB (1987): Biotechnology and hazardous waste disposal: An unfulfilled promise. ASM News 53:138–142.

Oldenhuis R, Kuijk L, Lammers A, Janssen DB, Witholt B (1989): Degradation of chlorinated and non-chlorinated aromatic solvents in soil suspension by pure bacterial cultures. Appl Microbiol Biotechnol 30:211–217.

Ottengraf SPP, Meesters JJP, van den Oever AHC, Rozema HR (1986): Biological elimination of volatile xenobiotic compounds in biofilters. Bioprocess Eng 1:61–69.

Ottengraf SPP, van den Oever AHC (1983): Kinetics of organic compound removal from waste gases with a biological filter. Bio/Technol Bio Eng 25:3089–3102.

Pritchard P (1991): Bioremediation as a technology: Experiences with the Exxon Valdez oil spill. J Hazard Mater 28:115–130.

Pritchard PH, Costa CF (1991): EPA's Alaska oil spill bioremediation project. Environ Sci Technol 25:372–379.

Ramanathan V (1975): Greenhouse effect due to chlorofluorocarbons. Climatic Implications. Science 190:50–52.

Sayles GD, Hinchee RE, Brenner RC, Vogel CM, Miller RN (1992): *In-situ* bioventing: Two U.S. EPA and Air Force sponsored field studies. In Proc. US EPA/A&WMA: *In Situ* Treatment of Contaminated Soil and Water. Cincinnati, OH: EPA.

Semprini L, McCarty PL (1992): Evaluation of enhanced *in situ* aerobic biodegradation of trichloroethylene and *cis*- and *trans*-1,2-dichloroethylene by phenol-utilizing bacteria. Abstr EPA's Biosys Technol Dev Prog Chicago, IL.

Semprini L, Roberts PV, Hopkins GD, McCarty PL (1990a): A field evaluation of in situ biodegrdation of chlorinated ethenes. Part 2. The results of biostimulation and biotransformation experiments. Ground Water 28:715–727.

Semprini L, Roberts P, Hopkins G, McCarty P (1990b): Enhanced *in situ* biotransformation of carbon tetrachloride under anoxic conditions. In EPA Symposium on Bioremediation of Hazardous Wastes: EPA's Biosys. Technol Develop Prog, p 27.

Sewell GW, Gibson SA (1991): Stimulation of the reductive dechlorination of tetrochloroethene in anaerobic aquifer microcosms by the addition of toluene. Environ Sci Technol 25:982–984.

Stormo KE, Crawford RL (1992): Preparation of encapsulated microbial cells for environmental applications. Appl Environ Microbiol 58:727–730.

Timmis KN, Rojo F, Ramos JL (1988): Prospects for laboratory engineering of bacteria to degrade pollutants in environmental biotechnology. In Omen GS (ed): Reducing Risks From Environmental Chemicals Through Biotechnology. New York: Plenum Press, pp 61–79.

Vanderlaan, M, Stanker LH, Watkins BE (1988): Improvement and application of an immunoassay for screening environmental samples for dioxin contamination. Environ Toxicol Chem 7:859–862.

Vogel TM, McCarty PL (1985): Biotransformation of tetrachloroethylene to trichloroethylene, dichloroethylene, vinyl chloride, and carbon dioxide under methanogenic conditions. Appl Environ Microbiol 49:1080–1983.

Wilson JT, Wilson BH (1985): Biotransformation of trichloroethylene in soil. Appl Environ Microbiol 49:242–243.

Winter RB, Yen KM, Ensley BD (1989): Efficient degradation of TCE by a recombinant *Escherichia coli*. Biol Technol 7:282–285.

Xun L, Orser CS (1991): Purification and properties of pentachlorophenol hydroxylase: A flavoprotein from *Flavobacterium* sp. strain ATCC 39723. J Bacteriol 173:4447–4453.

Xun L, Topp E, Orser CS (1992): Diverse substrate range of a flavobacterium pentochlorophenol hydroxylase and reaction stoichiometries. J Bacteriol 174:2829–2902.

Zurer PS (1992): Industry, consumers prepare for compliance with pending CFC ban. Chem Eng News.

INDEX

Acclimation, factors in, 564–565
Acclimation period in aryl reductive deha-
  logenation, 246, 252–253, 257–258
Acenaphthene
  enzymatic attack, 270
  metabolism, 273–275, 282
  structure, 46
Acenaphthylene
  enzymatic attack, 270
  metabolism, 274, 275
  structure, 69
Acetanilides, 38, 49
Acetate, 190–191
Acetone, 186, 190
Achromobacter sp., 323
Acids, 30
Acinetobacter calcoaceticus, 273
Acinetobacter sp., 295
Acrylonitrile, 43
Activated sludge, 436
  chlorinated organic compounds, 450
  continuous flow reactors, 374–377
  nitrogen cycle, 371–374
  requirements, 353
  sequencing batch reactors, 378–380
Aeration in bioremediation, 461–462, 612–
  613
  coal tar-contaminated soil, 503
  polycyclic aromatic hydrocarbon, 496
Aerobacter aerogenes, 6–7
Aerobic bioremediation
  BTEX compounds
    biofiltration, 531
    biosparging, 527–530
    bioventing, 525–527
    landfarming, 523–524
    oxygen sparging/hydrogen peroxide,
      524–525

chlorinated organic compounds, 450–
  451, 454
  fluidized–fixed bed reactors, 472–
  473
  natural, 467–468
  oxidation, 464
  sequential anaerobic–aerobic bio-
    reactors, 473, 474–475
  oxygen supplementation, 612–613
  systems, 353
  see also Aerobic metabolism
Aerobic metabolism
  aryl reductive dehalogenation, 244–246
  benzene, 308, 309
  chlorinated organic compounds, 439,
    443–444
  chlorophenols, 396–400, 413
  hydrocarbons, 87–88
  PCBs, 206
  see also Aerobic bioremediation
Agrochemical industry. *See* Pesticides
Airborne contaminants, 606–608
Air emissions, 511
Air samples, 488
Air sparging, 524–525, 527–530
Alachlor, 49
Alcaligenes denitrificans, 246, 277, 316
Alcaligenes eutrophus, 224, 275, 554, 555
Alcaligenes faecalis, 295
Alcaligenes paradoxus, 275
Alcaligenes sp.
  chlorophenol degradation, 416
  genetically engineered, 619, 620
Alcohols, 30
Aldehydes, 30
Aldicarb, 72
Aldolase, 289
Aldrin, 48

Algae
  hydrocarbon degradation in soils, 84–85
  polycyclic aromatic hydrocarbon metab-
    olism, 273–280
Aliphatic hydrocarbon degradation, 88–89;
  see also BTEX compounds
Alkyl halides, 438, 439
Amidase, 18
Amides, 38
Amines, 31
Aminoaromatics, 341
4-Amino-3,5-dichlorobenzoic acid, 249,
  251–252
Aminomethylphosphonic acid, 361–363
Amitraz, 72
Amitrole, 73
Ammonification, 371–374
Ammonium, 521–523
Amplification, genetic, 11–13, 540–550
Anabaena variabilis, 13
Anaerobic bioremediation
  BTEX compounds, 531–532
  denitrification, 532–533
  methanogenesis, 533–534
  sulfate reducing bacteria, 534
  transition metals, 534–535
  chlorinated organic compounds, 454
  natural, 466–467
  PCE, 465
  reductive, 464
  sequential anaerobic–aerobic bio-
    reactor processes, 473, 474–
    475
  TCE, 613
  see also Anaerobic metabolism
Anaerobic metabolism
  aryl reductive dehalogenation, 244–246
  chlorinated organic compounds, 439,
    444–446
  chlorophenols, 206, 400–401, 417
  homocyclic aromatic compounds, 307–
    308
    benzene, 308–310
    benzoate, 316–319
    cresols, 322–324
    ethylbenzene, 315–316
    phenol, 321
    phthalic acids, 319–321
    toluene, 310–313
    xylenes, 313–315
  hydrocarbons, 87–88
  Polycyclic aromatic hydrocarbons, 286
  see also Anaerobic bioremediation
Analog enrichment, 87

Analytical methods
  BTEX contamination and bioremedia-
    tion
    contaminant plume, 519
    nutrients, 523
    respirometry method, 527
    soil gas survey, 527
    soil permeability, 519–520
  coal tar–contaminated soil
    aeration, 503
    air samples, 488
    culturing and enumeration, 489
    excavation and application, 501–502
    humus percentage, 488
    leachate and run-off collection, 503
    metal analysis, 488
    mineralization of radiolabeled
      naphthalene, 489–491
    soil samples, 489, 503
    trapping and quantitation of vol-
      atilized hydrocarbons, 491
    volatile organic analysis, 488
    volatilization measurement, 497–500
  enumeration, 96–98, 363–366, 489,
    520–521
  hydrocarbons in soils
    indicators and measurement, 96–99
    methods, 93–96
  immunoassays, 562–563
  polychlorinated biphenyl dechlorination,
    133–135, 167–170
  site analysis, 623–625
  see also Risk assessment
Anilazine, 73
Anilines, 341
  immobilized bacteria in degradation,
    367–371
  structure, 43
Anisoles, 402
Anthracene
  enzymatic attack, 270
  fungal degradation, 335–337
  metabolism, 274, 276, 281, 282, 284,
    285, 295–296
  structure, 46
Anthraquinones, 335–337
Antibiotic-resistance markers, 10–11
Antibiotics
  gene amplification in production, 12
  resistance
    genetic vs. physicochemical causes,
      6–7
    mutations in, 8
    Pseudomonas aeruginosa, 17

Antinomycetes, 12
Aquifers. *See* Groundwater; Soil
Aquifer simulators, 453
D-Arabinose, 6–7
Arochlor. *See* Polychlorinated biphenyl dechlorination; Polychlorinated biphenyls
Aromatic compounds
anaerobic degradation, 307–308
benzene, 308–310
benzoate, 316–319
cresols, 322–324
ethylbenzene, 315–316
phenol, 321
phthalic acids, 319–321
toluene, 310–313
xylenes, 313–315
aryl reductive dehalogenation
anaerobic vs. aerobic, 244–246
chemical factors, 260–263
compounds involved, 243–244
environmental factors, 255–260
physiological factors, 246–255
chlorinated
bioremediation, 455–460
degradation, 443–444
in chlorophenol degradation, 401
degradation, 88–89
fungal degradation, 333–337, 341
ligninolysis, 332–333
in petrochemical industry, 28
polycyclic aromatic hydrocarbons
anaerobic metabolism, 286
carcinogenicity, 269–273
degradation to *cis*-dihydrodiols, phenols, and ring-fission products, 273–280
degradation to quinones, 284–286
degradation to *trans*-dihydrodiols, 281–284
enzymes in metabolism, 286–292
genetics of metabolism, 293–296
Pseudomonas in degradation, 17
structures and applications, 32, 41
see also Chlorophenols; BTEX compounds; *specific compounds*
Arthrobacter polychromogenes, 277
Arthrobacter sp., 393–394, 398–399, 404, 405, 413, 416
Arylglycerol-β-aryl ether, 332, 333
Aryl halides, 438, 439
Aryl reductive dehalogenation
acclimation period, 246
anaerobic vs. aerobic, 244–245

in bioremediation, 245–246
chemical factors, 260–263
compounds involved, 243
environmental factors, 255–260
PCBs, 202–204
physiological factors, 246–247
benefits to microorganisms, 252–253
catabolic enzyme derepression, 247–249
electron donors, 253–255
growth substrates or growth factors, 252
multiple substrate effects, 249–251
substrate concentration, 251–252
Aspergillus ochraceus, 284
Assays. *See* Analytical methods
Association for the Environmental Health of Soils, 100
Asulam, 72
Atrazine, 48
Autotrophy, 5
Azinphosmethyl, 73
Azo-adamantane, 50
Azoaromatics, 341
Azobenzene, 70
Azotobacter sp., 393, 398

Bacilli sporulation, 16
Bacillus subtilis, 16
Bacteria
anaerobic metabolism
benzoate, 316–317
cresol, 322–324
phenol, 321
phthalic acid, 319–321
toluene, 310–313
bioleaching, 615
bioremediation, 610
PCBs, 458–459
BTEX degradation, 520–521
chlorinated organic compounds
bioremediation, 456–457
degradation, 437–439
chlorophenol degradation, 392–401
soil treatment, 403–412
wastewater treatment, 413–417
genotypic surveys, 550–555
hydrocarbon degradation, 84–85, 86–87
immonoassays, 562–563
organohalide metabolism, 220–223
PCB dechlorination
isolation, 195–198
population H, 199–200
population M, 198–199

Bacteria (*Continued*)
  petroleum reservoir degradation, 78–79
  polycyclic aromatic hydrocarbon degra-
    dation, 334–335, 489
    anaerobic bacteria, 286
    to *cis*-dihydrodiols, phenols, and
      ring-fission products, 273–280
    to *trans*-dihydrodiols, 281–284
  versatility
    of biochemical pathways, 4–5
    consequences on survival, 19
    developmental cycles, 16
    environmental factors, 15, 18–19
    genome structure changes, 14–15
    mutational changes, 6, 7–8
    nutritional versatility of *Pseu-*
      domonas, 16–18
    vs. physicochemical causes, 5–7
    rearrangements, 8–14
Baffled reactor, 454
Barban, 72
Beijerinckia sp., 273–280
Bendiocarb, 72
Bentazon, 74
Benz[*a*]anthracene
  enzymatic attack, 270
  metabolism, 274, 280, 281, 283–285
Benzenes
  bioremediation
    aerobic, 523–531
    anaerobic, 531–535
    chemical/nutrient requirements,
      521–523
    microbial populations and enumera-
      tion, 520–521
    remedial options, 523
    requirements for *in situ*, 518–520
  degradation, 87–88
    anaerobic, 308–310, 311
    methylotrophic bacteria, 280
  production volume, 55, 57
  properties, 517–518
  regulations, 518
  sources, 515–517
  structures and applications, 28, 34
Benzidine, 69
Benzo[*a*]pyrene
  carcinogenesis, 271–273
  enzymatic attack, 270
  fungal degradation, 333–334, 335
  metabolism, 274, 280, 281, 284, 285–
    286, 292
Benzo(3,4)fluoranthene, 46
Benzoic acid and related compounds
  anaerobic degradation, 316–319

  phthalic acid metabolism, 319–321
  structures and applications, 36, 48
Benzonitriles, 36–37
Benzo(ghi)perylene, 69
4-Benzylbenzoic acid, 52
Benzylfumaric acid, 313
Benzylsuccinic acid, 313
Berkeley, M.J., 4
BESA, 175–180, 200
Bioavailability of substrate, 91–92
Biobenefication, 79
Biodegradation. *See* Degradation
Biofilms, 414, 450–451, 456–457
Biofiltration, 531
Bioleaching, 615
Biomass ratio, 355–357
Biopiles, 109–110
Bioreactors
  chlorinated organic compounds, 450–
    451, 461
  configurations, 454
  see also Bioremediation; *specific sys-*
    tems
Bioremediation
  aquifers, 608–610
  aryl reductive dehalogenation, 245–246
  bioleaching, 615
  BTEX compounds, 516–517
    aerobic, 523–531
    anaerobic, 531–535
    chemical/nutrient requirements,
      521–523
    microbial populations and enumera-
      tion, 520–521
    remedial options, 523
    requirements, 518–520
  chlorinated organic compounds
    aerobic fluidized–fixed bed reactors,
      472–473
    aerobic natural, 467–468
    anaerobic natural, 466–467
    bench-scale process designs, 450–
      454
    bioslurries, 468–471
    chlorobenzenes, 455–456
    chlorophenols and phenoxy her-
      bicides, 456–458
    composting, 471
    field-scale, 460–462
    fungal inoculation technology, 464–
      465
    in situ stimulation, 462–463
    in situ vs. *ex situ*, 473–476
    landfarming, 471–472
    liquid and gaseous wastes, 472–473

methanotrophic oxidation, 463–464
natural, 465–468
PCE, 465
reactor configurations, 454
requirements, 459–460
sequential anaerobic–aerobic bioreactor processes, 473
solid wastes, 468
solvents, 449, 455
trichloroethylene, 226
chlorophenols
soil treatment, 403–412
water treatment, 413–417
chlorophenyls
polychlorinated biphenyls, 205–209,
458–459, 464
coal tar–contaminated soil, 511–513
factors in, 436
genetically engineered microorganisms,
615–622
hydrocarbons in soils
bioventilation, 111–114
composting, 111
engineered biopiles, 109–110
land treatment, 105–109
regulations, 100–101
rhizosphere remediation, 114–115
strategies, 99–100
treatability studies, 101–105
in situ, 608–610
obstacles, 605–606
oxidative state, 612–613
polycyclic aromatic hydrocarbons, 296
Prince William Sound oil degradation,
604
site assessment, 623–625
soil
bioventing, 610–611
sorbed chemicals, 613–615
temperature, 611
trends, 603–604
volatile (airborne) contaminants, 606–
605
Biosensors, 557–558
Bioslurries, 453, 457, 461, 468–471
Biosparging, 524–525, 527–530
"Biotechnology and Hazardous Waste Disposal: An Unfulfilled Promise"
(Nicholas), 605
Bioventing, 610–611
BTEX compounds, 525–527
chlorinated organic compounds, 461,
462–463
hydrocarbons, 111–114
Biphenyls, 33

Bitertanol, 73
Bjerkandera sp., 284–285
Bleach plant effluents, 416–417
British Columbia, 392
Brodifacoum, 71
Bromacil, 72
Bromethalin, 71
3-Bromobenzoic acid, 249
Bromobiphenyls, 194
BTEX compounds (benzene, toluene, ethylbenzene, and xylene)
amount used, 55, 57
bioremediation
aerobic, 523–531
anaerobic, 531–535
chemical/nutrient requirements,
521–523
microbial populations and enumeration, 520–521
remedial options, 523
requirements for *in situ*, 518–520
degradation, 87–88
genetically engineered microorganisms, 621
genotypes in microorganisms, 551
properties, 517–518
regulations, 518
sources, 515–517
Bulk chemicals
amount produced, 55, 56
structure and applications, 29–32, 43
Butane and related compounds, 29–30

Candida sp., 395
CaO_2, 612
Capitella capitata, 85
Caprolactam, 69
Captan, 70
Carbamates, 39, 49, 51
Carbaryl, 72
Carbofuran, 49
Carbon
in chlorinated organic compounds, 438
PCB dechlorination, 173, 189–193
similarity of biochemical pathways in
living systems, 5
Carbon biomass ratio, 355–357
Carbon dioxide
BTEX degradation, 534
PCB dechlorination, 179
respirometric measurements, 98–99
Carboxin, 74
Carcinogenicity
dose–response models, 590–591
historical tumor rates, 586–587

Carcinogenicity (*Continued*)
  intermittent exposure in, 592
  multiple comparisons, 585–586
  multistage process, 597
  statistical tests, 580–584
  uncertainty in risk assessment, 598–599
Carcinogens
  chlorphenols, 392
  organohalides, 220
  PCBs, 204
  polycyclic aromatic hydrocarbons, 269–273
Catechol 2,3-dioxygenase, 290
Catechols, 308, 309, 396
catM gene, 558
Ceresan, 50
CFC-11, 44
CFC-113, 44
CFCs (chlorofluorocarbons), 607–608, 612–613
Chemical structure in environmental fate of compounds, 63
Chloral, 226
Chloramine-T, 45
Chlordane, 70
Chloridazon, 72
Chlorinated catechols, 52
Chlorinated compounds
  bioremediation, 437
    activated sludge system, 436
    aromatics, 455–459
    in situ vs. *ex situ*, 460–462, 473–476
    liquid and gaseous wastes, 472–473
    microbial stimulation, 462–465
    microorganisms, 449
    natural, 465–468
    requirements, 459–460
    solid wastes, 468–472
    solvents, 449, 455
    system comparisons, 450–454, 461
    two stage, 448–449
  chlorine to carbon ratios, 438
  degradation
    co-metabolic processes, 439–440
    environmental factors, 446–448
    fungal, 338–339
    microbial interactions, 440–441
    microorganism physiology, 437–439
    parameters, 435–436
  dehalogenation reactions, 262–263, 441–443
    aerobic, 443–444
    anaerobic, 444–446
  pump-and-vent systems, 436–437

sources, 436
see also Chlorophenols; Polychlorinated biphenyl dechlorination; Polychlorinated biphenyls; *specific compounds*
Chlorinated cymenes, 52
Chlorinated dioxins, 45–46
Chlorinated ethylenes, 225
Chlorinated guaiacols, 52
Chlorinated solvents, 55, 57
Chloroaniline, 341
Chlorobenzenes, 455–456
Chlorobenzilate, 70
3-Chlorobenzoate, 459
  aryl reductive dehalogenation, 249–251, 253, 255, 258–259, 261–262
  degradation, 220–221, 401, 551
Chlorobenzoic acid, 207–208
Chlorocatechols, 396–398
Chlorodecone, 70
Chloroethane, 466–467
1-Chloro-2-(2-[*p*-1,′3′-tetramethylbutylphenoxy]ethoxy)-ethane, 31, 44
Chlorofluorocarbons (CFCs), 607–608, 612–613
Chlorohydroquinones, 398–400
Chlorophacinone, 71
Chlorophenols
  bioremediation, 456–458
    soil treatment, 403–412
    water treatment, 413–417
  biotransformation, 402–403
  degradation
    aerobic, 396–400
    anaerobic, 400–401
    microbial isolates, 392–396
  fungal degradation, 337–340
  production, 389–390
  soil and groundwater contamination, 391–392
  sources, 389–390
4-Chlororesorcinol, 252
Chlorpyrifos, 73
Chromatography, 623
  air samples, 489
  hydrocarbons, 95–96
  polychlorinated biphenyl dechlorination, 133–134, 169–170
Chrysene
  enzymatic attack, 270
  metabolism, 280
  structure, 46
CL-20, 54
Classification, 16–18, 363–366

C/N ratio, 90
Coal tar–contaminated soil
  analytical methods, 488–489
  bioremediation
    alternatives, 512–513
    tar and land requirements, 512
    temperature, 511–512
  chemical content, 487–488
  field pilot demonstration
    air emissions, 511
    GC/FID fingerprints, 509–510
    leachate metals, 510–511
    leachate polycyclic aromatic hydro-
      carbons, 510
    materials and methods, 501–503
    nitrogen and phosphate, 504
    pH, 503–504
    polycyclic aromatic hydrocarbon
      concentration, 504–509
    run-off metals and polycyclic aro-
      matic hydrocarbon, 511
  laboratory investigation
    degradation, 491–501
    materials and methods, 489–491
Cochran-Armitage Test, 583–584
Coenzyme A
  chlorinated aromatic compounds, 444
  cresol, 322, 323
  phthalic acid, 320, 321
  toluene, 312–313
Coenzyme F_{430}, 231, 232
Colony hybridization, 548
Combustion of fossil fuels, 34, 38, 54; *see
  also* Petrochemical industry compounds
Co-metabolism
  chlorinated compounds, 439–440, 467–
    468
  hydrocarbons, 87
  organohalides, 221–223
    cooxidative, 223–229
    coreductive, 229–236
  PCBs, 458–459
Composting
  chlorinated organic compounds, 453,
    457, 461, 471
  chlorophenols, 407–412
  coal tar–contaminated soil, 513
  hydrocarbons, 111
Comprehensive Environmental Response,
  Compensation, and Liability Act
  (CERCLA), 100
Congeners, PCB, 128; *see also* Polychlorinated biphenyls; Polychlorinated biphenyl dechlorination

Continuous flow systems, 352, 374–377
Cosmetics industry compounds
  classes, 52
  structure and application, 36, 40
Cost
  biosparging, 530
  chemical wastewater treatment, 349–350
Coumachlor, 71
Coumafuryl, 71
C/P ratio in hydrocarbon degradation, 90
C230 protein, 563
Creosote, 409, 416
o-Cresol, 45
Cresols, 318–319, 322–324
Cross-acclimation, 87
Cubane, 54
Culturing
  PCB dechlorination, 185–187, 190–193
  polycyclic aromatic hydrocarbon-
    degrading bacteria, 489
Cunninghamella elegans, 282–284
Cunninghamella sp., 333
Cyanobacteria
  hydrocarbon degradation in soils, 84–85
  polycyclic aromatic hydrocarbon metab-
    olism, 281–286
Cyclodienes, 34, 48
Cyclohexanol, 321
Cyclohex-1-enecarboxylate, 319–321
Cycloisomerase, 396
Cyromazine, 73
Cytochrome P_{450}
  organohalide effects, 219–220
  organohalide metabolism, 232–236
  polycyclic aromatic hydrocarbon metab-
    olism, 271, 292

2,4-D. *See* 2,4-Dichlorophenoxyacetic acid
Data evaluation in treatability studies, 104–
  105
Dazomet, 69
DCB-1. *See Desulfomonile tiedjei*
DCE. *See* Dichloroethylene
4,4'-DDE, 70
DDT, 219
  structure, 48
  worldwide production, 56, 59
DDT "antiresistant," 70
DDT relatives, 34
Dechlorination of polychlorinated bi-
  phenyls. *See* Polychlorinatedbiphenyl
  dechlorination
Degradation
  anaerobic, 307–308
  benzene, 308–310

Degradation (*Continued*)
   benzoate, 316–319
   cresols, 322–324
   ethylbenzene, 315–316
   phenol, 321
   phthalic acids, 319–321
   toluene, 310–313
   xylenes, 313–315
  aryl reductive dehalogenation
   anaerobic vs. aerobic, 244–246
   chemical factors, 260–263
   environmental factors, 255–260
   physiological factors, 246–255
  BTEX compounds, 531–535
  chemical wastewater treatment
   activated sludge, 374–380
   elemental transformations, 371–374
   engineering design, 351–355
   glyphosate, 358–366
   microbial design, 355–357
  chlorinated organic compounds
   aerobic, 443–444
   anaerobic, 444–446
   co-metabolic processes, 439–440
   environmental factors, 446–448
   interactions of microorganisms, 440–441
   oxidative and reductive, 441–443
   physiology of microorganisms, 437–439
  chlorophenols
   aerobic, 396–400
   anaerobic, 400–401
   biotransformation, 402–403
   microbial isolates, 392–396
  genes in
   abundance and activity, 552–553
   acclimation of microorganisms, 564–565
   in environment, 550–552
   establishing in environment, 553–554
   hybridization, 547–550
   models, 545–547, 565–567
   protein detection, 562–564
   tracking in environment, 554–555
   transcript detection, 558–562
   transcription, 555–558
   translation, 562
  ligninolytic fungi
   amino-, nitro-, and azoaromatics, 341
   biochemical features, 332–333
   chlorinated phenols, 337–340
   discovery of, 331–332
   polycyclic aromatic hydrocarbons, 333–337
   polymerized organopollutants, 341–342
  organohalides
   challenges to, 220–223
   cooxidative, 223–229
   coreductive, 229–236
  plasmids in, 11
  polycyclic aromatic hydrocarbons
   anaerobic bacteria, 286
   carcinogenesis, 269–273
   to *trans*-dihydrodiols, 281–284
   to *cis*-dihydrodiols, phenols, and ring-fission products, 273–280
   environmental factors, 271
   enzymes, 286–292
   genetics, 293–296
   for naphthalene, 286–291
   to quinones, 284–286
   in soils, 491–501, 504–509
  Pseudomonas, 17–18
  schematic representation, 61–64
   complexity of compound in, 63
   toxic intermediates, 62–63
  see also Polychlorinated biphenyl dechlorination
Dehalobacter restrictus, 230
Dehalogenases, 443, 447
Dehalogenation. *See* Aryl reductive dehalogenation
Dehydrogenases, 287–288, 290
Deletions in microbial versatility, 11–13
Denitrification. *See* Nitrate
Denitrifying bacterium strain T1, 312–313
Desulfitobacterium dehalogens, 246
Desulfobacterium phenolicum, 321
Desulfomonile tiedjei, 132, 197, 201
  aryl reductive dehalogenation, 245–246, 247, 250, 252, 255, 258–262
  chlorophenol degradation, 400–401
  organohalide metabolism, 220–221, 230
Detection methods. *See* Analytical methods
Developmental cycles in gene expression, 16
Developmental risk
  bioassays, 587
  dose–response models, 593–594
Diallate, 74
Diazinon, 73
Dibenzofurans, 45–46, 412
Dicamba, 48, 262
Dichlone, 48
3,4-Dichloroaniline, 252
p-Dichloro-benzene, 50

3,5-Dichlorobenzoate, 249–250
Dichloroethylene (DCE), 609–610
    anaerobic bioremediation, 466–467
    bioremediation, 449, 455
    co-metabolism, 227, 235
    methanotrophic oxidation, 463–464
Dichlorofluoroethane, 44
Dichloromethane dehalogenase, 220
Dichlorophene, 70
2,4-Dichlorophenol, 340, 399–401
2,4-Dichlorophenoxyacetic acid (2,4-D)
    degradation
        genotypes in microorganisms, 551,
            552
        pathway, 390, 396, 398
    structure, 70
Dichlorvos, 73
Dicofol, 70
Dicumarol, 71
Dieldrin, 50
1,2-Diethoxyethane, 44
Dihydrodiol dehydrogenase, 287–288
cis-Dihydrodiols, 273
trans-Dihydrodiols, 281
Dihydroxynaphthalene dioxygenase, 288–
    289
Dimethirimol, 48
Dimethoate, 73
Dinocap, 70
Dinoseb, 48
Dioxane, 69
Dioxygenases
    chlorinated aromatic compounds, 444,
        447
    naphthalene, 286–287, 288, 290, 616
    organohalides, 227–229
Diphacinone, 71
Diphenamid, 71
Diphenatrile, 70
Diphenyl, 48
Diphenylamines, 38
Diphenyl ethers, 38
Diquat, 48
Disinfectants, 35, 36, 45
Disulfoton, 73
Dithiocarbamates, 41
Diuron, 72
DMC, 70
DNA (deoxyribonucleic acid)
    adducts, 596
    insertion sequences, 9
    supercoiling and looping, 14–15
    transcript measurement, 559–562
DNA–DNA hybridization, 547–548, 553–
    554

Dose
    ED_{10}, 594
    interspecies dose scaling, 596–597
    LED_{10}, 594
    low dose linearity, 595–596
    maximum tolerated dose, 597–598
Dose–response and risk estimation
    cancer, 590–591
    humans, 589–590
    intermittent exposure to carcinogens,
        592
    neurotoxic, 594
    nonparametric extrapolation, 592–593
    reproductive and developmental, 593–
        594
Downflow packed bed, 454

Earthworms, 402, 403
ED_{10} (effective dose), 594
Effective dose (ED_{10}), 594
Electron acceptors in degradation
    aryl reductive dehalogenation, 257
    BTEX compounds, 523, 532–535
    chlorinated aromatic compound, 445
    chlorophenol, 401
    PCB dechlorination, 173–180
Electron capture detector (ECD), 133–134,
    169–170
Electron donors, 253–255
Electronics industry compounds, 50–51
Elemental transformations, 371–374
ELISA (enzyme-linked immunosorbent as-
    say), 623
Endosulfan, 70
Endothall, 48
Engineered biopiles, 109–110
Enumeration techniques
    BTEX compound bioremediation, 520–
        521
    glyphosate degradation, 363–366
    hydrocarbon degradation in soils, 96–98
    polycyclic aromatic hydrocarbon-
        degrading bacteria, 489
Environmental factors
    aryl reductive dehalogenation, 255–260
    chlorinated organic compound degrada-
        tion, 439, 446–448
    chlorophenol bioremediation, 404
    DNA supercoiling, 14
    fate of compounds in, 61–64
    gene expression, 9, 11, 15, 18–19
    genetically engineered microorganisms,
        622
    hydrocarbon degradation, 92–93
    PCB dechlorination, 172–173

Environmental factors (*Continued*)
  polycyclic aromatic hydrocarbon metabolism, 271
Environmental Protection Agency (EPA), 474–476, 596; *see also* Regulations
Enzyme assays, 563–564
Enzyme-linked immunosorbent assay (ELISA), 623
Enzymes
  amplification, 11–13
  aryl reductive dehalogenation, 247–249
  broadened substrate specificity, 618–620
  chlorophenol degradation, 396–400
  fungal degradation
    amino-, nitro-, and azoaromatics, 341
    chlorinated phenols, 339–340
    polycyclic aromatic hydrocarbon, 335–337
  PCB dechlorination, 132–133, 201–204
  polycyclic aromatic hydrocarbon metabolism, 286–292
  see also specific enzymes
EPA (Environmental Protection Agency), 596; *see also* Regulations
EPA Bioremediation Field Initiative, 474–476
Epoxides, 31
EPTC, 74
Equal longevity, 582–584
Escambia Bay, 160
Escherichia coli, 616
  DNA supercoiling, 14
  enzyme amplification, 11
  lacZ gene, 10
  naphthalene metabolism, 289
  toluene dioxygenase purification, 228
Esters
  structures and applications, 31
Ethane and related compounds, 29
Ethers, 31
Ethion, 73
Ethylan, 70
Ethylbenzene
  anaerobic degradation, 311, 315–316
  structure, 43
Ethyl benzyl ether, 44
Ethylene and related compounds
  bioremediation
    aerobic, 523–531
    anaerobic, 531–535
    chemical/nutrient requirements, 521–523

  microbial populations and enumeration, 520–521
  remedial options, 523
  requirements for *in situ*, 518–520
  compounds and structure, 29, 44
  properties, 517–518
  regulations, 518
  sources, 515–517
  structure and applications, 29, 43
Eukaryotes, 13
Europe
  bulk chemical production, 55, 56–57
  chlorophenols, 391–392
  hazardous waste production, 59, 60
Expanded bed reactors, 454
Explosives
  classes, 53–54
  structure and application, 40
Expression. *See* Gene expression
Extraction in hydrocarbon analysis, 94
Extraction wells, 527

Facultative multiple-step resistance, 8
Famphur, 73
Fatal tumors, 581
FBR. *See* Fluidized bed reactors
Feasibility studies, 101–104
Federal regulations. *See* Regulations
Ferredoxin reductase, 286–287
Ferric oxyhydroxide, 179
Fertilizer. *See* Nutrients
Field application vectors, 459
Filtration, 414, 531
Finland, 391–392, 404, 409
Fish, 19
Fisher's Exact Test, 582–583
Fixed bed reactors, 472–473
Fixed-film reactors, 414
Flavobacterium sp., 230, 246, 610
  chlorinated compound bioremediation, 456–457
  chlorophenol degradation, 393–394, 398–399, 405, 407, 414
  wastewater bioremediation, 361, 362
Floc formation, 355–357
Fluazifop-butyl, 70
Fluctuation test, 8
Fluidized bed reactors (FBR), 454
  advantages, 352–353
  chlorophenols, 413, 417
Fluid thioglycolate medium with beef extract (FTMBE), 190
Fluometuron, 49

Fluoranthene
  enzymatic attack, 270
  metabolism, 277, 281, 283
  structure, 69
Fluorene
  enzymatic attack, 270
  metabolism, 274, 275–276, 282, 284
  structure, 69
Fluorenone, 334
3-Fluorobenzoate, 251–252
Fluridone, 72
Flushing/washing, 451–452, 461, 614
Folpet, 70
Fomesafen, 71
Food and Drug Administration, 596
Food contamination, 402
Formaldehyde, 43
Freon-11, 44
Fruit, 402
Fumigants. *See* Pesticides
Fungi in degradation
  amino-, nitro-, and azoaromatics, 341
  biochemical features, 332–333
  BTEX compounds, 520–521
  chlorinated organic compounds, 464–465
  chlorophenols, 337–340, 395, 396, 399–400, 402
  bioremediation, 412, 416–417, 456
  discovery of, 331–332
  hydrocarbons, 84–85, 86–87
  murrain agent, 4
  polycyclic aromatic hydrocarbons, 281–286, 333–337
  polymerized organopollutants, 341–342
Furalaxyl, 71
Furanes, 37–38
Furfural, 71

GA3, 71
Gas chromatography (GC), 623
  air samples, 489
  coal tar–contaminated soil, 509–510
  hydrocarbons, 95–96
  polychlorinated biphenyl dechlorination, 133–134, 169–170
Gaseous wastes, 472–473
Gasoline. *See* Petrochemical industry compounds
Gas tanks, 100, 515–516
GC. *See* Gas chromatography
GEMs. *See* Genetically engineered microorganisms

Gene conversion, 13–14
Gene dosage, amplifications in, 11–13
Gene exchange in acclimation, 565
Gene expression
  developmental cycles, 16
  environmental factors, 15
  genome structure, 14–15
  see also Gene rearrangements
Gene probes, 547–548, 554–555, 624
Gene rearrangements
  amplifications and deletions, 11–13
  classification, 8–9
  insertions, 9–10
  inversions and translocations, 13–14
  nutritional versatility in *Pseudomonas*, 16–18
  plasmids, 10–11
  see also Gene expression
Genetically engineered microorganisms (GEMs), 19, 615
  broadened substrate specificity, 618–620
  establishing, 553–554
  improving degradation, 616–618
  in situ, 622
  low-level contaminant degradation, 620–621
  regulations, 605–606
  resistance to environmental stress, 621–622
  tracking, 554–555
Genetic mutations
  vs. physicochemical changes, 6–7
  in *Pseudomonas aeruginosa*, 18
  in pure cultures, 7–8
  in resistance, 8
Genetics
  bacteria in wastewater treatment, 358
  chlorinated organic compound degradation, 441
  degradative genes
    abundance and activity, 552–553
    acclimation of microorganisms, 564–565
    in environment, 550–552
    establishing in environment, 553–554
    hybridization, 547–550
    models, 545–547, 565–567
    protein detection, 562–564
    tracking in environment, 554–555
    transcript detection, 558–562
    transcription, 555–558
    translation, 562

Genetics (*Continued*)
  field application vectors, 459
  naphthalene metabolism, 293–296
Genetic selection
  in microbial variation, 6–7
  in pure culture, 7–8
Genome structure, 14–15
Gentisate, 291
Geobacter metallireducens, 317, 321, 324
Germany, 60
Glucose utilization, 86
Glutathione transferases, 219
Glyphosate
  activated sludge
    continuous flow reactors, 374–377
    sequencing batch reactors, 378–380
  enumeration and numerical taxonomy,
    363–366
  research program, 358–359
  resistance/degradation correlation, 359–
    363
  structure, 73
Granular activated carbon anaerobic bio-
  reactors, 457
Green algae, 273–280
Groundwater
  bioremediation, 608–610
  BTEX contamination, 519–520
  chlorophenol contamination, 391–392,
    414
  see also Wastewater treatment
Growth factors, 252
Gyrase, 14

Halazone, 45
Half-life concept, 107–109
Halidohydrolases, 443
Halobenzenes, 260–263
Halogenated biphenyls, 193–195
Halogenated compounds. *See* Chlorinated
  compounds; Chlorophenols; Organ-
  ohalides; *specific compounds*
Hazard identification
  carcinogenicity, 580–584
  historical tumor rates, 586–587
  multiple comparisons, 584–586
  reproductive and developmental effects,
    587
Hazardous waste production, 59–60
HCCA isomerase, 288–289, 293
Heat shock proteins, 15
Heavy metal resistant genotypes, 550–552
Heavy metals, 185
Helicases, 15

Hematin, 231, 232
Heptachlor, 70
Herbicides. *See* Pesticides; *specific compounds*
Hexachlorobenzene (HCB)
  amount produced, 60
  bioremediation, 455
  metabolism, 232
  vitamin B_{12} in dechlorination, 202
Hexachlorophene, 45
High performance liquid chromatography
  (HPLC), 95–96
Hinshelwood, Cyril, 6–7
Historical tumor rates, 586–587
HMX, 54
Homocyclic aromatic compound degradation, 307–308
  aryl reductive dehalogenation, 245
  benzene, 308–310
  benzoate, 316–319
  cresols, 322–324
  ethylbenzene, 315–316
  phenol, 321
  phthalic acids, 319–321
  toluene, 310–313
  xylenes, 313–315
Homologs, PCB, 128
Hoosic River, 160–161
Household compounds
  classes, 50
  pesticides, 29–42
  structures, 35, 36
HPLC (high performance liquid chromatography), 95–96
Hudson Estuary and River. *See* Polychlorinated biphenyl dechlorination
Humans, polycyclic aromatic hydrocarbon
  effects, 271
Humus. *See* Soil
Hybridization, genetic, 547–550, 553–554
Hydratase-aldolase, 289
Hydraulic residence time, 355–356, 373–
  374, 380–382
Hydrocarbons
  bioremediation applications
    bioventilation, 111–114
    composting, 111
    engineered biopiles, 109–110
    land treatment, 105–109
    regulations, 100–101
    rhizosphere remediation, 114–115
    strategies, 99–100
    treatability studies, 101–105
  combustion products, 54

defined, 77–78
degradation
    environmental factors, 92–93
    measurement and indicators, 93–99
    pathways, 87–89
    physiological factors, 89–92
    in soil, 79–81, 83–84
    types and numbers of mediators, 84–
        87
    undesired, 78–79
    fate in soil, 79–81, 83–84
Hydrochlorofluorocarbons (HCFCs), 608
Hydrofluorocarbons (HFCs), 608
Hydrogen, 186
Hydrogen peroxide, 524–525
Hydrolase, 290
p-Hydroxybenzaldehyde, 322
p-Hydroxybenzoate, 322–324
4-Hydroxybenzoic acid, 321
p-Hydroxybenzylalcohol, 322
2-Hydroxychromene-2-carboxylate
        (HCCA), 288–289, 293
Hydroxycoumarins, 38
2-Hydroxycyclohexanecarboxylate, 319–
        321
Hydroxylase, 290
para-Hydroxylation, 398–399
2-Hydroxymuconic semialdehyde dehy-
        drogenase and hydrolase, 290
8-Hydroxyquinoline, 52

Imazapyr, 73
Imidazoles, 40
Iminodiacetic acid, 372–374
Immobilized cell systems
    chemical mixture degradation, 366–371
    chlorophenol degradation, 413–417
    tertiary biotreatment, 380–382
    types, 352, 355
Immunoassays, 562–563, 623
Incidental tumors, 581
Indandiones, 37
Indeno(123-cd)pyrene, 69
Injection wells, 527
Inoculation in bioremediation
    chlorinated organic compounds, 452–
        453, 461, 464–465
    chlorophenols, 404–408
Insecticides. See Pesticides; specific com-
        pounds
Insertions, gene, 9–10
In situ sparing, 530
Inversions, gene, 13–14
Invertases (gene product), 13

3-Iodobenzoate, 249
Iron in metabolism
    benzoate, 316–317
    BTEX, 524, 534–535
    cresol, 324
    phenol, 321
    toluene, 310–313, 315
Isocyanuric acid, 73
Isomerase, 289
Isomers, PCB, 128
Isophorone, 69
Isopropanol alcohol, 43
Isothiazolones, 41–42
Isothiocyanates, 31
ISP$_{NAP}$, 287

Japan, 55, 56–57
JP4 fuel, 314
JP4 fuels, 462

Ketelmeer, Lake, 161–162
β-Ketoadipate, 291
Ketones, 30
Kinetics of hydrocarbon degradation, 107–
        109
Kinetin, 73
Kluyver, A.J., 3

Laccases, 339, 402–403
Lactose gene (lacZ), 10
lacZ (lactose gene), 10
Lake Ketelmeer, 161–162
Lake Shinji, 162
Landfarming
    BTEX compounds, 523–524
    chlorinated organic compounds, 461,
        471–472
    hydrocarbons, 105–109
Leachate from coal tar-contaminated soil,
        503, 510–511
Leaching bed, 454
LED$_{10}$ (lowest effective dose), 594
LexA protein, 15
Lignin, 332
Lignin peroxidases
    biodegradation, 332–333
    fungal degradation
        amino-, nitro-, and azoaromatics,
            341
        fungal metabolism of chlorinated phe-
            nols, 339
        polycyclic aromatic hydrocarbon metab-
            olism, 284, 292, 335–337
Lignolytic fungi. See Fungi in degradation

Limonene, 44
Lindane, 50
Lindley, John, 4
Linearized multistage model, 591
Linuron, 72
Liquid waste bioremediation, 472–473
Loading rate in polycyclic aromatic hydrocarbon degradation, 496
Low dose linearity, 595–596
Lowest effective dose (LED$_{10}$), 594
Luciferase, 624
lux genes, 557–558, 559

Malathion, 48
Mammalian liver, 219–220
Manganese-dependent peroxidases, 333
    amino-, nitro-, and azoaromatic metabolism, 341
    chlorinated phenol metabolism, 340
Manure, 107
Mass balance in PCB dechlorination, 167–170, 209
Material characterization in treatability studies, 104
Maximum tolerated dose, 597–598
MCPA, structure, 70
Measurement. *See* Analytical methods
Mechanisms of toxicity
    interspecies dose scaling, 596–597
    mixtures, 597
    threshold doses, safety factors, and low dose linearity, 595–596
Medical industry compounds, 52
Membrane solids separation, 454
Mercury, 557–558, 560
mer gene, 550–552, 558, 560
mer–lux gene, 557–558
Metabolism. *See* Degradation
Metaldehyde, 69
Metalloenzyme catalysts, 221–222
Metals
    analysis, 488
    BTEX compound bioremediation, 534–535
    in coal tar–contaminated soils, 502, 510–511
    heavy metal resistant genotypes, 550–552
    PCB co-contaminants, 185
    trace metals in bioremediation, 173
Metals industry compounds, 42, 53
Methacrylate, 47
Methane, structure and applications, 29; *see also* Methanogenic conditions in degradation

Methane monooxygenase
    chlorinated organic compound degradation, 443, 444, 447
    organohalide metabolism, 223–227
Methanogenic conditions in degradation
    BTEX compounds, 533–534
    chlorinated compounds
        aromatic, 445, 449, 455, 463–464
        bioremediation, 467–468
        chlorophenols, 400–401, 412
        PCBs, 172, 173–180, 198–200
    organohalides, 220, 222, 223–227
    oxidation, 463–464
    polycyclic aromatic hydrocarbon metabolism, 280
Methanol, 191–192
Methiocarb, 72
Methoxychlor, 70
4-Methylbenzoate
    degradation, 551, 553
    structure, 52
Methylchloroisothiazolone, 53
Methyl ethyl ketone, 46
Methyl isobutyl ketone, 49
Methylisothiazolone, 53
Methyl methacrylate, 47
Methylocystis parvus OBBP, 224
Methylosinus trichlsporium, 224
Methylosinus trichosporium, 225
Methyl-parathion, 48
N-Methyl-2-pyrrolidone, 44
Methyl *tert*-butyl ether, 43
Mexacarbate, 72
MGK repellents, 71, 72
MgO$_2$, 612
The Microbe's Contribution to Biology (Kluyver and van Niel), 3
Microorganisms
    acclimation, 564–565
    anaerobic metabolism
        benzoate, 316–317
        cresol, 322–324
        phenol, 321
        phthalic acid, 319–321
        toluene, 310–313
    bioremediation, 610
        bioleaching, 615
        BTEX compounds, 520–521
        carbon biomass ratio, 355–357
        chlorinated organic compounds, 464–465
        community structure, 357
        glyphosate degradation, 358–366
        PCBs, 458–459
        regulations, 605–606

BTEX degradation, 520–521
chlorinated compounds
  bioremediation, 449, 450–453, 456–
    457, 464–465
  degradation, 437–439
chlorophenols
  bioremediation, 403–417, 456
  degradation, 337–340, 392–402,
    412, 416–417
degradation
  amino-, nitro-, and azoaromatics,
    341
  biochemical features, 332–333
  discovery of, 331–332
  polymerized organopollutants, 341–
    342
genetically engineered microorganisms
    (GEMs), 19, 615
  broadened substrate specificity, 618–
    620
  establishing, 553–554
  improving degradation, 616–618
  in situ, 622
  low-level contaminant degradation,
    620–621
  regulations, 605–606
  resistance to environmental stress,
    621–622
  tracking, 554–555
genotypic surveys, 550–553, 550–555
hydrocarbons
  degradation, 84–87
  spoilage, 78–79
immonoassays, 562–563
organohalide metabolism, 220–223
PCB dechlorination, 132–133
  isolation, 195–198
  population H, 199–200
  population M, 198–199
polycyclic aromatic hydrocarbon degra-
    dation, 333–337, 489
  anaerobic bacteria, 286
  to *trans*-dihydrodiols, 281–284
  to *cis*-dihydrodiols, phenols, and
    ring-fission products, 273–280
  enzymes, 286–292
  to quinones, 284–286
versatility
  of biochemical pathways, 4–5
  consequences on survival, 19
  genetics in
    developmental cycles, 16
    environmental factors, 15, 18–19
    genome structure changes, 14–15
    mutational changes, 6, 7–8

nutritional versatility of *Pseudo-
    monas*, 16–18
  vs. physicochemical causes, 5–7
  rearrangements, 8–14
  see also Bioremediation; Degradation;
    specific organisms
Mirex, 70
MnP. *See* Manganese-dependent perox-
    idases
Moisture in polycyclic aromatic hydrocar-
    bon degradation, 494–495
Monooxygenases in metabolism
  chlorinated compounds, 443–444, 447
  chlorophenols, 398–399
  organohalides, 221–227
    coreductive, 233–236
    methanotrophs, 223–227
    pathway, 221–223
Monsanto Company, 358
Moraxella sp., 316
Most probable number (MPN), 548
  hydrocarbon analysis, 97
  PCB dechlorination microorganisms,
    196
MPN. *See* Most probable number
Multistage model, 590–592
Murrain, 4
Mutations
  in acclimation, 565
  in bacteria, 6
  in microbial versatility
    vs. physicochemical causes, 6–7
    in pure cultures, 7–8
    resistance, 8
  in *Pseudomonas aeruginosa*, 18
Mycobacterium chlorophenolicum, 405–
    406, 407–409, 414–416
Mycobacterium sp. in degradation, 224
  chlorophenol, 394, 398–399
  naphthalene, 295
  polycyclic aromatic hydrocarbons, 273–
    277, 282–283, 334–335
  pyrene, 279

nahAc gene, 550–551, 553
nah gene, 293–296
NAH7 plasmids, 293–296
Naled, 73
Naphthalene dihydrodiol dehydrogenase,
    287–288
Naphthalene dioxygenase, 286–287
Naphthalenes
  metabolism
    to dihydrodiols, 273, 274, 281–282
    enzymes, 270, 286–292

Naphthalenes (*Continued*)
genetically engineered microorganisms, 616–617
genotypes in microorganisms, 551, 552
methylotrophic bacteria, 280
mineralization, 489–491, 497
structure, 43
Naphthenes, 81
Napropamide, 71
Neisseria gonorrhoeae, 13–14
Neurotoxic risk, 594
New Bedford Harbor, 141, 152–154, 173, 185, 200
Nicholas, Robert, 605
Nickel, 201–202
Niclosamide, 72
Nicotine, 72
nifD gene, 13
Nitrate in metabolism
benzene, 308–309
bioremediation, 371–374, 612
BTEX compounds, 521–522, 532–533
chlorophenols, 401
coal tar-contaminated soil, 504
cresol, 322–324
ethylbenzene, 315–316
hydrocarbons, 89–91, 107
PCBs, 179
phenols, 321
phthalic acid, 319–321
toluene, 310–313, 313–315
Nitriles, 31
Nitroanilines, 39
Nitroaromatics
fungal degradation, 341
production volume, 55, 58
Nitrobenzene, 43
Nitrofen, 71
Nitrogen cycle, 371–374
Nitrosomonas europaea, 224
NOAEL (no observed adverse effect levels), 594, 595
Nocardia sp., 287–288
NOCs (nonionic organic contaminants), 184
Nonionic organic contaminants (NOCs), 184
Nonparametric extrapolation, 592–593
No observed adverse effect levels (NOAEL), 594, 595
Nucleic acid hybridization, 547–549; *see also* DNA; RNA

Nutrients in metabolism
aryl reductive dehalogenation, 252
bioremediation
BTEX compounds, 521–523
Prince William Sound, 604
hydrocarbons, 89–91, 107
polycyclic aromatic hydrocarbons, 495
in *Pseudomonas* sp., 16–18
see also Nitrate; Sulfate

Obligatory multiple-step resistance, 8
Organic sulfur in crude oil, 80, 81
Organochloride. *See* Chlorinated compounds
Organohalides
chemistry and distribution, 217–219
microbial metabolism
challenges to, 220–223
cooxidative, 223–229
coreductive, 229–236
toxicology, 219–220
Organometals, 49
Organophosphates, 40–41, 48
Organosulfates and related compounds, 41
Organothiocyanates, 41–42
Oryzalin, 72
Otonabee River–Rice Lake, 162–163
OUR (oxygen uptake rate), 99, 113
Ovex, 73
Owasco Lake, NY, 148
Oxidases, 402–403
Oxidative reactions
chlorinated organic compounds, 438–440, 442–444, 463–464
methanotrophic, 463–464
organohalides, 223–229
polycyclic aromatic hydrocarbons, 335–337
Oxygenases in metabolism
benzene, 308
chlorinated organic compounds, 441, 443–444, 447
chlorophenols, 398–399
naphthalene, 286–286, 288–289, 290
organohalides
methane monooxygenase, 223–227
pathway, 221–223
toluene dioxygenase, 227–230
Oxygen in bioremediation. *See* Aeration in bioremediation
Oxygen sparging, 524–525, 527–530
Oxygen uptake rate (OUR), 99, 113
Oxyphenoxy acid esters, 37

Packed bed reactors, 355, 380–382
PAHs. *See* Polycyclic aromatic hydrocarbons
Paint industry compounds
    amounts produced, 59
    classes, 49–50, 53
    structure and application, 29–42
Paper industry. *See* Pulp and paper industry
Paracocccus denitrificans, 316
Paracoccus sp., 322
Paraquat, 72
Parathion, 48
Pasteur, Louis, 4
PBR. *See* Packed bed reactor
PCBs. *See* Polychlorinated biphenyls
PCE. *See* Perchloroethylene
PCP (pentachlorophenol). *See* Chlorophenols
PCR (polymerase chain reaction), 97, 549–550, 560–561
Penicillium sp., 282–283
Pentachlorobenzenes, 202
Pentachlorophenol. *See* Chlorophenols
Perchloroethylene (PCE)
    anaerobic degradation, 465
    bioremediation, 449–455, 465, 473
    coreductive metabolism, 229–230
Peroxidases in metabolism
    chlorinated compounds, 443–444, 447
    chlorophenols, 402–403
    fungal
        amino-, nitro-, and azoaromatics, 341
        chlorinated phenols, 339–340
        ligninolysis, 332–333
        polycyclic aromatic hydrocarbons, 335–337
Peroxide, 524–525
Perylene, 285
Pesticides
    amounts produced, 55–59
    aryl reductive dehalogenation, 253–255, 262
    classes, 47–49
    household use, 50
    organohalide toxicology, 219
    phenoxy bioremediation, 456–458
    structure and application, 29–42
Petrochemical industry compounds
    amounts produced, 55, 56–58
    bioremediation, 607
        bioventilation, 111–114
        composting, 111

engineered biopiles, 109–110
        land treatment, 105–109
        regulations, 100–101
        rhizosphere remediation, 114–115
        strategies, 99–100
        treatability studies, 101–105
    classes, 28, 43–46, 77–78
    hazardous waste output, 60
    microbial degradation
        environmental factors, 92–93
        measurement and indicators, 93–99
        pathways, 87–89
        physiological factors, 89–92
        in soil, 79–81, 83–84
        types and numbers of mediators, 84–87
    undesired, 78–79
    as PCB co-contaminant, 185
    physical fate of hydrocarbons, 81–83
    Prince William Sound, 604
    structure and application, 29–42
    see also BTEX compounds; Hydrocarbons; *specific compounds*
pH
    bioremediation, 521
    coal tar-contaminated soil, 503–504
    naphthalene metabolism, 289
    PCB dechlorination, 172
    polycyclic aromatic hydrocarbon degradation, 495–496
    wastewater treatment, 371
Phanerochaete chrysosporium metabolism, 282, 284–285
    aminoaromatics, 331–341
    chlorophenols, 338–340, 395, 399, 412, 416–417
    polycyclic aromatic hydrocarbons, 334–337
Phanerochaete sordida, 334, 395, 412
Phanerochaete sp., 406, 456
Pharmaceutical industry compounds, 52
Phenanthrene
    enzymatic attack, 270
    fungal degradation, 337
    metabolism, 274, 277–278, 281–285, 295–296
    structure, 46
Phenols
    metabolism, 280
        anaerobic, 318, 321
        fungal, 337–340
    structures and applications, 35–36, 45, 48
    see also Chlorophenols; *specific compounds*

Phenoxy acids, 37
2-Phenoxyethanol, 44
Phenoxy herbicides, 456–458
o-Phenylphenol, 45
Phenylureas, 39
Phorate, 73
Phosphonatase, 362–363
N-Phosphonomethylglycine. *See* Glyphosate
Phosphorus
  BTEX compound bioremediation, 521–523
  coal tar-contaminated soil, 504
  hydrocarbon degradation, 89–91, 107
Phthalic acid and related compounds
  anaerobic degradation, 318, 319–321
  fungal degradation, 337
  structures and applications, 37, 43, 48
pilE locus, 13
Pindone, 71
α-Pinene, 44
Plants, rhizosphere remediation, 114–115
Plasmids
  in microbial versatility, 10–11
  NAH, 293–296
  in *Pseudomonas*, 17
  toluene, 295
Plasticizer, 69
Plastics industry compounds
  classes, 46–47
  structure and application, 31, 35
Polybrominated biphenyls, 252
Polychaete, 85
Polychlorinated biphenyl dechlorination
  aryl reductive dehalogenation, 249, 252, 255, 256–257
  biological basis, 132–133
  bioremediation, 204–208, 458–460, 613–615
    aerobic oxidation, 464
    anaerobic natural, 467
    anaerobic reductive, 464
    anaerobic reductive, 464
    co-metabolism, 458–459
  degradation
    genetically engineered microorganisms, 616–618, 622
    genotype of microorganisms, 550–551
    oxygenases, 441
    reductive, 445
  discovery, 128–129
  enhancing
    carbon additions, 189–193
    halogenated biphenyls, 193–195
    surfactants, 193

in environment
    Escambia Bay, Hudson Estuary and River, and Hoosic River, 160–161
    Hudson River, 154–156
    Lake Ketelmeer, 161–162
    Lake Shinji, 162
    New Bedford Harbor, 152–154
    Otonabee River–Rice Lake, 162–163
    Sheboygan River and Harbor, 161
    Silver Lake, 158–160
    Waukegan Harbor, 161
    Woods Pond, 156–158
  enzymes and selective advantage in, 201–204
  factors in
    bioavailability of substrate, 184, 188
    co-contaminants, 184–185, 188–189
    concentration of substrate, 180–184, 188
    culture conditions, 185–187
    electron acceptors, 173–180, 187–188
    environmental, 172–173, 187
  laboratory studies, 129–131
  microbiology
    characterization, 198–201
    isolation attempts, 195–198
  patterns
    B, 155
    B,′ 155–156
    basis of, 135–136
    C, 156
    defined, 133
    discovery, 128–129
    E, 156
    F and G, 158–160
  processes (activities)
    C, 146–148, 149–152
    defined, 133
    experimental conditions, 136
    H, 141–144, 150–152
    H,′ 141, 150–152
    M, 137–139, 149–152
    M and H or H,′ 148–149
    N, 144–146, 149–152
    P, 144, 149–152
    Q, 139–141, 150–152
  pure congeners, 163–164
    mass balance and quantitation, 167–170, 209
    pathways and microbial succession, 164–165
    reactivity preference, 165–167
    unusual activities, 170–171

recognition and characterization in environment, 133–135
specificity, 136
terminology, 133
Polychlorinated biphenyls (PCBs)
compounds and structure, 45
formulations, 128
regulations, 474
structure, 128
toxicology, 204–205
uses, 127
Polychlorinated dioxins, 412
Polycyclic aromatic hydrocarbons (PAHs)
bioremediation, 410–412, 613
carcinogenesis, 269–273
in coal-tar contaminated soil
analytical techniques, 488–491
bioremediation, 511–513
concentration, 504–509
degradation, 491–501
field methods, 501–503
GC/FID fingerprints, 509–510
leachate, 510
nitrogen and phosphate, 504
pH, 503–504
run-off, 511
metabolism, 92
anaerobic bacteria, 286
to *trans*-dihydrodiols, 281–284
to *cis*-dihydrodiols, phenols, and ring-fission products, 273–280
enzymes, 286–292
fungal, 333–337
genetically engineered microorganisms, 616
genetics, 293–296
to naphthalene, 286–291
to quinones, 284–286
structures and applications, 32–33
in waste sludge, 46
see also Hydrocarbons
Polymerase chain reaction (PCR), 97, 549–550, 560–561
Polymerized organopollutants, 341–342
Polyurethane, 414–416
Preservatives
compounds and structures, 30, 36, 40
in cosmetics, 52
Primary treatment, 353
Prince William Sound, 604
Probability (*p*), 582–584
Probes, gene, 547–548, 554–555, 624
Process evaluation of hydrocarbon-contaminated soils, 104
Procymidone, 72

Prokaryotes in studies vs. eukaryotes, 6; *see also* Microorganisms
Promexal W50, 50
Pronamide, 71
Propane, 29–30
Propargite, 73
Propylene, 43
Propylene glycol, 43
Proteins, 562–564
Protozoa, 85
Pseudomonas aeruginosa
anthracene metabolism, 276
nutritional variability, 17
Pseudomonas cepacia, 224
chlorinated organic compounds, 457–458
genotype, 558, 563
insertion elements, 9–10
TCE degradation, 610
Pseudomonas mendocina, 224
Pseudomonas paucimobilis, 288–289, 622
Pseudomonas pseudoalcaligenes, 288–289
Pseudomonas putida, 224, 291
genetically engineered, 621
genotype, 17, 558, 563
naphthalene metabolism, 275
organohalide metabolism, 227–229
polycyclic aromatic hydrocarbon metabolism, 287–288
Pseudomonas sp. metabolism
benzoate, 316
chlorophenol, 393–398, 416
cresol, 322–323
genetically engineered, 619, 620
genotype, 553, 554
nutritional versatility, 16–18
phthalic acid, 319–321
polycyclic aromatic hydrocarbon, 273–280, 286–291
toluene metabolism, 310
wastewater treatment
chemical mixtures, 367–371
glyphosate degradation, 363
Pseudomonas stutzeri, 316
Pulp and paper industry
chlorophenols, 390, 416–417
classes of compounds, 51–52
hazardous waste output, 60
Pump-and-treat systems, 436–437, 530
Pyrazon, 72
Pyrene
metabolism, 274, 277, 279, 283, 285
bacterial, 334

Pyrene (*Continued*)
  enzymatic attack, 270
  fungal, 335, 336
  structure, 46
Pyrethrin I, 74
Pyrethroids, 42
Pyridines and related compounds, 39–40,
  48, 72
Pyrimidine, 48
Pyrithion, 52
Pyrocatechase, 396
Pyruvate, 174, 192–193, 197

Quinoline, 72
Quinones, 292, 335–340
  chlorophenol degradation, 398–400
  polycyclic aromatic hydrocarbon metab-
    olism, 284–286
  structures and applications, 37, 48
Quizalofop-ethyl, 70

RDX, structure, 54
Rearrangements. *See* Gene rearrangements
Recovery systems, 520
Red Squill, 74
Reductive metabolism
  chlorinated compounds, 439, 442–446,
    464
  chlorophenols, 400–401, 417
  homocyclic aromatic compounds, 307–
    308
  benzene, 308–310
  benzoate, 316–319
  cresols, 322–324
  ethylbenzene, 315–316
  phenol, 321
  phthalic acids, 319–321
  toluene, 310–313
  xylenes, 313–315
  organohalides, 229–230
Regulations
  bioremediation, 605–606
  BTEX and fuel oils, 518
  chemical wastewater treatment, 350–
    351
  hydrocarbon-contaminated soils, 104
  hydrocarbons, 100–101
  PCB contamination, 474
Regulatory genes, 294
Remediation. *See* Bioremediation
Reproductive risk
  bioassays, 587
  dose–response models, 593–594

Resistance to antibiotics
  facultative multiple-step, 8
  genetic vs. physicochemical causes,
    6–7
  mutations in, 8
  obligatory multiple-step, 8
  plasmid markers, 10–11
  Pseudomonas aeruginosa, 17
Resource Conservation and Recovery Act,
  100
Respirometric measurements, 98–99, 527
Rhizoctonia practicola, 402
Rhizosphere remediation, 114–115
Rhodococcus chlorophenolicus, 230, 246,
  457
Rhodococcus erythropolis, 224
Rhodococcus sp., 273, 276–277, 393, 396,
  398
Ribosomal RNA genes, 11–12
Ribulose-1,5-bisphosphate carboxylase
  (RuBisCOI), 559–560
Rice Lake, 162–163
Risk assessment, 579–580
  dose–response and risk estimation
    cancer, 590–591
    humans, 589–590
    intermittent exposure to carcinogens,
      592
    neurotoxic, 594
    nonparametric extrapolation, 592–
      593
    reproductive and developmental,
      593–594
  experimental design, 597–598
  exposure assessment, 587–589
  gene probes in, 554–555
  hazard identification
    carcinogenicity, 580–584
    historical tumor rates, 586–587
    multiple comparisons, 584–586
    reproductive and developmental ef-
      fects, 587
  mechanisms of toxicity
    interspecies dose scaling, 596–597
    mixtures, 597
    threshold doses, safety factors, and
      low dose linearity, 595–596
  uncertainty in cancer risk estimates,
    598–599
RNA, transcript measurement, 559–562
RNA polymerase, 555–556
RNA transcription, 555–558
Rotenone, 74

Run-off in coal tar–contaminated soil, 503
Run-off in coal tar–contaminated soils, 511

Saccharomyces cerevisiae, 292
Safety factors, 595–596
Salicylaldehyde dehydrogenase, 290
Salicylate
    polycyclic aromatic hydrocarbon metabolism, 294
    structure, 52
Salicylate hydroxylase, 290
Salinity in PCB dechlorination, 172
Salmonella sp., 13
Scilliroside, 74
Screening level analysis, 101–104
Secondary treatment, 353–355
Sediments. *See* Soil
Semicontinuous operations, 352, 353
Separation in hydrocarbon analysis, 94–95
Sequencing batch reactors, 378–380
Sequential anaerobic–aerobic bioreactor
    processes, 473, 474–475
Sheboygan River and Harbor, 161
Shinji, Lake, 162
Silver Lake, 137–138, 144, 158–160,
    165–166, 173, 185
Sludge enrichment, 406
Slurry-phase systems, 404, 409–412
Sodium dodecyl benzenesulfonate (SDBS),
    193
Soil
    bioremediation
        bioventing, 610–611
        genetically engineered microorganisms, 622
        PCBs, 458–459
        sorbed chemicals, 613–615
        temperature, 611
    BTEX contamination
        aerobic bioremediation, 523–531
        anaerobic, 531–535
        bioremediation, 516–517
        evaluation, 519–520
        nutrients in bioremediation, 521–523
        soil permeability, 519–520
    chlorinated organic compound bioremediation, 460–462, 468–475
    chlorophenols
        bioremediation, 403–412
        biotransformation, 402–403
        contamination, 391–392
    coal-tar contamination
        analytical methods, 488–489, 489–491

        bioremediation, 511–513
        chemical content, 487–488
        degradation, 491–501
        field methods, 501–503
        laboratory methods, 501–503
        leachate polycyclic aromatic hydrocarbons, 510
        nitrogen and phosphate, 504
        pH, 503–504
        polycyclic aromatic hydrocarbon
            concentration, 504–509
        run-off metals and polycyclic aromatic hydrocarbon, 511
    components, 461–462
    hydrocarbons degradation in
        biological fate of hydrocarbons, 83–84
        bioremediation strategies, 99–100
        bioventilation, 111–114
        composting, 111
        engineered biopiles, 109–110
        environmental factors, 92–93
        land treatment, 105–109
        measurement and indicators, 93–99
        microbial mediators, 84–87
        pathways, 87–89
        physical fate of hydrocarbons, 81–83
        physiological factors, 89–92
        regulations, 100–101
        rhizosphere remediation, 114–115
        substrate, 79–81
        treatability studies, 101–105
    nucleic acid extraction, 549
Soil vapor extraction (SVE)
    BTEX bioremediation, 525–526, 528
    hydrocarbon degradation, 111–113
    technologies, 472
Soil washing, 451–452, 461, 614
Solid-phase systems
    chlorophenol, 409–412
    hydrocarbon-contaminated soil, 106
Solid wastes, 468–472
Solvents
    bioremediation, 449, 455
    classes, 49
    compounds and structures, 30–32, 37,
        38, 44, 49
    production volume, 55, 57
Sorbic acid, 52
SOS functions, 15
Sources of synthetic organic compounds,
    27–28

Sources of synthetic organic compounds
(*Continued*)
    chemicals, structure, and application,
        29–42
    cosmetics and medical/pharmaceutical
        industry, 52
    electronics industry, 50–51
    energy industry, 54
    explosives industry, 53–54
    fate in the environment, 61–64
    hazardous waste, 59–60
    household use, 50
    metals industry, 53
    paint industry, 49–50, 59
    pesticide industry, 29–42, 47–49, 55–
        59
    petrochemical industry, 28, 43–46, 55,
        56–58
    plastics industry, 3146–47
    pulp and paper industry, 51–52
    textile industry, 51
    wood preservation, 53
Sparging, air, 524–525, 527–530
Sporulation, 16
Staphylococcus auriculans, 276
Staphylococcus epidermidis, 246
Statistical tests
    carcinogenicity, 580–584
    multiple comparisons, 584–586
    reproductive and developmental effects,
        587
Sta-Way, 69
Stirred tank reactors, 451
Strain T1, 312–313
Streptococcus sp., 10
Streptomyces glaucescens, 12–13
Streptomyces peuceticus, 12
Streptomycete sporulation, 16
Structure of chemical in environmental fate
    of compounds, 63
Strychnine, 74
Study conduct in treatability studies, 104
Styrene, 43
Sulfate in metabolism
    aryl reductive dehalogenation, 257–260
    benzene, 309–310
    benzoate, 317–319
    bioremediation, 615
    BTEX compounds, 534
    chlorinated aromatic compounds, 445
    chlorophenols, 401
    cresol, 324
    PCB dechlorination, 173–175, 178–
        180, 198–199, 200

petroleum reservoir degradation, 78–79
phenols, 321
toluene, 310–313, 315
Sulfide in benzene metabolism, 309
Sulfite
    aryl reductive dehalogenation, 257–260
    chlorophenol degradation, 401
Sulfur in crude oil, 80, 81
Surfactants, 193
Suspended bioreactors, 352, 450–451
Suspended culture reactions, 456–457
SVE. *See* Soil vapor extraction
Syncephalastrum racemosum, 282

2,4,5-T. *See* 2,4,5-Trichlorophenoxy-
    acetate
T1, strain, 312–313
Taxonomy
    glyphosate degradation, 363–366
    of *Pseudomonas*, nutrition in, 16–18
TBZ (thiabendazole), 73
TCA, anaerobic bioremediation, 466
TCE. *See* Trichloroethylene
TDE, structure, 70
TeCP (tetrachlorophenols). *See* Chloro-
    phenols
Temperature in bioremediation, 611
    aryl reductive dehalogenation, 256
    BTEX compounds, 526
    chlorophenols, 409, 415
    coal tar–contaminated soil, 511–512
    PCB dechlorination, 172
Terephthalic acid, 47
Tertiary treatment of wastewater, 380–382
2,3,4,6-Tetrachloroanisole, 402
Tetrachloroethylene, 44
Tetrachlorohydroquinone, 398
Tetrachlorophenols (TeCP). *See* Chloro-
    phenols
Tetrasul, 73
Textile industry compounds
    classes, 51
    structure and application, 39
tfd genes, 553
Thiabendazole (TBZ), 73
Thiobacillus ferrooxidans, 563
Thiobencarb, 74
Thiocarbamates, 41
Thiodicarb, 72
Thioureas, 41–42
Thiram, 74
Threshold doses, 595–596
TN9*16*, 10
TNAZ, 54

TNT, 54
TOL, plasmid, 17
Toluene dioxygenase, 227–229, 230
Toluene plasmids, 295
Toluene
  bioremediation
    aerobic, 523–531
    anaerobic, 531–535
    chemical/nutrient requirements,
      521–523
    microbial populations and enumera-
      tion, 520–521
    remedial options, 523
    requirements for *in situ*, 518–520
  degradation, 87–88
    anaerobic, 310–313
    genetically engineered microor-
      ganisms, 621
    genotypes in microorganisms, 551
  production volume, 55, 57
  properties, 517–518
  regulations, 518
  sources, 515–517
  structure, 43
  structure and applications, 35
Total Kjeldahl nitrogen (TKN), 519, 522
Total organic carbon, 519
Total petroleum hydrocarbons (TPH), 95
Toxaphene
  fate in environment, 63
  structure, 69
Toxic intermediates, 62–63
Toxicology, PCBs, 204–205
2,4,5-TP, 70
TPH (total petroleum hydrocarbons), 95
Trace metals, 173
Trametes hirsuta, 412
Trametes sp., 284–285
Trametes versicolor, 337, 402
Transcription, 555–556
Transformation. *See* Degradation
Translation, 562
Translocations, 13–14
Transposons, 9–10
Treatability studies, 101–105
Triazines, 40, 48
Triazoles, 40
2,4,6-Trichloroanisole, 402
1,1,1-Trichloroethane, 44
Trichloroethylene (TCE), 223–229
  bioremediation
    aerobic fluidized–fixed bed reactors,
      473
    bench-scale processes, 449–455

field-scale, 463–464, 607, 609–610,
    613
  degradation
    genetically engineered microor-
      ganisms, 620
    methanotrophs and methane mono-
      oxygenase, 223–227
    toluene dioxygenase from *Pseu-
      domonas putida*, 227–229,
      230
  structure, 44
  toxicology, 223
Trichloroethylene-epoxide, 226
Trichlorophenols, 398–401
  bioremediation, 417
  fungal degradation, 338, 339
  structure, 50
2,4,5-Trichlorophenoxyacetate (2,4,5-T),
    390, 398
  aryl reductive dehalogenation, 253–255,
    257, 258
  bioremediation, 457–458
  structure, 70
Trickling filter, 414
Tricyclazole, 74
Trifenmorph, 70
Trifluralin, 72
Trimetaphosphate, 525
1,2,4-Trimethylbenzene, 43
2,4,6-Trinitrotoluene, 341
Triphenyltin acetate, 49
Triton, 193
Tumors. *See* Carcinogenicity
Two-stage-proliferation model, 592

Underground storage tanks, 100, 515–
    516
United States
  bulk chemical production, 55, 56–57
  hexachlorobenzene production, 60
Upflow packed bed, 454
Upflow sludge blanket, 451, 454
Uracils, 39
Urea
  BTEX compound bioremediation, 522
  structure, 49, 51

Vadose zone, 82
Vancide BL, 70
van Niel, C.B., 3–4
Vapor phase contamination, 531
Vinylchloride
  bioremediation, 463
  structure, 28

Vinylchloride (*Continued*)
trichloroethylene metabolism product,
223
Vitamin B_{12}
organohalide metabolism, 231, 232
PCB dechlorination, 202–204
VOCs. *See* Volatile organic compounds
Volatile organic analysis, 488
Volatile organic compounds (VOCs)
bioremediation, 606–608
trapping and quantitation, 491
volume produced, 59
in waste sludge, 46
Volatilization, measuring, 497–500

Warfarin, 71
Washing, soil, 451–452, 461, 614
Waste sludge
al, 46
Wastewater treatment
chlorophenols, 391–392, 413–417
cost, 349–350
elemental transformations (ammonia),
371–374
glyphosate, 358–366
immobilized bacteria in, 366–371
regulations, 350–351
systems
activated sludge: continuous flow reactors, 374–377
activated sludge: sequencing batch
reactors, 378–380
engineering design, 351–355
immobilized bacteria: tertiary treatment, 380–382
microbial design, 355–357

Waukegan Harbor, 161
Well/trench sparging, 530
White-rot fungi. *See* Fungi in degradation
Wilson, E.O., 19
Windrow composting method, 409, 410
Winogradsky, Sergei, 4–5
Wood preservatives
amounts produced, 59
bioremediation, 409–412
chlorophenols, 391
classes, 49–50, 53
structure and application, 29–42
Woods Pond, 144, 156–158, 165–166,
193–194

Xylenes
bioremediation
aerobic, 523–531
anaerobic, 531–535
chemical/nutrient requirements,
521–523
microbial populations and enumeration, 520–521
remedial options, 523
requirements for *in situ*, 518–520
degradation
anaerobic, 311, 313–315
genetically engineered microorganisms, 621
production volume, 55, 57
properties, 517–518
regulations, 518
sources, 515–517
structure, 43
xyl genes, 558, 559
xyl genes, 296